Basic Exercise Physiology

Moran S. Saghiv • Michael S. Sagiv

Basic Exercise Physiology

Clinical and Laboratory Perspectives

Springer

Moran S. Saghiv
North Carolina State University
Jamestown, NC
USA

Michael S. Sagiv
Wingate Institute
Netanya
Israel

ISBN 978-3-030-48805-5 ISBN 978-3-030-48806-2 (eBook)
https://doi.org/10.1007/978-3-030-48806-2

This Springer imprint is published by the registered company Springer Nature Switzerland AG
The registered company address is: Gewerbestrasse 11, 6330 Cham, Switzerland

Moran Schiamama Saghiv

In presenting this book, I would like to dedicate this book to my deceased mother Amira Sciamama-Sagiv for her love and for being a constant resource of patience and understanding. I would also like to express my gratitude to Professor Ehud Goldhammer, a fine cardiologist, and Professor David Ben-Sira, a great statistician and biomechanist, who read parts of the written material and gave me constructive criticism and useful suggestions. Finally, yet importantly, I am thankful and blessed to have Aviv, Michael, Noam, and Romi my children in my life as well as Jayme my wife. All five are an immense source of happiness.

Michael S. Sagiv

To my deceased wife, Amira, who has been my guiding light, and to my deceased sister Lilyan Climor, for her unrestricted love and support.

Preface

Exercise requires the integration of physiological mechanisms to enable both the respiratory and cardiovascular systems to support the increased metabolic demands. During exercise, both pulmonary and cardiovascular systems are under stress. The ability to respond to this stress is a measure of physiological health and fitness. Exercise testing evaluates the respiratory and cardiovascular responses to increased gradual workloads in order to define the level of physical work capacity and to determine the limiting factors to the increased metabolic demands during exercise. Serial exercise studies help define the effects of treatment or fitness programs on the physiological response to physical work.

Basic Exercise Physiology: Clinical and Laboratory Perspectives is a comprehensive tool for assessing human performance and provides an excellent understanding of the role of the different modes of exercise. It offers comprehensive coverage of the basic laboratory procedures for exercise training in health and disease. For the last half century, the attention of exercise physiologists was in general devoted to maintaining the health of men and women, keeping cardiopulmonary and muscle dysfunction at bay. *Basic Exercise Physiology: Clinical and Laboratory Perspectives* is a comprehensive source of information for medical personnel, instructors, and students interested in practical laboratory experiences related to the field of exercise physiology. This book deals with noninvasive measurements and various physiological parameters as a part of the laboratory exercise program. *Basic Exercise Physiology: Clinical and Laboratory Perspectives* offers testing procedures used in the assessment of human performance in health and disease. The book explores testing for oxygen uptake, anaerobic and aerobic fitness, musculoskeletal fitness, pulmonary function, body composition, ECG, and other areas. Within this setting, *Basic Exercise Physiology: Clinical and Laboratory Perspectives* is incomparable with its approach. It covers comprehensively a wide range of exercise physiology issues and, in a way, combines the exercise physiologist approach and the clinical medical personal approach. *Basic Exercise Physiology: Clinical and Laboratory Perspectives* contains mainly laboratory procedures that detect physiological changes during exercise in wellness and disease including notes for dealing with aged subjects with a small portion devoted to physiology at rest. *Basic Exercise*

Physiology: Clinical and Laboratory Perspectives will guide us to analyze and synthesize data in a friendly way. Undergraduate students experience considerable difficulties in understanding various physiological processes due to insufficient knowledge of physics, both theoretical and practical. *Basic Exercise Physiology: Clinical and Laboratory Perspectives* is for undergraduate students studying exercise physiology laboratory courses; it is a supplemental text for other exercise science laboratory courses covering measurement and evaluation. *Basic Exercise Physiology: Clinical and Laboratory Perspectives* is also for medical and graduate students, physicians, exercise scientists, sports medicine specialists, athletic trainers, and sports and physical therapists. It will undoubtedly serve in courses dealing in depth with work capacity of the cardiopulmonary system in athletes, cardiac patients, and suspected healthy subjects. The text is research oriented but also deductive. This textbook is comprehensive: the measurements and analyses of the data will provide an excellent understanding of the role of exercise's different modes, for personal trainers and other fitness professionals preparing for certifications and licensure.

Jamestown, NC Moran S. Saghiv
Netanya, Israel Michael S. Sagiv

Contents

About the Authors

Moran Sciamama-Saghiv is an associate professor of clinical exercise physiology and department chairperson. Dr. Sciamama-Saghiv's research focuses on the influences of exercise modalities on S-Klotho in health and disease, exercise changes in left-ventricular function, and the influence of exercise on genetic polymorphism. Dr. Sciamama-Saghiv is well published, mentors students' theses, and is professionally active at the international, national, regional, and local levels.

Michael S. Sagiv received his PhD in exercise physiology from the University of Wisconsin, Madison WI, USA in 1982. Michael was a soccer player at the national and international level. Dr. Sagiv is a distinguished researcher and an emeritus professor and was the president of the Wingate College, Wingate, Israel, from 1999 to 2009. He established the Wingate College Human and Biogenetics Exercise Physiology Laboratory in 1982 and served as its head for over 17 years. In addition, from 1984 to 2007, Michael was a lecturer in the Department of Physiology and Pharmacology, Faculty of Medicine at the Tel Aviv University, Israel. Dr. Sagiv has written and coauthored more than 200 publications over the course of his career, including 2 books, 6 book chapters, peer-reviewed articles (129), and lay publication. He served as the editor in chief of the *European Review of Aging and Physical Activity* for the last 6 years. He was president of the European Group for Research into Aging and Physical Activity from 1998 to 2005, a member of its board of trustees for 15 years, and a recipient of ACSM Young Researcher Award. Many of his former students are now leaders in the field of exercise physiology. From 1984 to 2007, Michael was a lecturer in the Department of Physiology and Pharmacology, Faculty of Medicine at the University of Tel Aviv, Israel. As an exercise physiologist, Michael works closely with cardiac patients in the rehabilitation center that he established at the Wingate College in 1982 and served as its head for over 25 years.

Introduction

Chapter 1: This chapter entitled "Introduction to exercise physiology" gives an overview of the scientific wide range of exercise responses of the acute and chronic adaptations that can occur across time on the human body to different modes of exercise. No laboratory directions.

Chapter 2: This chapter entitled "Metabolism" deals with fuel consumption. The purpose of this chapter is to increase the reader's proficiency in the methodology of indirect calorimetry as a technique for investigating respiratory rates and to examine the relationship between metabolic rate and body size.

Chapter 3: This chapter is devoted to aerobic and anaerobic exercises, dealing with oxygen uptake and anaerobic utilization. The chapter also deals with prediction of oxygen uptake, anaerobic threshold, aerobic power measurements, and anaerobic power measurements.

Chapter 4: It presents pulmonary function at rest and during exercise. Lung volumes and lung capacities refer to the volume of air associated with different phases of the respiratory cycle. Lung volumes are directly measured. Lung capacities differ from lung volumes at rest and during exercise.

Chapter 5: This chapter deals with blood pressure responses at rest and during exercise. For example, during exercise the blood pressure and flow may increase. Blood pressure is an important physiological signal to measure and monitor over time.

Chapter 6: It contains a review of the cardiovascular function at rest and during exercise. Both nutrients and oxygen must be delivered by the blood, meaning that the myocardium has to work harder to pump more blood through the body. This means it has to increase the heart rate and stroke volume in order to achieve a higher cardiac output.

Chapter 7: "Exercise in the elderly" is an important part of healthy aging. No laboratory directions.

Chapter 8: It contains a review of the skeletal muscles, and how exercise training affects the muscular function, cardiovascular system, and bones. Exercise training forces the muscles to contract and thus increases strength, tone, muscle mass, and muscle aerobic capacity.

Chapter 9: This chapter discusses thermoregulation during exercise in clinical and exercise settings. The thermoregulatory mechanisms play an important role in maintaining physiological homeostasis during rest and physical exercise.

Chapter 10: The exact mechanism by which the immune system is affected by exercise is still not yet understood. However, several theories tried to give an answer, yet none of the suggested theories have been confirmed. No laboratory directions.

Chapter 11: "Exercising in hostile environment" is a condition in which mountain climbers, divers, and astronauts perform work tasks in ambient deep water and space, which restricts oxygen supply. No laboratory directions.

Chapter 12: Exercise epigenetics is the study of epigenetic modifications resulting from physical exercise to the genome of cells. No laboratory directions.

Chapter 13: "Exercise equipment" concerns the need of appropriate equipment, in order to define power, cardiorespiratory function, and energy utilization, namely stepwise bicycle and treadmill. In addition, this chapter provides equations for the determination of energy expenditure and discusses metabolic rates.

Appendix 1: Gives the measuring of functional capacity, energy expenditure, and power output which have applications in different areas of exercise. The measurements and analyses of the data will provide an excellent understanding of the role of exercise's different modes. No laboratory directions.

Appendix 2: This format guides the student how to fill a laboratory report.

Listed organizations

Chapter 1
Introduction to Exercise Physiology

Introduction

Physiology science investigates the function of a living organism, which can be explained through physics, mathematics, and chemistry. Physiology has three common characters:

(a) Coordination between the cell's structure and function,
(b) Material movement, and
(c) Energy, control and regulation.

Domains include:

1. General physiology—explains the basic processes existing in a living cell: osmosis, diffusion, and convection by carriers, electrical impulse and metabolism.
2. Physiology of systems that include cells and organs that have a common function.
3. Comparative physiology between humans and animals.
4. Fucose on the effect of the surrounding on the physiological function of humans and animals.
5. Exercise physiology, is the exchange of chemical energy into mechanical energy. The uniqueness of a muscle is its ability to increase significantly metabolic rate, compared to other body organs, by the ability to increase oxygen uptake (VO_2) 20 folds from resting values.

Exercise physiology is a growing area of study in both medical (exercise medicine) and fitness fields. It can be defined as the identification of physiological mechanisms underlying physical activity, the comprehensive delivery of treatment services concerned with the analysis, improvement, and maintenance of health and fitness, rehabilitation of heart disease and other chronic diseases and/or disabilities, and the professional guidance and counsel of athletes and others interested in athletics, sports training, and human adaptability to acute and chronic exercise. The knowledge of exercise physiology can help out patients to get in better shape for all

M. S. Saghiv, M. S. Sagiv, *Basic Exercise Physiology*,
https://doi.org/10.1007/978-3-030-48806-2_1

around good health or to treat patients with a medical condition through exercise. Exercise physiology is a field of study devoted to examining the body's response to physical activity. In a nutshell, its primary focus is medical exercise. Exercise physiology focuses on how the body reacts to different exercise modes that increase work intensities, and, thus, metabolic demand during acute and chronic exercises. Topics of interest for someone working in the field of exercise physiology may include human energy expenditure, human energy transfer, and environmental effects on physiology. Many exercise physiologists dedicate research time to define; exercise influences the cardiovascular system, endocrine system, and musculoskeletal system. The exercise physiology programs concerning healthy habits for patients are developed in order to enable them to manage their medical conditions in the most efficient possible way. Exercise physiology can be divided into two primary areas of specialty: fitness in health and rehabilitation in disease. Exercise physiologists who focus on fitness may work with professional athletes to improve their performance or counsel private clients who are trying to lose weight and get in shape. Exercise physiology professionals who specialize in rehabilitation often help people who are recovering from injuries or dealing with chronic conditions such as cardiovascular diseases by developing a customized exercise program that can be used as part of a comprehensive treatment strategy.

Exercise rehabilitation is a common form of this restoring process. Often utilized after major surgery, an accident, or any event that takes from the individual mobility or function, this form of rehabilitation braces the patient with trained personnel who help him to recover as much of his former physical competence as possible. Knowledge on the musculoskeletal system, cardiopulmonary system, and endocrine systems enables to prescribe different courses of exercise for either fitness or rehabilitation. It may help in behavioral modification programs. Understanding exercise physiology can assist medical conditions such as obesity, diabetes, cancer, osteoporosis, depression, asthma, and cardiovascular diseases, as well as many other conditions. Therefore, an exercise physiologist is an associated health professional who recommends exercise programs, also for cardiopulmonary patients. Exercise physiology deals with states of acute and chronic physiological adaptation to exercise, while in healthy subjects and athletes it improves and preserves their level of performance. Exercise is an extreme strain which the body can stand. The most important energy sources are carbohydrates and fatty acids for generating adenosine triphosphate (ATP). The ATP production at the muscle's cells depends mainly on oxygen. In the Krebs cycle, the electron that represents H+, is released from the breakdown of chemical molecule to form ATP. The anaerobic glycolytic pathway forms ATP and lactate.

Exercise can give an approximation of the practical and work ability of the body's structures. The capacity to perform physical task is distinct by the following aspects: (a) the capacity to deliver enough oxygen to the cells, (b) the ability of the metabolic pathways to use the oxygen to yield energy, (c) the neuromuscular drive comprises muscle power, (d) harmonization between muscle-nervous, (e) joint's variety of motion, and (f) inspiration strategies.

The knowledge about the broad range of exercise responses that can occur across time, intensities, and modes of exercise can improve work capacity and physical performance of humans in health and disease. Exercise alters physiological function in the organism, regardless of whether the effect is beneficial or harmful. Exercise physiology can be divided into two primary areas of specialty: fitness in health and rehabilitation in disease. It includes bioenergetics, biomechanics, biochemistry, cardiopulmonary function, skeletal muscle physiology, hematology, neuroendocrine function, and central and peripheral nervous system functions. Exercise is an extreme strain which the body can stand. Muscle contraction has both mechanical aspect that produces movement and metabolic properties producing energy.

Two typical efforts dominate exercise: isotonic-dynamic exercise that shortens the muscle fibers and static isometric exercise that is an effort without fiber shortening. However, most physical activities involve two kinds of contractions, dynamic and static, and two main energy pathways, aerobic metabolism and anaerobic metabolism. Thus, activities tend to be classified on the basis of their dominant mechanical and/or metabolic characteristics.

Exercise can give some estimation of the different body's systems range and capabilities. It is, however, important to understand that the body's performance depends on the mode of exercise and more important if oxygen is available (aerobic) such prolonged running or without enough oxygen (anaerobic) such as 100 m dash or isometric bout. Therefore, the physiological responses of exercise demand coordinate between the heart, lungs (cardiopulmonary), and skeletal muscle functions (mechanical and metabolic). This has to meet the increased metabolic demands in order to deliver oxygen to the working muscles and to eliminate CO_2 to the atmosphere. Since the responses of the cardiopulmonary system to the metabolic demands are accurate, it enables to predict the cardiopulmonary responses during aerobic exercise.

Cardiovascular adaptations to exercise have been systematically defined and differ according to the type of conditioning: aerobic or anaerobic. Endurance training is a kind of isotonic contraction and sometimes also described as dynamic contraction or aerobic exercise such as long-distance running and swimming. Strength training is also referred to as static, isometric, power, or anaerobic exercises such as wrestling, weightlifting, or throwing heavy objects. Long-term cardiovascular adaptation to chronic dynamic training produces increased VO_2max due to increased cardiac output (Q) and arteriovenous oxygen difference (a–v)O2. Long-term adaptation to isometric training produces an increase in strength and muscle mass, however, is not beneficial to the cardiovascular system.

Performing vigorous exercise involves a critical role in the coordination between multiorgan systems and the cardiovascular system. In order to respond to the increased oxygen demand by working muscles during repeated physical activity of high intensity following training, the myocardium undergoes profound morphological alterations, functional and electrophysiological alterations, sometimes known as left ventricular hypertrophy or athlete's heart. These changes are generally representing a physiological adaptation to systematic training [1].

Modes of Exercise

Modes of exercise can be an important part of sports performance [2], part of healthy lifestyle, physical therapy, diabetes, depression, rehabilitation, physiotherapy, and weight loss. There are four main modes of exercise though, this book focuses on two most important modes for fitness, health, and wellness: aerobic and anaerobic or isometric or resistance exercises.

(a) aerobic exercise utilizes that uses large muscle mass and causes increase significantly, oxygen demand by the contracting muscles and delivery by the cardiopulmonary system. The aim of aerobic exercise is to increase cardiopulmonary and oxygen extraction by the skeletal muscles. Running, cycling, swimming, and walking are examples for aerobic exercise mode,
(b) Anaerobic exercise, which includes strength and resistance training, increases strength and muscle mass, as well as improves bone density. Strength is improved by exercises that include push-ups, pull-ups, lunges, squats, and bench press. Anaerobic exercise also includes weightlifting training, isometric training, eccentric training, interval training, sprinting, and high-intensity interval training.

Cardiovascular mortality is tide to sedentary lifestyle, which is an independent risk factor for the development of coronary artery disease. Exercising regularly keeps body weight, increases cardiovascular functional capacity, and reduces all-cause mortality. Caloric expenditure needed to gain benefits from exercise is about 700–2000 kcal week^{-1}, by exercising 3 week^{-1} at intensities between 50% and 60% maximal oxygen uptake.

Energy for Exercise

Exercise requires energy, both during aerobic and anaerobic respirations, that include chemical reactions in the cell to produce ATP, which is needed for mechanical energy [3]. Aerobic respiration takes place in the mitochondria and requires oxygen, glucose, and fat. Therefore, low to moderate intensities can last for long periods. When workload intensity is above 80% of maximal oxygen uptake (VO_2max), lactate accumulates. Exercising aerobically at intensities below the anaerobic threshold (AT), blood lactate levels remain low [4], compared to that high intensity workload where anaerobic metabolism supplies ATP. Anaerobic pathway utilizes glucose and glycogen only [5]. But the best results are achieved when a workout includes both energy sources: aerobic and anaerobic [6]. During aerobic exercise with adequate metabolism and oxygen, muscle cells can produce an unlimited amount of ATP and, therefore, contract repeatedly without fatigue [7]. This is not the case with anaerobic exercise; during high-intensity exercise, muscle cells rely on other sources to produce ATP, namely, the breakdown of carbohydrates in the plasma cell that do not require oxygen [8]. Aerobic exercise means “with oxygen”, following long-lasting training it increases endurance and cardiac function while anaerobic exercise increases lean muscle mass. So, it is possible to summarize the difference between aerobic and anaerobic exercises in one word: oxygen. Each type of exercise is different in its energy expenditure and oxygen utilization. However, all activities are composed of more than one mode. For example, many endurance activities also build strength [9] however, not to the levels achieved following strength exercise training. Aerobic exercise utilizes the Krebs cycle to produce ATP, CO_2, and H_2O [10] and is related to cardiopulmonary and hemodynamic changes in addition to the low force opposing.

Myocardial Contractility

During static exercise or resistance exercise, muscle contraction is without or with minimal movement, while force is generated [11]. During anaerobic exercise, oxygen is not available; it is an exercise mode where the individual comes to be out of breath within a few minutes [12]. It consists of brief intense bursts of physical activity such as during heavy weightlifting loads or when performing all-out 200 m dash. Isometric exercise differs significantly from aerobic respiration, and muscles produce ATP by the anaerobic pathways without the presence of oxygen, where oxygen demand exceeds oxygen supply ability [5]. Therefore, during muscle contractions without oxygen, the cell produces a small amount of ATP and large amounts of lactate, which causes distress and fatigue [12], with minor changes in the cardiopulmonary system. For each molecule of glucose metabolized in the mitochondria through the Krebs cycle in the presence of oxygen (aerobic respiration), 38 ATP molecules are produced, while anaerobic respiration produces only two net ATP from four pyruvic acid molecules through the glycolytic pathway [3].

Anaerobic exercise is fueled in muscles by energy stored in phosphagen molecules and through a process called glycolysis [13].

Exercising 3–4 times week^{-1} 45 min^{-1} with an energy expenditure of 3500–4000 kcal week^{-1} will increase significantly individual’s physical fitness capacity [14]. These amounts of energy expenditure are sufficient to reduce significantly all-cause mortality and cardiovascular disease in middle-aged and elderly individuals [15]. On the other hand, physical inactivity is a leading cause of disease and disability, and it increases all causes of mortality, cardiovascular risks factors which may cause cardiovascular diseases, diabetes, obesity, high blood pressure (BP), and lipid disorders [16].

Exercise Benefits

Long-lasting exercise training has an advantageous effect through body organs and systems. Thus, it advances individual's well-being and maximal work capacity. Beneficial exercise is any physical activity that improves or keeps physical fitness and general health and work capacity. Exercising is done for numerous beneficial reasons, together with increasing growth and development, it slows down aging, increases muscular strength and cardiovascular system functional capacity, enhances athletic skills, facilitates weight loss or bodyweight maintenance, improves health, and also for leisure activity and fun. Exercise training increases physical fitness, which is defined as the work capacity to perform daily activities [17] and the ability to compete in different sports events of highest level intensities [18]. Physical fitness represents a set of qualities which individuals have or attain in association with the capacity to do physical activity tasks [19].

Physical fitness has four components that include the (a) cardiopulmonary functional capacity, (b) muscular power, (c) muscle strength. Any limitation in the production of muscle strength will interfire with the generation of muscular power [20] and (d) muscular endurance, which is the ability of a set of muscles to perform repeated contractions, against resistance for an extended period of time without tiring [21]. Additional components of physical fitness include body composition that describes the percentages of body's fat, bone, water, and muscle and, finally, flexibility (stretching) denotes the range of movement in a joint or series of joints and length in muscles that cross the joints to induce a bending movement or motion. Flexibility varies between individuals, particularly in terms of differences in muscle length of multijoint muscles.

Physical Work Capacity

Physical work capacity may be attained by changes in lifestyle that include proper nutrition [22] and moderate to vigorous exercise [23].

Physical fitness demands are considered as measures of the body functional capacity during sports and physical activities, in addition to being fit and healthy [24]. Functional capacity evaluation is a set of tests, practices, and observations that are combined to determine the ability of the evaluated individual to function in a variety of physical tasks under changing circumstances.

Table 1.1 [25–27] suggests that when an individual is physically active, he preserves and improves its physical work capability by increasing its VO_2max. Therefore, contributing to entire health maintains bone modification and conservation, muscular function and mass, joint elasticity, and thus, power, by fiber hypertrophy, neural changes, and intensification of the immune system reaction. It has been previously suggested that exercise training may increase longevity and quality

Table 1.1 Hemodynamic and muscular function changes following exercise training

Variables	Aerobic exercise	Resistance exercise
Maximal oxygen uptake	Increase	Unchanged
Maximal minute ventilation	Increase	unchanged
Cardiac output	Increase	Unchanged
Heart rate rest and maximal	Decrease	Increase or unchanged
Stroke volume rest and Maximal	Increase	Unchanged
Systolic blood pressure rest	Unchanged or decrease	Increase
Diastolic blood pressure	Unchanged or decrease	Increase
Mitochondrial capacity	Increase	Unchanged
Mitochondrial size and number	Increase	Unchanged
Basal metabolic rate	Increase	Unchanged
Anaerobic Capacity	Increase	Increase
Lactate tolerance	Increase	Increase
Capillary density	Increase	Unchanged
Muscle mass	Unchanged or decrease	Increase
Strength	Unchanged	Increase
Skeletal muscle hypertrophy	Unchanged or decrease	Increase
Circulating s-Klotho	Increase	Unchanged

of life by increasing blood circulating s-Klotho [28, 29]. Furthermore, a lesser death rate was found in individuals who participate in moderate to high exercise intensities compared to individuals who by comparison were sedentary [30]. Following moderate intensity of lasting exercise training decreased inflammatory incidents thus reducing aging rate [31] and improves physical work capacity [26], increases muscle size or hypertrophy from resistance workloads [25, 27]. The epigenetic response to exercise is an important physiological key in explaining the exercise effects [32]. Studies have shown that exercising in middle age improves significantly physical work capacity in later age [33]. In addition, it has been suggested that trained individuals have faster regulation of post-exercise metabolism when exercising at either the same relative or absolute work rate [34].

Exercise and Cardiovascular Disease

Cardiovascular mortality and diseases are linearly related to the independent risk factor of physical inactivity or low levels of exercise intensities [35, 36]. Studies have shown that heart disease is the principal reason for death for women [37] however, regular aerobic exercise in aging women leads to healthier cardiovascular profiles [38]. Exercising aerobically at intensities above the AT, usually above 55% of VO_2max at any age, produces beneficial effects on cardiovascular disease and mortality [39]. Individuals who modify their behavior after myocardial infarction to include regular exercise have improved rates of survival [40], opposite from those inactive individuals, with the highest probability for cardiovascular disease mortality [41]. According to the American Heart Association, exercise reduces the risk of cardiovascular diseases, including myocardial infraction and stroke [42].

Pressure Overload and Volume Overload

Left ventricular hypertrophy might occur following hemodynamic alterations and forces opposing left ventricular ejection, during both aerobic and resistive exercises. Increased afterload or pressure overload, developed following chronic isometric or resistance training, induces thickening of the ventricular walls with unchanged or reduced internal dimension (concentric left ventricular hypertrophy). It has been suggested that these myocardial adaptations serve to normalize wall stress [43, 44], along with significant increase in wall thickness [45]. In addition to the high force opposing myocardial ejection, compared to resting values, during isometric maneuver, the hemodynamic variations include a slight elevation of minute ventilation (VE), oxygen uptake (VO_2), cardiac output (Q), due to the increase of heart rate (HR), while stroke volume (SV) decreases, increases, or remains unchanged [46]. However, a more pronounced rise in SBP, DBP, and thus, MABP [47] are seen, resulting in a significant increase in TPR to levels almost four times that recorded during aerobic exercise. Therefore, during isometric or resistance exercise, the left ventricular ejects against high opposing force meaning, pressure overload [48].

This is not the case during aerobic exercise where following long-lasting aerobic exercise training, volume overload may lead to expansion of the left ventricular diameter cavities, internal diameters, and a proportional increase of wall thickness [45]. This type of adaptation is termed eccentric left ventricular hypertrophy [44]. On the other hand, during aerobic exercise force development is low while the main objective is movement [49]. During aerobic bout, compared to resting values, there are increases in VA, VO_2, and SBP, while DBP remains unchanged or slightly decreases. In addition, a significant increase in Q is noted following the increase in SV and maximal HR, while TPR drops due to the BP response [5]. Therefore, during aerobic exercise left ventricular ejects against low forces meaning, volume overload [48].

Muscular Function

During aerobic exercise, the required energy supply is significantly above that of the resting values. This requirement by the working muscles is provided through the aerobic pathway [50]. Muscle contraction has both kind of energy: mechanical that produces movement and chemical reactions generating energy (ATP). Muscle cells comprise actin and myosin which produce contraction by sliding past one another, and thus, fiber's length and form of the myofibril change. This mechanical action follows chemical reactions that split ATP molecules at the myosin heads to produce force and motion [51].

During long-lasting aerobic (isotonic) exercise, individuals rely mainly on aerobic energy metabolism [52]. While static depends mainly on glycolysis. Isometric exercises are characterized by the ability of the muscle to oppose equally loads without net movement during contraction [53]. However, most physical activities involve both dynamic (aerobic metabolism) and static (anaerobic metabolism) contractions [54].

Oxygen Uptake and Muscle Fibers

VO_2 is the amount of oxygen extracted from the blood by the muscles, which enters the working muscle cells. At rest with under normal conditions and Q, oxygen availability is more than satisfactory to meet cell's requirements to produce ATP by aerobic metabolism. Aerobic exercise utilizes oxygen during small (arm cranking) and large muscle mass (running, walking, and swimming) with low to moderate intensity and may last for an extensive period [52]. Aerobic exercise lasts 20–40 minutes at least at intensities between 60 and 80% of VO_2max [55]. This is achieved by increasing oxygen delivery to the working muscles, by the circulatory system, namely, Q and the increased ability of the working muscles to extract oxygen from the delivered blood [56]. Main aerobic activities include walking, biking, jogging, and swimming.

Figure 1.1 describes the linear relationship between workload and oxygen uptake during maximal work capacity test (VO_2max). A subject goes gradually, from a calm exercise on an ergometer mainly treadmill or bicycle ergometer, through to fatigue [57]. The increase in VO_2 during exercise is due to the increase in demand for ATP, mainly, for muscle contraction and the movement of numerous materials into and out of the cells.

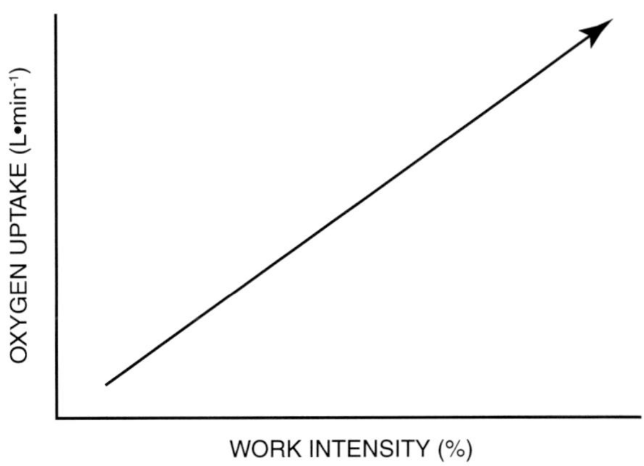

Fig. 1.1 Linear relationship between workload and oxygen uptake. With an increase in exercise intensity, the oxygen uptake increases equivalently

The increase in VO_2 during exercise is due to the increase in demand of ATP mainly, for muscle contraction and the movement of numerous materials into and out of the cells.

Muscle Fibers

There are numerous muscle fiber types. Between these fiber muscles, some of them are voluntary while others are involuntary. Skeletal muscles have the voluntary contracting muscles which in general changes body position and motion. These skeletal muscles have two types of fibers, namely, slow twitch fiber (type I) and fast twitch fiber (type II). Usually, a human body comprises a relatively identical quantity of these two fibers. Human muscles contain a genetically determined mixture of both slow and fast fiber types. In the general population, slow twitch and fast twitch fibers are utilized for movement. In skeletal muscles, both types are almost equal in their quantity used for movement. But, 100, 200, and 400 m runners have significantly higher fast twitch fiber content, while the opposite is found in long-distance runners that have a very high content of slow twitch fibers. This difference between the two fiber kinds is genetically determined [58]. Thus, genetics has a high impact on individuals to choose sports disciplines that suit their muscular fiber content [59]. The nature of metabolic pathway utilized and exercise intensity affect the extent of an exercise activity. During anaerobic exercise, the breakdown of phosphagen molecules and glycolysis produce little amount of ATP. Therefore, the extent of anaerobic bouts is short in length [60]. During this workout the individual relies on fast-twitch muscle fibers to perform high intensity activity. In contrast, aerobic exercise produces significantly higher amount of ATP based on slow-twitch muscle fibers. This enables rhythmic aerobic exercise utilizing large muscle mass for long periods of activity. Thus, the unlike metabolic pathway determines the variances between anaerobic and aerobic activities. Aerobic exercise does not accumulate lactate using muscle slow twitch fiber (red fibers or fiber I) that allows lengthy muscle contractions long-lasting duration [61]. The slow twitch fiber impulse progresses slowly compared to the fast twitch fiber (white fiber, or fiber II) and can go for a long time before fatigue is reached. The main difference between both kinds of fibers is that the slow twitch is mostly aerobically and used for long-lasting endurance activity, such as long-distance activities [62]. Slow twitch fibers do not become worn-out easily compared to fast twitch fibers that are used anaerobically, for a powerful burst of movements such as weightlifting and sprinting. However, fast twitch fibers cannot continue contractions for long periods since they become exhausted rapidly [63]. Therefore, slow twitch fibers help athletes to run marathons and bike for hours. Aerobic capacity defines the practical capacity of the cardiopulmonary system [64], which is the maximal oxygen uptake (VO_2max) typically achieved, at peak strenuous exercise [65].

Aerobic and Anaerobic Capacities

Aerobic capacity is defined by three physiological factors: VO_2max, AT and the efficient use of the three energy systems. Profits of aerobic exercise are as follows: reduced body's lipids thus increases lean body mass, delays fatigue stage, increases VO_2max, increases red blood cell number and blood volume, reduces cardiac risk factors, and alleviates cardiovascular disease problems.

In addition, it reduces daily tension, depression, and anxiety and improves general stamina and mood.

Anaerobic exercise is strenuous and short exercise bout lasting from seconds up to about 2 min, sufficient to cause anaerobic metabolism [66], with none to slight positive effect on the cardiovascular system [67]. Anaerobic pathway, or anaerobically energy spending is a normal share of whole-body metabolic energy expenses, any additional enrollment of muscle fiber II ends up in augmentation of anaerobic energy cost. Because fast twitch fibers use anaerobic metabolism to produce ATP, however, they reach fatigue stage faster than fiber I. Fiber II usually produces the similar quantity of power per contraction to fiber I, but it is able to react faster. Having faster twitch fibers can be an advantage to athletes since they want to rapidly produce high level of strength [68].

During anaerobic and resistance exercises, oxygen is not important since ATP production is achieved through the glycolytic pathway. To perform strenuous bout, ATP is produced anaerobically in the cell's cytosol, thus, it differs from aerobic exercise in which ATP is produced in the Krebs cycle that occurs in the mitochondria [69]. Yet, it increases muscle and bone strength and muscle mass. This is very vital for elderly individuals with regard to sarcopenia and osteoporosis [70, 71]. Weightlifting is an anaerobic exercise where the muscles use the glycolytic pathway; therefore, it does not utilize large quantities of oxygen due to the use of white muscle fibers (fast twist or fiber II) and has a very brief maneuvering time [72]. Weightlifting training augments daily energy expenditure throughout the exercise bouts by the large muscle mass involved [73]. Aerobic and resistance exercises increase metabolism and aerobic bout increases calories utilized, while weightlifting training increases lean body mass which in turn increases caloric expenditure at rest [74].

Strenuous Exercise

Strenuous rhythm exercise increases significantly blood lactate; thus, blood pH decreases from 7.4 (alkaline) value at rest to acidic 6.5 and, thus, becomes too acidic due to muscle metabolism that is deprived of oxygen, leading to cell hypoxia and lactic acidosis. This is primarily due to reliance on the glycolytic pathway systems and the hydrogen ion (H+) accumulation. In addition, anaerobic activities such as weightlifting and strenuous running interval training have less or none effect on oxidative energy metabolism and endurance capacity [75].

Exercise that changes physiological function in humans can have beneficial or damaging effects as in the case of overtraining, arising when an individual top its work capacity to recover from long strenuous exercise performance. At this point, the individual experience decreased work performance a workload above his recovery capacity. Individual who is overstrained stops increasing work capacity and sometimes loses strength and fitness and suffers from chronic fatigue [76].

Vigorous exercise has acute and chronic impacts on the immune system. In the resting state, the immune systems of athletes and sedentary individuals are more alike than, it become dissimilar with the exclusion of NK cell activity, which tends to be elevated in athletes.

Energy Interplay

When oxygen is present, the muscle produces chemical energy in the form of ATP molecules, used for muscle contraction (mechanical energy). During aerobic or anaerobic exercise bouts, both energy pathways donate to yield ATP molecules; however, one pathway regularly dominates. Thus, it is suitable to consider the metabolic pathways as a constant mixture of energy source [77]. Figure 1.2 points out the donation portion (percentage) of each metabolic pathway during various physical exercises

It is important to understand that the aerobic performance depends on oxygen availability and mode of exercise [78]. If the exercise bout lacks oxygen, then the activity is defined as anaerobic or resistance bout. Subsequently, there is a need for coordination between the cardiopulmonary system delivery and skeletal muscle

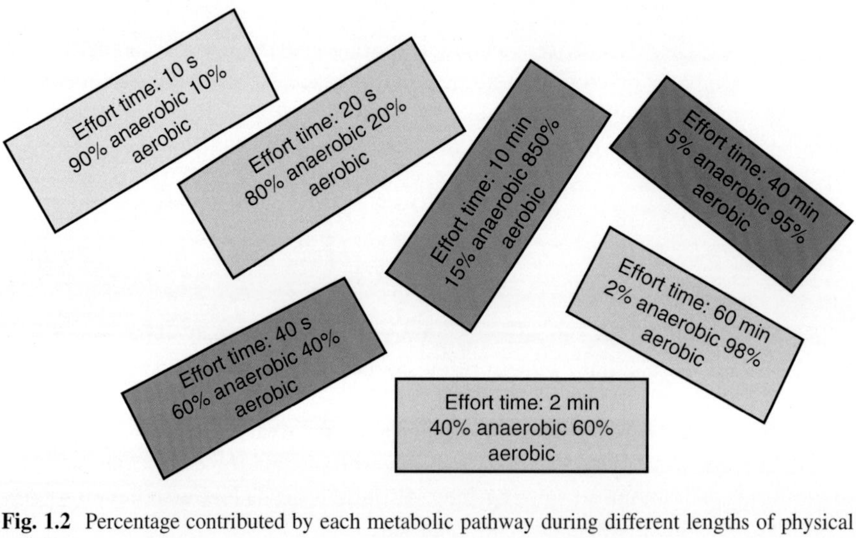

Fig. 1.2 Percentage contributed by each metabolic pathway during different lengths of physical activity

demand. Strenuous exercise involves multiorgan system synchronization, mainly by the circulatory system, to increase oxygen delivery to the working muscles [79]. In the case of exercising aerobically, increased metabolic demands, oxygen delivery extraction has to be balanced and eliminate or reduce CO_2 production. The cardiopulmonary response to increased metabolic demand is precise and predictable to be accurate during aerobic exercise. In sedentary individuals, attempting to perform strenuous anaerobic bout has a significant essential risk for damage [80].

The Pulmonary Function

Two most important functions of the pulmonary during exercise are oxygen delivery to the working muscles and CO_2 removal from the tissues. In addition, the pulmonary maintains arterial oxygen saturation, acid–base balance, fluid homeostasis, temperature balance, and even hormonal regulation. Lung's minute ventilation (VE) that includes the dead space (VD) and alveoli ventilation (VA) diffuses oxygen to the blood, which is propelled to the left atrial and left ventricle, while pulling by diffusion CO_2 from the venous blood to the atmosphere [5]. When exercising at maximal intensity VE increases to meet the exercise demands by an increase in VD and an increase in breathing rate (f). VO_2, CO_2 production, and VA increase by about 20–30-fold compared to the resting values [81]. A curvilinear relationship exists between VO_2 and VE (Fig. 1.3), while a linear relationship exists between CO_2 and VE (Fig. 1.5).

At maximal exercise, untrained individuals' breath frequency is up to 60 breaths min^{-1} (f), VE is 100–130 L min^{-1}, while maximal VE capacity is 150–180 L min^{-1}. Elite athletes can reach VE up to 220 L min^{-1} (Fig. 1.4). VO_2max is the rate of VO_2 at peak all-out strenuous exercise. VO_2max is determined by genetics [82] and long-lasting aerobic training [83].

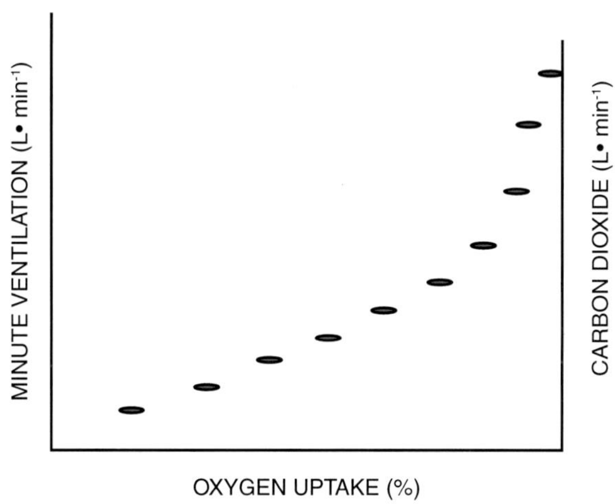

Fig. 1.3 Curvilinear relationship between minute ventilation and oxygen uptake. With increase in exercise intensity, minute ventilation increases inequivalently

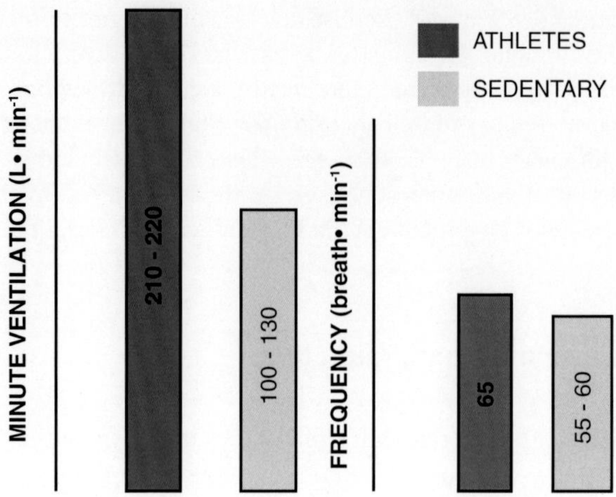

Fig. 1.4 Maximal minute ventilation and breathing frequency of sedentary and elite athletes at peak aerobic exercise

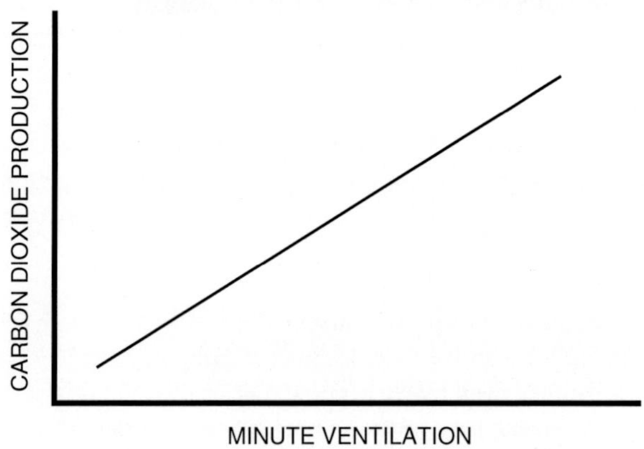

Fig. 1.5 Linear relationship between minute ventilation and carbon dioxide. With increase in exercise intensity, minute ventilation increases equivalently

During aerobic exercise, increased blood flow through the lungs causes more pulmonary capillaries to be opened, providing a larger surface for diffusion into and out of the pulmonary blood capillary [84]. VE increases linearly with CO_2 production (Fig. 1.5) controlled by an electrical impulse from the pulmonary nervous system [85].

Maximal VO_2 Values

Usually body VO_2 is measured in L min^{-1}, however, this value does not allow comparisons of individuals' body size, gender and age to be accurate. Thus, measure by kg body weight per minute (mLO_2 kg^{-1} min^{-1}) is used. The values at peak aerobic exercise, for the average young sedentary male, are approximately 35–50 mLO_2 kg^{-1} min^{-1}. World-class trained male athletes have values of 80 to 90 mLO_2 kg^{-1} min^{-1}. VO_2max is affected mainly by Q during aging, and VO_2max decreases due to the biological process that reduces maximal HR. Muscle is the second aspect that also determines VO_2 by the amount of oxygen extracted (a–v)O2 by the cells [56].

The Cardiovascular System

The cardiovascular system is a closed-loop system, where flow results from pressure gradient between the left ventricle pressure of 90 mmHg and right atrium pressure that is around 0 mmHg [86]. The myocardium alternates periods of relaxation and contraction by precise timing with the opening and closing of its valves. Relationship between systemic vascular resistance and left ventricular contractility ability affects blood flow [87] and, thus, oxygen to the cells. Reduced systemic vascular resistance during aerobic exercise facilitates oxygen delivery and blood flow to the working muscle.

Cardiovascular adaptations to exercise have been methodically well-defined and vary based on the type of training: (a) aerobic training occasionally labeled as dynamic or isotonic exercise such as long-distance running and swimming and (b) strength training also referred to as static, isometric, power, or anaerobic such as wrestling, weightlifting, or throwing heavy objects. Aerobic training results in functional and electrophysiological alterations, increases myocardial efficient that in turn increases SV to its maximum, and thus, increases VO_2 for a given workload [88]. When maximum SV is increased, it allows to have a lower HR for a given workload. The reduced HR increases ventricles filling time and once more keeps large SV. This results in an additional ability to increase Q with an increase in demand for oxygen by increasing HR [27]. In addition, the adaptation of the myocardium includes the reduction in resting HR as well during submaximal aerobic exercise. In addition, long-term cardiovascular adaptation to dynamic training produces increased VO_2max due to increased Q and (a–v)O2 [89]. Long-term cardiovascular adaptation to isometric training produces an increase in strength and skeletal muscle mass [90] without influence on SV, HR, Q, and (a–v)O2.

Following aerobic training, sedentary individuals increase significantly oxygen delivery (Q) from 5–6 to 24–25 L min^{-1} during exercise to working muscles, due to a substantial increase from resting to peak exercise values in SV from 70–80 to about 110–120 mL $beat^{-1}$, HR from 60–70 to 220 beats min^{-1} [91], SBP from 120 to 210 mmHg, and DBP remains unchanged or slightly reduced. Oxygen (a–v)O2 by the working muscle increases from 40 mLO_2 L^{-1} of blood at rest to about 180 mLO_2 L^{-1} of blood at peak strenuous aerobic bout [92].

Increase in the efficiency of cardiovascular system by delivering more oxygen to the working muscles is due to catecholamine release that increases significantly SV and HR [93]. In addition, modes of exercise, intensity, and duration affect cardiovascular adaptations [94]. Aerobic training such as walking, cycling, swimming, and jogging has different effects on the cardiovascular system [95] compared to strength (weightlifting) training [96] or anaerobic bouts such as 100 and 200 m dash.

Long-term cardiovascular adaptation to dynamic aerobic exercise training may produce eccentric left ventricle hypertrophy, increase left ventricular contractility and filling, and thus, increase end diastolic volume (EDV). These changes enable to increase SV at rest and during exercise and decrease HR, as Q at rest is quiet constant. During aerobic exercise in order to meet the increased muscular demand for oxygen, Q increases significantly during maximal exercise effort due to the increase in SV and HR. In addition, due to the increase in oxygen delivery, (a–v)O2 is increased at the working muscles and, consequently, VO_2max. Individuals with BP in the normal ranges experience reduced values in SBP and DBP at rest and at peak exercise [5]. Hypertensive responders show similar responses to those who do aerobic exercise training [97]. This BP response is due to the autoregulation mechanism that reduces TPR within the cardiovascular system. During running or cycling efforts that utilize large muscle mass, the metabolic costs of contracting muscles can be high and long lasting. Therefore, blood flow to the working muscles need to be increased and coordinated to the metabolic demand. In addition, the autoregulation mechanism (affected by metabolites in oxidative slow fiber muscle) increases vessel diameters, in order to avoid mismatches between Q and vascular conductance which is essential to increase Q [98].

Resistance training increases muscle strength and tone, lean body mass and muscle mass (hypertrophy), and therefore, increases basal metabolic rate, reduces risk for osteoporosis improve mobility, and balances by working against external weight or force. VO_2, BP, SV, HR, and Q differ significantly when compared to aerobic exercise. VO_2 and (a–v)O2 are not affected at all, while long-term cardiovascular adaptation to isometric training produces concentric left ventricular hypertrophy, skeletal muscle hypertrophy, and increased strength.

Anaerobic exercise is different from the term AT, which is the measurement of aerobic capacity. It is worth to point out that exercise imposes strenuous strains on the body that can stand it. For example, during marathon race, runner will increase its energy cost to a very high level. When VO_2 reaches steady state at a given workload, it echoes the energy cost of the physical activity at a constant submaximal load. Athletes with efficient motion utilize less oxygen than athletes with low efficient at the same given VO_2 steady-state rate. It has been reported that differences in movement efficiency can differ by 25–30% among trained athletes at comparable VO_2max.

Gender Effects on Physiological Differences

No gender differences exist in exercise physiology principles and measurements. However, there are important biological and behavioral differences between the two genders [99]. Differences are in body size, body composition, and hormonal responses [100, 101]. All physiological variables recorded significantly reduce in females compared to the recorded values for males in the range of 60–75% of work capacity, and when measured in terms of strength per square centimeter, female's muscle can generate the same force of contraction as that of a male.

Thermoregulation

During exercise, some influential active physiological mechanisms of heat loss prevent an excessive increase in body core temperature [102]. Because of a metabolic inadequacy in the muscles, very high portion of the energy devoted ATP production is converted to heat [103]. Adding to that heat produced, ambient temperature, and humid, it may significantly result in an imbalance between heat production and heat loss. Under these conditions, physical performance will be significantly impaired along with the risk of having heat illness [104]. The main mechanism to reduce core body temperature, with increased metabolic demand, during exercise in a hot environment, is by increasing skin sweat rate, which vaporizes from the surface of the skin to the atmosphere.

Physiological Responses During Diving

Following diving adaptations, the ability of mammals to function at depth on a breath hold relies on gas delivery (cardiopulmonary) and extraction (cell level). The outstanding breath-hold capacities of these mammals result from anatomical and physiological adaptations. These adaptations increase breathing efficiency, oxygen storage, hypoxemic, ischemic, and pressure tolerance [105].

However, humans descend underwater to relate to the environment in sport and professional activities and thus associate with the environment. Humans are not physiologically and anatomically well adapted to the ambient conditions of diving which during submersion to elevated ambient pressure (hyperbaric conditions) both, breath-hold dives and with SCUBA (breathing gas supply), significant physiological variations happened due to the change in ambient pressure and oxygen availability [106]. In addition to the increased work of breathing, during hyperbaric conditions, ratio of VE over physiological VD ratio is reduced, while gas mass and partial pressures are increased [107].

The physiological effects of the underwater ambient increase the amount of efforts used to expand the lungs, mainly the energy utilized to inhale. Suitable exchange of O_2 and CO_2 during diving needs to maintain lung ventilation due to the increase in gas density, and thus, resistance and elastic loads are augmented [108]. Increased breathing work is due to the density of the gas inhaled, problems with the breathing apparatus, and hydrostatic pressure variations due to posture in the water.

Breath-hold dives is a form of underwater diving that relies on breath holding without breathing apparatus. However, self-contained underwater breathing apparatus (SCUBA) is a suitable breathing gas supply at ambient pressure, achieved at pressure up to 71 atmospheres underwater [108]. Some aspects affect the diver, including a. immersion that distresses fluid balance, oxygen delivery and work of breathing, b. core temperature reduction results from excessive loss of body warmness, c. reduced oxygen availability, and thus, hypoxic stage. Significant increase in underwater environment pressure may result in barotrauma, that is, the most common problem for divers. It is a medical problem affected by water pressure when a scuba diver is underwater. As air molecules are significantly lighter than water molecules, the outcome of any additional descending will increase the ambient pressure. In general, barotrauma shows sinus or middle ear effects, decompression sickness, lung overpressure injuries, and external squeeze injuries [109].

The increased pressure on the pulmonary function results in gases captivated by the tissues bringing unwanted consequences, such as the formation of bubbles during decompression. It is caused by bubbles which expend in the body during or after a reduction in environment pressure [110]. In addition, the underwater ambient affects the nerve input that influences senses ability that may have consequences on diver's safety. An unexpected severe nervous system problem can cause many different symptoms. Some serious circumstances, sicknesses, and damages that can cause nervous system problems include mental health, such as anxiety disorders, depression, or psychosis [111].

Physiological Response Altitude

Height usually refers to elevations of 1000 m (3000 feet) above the sea level. The ambient conditions at high altitude are signs that differ significantly from those at the sea level by presenting physiological stress, and tolerance of modest altitude is familial for humans [112, 113]. The progress of high elevation physiology depends on the balance of oxygen demand delivery and on the reduced (as much as possible) dependency on anaerobic metabolism [114]. Ascending to high elevations drops air pressure and moisture levels, which changes the body water content by losing more wetness to the surroundings compared to the sea level [115]. When exercising under these conditions, it is very important to keep adequate hydration.

On arrival to the high elevations, individuals that perform physical work, inhale and exhale more deeply, indicating low oxygen content in the blood. With time blood plasma, total blood volume and Q are all decreased, while fluid loss, resting HR, and resting metabolic rate are all increased [116]. At high altitude, barometric pressure falls, compared to that at the sea level, and reduces oxygen partial pressure in lungs' inhaled air and, thus, resulting in oxygen deficiency in arterial blood that causes tissue hypoxia [117] that continues even following an acclimatization period. Even though hypoxia is often a pathological condition, differences in arterial oxygen levels can be normal physiological response, such as during all-out dynamic exercise. Under this condition, VE increases to introduce more molecules of oxygen in the lungs. In addition, altitude experience is associated significantly with alterations in cardiac and skeletal muscle energy metabolism: there is an increase in BP,

Q, HR, and imbalance changes in energy supply and demand, while SV remains unchanged [118]. Human inhabitants at high altitude experience hypoxia, which points out on the ability of humans to adapt to this hostile environment. Studying native Tibetan, Andean, or Ethiopian populations suggests that different genetic factors contribute to adaptive differences in these populations related mainly to hemoglobin concentration [119, 120]. In general, long-term physiological responses to high-altitude acclimatization is related to heritable behavioral and genetic changes. Inhabitants living at high altitude undergone extensive physiological and genetic changes in the regulatory systems of oxygen respiration and blood circulation [121].

Physiological and Medical Responses in Space

Medical exercise in space is the practice of medicine on astronauts in outer space through physical activity. Healthy individuals in microgravity of space have negative adaptive physiological alterations similar to accelerated aging. The atrophy of the cardiovascular system in space and continuous bed rest is 10 fold sooner than in the natural aging progression [122]. Long-duration head-down tilt bed rest is analogous to microgravity, affecting biosystems such as bone damage, muscle atrophy myocardial dysfunction, nervous system function, coordination deterioration, orthostatic intolerance, and low physical work and motion capacity [123]. In addition, the effects of 6 months space flight on astronauts boarding the international Space Station may develop neuro-ophthalmic and cerebral abnormalities [124]. Yet, those developed abnormalities are reversible; however, the question is what will happen following longer exposures. The dislocation of humans from earth to interplanetary results in deconditioning due to transitional gravitational forces [125]. This space deconditioning due to microgravity has applications as a medical risk when returning to Earth similar to inactive elderly [122]. Therefore, astronauts on board the space station exercise between 2 and 2.5 hours $\cdot$ day^{-1} to prevent bone and muscle loss occupying them a significant part of their daily routine. In space, astronauts use two modes of exercise: aerobic exercise and resistance training, utilizing three different equipment: cycle ergometer, treadmill (both aerobic), and resistance exercise device. The exercise apparatus is attached on an elevated platform for the lack of gravity which have the tendency to bring astronauts to float, and harnesses are attached to the astronauts to hold them to the bicycle and treadmill to avoid the effect of microgravity and to reduce the noise produced by the equipment.

Epigenetic Responses

In recent years, genetic research driven by genomic scans in humans and animal models has evaluated the impact of heritable factors on physical performance, health-related fitness phenotypes, aging, and individuals' responses to regular exercise. Although there is no uniform definition of epigenetics, it has the potential to explain various biological phenomena that have before now challenged whole clarification. It is a series of chemical tags that modify nicotinamide adenine dinucleotide (DNA) and its associated structures constitute the epigenome and include any genetic expression modifier independent of the DNA sequence of a gene. However, it does not involve changes in the nucleotide sequences [126].

Epigenetic factors include DNA methylations, histone modifications, and microRNAs, which can help to explain how cells with identical DNA can differentiate into different cell types with different phenotypes. Some interfaces of aging and energetic were identified, and these included molecular mechanisms that control the rate of epigenetic aging, H3K36 methylation machinery, or in contrast promote longevity such as Klotho gene [127]. The interaction of genes and exercise in modifying health status occurs at multiple levels. At the molecular level, the direct mechanism by which exercise alters gene expression involves activation of signal transduction pathways resulting in enhanced transcription of messenger RNA and subsequently increased translation into protein [128, 129].

Environmental factors including physical exercise have been shown to have a beneficial influence on epigenetic modifications, such as histone modifications and DNA methylation, which alter the accessibility to DNA and change chromatin structure, thereby regulating patterns of gene expression [126].

A common epigenetic tag found in DNA is the covalent attachment of a methyl group to the C5 position of the cytosine found in CpG dinucleotide sequences [130]. A major component of aging is significant loss of DNA methylation over time [131]. However, genome with DNA methylation in adult individuals who obtained 30 or more minutes of exercise a day had significantly more DNA methylation as compared to sedentary individuals [132]. Exercise has been shown to induce positive changes in DNA methylation within adipose tissue and regulate metabolism in both healthy and diseased individuals [132]. Exercise has been shown to induce positive changes in DNA methylation within adipose tissue and regulate metabolism in both healthy and diseased individuals [132, 133]. Increased DNA methylation of genes Hdac4 and Ncor2 has also shown to increase lipogenesis following exercise [132].

Due to the proportion of older adults expanding in the overall population, it is important to understand the competence of an individual to have the physiological capacity to perform normal everyday activities safely and independently without excessive fatigue [134].

Aging Effects on Physiological Responses

Aging is a complex multifactorial process that not only involves the natural processes of aging but also the increased risk of diseases—coronary heart disease, diabetes, and cancer [135]. Aging-related changes occur mainly in the cardiopulmonary and skeletal muscles, bringing about a reduction in physical performance [136]. Consequently, the VO_2max decreases. Such consequences contribute to the geriatric syndrome of frailty, thereby severely limiting the function, quality of life, and longevity [137].

Successful aging is a function of both genetic and environmental factors [138]. The primary aging process, itself genetically associated, occurs both independently of lifestyle and in the absence of disease [139]. An alteration may contribute to the accumulation of deleterious macromolecules and altered membranes and organelles in cells, leading to the progressive age-associated decline in the function of most physiological organs and systems including anti-aging cell repair mechanism that starts a vicious cycle [140].

Maximal work capacity is decreased with aging irrespective of lifestyle because of genetic factors. The basic mechanism by which exercise activates genes involves a stimulus signal to the DNA, then transcription via messenger RNA, and finally translation into protein [141]. Data further suggest that the epigenetic pattern may change during the aging, affecting key genes in the respiratory chain [142]. Thus, age can influence DNA methylation, gene expression, and subsequently in vivo metabolism. In contrast, chronic aerobic exercise significantly impacts positively DNA methylation, in a muscle tissue and gene-specific manner.

Reduced blood flow to the working muscles is the second factor associated with aging and thus VO_2max [143]. Untrained and trained elderly can increase the response of the central factors, i.e., cardiopulmonary without a significant reduction in (a–v)O2. It was found that in elderly subjects, skeletal muscle, mitochondrial tissue, blood flow, and oxygen exchange capacities appear to be well matched. It seems that intrinsic mitochondrial function and regulation are not altered significantly [144]. Much of the age-associated decline in oxidative function is related to the reductions in fitness and/or habitual physical activity that typically occur in this population [144]. Apart from genetic endowment, an individual must also interact with environmental factors associated with longevity. One of these factors includes maintaining high level of physical activity [145] Chronic endurance training will attenuate the decline in VO_2max associated with age [139].

Evidence from animal models has indicated a role for exercise in angiogenesis, neurogenesis, learning, and cognition [146].

Functional Capacity in Health and Disease

In healthy subjects, peak oxygen uptake or VO_2max can be limited by oxygen delivery, arterial oxygen content (hemoglobin), and/or oxygen extraction [147, 148]. During strenuous exercise that engenders a significant lactic acidosis, oxygen uptake continues to increase until exercise is terminated or exhaustion ensues [149]. The nature of the contractile and metabolic properties of skeletal muscle suggests that the speed of shortening will influence energy turnover during contractions. Indeed, it has been observed in vitro using isolated muscles and single muscle fibers that energy varies with shortening velocity [150].

In healthy subjects, it has been found that during aerobic exercises: VO_2 of male exceeds that of female due to gender-related variance in heart size [151]. Mean values for VO_2max in children, adolescents, and adults are consistently greater in males than in females, whether VO_2max is expressed in absolute terms or relative to body mass [152]. Factors linked mainly to oxygen-carrying capacity and cardiac size have been shown to contribute significantly to this gender-related difference [153].

The factor that facilitates oxygen extraction by males is the different muscle fiber recruitment patterns [154]. Males have larger muscle fibers than females but similar fiber composition, influencing metabolism and the adaptive response to exercise [155].

The mechanism that enables the females to increase their oxygen extraction during exercise is the lower velocity of contraction by skeletal muscles [156]. From in vitro studies, it has been observed that at low contraction velocities the efficiency of slow twitch fibers is higher than that of fast twitch fibers, and the reverse is observed at high speeds [157].

Left ventricular responses (delivery) and metabolic (oxygen extraction) during incremental exercise differ among men, women, adolescents, and elderly despite a similar fitness [158].

During the period of adolescence, many structural, hormonal, biochemical, and physiological changes take place that interfere with the oxygen delivery and extraction relationship. As a consequence, at peak aerobic test that creates a significant lactic acidosis [159], left ventricular contractility and function in healthy adolescent subjects may be altered. Namely, exercise may induce a myocardial phenotype that reduces Ca^{++} responsiveness during acidosis [160].

Aging-related changes occur mainly in the cardiopulmonary and skeletal muscles, bringing about a reduction in physical performance [161] Such myocardial and peripheral functional changes include a decline in the maximum HR, SV, and contractility and an increase in peripheral vascular resistance. Consequently, the VO_2max decreases.

The primary aging process, itself genetically associated, occurs both independent of lifestyle and in the absence of disease [162]. Accordingly, one may expect maximal Q to decrease with aging irrespective of lifestyle because of genetic factors. Reduced arteriovenous oxygen difference at maximal effort (Shephard 1987) is the second factor associated with decrease with aging of VO_2max.

Incremental exercise is characterized by exposing the subjects to a high degree of load which may alter the left ventricular contractility and function [163].

This has the effect of placing a large load on the left ventricular which might have significant effects on oxygen delivery to the working muscles. Thus, oxygen delivery to the working muscle may be reduced, and since metabolic demand during incremental exercise is increased over time to maximum, elderly subjects may maintain the energy supply due to the balance between oxygen delivery and extraction. Since metabolic demands during incremental exercise is increased over time, diverse populations may have different rates of reaction for energy supply between oxygen delivery and extraction [163].

Left ventricular responses (delivery) and metabolic (oxygen extraction) during incremental exercise differ among health, disease, men, and women in the life cycle, despite a similar fitness [164]. In healthy subjects, it has been found that during aerobic exercises: VO_2 of male exceeds that of female due to gender-related variance in heart size [164].

Since metabolic demands during incremental exercise are increased over time, diverse populations may maintain the energy supply in an imbalance between oxygen delivery and extraction and hence different VO_2 rates of reaction. It has also been suggested that under elevated pressure loads, a failure in the performance of the left ventricular may be affected such as reduced SV and left ventricular contractility and thus, oxygen delivery to the working muscle may be reduced [163].

Patients on dialysis have extremely limited exercise capacity, and poor physical functioning has been linked to low quality of life and high mortality in this population. The reason for the disability of patients on dialysis is far from clear despite years of study. The anemia of chronic renal disease is clearly a contributing factor, but uremic myopathy and resulting decreased muscle oxygen utilization have a significant impact on the physical functioning of patients on dialysis as well. Although it is likely that factors related to uremia adversely affect muscle function, some of the abnormalities demonstrated in uremic muscle are consistent with disuse atrophy. The clear contribution of anemia and the possible role of limited physical activity have led to studies of the effects of erythropoietin and aerobic exercise training on exercise capacity in end-stage renal disease patients. Both of these interventions result in increased exercise capacity. Thus, vigorous treatment of anemia and uremia and encouragement of physical activity are important interventions to maximize the physical functioning of patients on dialysis. In addition, more studies are needed to clarify the causes of debility in this population and the impact of interventions on physical functioning, quality of life, and mortality [165].

The mechanisms of exercise intolerance in heart failure in comparison with rest indices of cardiac performance, measures of pulmonary function, and peripheral vasodilator capacity were more closely associated with peak exercise VO_2. Furthermore, the associations were independent of each other and together accounted for 50% of the variance in peak oxygen uptake. This suggests that pulmonary and peripheral vascular adaptations may be important determinants of exercise intolerance in heart failure [166].

Exercise intolerance in patients with chronic heart failure shows no correlation with the degree of left ventricular dysfunction. This surprising finding has directed attention to peripheral changes in chronic heart failure patients. During the last few years, several different peripheral factors as determinants of exercise intolerance have been defined, i.e., abnormalities in ventilation, reduced endothelium-dependent vasodilatation of peripheral conduit and resistance vessels, and altered skeletal muscle metabolism. Skeletal muscle alterations are characterized by a reduced oxidative capacity, a catabolic state with reduced local IGF-I expression and muscle atrophy, chronic inflammation with local expression of the inducible isoform of nitric oxide synthase, and an accelerated rate of programmed cell death (apoptosis). Physical exercise training has evolved as an important therapeutic approach to influence these non-cardiac causes of exercise intolerance. After the first studies documenting the effect of aerobic training on the peripheral causes of exercise intolerance in chronic heart failure patients, the following question was asked: Should we treat the heart or the periphery to improve exercise intolerance in chronic heart failure patients? Today, we have come closer to the answer: It is now clear that these two systems are not mutually exclusive. Exercise training in chronic heart failure patients has been shown to improve skeletal muscle metabolism and function, to avert muscle catabolism, to reduce neurohumoral overactivation, to reverse endothelial dysfunction, and to contribute to the prevention of pathologic left ventricular remodeling. After 6 months of regular exercise training oxidative capacity of the working skeletal muscle increases by approximately 40%. Regular exercise training leads to a significant improvement in endothelium-dependent vasodilatory capacity of peripheral resistance vessels, thereby reducing peripheral resistance in particular during exercise. These beneficial training effects result in a small, but significant improvement of SV and reduction in cardiomegaly [166].

References

1. Maron BJ, Pelliccia A. The heart of trained athletes: cardiac remodeling and the risks of sports, including sudden death. Circulation. 2006;114:1633–44.
2. Stone M, Plisk S, Collins D. Training principles: evaluation of modes and methods of resistance training-a coaching perspective. Sports Biomech. 2002;1:79–103.
3. Rich PR. The molecular machinery of Keilin's respiratory chain. Biochem Soc Trans. 2003;31:1095–105.
4. MacRae HS, Dennis SC, Bosch AN, et al. Effects of training on lactate production and removal during progressive exercise in humans. J Appl Physiol (1985). 1992;72:1649–56.
5. Sagiv M. Exercise cardiopulmonary function in cardiac patients. London: Springer; 2012.
6. Lukacs A, Barkai L. Effect of aerobic and anaerobic exercises on glycemic control in type 1 diabetic youths. World J Diabetes. 2015;6:534–42.
7. Wells GD, Selvadurai H, Tein I. Bioenergetic provision of energy for muscular activity. Paediatr Respir Rev. 2009;10:83–90.
8. Baker JS, McCormick MC, Robergs RA. Interaction among skeletal muscle metabolic energy systems during Intense exercise. J Nutr Metab. 2010;2010:905612. https://doi.org/10.1155/2010/905612.
9. Goulet EDB, Melancon MO, Lafreniere D, et al. Impact of mild hypohydration on muscle endurance, power, and strength in healthy, active older men. J Strength Cond Res. 2018;32:3405–15.
10. Dashty M. A quick look at biochemistry: carbohydrate metabolism. Clin Biochem. 2013;46:1339–52.
11. Malisoux L, Francaux M, Nielens H, et al. Stretch-shortening cycle exercises: an effective training paradigm to enhance power output of human single muscle fibers. J Appl Physiol (1985). 2006;100:771–9.
12. Cox MM, Nelson David L Chapter 14: Glycolysis, gluconeogenesis, and the pentose phosphate pathway. Lehninger principles of biochemistry (5 ed.). W. H. Freeman & Co; 2008, pp. 527–568.
13. Barclay CJ. Energy demand and supply in human skeletal muscle. J Muscle Res Cell Motil. 2017;38:143–55.

14. Ades PA, Savage PD, Toth MJ, et al. High-caloric expenditure exercise: a new approach to cardiac rehabilitation for overweight coronary patients ades: high-caloric exercise overweight coronary patients. Circulation. 2009;119:2671–8.
15. Thompson PD, Bucner D, Pina IL, et al. Exercise and physical activity in the prevention and treatment of atherosclerotic cardiovascular disease. Arterioscler Thromb Vasc Biol. 2003;23:1319–21.
16. Knight JA. Physical inactivity: associated diseases and disorders. Ann Clin Lab Sci Summer. 2012;42:320–37.
17. Kylasov A, Gavrov S. Diversity of sport: non-destructive evaluation. Paris: UNESCo: Encyclopedia of Life Support Systems; 2011. p. 462–91.
18. Iaia FM, Bangsbo J. Speed endurance training is a powerful stimulus for physiological adaptations and performance improvements of athletes. Scand J Med Sci Sports. 2010;20(Suppl 2):11–23.
19. Nieuwenburg-van Tilborg EM, Horstman AM, Zwarts B, et al. Physical strain during activities of daily living of patients with coronary artery disease. Clin Physiol Funct Imaging. 2014;34:83–9.
20. Macaluso A, De Vito G. Muscle strength, power and adaptations to resistance training in older people. Eur J Appl Physiol. 2004;91:450–72.
21. Hughes DC, Ellefsen S, Baar K. Adaptations to endurance and strength training. Cold Spring Harb Perspect Med. https://doi.org/10.1101/cshperspect.a029769.
22. Despres JP. Physical activity, sedentary Behaviours, and cardiovascular health: when will cardiorespiratory fitness become a vital sign? Can J Cardiol. 2016;32:505–13.
23. So C, Pierluissi E. Attitudes and expectations regarding exercise in the hospital of hospitalized older adults: a qualitative study. J Am Geriatr Soc. 2012;60:713–8.
24. Gault ML, Willems MET. Aging, functional capacity and eccentric exercise training. Aging Dis. 2013;4:351–63.
25. Cholewa J, Guimaraes-Ferreira L, da Silva TT, et al. Basic models modeling resistance training: an update for basic scientists interested in study skeletal muscle hypertrophy. J Cell Physiol. 2014;229:1148–56.
26. Konopka AR, Harber MP. Skeletal muscle hypertrophy after aerobic exercise training. Exerc Sport Sci Rev. 2014;42:53–61.
27. Gielen S, Schuler G, Adams V. Cardiovascular effects of exercise training molecular mechanisms. Circulation. 2010;122:1221–38.
28. Manini TM, Pahor M. Physical activity and maintaining physical function in older adults. Br J Sports Med. 2009;43:28–31.
29. Saghiv MS, Goldhammer E, Sagiv M, et al. Effects of aerobic exercise training on S-Klotho in young and elderly. J J Physiology. 2015;1(1):001.
30. Kokkinos P. Physical activity, health benefits, and mortality risk. https://doi.org/10.5402/2012/718789.
31. Garatachea N, Pareja-Galeano H, Sanchis-Gomar F, et al. Exercise attenuates the major hallmarks of aging. Rejuvenation Res. 2015;18:57–89.
32. Brown WM. Exercise-associated DNA methylation change in skeletal muscle and the importance of imprinted genes: a bioinformatics meta-analysis. Br J Sports Med. 2015;49:1567–78.
33. Laine MK, Johan G, Eriksson JG, Kujala UM. Effect of intensive exercise in early adult life on telomere length in later life in men. J Sports Sci Med. 2015;14:239–45.
34. Short KR, Sedlock DA. Excess postexercise oxygen uptake and recovery rate in trained and untrained subjects. J Appl Physiol. 1997;83:153–9.
35. Lyerly GW, Sui X, Lavie CJ, et al. The association between cardiorespiratory fitness and risk of all-cause mortality among women with impaired fasting glucose or undiagnosed diabetes mellitus. Mayo Clin Proc. 2009;84:780–6.
36. Booth FW, Roberts CK, Thyfault JP, et al. Role of inactivity in chronic diseases: evolutionary insight and pathophysiological mechanisms. Physiol Rev. 2017;97:1351–402.

37. Xu JQ, Murphy SL, Kochanek KD, et al. Deaths: final data for 2013 Cdc-pdf [PDF-7.3M]. National vital statistics report. 2016:64(2).
38. Piercy KL, Troiano RP, Ballard RM, et al. The physical activity guidelines for Americans. JAMA. 2018;320:2020–8.
39. Witvrouwen I, Van Craenenbroeck EM, Abreu A, et al. Exercise training in women with cardiovascular disease: differential response and barriers – review and perspective. Eur J Prev Cardiol. 2019; https://doi.org/10.1177/2047487319838221.
40. Chang P, Nead KT, Olin JW, et al. Effect of physical activity assessment on prognostication for peripheral artery disease and mortality. Mayo Clin Proc. 2015;90:339–45.
41. Martinez-Gomez D, Guallar-Castillon P, Higueras-Fresnillo S, et al. Physical activity attenuates total and cardiovascular mortality associated with physical disability: a national cohort of older adults. J Gerontol A Biol Sci Med Sci. 2018;73:240–7.
42. Myers J. Exercise and cardiovascular health. https://doi.org/10.1161/01.CIR.0000048890.59383.8DCirculation
43. Fagard RH. The ALLHAT trial: strengths and limitations. J Hypertens. 2003;21:229–32.
44. Mihl C, Dassen WRM, Kuipers H. Cardiac remodelling: concentric versus eccentric hypertrophy in strength and endurance athletes. Neth Heart J. 2008;16:129–33.
45. Rawlins J, Bhan A, Sharma S. Left ventricular hypertrophy in athletes. Eur J Echocardiogr. 2009;10:350–6.
46. Runte K, Brosien K, Salcher-Konrad M, et al. Hemodynamic changes during physiological and pharmacological stress testing in healthy subjects, aortic stenosis and aortic coarctation patients – a systematic review and meta-analysis. Front Cardiovasc Med. 2019;6:43. https://doi.org/10.3389/fcvm.2019.00043.
47. Sagiv M, Hanson PG, Ben-Sira D, et al. Direct vs indirect blood pressure at rest and during isometric exercise in normal subjects. Int J Sports Med. 1995;16:514–8.
48. Carabello B, Zile MR, Tanaka R, et al. Left ventricular hypertrophy due to volume overload versus pressure overload. Am J Phys. 1992;263(4 Pt 2):H1137–44.
49. Ertelt T. Walking with chronic non-specific low back pain – a failed strategy: what can we learn from sports? Med Hypotheses. 2014;82:601–5.
50. Voet D, Voet JG, Pratt CW. Fundamentals of biochemistry: life at the molecular level. 4th ed. Hoboken, NJ: Wiley; 2013.
51. Nishikawa KC, Lindstedt SL, LaStayo CP. Basic science and clinical use of eccentric contractions: history and uncertainties. J Sport Health Sci. 2018;7:265–74.
52. Plowman SA, Smith DL. Exercise physiology for health, fitness, and performance: Lippincott Williams & Wilkins; 2007. p. 61.
53. Oranchuk DJ, Storey AG, Nelson AR, et al. Isometric training and long-term adaptations: effects of muscle length, intensity, and intent: a systematic review. Scand J Med Sci Sports. 2019;29:484–503.
54. Ehrsam JK, Kerrigan DJ, Keteyian SJ. Advanced exercise physiology: essential concepts and applications. Human Kinetics: Champaign IL; 2018.
55. Schmolesky MT, Webb DL, Hansen RA. The effects of aerobic exercise intensity and duration on levels of brain-derived neurotrophic factor in healthy men. J Sports Sci Med. 2013;12:502–11.
56. Montero D, Díaz-Canestro C. Endurance training and maximal oxygen consumption with ageing: role of maximal cardiac output and oxygen extraction. Eur J Prev Cardiol. 2016;23:733–43.
57. Gronwald T, Hoos O, Ludyga S, et al. Non-linear dynamics of heart rate variability during incremental cycling exercise. Res Sports Med. 2018;24:1–11. https://doi.org/10.1080/15438627.2018.1502182.
58. Baguet A, Everaert I, Hespel P, et al. A new method for non-invasive estimation of human muscle fiber type composition. PLoS One. 2011;6(7):e21956. https://doi.org/10.1371/journal.pone.0021956.

59. Fulda KG, Lykens K. Ethical issues in predictive genetic testing: a public health perspective. J Med Ethics. 2006;32:143–7.
60. Ben-Sira D, Goldhammer E, Saghiv M, et al. Effects of all out-cycling bouts on left ventricular function in master cyclists. J J Geronto. 2017;3(1):025.
61. Goodwin ML, Harris JE, Hernandez A, Gladden LB. Blood lactate measurements and analysis during exercise: a guide for clinicians. J Diabetes Sci Technol. 2007;1:558–69.
62. Wang YX, Zhang CL, Yu RT, et al. Regulation of muscle fiber type and running endurance by PPARδ. PLoS Biol. 2004;2(10):e294.
63. Bogdanis GC. Effects of physical activity and inactivity on muscle fatigue. Front Physiol. 2012;3:142. https://doi.org/10.3389/fphys.2012.00142.
64. Khan AM, Paridon SM, Kim YY. Cardiopulmonary exercise testing in adults with congenital heart disease. Expert Rev Cardiovasc Ther. 2014;12:863–72.
65. Kaminsky DA, Knyazhitskiy A, Sadeghi A, et al. Assessing maximal exercise capacity: peak work or peak oxygen consumption? Respir Care. 2014;59:90–6.
66. Saghiv M, Sagiv M, Ben-Sira D, et al. What maintains total energy release at peak anaerobic effort in young and old men? ARC J Res Sports Med. 2016;1:4–8.
67. Golbidi S, Laher I. Exercise and the cardiovascular system. Cardiol Res Pract. 2012;2012:210852. https://doi.org/10.1155/2012/210852.
68. Trappe S, Luden N, Minchev K, et al. Skeletal muscle signature of a champion sprint runner. J Appl Physiol (1985). 2015;118:1460–6.
69. Romano AH, Conway T. Evolution of carbohydrate metabolic pathways. Res Microbiol. 1996;147:448–55.
70. Hurley B, Armstrong TJ. Bisphosphonates vs exercise for the prevention and treatment of osteoporosis. J Nurse Pract. 2012;8:217–24.
71. Steffl M, Bohannon RW, Sontakova L, et al. Relationship between sarcopenia and physical activity in older people: a systematic review and meta-analysis. Clin Interv Aging. 2017;12:835–45.
72. Slade JM, Miszko TA, Laity JH, et al. Anaerobic power and physical function in strength-trained and non-strength-trained older adults. J Gerontol A Biol Sci Med Sci. 2002;57:M168–72.
73. Robinson SL, Lambeth-Mansell A, et al. A nutrition and conditioning intervention for natural bodybuilding contest preparation: case study. J Int Soc Sports Nutr. 2015;12:20. https://doi.org/10.1186/s12970-015-0083-x.
74. Ferrara CM, Oldberg AP, Ortmeyer HK, et al. Effects of aerobic and resistanceexercise training on glucose disposal and skeletal muscle metabolism in older men. J Gerontol A Biol Sci Med Sci. 2006;61:480–7.
75. Foster C, Farland CV, Guidotti F, et al. The effects of high intensity interval training vs steady state training on aerobic and anaerobic capacity. J Sports Sci Med. 2015;14:747–55.
76. Carfagno D, Hendrix J. Overtraining syndrome in the athlete. Curr Sports Med Rep. 2014;13:45–51.
77. Skoog E, Shin JH, Saez-Jimenez V, et al. Biobased adipic acid – the challenge of developing the production host. Biotechnol Adv. 2018;36:2248–63.
78. Robineau J, Babault N, Piscione J, et al. Specific training effects of concurrent aerobic and strength exercises depend on recovery duration. J Strength Cond Res. 2016;30:672–83.
79. Burnside WS, Snowden C. Physiological basis of preoperative cardiopulmonary exercise testing. Surgery (Oxford). 2014;32:59–62.
80. Goodenkuf R, Heesch M, Hassenstab B, et al. Acute high intensity anaerobic training and rhabdomyolysis. Int J Exerc Sci. 2015;8:65–74.
81. Amann M. Pulmonary system limitations to endurance exercise performance in humans. Exp Physiol. 2012;97:311–8.
82. Eroglu O, Zileli R, Nalbant MA, et al. Prevalence of alpha actinin-3 gene (ACTN3) R577X and angiotensin converting enzyme (ACE) insertion/deletion gene polymorphisms in national and amateur Turkish athletes. Cell Mol Biol (Noisy-le-Grand). 2018;64:24–8.

83. Saghiv M, Goldhammer E, Sagiv M, et al. Klotho gene expression responses to long lasting aerobic training and aging. J J Geronto. 2015;1(2):009.
84. West JB. Vulnerability of pulmonary capillaries during exercise. Exerc Sport Sci Rev. 2004;32:24–30.
85. Clark AL, Poole-Wilson PA, Coats AJ. Relation between ventilation and carbon dioxide production in patients with chronic heart failure. J Am Coll Cardiol. 1992;20:1326–32.
86. Sabbah HN, Anbe DT, Stein PD. Negative intraventricular diastolic pressure in patients with mitral stenosis: evidence of left ventricular diastolic suction. Am J Cardiol. 1980;45:562–6.
87. Chirinos JA, Segers P. Noninvasive evaluation of left ventricular afterload: part 1: pressure and flow measurements and basic principles of wave conduction and reflection. Hypertension. 2010;56:555–62.
88. Wilmore JH, Costill DL, Kenney WL. Physiology of sport and exercise. 4th ed. Champaign, IL: Human Kinetics; 2008.
89. Duncker DJ, Bache RJ. Regulation of coronary blood flow during exercise. Physiol Rev. 2008;88:1009–86.
90. Rivera-Brown AM, Frontera WR. Principles of exercise physiology: responses to acute exercise and long-term adaptations to training. PM&R. 2012;4:797–804.
91. Tucker WJ, Lijauco CC, Hearon CM Jr, et al. Mechanisms of the improvement in peak VO2 with exercise training in heart failure with reduced or preserved ejection fraction. Heart Lung Circul. 2018;27:9–21.
92. Garber CE, Blissmer B, Deschenes MR, Franklin BA, et al. American College of Sports Medicine position stand. Quantity and quality of exercise for developing and maintaining cardiorespiratory, musculoskeletal, and neuromotor fitness in apparently healthy adults: guidance for prescribing exercise. Med Sci Sports Exerc. 2011 Jul;43(7):1334–59.
93. Lameris TW, de Zeeuw S, Alberts G, et al. Time course and mechanism of myocardial catecholamine release during transient ischemia in vivo. Circulation. 2000;101:2645–50.
94. Martin J, MacInnis MJ, Gibala MJ. Physiological adaptations to interval training and the role of exercise intensity. J Physiol. 2017;595:2915–30.
95. Lindgren M, Alex C, Shapiro PA, et al. Effects of aerobic conditioning on cardiovascular sympathetic response to and recovery from challenge. Psychophysiology. 2013;50:963–73.
96. Kingsley JD, Figueroa A. Acute and training effects of resistance exercise on heart rate variability. Clin Physiol Funct Imaging. 2016;36:179–87.
97. Lakin R, Notarius C, Thomas S, Goodman J. Effects of moderate-intensity aerobic cycling and swim exercise on post-exertional blood pressure in healthy young untrained and triathlon-trained men and women. Clin Sci (Lond). 2013;125:543–53.
98. Joyner MJ, Casey DP. Regulation of increased blood flow (hyperemia) to muscles during exercise: a hierarchy of competing physiological needs. Physiol Rev. 2015;95:549–601.
99. Regitz-Zagrosek V. Sex and gender differences in health science and society series on sex and science. EMBO Rep. 2012;13:596–603.
100. Nindl BC, Scoville CR, Sheehan KM, et al. Gender differences in regional body composition and somatotrophic influences of IGF-I and leptin. J Appl Physiol. 2002;92:1611–8.
101. Geer EB, Shen W. Gender differences in insulin resistance, body composition, and energy balance. Gend Med. 2009;6(Suppl 1):60–75.
102. Alhadad SB, Tan PMS, Lee JKW. Efficacy of heat mitigation strategies on core temperature and endurance exercise: a meta-analysis. Front Physiol. 2019; https://doi.org/10.3389/fphys.2019.00071.
103. Rolfe DFS, Brown GC. Cellular energy utilization and molecular origin of standard metabolic rate in mammals. Physiol Rev. 1997;77:731–58.
104. Hanna EG, Tait PW. Limitations to thermoregulation and acclimatization challenge human adaptation to global warming. Int J Environ Res Public Health. 2015;12:8034–74.
105. Ponganis PJ. Diving mammals. Compr Physiol. 2011;1:447–65.
106. New K, Bailey DM, Campbell I. Extremes of barometric pressure. Anaesth Intens Care Med. 2005;6:376–9.

107. Doolette DJ, Mitchell SJ. Hyperbaric conditions. Compr Physiol. 2011:163–201.
108. Moon RE, Cherry AD, Stolp BW, et al. Pulmonary gas exchange in diving. J Appl Physiol. 2009;106:668–77.
109. Brubakk AO, Neuman TS. Bennett and Elliott's physiology and medicine of diving, 5th Rev ed. Philadelphia, PA: Saunders; 2003. p. 800.
110. Papadopoulou V, Eckersley RJ, Balestra C, Karapantsios TD, Tang MX. A critical review of physiological bubble formation in hyperbaric decompression. Adv Colloid Interface Sci. 2013;191–192:22–30.
111. Grover I, Reed W, Neuman T. The SANDHoG criteria and its validation for the diagnosis of DCS arising from bounce diving. Undersea Hyperb Med. 2007;34:199–210.
112. Young AJ, Berryman CE, Kenefick RW, et al. Altitude acclimatization alleviates the hypoxia-induced suppression of exogenous glucose oxidation during steady-state aerobic exercise. Front Physiol. 2018;9:830. https://doi.org/10.3389/fphys.2018.00830.
113. Ruggiero L, Hoiland RL, Hansen AB, Ainslie PN, McNeil CJ. UBC-Nepal expedition: peripheral fatigue recovers faster in Sherpa than lowlanders at high altitude. J Physiol. 2018;596:5365–77.
114. Storz JF, Cheviron ZA, McClelland GB, Scott GR. Evolution of physiological performance capacities and environmental adaptation: insights from high-elevation deer mice (Peromyscus maniculatus). J Mammal. 2019;100(3):910–22.
115. Taylor AT. Risk factors, prevention, and treatment. Rambam Maimonides Med J. https://doi.org/10.5041/RMMJ.10022.
116. Moore LG. Measuring high-altitude adaptation. J Appl Physiol (1985). 2017;123:1371–85.
117. Eckman M. Professional guide topathophysiology. 3rd ed. Philadelphia: Wolters Kluwer/Lippincott Williams & Wilkins; 2010. p. 208.
118. Murray AJ. Energy metabolism and the high-altitude environment. Exp Physiol. 2016;101:23–7.
119. Alkorta-Aranburu G, Beall CM, Witonsky DB, Gebremedhin A, et al. The genetic architecture of adaptations to high altitude in Ethiopia. PLoS Genet. 2012;8(12):e1003110. https://doi.org/10.1371/journal.pgen.1003110.
120. Simonson TS, Huff CD, Witherspoon DJ, et al. Adaptive genetic changes related to haemoglobin concentration in native high-altitude Tibetans. Exp Physiol. 2015;100:1263–8.
121. Bigham A, Bauchet M, Pinto D, et al. Identifying signatures of natural selection in Tibetan and Andean populations using dense genome scan data. PLoS Genetics. 2010;6(9):e1001116. https://doi.org/10.1371/journal.pgen.1001116.
122. Vernikos J, Schneider VS. Space, gravity and the physiology of aging: parallel or convergent disciplines? A mini-review. Gerontology. 2010;56:157–66.
123. Hargens AR, Vico L. Long-duration bed rest as an analog to microgravity. J Appl Physiol (1985). 2016;120:891–903.
124. Marshall-Goebel K, Mulder E, Donoviel D, et al. An international collaboration studying the physiological and anatomical cerebral effects of carbon dioxide during head-down tilt bed rest: the SPACECoT study. J Appl Physiol (1985). 2017;122:1398–405.
125. Cardus D, McTaggart WG. Artificial gravity as a countermeasure of physiological deconditioning in space. Adv Space Res. 1994;14:409–14.
126. Kanherkar RR, Bhatia-Dey N, Csoka AB. Epigenetics across the human lifespan. Front Cell Dev Biol. 2014;2:49.
127. Blagosklonny MV, Campisi J, Sinclair DA, et al. Impact papers on aging in 2009. Aging (Albany NY). 2010;2:111–21.
128. Koopman R. Role of amino acids and peptides in the molecular signaling in skeletal muscle after resistance exercise. Int J Sport Nutr Exerc Metab. 2007;17(Suppl):S47–57.
129. Churchward-Venne TA, Murphy CH, Longland TM, et al. Role of protein and amino acids in promoting lean mass accretion with resistance exercise and attenuating lean mass loss during energy deficit in humans. Amino Acids. 2013;45:231–40.

130. Handy DE, Castro R, Loscalzo J. Epigenetic modifications: basic mechanisms and role in cardiovascular disease. Circulation. 2011;123:2145–56.
131. Walter CA, Zhou ZQ, Manguino D, et al. Health span and life span in transgenic mice with modulated DNA repair. Ann N Y Acad Sci. 2001;928:132–40.
132. Ronn T, Volkov P, Davegardh C, et al. Six months exercise intervention influences the genome-wide DNA methylation pattern in human adipose tissue. PLoS Genet. 2013;9:e1003572. https://doi.org/10.1371/journal.pgen.1003572.
133. Ntanasis-Stathopoulos J, Tzanninis JG, Philippou A, et al. Epigenetic regulation on gene expression induced by physical exercise. J Musculoskelet Neuronal Interact. 2013;13:133–46.
134. Rikli RE, Jones CJ. Development and validation of criterion-referenced clinically relevant fitness standards for maintaining physical independence in later years. Gerontologist. 2013;53:255–67.
135. Lim JY. Therapeutic potential of eccentric exercises for age-related muscle atrophy. Integr Med Res. 2016;5:176–81.
136. Sagiv M, Goldhammer E, Ben-Sira D, et al. What maintains energy supply at peak aerobic exercise in trained and untrained older men? Gerontology. 2007;53:357–61.
137. Strandberg TE, Sirola J, Pitkala KH, et al. Association of midlife obesity and cardiovascular risk with old age frailty: a 26-year follow-up of initially healthy men. Int J Obes (Lond). 2012;36:1153–7.
138. Vina J, Tarazona-Santabalbina FJ, Perez-Ros P, et al. Biology of frailty: modulation of ageing genes and its importance to prevent age-associated loss of function. Mol Asp Med. 2016;50:88–108.
139. Hawkins S, Wiswell R. Rate and mechanism of maximal oxygen consumption decline with aging: implications for exercise training. Sports Med. 2003;33:877–88.
140. Feng Z, Hanson RW, Berger NA, et al. Reprogramming of energy metabolism as a driver of aging. Oncotarget. 2016;7:15410–20.
141. Vellers HL, Kleeberger SR, Lightfoot JT. Inter-individual variation in adaptations to endurance and resistance exercise training: genetic approaches towards understanding a complex phenotype. Mamm Genome. 2018;29:48–62.
142. Ling C, Groop L. Epigenetics: a molecular link between environmental factors and type 2 diabetes. Diabetes. 2009;58:2718–25.
143. Proctor DN, Parker BA. Vasodilation and vascular control in contracting muscle of the aging human. Microcirculation. 2006;13:315–27.
144. Hart CR, Layec G, Trinity JD, et al. Evidence of preserved oxidative capacity and oxygen delivery in the plantar flexor muscles with age. J Gerontol A Biol Sci Med Sci. 2015;70:1067–76.
145. Baldwin KM, Haddad F. Research in the exercise sciences: where we are and where do we go from here – Part II. Exerc Sport Sci Rev. 2010;38:42–50.
146. Ding Q, Vaynman S, Akhavan M, et al. Insulin-like growth factor I interfaces with brain-derived neurotrophic factor-mediated synaptic plasticity to modulate aspects of exercise-induced cognitive function. Neuroscience. 2006;140:823–33.
147. Basset DR Jr, Howley ET. Limiting factors for maximum oxygen uptake and determinants of endurance performance. Med Sci Sports Exerc. 2000;32:70–84.
148. Tolle J, Waxman A, Systrom D. Impaired systemic oxygen extraction at maximum exercise in pulmonary hypertension. Med Sci Sports Exec. 2008;40:3–8.
149. Poole DC, Jones AM. Measurement of the maximum oxygen uptake Vo2max: Vo2peak is no longer acceptable. J Appl Physiol (1985). 2017;122:997–1002.
150. Lodder MAN, de Haan A, Sargeant AJ. Effect of shortening velocity on work output and energy cost during repeated contractions of the rat EDL muscle. Eur J Appl Physiol. 1991;62:430–5.
151. Sagiv M, Metrany R, Fisher N, et al. Comparison of hemodynamic and left ventricular responses to increased after-load in healthy males and females. Int J Sports Med. 1991;12:41–5.

152. Ma Y, Chen Y, Li X, et al. Aerobic power of students aged 13-15 years in Wuhan City. (Article in Chinese). Wei Sheng Yan Jiu. 2016;45:608–13.
153. Lolli L, Batterham AM, Weston KL, et al. Size exponents for scaling maximal oxygen uptake in over 6500 humans: a systematic review and meta-analysis. Sports Med. 2017;47:1405–19.
154. Gouzi F, Maury J, Molinari N, et al. Reference values for vastus lateralis fiber size and type in healthy subjects over 40 years old: a systematic review and metaanalysis. J Appl Physiol (1985). 2013;115:346–54.
155. Lenell C, Johnson AM. Sexual dimorphism in laryngeal muscle fibers and ultrasonic vocalizations in the adult rat. Laryngoscope. 2017;127(8):E270–6.
156. Mierzejewska-Krzyżowska B, Drzymała-Celichowska H, Bukowska D, et al. Gender differences in morphometric properties of muscle fibres measured on cross-sections of rat hindlimb muscles. Anat Histol Embryol. 2012;41:122–9.
157. Kang J, Hoffman JR, Chaloupka EC, et al. Gender differences in the progression of metabolic responses during incremental exercise. J Sports Med Phys Fitness. 2006;46:71–8.
158. Terjung RL, Zarzeczny R, Yang HT. Muscle blood flow and mitochondrial function: influence of aging. Int J Sport Nutr Exerc Metab. 2002;12:368–78.
159. Renna BF, Kubo H, MacDonnell SM, et al. Enhanced acidotic myocardial Ca2+ responsiveness with training in hypertension. Med Sci Sports Exerc. 2006;38:847–55.
160. Thomas L, Levett K, Boyd A, Leung DY, et al. Changes in regional left atrial function with aging: evaluation by Doppler tissue imaging. Eur J Echocardiogr. 2003;4:92–100.
161. Zeiher J, Ombrellaro KJ, Perumal N, et al. Correlates and determinants of cardiorespiratory fitness in adults: a systematic review. Sports Med Open. 2019;5(1):39. https://doi.org/10.1186/s40798-019-0211-2.
162. Sagiv M, Ben-Sira D, Goldhammer E, et al. Left ventricular contractility and function at peak aerobic and anaerobic exercises. Med Sci Sports Exerc. 2000;32:1197–201.
163. Scott JM, Esch BT, Haykowsky MJ, et al. Sex differences in left ventricular function and beta-receptor responsiveness following prolonged strenuous exercise. J Appl Physiol (1985). 2007;102:681–7.
164. Morishita S, Tsubaki A, Shirai N. Physical function was related to mortality in patients with chronic kidney disease and dialysis. Hemodial Int. 2017;21:483–9.
165. Aoike DT, Baria F, Kamimura MA, et al. Impact of home-based aerobic exercise on the physical capacity of overweight patients with chronic kidney disease. Int Urol Nephrol. 2015;47:359–67.
166. Hambrecht R. Stress and stress tolerance in chronic heart failure. Herz. 2002;27:179–86.

Chapter 2
Metabolism

Anabolism vs. Catabolism

Cells in human are very dynamic with multifaceted metabolic pathways, where enzymes do not function in separation form of large molecules structures [1]. In humans, there are significant numbers of metabolic pathways in order to obtain ATP, electrons removal, and the elimination of ammonia (NH_4^+) through the urea cycle [2]. The utmost imperative metabolic pathways are (a) carbohydrates oxidation namely glycolysis, (b) acetyl-CoA oxidation and fatty acid β-oxidation to acetyl-CoA in the citric acid cycle or Krebs cycle, (c) oxidative phosphorylation, electrons removal released during the glycolysis and citric acid cycle, (d) pentose phosphate pathway, synthesizes pentoses required for anabolic reactions, and (e) glucose synthesis (gluconeogenesis) from smaller precursors and removal of NH_4^+ to the urea cycle.

Metabolism means change that comes from the Greek word metabole, it includes all the chemical reactions of an organism. A metabolic pathway alters particles or substrates, such as fat, into other compounds, to a more usable resources to generate energy through metabolic pathways. Metabolic reactions occur inside the cells, where anabolism and catabolism take place in order to break down or build up molecules. Anabolism and catabolism occur at the same time throughout the body. However, at the same cell, when catabolism occurs there is no anabolism activity and vice versa. Catabolism is the degradation of substrates; it typically releases energy for the cell to use. This generates the metabolic cycle, where anabolism then produces other particles that catabolism breakdown, numerous of which keep on in the organism to be utilized again. The main catabolic course is digestion, where nutrient ingredients are consumed and broken into simpler molecules for the cells to use. By breaking-down polysaccharides (starch) within the cells glycogen, glucose, ribose, and fructose, the cells can produce energy. Same with proteins which are broken-down to amino acids.

M. S. Saghiv, M. S. Sagiv, *Basic Exercise Physiology*,
https://doi.org/10.1007/978-3-030-48806-2_2

Anabolism, also called biosynthesis, is the construction of molecules; it typically needs energy. At the cellular level, in order to build complex molecules, anabolic processes use amino acids to synthesize proteins. These reactions are under tight enzymatic control; they can reduce, increase speediness, and halt metabolic procedures [3]. Anabolism is a productive metabolic process in which a cell utilizes energy to construct particles such as enzymes nucleic acids, amino acids, monosaccharides nucleotides, and the assembly of these precursors into complex molecules [4]. Catabolism is a reaction that releases heat (exothermic reaction) where the cell breakdowns multipart molecules into carbohydrate, fat, and protein catabolism [4]. Anaerobic exercise typically are resistance exercises such as weightlifting, isometrics bouts, sprinting and high intensity for short time are anabolic in their nature, commonly increasing, and shaping muscle mass. Catabolic exercise is aerobic physical activity where oxygen is present and consumed (aerobic exercise) mostly utilizes fat. Thus, oxygen utilization is an important aspect in catabolism, since oxygen is a reducing mediator in chemical reactions. Usual catabolic exercises are aerobically lasting minimally above 20 min, activities such as running, cycling, walking, and swimming.

The net result of the anabolism and catabolism reactions is metabolism, connected also to biological chemistry, which is the learning of the assembly, arrangement, and biochemical responses of substances in living systems. It is also the study of substances found in organisms such as deoxyribonucleic acid (DNA) and proteins. Chain of biochemical responses that occur in living cells is cellular metabolism, which includes multifaceted arrangements of precise organic reactions, namely, metabolic pathways, a connected sequence of chemical reactions happening within a cell. These metabolic pathways are biological responses, where chemical particles with the enzymatic effects are altered into different molecules.

These progressions allow organisms to grow and reproduce, maintain their structures, enable to keep living, growing cell dividing, and respond to environmental changes. The products and in between of an enzymatic response are identified as metabolites that are end product of metabolism [2], modified by a sequence of chemical reactions catalyzed by enzymes [5]. Some mechanisms control the changes of metabolites over metabolic pathways to guarantee that pathways' product meets biological demand and that energy stored in ATP is not lost by having in the same cell opposite pathways activity. Usually, in a metabolic pathway, the product of one enzyme acts as the substrate for the next product. However, some products are removed from the cell since they are considered leftover [6]. Enzymes often require dietary supplementation of minerals, vitamins, and other cofactors to function [7]. The metabolism process needs vigorously enzymes for the satisfactory metabolic pathways regulation in the cell's environment [8], a procedure identified as homeostasis. Enzyme kinetics describes the chemical reactions that are catalyzed by enzymes. Numerous aspects affect the degree of enzymatic reactions advance [9]. These include altering enzyme's activity and substrate concentration, temperature, pH value, and alternation in the transcription of enzyme's gene [10]. Almost all substrates occupy most of the time the enzyme active, thus reaction rate is influenced by the characteristic turnover rate of the enzyme [11]. Enzymes control the reactions through a typical attraction (dissociation constant) and a kinetic parameter called Km which is a unit of molarity representing substrate concentration where the response rate is halfway to maximal velocity [11]. Enzyme has numerous kinetic properties yet the two most important are (a) how simply a specific substrate saturates the enzyme and (b) what is the maximal reaction rate that can be attained. As an example, during exercise, ATP production through glycolysis last 100–120 s along with the production of lactate that prevents continual energy generation before cell's enzymes and substrate amounts are reduced. Most enzymes are protein molecules that activate other molecules by binding to the enzyme's active site and by series of steps they are altered into other products through enzymatic mechanism. There are two kinds of enzymatic mechanisms: single-substrate that binds only one substrate and multiple-substrate that bind multiple substrates. Mostly a single substrate is involved which is turned into a single product by an enzyme. Hexokinase is an example of a single-substrate enzyme found in the glycolytic pathway. There are four isozymes of hexokinase: types 1–4 (glucokinase). During glycolysis, they differ in their catalysis, localization, and regulation, and thus influence the different outlines of glucose and glycogen metabolism. While an example for multiple-substrate enzymatic mechanism is the cleavage of one protein substrate into two polypeptide products. Yet, the overall kinetics determination is a chemical reaction that changes the enzyme or substrate characteristically one rate-determining step [12].

There are two main forms of metabolic pathways to produce ATP that are considered by their capability to build up molecules with the use of energy (anabolic pathway) or break down of large molecules that release energy in a process (catabolic pathway). Thus, the two pathways balance each other, where energy is released by the degradative course of a catabolic pathway used for molecules build up. Catabolic responses hold and yield energy from nutrients and reduce larger molecules into lesser molecular raw materials for reutilize by the cell, where the energy is kept in the form of ATP, that influences the reactions of anabolism. Enzymes in anabolic pathways need energy to alter or build molecules into more complex molecules or macromolecules. Enzymes in catabolic pathways degrade molecules and liberate energy. Thus, catabolic pathways liberate energy while breaking down molecules into lower molecules. Cellular respiration is one example of a catabolic pathway, glucose is degraded in the cell to set free energy for the cell's needs during cellular respiration. Additional kind of catabolic pathways contains the citric cycle or Krebs cycle, where protein, fat, and glucose molecules undergo oxidation. The end result is the chemical CO_2 molecules. The aerobic metabolic pathway starts in the cytosol and goes on into the mitochondria, where the electron transport chain and oxidative phosphorylation take place [13]. In contrast, glycolysis, pentose phosphagen pathway, and fatty acid biosynthesis all occur in the cytosol of a cell [14]. Glycolysis is an additional type of catabolic pathway, in which the cell store and breakdown glucose and other carbohydrates energy for the renewal of high-energy molecule ATP.

Energy Production

The energy that the body needs to function arises from the eaten food. However, energy cannot be used directly from food breakdown. Therefore, energy is transformed into ATP, a useable form of chemical energy utilized for all cellular function. ATP amount stored within the muscles is nominal, however, it is resynthesized rapidly using the phosphagen, carbohydrates, fats, and proteins energy. Every metabolic pathway involves a series of biochemical reactions that are linked by their intermediates to produce new substrates intended for following chemical reactions and so on. The metabolic pathway's flux direction is one way since there is a relationship between glucose uptake rate and the present of oxygen [12], yet, these reactions are sometimes reversible [2], such as during glycolysis pathway where reactions are reversible such as gluconeogenesis [2]. Conditions in the cell during energy production regulate metabolic pathway reactions flux in one direction [14] or often by feedback inhibition. When excess fat or protein are presented in the cell, reversible reactions may take place during glycolysis to produce glucose 6-phosphate, which is then converted to glucose in the liver (Cori cycle).

The Krebs cycle flow is in a cycle structure in which each previous substrate is a precursor for the next substrate reaction in the cycle. All metabolic pathways are transformed and degraded by chemical reactions substrates into different particles, which are oxidized in the end leading, to CO_2 production. Yet, particles oxidation may be interrupted for some reasons, mainly, oxygen supply [15]. Glucose and free fatty acids are the main precursors for the aerobically ATP synthesis in skeletal muscles [16]. The interplay, between fat and glucose consumption during exercise, can differ extremely when it rests on oxygen availability or exercise intensity [17]. During low-to-moderate aerobic exercise intensities, the muscle utilizes more fats and less glucose [18]. However, at high exercise intensities, lack of oxygen obliges the muscle to utilize glucose and glycogen as favorite substrates for muscle contraction [19]. Thus, free fatty acid oxidation rates decrease during high-intensity exercise due to decrease in carnitine levels that reduce transportation of fat into mitochondria [20]. Carnitine is an ammonium compound containing a cation of the form NR^{4+}, where R represents organic groups or atoms other than hydrogen [21], involved in metabolism in most mammals [22].

Metabolic Pathways

Aerobic respiration is the course by which oxygen-breathing turn fats and carbohydrates into energy. It is used by all cells to produce ATP, that energy is stored in its phosphate bonds. The aerobic pathway that uses oxygen is considerably more efficient and produces larger ATP amount, compared to anaerobic respiration pathway, since oxygen is an exceptional electron acceptor. Through the aerobic pathway, energy released is achieved by catabolic reactions.

Muscle cell is exceptional in its ability to fluctuate metabolic frequency to a larger degree than any other tissue depending on the strains loaded on [23]. Any exercise performance depends on the hydrolysis of ATP for muscle contraction and the connected release of free energy [24]. During high-intensity exercise bout, ATP consumption in human skeletal muscle may increase by 1000-folds, compared to resting levels [25], pointing out about the potential for skeletal muscle to induce metabolic stress. At rest, ATP concentration in muscle is approximately 8 mmol kg^{-1} dry muscle, during intense bout, it may decrease by about 1–2 mmol kg^{-1} dry muscle, yet, it is not lower than 5 mmol kg^{-1} dry muscle [26]. This is possible due to the ATP, ADP, AMP, and Pi concentrations in muscles that optimally support energy transference to from ATP. Any additional decrease in muscle ATP concentration is related to fast development of fatigue, expressed in muscle strength reduction [27]. Below 5 mmol kg^{-1} dry muscle, fatigue ensues (Fig. 2.1). Fatigue due to reduced ATP levels is a physiological warning to avoid irreversible muscle impairment [28].

Therefore, to tolerate muscle contraction at high intensity a precise balance, between rates of ATP hydrolysis and ATP synthesis is kept [29]. As a result, cells depend on energy mechanisms to immediately supply ATP for contraction, stored in glycogen and fats. This balance is achieved through all three different energy systems: (a) phosphagen, (b) anaerobic pathway, and (c) aerobic pathway [25]. A well-coordinated interaction between anaerobic and aerobic pathways facilitates the regeneration of ATP during intense exercise [30]. The proportional contribution of each pathway for ATP regeneration depends on the exercise bout intensity and the time it last. In addition, during high-contraction rate bout, ATP needed is minimal for the myosin rotation at the cross-bridge, resulting in a shorter actin sliding over the myosin for each contraction [24].

Since there are no anaerobic or aerobic efforts per se, the aerobic and anaerobic energy pathways are independent as physiological systems, however, they are in close relationships concerning energy supply [31]. Therefore, during different

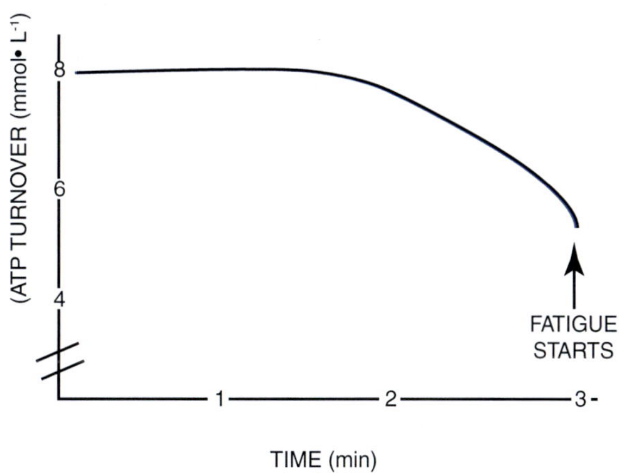

Fig. 2.1 ATP, ADP, AMP, and Pi concentrations enable energy transference to and from ATP. However, under 5 mmol kg^{-1} dry muscle, fatigue ensues

exercise activities, these two metabolic pathways are well united and synchronized to supply the needed energy amount, due to variations in ATP cell's levels. Nevertheless, during exercise regeneration of ATP is immediately [25].

The Phosphocreatine System

Energy stored in any molecule that contains phosphates (phosphagen molecules) is liberated when the inorganic phosphate (Pi) splits from the molecule by means of enzymes or water [2].

Since ATP is not kept in any substantial amount, phosphagens provide the energy that could not be instantly supplied by oxidative phosphorylation or glycolysis. Phosphagens are found in tissues that experience rapid changes in energy needs; muscles and nerves. Muscle cells have small amount of ATP stored; therefore, when exercise commences, this amount enables the muscle to work immediately at high intensity for 3 s. The cell turns ATP into ADP, then the inorganic phosphate is removed from creatine phosphate (CP) by creatine kinase enzyme, in order to rapidly turn the ADP back into ATP. Both muscle-stored ATP and CP are named the phosphagen system, which supplies the energy needed for 8–10 s at very high exercise intensity. For an individual weighing 70 kg creatine, turnover is about 2 g every day. CP-ATP system or phosphagen system is the fastest way to turn ADP and Pi to ATP. When exercise commences, rapid ATP regeneration takes place by the interaction of CP with ADP.

However, CP is limited in its quantities, and thus in cases of all-out exercise intensity, two molecules of ADP attach to yield one ATP molecule and one AMP molecule. The phosphagen system that supports rapid ATP production does it by three reactions. The following equations summarize the ATP's regeneration sources:

$$CP + ADP + H+ \rightarrow\rightarrow ATP + HPi,$$

$$ADP + ADP \rightarrow\rightarrow ATP + AMP,$$

and,

$$AMP + H+ \rightarrow\rightarrow IMP + NH_4^+$$

It seems that during the first 15 s of a severe exercise, CP is the only source for ATP renewal [32], a reaction that does not need the presence of oxygen [33]. Coaches turn to supplements creatine as a way to improve explosive sports performance such as 100, 200, and 400 m dashes. Creatine increases muscle mass and thus increases power. However, not all athletes who used creatine benefited from that. Creatine found in muscles is an amino acid produced in the liver, kidney, and pancreas. It stores energy in the muscle for high-intensity exercise. At the onset of exercise, CP, stored in the cytosol, signifies the extremely rapid rates for ATP replacement, by transferring the energy and inorganic phosphate from CP to ADP. The stockpile of CP in the muscle at rest is approximately 26 mmol kg^{-1} dry muscle. Hydrogen (proton) from the creatine kinase reaction changes the muscle to be alkalized. However, due to the intense effort, metabolic acidosis activates the enzyme AMP deaminase that catalyzes the alteration of AMP to inosine monophosphate (IMP) to produce in the end NH_4^+ (ammonia) which is converted to urea $(NH_2)_2Co)$, by the liver [25]. Additionally, AMP is an effective allosteric activator during glycolysis, it increases glucose-6-phosphate production and activates phosphofructokinase enzyme that result in higher rates of ATP renewal [34]. Following an exercise bout, during a passive recovery, athletes' ability to frequently regenerate muscle CP levels may improve significantly their performance.

Creatine transports high-energy phosphate from mitochondria to actin–myosin sites where muscle explosive contraction occur. CP can be a backing source for vigor muscle contraction lasting around 6–8 s, which thereafter, glycolysis ensues. Following aerobic training, power capacity generated by ATP-PC and glycolytic pathways may rise 10–20%, while the aerobic system is more increased with the genetic limitation.

The Anaerobic Pathway

When the ratio ATP/ADP declines, catabolism ensues to form ATP from ADP + Pi + energy. When the ratio ATP/ADP increases; anabolism of glucose and fat starts (Fig. 2.2). During intense aerobic bout, the main source for energy is the breakdown of carbohydrates due to its low oxygen cost.

Anaerobic glycolysis generates energy only half as quickly (about 16 kcal min^{-1}) compared to the ATP-PC system, while the aerobic system has the lowest rate (about 10 kcal min^{-1}). It is the production of energy through the carbohydrate's

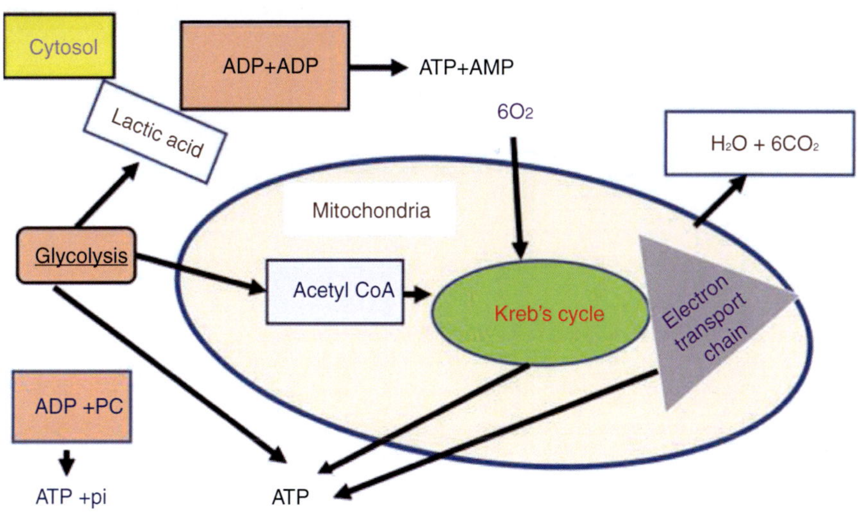

Fig. 2.2 Summary of the anaerobic and aerobic phases to produce ATP

breakdown, in the absence of oxygen [35]. Anaerobic reactions take place in the cytoplasm, while aerobic reactions happen in the Krebs cycle of the cell's mitochondria. Since carbon and oxygen are present in carbohydrate, its breakdown cost less oxygen compared to the breakdown of fat and protein which are built differently. That is why muscles utilize carbohydrates and not fat when oxygen is in short supply. However, when oxygen is available (during submaximal aerobic exercise), the utilization of fat become the major source to produce ATP [36]. This occurs during intense exercise when an imbalance between muscle oxygen demand supply is extreme. The myocardium is a muscle that differs from skeletal muscles for having a large number of mitochondria [37], enabling continuous aerobic respiration through oxidative phosphorylation. When exercising aerobically, lactate produced by the working skeletal muscles can be recycled and consumed, aerobically by the liver [38], and Krebs cycle of the myocardium due to its large mitochondria numbers [39]. Thus, under ischemic conditions (severe hypoxic), the left ventricular is not able to eject blood into the aorta adequately. When oxygen demand supply of the myocardium is balanced, the breakdown of free fatty acids and triglycerides is the main source for regenerating ATP, while carbohydrates donation is minor.

Usually the glycolytic pathway lasts up to 20 s of vigorous exercises bout, such as when performing 100 or 200 m dashes. When oxygen demand supply imbalance is reached, glucose and glycogen cannot be fully degraded to CO_2 and H_2O. Then, pyruvate is converted to lactate by lactate dehydrogenase catalyzes. Lactate, the final glycolysis product, is formed, which increases its muscle concentration to the point of acidosis onset, usually attributed to the buildup of lactate. After this, it has been suggested that lactate production is the main reason for muscle fatigue during exercise.

Glycolysis or anaerobic pathway, an oxygen-independent metabolic pathway, changes glucose into pyruvate and H+. The free energy used in the process comes from two contributed ATP molecules and NADH.

In glycolysis, glucose with its six carbon-sugar is divided into 2 molecules of a three-carbon sugar. During glycolysis 2 ATP molecules are contributed, 4 ATP molecules are extracted, and thus the net gain is 2 ATP molecules, 2 molecules of pyruvic acid and 2 electron carrying molecules of NADH. Conversion of glucose to lactate occurs when oxygen is not obtainable, along with the production of small amounts of ATP. This process is called fermentation.

Pyruvate is used as an electron acceptor to regenerate NAD+ from NADH. Pyruvate enters the mitochondria in the presence of oxygen (aerobic condition) then it is converted to acetyl-CoA by the enzyme pyruvate dehydrogenase which enters the Krebs cycle, where a large amount of ATP molecules is generated, CO_2 and H_2O (Fig. 2.3). This is anaerobic glycolysis which means that molecular oxygen was not consumed during the process. In the presence of oxygen, glycolysis is the first stage of cellular respiration.

Glycolysis is a pathway that does not involve the mitochondria and occur in the cytoplasm with or without oxygen. The glycolytic pathway can be divided into two phases: (a) during this first phase, glucose is converted to glucose 6,1-phosphate and

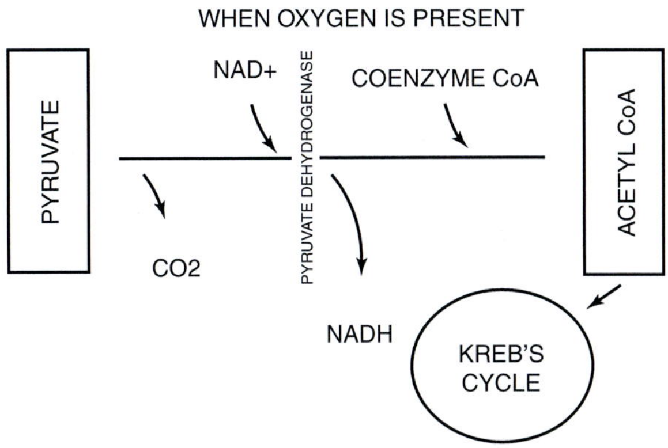

Fig. 2.3 Pyruvate conversion to acetyl-CoA when oxygen is present

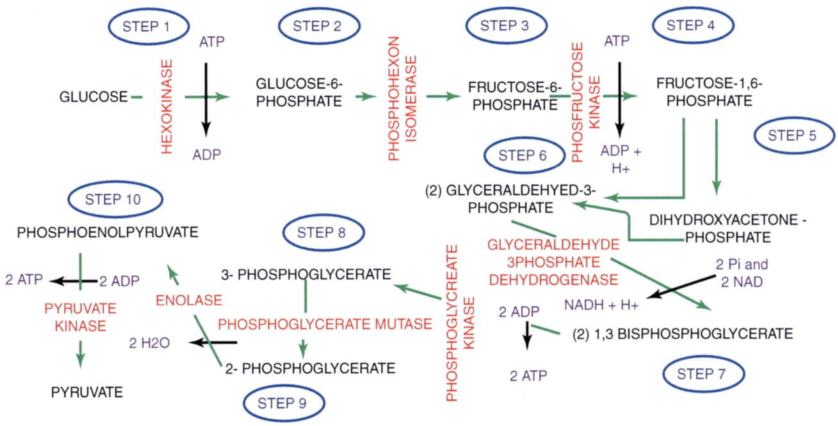

Fig. 2.4 The anaerobic pathway from glucose to pyruvate

two molecules of ATP are contributed and (b) the hexose chain is cleaved into two triose phosphates and the conversion of glyceraldehyde-3-phophate to pyruvate where the production of four ATP take place.

The anaerobic pathway is a serial of ten reactions to produce pyruvate in 10 steps by glycolysis (Fig. 2.4).

Step 1. This reaction traps glucose within the cell, since the negatively charged phosphate group prevents glucose -6-phosphate from crossing the membrane out of the cell. So, cells can stock up glucose. However, regulation of cell glucose levels is controlled by hexokinase reaction providing a feedback inhibition of the enzyme. Glucose is phosphorylated to form glucose-6-phosphate, a main metabolic fuel in most cells, while glycogen is stored energy within the cells. Catalyzed by the enzyme glucokinase which splits the ATP (first donation) into ADP and the Pi is added to the sixth carbon of the glucose. Hexokinase catalyzes a regulatory step in glycolysis that is irreversible. Hexokinase, as other kinas enzymes, needs Mg^{2+} to improve his active abilities.

Step 2. Glucose-6-phosphate is isomerized to fructose-6-phosphate by the enzyme phosphohexose isomerase.

Step 3. Fructose-6-phosphate is catalyzed by the enzyme phosphofructokinase to phosphorylate and then to fructose 1,6-bisphosphate. The source of the phosphate to carbon 1 comes from ATP (second donation) to fructose-6-phosphate. This reaction is irreversible since glucose 6 carbons are split into two molecules of 3 carbons each. Therefore, the enzyme phosphofructokinase plays a key role in the regulation of glucose breakdown.

Step 4. Fructose-1,6-bisphosphate (6 carbons) is cleaved by the enzyme aldolase into two molecules with 3 carbons each; one glyceraldehyde-3-phosphate (GAP) and another molecule of dihydroxyacetone phosphate (DHAP). This reaction is reversible.

Step 5. The dihydroxyacetone phosphate chemical structure does not fit the glycolysis pathway. Therefore, it is transformed into glyceraldehyde-3-phosphate which suits the pathway. Thus, the outcome is that glucose is cleaved into 2 glyceraldehyde-3-phosphate molecules. Still, this reaction is fast and reversible.

Step 6. Glyceraldehyde 3-phosphate is catalyzed by glyceraldehyde 3-phosphate dehydrogenase to1,3-bisphosphoglycerate and oxidized. This reaction yields energy where NAD+ is reduced to NADH.

Step 7. Oxidation and phosphorylation of 1,3-biphosphoglycerate are transformed to 3-phosphoglycerate by the enzyme phosphoglycerate kinase to produce ATP by adding Pi to ADP. The ATP produced in this reaction is by a substrate degradation (substrate-level phosphorylation) without the involvement of electron transport.

Step 8. A reversible reaction where 3-phosphoglycerate is isomerized by the enzyme phosphoglucose mutase and Mg^{2+} to 2-phosphoglycerate by moving the phosphate from carbon 3 to carbon 2.

Step 9. A reversible reaction where 2-phosphoglycerate is converted to phosphoenolpyruvate by the enzyme enolase which requires Mg^{2+} and the removal of molecule water.

Step 10. pyruvate kinase converts phosphoenolpyruvate into dephosphorylated to pyruvate. One molecule of ATP is produced out of this reaction. As in step 7, this reaction is substrate-level phosphorylation. This step is irreversible.

Note that most of the reactions of the glycolytic pathway are reversible, used for gluconeogenesis.

When oxygen is not presented in the cell, in order to regenerate NAD, pyruvate oxidizes NADH produced in glycolysis and is reduced to lactate, this reaction is catalyzed by lactate dehydrogenase (Fig. 2.5).

During all-out effort, the anaerobic glycolysis pathway is the main source for the renewal of ATP lasting about 10–30 s. It regenerates ATP rapidly over this time and produces 2 ATP per glucose molecule (Fig. 2.6), compared to potential of 38 ATP molecules produced by the aerobic pathway.

However, during all-out exercise intensity glycogen is utilized during very intense exercise bout glycolysis with significant insufficient oxygen, 3 ATP (not 2 ATP) are yield since glycogen contains phosphate; therefore, there is no need to donate second phosphate in step 1 (Fig. 2.7). Primary sites where glycogen can be

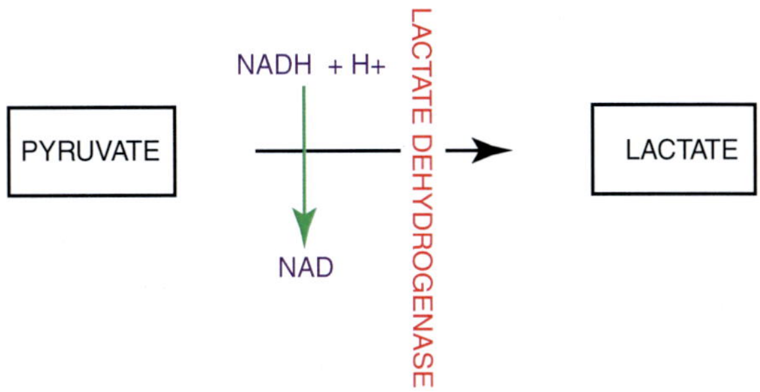

Fig. 2.5 The reduction of pyruvate is catalyzed by lactate dehydrogenase

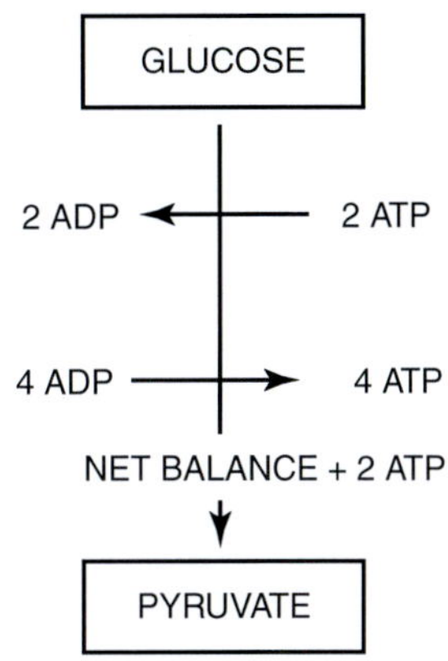

Fig. 2.6 Energy yield in glycolysis pathway from glucose

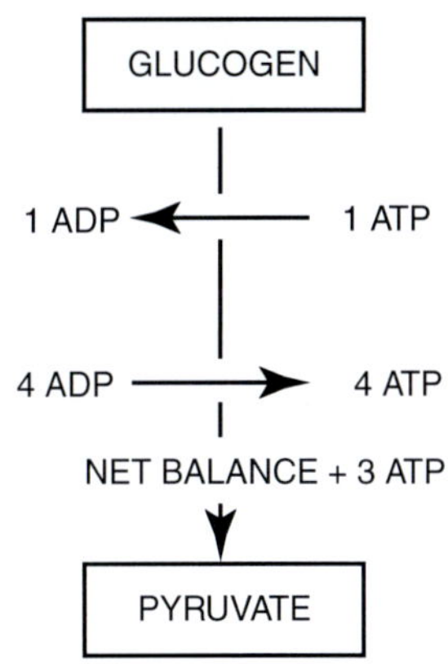

Fig. 2.7 Glycogen turn to glucose 6,1-phosphate by donating only 1 ATP

found are liver and skeletal muscle. Glycogen is a long polymer of glucose up to120,000 deposits and is a main carbohydrate storage form in humans, and thus it serves as an energy backup. The polymer is composed of units of glucose approximately every 8–12 residues. In addition, glucose produced through gluconeogenesis glycogen is an extra source of glucose. Glycogen is a fast glucose supplier when needed and serve to store excess glucose from the blood.

The branching of glycogen is an important feature of the molecule metabolically as well. Since glycogen from the peripheral boundaries is cleaved down, more glucose can be released at the same time. Compared to the 2 ATP molecules produced by the anaerobic pathway, net ATP molecules yield in aerobic glycolysis equals to 8 molecules. During glycolysis, four hydrogens (protons H^+) and four electrons (e^-) are formed following the conversion of glucose to pyruvate. The electrons and 2 of the protons are transfered to 2 NAD molecules to produce NADH. The instant energy source for ATP production in the outer and inner membranes (chloroplasts) and mitochondria is completed by the proton pump across a membrane. Cells also can yield ATP in the cytosol by a course of substrate-level phosphorylation, catalyzed by water-soluble enzymes, without the involvement of membranes or ion gradients. This substrate-level phosphorylation happens two times in the glycolytic pathway. The first occur with the glyceraldehyde 3-phosphate alteration to 3-phosphoglycerate (Fig. 2.4), steps 5 and 6 are assisted by the enzyme, glyceraldehyde 3-phosphate dehydrogenase. The Pi of 1,3-bisphosphoglycerate is then transported to ADP, generating ATP and 3-phosphoglycerate, a reaction catalyzed by phosphoglycerate.

The Aerobic Pathway

The aerobic pathway is a metabolic pathway to produce energy through combustion of fat, carbohydrates, and amino acids in the presence of oxygen. During low-to-moderate intensities of exercise such as walking, running, or cycling with continued effort, aerobic metabolism is used. However, it is important to note that during exercise athletes often switch between aerobic and anaerobic metabolism depends if it requires short vigorous bout of short dash or a long-lasting jogging, walking, or game sports activities such as in basketball, football, and basketball.

The oxidation of pyruvate through the Krebs cycle in the mitochondria produces large amount of ATP. Four molecules of ATP are formed by substrate-level phosphorylation in glycolysis: (a) two during dephosphorylation of two molecules of 1,3-diphosphoglyceric acid producing two molecules of 3-phosphoglyceric acids and (b) two during dephosphorylation of two molecules of phosphoenolpyruvate forming two molecules of pyruvate. Pyruvate formed following glycolysis, with oxygen presence, enters the Krebs cycle in the mitochondria, which finally generates 34 of the 36 ATP molecules, H_2O and CO_2. Mitochondria is made of two membranes, resulting in two submitochondrial compartments (Fig. 2.8).

Endurance training causes increase whole-body aerobic and general muscle capacities, especially, adaptations in the mitochondria. Mitochondrial biogenesis is an important metabolic adaptation to aerobic exercise training, resulting in greater mitochondrial content, density, size, number, and enzymatic actions during fatty oxidation [40, 41].

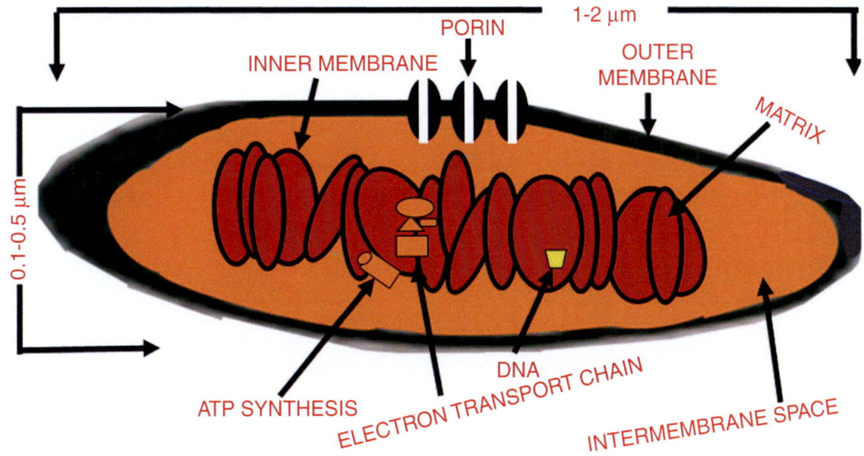

Fig. 2.8 Mitochondrial structure

The Krebs cycle is the second stage of aerobic respiration, the first being glycolysis and last being the electron transport chain. The outer membrane covers the organelle, while the inner membrane wrinkles over many times and generates layered structures called cristae. In normal cells, mitochondria's volume is almost 25% of the cytoplasm volume. Nutrients are degraded in the mitochondrial matrix which is a fluid to produce energy stored in ATP molecules. This biochemical process is aerobic cellular respiration or cellular respiration. The place where ATP is synthesized is the intramembrane particles that overhang from the inner membrane into the matrix. In addition, the matrix contains the mitochondrial deoxyribonucleic acid (DNA). The outer membrane defines the smooth external edge of the mitochondrion and includes mitochondrial transmembrane porins (Fig. 2.9). However, mitochondrial oxidation rate is influenced by metabolites movement that crossways the outer membrane. The matrix and the cytosol portions are separated by the inner membrane which is a key penetrable barrier. In the order of each step of the Krebs cycle course to occur, all the necessary enzymes are situated in the mitochondria. Krebs cycle or tricarboxylic acid cycle is also known as the citric acid cycle because citric acid is the very first product generated and it is also regenerated at the end of the cycle, without any ATP molecule consumed. The Krebs cycle stage follows glycolysis which is the first stage of aerobic respiration. Therefore, Krebs cycle is the second stage while the last being the electron transport chain, linked to Krebs cycle. The cycle refers to a complex series of chemical reactions that produce ATP required for numerous activities such as muscle contraction protein synthesis DNA duplication, etc. Enzymes of Krebs cycle are located both in the matrix and inner membrane of mitochondrion.

Krebs cycle exists in all organisms that utilize oxygen to breath and has a key role in glucose oxidation by the aerobic pathway. It is also the precursor of numerous particles; therefore, it is both anabolic and catabolic (amphibolic pathway). It is

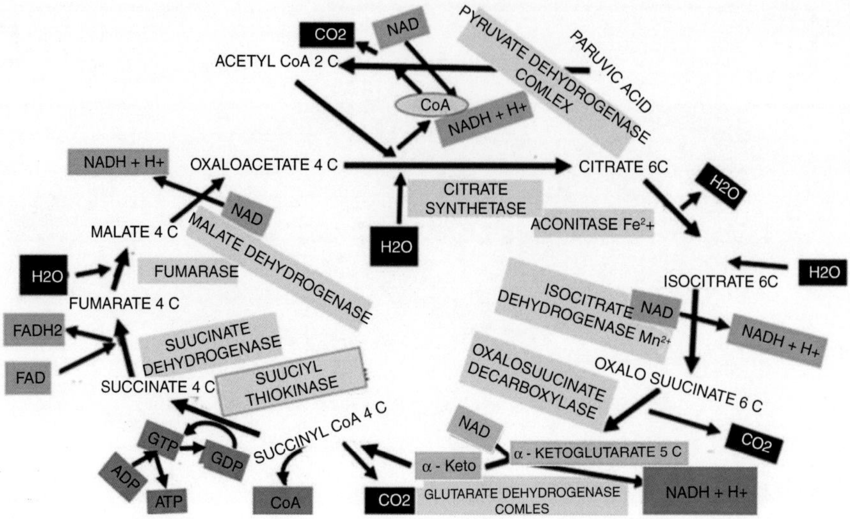

Fig. 2.9 The Krebs cycle refers to a complex series of chemical reactions that produce ATP

located in the matrix of mitochondria and is an important metabolic pathway that attaches reactions of glucose, fat, and protein metabolism. It releases stored energy through the oxidation of acetyl-CoA derived from pyruvate, fats, and proteins. The energy released in the cycle enables to produce ATP and CO_2. In the Krebs cycle, eight enzymes activate the cycle reactions to oxidize pyruvate formed during the glycolytic breakdown of glucose, to form CO_2 and H_2O.

Krebs cycle actually begins when acetyl-CoA which rises from breakdown of glucose, fat, and protein enters into a reaction to form citric acid, the first product of the cycle, while oxaloacetate molecule is the last in the cycle and serves as an acceptor. Krebs cycle is a step-by-step cyclic oxidation course that include four dehydrogenation steps and two decarboxylation stages where acetate group produce reduced CoA and CO_2. The following steps occur in the overall Krebs cycle:

- Step 1: combines two-carbon acetyl-CoA with 4-carbon oxaloacetate (condensation), with enzyme citrate synthetase, to form citrate a 6-carbon molecule. CoA spreads away to ultimately combine with another acetyl group. This step is an irreversible reaction for it is very exergonic, controlled by negative feedback and ATP amount existing. When ATP rises, reaction rate declines and the other way around.
- Step 2: in a two-step reaction with the help of the enzyme, aconitase citrate loses one H_2O molecule and gains another when citrate is converted to isocitrate: (a) water is freed thus citric acid is altered to cis-aconitate and (b) water is recaptured, cis-aconitase associates with water to form isocitrate.
- Step 3: isocitrate undergoes dehydrogenation with the enzyme isocitrate dehydrogenase, Mn^{2+} ion which is essential for the enzyme to function to form oxalosuccinate 6-carbon molecule. Oxalosuccinate undergoes decarboxylation with the presence of oxalosuccinate decarboxylase enzyme.

- Step 4: oxalosuccinate is altered to α-ketoglutarate (5-carbon molecule), CO_2, and two electrons, which reduce NAD^+ to $NADH_2$. This reaction is controlled by ATP and NADH (negative feedback) and by ADP (positive effect). Following the loss of hydrogen, isocitrate is altered into 6-carbon oxalosuccinate molecule.
- Step 5: a phosphate group is replaced for CoA; α-ketoglutarate undergoes simultaneous dehydrogenation and decarboxylation with the enzyme α-ketoglutarate dehydrogenase complex. This energy is utilized in substrate-level phosphorylation to form either guanine triphosphate (GTP) or ATP, the energy amount stored in GTP and ATP is the same. However, GTP is used only for protein synthesis, where NAD^+ and CoA are required, to form succinyl-CoA (4-carbon molecule), $NADH_2$, and CO_2.
- Step 6: a dehydration process that converts succinyl-CoA to succinate with the enzyme succinyl thiokinase and then into fumarate a 4-carbon molecule with the assistance of enzyme succinate dehydrogenase attached to inner mitochondrial membrane and FAD as hydrogen acceptor producing $FADH_2$. The enzyme succinate dehydrogenase has a non-hem iron (Fe-S) protein, which facilitates the link between the enzyme and the electron transport chain. The reason for utilizing FAD instead of NAD is that energy stored in the electrons is inadequate to reduce NAD, nonetheless, enough to reduce FAD, located in the inner membrane of the mitochondrion. Unlike NADH, $FADH_2$ carries the electrons directly to the electron transport chain. Energy liberated is used for the synthesis of GTP, with CoA release.
- Step 7: water is added and with the enzyme malate dehydrogenase, fumarate is altered to another 4-carbon dicarboxylic acid called Malate. The last step in the Krebs cycle regenerates oxaloacetate by oxidizing malate and another molecule of NADH is produced.
- Step 8: with the help of the enzyme malate dehydrogenase, malate formed in step 7 is altered to oxaloacetate, combined again with acetyl-CoA to form citrate, while NAD^+ is reduced to $NADH_2$.

In summary, the following reactions occur in the Krebs cycle: condensation, isomerization, dehydrogenation, decarboxylation, oxidative decarboxylation, substrate-level ATP/GTP synthesis, dehydrogenation (oxidation) of succinate, hydration, and dehydrogenation (oxidation) of malate. Subsequent to glycolysis, pyruvate (3-carbon molecule) is transformed into acetyl-CoA (2-carbon molecule) as the first stage of the Krebs cycle. During this process, one molecule of CO_2 and one molecule of the electron carrier NADH and acetyl-CoA molecule are produced. Acetyl-CoA comes into the Krebs cycle from each acetyl group, representing four of the six carbons of one glucose molecule, the other two carbons were released during each cycle as 2 CO_2 molecules. The Krebs cycle involves converting acetyl-CoA into CO_2. The two acetyl carbon atoms will ultimately be incorporated into 2 CO_2 released along the cycle; consequently, all 6 glucose carbon atoms are finally combined into CO_2. Accordingly, for every pyruvate molecule entering the Krebs cycle, the following molecules are produced: 2 CO_2, 3 NADH, 1 $FADH_2$, and 1 GTP or ATP.

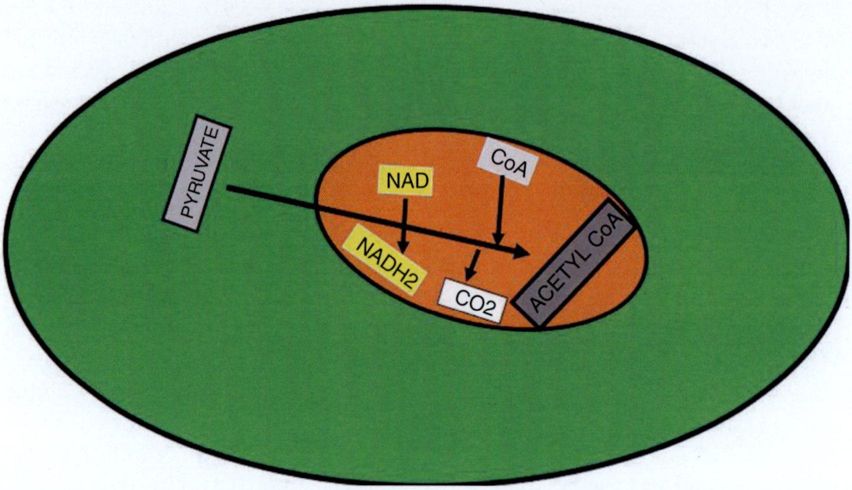

Fig. 2.10 Pyruvate loses carbon to form CO_2. The other two carbons are linked to the enzyme CoA to produce acetyl-CoA

In the conversion of pyruvate to acetyl-CoA, each pyruvate molecule loses one carbon atom with the release of CO_2 (Fig. 2.10). During the breakdown of pyruvate, electrons are transferred to NAD to produce NADH, used later by the cell to produce ATP. In the final step of the breakdown of pyruvate, an acetyl group is transferred to CoA to produce again acetyl-CoA.

Thus, the overall calculation of the oxidative catabolism of pyruvate can be summarized as follows:

$$\text{Pyruvate} + 4\,\text{NAD} + 2\,\text{H}_2\text{O} \rightarrow\rightarrow 3\,\text{CO}_2 + 4\,\text{NADH}_2 + \text{ATP}$$

$NADH_2$ and $FADH_2$ are linked to electron transport system and the formation of ATP by oxidative phosphorylation. Glycolysis do not produce CO_2 molecules; however, three molecules of CO_2 are developed during comprehensive oxidation of each of the two pyruvates. One molecule is produced when oxidative decarboxylation of pyruvate is transformed to acetyl-CoA. Two molecules are produced during Krebs cycle: one during decarboxylation of oxalosuccinate to α-ketoglutarate and another during decarboxylation of α-ketoglutarate to succinyl-CoA. Two molecules of GTP or ATP are formed through substrate-level phosphorylation which is linked to the released energy at the time of transforming two molecules of succinyl-CoA to succinate state.

Two molecules of NADH + H+ are formed in glycolysis during oxidation of two molecules of glyceraldehyde-3-phosphate to 1,3-diphosphoglycerate state, two molecules of NADH +H+ are produced in link reaction when two pyruvate molecules are oxidatively decarboxylated to acetyl-CoA. In Krebs cycle, six molecules of NADH +H+ and two molecules of $FADH_2$ are formed: oxidation of isocitrate to oxalosuccinate yields two NADH +H+, oxidative decarboxylation of α-ketoglutarate to form succinyl-CoA yield two NADH +H+, dehydrogenation of succinate to form fumarate yields $FADH_2$ and dehydrogenation of malate to oxaloacetate yields two NADH +H+. The net number of coenzymes formed from one molecule of glucose is 10 NADH +H+, 2 $FADH_2$, and 4 ATP molecules. Krebs cycle produces 6 NADH +H+ and 2 FADH2 molecules. 2 NADH +H+ are additionally formed through link reaction.

Electron Transport-Linked Phosphorylation

Oxidative phosphorylation is a highly efficient metabolic pathway in which oxidize fuel, so, liberating energy which is utilized to yield ATP, happens in the mitochondria, where oxidative phosphorylation is carried out. During oxidative phosphorylation, redox reactions that release energy used to produce ATP take place; electrons are relocated from electron donors to electron acceptors such as oxygen [42]. Electron transport chain, composed of 4 complexes, all take place in the mitochondrial membrane [43]. Energy produced through the electron transport chain is then used to tie an inorganic phosphate to ADP molecule, generating ATP. This concludes the course of oxidative phosphorylation.

Electrons and protons produced during glycolytic process, oxidation of fatty acid beta metabolism, and amino acids oxidation are transferred to coenzymes NAD and FAD in the Krebs cycle. Both coenzymes are reduced to NADH and $FADH_2$ with electrons and protons [44].

The Krebs cycle brings together eight high-energy electrons formed by oxidizing glucose, fat, and protein. Following the binding of NADH and $FADH_2$ carriers to the proteins of the electron transport chain, their electrons and protons are transported to the electron transport chain. This becomes the first stage in the electron transport chain. As the proteins of the electron transport chain passes from one complex to another, the electrons lose energy, however, portion of that energy is utilized to move hydrogen ions from the mitochondrial matrix into the intermembrane space. Once the electrons and protons are transferred within the electron transport chain, it is the last stage of cellular respiration. An electron transport chain is a sequence of electron transporters in the inner mitochondrial membrane that move electrons from electron giver molecules to electron acceptors by redox reactions, along this, transfer of H^+ ions (protons) across a membrane occurs (Fig. 2.11). Electron transporters are reduced during glycolysis and the Krebs cycle to NADH + H+ and $FADH_2$. These transporters then donate electrons and protons to the electron carrier proteins of the electron transport chain. The final electron acceptor is oxygen.

Both reduction and oxidation occur at the same time. In this process, protons are driven from the mitochondrial matrix to the intermembrane space [44]. The electron transport chain, which is the latter element of aerobic respiration, is the only part of glucose metabolism that uses atmospheric oxygen.

The respiratory system supplies oxygen unceasingly, it diffuses into the body cells which is the endpoint of the electron transport chain, where the electrons

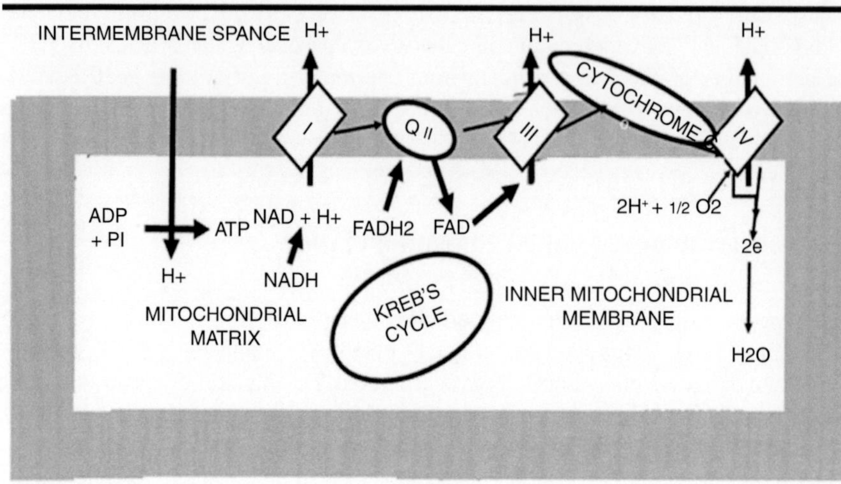

Fig. 2.11 The electron transport chain in the inner mitochondrial membrane transports electrons from the coenzymes NADH and $FADH_2$ to molecular oxygen to form water. In this process, protons are driven from the mitochondrial matrix to the intermembrane space

pass the sequence of redox reactions, meaning, the electrons react with oxygen in the fourth protein complex to produce water [45]. Electron transport chain is made of four complexes composed of proteins, labeled I–IV (Fig. 2.10) [45]. Each of these complex acts as an H^+ pump. Complexes I, III, and IV are proton pumps, where electrons transfer actively downhill, the released energy from this action, is used by the complexes to pump H^+ ions from the matrix to the intermembrane space. The common feature of all electron transport chains is the presence of a proton pump to create an electrochemical proton- gradient crossway in the membrane. As electrons move from a higher to a lower energy level, they liberate energy. Each complex receives electrons from a NADH and $FADH_2$ or from other complexes in the chain [2]. Complex I accepts hydrogens and electrons from NADH. The energy released produced by the hydrogens and protons facilitates their passage to the inner mitochondrial membrane by complex I. The hydrogens and electrons in the inner mitochondrial membrane increase hydrogens level in the intermembrane space. Then hydrogens and protons from complex I, $FADH_2$, and Krebs cycle are driven to complex II (coenzyme Q). Thus, complex II actually relieves $FADH_2$ from its electrons and passes them to coenzyme Q. The coenzyme passes on the electrons to complex III, which now receives electrons and their energy from two sources (a) the large gradient in redox potential across the three enzyme complexes and (b) electron transport can also pump H^+. Thus, large hydrogen numbers are transferred across the membrane by complex III with the support of cytochrome c, hydrogens move to complex IV. At that point hydrogen binds to oxygen to produce water [46]. The following highlights the electron transfer chain (Fig. 2.12).

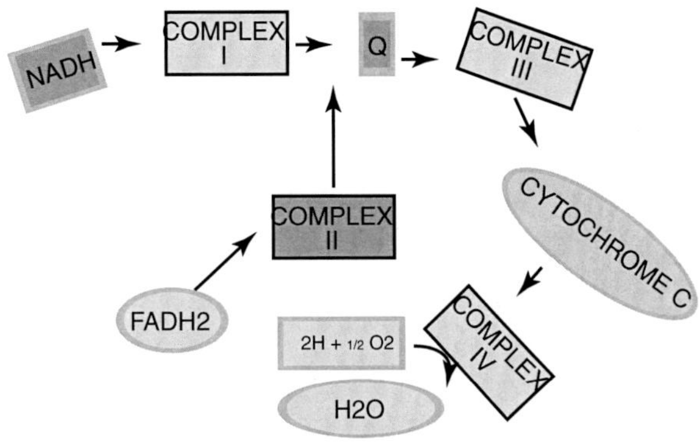

Fig. 2.12 Summary of the respiratory or electron transfer chain steps

- The inner mitochondrial membrane is the location of the electron transfer chain.
- Electrons are released following substrate oxidation which is moved to complex I with low redox potential to a high redox potential; complex IV.
- Electrons reaching complex IV are ultimately added to oxygen to form H_2O.
- Energy made at complexes I, III, and IV is utilized for proton pump (H+) in the inner mitochondrial membrane.
- ATP generation from ADP + Pi: 3 ATP are driven from each NADH while each $FADH_2$ gives 2 ATP.

The stage of oxidative phosphorylation produces ATP by the proton force and the enzyme ATP synthase. The proton force or protons gradient occurs in the mitochondrial intermembrane space. During cellular respiration, potential for ATP production is 38 molecules per glucose. The equation for glycolysis is

$$\text{Glucose} + 2\,\text{HPo}_4^{\;2-} + 2\,\text{ADP}^{3-} + 2\,\text{NAD}^{+} \rightarrow\rightarrow$$
$$2\,\text{Pyruvate}^{-} + 2\,\text{ATP}^{4-} + 2\,\text{NADH} + 2\,\text{H}^{+} + 2\,\text{H}_2\text{O}$$

The equation for aerobic respiration defines the reactants and products of all of its steps, including glycolysis:

$$1\,\text{glucose} + 6\text{O}_2 \rightarrow\rightarrow 6\text{CO}_2 + 6\,\text{H}_2\text{O} + 38\,\text{ATP}$$

However, the 38 ATP molecules are not attained due to the energy needed for active transports of: (a) pyruvate into the mitochondrial matrix, (b) phosphate exchange of phosphate ($H_2PO_4^-$; Pi) for O H^-, and (c) exchanges ADP and ATP across the inner membrane. The outcome of these is that additional 3 H^+ are needed to produce 1 ATP. Figure 2.13 elucidates this energy cost of the entire process in the mitochondria ending up closer to 28–30 ATP molecules [47]. Adding to that are the two net ATP produced from glucose glycolysis in the cytoplasm. Thus, total ATP net produced through the aerobic pathway adds-up to 30–32 molecules.

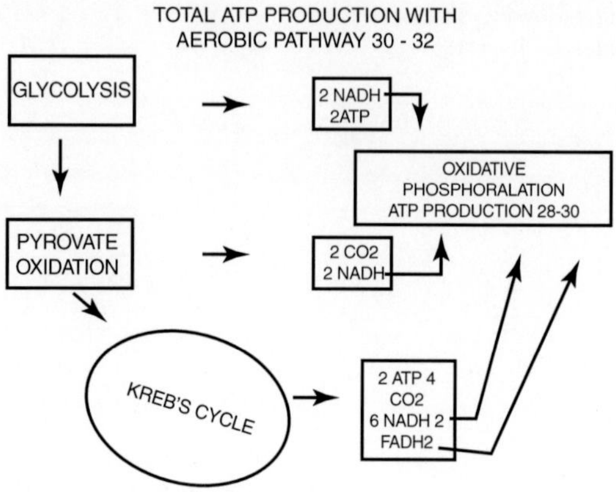

Fig. 2.13 Total ATP production through oxidative phosphorylation

Lactate Production and Utilization

Anaerobic metabolism can only use glucose and glycogen, while aerobic metabolism can also breakdown fats and protein. Intense exercise, at 90% VO_{2max}, results in a significant increase in muscle lactate concentration due to the shortage in oxygen, and thus the muscle utilizes the glycolytic pathwayin order to produce the needed ATP. Lactate is produced when oxygen supply is insufficient during all-out exercise. Although lactate is used later by the myocardium and liver to produce ATP, an extreme amount of lactate in the working muscles will restrict muscle contraction and strength output. Reducing work intensity will allow the muscles to recover and let lactate to diffuse out of the cells to the bloodstream. During moderate exercise intensity with adequate oxygen supply, part of lactate is used in the cell as an energy resource, while the remaining diffuse out of the cells which is utilized by the myocardium to be converted into acetyl-CoA [48] and Cori cycle in the liver to reproduce glucose. The following exercise activities use anaerobic metabolism during effort:

1. Resistance maneuvers such heavy-load weightlifting and other efforts where strength is applied.
2. Running or cycling at high speeds above 90% of VO_{2max} turning to anaerobic metabolism to regenerate ATP at the amount needed.
3. Short burst of speed, dashes such as 100, 200, and 400 m.
4. Interval training that relies more on anaerobic metabolism for rapid fast muscle fibers contractions but fatigue fast as well.

Intensity of an aerobic exercise training for healthy purposes is between 60 and 75% of VO_{2max}, without higher intensities. This enables the breaking down of carbohydrates and fats through the aerobic metabolic pathway. During exercise, using aerobic metabolism where oxygen is enough for energy results in well-controlled low lactate levels, less muscle distress and fatigue compare to anaerobic bout.

The following exercise activities use aerobic metabolism during effort: Walking, one of the easiest aerobic exercises is walking at a brisk pace, for about 30 min per day. Running, cycling, rowing, swimming, cross-country skiing, and laboratory cardiopulmonary exercise machines exercising at moderate-intensity, 3–5 times per week for 30–45 min each session. In contrast, glycolysis, pentose phosphate pathway, and fatty acid biosynthesis all occur in the cytosol of a cell [49].

Fat Metabolism

Fatty acids and fats converted to acetyl-CoA can be utilized only in the mitochondria, with the presence of oxygen through the Krebs cycle for ATP synthesis [50]. Fatty acids are a significant storage of energy utilized during exercise, mainly during long-lasting exercise at mild-moderate intensity [51, 52]. In humans, the majority of the fatty acids are placed in dermal and deep visceral adipose tissue with small amount in muscles (around 300–350 g). In addition, fatty acids are vital parts of the phospholipids forming phospholipid membranes bilayers of the cells and numerous organelles within the cells. Energy per gram of fatty acid produce most ATP when they are entirely oxidized to CO_2 and H_2O by β-oxidation and the citric acid cycle [53]. Whole oxidation fatty acid containing 16 carbons such as palmitate molecule generates 129 ATP molecules when comparing energy equivalent of one-gram carbohydrates to one-gram fat, it reveals that carbohydrate contains 4 kcal and fat 9 kcal. In fact, fat energy stored in a person weighing 70 kg equals between 75,600 to 85,500 kcal (8.5–9.5 kg fat), while in carbohydrate, it equals to 1600 kcal (400 g). This suggests that physical performance length is very limited if it relies only on carbohydrates. During aerobic respiration, fatty acid (Fig. 2.14), mainly in the form of triglycerides, is the primary stored form of energy in the body.

However, since most of the body's fat is stored in inaccessible adipose tissue stores, its availability for immediate use does not exist. Subsequently, it is essential to free fat stored from adipose tissue, mainly as triacyclglycerols. The breakdown of

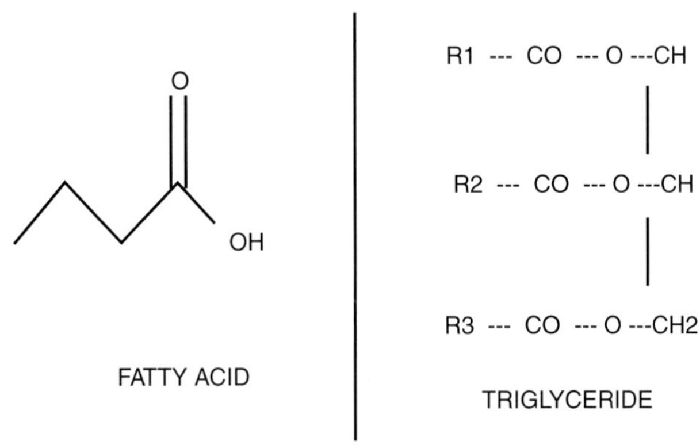

Fig. 2.14 Triglyceride is an ester derived from glycerol and three fatty acids, while fatty acid is a carboxylic acid with a long aliphatic chain, which is either saturated or unsaturated

triacyclglycerols gives free fatty acids (released to the plasma) and glycerol. At that point, free fatty acids are delivered by the bloodstream to the muscles. Thus, fat can be used during low-to-moderate exercise intensity due to its slow course. Fat oxidation seems to be diminished when exercise intensities are above 80% of VO_{2max}. Exercising at intensities above 90% of VO_{2max}, ATP production by fat oxidation turn out to be very small [54]. At rest and during long-lasting exercise, dependent on the mode, intensity, and duration of exercise and on the training status, the greatest portion of energy supplied during aerobic exercise comes from triglycerides [55]. Long-lasting aerobic exercise training significantly improves fatty acid oxidative volume and for ATP synthesis, and thus fatty acid proportion of total energy increases during exercise.

At the onset of exercise, sympathetic nervous system impulse releases epinephrine and norepinephrine, which stimulate beta-adrenergic receptors which activate lipolytic enzyme, thus starting lipolysis to separate fatty acids from triglycerides in the adipose tissue (Fig. 2.15).

In addition, in trained subjects, due to increased bloodstream catecholamines concentration, lipolytic activity improves [56].

When glucose and glycogen stores in the muscles and liver are depleted, amino acids (protein) released into the bloodstream are used for ATP synthesis. In the liver, through the Cori cycle (gluconeogenesis pathway), amino acids are converted into glucose. The glucose, when needed, is released from the liver into the bloodstream and then to the muscles. Triglycerides are broken-down by most lipases at a specific position on the glycerol backbone (Fig. 2.14). Reduced blood glucose levels lower bloodstream insulin levels, simultaneously to that, bloodstream glucagon, epinephrine, norepinephrine, growth hormone, and cortisol levels increase, and two enzymes hormone-sensitive lipase and lipoprotein lipase both fat-splitting enzymes found in the blood are activated. Additional aspects that can affect lipolysis may be the effects of long-lasting aerobic exercise training on VO_{2max} and gender.

Trained individuals with higher VO_{2max} improve significantly their capacity to oxidize larger amount of free fatty acids compared to untrained individual [57], and it also appears that males have a lower rate of fat oxidation when compared to

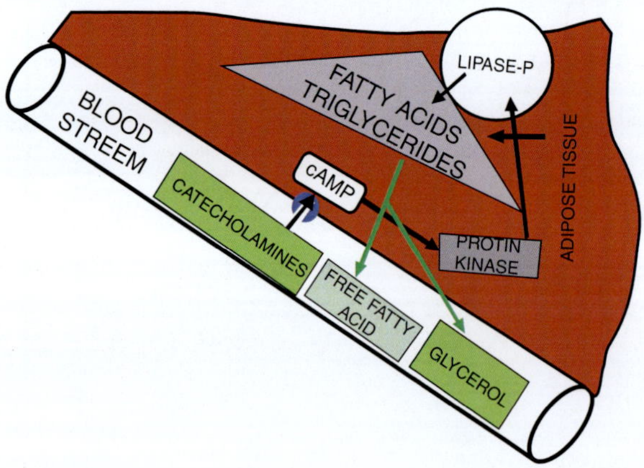

Fig. 2.15 Lipolysis releases triglycerides from the bounded glycerol to produce free fatty acid and glycerol

females during moderate exercise intensity [58]. It seems that lipolysis in females is affected by estrogen, through inhibiting lipoprotein lipase enzyme and by triggering the beta-adrenergic receptors in fat cells affecting lipolysis.

Plasma free fatty acid oxidation is directly related to the rate of lipolysis in adipose tissue. However, triglyceride hydrolysis within the muscle seems to be facilitated completely by beta 2-adrenergic stimulus. Ninety nine percent of long-chain free fatty acid are transported by plasma albumin to muscles, where they enter into the cells by specific protein-mediated transport SLC27 [59]. In the cell, the enzyme, long-chain-fatty acid CoA ligase, accelerates the energetic reaction between ATP and fatty acid to fatty acyl-adenylate, which latter is converted to fatty acyl-CoA molecule. Mitochondrial oxidation of long-chain fatty acid delivers a significant basis of energy for the skeletal muscle during long-lasting aerobic exercise training. Acyl-CoA enters the mitochondrion by the carnitine shuttle that is [60] responsible to move long-chain fatty acids over the inner mitochondrial membrane to be catalyzed by β-oxidation enzymes [61]. β-Oxidation is named as such for the reason that the beta carbon of the fatty acid undergoes oxidation to a carbonyl group. β-Oxidation, in the mitochondrial matrix, converts acyl-CoA molecules into acetate (2 carbons molecule), which combine with CoA to form acetyl-CoA, NADH, and $FADH_2$, used in the electron transport chain. At the starting point of the Krebs cycle, acetyl-CoA by combining with oxaloacetate forms citrate. This results in complete transformation of the acetyl-CoA to CO_2 and H_2O. From one oxidized acetyl-CoA where β-oxidation reaction of fatty acids takes place, energy yield adds up to1 GTP and 11 ATP molecules [62]. The glycerol is converted to glucose in the liver through gluconeogenesis.

The Cori Cycles

The Cori cycle is located in the liver and is enzymatically able to add two lactate molecules that each of them have three carbons to produce glucose that have six carbons. In addition, the Cori cycle is also able to convert amino acids to glucose. Therefore, the Cori cycle produces from two different substrates; lactate and proteins, glucose. During strenuous aerobic exercise, lactate produced is increased due to the utilization of the glycolytic metabolic pathway in order to reduce muscle lactate levels, it diffuses to the bloodstream and then to the liver. Lactate accumulated in the bloodstream is taken up by the liver where chemical process known as gluconeogenesis takes place to convert 20% of the produced lactate back to glucose. This glucose is then presented into the circulation, which transports back to the working muscles, where it is utilized to feed the additional energy demands of the muscles (Fig. 2.16).

During supramaximal exercise, the produced lactate is converted to glucose during recovery, due to the short-lasting bout. The proteins breakdown takes place only during long-lasting intense aerobic exercise (above 4 h), where muscle's glycogen levels are significantly low. Under this condition, only 3% of the total body's proteins can be converted to glucose. Figure 2.17 reveals the protein breakdown in the Cori cycle using the amino acid alanine to produce glucose.

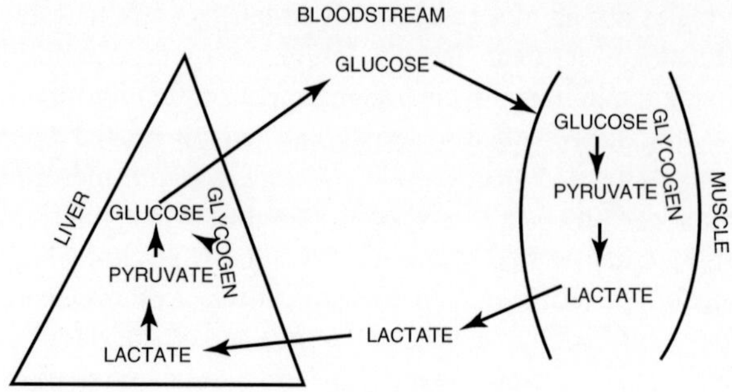

Fig. 2.16 Glucose production from lactate in the Cori cycle

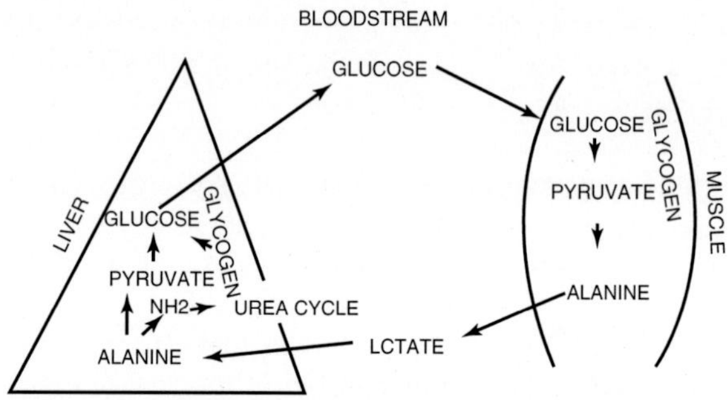

Fig. 2.17 Glucose production from alanine in the Cori cycle

Protein cannot be converted into acetyl-CoA in the muscle, consequently, it must first be converted into glucose from the amino acid alanine, the only amino acid that can enter the liver, however, alanine is limited in its amount in the bloodstream. Therefore, muscles' amino acids are broken-down, the NH_2 is connected to pyruvate through an enzymatic process to form alanine. This alanine is then presented into the circulation, which transports it to the liver. In the liver again, through an enzymatic process alanine molecule is broken-down to NH_2 and pyruvate. NH_2 is recycled to amino acid or exerted by the kidney while pyruvate is recycled to glucose in an opposite direction of glycolysis pathway.

Metabolic Rates

Metabolism includes the chemical reactions that create energy in order for the body to function. Rate of metabolism or metabolic rate is the total energy utilized per unit of time. About two-thirds of total daily energy applied is payable to the basal life processes, and one-third is for food digestion and routine physical efforts. These processes need mainly oxygen, fats, and carbohydrates. The effects of aerobic training on total daily energy utilizes are comparatively minor [63]. The term metabolism defines body's energy alteration in order to perform physical activity tasks, while the metabolic rate is the calories amount that the body utilizes. Caloric utilization is necessary to maintain cardiopulmonary function, blood circulation, body temperature, nervous system function, and internal organs. The amount of energy needed at rest to maintain body's systems is resting metabolic rate (RMR) measured with less strict criteria and less controlled conditions than BMR. The larger portion of the daily energy expenditure is RMR, and it increases in response to exercise in spite of reductions in muscle mass and fat mass [64]. However, countless additional researches were unsuccessful to confirm the effects of exercise on RMR [65].

BMR energy accounts for basal life processes and it is the amount of energy used daily at rest, per unit time expressed as, calories, watts, VO_2 kg^{-1} min^{-1} or J h^{-1} kg^{-1}). Oxygen uptake is substitutable to calories and therefore, it is suitable for measuring BMR. VO_2 of one-liter equals to 4.83 kcal of energy, an adult at rest consumes about 14.4 VO_2 h^{-1} or 69.55 kcal h^{-1}, which is also 78.3 W. A doctor or nutritionist can devise a plan specifically for weight loss, maintenance, or weight gain goals through diet and exercise. In order to be precise in the determination of the metabolic rate, the metabolic cart is utilized. Although this method is not precise still, it gives some indication for therapy of obesity and metabolic diseases.

RMR differ significantly between individuals owed to age, gender, body composition, and health status. Knowing individual's BMR is an important variable in the ability of health professional staff to plan a personal course to control body weight; lose, gain, or maintain, by means of diet and exercise. It is measured under very restrictive conditions; when a person is awake at complete rest in a reclining position, following 8 h of sleep in a darkened room and when the digestive system is inactive. VO_2 is measured during the 20 min test along with CO_2 production and HR with the subject sitting in a lie-down, relaxed position. In addition, an accurate BMR measurement requires that the person's sympathetic nervous system is not stimulated. BMR accounts for almost 60–75% of the daily calorie expenditure by individuals, about 1200–1400 calories for females and 1400–1800 calories for males, 10% is the cost of the digestive activity and the rest comes from daily physical activity [66]. The aging process decreases energy expenditure, BMR declines by 1–2% per decade after age 20, mostly due to reduced lean body mass, while increasing muscle mass has the effect of increasing BMR. Resistance training that increases muscle mass and aerobic exercise with intensities above 60% VO_{2max} maintain metabolically active muscle and thus avoid increased fat mass which are the key to control age-associated weight and decreased metabolic rate [67]. However, variability between individuals is high [68].

Knowledge of the metabolic rate may point on the relationships between fat and lean body mass [69]. It is reasonable to assume that low metabolic rate is a result of small muscle mass.

Aerobic exercise increases metabolism during exercise at the working muscles [70] and remains slightly increased following exercise for a few hours [71]. It equals up to15% of calories expenditure in 48 h following exercise. A man weighing 70 kg who exercises aerobically 4 times $week^{-1}$ 30 min^{-1} his caloric expenditure equals almost 4000 kcal on top of the regular weekly caloric intake requirements, this includes the 15% of the recovery following the aerobic bout. Weightlifting or resistance training adds additional calories on top of the daily requirements during the workout by increasing muscle mass involved in the bout [72]. When performing regular strength training, BMR increases due to the increase in muscle mass [73]. Both aerobic and resistance exercise training are useful for increasing metabolism. Regular aerobic exercise and resistance training increase lean body mass [74] and so increase the number of calories utilized at rest.

An operational definition of metabolism is the rate of heat produced that describes the metabolic rate. The metabolic rate varies between individuals due to body size, gender, and fat percentage. When comparing two individuals with the same body weight and fat percentage, metabolic rate is different. Some studies have suggested that minimum of 20–25 min of cardiovascular training per day can temporarily increase BMR by around 10%, owing to an increase in the metabolism of the working muscles required for recovery as well as glycogen and other fuel sources used by the body like ATP and CP restoration. However, even a substantial increase in muscle mass would make only a minor impact on BMR.

BMR can be measured in units ranging from watt (J/s) to $mLO_2\ min^{-1}$ or joule hour per kg body mass [J/(h kg^{-1})] and, calories per unit of time, however, most often given in calories per day. In order to standardize BMR, it requires the measurement of the temperature at which the metabolic rate was measured. It means that basal rate of heat production is equal to the rate of heat loss to the environment. One characteristic of living animals is that they all give off heat produced by cellular respiration and cellular work (Fig. 2.18). Chemical reactions of the catabolic process increase body heat in warm-blooded animals. The first law of thermodynamics circumstances that energy cannot be abolished nor created but only changed in form. Is this the theoretical basis behind open-circuit spirometry? BMR may be reported in units ranging from watts h^{-1} to $mLO_2\ min^{-1}$ or kJ $kg^{-1}\ h^{-1}$. Energy quantity needs a firm set of criteria to be met. These standards include person's sympathetic nervous system not being stimulated, a condition which requires complete rest, being in a physically and psychologically undisturbed state, in a thermally neutral environment and, while in the postabsorptive state i.e., not actively digesting food. Thus, it is quite restrictive and only used in clinical or laboratory settings.

Fig. 2.18 From respiration to mechanical work and the heat production

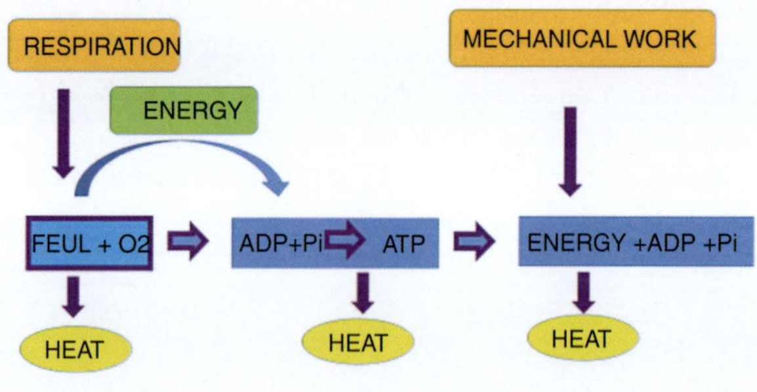

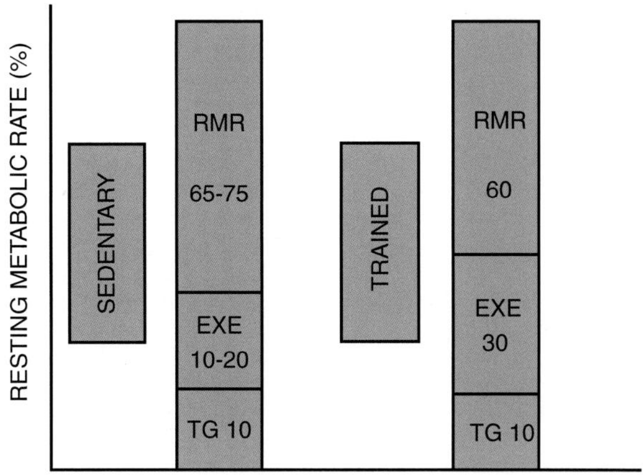

Fig. 2.19 Total resting metabolic rate (RMR) in sedentary and athlete's *TG* thermogenesis and *EXE* exercise

Aging, gender, and body composition are factors that define BMR levels. Since there is not much one can do regarding genetics, age, or gender, what is left is increased lean body mass namely muscle mass that increases metabolic rate. In addition, previous studies suggested that exercise training influences energy turnover during resting as other factors such as lean body mass, hormonal activity, genetics, and energy balance (Fig. 2.19).

The increased metabolic rate consumes more energy from fat as a main fuel source. However, even a substantial increase in muscle mass would make only a minor impact on BMR. Although variableness among individuals is wide, daily BMR expenditure until the age of 20 is 60–75% of the total daily calorie utilized by the individual. This value declines by 1–2% per decade thereafter due to aging, namely, loss of muscle mass. Thermogenesis fluctuations are influenced mainly by alterations in BMR, response to variations in diet intake, environmental temperature, and emotional tension. For example, under cold conditions, the adaptation is to rely on increased heat production.

The terms BMR and resting metabolic rate (RMR) are interchangeably used, however, these two terms although alike they differ. Common measurement that uses less strict criteria is for RMR. RMR is the energy expended while an individual is resting quietly in a supine position. Time of measuring RMR is regularly during the morning prior to eating or exercise and following a full night sleep.

The open-circuit spirometry is an indirect technique utilized to evaluation RMR. Measuring volume, composition of expired air, and converting oxygen consumption to kcal determine energy expenditure. Aerobic and anaerobic fitness level of an individual does not have any influence on resting metabolism. These modes of exercises cannot be considered as a factor for RMR of lean body mass. However, some studies suggest that a minimum of 25–30 min 3–4 week^{-1} of aerobic training can increase the BMR by around 10–50% due to the increase in the metabolism of the working muscles which depend on the training intensity and active muscle mass involved (as long as the individual exercises).

Weightlifting is an anaerobic mood of exercise that increases body muscle mass and contributes to the lean body mass of an individual. Therefore, the expected effective results following weightlifting exercise training will be an increased BMR. However, the actual effect of weightlifting exercise training on BMR is debated and difficult to count. It has been suggested that the RMR of trained muscle is around 55–60 kJ kg^{-1} d^{-1}, so, even a considerable increase in muscle mass would make only a slight influence on BMR.

Gender differences during exercise reveal that women depend on fat as the main source for energy. Consequently, since women utilize smaller muscle mass compared to males, they increase their storages of glycogen, protein, ATP, and CP availability. BMR can be estimate as suggested by the Harris–Benedict equation:

For men:

$$P\left(\mathrm{kcal\,day}^{-1}\right) = 13.751\mathrm{m} + 5.0033\mathrm{h} - 6.755\mathrm{a} + 66.473$$

For women:

$$P\left(\mathrm{kcal\,day}^{-1}\right) = 9.5634\mathrm{m} + 1.8496\mathrm{h} - 4.6756\mathrm{a} + 655/0955$$

where P = total heat production at complete rest; m = mass (kg); h = height (cm); a = age (years).

RMR includes the thermal effect of substrate metabolism and heightened metabolic activity due to prior physical or mental activity and/or stress. These factors, collectively known as facultative thermogenesis, may be consider as components of a person's RMR and are not part of the BMR. RMR is relatively constant, with minor fluctuations due to muscle mass, fitness level, and to dietary shifts. The exercise components have large variability and can greatly influence individual's total caloric expenditure. There is a possibility to determine how many calories an individual consume in a given day. A relatively accurate (but more difficult) method is the indirect calorimetry (metabolic carte), which uses expired gases to calculate the amounts and types of fuel being utilized. Energy balance aim in living systems is to reach homeostasis of energy. It is measured as follows:

Energy intake = internal heat produced + external work + storage

To define the amount of energy expenditure the unit kilogram calorie (kcal) is used, which equals the energy needed to increase the temperature of one liter water by 1 °C, from 16 to17°C, and equals to about 4.184 kJ. Each minute during repetitive knee extension muscles utilizes 900.5 mmol of glucose, only 22–26% (≈24 watts) is converted to mechanical energy, and ≈76 watts are heat energy. At rest, skeletal muscle has a BMR of 0.63 watt kg^{-1}, during exercise it increases by 23-fold. During short strenuous exertion, energy cost is even higher up to 314 watts kg^{-1} in young male when jumps up from a squat mechanically. Total daily energy expenditure is significantly higher than BMR and it depends on physical activity intensity, exercise lasting time, and mode (aerobic or anaerobic). During strenuous exercise, the capacity to maintain cell's ATP levels depends mainly on

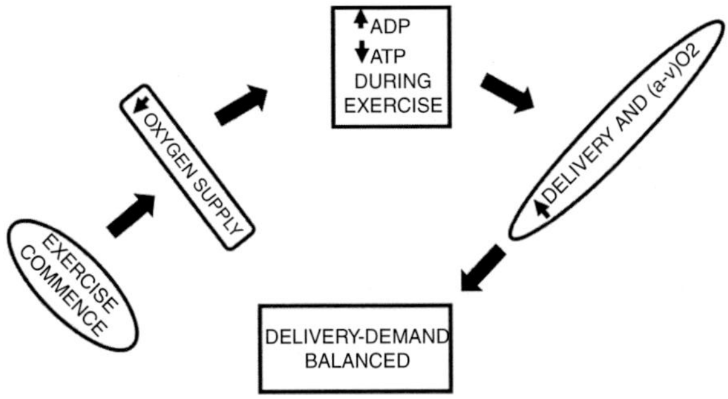

Fig. 2.20 Describes the changes in oxygen delivery-extraction during exercise. With increase in energy demand, consequently, an imbalance ATP/ADP occurs. This in turn will increase oxygen delivery and extraction to keep balanced ATP/ADP

the pulmonary function, heart's Q to increase oxygen delivery and, $(a–v)O_2$ at the tissue level (Fig. 2.20).

Consequently, the biological variations desirable through exercise are essential to be synchronized. Accordingly, during exercise, substantial increases in cardio-pulmonary system function and muscle efficient are seem to balance the increased metabolic demands and eliminate CO_2.

At rest, oxygen demand averages in males 250 and 200–210 $mLO_2\ min^{-1}$ in females. During aerobic bout, the cardiopulmonary increases its oxygen delivery up to 20–25-fold from resting values.

The following calculates total daily energy expenditure based on kcal per hour, with energy value of 1 kcal $kg^{-1}\ h^{-1}$, thus calculating total daily energy expenditure for a person weighing 70 kg for 24 h is as follows:

$$\text{Daily energy expenditure} = 70\,\text{kg}\,\text{kcal}^{-1}\,\text{kg}^{-1}\,24\,\text{h}^{-1} = 1680\,\text{kcal}.$$

The second possibility to calculate total daily energy expenditure is by means of METs, for a person weighing 70 kg for 24 h is as follows:

Average person consumes for each 1 K of body weight 3.5 $mLO_2\ min^{-1}$.

Calculating energy expenditure for whole body per minute is

$$VO_2\ \text{whole body} = 70\,\text{kg}\cdot 3.5\,\text{mLO}_2\,\text{kg}^{-1}\,\text{min}^{-1} = 245\,\text{mLO}_2\,\text{min}^{-1}$$

Therefore, energy expenditure for 1 h is

$$VO_2\ \text{whole body for hour} = 245\,\text{mLO}_2\,\text{min}^{-1}\,60\,\text{min} = 14.7\,\text{LO}_2$$

Thus, total energy expenditure for 24 h is

$$VO_2\ 24\,\text{h} = 14.4\,\text{LO}_2\,24\,\text{h} = 352.8\,\text{LO}_2$$

Since the energy value of 1-liter oxygen is 4.83 kcal, total energy expenditure equals to:

$$\text{Total energy expenditure} = 352.8\,\text{LO}_2\,4.83\,\text{kcal} = 1700\,\text{kcal}$$

Factors Affecting BMR and RMR

Body size and body composition are important determinants of BMR and RMR. A larger individual will generally have a higher BMR than a smaller individual. Larger body will require a greater amount of energy at rest to support larger organ systems (ventilatory, circulatory, etc.) and a greater muscle mass to maintain thermal equilibrium. BMR and RMR often expressed relative to an individual's estimated body surface area in order to allow a comparison between individuals of different body sizes. Body composition is also a major determinant of BMR and RMR. For two individuals of a given size, the individual with the greater lean body mass will generally have a higher BMR and RMR. This is attributed to the greater metabolic activity of muscle as compared to fat. Differences in BMR and RMR between men and women (males usually have larger muscle mass than women by 5–10%) are as a result of differences in body composition. If expressed BMR and RMR per unit of lean body mass, the difference between male and female in metabolic rates disappears. Most of the decrease in RMR associated with aging is generally associated with body composition changes (i.e., a decrease in lean body mass). The amount and type of food ingested by an individual affect RMR. Severe caloric restriction decreases RMR. It represents the greatest percentage of daily caloric expenditure in sedentary individuals; therefore, weight loss slows down. This situation slows the expected weight loss due to the amount of dietary restriction and negative caloric balance. In addition, if severe caloric restriction continues, the weight loss will become progressively slower. To obtain a true RMR value, the subject must be in a completely rested state and in the supine position. Use of leg and core postural muscles when the subject is sitting or standing will result in greater oxygen uptake and cause the RMR to be higher. Additionally, a fasted condition is required to minimize the thermic effect of food. Thus, for this experiment, the preexercise energy expenditure is not a true RMR but rather a pseudoresting measure, with the subject in a standing position.

The total amount of calories an individual burn in a given day is known as total daily energy expenditure, and it comes from several sources including RMR, increasing muscle mass increases BMR, although the effect is not significant enough to act as a weight-loss method.

Increase aerobic fitness level does not increase BMR when adjusted for fat-free body mass. Illness, previously consumed food and beverages, environmental temperature, stress levels and, exercise energy cost can affect individual's overall energy expenditure and its BMR measured by gas analysis through either direct (Bomb-calorimeter) or indirect calorimetry (metabolic carte, Fig. 2.21).

BMR is interchangeable with basal energy expenditure (BEE), while RMR is the same as resting energy expenditure (REE).

Predictive equations for basal energy expenditure (BBE) could be derived using weight, age, and gender as follows:

For females

BEE [kcal day^{-1}] = 10 body weight (kg) − 3 age (years) + 125.

For males

BEE [kcal day^{-1}] = 10 body weight (kg) − 3 age (years) + 750.

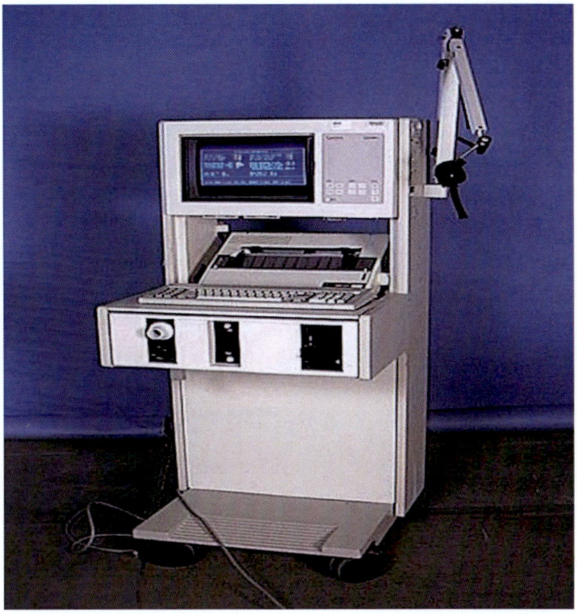

Fig. 2.21 Open circuit spirometry metabolic cart to define oxygen uptake in a direct method on line in real time

Direct vs Indirect Calorimetry

Bomb-calorimeter uses expired gases to calculate the amounts and types of fuel being utilized and measures the combustion's heat of a particular reaction. It serves as a direct measure of human heat production. Bomb-Calorimeter is an object used for calorimetry or the process of measuring the heat of chemical reactions or physical changes as well as heat capacity. This approach is difficult to operate, and therefore, the indirect calorimetry is used. Resting energy expenditure can be measure with indirect calorimetry using a metabolic cart to measure VO_2 and VCO_2 production.

Every liter of oxygen consumed is equivalent to the energy cost of 4.83 kcal. A metabolic cart can assess the energy requirements at rest and during exercise and determine work capacity. The indirect calorimetry method measures the volume of oxygen consumed or utilized by the body over CO_2 produced.

During metabolic energy transformations, oxygen is consumed and heat is produced. Either of these variables can therefore be used to estimate energy expenditure. Using the fact that one-liter of oxygen liberates 4.83 kcal of heat energy, a highly accurate indirect measure of energy production is possible when a mixture of carbohydrate, fat, and protein is burned in a bomb calorimeter. Two procedures of indirect calorimetry are closed circuit and open circuit spirometry. Open circuit spirometry is utilized for this laboratory experiment. In open-circuit method, subject breathes ambient room air, expired air passes directly through a gas meter. Both the volume of air inhaled per unit time and the percentage of oxygen and VCO_2 exhaled are important for determining VO_2 and energy expenditure.

Respiratory Quotient and RER

Comparisons of energy breakdown using both means offer substantial indication for the validity of the respiratory quotient (RQ). This ratio measures the number of oxygen units used by glucose, fats, and proteins and CO_2 production at rest and during exercise. At rest, individual's RQ values are usually in the mid-seventies (0.73) and will increase during exercise up to1.00–1.15 value as workload increases.

Although with reduced precision, compare to the RQ method, respiratory exchange ratio (RER) approach RQ values, however, it is far comfortable. The RER is the ratio of the CO_2 volume produced to the oxygen volume consumed at the body. RER is a measure determining VO_{2max}, however, its values has to be higher than 1.15. Because of inherent chemical differences in the composition of carbohydrates, fats, and proteins, different amounts of oxygen are required to oxidize completely the carbon and hydrogen atoms in carbohydrates, fats, and protein into VCO_2 and water. Thus, the quantity of VCO_2 produced relative to VO_2 vary depending on the proportional mix of energy nutrients (carbohydrate, fat, or protein) metabolized. Therefore,

$$RER = VCO_2 \,/\, VO_2$$

The RER ratio value is important not only in determining the body's rate of energy expenditure but also enables to determine the nutrient mixture being metabolize during rest or exercise.

RER for Carbohydrate, Fat, and Protein

The ratio of hydrogen to carbon and oxygen atoms in carbohydrates is 2:1. Therefore, all of the oxygen consumed is used to oxidize the carbon into VCO_2. Consequently, during the complete oxidation of a glucose molecule, six-molecules of oxygen need to oxidize six molecules of carbon into CO_2 molecule, and thus the overall reaction for carbohydrate RER will be:

$$\text{Glucose} + 6O_2 \rightarrow 6CO_2 + \text{water}$$

or;

$$C_6H_{12}O_6 + 6O_2 \rightarrow 6CO_2 + 6H_2O$$

Thus,

$$RER = 6CO_2 \,/\, 6O_2 = 1.0$$

The chemical composition of fat molecules differs from carbohydrates composition specifically; fat molecules contain considerably less oxygen atoms in proportion to atoms of carbon and hydrogen. Consequently, when fat is oxidized, more oxygen is required to convert fat into CO_2 and H_2O. For example, when palmitic acid is oxidized, 16 CO_2 molecules are produced for every 23 oxygen molecules consumed:

$$C_{16}H_{32}O_2 + 23O_2 \rightarrow 16CO_2 + 16H_2O$$

Therefore, the RER for this fatty acid is

$$\left(16CO_2 / 23O_2\right) = 0.696.$$

Note: RER value for fat in general is consider 0.70.

Protein is first determined (the amino or nitrogen groups are removed) in the liver and the nitrogen and sulfur fragments are released and excreted. The remaining fragments (ketoacids) are then oxidized to CO_2 and water. As with fats, more oxygen is needed for complete protein combustion. For example, the protein albumin is oxidized as follows:

$$C_{72}H_{112}N_2O_{22}S + 77O_2 \rightarrow 63CO_2 + 38H_2O + SO_3 + 9CO\left(NH_2\right)_2$$

The RER for this protein would be ($63CO_2/77O_2$) = 0.818.

The RER for protein in general is 0.82.

Once the nitrogen atoms and atoms other than carbon and oxygen are removed from protein, the protein expressions are similar to that of carbohydrate or fat. Therefore, it is not possible to separate out the contribution of protein in substrate utilization using expired air. Therefore, the RER is called a non-protein RER, since it is impossible to determine the contribution of proteins to energy metabolism. One-way to estimate the contribution of protein to energy metabolism is to measure urinary nitrogen. It has been determined that approximately one gram of urinary nitrogen is excreted for every 6.25 g of protein metabolized, and every one gram of nitrogen results in approximately 4.8 l of CO_2 produced with 6.0 l of O_2 consumed. With this information, it would be possible to estimate the contribution of protein to energy metabolism.

Rate of Energy Expenditure

The rate of energy expenditure of an individual during rest or activity is the amount of energy divided by heat liberated over time. It is calculated by taking VO_2 times the caloric equivalent per liter of oxygen utilized (4.83 kcal) for the RER value obtained.

Rate of Energy Expenditure = VO_2 caloric equivalent (kcal min^{-1}) = (LO_2 min^{-1}) (kcal L^{-1})

For example, using the mixed-diet RER of 0.82, the amount of energy liberated per liter of oxygen consumed is 4.83 kcal. If the rate of VO_2 is 0.25 l per minute, the RMR is

$$4.83\,\text{kcal}\,L^{-1}\,0.25\,L\,\text{min}^{-1} = 1.2\,\text{kcal}\,\text{min}^{-1}$$

If the above data (1.2 kcal min^{-1}) were obtain at rest from an individual (weighing 70 kg), his RMR for 1 min would be equal to 1.2 kcal min^{-1} and for 1 h it is 72 kcal h^{-1}.

Table 2.1 summarizes the RQ values of different energy sources

Energy sources	RQ values
Ketosis	<0.70
Fat	0.70
Protein	0.80
Mixed energy	0.85
Carbohydrate	1.00

Additional way to calculate RMR is by respiratory quotient (RQ) ratio, which is useful for understanding human's metabolism, while indicating which fuel is metabolize. Calculations of the RER and RQ are almost the same, namely from the production of VCO_2 and VO_2 volumes. The RQ calculated at rest and during exercise ranges from value of 0.7 for fat to1.0 for glucose, pointing about the metabolic pathway applied by the body. At peak all-out strenuous exercise, RER ratio exceeds that of 1 value up to1.4 value, due to hyperventilation. At this point of hyperventilation, high lactate and VCO_2 production intermixes the different metabolic pathways and therefore RER is not reliable. On the other hand, during moderate-intensity of aerobic exercise, the RQ and RER values will be 0.8 to 0.9, suggesting that the main energy source is fat and to the less extent glucose. Above that ratio of 0.9, carbohydrates are the main energy source (Table 2.1).

Hormonal Regulation of Metabolism

Hormonal glands, such as the pituitary, testes, ovaries, pancreas, thyroid, and the adrenal cortex, release and store body's hormones.

They are secreted to the bloodstream following stimulation of the autonomic nervous system. Hormones affect metabolism in the targeted cells and energy production in the muscles. Hormone acts as an accelerator or decelerator of muscle's energy production rate and facilitates fuel substrates to enter into the cell. The secreted hormone is attached to the cell's membrane sensor that fits together as a key and lock; in other words, the hormone has a specific sensor that it can only can combine. Hormones initiate increase in cell enzymes' amount, and by that, they increase simultaneously chemical reactions number. However, sometimes, in-spite of the above mentioned, hormones may significantly increase enzymatic activity, without increase in their cell amount. These hormonal effects, within the cell, affect the enzymatic activity in the long run to produce large energy quantities at a given time [75]. Other hormones affect the substrate entry to the cell by binding to the cell membrane, as an example, glucose enters the muscle cell by the action of the insulin hormone, while stress hormone such catecholamines, when in the presence of oxygen, slows cell glucose entry. The endocrinology system is involved in the regulation of metabolism in an extensive inconsistency of physiological responses and pathophysiological states. The endocrine system acts in the same way as the sympathetic nervous system [76], but the endocrinology system is very slow acting and hormone release takes time.

Following the activation of the autonomic nervous system, hormones are secreted where there are capillaries, enter bloodstream fast, circulate freely in bloodstream, or bind to a carrier protein freely. In addition, sympathetic nervous system has a key role in the regulation of hepatic metabolism [86]. Hormones are regulators that influence cell's metabolic rate [77] and energy production, by increase or reduce metabolic rate, and enable substrate entry to the cell of fat, glucose, and proteins. The hemodynamic effects of catecholamines simultaneously serve to distribute the mobilized substrates according to the various tissues needs; thus, they determine the metabolic rate.

The Endocrine System

The nervous system and endocrine system have an automatic regulation of the body's internal environment. During exercise, endocrine system's roles are increase in hormonal sensitivity, maintain homeostasis, control substrate metabolism, and regulates fluid and electrolyte balance. Hormones are secreted and regulated through negative feedback. Hormones have very specific functions when they reach nearly all parts of the body through blood. Each cell has specific hormone receptors influenced by the target tissues so they can bind only to specific hormones. When hormone is deficient, target cells become more sensitive and receptors are added, while when hormone is in excess, target cells becomes less sensitive and the number of receptors decreases. Hormones that float in the blood freely do not have any effect.

The interplay between the glands, hormones and other target organs is referred to as the endocrine system. Endocrine glands provide the basic case, but specialized cells in various other organs also secrete hormones. Hormone is a controlling ingredient that is delivered by the circulatory system or fluid, to send a message to another part of the body, and to excite specific cells or tissues into action. Hormonal effects depend on where they are released, as they can be released in different manners [78]. Not all hormones are released from a cell and into the bloodstream until, it binds to a receptor on a target. The major types of hormone signaling are (a) endocrine that acts on the target cell after being released into the bloodstream, (b) paracrine acting on a nearby cell and does not enter general circulation, (c) autocrine that affects the cell type that secreted it and causes a biological effect, and (d) intracrine acts intracellularly on the cell that synthesized it.

The majority of hormones start a cellular activity by primarily binding to either cell membrane related or intracellular receptors. Cell may have numerous diverse receptors, sorts that identify the different hormone but then activate different signal transduction pathways or the opposite may happen, a cell may have numerous dissimilar receptors that identify unlike hormones and activate the same biochemical pathway [79].

Occasionally, the term hormone is used to describe chemicals molecules produced by cells that affect the same cell (autocrine or intracrine signaling) or nearby cells (paracrine signaling). Hormone is a chemical signaling molecule, made by specialist cells (glands), found in multicellular organisms, targeting organs to afford an interior communication arrangement between cells located in different parts of the body, in order, to regulate physiology and behavior variables [80]. Hormones in humans serve two sorts of communication: (a) communiqué among two endocrine glands, in which, one gland is affected by the release of hormone, by the other gland, and (b) communication, between an endocrine gland and a target cells, for example, when the pancreas releases insulin which causes muscle and fat cells to take up glucose from the bloodstream.

Hormones have diverse chemical structures, mainly of three classes: (a) eicosanoids, (b) steroids, and (c) amino acid/protein derivatives (amines, peptides, and proteins). The glands that secrete hormones contain the endocrine signaling system.

Hormones aid to connect among tissues for physiological control and communication activities such as metabolism, respiration, growth, movement, appetite, puberty, and fertility [81]. Hormones affect targeted tissues by binding to specific receptor proteins, leading to change in cell function. When a hormone binds to the receptor, it results in the activation of a signal transduction pathway that typically activates gene transcription, consequential in increased expression of target proteins; nongenomic effects are faster, and can be synergistic with genomic effects [82]. Amino acid-based hormones are water-soluble and act on the surface of target cells, by second messengers steroid hormones, being lipid-soluble, move through the plasma membranes of target cells (both cytoplasmic and nuclear) to act within their nuclei. Hormone secretion may occur in many tissues.

Hormone secretion occurs in response to specific biochemical signals from a wide range of regulatory systems. For instance, serum calcium concentration affects parathyroid hormone synthesis; blood sugar (serum glucose concentration) affects insulin synthesis. Regulation of hormone synthesis of gonadal hormones, adrenocortical hormones, and thyroid hormones often depends on complex sets of direct-influence and feedback interactions involving the hypothalamic–pituitary–adrenal, gonadal, and thyroid axes.

Upon secretion, certain hormones, including protein hormones and catecholamines, are water-soluble and are thus readily delivered via the circulatory system. Other hormones, including steroid and thyroid hormones, are lipid-soluble; to achieve widespread distribution, these hormones must bond to carrier plasma glycoproteins: thyroxine-binding globulin to form ligand–protein complexes. Some hormones are completely active when released into the bloodstream such as the case for insulin and growth hormones, while others are prohormones that must be activated in specific cells through a series of activation steps that are commonly highly controlled. The endocrine system secretes hormones directly into the bloodstream, typically via fenestrated capillaries, whereas the exocrine system secretes its hormones indirectly using ducts. Hormones with paracrine function diffuse through the interstitial spaces to nearby target tissue.

Hormone can control production and release of other hormones. Hormone signals control the internal environment of the body through homeostasis.

Hormones have the following effects on the body [83]:

- stimulation or inhibition of growth,
- wake–sleep cycle and other circadian rhythms,
- mood swings,
- induction or suppression of apoptosis (programmed cell death),
- activation or inhibition of the immune system,
- regulation of metabolism,
- preparation of the body for mating, fighting, fleeing, and other activity,
- preparation of the body for a new phase of life, such as puberty, parenting, and menopause,
- control of the reproductive cycle, and
- hunger cravings.

The rate of hormone biosynthesis and hormone secretion is mostly controlled by a homeostatic negative-feedback control mechanism. The mechanism depends on factors that influence metabolism and elimination of hormones. Therefore, increased hormone concentration levels alone is not enough to activate the negative-feedback mechanism. Consequently, there is a need for hormonal overmuch in order to activate the negative response [84]. Hormone effects can be inhibited and thus regulated by competing substances that bind to the same target receptor as the hormone in question. When a competing substance is bound to the receptor site, the hormone is unable to bind to that site and is unable to elicit a response from the target cell, these are antagonists of the hormone.

Anabolic and Catabolic Hormones

Many of the metabolic procedures are controlled by hormones. Overall, hormones can be labeled as anabolic or catabolic. Usually hormones were classified as anabolic or conditional on which fragment of metabolism they act. The classic anabolic hormones are the anabolic steroids, involved in the stimulation of protein synthesis, thus increasing muscle mass. The second well-known hormone is insulin. It is considered to be the most important anabolic hormone of the body, promoting the change of small molecules in the blood into large molecules inside the cells. Insulin deficiency may lead to diabetes. Estrogen, a key female sex hormone, regulates and backs female's progress reproductive structure system, and to the lesser sex features. Testosterone, present in males and females, regulates some male sexual features, strengthens bones, and muscle mass. Growth hormone stimulates and regulates growth regulate bone restoration. Catabolic hormones are involved in metabolism process and numerous signals that control catabolism. The well-known catabolic hormones are cortisol, glucagon, catecholamines, cytokines, orexin, and melatonin. Catecholamines are the key factor that increases HR, affects lungs bronchioles, and increases body's glucose used for rapid energy. Cortisol is a stress hormone which increases blood pressure, blood glucose concentration, and overwhelms the immune system function. Glucagon stimulates the breakdown of glycogen during glycolysis into glucose-6-phosphate. Glucagon, in the liver, stimulates the transformation of glycogen to glucose which enters the blood. Cytokines is a small protein that regulates communication and interactions between cells. Cytokines are both proinflammatory and antiinflammatory and have parallel functions stimulated by dissimilar cytokines. When bacteria or virus invade the body, interleukin and lymphokines (two kinds of cytokines) are released as an immune response to the invasion.

In the muscle, during ATP hydrolysis, the released energy is utilized for contraction by rotating actin over myosin to generate force [85]. Force generation depends on loads applied upon the muscle to shorten in length (concentric contraction), unchanged fibers length (isometric contraction) or instead, lengthening of muscle fibers (eccentric contraction). Work-capacity of a skeletal muscle measured after the 30 s equals to 24kJ mol^{-1} of ATP [86].

G Protein-Coupled Receptors

Also known as seven-transmembrane domain receptors respond to a variety of hormonal stimuli by binding to variety hormones and neurotransmitters [87]. G protein-linked receptors create a large receptor that are sensitive to external effectors and activate the cells inside (Fig. 2.22). When a G protein-coupled receptor is activated by its extracellular ligand, the Gs alpha subunit of the stimulated G protein complex converts GDP for GTP released from the complex [87]. In a cyclic adenosine 3,5′-monophosphate (cAMP)-dependent pathway, the activated Gs alpha subunit binds to adenylyl cyclase enzyme, a key regulator enzyme found in all cells. Adenylyl cyclase catalyzes the conversion of ATP into cAMP [88].

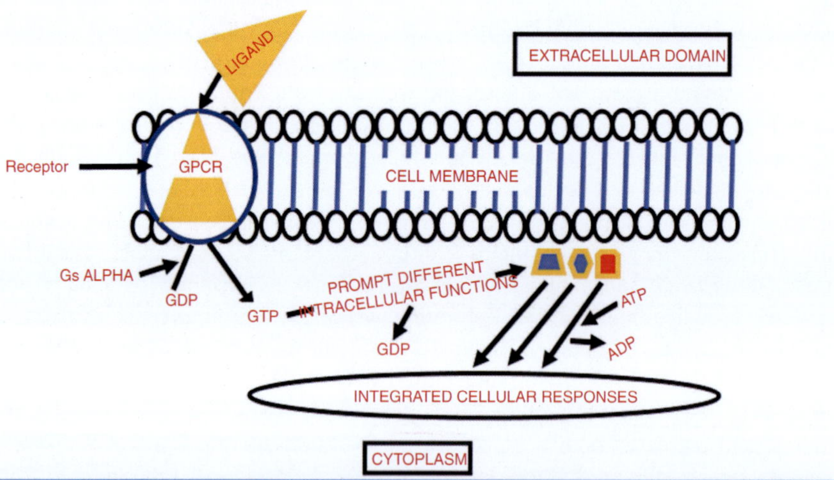

Fig. 2.22 The basic concept of GPCR (G protein-coupled receptors). It reorganizes to allow activation of G-alpha proteins in the plasma membrane

Mitochondrial cAMP controls the function of mitochondrial enzymes, regulating oxidative phosphorylation through intramitochondrial protein kinase A [89]. Increase in the concentration of cAMP can activate the central nervous system [90], causing a combined sympathoadrenal, cardiopulmonary, hepatic and adipocyte response to increase oxygen, and substrate supply for increase energy production. The sympathoadrenal system releases the hormones epinephrine, norepinephrine, and cortisol, although norepinephrine acts as a neurotransmitter.

Norepinephrine functions in the brain and body as a hormone and neurotransmitter by binding to alpha and beta receptors on the cell membrane [91]. Catecholamines simultaneously breakdown and distribute rapidly local fuel stocked according to the needs of the working muscles. Facilitate transfer of free fatty acids from triglyceride stored in adipose tissue [92]. Catecholamines does it by insulin suppression and the release of glucagon. Circulating epinephrine, if present in high concentration, can definitely stimulate glycolysis, gluconeogenesis, and lipolysis. Low blood glucose level (hypoglycemia), though is the only situation in which epinephrine appears to be preferentially and predominantly secreted.

Stress hormones break down triglyceride molecules into glycerol and free fatty acids and convert glycogen to glucose. Triglycerides are a type of fat (lipid) found in the blood. Triglycerides are stored in fat cells. Later, stress hormones release triglycerides form that storage. In addition, stress hormones increase liver glucose production which later is diffused to the bloodstream, thus, again glucose levels rise in the blood. Along with it, the cell increases its utilization of glycerol and free fatty acid to produce energy. Free fatty acids enter the cell facilitated by growth hormone and catecholamine to the Krebs cycle as acetyl-CoA, while glycerol enters the cell by diffusion utilized in the glycolytic pathway.

Insulin is formed in the pancreas by the beta cells and is secretes into the bloodstream to regulate body's glucose or sugar level [93] and to affect protein synthesis in a wide variety of tissues. Insulin is an anabolic hormone that promotes glucose transport (Fig. 2.23) by glucose transporter (GLUT4). It is the key hormone of carbohydrate metabolism; it also influences the metabolism of fat and proteins [94]. It lowers blood glucose by increasing glucose uptake, glycogen synthesis, and stimulates fat and protein [94]. In the cells, glucose is converted into either glycogen via glycogenesis or triglycerides through lipogenesis, in the liver, glycogenesis and lipogenesis take place [95]. High blood insulin levels, also, controls and inhibits liver's glucose production and secretion [96]. Reduced blood insulin levels in the blood have the opposite effect on its anabolic activity.

Substrate Oxidation

During exercise, fatty acid (FA) oxidation is regulated in skeletal muscle during exercise. From the available evidence, it seems that acetyl-CoA availability in the mitochondrial matrix adjusts fatty acid oxidation to exercise intensity and duration. This is performed at the step of mitochondrial fatty acyl import (Fig. 2.24), as the extent of acetyl group sequestration by carnitine determines the availability of

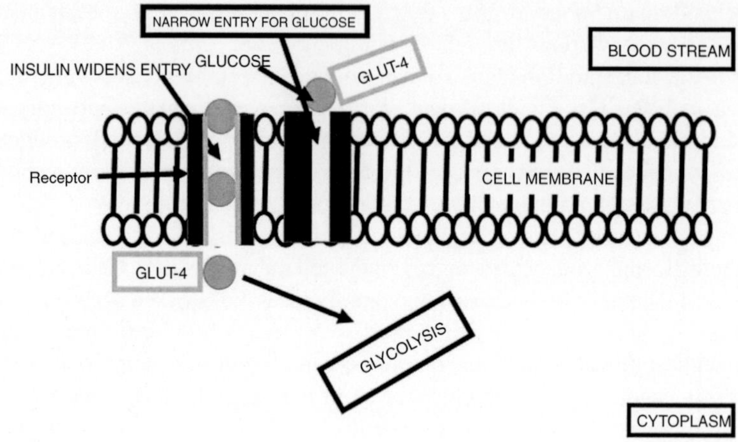

Fig. 2.23 Insulin action on its receptor widen glucose entry channel

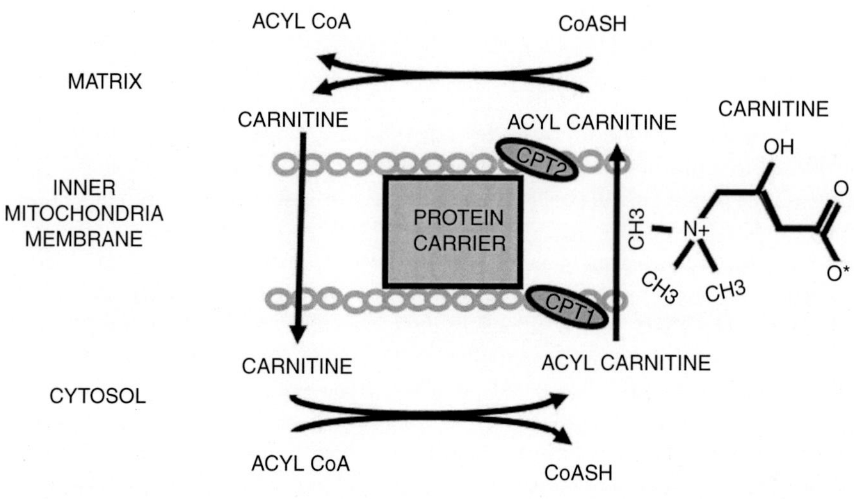

Fig. 2.24 Isoforms carnitine palmitoyltransferase (CPT1 and CPT2) mechanism. Asterisk, carnitine bound in the catalytic site of carnitine acetyltransferase, an enzyme homologous to CPT1

carnitine for the carnitine palmitoyltransferase 1 (CPT1) reaction [97]. In humans, carnitine is arranged in sets of four ammonium compound [98], involved in metabolism [48]. The function of carnitine is the transportation of fatty acids and acetate through the inner mitochondrial membrane [99].

Carnitine may exist in two isomers, labeled D-carnitine and L-carnitine, as they have the same molecular formula but a different arrangement of atoms in space. The rate of glycolysis seems therefore to be central to the amount of β-oxidation-derived acetyl-CoA that is oxidized in the tricarboxylic acid cycle.

Fatty acid oxidation during exercise is also determined by fatty acid availability to mitochondria, dependent on trans-sarcolemma fatty acid uptake via cluster of differentiation 36/SR-B2 (CD36), and fatty acids mobilized from myocellular lipid droplets.

Insulin is a peptide hormone that is critical for managing the body's metabolism. Insulin is released by the pancreas when blood sugar levels rise, additional to many effects it stimulates lipogenesis.

Insulin stimulates lipogenesis primarily by activating two enzymatic pathways. Pyruvate dehydrogenase converts pyruvate into acetyl-CoA. Acetyl-CoA carboxylase converts acetyl-CoA produced by pyruvate dehydrogenase into malonyl-CoA. Malonyl-CoA provides the two-carbon building blocks that are used to create larger fatty acids. Insulin stimulation of lipogenesis also occurs through the promotion of glucose uptake by adipose tissue [100]. Lipogenesis stimulation increases glucose consumption by activating lipogenic and glycolytic enzymes through covalent alteration [101].

Glucagon is a peptide hormone produced by alpha cells of the pancreas. It increases the concentration of glucose and fatty acids in the bloodstream and is considered to be the main catabolic hormone of the body [93]. Its effect is opposite to that of insulin, which lowers the extracellular glucose [102].

Cortisol is a catabolic steroid hormone produced by the adrenal gland in response to low blood sugar and exercise. It supports energy metabolism during long periods of exercise by facilitating the breakdown of triglyceride (use glycerol) and protein to create the glucose necessary to help fuel exercise. Cortisol is released when the body experiences too much physical stress or is not sufficiently recovered from a previous workout. While cortisol helps promote fat metabolism, when carbohydrate level is low, elevated levels of cortisol catabolize muscle protein for fuel. Proteins are composed by amino acids and are the main structural component of muscles, hormones, enzymes, and hemoglobin. Proteins serve as energy source; however, proteins are not the primary choice as an energy fuel source. To utilize proteins as energy source, they are metabolized into amino acids [103].

Testosterone is a steroid hormone produced by the Leydig cells of the testes in males and the ovaries of females, with small amounts produced by the adrenal glands of both genders. Testosterone is responsible for muscle protein resynthesize and the repair of muscle proteins damaged by exercise and plays a significant role in helping grow skeletal muscle. Testosterone works with specific receptor sights and is produced in response to exercise that damages muscle proteins.

Human growth hormone is an anabolic peptide hormone secreted by the anterior pituitary gland that stimulates cellular growth [104]. Similar to all hormones, growth hormone works with specific receptor sites and can produce a number of responses, including increasing muscle protein synthesis responsible for muscle growth [105], increasing bone mineralization, supporting immune system function and promoting lipolysis, or fat metabolism. Growth hormone is stimulated by high-intensity exercise during resistance bouts and aerobic exercise.

Insulin-like growth factor (IGF) has a similar molecular structure as insulin and is stimulated by the same mechanisms that produce HGH [48, 60]. IGF is a peptide hormone produced in the liver and supports the function of HGH to repair protein damaged during exercise, which makes it an important hormone for promoting muscle growth.

Hormonal Regulation During Anaerobic Exercise

Increased energy needs during exercise press the nervous, muscular, cardiovascular, metabolic, and respiratory systems. These trigger responses and adaptations to exercise, through a number of multipart hormonal interactions happening at the same time. For example, hormones can increase blood pressure, stimulate protein synthesis, and increase the body's metabolic rate.

During vigorous exercise, an oxygen delivery supply imbalance is reached, consequently, glucose is the only muscle fuel. Glucose production increases 7–8-folds while its consumption increases 3–4-folds; so, blood glucose increases and plasma insulin stay stable or decreases nonsignificantly [106]. Catecholamine responses to this kind of exercise are in control for glucose production increment and controlled increase in glucose consumption [107]. Catecholamines, via α2-adrenoceptors, usually inhibit insulin release during exercise [108]. However, it does not restrict glucose entry to the working muscles, and, do not suppress insulin activity and so glycolytic metabolism in the cell is possible along with reduction in bloodstream glucose.

During vigorous interval exercise, training responses differ; sessions lasting 25–30 min improve blood glucose better than exercise volume [109].

Hormonal Regulation During Aerobic Exercise

Long-lasting aerobic exercise training with intensities up to 85% VO_{2max} increases skeletal muscle work capacity and fat oxidation by increasing mitochondrial number and density and improves enzymes involved in β-oxidation and muscle oxygen uptake. In addition, aerobic training increases the number of several free fatty acid transporters, consequently increases fat uptake in cell [71, 97]. Chronic moderate-intensity aerobic exercise training improves blood glucose and increases insulin sensitivity and efficiency [110]. During long-lasting aerobic exercise, there is a significant increase in fat intake by the muscle and less carbohydrates utilization accordingly, fat become the main muscle energy source. Stress hormones, epinephrine, norepinephrine, glucagon, and cortisol, and growth hormone [46], restrain insulin action, causing a significant increase in bloodstream glucose. High blood catecholamines affects metabolic level associated with enhanced rates of aerobic glycolysis, by increasing glucose blood and lactate levels [111].

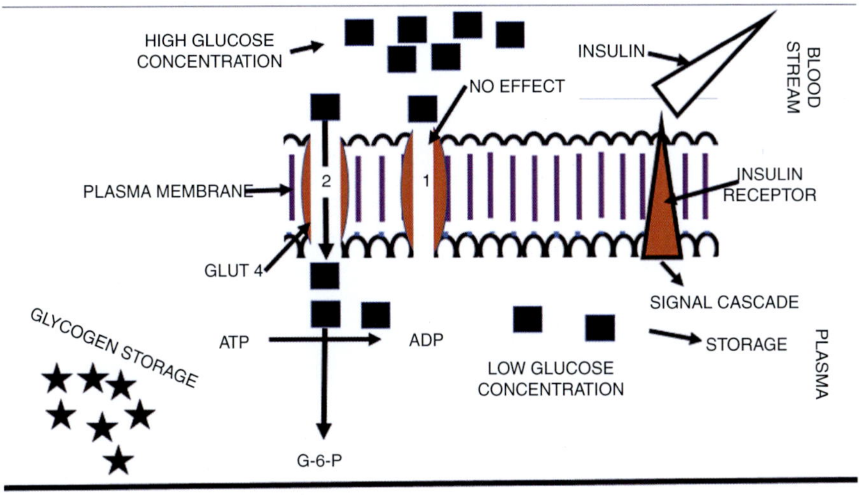

Fig. 2.25 Exercise stimulates GLUT4, translocates, and increases glucose transport through an insulin-independent pathway. Low intracellular glucose increases insulin sensitivity

Though more free fatty acid enters the cell as a main fuel for energy production, catecholamine simultaneously suppresses insulin activity and epinephrine increases blood flow to working muscles, cardiac output, and blood glucose concentration [112]. During low intensities of aerobic exercise, when there is no limit of oxygen, insulin activity is suppressed by β-cell α-adrenergic receptor activation [106].

Substrate utilization is a product of both exercise intensity and duration, with a key role to the sympathetic nerves, which mobilize free fatty acids from adipose tissues [113].

At rest and during aerobic exercise, insulin hormone regulates the blood glucose level by stimulating several functional responses in working muscle and other target cells. Insulin helps GLUT4 to transfer glucose across cell membrane (Fig. 2.25) to the plasma membrane, in skeletal muscle, and adipose tissue. Thus, insulin facilitates diffusion and consumption of glucose by the cells from the bloodstream, in addition to blood glucose increases. Glucose transport is achieved by promoting translocation of intracellular vesicles that contain the GLUT4 and GLUT1 glucose transporters to the plasma membrane. During aerobic exercise when oxygen demand supply is balanced, fat is the main muscle energy source, stored out of the muscle, while carbohydrates with small quantities (400 g body weight) are stored largely in the muscles. Stress hormones breakdown triglyceride molecules into glycerol and free fatty acids and convert glycogen to glucose.

Exercising aerobically at intensities of 55–75% VO_2 (moderate) increases blood glucose levels, however, it depends on exercise length and volume. The majority of references are for 150–210 min week^{-1} at the abovementioned intensities [114]. In addition, it is recommended to add resistance exercise 3–4 times week^{-1} [115].

Aerobic and resistance exercises training increase GLUT4 amount and transfer, thus, blood glucose uptake which increases insulin independently [115].

At rest and during aerobic exercise, insulin hormone regulates the blood glucose level by stimulating several functional responses in working muscle and other target cells. The sympathoadrenal system releases the hormones epinephrine, norepinephrine, and cortisol, although norepinephrine acts as a neurotransmitter. An increase in exercise intensity from rest to 55% VO_{2max} and on to 85% and above give rise of 200 and 400% in the bloodstream of norepinephrine and epinephrine level, respectively. These exercise intensity increments are key factors in the mobilization of substrates required by cells to supply the ATP needed by the working muscle [116]. During aerobic exercise, catecholamines increase lipolytic rate to release free fatty acids from stored triacylglycerol [117], with peak oxidation of fatty acids reached at 65% of VO_{2max} [118], a response matching the increased bloodstream epinephrine level [119]. However, at 85% VO_{2max} and above, epinephrine level in bloodstream increases, but then utilization of fatty acid in cells drops, since free fatty acid mobilization fails to keep up [120]. Another factor that keeps a normal level of glucose in the blood during steady-state aerobic exercise, is the stable hepatic yield of glucose. This response is facilitated by glucagon. It influences glycolysis and gluconeogenesis. Due to the interplay between insulin and glucagon release, glucagon level increases during all-out exercise bout, however, during moderate exercise, in dogs, glucagon does not increase [121]. Glucagon level increases during exercise, almost by 35% compared to its levels during low to maximal intensities.

Hormonal Regulation During Resistance Exercise

With regard to gender differences, the most important differences between male and female endocrinologically end up in the testes versus ovaries differences. Females produce low levels of testosterone, higher levels of estrogen and progesterone, while males have higher levels of testosterone [122].

Resistance exercise and aerobic exercise training above 150 min $week^{-1}$ improve HbA1c, while in those who train less time per week there is no change [123]. Resistance exercise sets have been shown to cause an immediate increase in epinephrine, norepinephrine, growth hormone, estradiol, and cortisol [124], these increases are important for best force production and energy utilization. Testosterone and growth hormone (anabolic hormones) were elevated during 15–30 min following resistance exercise bout. Testosterone, growth hormone, and cortisol (catabolic hormones) levels are highest when utilize large muscle mass during high-intensity resistance training with short rest intervals [125]. Thus, high-intensity resistance training is superior for enhancements in power, strength, and hypertrophy [125].

Insulin and insulin-like growth factor-1 (IGF-1) (anabolic hormones) are important to skeletal muscle growth [126]. Insulin is regulated by blood glucose and amino acid levels. Though IGF-1 increases in circulating blood probably as counteraction for the growth hormone action on hepatic secretion, a significant rise in blood catecholamine increases muscle force generation by supporting membrane excitability and, energy production and muscle contraction rate by stimulating electrogenic Na + -K + -transport [127]. With resistance exercise there is an immediate increase in epinephrine and norepinephrine [128]. Hormonal rise happens prior to the exercise bout itself; it rises during the anticipation for the challenging exercise to follow [127]. Glucose uptake by the cell is increased, owing to it the increased glucose transporters function.

The acute effects of different set loads during resistance bouts on hormonal profile revealed that testosterone is not affected during all different loads, while cortisol and growth hormone increased after a single bout of low intensity [129]. However, different study suggested that high training volume produce hormonal increases in testosterone, as well as cortisol, and growth hormone [129].

Resistance bouts has an acute effect of hormonal responses following training. It increases strength, power, and muscle mass, in addition to body figure and evolution. High volume resistance exercise training that utilize large muscle mass and intensities from moderate to high, demonstrate the highest hormonal rises [130].

The American College of Sports Medicine recommends that most adults engage in moderate-intensity cardiorespiratory exercise training for 30 min 5 week^{-1} for a total of at least 150 min week^{-1}, vigorous-intensity cardiorespiratory exercise training for minimum 20 min 3 week^{-1}, or a combination of moderate- and vigorous-intensity exercises to achieve a total energy expenditure of ≥1750–3500 kcal week^{-1} [131].

Hypothalamic Effect on Metabolic Rate Regulation

The primary organ responsible for regulating metabolism is the hypothalamus. The elementary procedure of metabolic regulation happens once genes affect cells to produce enzymes and to control their amount. Moreover, in a metabolic pathway, substrates are dynamic effectors of metabolic regulation over their level. Rate of a metabolic path will be altered, due to the obtainability and level of a substrate, and its ability to bind enzyme. Furthermore, hormones control metabolism by chemical signals which control enzyme activity by influencing genes expression which regulate production. Some forms of metabolic regulation affects biochemical process rate of time, other forms activate or inhibit commencement. The hypothalamus which is a part of the limbic system is crucial in regulating metabolism. It connects between the nervous system and the endocrine system through the pituitary gland [132]. The hypothalamus is highly involved in pituitary gland function. When the hypothalamus is stimulated by the nervous system, it secretes neurohormones that stimulate or dissimulate the secretion of pituitary hormones [133].

Primary hormones secreted by the hypothalamus which help regulate metabolism include:

- Antidiuretic hormone (ADH): this hormone increases water absorption into the blood by the kidneys.
- Corticotropin-releasing hormone: sends a message to the anterior pituitary gland to stimulate the adrenal glands to release corticosteroids, which help immune response.

- Thyrotropin-releasing hormone: triggers the release of thyroid hormones, which regulate energy, growth, and development.
- Somatostatin: prompts the anterior pituitary to release growth hormone.

AMP-activated protein kinase, occupied in the hypothalamus, is activated in circumstances of low energy, and thus, it has a key role in the variation of energy balance. It increases energy production and reducing energy consumption and therefore keeps energy homeostasis. AMP-activated protein kinase does it by hormones and metabolites, with the nervous system control [134].

Metabolic syndrome components result in disturbance of the hypothalamus function that avoid secretion of hormones required to regulate obesity and glucose intolerance to cardiovascular dysfunctions [135]. Since the pituitary gland is controlled by hypothalamic hormones, any impairment of the hypothalamus affects organs that are under pituitary regulation, such as the adrenal glands, gonads, and thyroid gland [136]. Hypothalamus disturbance include lacking pituitary hormone yield (hypopituitarism), insufficient thyroid hormone yield (hypothyroidism), and sexual development disturbance [137]. Hypothalamic disease is usually triggered by brain damage, surgical procedure, undernourishment, irritation, and cancers [138].

Myocardial Metabolism

In the healthy myocardium, free fatty acids and glucose are the main substrates of myocardial energy metabolism. In addition, lactate can be used since the myocardium is metabolically aerobic. Under normal conditions, 70–90% of ATP production comes from oxidation of free fatty acids and triglycerides (Fig. 2.26) and to a smaller extent 10–30% from glucose and lactate oxidation [139]. The myocardium uses around 60–70% of produced ATP for contraction, while the remaining is utilized for several ion pumps, especially the Ca^{2+}—ATPase in the sarcoplasmic reticulum [140]. ATP levels of the myocardium are around 5 μmol g^{-1} wet weight, and phosphocreatine about 8 μmol g^{-1} wet weight with phosphocreatine acts as an ATP transport and buffer system, with rate of ATP hydrolysis around 0.5 μmol g^{-1} wet weight. Consequently, since myocardial ATP storage is slight, without appropriate ATP renewal, it significantly reduces left ventricular contractility.

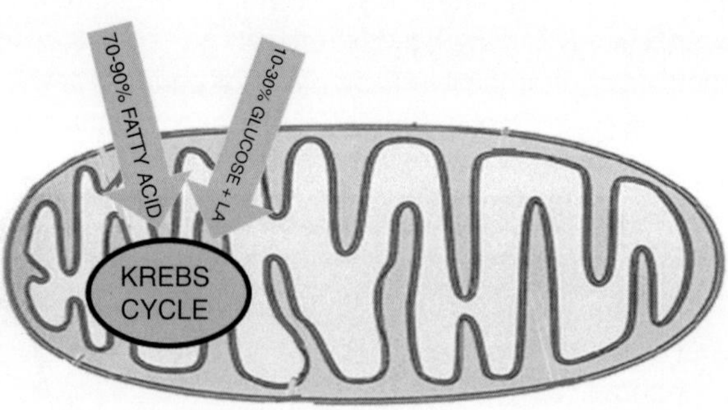

Fig. 2.26 Fuel sources for the production of ATP in cardiomyocyte, *LA* lactate

The myocardium is an organ that highly resist fatigue, it contains great number of mitochondria, and therefore, allowing continuous aerobic respiration via oxidative phosphorylation, due to large myoglobin number, adequate oxygen delivery, and nutrients [141]. On the other hand, glycolysis metabolism in myocardium muscle conserves of myocardial function only during reduced oxygen supply [142].

Free fatty acids consumption is determined mainly by the circulating free fatty acids levels. It can cross the cell membrane by passive diffusion or by transport proteins. Once inside the cardiomyocyte, free fatty acids enter the mitochondria by carnitine palmitoyltransferase 1 (CPT1) and undergo beta-oxidation. The precise mechanism of CPT1 is not presently known. CPT1 is a part of a family of enzymes called carnitine acyltransferases [143]. It moves acyl carnitine from the cytosol into the intermembrane space of mitochondria.

At basal metabolic rates, between 1 and 10% of myocardial, energy needed may be supplied by anaerobic metabolism. However, above 10%, that is, severe hypoxic conditions reduce left ventricular contractility significantly [144]. In diabetic patients, the myocardium utilizes more free fatty acid and less glucose, due to the reduced initiation of GLUT4 glucose transporters to the plasma membrane. However, heart's contraction partially affects GLUT4 transporters to the plasma membrane [145]. Although this mechanism is skeletal muscle, it is very relevant to the heart for supporting GLUT4 action, due to myocardial unceasing contractions.

The Metabolic Syndrome

The metabolic syndrome is a collection of conditions that result in metabolic disorders that describe a mixture of metabolic and cardiovascular risk factors, it is related to unsuitable initiation of the renin–angiotensin–aldosterone system in the cardiovascular system and increased cardiovascular morbidity and mortality [146]. Collection of conditions, include hypertension, high blood sugar, excess body fat around the waist, and abnormal cholesterol or triglyceride levels (Fig. 2.27).

The risks factors include insulin resistance, glucose intolerance, atherosclerosis, oxidative stress, polycystic ovary syndrome, hyperinsulinemia, dyslipidemia, proinflammatory status, and microalbuminemia. Angiotensin II action results in the

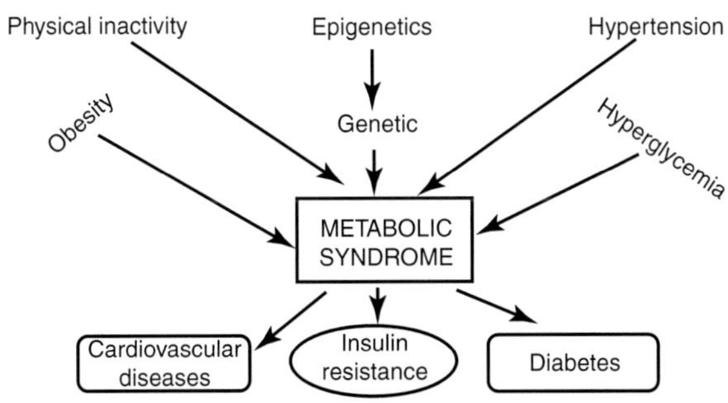

Fig. 2.27 Collection of conditions that result in metabolic disorders

impairment of endothelial nitric oxide production and Glut-4 translocation causing cardiovascular insulin resistance [147]. Of specific note is the usage of waist boundary as a pointer of obesity. Obesity turns to be one of the leading factors in the occurrence of metabolic syndrome and type 2 diabetes. Obesity is the primary risk factor for developing the metabolic syndrome, a group of metabolic abnormalities which include insulin resistance, fatty liver disease, dyslipidemia, and hypertension. Thus, both obesity and insulin resistance are thought to be the most important risk factors for metabolic syndrome [148]. Frequency of metabolic syndrome increases with obesity severity. One of the two obese adolescents advance to have metabolic syndrome [149]. Compared to individuals of normal weight, obese individuals with BMI above 30 kg m^{2-1} have 5–6 times higher chances to develop type 2 diabetes.

Most of the disorders associated with metabolic syndrome do not have noticeable signs or indications. However, some signs such as increased thirst, urination, fatigue, and hazy vision large waist circumference, that is evident and increased blood sugar might suggest diabetes (Fig. 2.28).

Although metabolic syndrome is not completely related to type 2 diabetes and the linked insulin resistance, still, based on different evidences on the increasing occurrence of obesity and related progress of type 2 diabetes set insulin resistance as a key donor of the syndrome.

Leptin signaling cascades and components within the insulin is suggested as mechanisms of insulin resistance and metabolic syndrome [146]. In addition, a combination of genetic and lifestyle factors may result in insulin resistance [150].

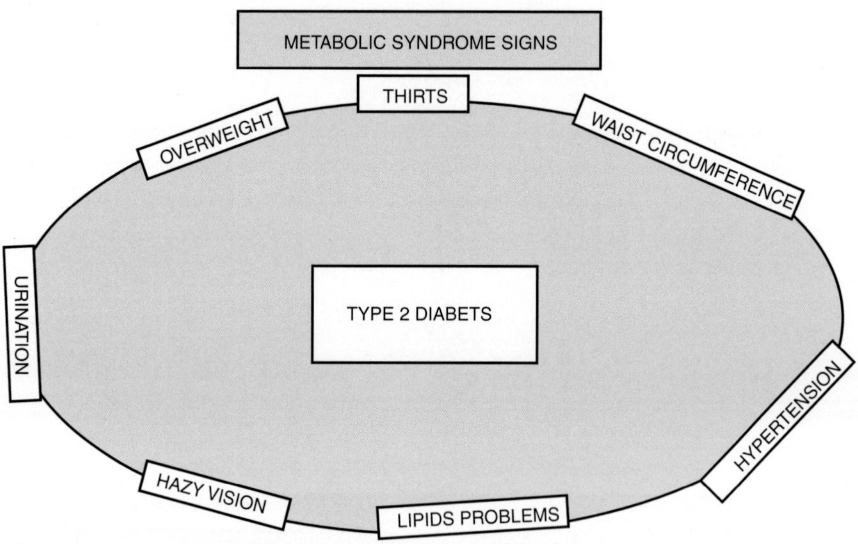

Fig. 2.28 Metabolic syndrome signs might suggest diabetes

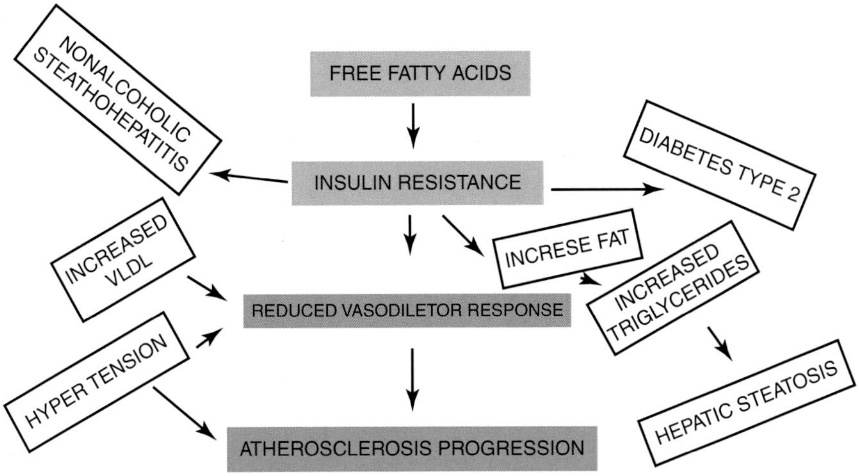

Fig. 2.29 Insulin resistance impact on metabolic syndrome

With insulin resistance, cells do not respond normally to insulin and thus glucose cannot be utilized in cells without difficulty. Therefore, it is believed that insulin resistance causes the cardiovascular pathologies of the metabolic syndrome (Fig. 2.29). Insulin resistance increases bloodstream triglycerides (dyslipidemia). Consequently, elevated plasma concentrations of glucose and fatty acids may improve hepatic fatty acid and triglyceride uptake and production and impair β-oxidation [151].

As a result, an accumulation of triglycerides in the liver occurs, leading to hepatic steatosis. From triglycerides, lipoprotein particles (very low-density lipoproteins) are produced, which are given back to the bloodstream [152]. As stated above, insulin resistance is the symbol appearance of metabolic syndrome.

Other related role of insulin resistance associated with metabolic syndrome is its role in platelet function. Insulin action increases endothelial nitric oxide action on platelet that is due to its phosphorylation by AMPK. Activation of nitric oxide synthesis in platelets leads to a decrease in thrombin-induced aggregation, so, preventing the procoagulant properties of platelet stimulation. This response shows why disturbance in insulin action is a main causal factor in the progress of the metabolic syndrome. Insulin resistance negative effects, combined with coagulation and fibrinolysis pathologies, come up with the evolution of atherosclerosis.

The next factors intensify the probabilities of having metabolic syndrome: (a) biological changes due to aging process, (b) ethnicity, as an example, Hispanic women seem to be at the highest risk compared to others in the United States, (c) obesity, and (d) diabetes.

Diabetes

Diabetes is a disease in which the body's ability to produce or respond to insulin is lessened, resulting in irregular metabolism of carbohydrates characterized by raised levels of bloodstream glucose. Since insulin has a key role in glucose utilization and metabolism, therefore, glucose significantly affects insulin biosynthesis and secretion [153]. This impairment is key to the development of type 2 diabetes. Diabetes mellitus is a diverse metabolic disorder. Increased bloodstream glucose is the primary stimuli for insulin release [154], from secretory granules in the pancreatic β cell, referred to as glucose-stimulated insulin secretion [155]. Normal blood insulin level is 20–30 pmol L^{-1}, oral ingestion of 75 g of glucose will increase insulin level up to 250–300 pmol L^{-1} [156].

Diabetes have two foremost types: type 1 and type 2 (diabetes mellitus). Both kinds of diabetes are chronic diseases, affecting blood glucose level [157]. Both diabetic types are characterized by having high than normal blood glucose levels [158]. However, the reason and progress of each diabetic type are different.

There is no mutual diabetes origin that fits each type of diabetes, due to numerous variations depending mainly on the individual and the diabetes type. There are a number of risk factors that are closely linked to type 2 diabetes, obesity or bad diet, sedentary lifestyle, aging, high BP, high cholesterol levels, and family history which is the most devastating factor. Type 2 is by far the most common and is generally related to insulin resistance [159], which is a failure of the body to respond adequately to the blood sugar-lowering effects of insulin. An important risk factor for cardiovascular disease and diabetes is glucose tolerance [160]. High blood glucose and high insulin levels can also have a deleterious effect on hypertension and blood lipids (Fig. 2.30) [161]. Circulatory disorders associated with diabetes include coronary heart disease, stroke, peripheral arterial disease, cardiomyopathy, and congestive heart failure.

Compare to type 1, type 2 diabetes is a milder disease, it is the most communal form of diabetes mainly in adults and it accounts for above 90% of type 2 diabetes [162] and is a condition problematic to treat and costly to handle [163]. Yet, type 2 diabetes can cause main health difficulties [164], mainly in the smallest blood vessels that deliver blood to the kidneys, nerves, and eyes, it also increases the risk of heart disease and stroke.

Type 1 diabetes or juvenile-onset diabetes, is an insulin-dependent and is an autoimmune disorder (Fig. 2.31).

This can happen at any age but is greatest usually identified from infancy to the late 30s [165]. Its pancreas is attacked by its immune system by antibodies, and therefore, insulin synthesis stops. Genetic tendency may be the source of this

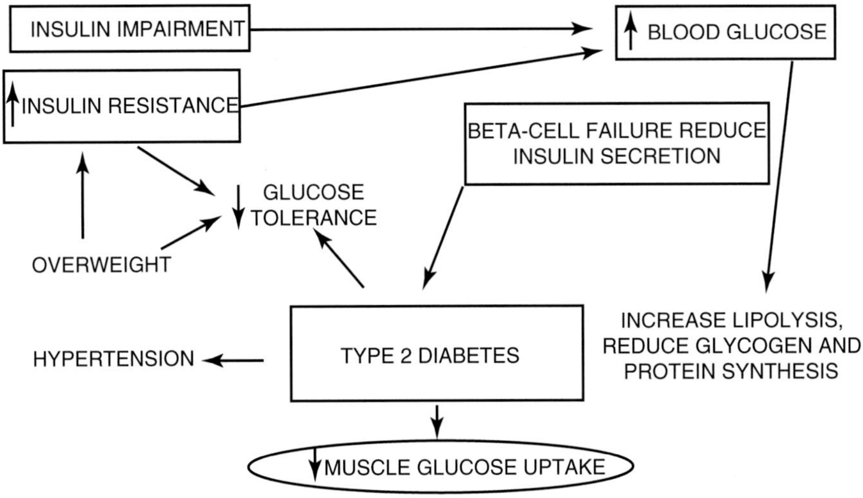

Fig. 2.30 Causes of diabetes type 2

Table 2.2 Summarizes the main differences between the two forms of diabetes

Type 1 diabetes	Type 2 diabetes
Diagnosed in childhood	Diagnosed at age 30-years and over
Individuals have high ketone Levels	Overweight individuals
Treated with insulin injections or insulin pump	High blood pressure and/or cholesterol
Glucose regulation is only by taking insulin	Treated tablets medication
Normal body weight	Individuals sometimes following exercise can be off medications

diabetic disease, in addition, it may also result from beta cells dysfunction [166]. Insulin injection through the skin into the fatty tissue below is the treatment for type 1 diabetes. Glycated hemoglobin (HbA1C) blood periodic test evaluates bloodstream glucose levels in the last 3 months. The importance of this test is the identification ahead a time of abnormal glucose levels and to estimate the risk of difficulties from diabetes. Daily exercise is recommend in addition to frequent blood testing, diet changes, and most important, taking insulin and other medications as needed.

In type 2 diabetes, pancreas synthesis of insulin is reduced to an amount that is not sufficient to influence glucose entry to the cells [167] (Table 2.2).

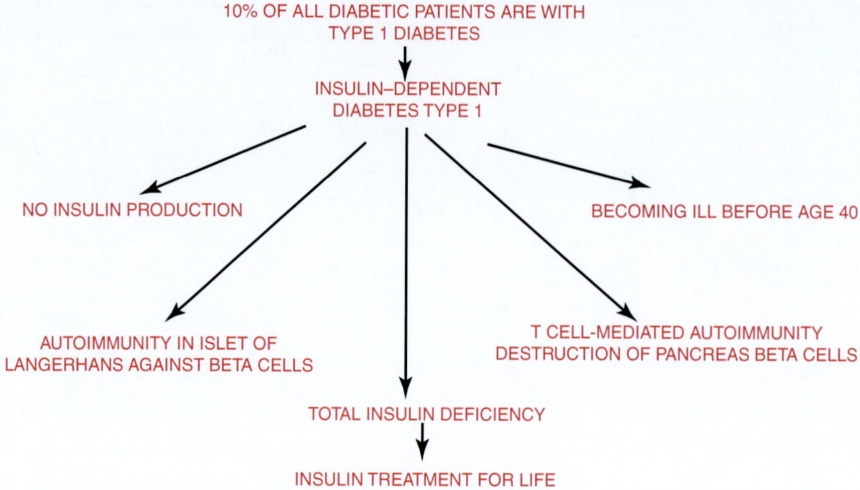

Fig. 2.31 Diabetes type 1 is related to many medical problems

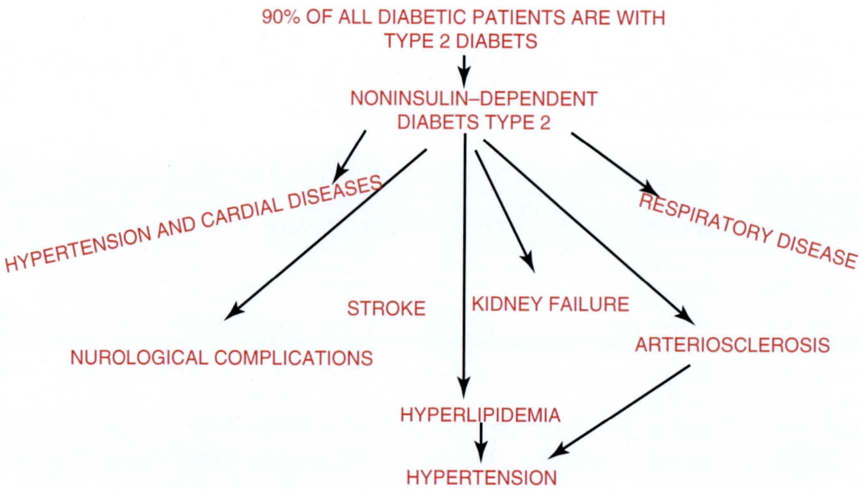

Fig. 2.32 Diabetes type 2 and its medical impacts

Increased blood glucose levels cause hypertension by reducing blood vessels stretching ability, with diabetes, kidneys retain water and, insulin resistance increase the risk of hypertension [168]. Circulatory disorders; coronary heart disease, stroke, peripheral arterial disease, cardiomyopathy, and congestive heart failure are associated with diabetes (Fig. 2.32). Diabetes generally results in early death from cardiovascular diseases [169].

Aging process is a factor that increases the occurrence of type 2 diabetes [170]. In addition, also muscle strength, muscle mass, and metabolic control decline [171].

Insulin Resistance

Insulin resistance causes impaired glycogen production and protein breakdown in skeletal muscles [172], and prevents lipoprotein lipase activity in adipocytes [174], as well as being a key to a rise in delivering free fatty acids and inflammatory cytokines. Insulin resistance reduces nitric oxide production, resulting in endothelial cell dysfunction, bringing about platelet aggregation [173], leading to atherosclerosis [174].

Insulin resistance (Fig. 2.33) is a pathological condition in which cells fail to respond normally to the insulin hormone [175] and is the best predictor to have diabetes in the future [176]. Usually, it means resistance to the effects of insulin on glucose consumption, metabolism, or storage, it happens mainly in fat, liver, and muscle cells [177].

This vision is due to obvious faults in the insulin-stimulated glucose consumption, mainly, in glycogen production and, to a smaller degree, glucose oxidation. Insulin resistance is answered by compensatory hyperinsulinemia which reduces glucose transport and diminished destruction of hepatic glucose production [178]. Insulin resistance affects differently dissimilar tissues, rest on the biological and metabolic function of the tissues. Since muscles have a high metabolic rate and thus, metabolic demand, insulin resistance has significant effects on skeletal muscle, adipocytes, and liver tissue. These organs are the focal targets of intracellular glucose transport, glucose, and lipid metabolism [179].

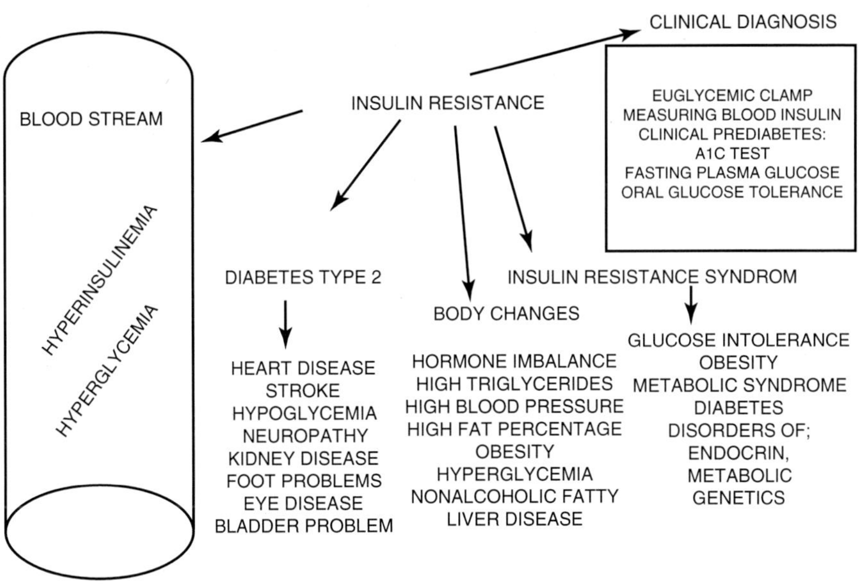

Fig. 2.33 Risk indices related to insulin resistance syndrome

Obese individuals are at high risk of developing type 2 diabetes and its related medical complications [163]. Obese individuals have insulin resistance, thus, the workload on the pancreas is exaggeratedly hard to produce more insulin [180]. However, there is no adequate insulin to control blood glucose at a normal level. Additionally, obesity is related to inflammatory features which surge adipose tissue macrophages [181]. The above factors intensify lipolysis, boost up hepatic triglyceride production, and hyperlipidemia due to the increased esterification of fatty acid.

Glycated Hemoglobin

HbA1c or A1c is considered a standard test. It is of specific interest because it is easy to notice by simple blood test [182]. The connection between glucose and hemoglobin (glucose-hemoglobin), often reveals the presence of diabetes (Fig. 2.34).

At first, an unstable bond is formed between hemoglobin and glucose, with time, arranges a steadier molecule composition where glucose is bound covalently to hemoglobin. When blood glucose is high, it accelerates the quantity of the unstable compound. This, where the stable form changes slowly, gives life span of the red blood cell. Thus, HbA1c is a significant pointer of long-term glycemic regulator with glycemic history in the last 2–3 months [183].

Levels of hyperglycemia and HbA1c are increased, which is a form of hemoglobin chemically linked to glucose to form nonenzymatically within the cell [184]. HbA1c observed with these disorders are known to increase the risk of microvascular and macrovascular complications that can cause neuropathy, end-stage renal

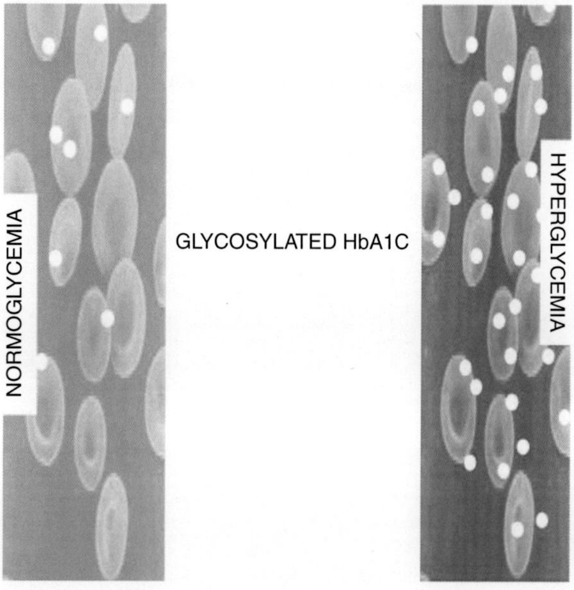

Fig. 2.34 Description of normoglycemic bloodstream and hyperglycemic

disease, vision loss [185], and other health matters, from hypertension to kidney failure and heart disease [186]. HbA1c causes an increase in highly reactive free radicals inside blood cells, that, change blood cell membrane properties. Thus, blood cell aggregation occurs which augment blood viscidness, consequently, blood flow is impaired [187].

Effects of Diabetes on Body's Systems

The pathological processes and risk factors associated with cardiovascular disease begin as early as during childhood. Diabetes, first thought is the connection to high blood glucose, affects the body's ability to produce or use insulin, thus, cells are deprived of the needed energy. Diabetes can be efficiently taken care when detected early. However, when it is untreated, it led to potential dysfunction of almost each main body system: diabetic ketoacidosis, kidney dysfunction, cardiovascular system, and central nervous system (Fig. 2.35).

Diabetic Ketoacidosis

Diabetic ketoacidosis is a serious complication of diabetes mellitus [188]. Individuals without an earlier finding of diabetes can develop diabetic ketoacidosis as symptom suggesting of the disease [188]. Diabetic ketoacidosis usually found in individuals with type 1 diabetes however, it may be found in patients

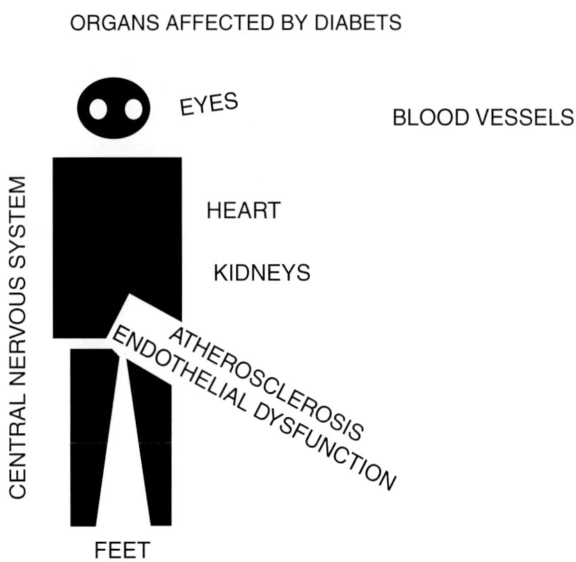

Fig. 2.35 Main body organs affected by diabetes

with different types of diabetes in specific cases. Triggers may include infection, not taking insulin correctly, stroke, and certain medications such as steroids. It may happen due to low insulin level, consequently, fatty acids are utilized, producing acidic ketone bodies [189]. Diabetic ketoacidosis is usually detecting in the blood or urine, high blood sugar between 300 and 500 mg dl^{-1}, positive ketones, potassium, anion gap, and low blood pH below 7.3, total CO_2, HCO_3 below 15 meq L^{-1} and sodium [190].

Treatment includes: intravenous fluids replacement and electrolyte deficits, insulin administration, and inspection of hypokalemia, hypoglycemia, and cerebral edema [190]. Depending on the severity of the ketoacidosis, insulin may be given intravenously or by injection subcutaneous [189], and potassium needed in the prevention of low blood potassium development [188]. In addition, diabetic ketoacidosis may need antibiotics treatment in individuals with an increased incidence of infection [191]. In diabetic ketoacidosis patients, with blood pH less than 7.30, due to metabolic acidosis, may be treated with sodium bicarbonate, however, it is usually not suggested [192].

When blood insulin levels are low, counter-regulatory hormones, glucagon, epinephrine, cortisol, and growth hormone (Fig. 2.36), stimulate the utilization of stored body fat for energy.

It stimulates fatty acid released from triglycerides kept in adipocyte fat droplets, and then on, to the production of ketone bodies [193], which are energy-rich compounds that transport energy from the liver to other tissues [194]. Ketone bodies production is possibly mediated by insulin resistance that suppresses the activity of the branched-chain 2-oxo acid dehydrogenase, often related to positive effects on the regulation of body weight, muscle protein synthesis, and glucose homeostasis [195]. Epinephrine and glucagon stimulation may generate high levels of toxic

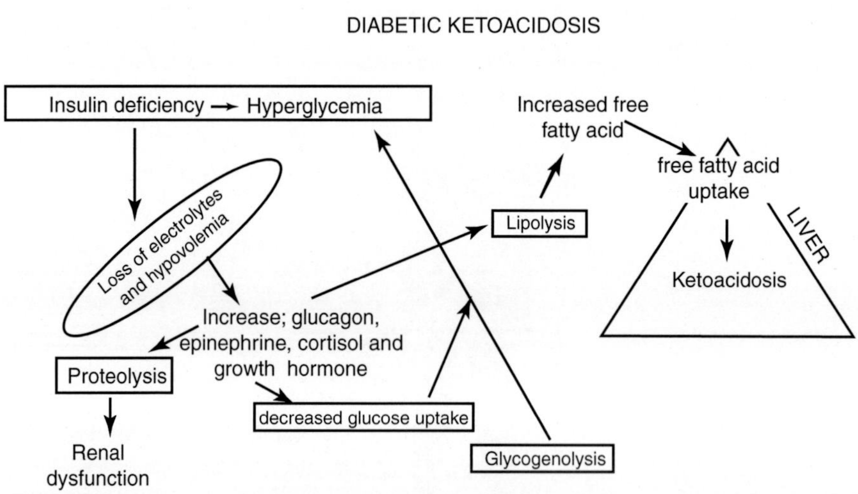

Fig. 2.36 Diabetic ketoacidosis is a state of metabolic acidosis, ketonemia, and hyperglycemia

chemicals, with acids as well as ketone bodies, leading a serious complicated diabetic ketoacidosis disease [196]. Symptoms are significant thirst, extreme urination, and exhaustion. High blood glucose levels and additional ketones in urine point on diabetic ketoacidosis. If diabetic ketoacidosis is untreated, it may lead to loss of awareness and sometimes death.

Kidney Dysfunction in Diabetes

Both types of diabetes are related to increased rapidity of quite a few incapacitating microvascular and macrovascular difficulties, such as nephropathy, retinopathy, neuropathy, and cardiovascular diseases. Atherosclerosis and stroke [197] resulting in end-stage renal diseases. Diabetic kidney disease is a problem rising in some individuals with diabetes. It may result in some instances of kidney failure. Twenty percent of kidney failure patients, due to diabetic kidney disease, are in need of dialysis. Overweight, high HbA1c level, being male and length of time that the diabetic patient has had diabetes, are factors that increase the risk of developing diabetic kidney disease. Treatment is given in order to avoid the development of diabetic kidney disease, myocardial infarction, and stroke. Individuals with type 1 diabetes when discovered, microalbuminuria or proteinuria are seldom presented without symptoms. However, 5 years following the disease discovery, about 14% of all diseased individuals will have advanced microalbuminuria. Following 30 years, it will be in 40% of the individuals. Some individuals with microalbuminuria progress to proteinuria and kidney failure, or diabetic nephropathy occurs in individuals with type 2 diabetes. Blood chemicals scare the glomeruli resulting in glomerulosclerosis, a condition that increases unwanted glomeruli removal of albumin (protein) through the urine. Diabetic kidney disease has two main categories depending on how much albumin is lost through the kidneys: microalbuminuria and proteinuria. Microalbuminuria is a condition that albumin up to 300 mg day^{-1} is in the urine, while levels of albumin in proteinuria that is irreversible are above 300 mg day^{-1}. This compared to healthy individuals that albumin quantity found in the urine is neglectable.

At the time that proteinuria was diagnosed, about 12.5% of them were with microalbuminuria and 2% already had proteinuria. The reason for these figures is that it takes time until type 2 diabetes is diagnosed after the disease had begun. Higher kidney disease prevalence is found in Asians and black individuals compared to white individuals. When kidney dysfunction occurs, problems such as anemia that cause fatigue, blood calcium imbalance, and variations in phosphate level. Calcium and phosphate imbalance cause bone cracks and breaks. Right treatment may reverse end-stage kidney failure that is ultimately fatal if not treated (Fig. 2.37). Individuals having proteinuria for 10 years, end-stage kidney failure advances in about 12% of individuals, which after that they need kidney dialysis or a kidney transplant.

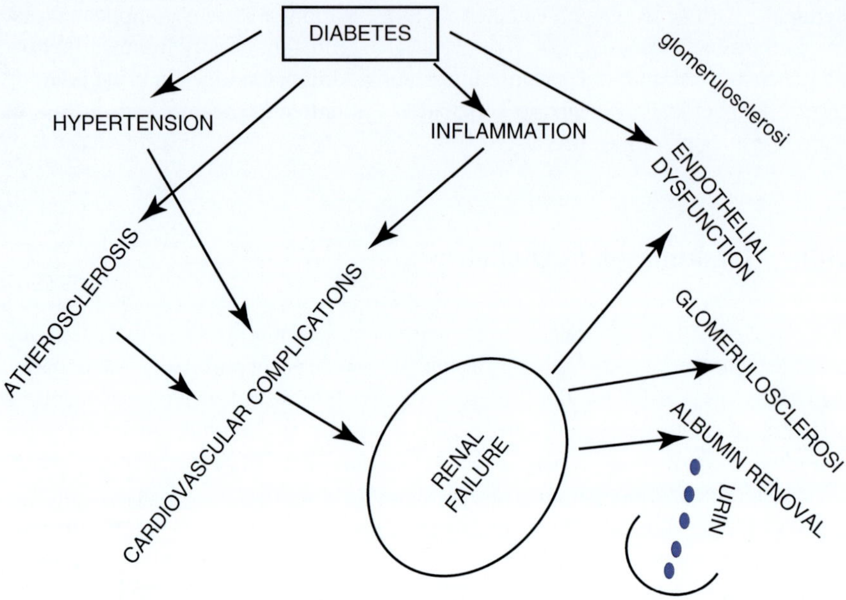

Fig. 2.37 Diabetic kidney disease is a leading cause of kidney disease, figure perceiving the primary causes of kidney failure

The Cardiovascular System in Diabetes

Diabetes type 2, glucose intolerance and insulin resistance are essential characteristics that increase risk for CAD associated with high BP (hypertension), irregular quantity of lipids (dyslipidemia), and endothelial dysfunction [198]. These anomalies increase during aging and signify the initial phases of circulatory disease that come first to the clinical appearances of cardiovascular disease. Regardless of age, due to diabetes, cardiovascular disease remains the greatest mutual cause of death [199]. The pathophysiology state of the connection between diabetes and cardiovascular disease involves numerous factors (Fig. 2.38).

Suggested factor for the etiology of type 1 and 2 diabetes is epigenetic [201]. Hyperglycemia affects gene expression regulation (epigenetic alterations), which happen even following glucose stabilization. Epigenetic mechanisms may play a key role in the pathophysiology of diabetes and its associated, atherosclerosis, cardiomyopathy, nephropathy, and retinopathy [200]. Although diabetes is curable, nonetheless, when glucose levels are under control it still increases the risk of myocardial disease and stroke [201]. In addition, an indication that environmental factors control phenotypic expression arisen from the Diabetes Prevention Program [202]. In type 1 and type 2 diabetic individuals, an increment in the risk for bone break was found to be widespread. Hyperglycemia alters bone matrix proteins such as collagen I through nonenzymatic glycation [203]. The relationship between

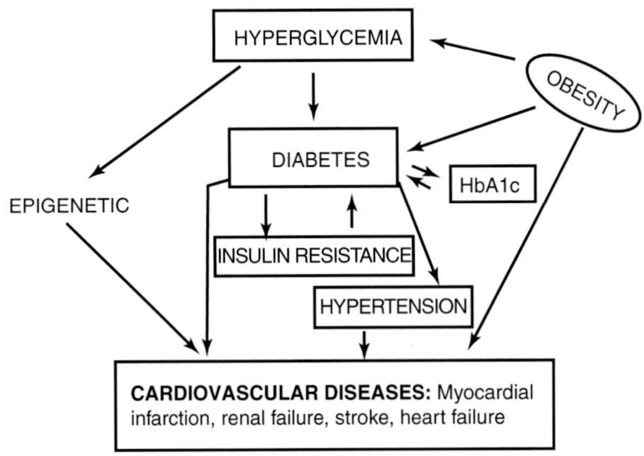

Fig. 2.38 Diabetic factors affecting cardiovascular diseases

diabetes and cardiovascular diseases is based on biological and molecular mechanisms from hyperglycemia, insulin-resistant, and vessels, which supply blood to the myocardium and the brain begin to stiffen and occlude [204]. Diabetic polyneuropathy is a complex and multifactorial entity in which various factors besides hyperglycemia play an important role [205]. Diabetes is a major risk factor for both cardiac and vascular disease. Hyperglycemia causes myocardial damage and therefore left ventricular systolic and diastolic dysfunction and thus cardiac failure [206]. Vascular disorders include increasing atherosclerosis [199], retinopathy, nephropathy, peripheral vascular disease, stroke, and CAD [207]. The course development of diabetes effects on the cardiovascular system is slow. However, the course of the disease development is quicker, in aged individuals with type 2 diabetes who are exposed to a multifactorial metabolic abnormality consisting of insulin resistance with compensatory hyperinsulinemia, essential hypertension, overweight, high triglycerides, and hypercholesterolemia [208]. This abnormal lipid levels regularly happens in individuals with early coronary heart disease. Additionally, lipid disorder is typically related to insulin resistance (atherogenic dyslipidemia or diabetic dyslipidemia) in those individuals with diabetes. Lipid abnormalities in the course of diabetes triggered dyslipidemia as hypercholesterolemia, hypertriglyceridemia, elevated LDL, and decreased HDL [209].

High blood pressure has long been documented as a main risk factor for cardiovascular disease. Two foremost risk factors that are accountable for the development of atherosclerosis are diabetes and hypertension, and then myocardial infarction and strokes [210]. High blood glucose and insulin levels can also have a harmful outcome on hypertension and blood lipids. A positive relation exists between hypertension and insulin resistance which doubles the risk for cardiovascular disease [211].

About half of hypertensive subjects also have hyperinsulinemia or glucose intolerance, while above three-quarters of people with type 2 diabetes also have hypertension [212]. This relationship between hypertension in diabetic patients significantly increases the probabilities of these individuals to develop cardiovascular disease. It seems that abnormalities in the renin–angiotensin-aldosterone system can be related to hypertension and insulin resistance [213]. Neuronal vascular tone regulation is altered due to insulin-resistant which is another cause of hypertension [214]. The renin–angiotensin-aldosterone system effect is an increment of blood pressure by initiating vasoconstriction along with increased cardiac contractility and cardiac output [215]. Increase in aldosterone and antidiuretic hormone levels causes sodium to be absorbed and potassium to be excreted [216]. These two hormones regulate sodium and water homeostasis, which cause an increase in water reabsorption, and thus, increasing BP [217]. Aldosterone, also, produce mortal structural changes in kidney tissues (hypertrophy), dysregulation of proliferation and apoptosis, leading to fibrosis and tissue remodeling; in addition, its effects on blood vessels and the myocardium can have pathophysiological consequences [218].

Nervous System and Epigenetic Role in Diabetes

Diabetic neuropathy is the greatest problem of diabetes mellitus, it is a diverse group of disorders that affects both somatic and autonomic parts of the nervous system [219]. Neuropathy is believed to be existing in up to half of the individuals with diabetes. The nervous system permits the body to take actions, automatically, or by reflex responses, while other actions will come right by the individual's choice. There are two main types of nerves, nerves that transmit impulses from the organs to the brain for taking the appropriate action and motor nerves that transmit impulses from the brain to organs and muscles. Increased blood glucose levels that last for over a number of years damage the nerves (neuropathy). Nerve damage occurs when blood sugars increase over 7.8 mmol L^{-1} on glucose tolerance tests [220]. Previously, it has been suggested that keeping blood sugars below 140 mg dL^{-1} (7.8 mmol L^{-1}) at all times increases odds for avoiding neuropathy and survive diabetes [221]. Nerve damage can take the form of sensory neuropathy, motor neuropathy, or autonomic neuropathy. Hyperglycemia plays a key role in the diabetic neuropathy progress, resulting, from metabolic disturbances such as protein kinase C activity alteration, increased polyol pathway activity in neurons, and Schwann cells. Initiation of these metabolic paths leads to hyperglycemia by producing oxidative stress [219]. Prolonged exposure to blood glucose over 140 mg dL^{-1} (7.8 mmol L^{-1}) kills pancreas beta cells [222].

Healing involvement of the abovementioned metabolic pathways should include all metabolic paths for the improvement in diabetic neuropathy [223]. This may lead to loss of feeling in the upper and lower extremities. Nerve damage in the feet is particularly dangerous for patients with diabetes, push them to a higher risk of foot damage which could lead to amputation. Thus, it is advisable to check the feet regularly, since neuropathy is common in diabetes. Nerve damage in the feet in patients with diabetes may have foot damage, which could lead to amputation [224]. In addition, sustained damaged nerves can affect the functional capacity of different organs such as the cardiopulmonary system and motion ability.

Gene expression regulation depends on the availability of DNA to several transcription factors. Gene transcript and initiation rest on a chromatin assembly that is very active reliant on a multitude of posttranslational alterations of histones that permit for the change of unreachable, compacted, or repressive heterochromatin to the available or dynamic euchromatin state of DNA. Changes happening on the histone tails (acetylation, methylation, and phosphorylation), these epigenetic modifications constitute gene regulation that can be altered without changing the original DNA sequence [225].

Although a significant genetic tendency for diabetes exists, the precise genetic faults are not yet well-defined. Both genetic and environmental factors are involved in these pathologies. Yet, the complete risk elucidated by genetic distinction is actually uncertain. Interplay between genes, such as DNA methylation, histone post-translational modifications in chromatin, non-coding RNAs, and the environment, affects diabetes and its effects on health [197, 226]. Hyperglycemia, oxidant stress, and inflammatory growth factors are related to diabetic pathology difficulties which may lead to alteration in epigenetic mechanisms.

Exercise's Effects on Diabetic Patients

Exercise training is a vital element in the therapeutic management of individuals with type 2 diabetes and hypertension. Exercise is a basis for actual diabetes prevention and control [227]. Exercise training helps regulate different health problems such as cardiovascular diseases, obesity, insulin sensitivity increments, and increase glucose utilization (Fig. 2.39). The target of exercise training, in type 2 diabetes patients, is to

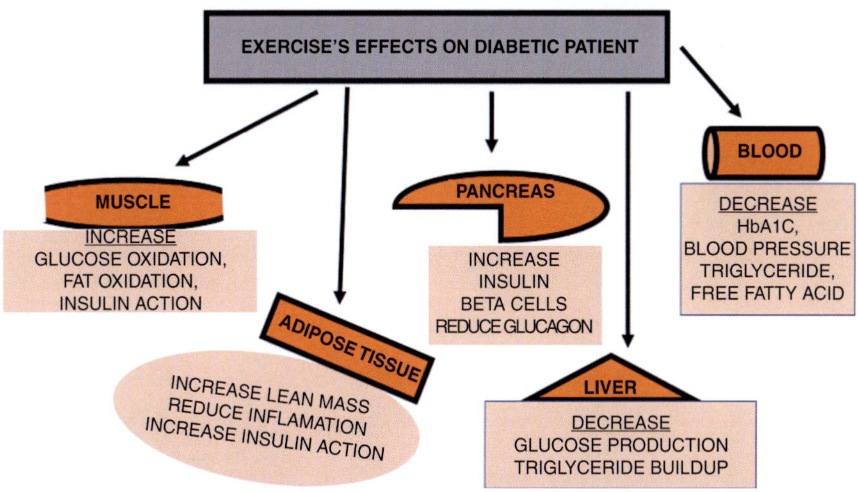

Fig. 2.39 Effects of adding long-lasting aerobic exercise training on diabetic patients

maintain normoglycemia, control lipid levels, to avoid or delay the microvascular, macrovascular, and neural complications. This protective effect can be following exercise training, the most of the individuals at highest risk for developing type 2 diabetes. When combining exercise and controlled diet, results are very positive [228].

Aerobic exercise may be a beneficial tool in a variety of diabetic patients [227]. Individuals who have metabolic syndrome or are in a prediabetic stage may anticipate that, following long-lasting exercise training, progression of diabetes is averted, or slowing down its progression. Clinicians need to make clear to diabetic patients of the exercise benefits in order to control their type 2 diabetes. Cardiac rehabilitation centers with clinical exercise training programs can guide those patients with type 2 diabetes and hypertension [229]. From an applied point of view, understanding this tool by the health care team, patient will gain from the knowledge about the benefits of exercise efforts. Therefore, indorsing exercise as an important part of the prevention and management of type 2 diabetes must be given top importance. In nondiabetic individuals, during aerobic exercise, VO_{2max} increase from resting values, by as much as 20–25 folds, at peak exercise [230]. Under these conditions, skeletal muscles utilize glycogen and glucose [231], while triglycerides and glucose released from the liver and free fatty acids are utilized at intensities between 50% and 80% VO_{2max} [232].

Aerobic exercise lasting 150 min $week^{-1}$ or 500 kcal $week^{-1}$ in energy expenditure of moderate-intensity or at least 90 min $week^{-1}$ of vigorous aerobic exercise is associated with a 6% decreased incidence of type 2 diabetes [233, 234]. Moderate-intensity aerobic exercise increases glucose uptake by 2–3 mg kg^{-1} min^{-1} above usual requirements (70-kg person: 8.4–12.6 g h^{-1} of exercise). During high-intensity aerobic exercise, glucose uptake increases by 5–6 mg kg^{-1} min^{-1}; however, exercise of this intensity cannot usually be sustained for long intervals. Traditionally, diabetic patients have been encouraged to train aerobic exercises such as walking, running, cycling, and swimming.

Long-lasting aerobic activity, utilizing large muscle groups, stimulates cardiovascular adaptations, VO_{2max}, and skeletal muscle strength, thus, increases work capacity. Chronic exercise training, also, prevents the progress of coronary artery disease and reduces symptoms in patients with diabetes and cardiovascular diseases [235].

Exercise training benefits for diabetic patients are significant for enhancements in endothelial vasodilator function [236] and left ventricular global function [237]. Chronic exercise training expands the cardiovascular health of diabetic patients, through numerous mechanisms. These mechanisms consist of arterial stiffness and endothelial vasodilator function improvement, left ventricular diastolic and systolic function, systematic inflammation, and reducing left ventricular risk for hypertrophy. Exercise training also reduces total and abdominal fat, which mediate insulin sensitivity, enhancements, and hypertension.

To keep normal central nervous system function, blood glucose hemostasis (normoglycemia) is soundly sustained during exercise, thus, hypoglycemia during exercise rarely occurs [238]. Normoglycemia regulations that preserve its hemostasis during aerobic exercise is hormonally facilitated [239]. In the liver, the metabolic pathway gluconeogenesis generates glucose from the breakdown of proteins,

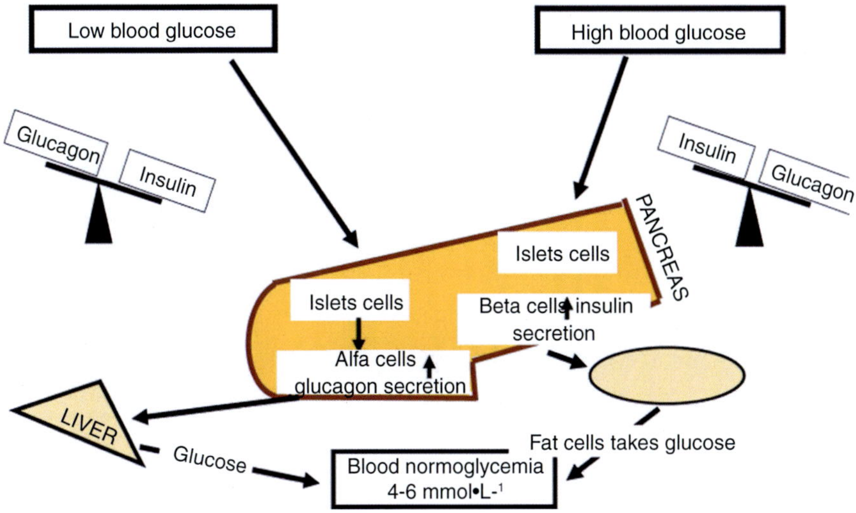

Fig. 2.40 Insulin and glucagon work together to maintain normoglycemia

glycerol, and from pyruvate and lactate (Cori cycle). Thus, during aerobic exercise, in order to increase hepatic glucose production, insulin plasma levels have to decrease, along, with the secretion of glucagon and catecholamines [240]. Glucagon–insulin ratio is in an opposing action (Fig. 2.40) to preserve a normal fuel balance [241]. Glucagon response is seen when blood glucose levels are below 3.8 mmol L^{-1} [242]. However, these hormonal responses are basically missing in patients with type 1 diabetes [243].

Consequently, due to low blood insulin, the counter insulin hormones may increase already high levels of glucose and ketone bodies to result in diabetic ketoacidosis. Yet, in patients with type 2 diabetes, hypoglycemia during exercise is less of a problem. Indeed, in these patients, exercise may have normal blood glucose levels.

Exercise improves regulation of blood glucose homeostasis and decreases the risk of cardiovascular disease and mortality in patients with type 2 diabetes and healthy subjects [244].

It is assumed that following 15 days of aerobic exercise training, advances glycemic regulation in type 2 diabetes patients can be seen [245], mainly by increasing greater reductions in HbA1c, greater increase in VO_{2max}, and greater increase in insulin sensitivity [246].

Intensities of moderate to forceful aerobic exercise are suggested to reduce type 2 diabetes complications [247]. Still, at the beginning of training, patients with type 2 diabetes cannot start with the abovementioned intensity, as any healthy untrained subject. For beginners, workload (walking, swimming) should be between 50 and 60% of VO_{2max} and with time to exercise at the abovementioned intensities of 60% up to 70% [248]. Working at 60% up to 70% intensities increases insulin sensitivity, pointing out the relations between profits of aerobic exercise and type 2 diabetes [246].

The increased metabolic demands during aerobic exercise increase vigorously liver production of glucose to increase blood glucose level. Without this liver response, during exercise, hypoglycemic stage would be reached, and therefore, long-lasting exercise would be impossible. The liver, when metabolic demand is high, it increases blood glucose level, while at rest following meal its energy stores are replenished. Liver functional capacity is especially relevant to diabetes mellitus patients. Worries of hypoglycemia have been recognized as the highest blockade to exercise in individuals with type 1 diabetes [249], it seems that the same worries holds for individuals with type 2 diabetes [250].

Endothelial Dysfunction and Exercise

Alterations in blood glucose levels have a reflective effect on vascular endothelium which is very sensitive to these variations. So, possible primary effect of hyperglycemic damage is the vascular endothelium [251]. The effect of reactive oxygen species (ROS) on the vascular endothelium is the basis of microvascular and macrovascular brought about by hyperglycemia [252]. Arterial stiffens describes the inflexibility of the arterial wall that occurs as a result of the aging process and arteriosclerosis. Inflammation following hyperglycemia has a major part in arteriosclerosis progress, so, it is the main donor in large arterial stiffening [253]. In brief, atherogenesis results from hyperglycemia and lipid peroxidation-derived aldehydes oxidized to carboxylic acids. The proinflammatory oxidized phospholipids, resulting from the oxidation of low-density lipoprotein and phospholipids including arachidonic acid, formed in the lipoxygenase and myeloperoxidase pathways, attract and trigger inflammatory cells, such as monocytes, T cells, and macrophages. Matrix degradation brings about atherosclerotic as a result of macrophages activation through cytokines, ROS, and proteolytic enzymes. Oxidative stress is also considered to be a key factor in the mechanisms of changes in cell function [254]. In addition, indications suggest that the aging process is in oxidative stress condition [255]. When oxygen is present in large amounts in the cell, such as during exercise, ROS production is increased, therefore, the cell needs to eliminate those molecules. ROS molecules have an important part in vascular disease and are known to be a key donor to cardiovascular dysfunction. ROS is produced in large amounts by the enzyme nicotinamide adenine dinucleotide phosphate (NADPH) oxidases [256]. ROS affect other enzymes, such as nitric oxide synthase and endothelial nitric oxide synthase and thus, it is a risk factor for atherosclerosis, cardiac hypertrophy, and heart failure. NADPH oxidase activation for the ROS production is facilitated by an extensive number of factors such as hypoxia, cytokines, mechanical forces, and hormones [257].

Previously it has been reported that oxidizing ROS are generated during moderate and high aerobic bouts [258]. Skeletal muscle generates superoxide and nitric oxide during aerobic exercise in intensities above the anaerobic threshold [259]. ROS is essential for skeletal muscle force generation, however, ROS in high values

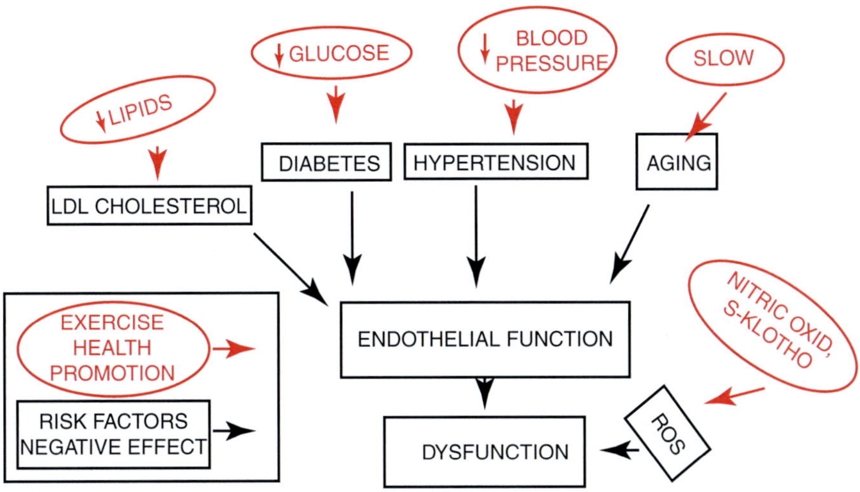

Fig. 2.41 Causes of endothelial dysfunction (black frame) and role of exercise in the attenuation of these factors

may reduce muscle contraction properties and thus bring about early exhaustion [260]. Following long-lasting aerobic exercise training, blood s-Klotho concentration increases as a response to ROS that increases in muscle cells as a result of aerobic training. S-Klotho reduces apoptosis through the nitric oxide production (Fig. 2.41) and thus suppresses oxidative stress [261]. Thus, an increase in Klotho gene expression reduces ROS damages to the vascular endothelial cell. Nitric oxide acts as an endothelium-derived relaxing factor [262], released from endothelial cells and acts as an inhibitor of ROS production, by a decrease in L-arginine endogenous asymmetric dimethyl-L-arginine ratio connected with endothelial dysfunction [263]. Nitric oxide stimulates phospholipase A2 and inhibitors of lysolecithin acyltransferase inducing smooth muscle tone relaxation by inhibiting low-density lipoprotein oxidation [264]. In addition, nitric oxide protects endothelial penetrability, smooth muscles' contraction by calcium homeostasis, and inhibits insulin-like growth factor-1 signaling. [265]

Effect of Exercise on Calmodulin Signaling Pathway

Epigenetic modifications on GLUT4 and mitochondrial proteins reduces skeletal muscle glucose consumption and type 2 diabetes progression [266]. In diabetic patients, calcium reliant on signing path and subsequently glucose uptake is diminished [267]. While in nondiabetic individuals, activation of signaling cascades, during exercise, changes in skeletal muscle cells, calcium concentration and thus influence cellular metabolism including glucose uptake [267], consequently, number

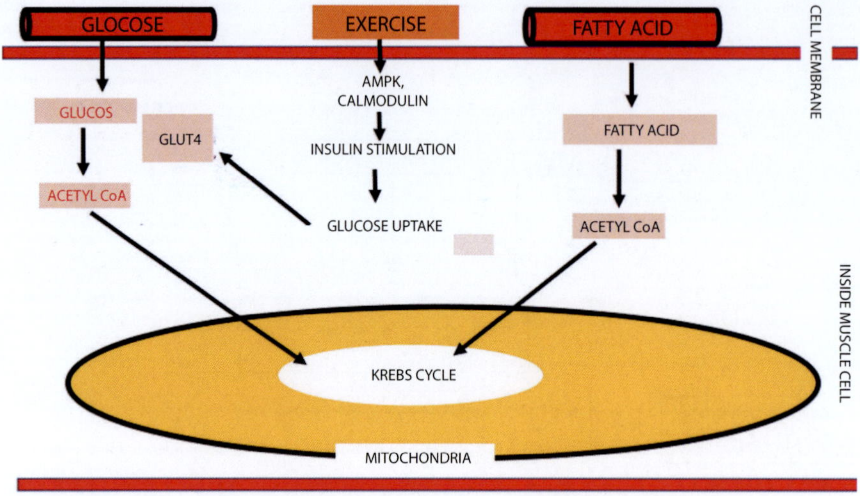

Fig. 2.42 Glucose uptake into muscle cell facilitated by calmodulin

of Ca^{2+}/calmodulin increase. Calmodulin is a multifunctional intermediate calcium-binding messenger protein expressed in all eukaryotic cells. Signaling pathway Ca^{2+}/calmodulin-dependent protein kinases (Fig. 2.42) have an important role in the Ca^{2+}/calmodulin to induce glucose uptake [268]. Calmodulin has a key role in the phosphorylase kinase activation, which eventually cause the cleavage of glycogen by glycogen phosphorylase to produce glucose being from glycogen [269]. Calmodulin plays a crucial role in lipid metabolism by activating calcitonin hormone, which lowers blood Ca^{2+} levels and activates G protein that generates cAMP.

The impact of calcitonin can be inhibited by blocking calmodulin actions, showing the important role of calmodulin in the activation of calcitonin [270]. In mouse muscle, the mechanism for the calmodulin to affect muscle glucose uptake is through AS160 CBD, which adjusts contraction to increase glucose uptake. This discovery defines AS160 signaling factor as an exclusive to muscle contraction and not insulin, leading to glucose uptake in skeletal muscle [271].

Regular exercise modifies numerous intracellular pathways, alleviates insulin resistance effects, and increases glucose uptake in skeletal muscle [272], leading to an improved expression frequency of GLUT4 [273]. Following 6 months of aerobic training, DNA methylation is reduced [274]. As a result of skeletal muscle contraction, cytosolic Ca^{2+} concentration and consequently the number of Ca^{2+}/calmodulin complexes increase. More important key players in the Ca^{2+}/calmodulin signaling pathway are the Ca^{2+}/calmodulin-dependent protein kinases (CaMKs). Following an acute endurance cycling exercise at an intensity of 80% of VO_{2max}, temporary postexercise bout gene transcription increases Ca^{2+} release and phosphorylation of CaMKII. It signifies that high-intensity exercise results in enhanced Ca^{2+}/calmodulin signaling [275]. Moreover, long-last aerobic exercise increases CaMKII, which drives higher Ca^{2+}/calmodulin signaling pathway activity [276]. Other ways in the framework of glucose uptake-related signaling are the Ca^{2+}/calmodulin signaling pathway and the AMPK-signaling pathway and Akt2-dependent TBC1D4 phosphorylation [277].

Lipids Metabolism

Training may elicit favorable changes in lipid metabolism and can bring about improvements in the regulation of hepatic glucose output, which is especially relevant to diabetes mellitus [278]. Meta-analysis examining the effects of aerobic and resistance exercise training on lipid profile in individuals with type 2 diabetes suggested that following endurance training, there was a 5% decrease in low-density lipoprotein cholesterol, without effect on high-density lipoprotein cholesterol, total cholesterol, or triglycerides [279], while other did not find any improvements [280]. Aerobic exercise is proving to have significant benefits for people with both type 1 and type 2. Aerobic exercise has significant and particular benefits for people with type 1 diabetes. It increases sensitivity to insulin, lowers blood pressure, improves cholesterol levels, and decreases body fat. In type 2 diabetes, moderate aerobic exercise can lower blood sugar levels and improve insulin sensitivity. Strength training, which increases muscle and reduces fat, may be particularly helpful for people with diabetes.

Notification of the differences between the acute effects of exercise and chronic training effects should be made. Exercise, irregular insulin secretion, and peripheral insulin resistance are key factors that affect the acute effects of exercise on metabolic responses in type 2 diabetes patients. Following a single acute bout of aerobic exercise, diabetic patient and healthy individual increase insulin sensitivity for at least 16 h postexercise [281]. Additionally, glucose uptake is still elevated owing to insulin-independent mechanisms, calmodulin effect, muscle contraction-induced stimulated GLUT4 translocation, and increased quantity of muscle GLUT4. However, insulin sensitivity is gone within a few days.

Fat following aerobic exercise training results in loss from the central regions of the body and should, consequently, contribute significantly in avoiding or alleviating insulin resistance [282]. Mechanism underlining for these improvements is that exercising muscles become more sensitive to circulating insulin. Blood sugar uptake turns out to be more easily and sugar utilization is more effective. Additionally, training may cause promising changes in lipid metabolism bringing about enhancements in the control of hepatic glucose output, relevant to diabetes mellitus [281]. In addition, physical training potentiates the effect of aerobic exercise on insulin sensitivity through multiple adaptations in glucose transport and metabolism. Blood glucose reduction during physical activity is attributed to an attenuation of hepatic glucose production, whereas muscle glucose utilization increases normally.

Aerobic exercise also improves blood cholesterol levels and blood pressure control. This benefit occurs regardless of weight loss. Exercise program reduces triglyceride levels by 20%, increases high-density lipoprotein cholesterol significantly, and decreases BP.

Aerobic exercise, at moderate-intensity aerobic exercise, increases glucose uptake by 2–3 mg kg^{-1} min^{-1} above usual requirements (70-kg person: 8.4–12.6 g h^{-1} of exercise). During high-intensity aerobic exercise, glucose uptake increases by 5–6 mg kg^{-1} min^{-1}; however, exercise of this intensity cannot usually be sustained for long intervals. Traditionally, diabetic patients have been encouraged to train aerobic exercises such as walking, running, cycling, and swimming.

Resistance Training in Diabetics

Mixture of aerobic and resistance modes of exercise is suggested, however, the benefit of this approach is still not well defined [283]. The American Diabetes Association recommends resistive training for type 2 diabetes patients, however, with the condition that in contraindications for exercising are not present [284].

Resistance exercise improves general metabolic function in individuals with diabetes, by enhancements muscle mitochondrial function and increases in muscle mass which may have an optimistic influence on insulin sensitivity and glucose regulation [285]. Resistance training is a method that may avoid muscle atrophy and diminished physical work capacity. It improves whole body metabolism of glucose, glycogen, and insulin action in the life cycle, in man and women. Weight training is usually not recommended much for diabetic patients. Due to the limited information on the role of resistance training for older patients with type 2 diabetes, and, since resistance training increase muscle mass, it decreases insulin resistance [286], it has been recommended that patients utilize low-moderate weights with high repetitions [287]. Consequently, strength muscle contraction increases glucose uptake in skeletal muscle [288], thereby forming the basis for recommending resistance training for individuals with irregular glucose metabolism. In the absence of contraindications, the American Diabetes Association recommends resistance training for those with type 2 diabetes [289]. Resistance training is an effective way for an integral element in the therapeutic management of glycemic regulation in both young and older type 2 diabetics patients [290]. The American Diabetes Association recommends performing resistance exercise 8–10 reps 3 $times^{-1}$ $week^{-1}$, without developing fatigue, while the American College of Sports Medicine suggests practicing progressive resistance training in individuals with type 2 diabetes [291]. Increased muscle mass following resistance training enhanced insulin sensitivity owing it to muscle contractions, which increase glucose consumption through insulin-independent mechanisms [292]. Muscle mass recruitment during resistance bouts may provide equally high or higher muscle mass and similar lasting training time, compared to aerobic exercise, still, effects of resistance training are controversial. Weightlifting training of several months significantly lowers insulin response and is without effect on glucose tolerance and to increase the rate of glucose clearance during a euglycemic clamp [293]. Compared to aerobic, resistance training may improve more glycemic control [285].

In rodents, resistance training enhanced insulin action in skeletal muscle, GLUT4 phosphorylation, and glucose consumption. Two pathways were designated which play a dominant part in cellular growth, developed in response to resistance exercise training. Protein kinase B PKB or, Akt (intracellular), signaling pathways that play a key role in glucose metabolism inhibit the apoptotic process, cell proliferation, transcription, cell migration [293], and ERK1/2 regulated kinase.

High muscle glycogen storage and Akt phosphorylation were present after 30 min of resistance exercise [294]. These are the key features that decrease the blood glucose level and avoid muscle atrophy in diabetic patients.

Resistance training differs significantly from aerobic exercise training for healthy subjects and patients with type 2 diabetes. Following strength training for 30 min 3 times week^{-1} 6 weeks^{-1}, insulin action in skeletal muscle in healthy subjects and patients was increased. The adaptation for that improvement is probably due to local contraction-mediated mechanisms involving key proteins in the insulin signaling cascade [295]. Following long-lasting, chronic resistance exercise training, insulin resistance is decreased, mechanisms of the insulin signaling pathway in skeletal muscles are altered and improve insulin sensitivity by reducing insulin amount needed to elevated rates of glucose transport [296]. Resistance exercise training, also known as strength or weightlift training, is a method of therapy [297], by increasing lean body mass, muscle mass, and muscle fiber-type alteration. Following resistance training, body sensitivity improves to insulin and insulin resistances decrease (Fig. 2.43); these changes are related to an increase in lean body and muscle mass, thus, enabling more glucose to enter the muscle for metabolism and storage [298].

Muscle fiber types have usually been categorized as fast twist fiber (types IIb, IIc, and IIa) and slow twist fiber (type I). Fiber types differ in their myosin heavy

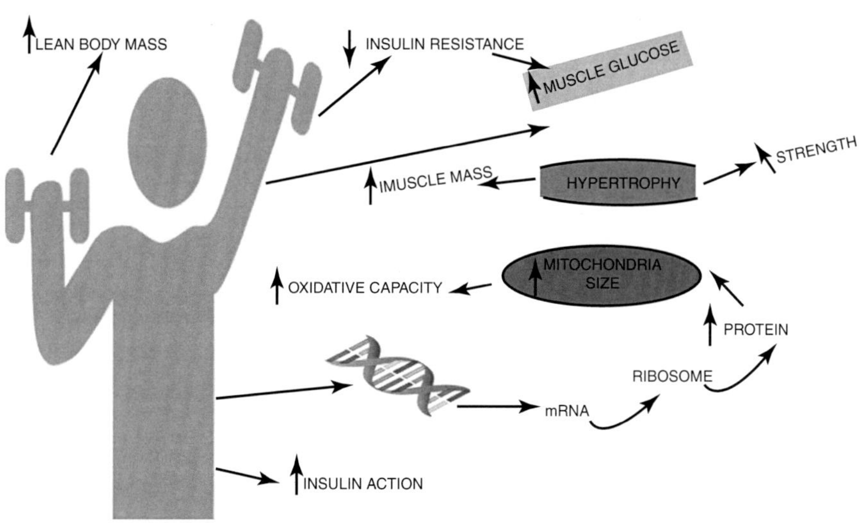

Fig. 2.43 Resistive training effect on diabetic patients

chain isoforms and consequently in their strength, oxygen utilization, and reaction. Since relation exists between fiber type and function, it affects mitochondrial number and content, therefore, type I is reach with mitochondria, and the opposite is true for type II fiber time [299]. In diabetic patients, following long-lasting exercise, muscle fiber type alteration, enhance skeletal muscle insulin sensitivity and decreases insulin resistance, this is mainly achieved by altering muscle fast twist fiber IIb to fast twist fiber IIa and times to type I [299].

Effect of Aerobic Exercise on GLUT4 Translocation

Glucose isoform transporter 4 (GLUT4) is found in skeletal muscles and fat cells, it is the insulin controller, GLUT4 permits the glucose diffusion from bloodstream into muscle and fat cells. GLUT4 is a protein and an essential adaptation to exercise training in skeletal muscles [267]. Glucose muscle uptake during exercise is facilitated by intramuscular breakdown of glycogen to glucose-1-phosphate and glycogen (glycogenolysis) by increased glucose uptake. Both aerobic and resistance exercises increase GLUT4 significantly and translocation, and consequently, blood glucose consumption, without the presence of insulin, even in type 2 diabetes mellitus [115]. At rest, replenishment of muscle glycogen stores depends on insulin [115]. Single bout of exercise produces different metabolic responses compared to those made by chronic exercise training. A single bout of exercise increases skeletal muscle GLUT4 and mRNA expression in patients with type 2 diabetes to a similar extent as in healthy subjects [300]. An acute bout of exercise requires the utilization of glycogen to produce ATP. Consequently, acute exercise changes glucose consumption, occurring by insulin-independent pathways [301] to guarantee glucose source to the working muscles. Additionally, acute exercise produces alterations in the GLUT4 cycling, as GLUT4 transportation is increased into the cell (endocytosis) or, GLUT4 transportation of membrane-bound vesicles containing cellular molecules to the cell membrane (exocytosis). Thus, exocytosis is an important process in cells, as it performs the opposite function of endocytosis [302].

In sled dogs, acute bout of exercise at 90 of VO_{2max} increases surface GLUT4 levels, this results is in agreement with findings in skeletal muscle of humans [303]. Findings suggested the existence of two different intracellular sources of GLUT4, one affected by insulin and the other is affected by exercise [304]. In mice, with impaired GLUT4-dependent glucose uptake, glucose consumption increased during exercise. This signifies that augmented GLUT4-dependent glucose flux reduces nutrient strain by changing in skeletal muscle, lipid, and amino acid metabolism [305]. This supports the notion that acute exercise can activate molecular pathways of storage vesicles mobilization that are not insulin-dependent. In insulin-independent pathways, two proteins have a major role in the mobilization of GLUT4: 5′ adenosine monophosphate-activated protein kinase (AMPK), that activates glucose and fatty acid uptake and Ca^{2+}/calmodulin-dependent protein kinase II (CaMKII) is controlled by the Ca^{2+}/calmodulin compound. CaMKII is involved in many signaling cascades.

Chronic exercise training is the utmost active stimulus to increase skeletal muscle GLUT4 expression, along with enhanced insulin effect, glucose disposal, and improved muscle glycogen storage following.

Long-lasting aerobic exercise training increases GLUT4 mainly through exocytosis and thus intracellular stores. Also, this adaptation regulates variations in molecular pathways that stimulate GLUT4 translocation, and GLUT4 expression at cell surface augmenting microsomal fraction [305]. During exercise, glucose transport is rate limiting for glucose utilization in muscle. Subsequent to exercise effort, glucose consumption and the contraction-mediated pathway stay elevated for a few hours [109].

The mechanism for improving glycemic regulation during aerobic exercise is the GLUT4 translocation and therefore glucose uptake by the muscle. Skeletal muscle expresses multiple isoforms of GLUT4 [306]. During exercise, the most important of these is GLUT4 because systemic [307] and muscle-specific GLUT4 knockout abolishes contraction-stimulated glucose uptake [308]. Validation of molecular machinery accountable for the increment in glucose transport and GLUT4 translocation during exercise is still known. It is generally accepted that GLUT4 arises from local factors within the contracting skeletal muscles [306], such as activation of AMP-activated protein kinase (AMPK). During exercise, translocation of GLUT4 to surface membranes in skeletal muscle bring about an increase in glucose consumption and 5′ AMP-activated protein kinase activity [309]. Different possible mechanisms are the signaling via nitric oxide, ROS, and by the contraction itself. The major glucose transporter protein expressed in skeletal muscle is GLUT4.

Combined Aerobic and Resistance Exercises in Diabetics

Improved glycemic control and decreased HbA1c levels are important for reducing the microvascular and macrovascular complications of diabetes, each percentage point reduction in HbA1c was associated with a 35% reduction in microvascular complications [169]. In brief, chronic resistance training, at moderate intensities produces the greatest metabolic functional responses in relations to HbA1c and insulin sensitivity in individuals with type 2 diabetes [285].

Combined aerobic and resistance training were tested in a group of type 2 diabetics, it revealed very encouraging results [310]. Combined of both modes of exercise and modest weight loss resulted in 58% reduction in type 2 diabetes risks in high-risk populations. Both aerobic and resistance training exercises improve acute and chronic insulin action [311]. Although both aerobic and resistance types of exercise have been shown to be related to glycemic improvement, significant improvement is when both aerobic and resistance exercises training are combined [312].

Recommended Exercise Training Intensities

Before beginning a program of exercise training, diabetic patients should be assessed for current work capacity that might restrict certain exercise work intensities [313]. For improving glycemic control and reducing the risk of cardiovascular disease, the American Diabetes Association recommends to exercise at least 3 days a week, without allowing more than 2 days to elapse between exercise sessions, at least 150 min $week^{-1}$, at intensities of 60–70% of reserved HR), or at least 90 min $week^{-1}$ at vigorous intensity above 70% of reserved HR) [314]. So, patients with diabetes should be stimulated to perform 30–60 min of moderate-intensity aerobic activity such as brisk walking on most (preferably all) days of the week.

Resistance training-induced enhancements in glycemic control appear to be intensity-dependent, with beneficial effects occurring when subjects train at 70% to 90% of the 1-repetition maximum strength (1-RM). Heavy-resistance strength training may be related to a decrease in HbA1c and cholesterol in type I diabetic men and women regardless of age following training in addition to an increase in overall strength [312].

Obesity

Obesity is a medical problem; it is a multifactorial disease connected to an extreme quantity of body fat. Obesity is a genetic trait that gets up from the connections of numerous genes and lifestyle influences. Obesity is the results of energy imbalance between caloric consumption and calories utilization. Obesity is a serious public health problem because it is associated with some of the leading causes of death in the western population. Obesity is associated with a variety of chronic diseases and increases the risk of other diseases [315]. Obesity is an independent risk factor strongly related to coronary artery disease, however, partial data are accessible on obesity involvement in CAD patients. There are many reasons why some people have difficulty avoiding obesity. Usually, obesity results from a combination of genetic factors, combined with the environment and personal behaviors such as diet (Fig. 2.44) and physical inactivity [316].

Almost, two-thirds of western population is overweight or obese, the occurrence of obesity stands at one-third and continues to increase [317].

Tide relationship exists between human obesity and cardiovascular disease [318], as well as with diabetes [319]. In addition, bigger body mass index (BMI) is a better predictor of recognized cardiovascular disease [320]. BMI (Fig. 2.45) is a value resulting from the body mass and height of an individual.

Although obesity is seen as an epidemiological result, it is also a medical-metabolically healthy problem which results in adipocyte dysfunction, however, this can be helped by exercise. In addition to exercise, weight loss can advance or avoid the health complications that are related to obesity. Medications and behavioral changes are other choices for treating obesity [321].

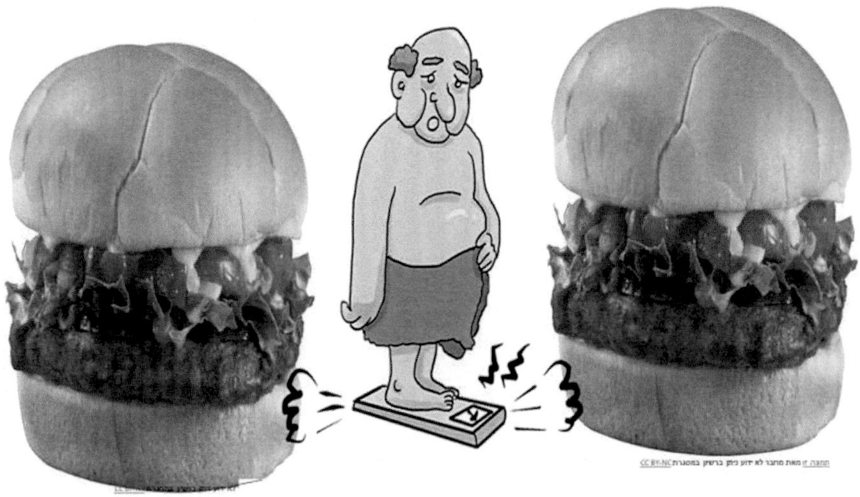

Fig. 2.44 Wrong diet, physical inactivity, and undesirable lifestyle result in human obesity

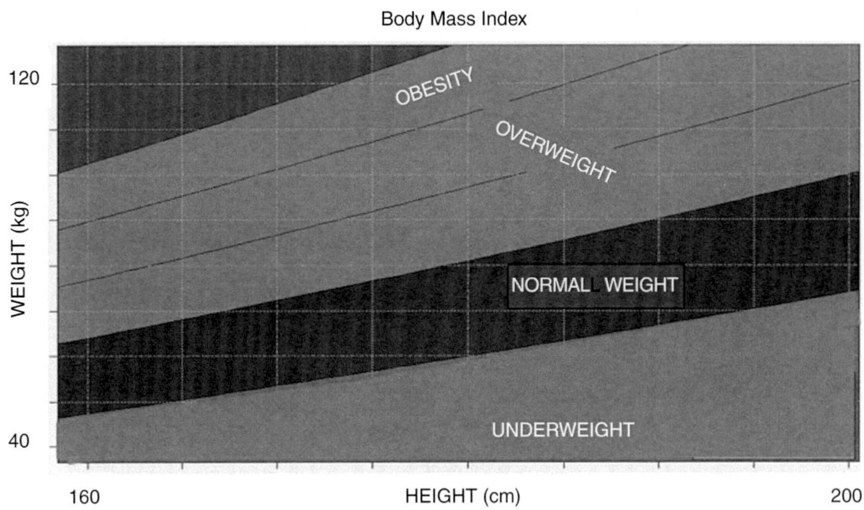

Fig. 2.45 Body mass index (BMI) is a person's weight in kilograms divided by the square of height in meters. A high BMI can be an indicator of high body fatness

Obesity is identified when BMI is 30 or higher. The equation for calculating BMI is weight (in kilograms) divided by height (in meters squared) to have kg m$^{2-1}$. When using pounds and inches measurements, pounds should be divided by inches squared. This should then be multiplied by 703 to convert it from lbs inches 2–1 to kg m$^{2-1}$. If BMI is below 18.5, it is considered underweight, 18.6–24.9 is considered normal, above 25–29.9 is considered overweight, and higher than 30 is considered obesity.

Obesity is frequently subdivided into three categories:

- Class 1: BMI from 30 to <35.
- Class 2: BMI from 35 to <40.
- Class 3: BMI from 40 or higher, it is life-threatening [324]. BMI in general gives a practical assessment of body fat, still, BMI is not a direct measurement of body fat percentage, thus, sometimes, BMI especially, in muscular athletes, might result as obesity category even though they have high lean body mass.

Body Composition Measurements

In adults, the determination of obesity and overweight standing is by calculating body mass index (BMI). This measurement is a good indicator of body fat or adipose, which is a value derived from the height (m) and weight (kg) that applies to adult men and women. Storage body fat consists of fat accumulation in adipose tissue, part of which protects internal organs. Health risk factors are related to hypertension, coronary artery disease, diabetes, and respiratory apnea. However, essential body fat is necessary to maintain life and reproductive functions. Adipose tissue is an anatomical term for moveable connective tissue composed of adipocytes. Its main role is to store energy in the form of fat, in addition, it also cushions and insulates the body. Adipose tissue is found under the skin (subcutaneous), kidneys, liver, and minor quantity in muscles. Gender is the main factor for defining where fat will accumulate. Males have a tendency to store fat in their chest, abdomen, and backsides. In women, fat accumulates in the breasts, hips, waist, and behinds. Differences of fat distribution between males and females are essential, body fat is attributed to women's sex hormones and to the burdens of child bearing. Fat cells are formed in the fetus during the third trimester of pregnancy and later at the onset of puberty. In general, fat cells usually do not generate after puberty; thus, the fat cell's number remains unchanged. However, gaining weight is accomplished by the swell of fat cell's size. Body composition is one variable related to the figure. Body composition has two components: lean body mass and fat tissue. In men, essential percentage of fat is 3–5%, while in women is 8–12%. Body composition concerted with bodyweight often is requested in sport's competitions for appropriate playing weight.

Obesity as a Risk Factor

Weight reduction, reduction programs is needed in preventive cardiology. In addition, obesity has significant adverse effects on various coronary heart disease risk factors, including increasing arterial pressure, promoting the development of left ventricular hypertrophy, worsening insulin resistance, adversely affecting plasma lipids particularly increasing levels of triglycerides, and reducing levels of high-density lipoprotein cholesterol, and leading to a more sedentary lifestyle [322]. In addition, obesity has substantial opposing effects on numerous coronary heart disease risk factors, including hypertension, development of left ventricular hypertrophy with small left ventricular volumes, insulin resistance and thus, diabetes, triglycerides increment and lowering levels of high-density lipoprotein cholesterol [323].

Lavie et al. [324] demonstrated the benefits of cardiac rehabilitation and exercise training programs in obesity indices, exercise capacity, plasma lipids, behavioral characteristics, and quality of life in obese patients with coronary artery disease, which were particularly noted in a subgroup of obese patients who were more successful with.

Adult obesity is a common health problem associated with significant adverse health outcomes. Overweight and obesity are highly prevalent in cardiac rehabilitation. Overweight and obese patients had a greater adverse cardiovascular risk profile, including a lower exercise capacity in the latter. Thus, targeted interventions toward weight management in contemporary cardiac rehabilitation programs are important. Although short-term outcomes appear promising, greater efforts to improve these outcomes and to support long-term management are needed [325]. However, pharmacotherapy agents are available for long-term use to enhance weight loss efforts for some patients [326]. Weight reduction in obese persons will reduce all of the coronary vascular disease risk factors associated with type 2 diabetes mellitus and will improve hyperglycemia. Despite adversely affecting all of these risk factors, physical inactivity seems to have an independent effect on cardiovascular disease risk, whereas obesity increases the risk partly through the modification of other risk factors [327]. Although data are available to indicate that weight reduction with dietary therapy or exercise improves lipid profiles and leading to increases in high-density lipoprotein cholesterol, along with a reduction in insulin resistance, and that weight reduction in severe obesity reduces both arterial pressure and left ventricular hypertrophy, limited data are available on the effects of vigorous risk factor intervention in obese patients with known coronary artery disease. Evidence-based guidelines support intensive nutrition and behavioral counseling and moderate aerobic exercise [328]. A growing body of literature demonstrates that aerobic exercise, accompanied with or without weight loss, can lead to favorable changes in body composition including a reduction in abdominal adiposity [329].

Obesity and Cardiovascular Disease

Increase morbidity is mainly caused by hypertension and cardiovascular disease and correlates directly with obesity. Yet, since type 2 diabetes also correlates highly with obesity, the pathophysiological problems related to diabetes add as well to mortality rate related to obesity (Fig. 2.46). As adipose accumulates in tissues, due to excess triglyceride amass, physiological changes happen that produce an

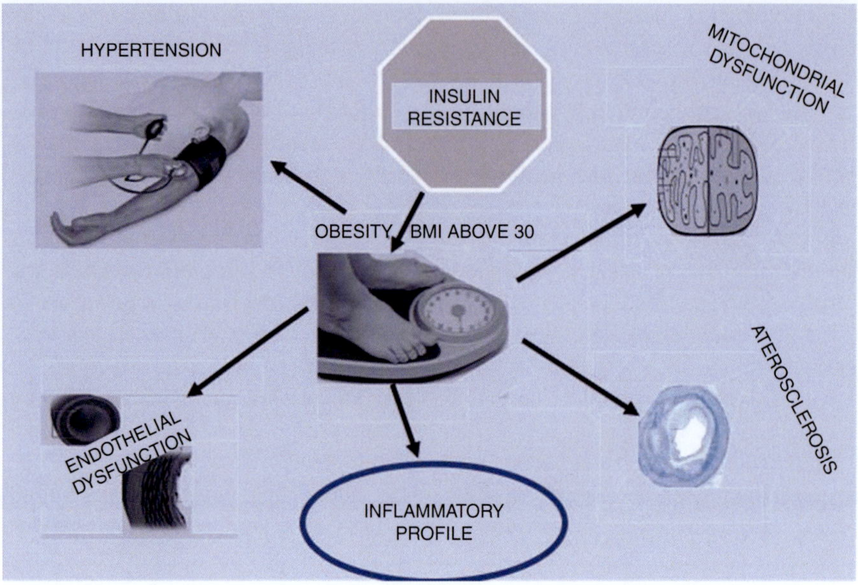

Fig. 2.46 Obesity effects on risk factors for diabetes and cardiovascular diseases

increased tissue inflammatory response [330]. These disturbances also happen in the liver, consequently, both liver inflammatory and adipose tissue in obesity cause insulin resistance [331].

It turns out, that insulin resistance due to obesity is a key factor in the progress of cardiovascular disease, by exerting vasodilatation due to increased nitric oxide levels.

Atherosclerosis is a result of proinflammatory state where leukocyte infiltration into the vascular endothelium increases significantly which increases proinflammatory cytokines secretion which increases in obesity and has a significant role in the expansion of atherosclerosis [332]. Mounting adipose tissue increases quantities of TNF-α and other proatherogenic factors [333]. Augmented circulating TNF-α improves grip of leukocytes to the endothelium [334]. Adherent monocytes and additional lipoprotein particles in the vasculature infiltrate the endothelial layer which becomes foam cells. Foam cell secretion of cytokines increases extracellular matrix, thus, decreases vasculature fluidity [335].

Mitochondrial Dysfunction in Obesity

Significant increase in caloric intake causes mitochondrial dysfunction, and then obesity-related pathologies, partially because of the ROS possessions. Mitochondrial dysfunction results in increased production of ROS which activates stress responses, leading to increased activity of MAPK and c-Jun N-terminal kinases (LNK) decrease insulin receptor signaling. In addition, mitochondrial dysfunction contributes to the pathogenesis of metabolic disorders [336]. Mitochondrial dysfunction changes mitochondrial dynamics, decreased activity nuclear respiratory factor-1 (NRF1), and impaired ability to produce ATP [337]. Mitochondrial dysfunction is mainly associated with progressions of oxidative phosphorylation (morphos), it is causative to the progress of brain disease, particularly chronic degenerative conditions (encephalomyopathy) [338], mitochondrial myopathy, and aging-related disorders such as neurodegenerative diseases, metabolic syndrome, and diabetes [339]. During adipocyte disparity, mitochondrial biogenetics and activity increase significantly [340]. Furthermore, in mature adipocytes, mitochondrial dysfunction is related to imperfections in fatty acid oxidation [341], excretion of adipokines [342], and decontrol of glucose balance [343]. This proposes that decreased adipose tissue mitochondrial content in mouse is related to the manner of the disease development, and the development of imbalanced glucose homeostasis. This suggests that decrement in adipose tissue mitochondrial proteins is not a causal factor, in the progress of glucose homeostasis imbalance [344].

Normal production of mitochondria is activated in response to variations in the ATP/ADP ratio and AMPK activation, subsequently, the transcriptional coactivator peroxisome proliferator-activated receptor gamma coactivator-1 alpha (PGC-1α), and NRF1 expression are increased [345]. PGC-1α's have the capacity to coactivate several nuclear transcription factors that increase conversion of mitochondrial gene information encoded, as well as, PGC-1α through edge with mitochondrial transcription factor A, inside mitochondria to intensify the mitochondrial DNA transcription [346].

The NRF1 gene is required in regulating the production of nuclear-encoded parts of the mitochondrial respiratory chain and heme biosynthesis. It controls the expression of mitochondrial transcription factor A (TFAM, for transcription factor A, mitochondrial, designated mtTFA) which is a nuclear transcription factor essential for replication, maintenance, and transcription of mitochondrial DNA [347]. Mitochondrial dysfunction results in decreased activation of PI3K, which participate in the translocation of GLUT4 to the plasma membrane thus glucose uptake is reduced [348]. Therefore, inhibition of PI3K results in a reduction in glucose consumption in skeletal muscle and adipose tissue.

The effect of enzymes on lipid oxidation occurring in the mitochondria, through β-oxidation, which in turn, increases intramyocellular lipid content. There are four major enzyme activities in the β-oxidation pathway: acyl-CoA dehydrogenases, enoyl-CoA hydrateses, 3-hydroxyacyl-CoA dehydrogenases, and 3-ketoacyl-CoA thiolates [349]. In diabetes type 2 patients, skeletal muscle lipids metabolism is impaired. Reduced mitochondrial oxidation and augmentation of fatty acids to skeletal muscle results in augmentation of intracellular fatty acid metabolites. These metabolites stimulate protein kinase C isoforms, resulting in reduced insulin signaling of the insulin receptor.

Effects of Exercise on Obesity

Weight loss decrease the multifactorial disease risks related to obesity, such as diabetes mellitus, hypertension, cancer, hyperlipidemia, and cardiovascular disease. When exercise is included in a weight loss therapy, it signifies a positive approach to improve weight loss and in obese patients, along with, greater enhancements in body composition and metabolic outcomes [350]. In the literature, a debate exists over the independent and combined effects of dieting and increased physical activity on improving metabolic risk factors. Recently, it has been suggested by Redman et al. that exercise plays an equivalent role to caloric restriction in terms of energy balance [351]. Therefore, it has been suggested that exercise is an important adjunct to caloric restriction for achieving and maintaining weight loss. The National Weight Control Registry enrolled 3000 individuals who lost >10% of their body weight and maintained this weight loss for at least 1 year [352].

Yet, exercise alone, without caloric intake restraint, is less efficient, while adding exercise to caloric intake restraint does not make big different in fat-mass loss compared to dietary restriction alone [353]. Although, sometimes, exercise may even act as a mild appetite.

Suppressant

Losing weight is a combination of a process that includes a decrease in caloric dietary intake, an obese person may reduce energy intake 1500 kcal d^{-1} and an increase in exercise intensity that adds few hundred kcal d^{-1} [354]. Even so, adding exercise training is vital to avoid a decrease in muscle mass and augment relative visceral fat-mass decrement [348]. At rest and during exercise, skeletal muscles are the main site of oxidation of free fatty acid. At rest, when oxygen supply is adequate for muscle contraction, free fatty acids are the predominant fuel used. During low-intensity exercise (45–50% of VO_{2max}), fat oxidation is increased by 6–7-folds than resting conditions. Endurance exercise training elicits many skeletal muscle adaptations, including increased capacity for the oxidative metabolism of fatty acids and carbohydrates. When the exercise intensity increases, fat oxidation increases further, until exercise intensities of about 65% VO_{2max}, after which a marked drop in the rate of fat oxidation is observed [355]. This is in contrast to carbohydrate metabolism, which increases as a function of the aerobic work rate. Fat oxidation is reduced at high exercise intensities. At around 85–90% VO_{2max}, the majority of energy needed is produced from carbohydrates metabolism only without the utilization of fatty acid (Fig. 2.47). An increase in fatty acids oxidation is facilitated by increased capacities for fatty acids uptake into the myocyte [356], their subsequent mitochondrial transport and β-oxidation [357].

Figure 2.48 Explains the fatty acids utilization as an energy source, lipolysis of triglycerides is needed to produce an increase in free fatty acids and glycerin levels, from adipose tissue to be oxidized in the mitochondria for ATP synthesis [358].

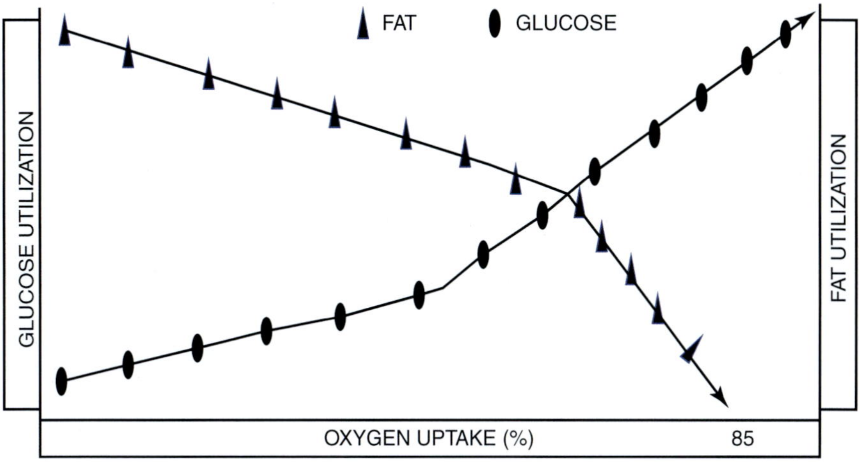

Fig. 2.47 Fat oxidation and glucose utilization as a function of aerobic exercise intensity. Fat oxidation increases at 65% VO_{2max} and entirely decreases around 85% VO_{2max}

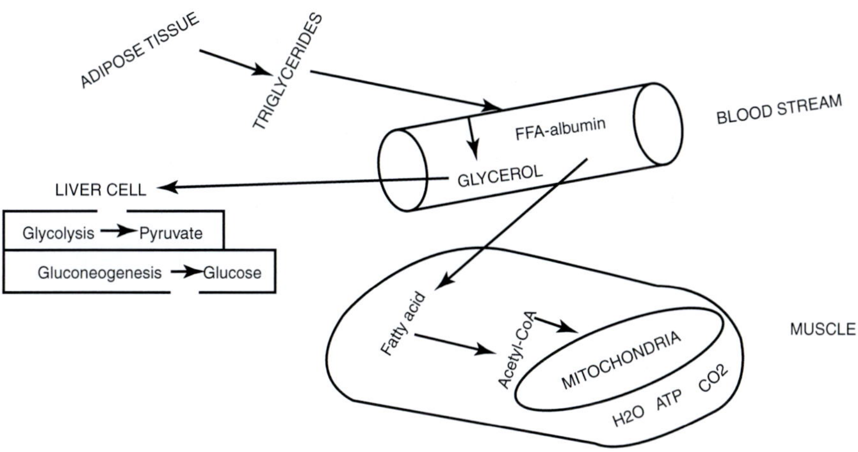

Fig. 2.48 Stored triglyceride is mobilized from adipose tissue, can be broken down to glycerol and free fatty acid, which is transported in the bloodstream, bounded to albumin, to the working muscles. Triglyceride is also broken to glycerol and fatty acid, which enter the mitochondria for oxidation during exercise

Lipids oxidation are the energy source mostly accountable for energy supply during submaximal exercise. At the onset of aerobic exercise, there is a large increase in oxygen consumption, making possible in skeletal muscle, to increase uptake and oxidation of fatty acids, the largest fuel reservoir in the body. Intensities of 60–65% during aerobic exercise, fatty acid levels rise by 2–3 above resting values [359].

It is also well recognized that prolonged exercise for several hours at a low intensity induces gradual decrease in the respiratory quotient (RQ) and hence, an improved lipid utilization at the conservation of carbohydrates as energy fuel [360].

Thus, fatty acid oxidation during aerobic exercise allows long-lasting effort and delays glycogen depletion and bloodstream hypoglycemia. With increased workload intensity, triacylglycerol utilization also increases gradually. Energy molecules, such as non-esterified fatty acids from the bloodstream, are vital for the muscle metabolism during moderate exercise intensities [360]. The nonesterified fatty acids are made through lipolysis of the intramuscular triacylglycerol [361].

During lasting moderate-intensity exercise, there is a direct relationship between the nonesterified fatty acids of exercising muscle transport and consuming. The rise in nonesterified fatty acids distribution is due to increase in capillary enrolment, for the increase in Q, where a greater share of the total accessible nonesterified fatty acids delivery turn out to be available [362].

At any intensity, the energy needs of working muscles depends on oxygen availability; at intensities up to 65% the main energy source is fatty acid transported to muscle mitochondria, to be oxides [363]. Above this intensity, more and more the muscle utilizes carbohydrates, increases in exercise intensity beyond 65 VO_{2max}, is a point where maximal lipid is oxidized is maximal [364]. When reaching VO_{2max}, cells energy sources are glycogen and glucose. In other words, as exercise workload is increased, the involvement of glucose oxidation rises in percentage to the reduction in lipid oxidation [365].

During an aerobic bout, of an untrained obese individual, at the anaerobic threshold intensity (50–55% of VO_{2max}), triacylglycerol utilization is lower than that at 70%. Thus, increasing exercise above the 75% intensity changes fuel utilization toward an increase in carbohydrate and decreases in fat utilization. Aerobic exercise training improves VO_{2max}, which has implications on cardiovascular and metabolic responses [351]. Subsequent to aerobic exercise training, with increased VO_{2max}, the anaerobic threshold also increases, and consequently, an upward change in the improved lipid utilization is obvious [366]. The control mechanisms regulating fatty acid consumption and oxidation depend on the accessibility of oxygen. Exercise intervention can enhance glucose utilization in skeletal muscle and affect the metabolic pathways supplying ATP to the exercising muscles, though exercise-mediated lipolysis in adipose tissue and mitochondrial biogenesis.

Subjects who habitually exercise aerobically 176 min 3–4 $week^{-1}$ loses significantly more weight than those who did not exercise regularly. Subjects who exercised more than 195 min 3–4 $week^{-1}$ may lose nearly 7% of its abdominal fat. Aerobic exercise lasting at 60–75% maximal work capacity at least 30 min 3–4 $week^{-1}$ may be adequate to maintain cardiovascular health, but it may not be enough to prevent unhealthful weight gain. Recommendations published in 2003 and 2004 suggest that 45–60 min day^{-1} is necessary to prevent weight gain, children may need more activity. Studies suggest that people who have trained for a long time develop more efficient mechanisms for burning fat and are able to stay leaner. Following exercise training, these increments have been associated with the upregulation of membrane-associated fatty acid transport proteins, fatty acid-binding protein [367], fatty acid translocase [368], the mitochondrial transporter carnitine palmitoyltransferase I [369], and a key enzyme of ß-oxidation and ß-hydroxyacyl-CoA dehydrogenase [370].

Resistive Exercise Training Effect on Obesity

Resistance training has been regarded as a lesser tool of exercise in relations of weight loss due to: (a) the less caloric expenditure during each bout compared to aerobic bout [371] and (b) during resistive bout, muscles utilize mainly anaerobic pathways that rely on ATP, CP, and stored glycogen [372]. However, adding resistance exercise training as an essential part of an exercise therapy program has been recommended by the American Heart Association [373], the American College of Sports Medicine [374], and the American Diabetes Association [375]. Resistance training programs can increase lean body mass (or fat-free mass) and decrease percentage of body fat. Indications propose that the preservation of a large muscle mass can decrease metabolic risk factors, namely, obesity, dyslipidemia, and type 2 diabetes mellitus related to cardiovascular disease [376]. One of the outstanding benefits of resistance exercise, as it relates to weight loss, is the positive impact of increasing energy expenditure, during the exercise session and somewhat during recovery, and on maintaining or increasing fat-free body mass, while encouraging the loss of fat body weight. It is more likely that body composition is affected and controlled by resistance training programs using the larger muscle mass and greater total training volume [377].

Although it is obvious that aerobic exercise training is related to a higher energy expenditure compared to resistance training, yet, it has been suggested that chronic resistive training is a better mean for weight loss in obese individuals [378]. Resistance training increases lean body mass, consequently, it does not affect total body weight [245]. The application of resistive training combined with dietary intake restraint avoids the loss of lean body mass, lesser to dietary restraint [379]. Resistive training can avoid age-related reduction in lean body mass, and RMR, two variables that have a direct correlation [380]. Resistive training elevates RMR due to muscle hypertrophy and muscle protein turnover [381]. Change in lean body mass following resistive training depends on exercise intensity, that has to be sufficient to increase in lean body mass [382].

In general, relative intensity appears to account for 18–35% of the variance for muscle hypertrophy response to resistance exercise [383]. Volume in resistance training is equal to the total workload, which is directly proportional to the energy expenditure of the work bout. Total volume is determined by the total number of repetitions (repetitions × sets) performed times the weight of the load (total repetitions × weight). Often total volume calculated multiplying the number of sets × repetitions × load. For example, four sets of 10 repetitions with 30 kg would be expressed, $4 \times 10 \times 30 = 1200$ kg of volume. An impressive finding to highlight with resistance training is that the energy expenditure following the higher total volume workouts appears to be elevated, compared to other forms of exercise, and thus, further contributes to weight loss objectives. Furthermore, studies evaluating the impact of high-intensity resistance exercise on body composition and other physiological adaptations during weight loss have reported inconsistent findings [384]. However, one has to keep in mind that there are limitations with research examining

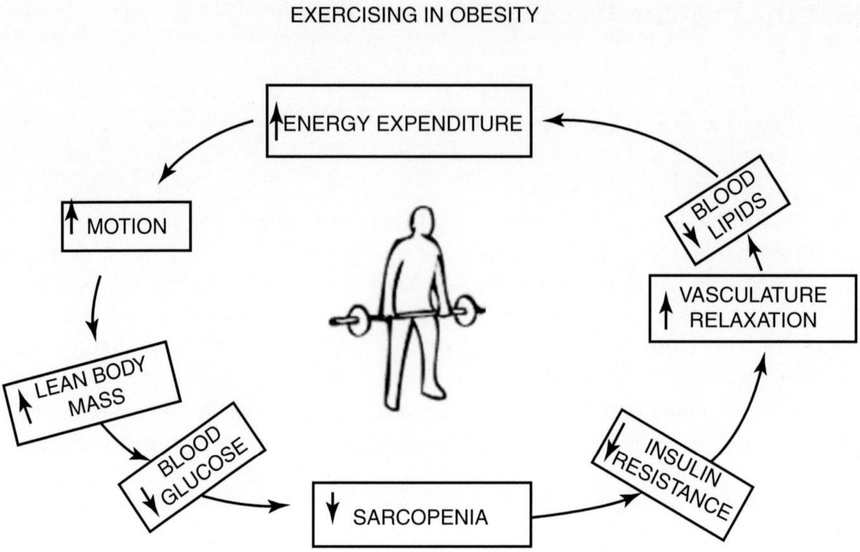

Fig. 2.49 Resistive training effect on different risk factors in obesity

the physiological effects of resistive training (Fig. 2.49) associated with various forms of weight loss. Only a few studies have simultaneously reported data on body composition, exercise performance (muscular strength, endurance, power), metabolic rate, and blood lipid profiles, and no studies have examined all of these variables in response to resistance training [384].

The Endocrine System in Fat Metabolism

Findings reveal that obesity progression includes hormones and neurotransmitters that control hunger and energy expenditure [385]. These hormones control satisfaction feeling in certain centers of the brain. In case of mutations in these hormones or their receptors can lead to obesity [386].

Insulin and catecholamine have pronounced metabolic effects on human adipose tissue metabolism. Insulin stimulates lipoprotein lipase and inhibits hormone-sensitive lipase; the opposite is true for catecholamine in which its level is increased during exercise. Catecholamine induced lipolysis in visceral fat but decreased in subcutaneous fat. The endocrinology system during aerobic exercise controls lipolysis of adipose tissue triacylglycerols and, seemingly, intramuscular triacylglycerols [387]. Lipolysis stimulated by epinephrine augment serum free fatty acid levels. At rest, blood epinephrine level is about 10 ng L^{-1} but, it may increase by 15-fold during aerobic exercise [388]. Plasma epinephrine levels also increased progressively with the graded exercise increments. The increased plasma epinephrine level, with increased exercise intensity, serves numerous metabolic and cardiovascular

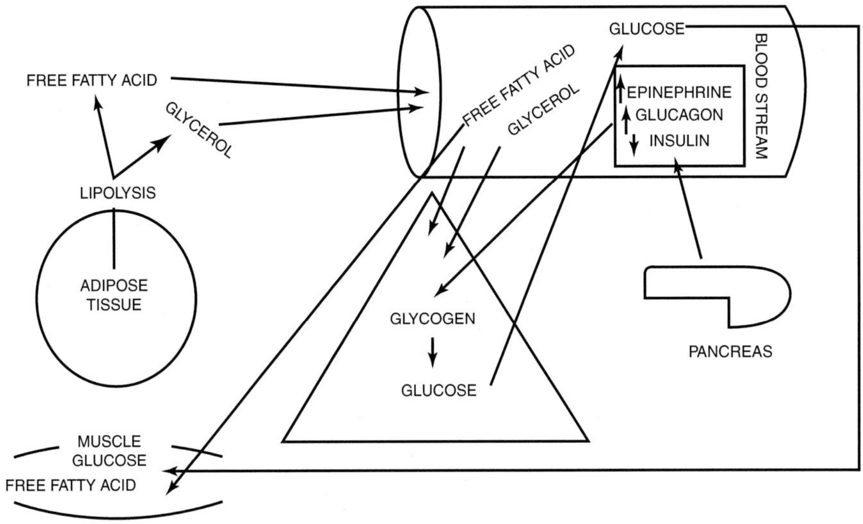

Fig. 2.50 Epinephrine has metabolic actions on adipose tissue, muscle, and liver

functions such as lipolysis, elevate Q, and liver gluconeogenesis. Epinephrine that binds to the β-adrenergic receptor on adipose cell membranes [389] activates a waterfall of processes, which start with the phosphorylation of adipose triglyceride lipase. Secretion of epinephrine from the adrenal medulla is triggered by impulses from preganglionic sympathetic nerves in response to stress, prolonged exercise and hypoglycemia [390].

Epinephrine has metabolic actions on adipose tissue, muscle, and liver (Fig. 2.50). This consists of fat mobilization in adipose tissue and, glycogenolysis in muscle and the liver [390].

Epinephrine receptors in muscle and adipose tissue are β-adrenergic, leading to the mobilization of glucose from glycogen for energy in muscle, and the mobilization of free fatty acids from adipose tissue, in both the muscle and the liver. In the liver, epinephrine receptors are different and are α1-adrenergic, leading to glycogen phosphorylase activation, as seen with glucagon. Insulin roles on metabolism are in the liver, muscle, and adipose tissue (Fig. 2.51), which increase in general fuel storing. Building stock includes the production of glycogen in liver and muscle and production of triglycerides mainly in liver and also in adipose tissue.

Numerous adaptive responses take place with physical training. These adaptations result in a more efficient system for oxygen transfer to muscle, which is now able to better utilize the unlimited lipid stores instead of the limited carbohydrate reserves available. To generate beneficial metabolic effects, the recommendation for physical activity of moderate intensity is 30–45 min performed 3–5 times per week. Walking is likely to be the most successful because it is safe and accessible. It appears that as long as the increase in energy expenditure is sufficient, low-to-moderate intensity endurance exercise is likely to reach the aim [391].

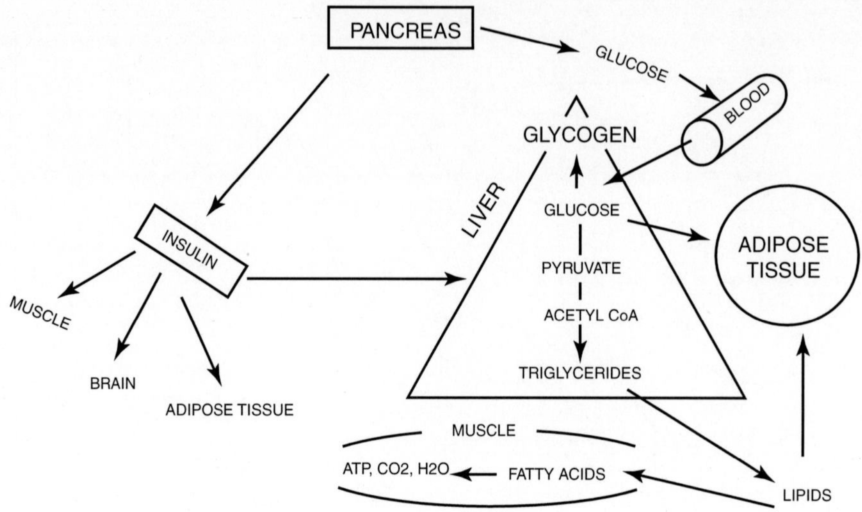

Fig. 2.51 Metabolic effects of insulin in liver, adipose, and muscle tissue

Exercise Epigenetics in Obesity

Obesity seems to alter genetic expression, to progress diabetes disease process, and other diseases, alterations, that may affect the subsequent generation. Genetic tendency and environmental factors suggest a dynamic part in the progress of obesity, Twin and approved studies suggest that up to 70% of the interindividual alteration in fat mass is controlled by genetic factors [392]. Interactions between environment and epigenetic mechanisms facilitate the expression of genes related to BMI increment and adiposity [393]. Variances in BMI in different populations are due to genetic factors that may explain up to 40% these variations in BMI [394]. Thus, Obesity and common overweight share a genetic basis. Epidemiological approaches utilized numerous genetic to categorize genetic loci for obesity. Loci in genetics is a fixed position on a chromosome, where a particular gene or genetic marker is located. A variant of the similar DNA sequence located at a given locus is called an allele. The ordered list of loci known for a particular genome is called a gene map.

Changes in epigenetic signs, with DNA methylation and histone alterations, are related to the skeletal muscle transcriptional responses to exercise that ultimately facilitate the exercise variations [395]. Exercise controls the activity of epigenetic modifying enzymes that are controlled and activated, by signaling pathways, consequently, exercise affect transcriptional response. Alterations in the DNA methylation outline in adipose tissue take place on CpG sites all over the genome following exercise training program [396]. Key metabolic and myogenic genes respond early to exercise by increasing transcription are significant in facilitating the following adaptations in skeletal muscle [397]. It is estimated that exercise training program consistent exclusive genes almost roughly equal to 33% of all genes in the human genome. It seems that enhanced fat oxidation after exercise training is most closely associated with the genes involved in regulating fatty acid uptake across the plasma membrane and across the mitochondrial membrane.

Effects of Exercise on Lipids

Aerobic exercise is widely recommended in the treatment model of various hyperlipoproteinemia for its putative antiatherogenic action on circulating lipids and apolipoproteins and to improve cardiorespiratory fitness, which may lower atherosclerotic risk, independent of lipids, and other risk factors [398]. However, training studies in hypercholesterolemic subjects are rare, and lipid results are inconclusive [399]. Aerobic fitness and exercise programs such as walking, jogging, and aerobics have been encouraged, as a means to reduce total cholesterol, low-density lipoprotein cholesterol (LDL-C), and triglycerides, while elevating high-density lipoprotein cholesterol (HDL-C). The association between serum cholesterol levels and the incidence of ischemic heart disease has been demonstrated by experimental and epidemiological studies [400]. The relationship between cholesterol and ischemic heart disease is continuous, gradual, and highly intense. The predictive value of the cholesterol level decreases with age, and actually it is low, from the sixth decade of life on. The risk attributed to hypercholesterolemia is due to LDL-C. When it involves large- and medium-sized arteries, it is referred to as atherosclerosis. The hypothesis that physical training programs are beneficial to the myocardium is based on evidence that chronic repeated exercise may reduce the risk of developing CAD [401]. Studies have established that HDL-C blood level is inversely related to the incidence of CAD. On the other hand, reductions in low LDL-C produce definite further reductions in the incidence of myocardial infarction, revascularization, and ischemic stroke. Number of intervention studies have demonstrated that lowering LDL-C by means of hypolipidemic agents is accompanied by significant reductions in cardiovascular morbidity and mortality, both in primary and secondary care [402].

The first indication of the HDL-C raising effect comes from the fact that highly trained athletes have lower levels of LDL-C and higher level of HDL-C than, do sedentary persons matched to age and sex. This is also true following aerobic exercise training programs [403]. Disabled individuals, whose movement is further restricted, exhibit even lower levels of HDL-C [404]. Running distance per week was found to be the best predictor of HDL-C levels among runners [405]. Rotkis et al. found in 90 runners reporting mean weekly distance run of 13, 27 and 58 miles and 19 non-runners, in runners, mean HDL-C was 47 mg dL^{-1} in 28 runners, 53 mg dL^{-1} in 30 runners, and 60 mg dL^{-1} in 32 runners. Nineteen nonrunners had a mean HDL-C level of 34 mg dL^{-1} [406]. Athletes mainly engaged in anaerobic training have lower HDL-C and high-density lipoprotein cholesterol/total cholesterol ratios when compared to endurance-trained subjects such as long-distance runners [407].

Some investigators reported HDL-C values in male strength-trained athletes similar to those of endurance-trained athletes [408]. Others questioned of strength training role for improving HDL-C and triglycerides in adults. Tucker and Silvester [409] found a reduced risk of hypercholesterolemia among individuals participating above 4 h in resistive training programs. In contrast, Kohl et al. [410] reported no significant relation, between muscle strength and total or LDL-C for either gender.

Nevertheless, one published and well-designed study indicated that 5 months of sustained resistance training significantly decreased total cholesterol and low-density lipoprotein cholesterol in women [411].

The evidence that blood levels of lipoprotein not only differ between persons of different activity levels but are also affected by physical activity is substantial [412]. Physical training is thus thought to be a stimulus for incrementing the level of HDL-C and reducing triglycerides in the blood. The response of different lipoproteins to aerobic training is well-established, at least in males [413]. Aerobic training is associated with two of the major lipoproteins that transport cholesterol in plasma: HDL-C, very low-density lipoprotein (LVDL) cholesterol, and possible LDL-C. Moreover, the cholesterol composition of high-density lipoprotein cholesterol can vary in response to different modes of exercise [414]. Thompson et al. [415] have suggested that any level of continued running exercise may have a favorable effect on HDL-C in adults. Subjects with greater VO_{2max} values are less likely to show a change in serum lipids and lipoproteins with training than individuals with lower initial VO_{2max}. World-class marathon runners are known to have exceptional high plasma HDL-C and relatively low total cholesterol, LDL-C, and triglycerides concentration, suggesting that following long-lasting strenuous aerobic exercise serum lipids and lipoproteins were altered [413].

Angelopoulos et al. [416] have reported that previously sedentary men can expect an increase in high-density lipoprotein cholesterol and a reduction in low-density lipoprotein cholesterol and triglycerides as a result of an exercise program [417].

The dependency of lipoprotein cholesterol fraction changes on the intensity of training revealed that moderate exercise intensity is required to increase HDL-C levels [413] This is important since it has implications in the management of patients with proven CAD, who cannot perform strenuous exercise. Kodama et al. [418] concluded by meta-analysis review that the minimum training session needed to raise HDL-C cholesterol is 120 min $week^{-1}$. Greater duration of cardiovascular exercise elicits greater improvements in HDL-C levels although at this time an optimal training dose in minutes or kcal $week^{-1}$ has not been identified.

Eapen et al. [419] demonstrated that LDL-C cholesterol levels are actually the best predictor of CAD risk factor, in women, regardless of whether LDL-C levels and triglycerides are elevated or not. Estrogen, which reaches its peak during the childbearing years, usually raises HDL-C. In women, moderate exercise may not have an identical effect on HDL-C to that seen in men [420].

The primary explanation for the increase in HDL-C levels by exercise focuses on the increase in lipoprotein lipase, the key enzyme in triglyceride hydrolysis. Lipoprotein lipase is synthesized in adipocytes and muscle cells. It is released and transported to the capillary endothelium where it binds to the luminal surface [414]. The accelerated catabolism of triglycerides-rich lipoproteins results in an increased transfer of surface components to high-density lipoprotein cholesterol, thus increasing HDL-C mass. Other enzyme that might contribute to the rise in HDL-C, following exercise, includes the decrease in hepatic lipase and the increase in lecithin cholesterol acyl transferase. The hepatic lipase activity correlates inversely with exercise and with HDL-C-2 [421]. Lecithin cholesterol acyl transferase activity is increased after a period of training in relation to changes in HDL-C. The changes described in hepatic lipase and lecithin cholesterol acyl transferase activities result in increased HDL-C-2. The finding that the HDL-C-2 level rather than HDL-C-3. HDL-C is affected by aerobic training supports this contention [414].

Increased synthesis of nascent HDL-C might occur, of the triglyceride-rich lipoprotein, implicated in forming nascent HDL-C [422]. Alternatively, secretion of nascent HDL-C by the intestine could be elevated. Ben Ounis et al. [423] reported an increase in apoprotein AI (APO A-I) synthesis following a training period of 8 weeks in obese children. Morris et al. [424] reported on 142 cardiac rehabilitation patients on a progressive increase in serum levels of apoprotein AI, over a training period of 6 months. These observations demonstrate the key role of lipoprotein lipase in high-density lipoprotein cholesterol and very low-density lipoprotein cholesterol regulation and how exercise affects this regulation.

Laboratory Assignment on Obesity

The optimal method to be utilized depends on the individuals' number to be measured, equipment availability, and characteristic of the measurement; testing in laboratory setting or at the field. Currently, several laboratory methods for determining body fat percentage are available, with technical or biological limitations. Methods include calipers, bioelectrical impedance analysis, and water weighing. Best means for measuring percent body fat is the relationship between body mass and percent body fat. The skinfold thickness approach done by caliper allowing for the assessment of skin-fold thicknesses at various sites is routinely used to estimate body composition in a field setting and laboratories. It correlates well with percentage body fat determined by hydrostatic weighing. This is an inexpensive method and shown to be accurate when taken by an experienced technician. The second method is the densitometry, using underwater, or hydrostatic, weighing. Hydrostatic weighing, which is the gold standard for other estimation methods, requires correction for lung volume. In addition, it is an expensive method, complicated inconvenient, and very time consuming. Currently, comfort to measure, reliability, low time consuming, makes the bioelectric impedance analyzers an attractive method. However, the fact that equations given by the producers were not performed well and therefore, impedance analyzers became insensitive at extreme body structures. Of the three estimates of body composition, skin-fold measurements provided the

Fig. 2.52 Percentage body fat, by gender and age groups

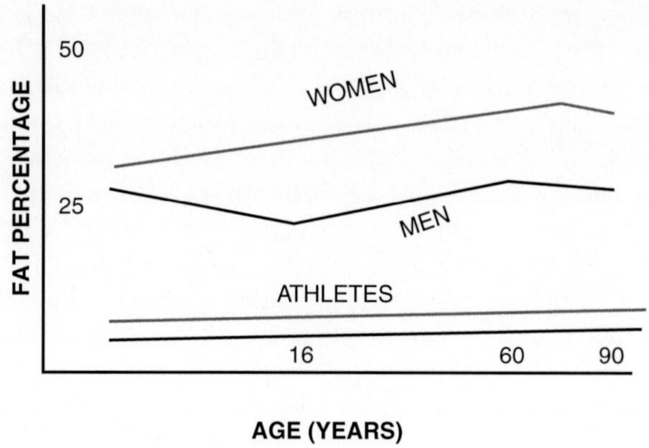

best estimate with the least amount of error. It should be noted, however, that there are several underlying assumptions made when using skin-fold determine percentage body fat. Average percent body fat values for 20–29-year-old individuals in the general population are 15.9% for men and 22.1% for women. In athletes, fat percentage is significantly lower than the general population in men it is 4–10% and 7–15% in women (Fig. 2.52).

Controversy exists as to whether a particular body fat percentage is better for one's health; athletic performance may also be affected. The leanest athletes typically compete at levels of about 6–13% for men or 14–20% for women. Bodybuilders may compete at ranges even lower than these levels. Certified personal trainers will suggest to male bodybuilders that they aim for a body fat percentage between 2% and 4% by contest time. However, it is unclear if such levels are ever actually attained since; (a) the means to measure such levels are, as noted below, lacking in principle, are inaccurate and (b) 4–6% is generally considered a physiological minimum for human males.

Measurement Techniques

Underwater Weighing

Irrespective of the location from which they are obtained, the fat cells in humans are composed almost entirely of pure triglycerides with an average density of about 0.9 kg per liter. These days laboratories use the value of 1.1 kg per liter for the density of the fat free mass, a theoretical tissue composed of 72% water (density = 0.993), 21% protein (density = 1.340) and 7% mineral (density = 3.000) by weight. With a well-engineered weighing system in a dunk tank, body density can be determined with great accuracy by completely submerging an individual in water and calculating the volume of the displaced water from the weight of the

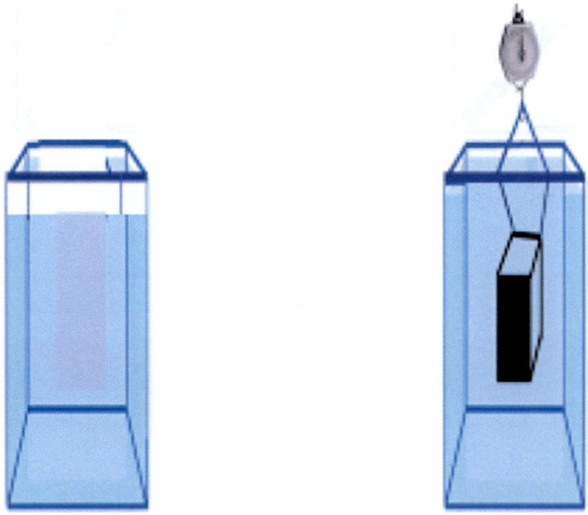

Fig. 2.53 Measure body fat with hydrostatic underwater weighing

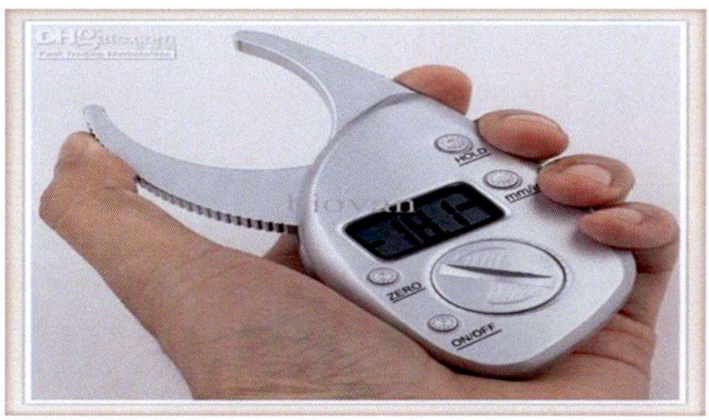

Fig. 2.54 Digital LCD body fat caliper skin fold analyzer for fat measurement. From, DHgate China

displaced water (Fig. 2.53). A correction is made for the buoyancy of air in the lungs and other gases in the body spaces. If there were no error whatsoever in measuring body density, the uncertainty in fat estimation would be about ±3.8% of the body weight, primarily because of normal variability in body components.

Skinfold Method

The skinfold estimation method is based on a skinfold test, measured by calipers (Fig. 2.54) at several standardized points on the body to determine the subcutaneous fat layer thickness. These measurements are converted to an estimated body fat percentage by an equation. Some formulas require as few as three measurements, others as many as seven. The accuracy of these estimates is more dependent on an individual's unique body fat distribution than on the number of sites measured.

As well, it is of utmost importance to test in a precise location with a fixed pressure. Although it may not give an accurate reading of real body fat percentage, it is a reliable measure of body composition change over a period of time, provided the test is carried out by the same person with the same technique.

Skinfold-based body fat estimation is sensitive to the type of caliper used and technique. This method also only measures one type of fat: subcutaneous adipose tissue (fat under the skin). Two individuals might have nearly identical measurements at all of the skin fold sites, yet differ greatly in their body fat levels due to differences in other body fat deposits such as visceral adipose tissue: fat in the abdominal cavity. Some models partially address this problem by including age as a variable in the statistics and the resulting formula. Older individuals are found to have a lower body density for the same skinfold measurements, which is assumed to signify a higher body fat percentage. However, older, highly athletic individuals might not fit this assumption, causing the formulas to underestimate their body density.

To take a skin-fold measurement, first it must determine the correct measurement site. Grab the skin with the thumb and forefinger about 0.5 inch from the measurement site following the natural fold of the skin. Lift the skin up from the muscle, apply the calipers and wait for 4 seconds before reading the calipers. Fat is compressible, so reading the scale before or after the 4-s delay may affect the results.

Measurement Sites (Figs. 2.55, 2.56, 2.57, 2.58, and 2.59)

Along the midline on the back of the triceps of the right arm determine the midpoint located between the top of the acromial process (top of the shoulder) to the bottom of the olecranon process of the ulna (elbow). Pinch the skin so that the fold is running vertically.

Using a line from the fold of the axillary (armpit) to the nipple, determine the midpoint. Pinch the skin with the fold running in the same direction of the line. Select a site on the right side about 1 inch lateral from and 0.5 inch below the umbilicus (bellybutton). Lift a horizontal fold of skin for the measurement. Note: The Jackson and Pollock (1978) and Jackson, Pollock, and Ward (1980) abdominal skinfolds were made using a vertical fold. Consequently, one should stay consistent with how the original equations were validated when using these equations.

Determine the midaxillary line and palpate for the iliac crest (top of the hip bone). Grasp the skin that follows the natural fold which will follow a line of approximately from the suprailiac to the umbilicus (bellybutton), an angle of approximately 30°.

Determine the midline of the front of the thigh and measure midway between the inguinal crease (the natural crease between the thigh and hip which is at an approximate 45°) and the top of the patella (kneecap). Grasp a vertical fold of skin for the measurement.

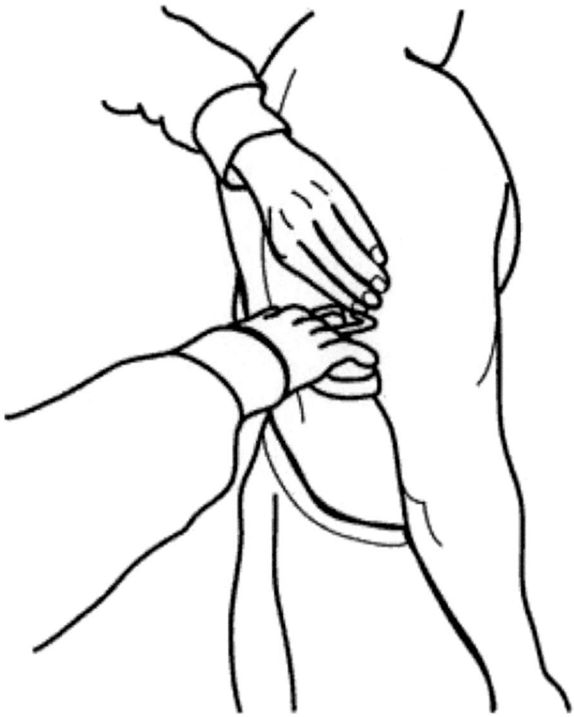

Fig. 2.55 Triceps skinfold

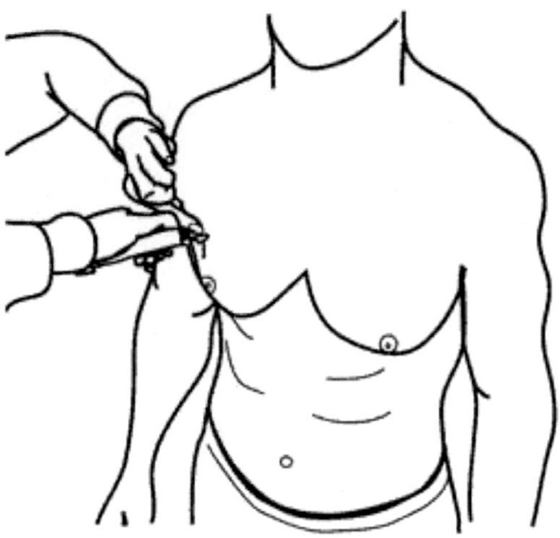

Fig. 2.56 Pectoral (chest) skinfold

Estimation of Body Density

There are two calculation steps for determining percent fat: determining body density and then using body density to estimate percent fat. All the skinfold equations estimate body density from the measurements which are then used to estimate

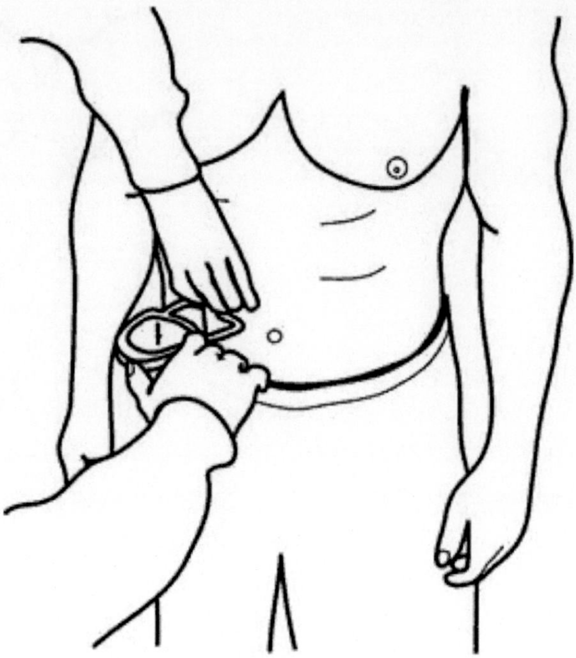

Fig. 2.57 Abdominal skinfold

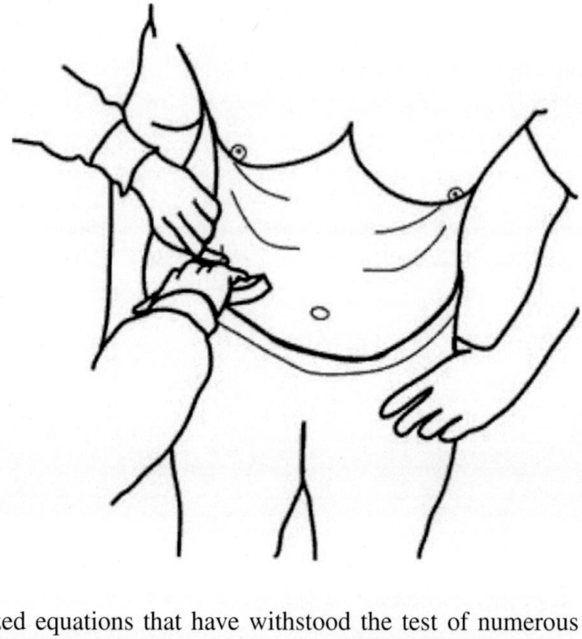

Fig. 2.58 Suprailiac skinfold

percent fat. Two generalized equations that have withstood the test of numerous cross-validation studies are ones published by Jackson and Pollock (1978) for men and Jackson, Pollock, and Ward (1980) for women. Each study developed numerous equations based on different skinfold measurements, although they all were similar

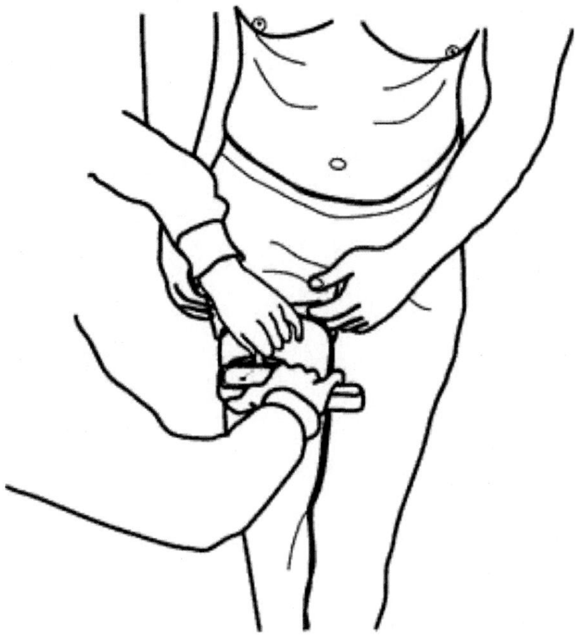

Fig. 2.59 Thigh skinfold

in prediction error. Listed below are a single equation from each study based on the sum of three skinfold measurements. Standard errors of their formulae ranged from 3.6% to3.8%. The potential errors are increased if these equations are used on the young or old, very lean and muscular or obese. Other, population specific equations are available and more suitable for these groups.

From 3 sites measured: (triceps, suprailiac, and thigh skinfold measurements).

Equations:

For Females:

$$\text{Body density} = 1.0994921 - 0.0008267 \circ \text{sum} + 0.0000023 \circ \text{sum2} - 0.0001392 \circ \text{age}$$

For males:

$$\text{Body density} = 1.1093800 - 0.0009929 \circ \text{sum} + 0.0000016 \circ \text{sum2} - 0.0002574 \circ \text{age}$$

Estimation of Percent Fat

The two most commonly used equations for estimating percent fat from body density are the Siri (1961) and Brozek (1963) formulae. A limitation to these formulae is that they assume the density of fat-free mass to remain a constant across the population when in fact is varies. Thus, the actual percent fat tends to be slightly higher than the measured percent fat in the lean, muscular individual, and the opposite effect in obese individuals.

$$\text{Siri : Percent fat} = \left[\left(495 / \text{body density}\right) - 450\right] \cdot 100.\text{Broze : Percent fat}$$
$$= \left[4.570 / \text{body density}\right) - 4.142] \cdot 100$$

Laboratory on Metabolism

For this laboratory, each group will calculate and compare RER and energy expenditure values of two subjects (one trained and one untrained) at rest and during aerobic exercise. Through this experiment, it will:

(a) use RER data to determine the probable fuel source used at rest and during exercise,
(b) observe how energy expenditure changes from rest to exercise and throughout exercise, and
(c) compare resting and exercise energy expenditures in trained and untrained subjects. Make hypotheses reflecting each of these goals.

Equipment required:

(a) open circuit spirometry apparatus—computer analysis (metabolic cart),
(b) stopwatch,
(c) treadmill, and
(d) heart rate monitor.

Experimental procedure:
In this laboratory, each group will test two subjects.
Pretest instructions for subjects:

(a) Subjects must refrain from exercise 24 h prior to test.
(b) Subjects should minimize activity on the day of test.
(c) Subjects must fast (including all food and beverages, except water) for 12 h prior to the test.

Test procedures:

(a) Obtain subject data and environmental conditions (temperature, barometric pressure, and humidity). Record it on laboratory sheet.
(b) Secure the HR monitor and prepare the receiver to record HR.
(c) Instruct subject to lay quietly on the patient bed or a mat on the floor. Dim the lights and try to maintain a low-stimulus environment.
(d) Calibrate the metabolic cart.
(e) Input subject data into computer.
(f) Secure the mouthpiece and headgear.

(g) Begin collecting metabolic data.
(h) Allow subject to rest quietly for 10 min. Record resting metabolic data.
(i) Instruct the subject to stand quietly on the treadmill for 3 min. Record preexercise metabolic data.
(j) At the end of the 3 min, record resting heart rate, VO_2 and VCO_2.
(k) Begin exercise. The subject will walk on the treadmill for 8 min at a speed of 5.7 km h^{-1} (3.5 mph) and a 7% grade.
(l) Record heart rate (HR), VO_2, and VCO_2 at the end of each minute.

Data analysis:

1. Calculate the RER at rest and for at each minute of exercise. Use the final minute of rest for resting calculations. (See sample calculations below.)
2. Use the caloric equivalents from Table 2.1 and the RER values calculated to estimate energy expenditure at rest and during each minute of exercise. (See sample calculations below.)
3. Enter the descriptive data, RER values, and energy expenditure values for the two subjects.
4. Create a table that shows the subject descriptive data. Be sure that the table reflects a comparison between the trained and untrained subjects.
5. Create the following two graphs, using the averages from the trained and untrained subjects:

   (a) Graph that compares resting and exercise RER in trained and untrained subjects. (Use the RER during the last minute of exercise for the exercise value.)
   (b) Draw a graph to show energy expenditure at rest and during 8 min of exercise in trained and untrained subjects.
   (c) Be sure that, graphs include a title and descriptive caption.
   (d) Attach data collection tables and sample calculations to the end of your laboratory report. Sample calculations:

      Determine the Respiratory Exchange Ratio (RER).
      $VO_2 = 0.17$ L min^{-1} $VCO_2 = 0.15$ L min^{-1}
      $RER = VCO_2/VO_2 = 0.15/0.17 = 0.8947$ 2.

Laboratory Report Instructions

(a) Should write up a report with introduction, methods, results, and discussion that includes the information requested in the data analysis section above. The discussion section summarizes the data and address the following questions in paragraph format:
(b) What was the probable fuel source during rest? Is this what you expected? Explain.
(c) What was the probable fuel source during the steady-state exercise? Is this what you expected? Explain.

(d) How did energy expenditure change from rest to exercise? What happened to energy expenditure during the 8 min of exercise? Why does this happen?
(e) How RMR values are compared to the exercise energy expenditure in the trained and untrained subjects? Discuss the physiological factors that might explain inter-subject differences in energy expenditure.
(f) For this experiment, you compared energy expenditure in trained and untrained subjects. What would you expect to happen if, instead, you compared males and females? How would you expect energy expenditure to compare in males and females and why? Be sure to reference material properly if you used it to obtain answers to Laboratory report questions.
(g) Be sure to reference material properly if you used it to obtain answers to laboratory report questions.

Collection Sheet Note:
Note: Must calculate RER.
Name ______________ Height______________
Weight ______________ Date______________

Time (min)	V02 (L min^{-1})	VC02 (L min^{-1})	RER	HR (beats min^{-1})
Rest				
1				
2				
3				
Exercise				
1				
2				
3				
4				
5				
6				
7				
8				

Laboratory on body mass index and body composition
Name_______________ Date________________
Equipment:

(a) Weight scale.
(b) Height scale.

Instructions:
Measure subjects;
Height (cm) _____________ Weight (kg) ______________
Calculate BMI __________

Table 2.3 Rating subject's BMI

Classification	BMI (kg m^{-2})
Underweight	18.5
Normal	18.5–24.9
Overweight	25.0–29.9
Obesity (I)	30.0–34.9
Obesity (II)	35.0–39.9
Extreme obesity (III)	≥40.0

(a) Convert subject's body weight to kilograms by dividing weight in pounds by 2.2.

Body weight (lb) ________ ÷ 2.2 lb.·kg-[1] = Body weight (kg) _______.

(b) Convert height measured to meters by multiplying height in inches by 2.54 cm.

Height _________(in) × 2.54 cm height_________(m).

(c) Square your height measurement.

Height________(m) × height________(m) = height_______ (m).

(d) BMI equals body weight in kilograms divided by height in meters squared (kg m^{-2}).

Body weight ______(kg) ÷ height ______ (cm) = BMI (kg m^{-2}) ______.
Refer to the enclosed table, for rating subject's BMI. Record the results below and on the final page of this laboratory (Table 2.3).

References

1. Jorgensen K, Rasmussen AV, Morant M, et al. Metabolon formation and metabolic channeling in the biosynthesis of plant natural products. Curr Opin Plant Biol. 2005;8:280–91.
2. Nelson DL, Cox MM. Lehninger principles of biochemistry. New York: Freeman; 2013.
3. Lee H, DeLoache WC, Dueber JE. Spatial organization of enzymes for metabolic engineering. Metab Eng. 2012;14:242251. https://doi.org/10.1016/j.ymben.2011.09.003.
4. Ventura SPM, Silva FA, Quental MV, et al. Ionic-liquid-mediated extraction and separation processes for bioactive compounds: past, present, and future trends. Chem Rev. 2017;117:6984–7052.
5. Kauffman GB, Chooljian SH. Friedrich Wohler (1800–1882), on the bicentennial of his birth. Chem Educ. 2001;6:121–33.
6. Voet D, Voet JG. Biochemistry. 3rd ed. Hoboken, NJ: Wiley; 2005.
7. Huskisson E, Maggini S, Ruf M. The role of vitamins and minerals in energy metabolism and well-being. J Int Med Res. 2007;35:277–89.
8. Aughey GN, Liu JL. Metabolic regulation via enzyme filamentation. Crit Rev Biochem Mol Biol. 2016;51:282–93.
9. Mahankali M, Alter G, Gomez-Cambronero J. Mechanism of enzymatic reaction and protein-protein interactions of PLD from a 3D structural. Cell Signal. 2015;27:69–81.
10. Singh RK, Fan J, Gioacchini N, et al. Transient kinetic analysis of SWR1C-catalyzed H2A.Z deposition unravels the impact of nucleosome dynamics and the asymmetry of histone exchange. Cell Rep. 2019;27:374–86.

11. Zhang R, Wong K. High performance enzyme kinetics of turnover, activation and inhibition for translational drug discovery. Expert opin Drug Discov. 2017;12:17–37.
12. Fromm HJ, Hargrove MS. Enzyme kinetics. In: Essentials of biochemistry: Berlin, Springer; 2012.
13. Papachristodoulou DK, Elliott WH, Elliott DC. Alison and Snape. Biochemistry and molecular biology. 5th ed. Oxford university press; 2018.
14. Smith RL, Soeters MR, Wüst RCI, et al. Metabolic flexibility as an adaptation to energy resources and requirements in health and disease. Endocr Rev. 2018;39:489–517.
15. Huberts DH, Niebel B, Heinemann M. A flux-sensing mechanism could regulate the switch between respiration and fermentation. FEMS Yeast Res. 2012;12:118–28.
16. Nelson DL, Cox MM. Lehninger principles of biochemistry: Freeman; 2008.
17. Luo L, Molnar J, Ding H, et al. Physicochemical attack against solid tumors based on the reversal of direction of entropy flow: an attempt to introduce thermodynamics in anticancer therapy. Diagn Pathol. 2006;1:43. https://doi.org/10.1186/1746-1596-1-43.
18. Werner C, Doenst D. Metabolic pathways and cycles. In: Schwarzer M, Doenst D, editors. The scientist's guide to cardiac metabolism: Elsevier; 2016. p. 39–55.
19. Spriet LL. The metabolic systems: lipid metabolism. Advanced exercise physiology. 2nd ed. Philadelphia: Lippincott, Williams and Wilkins; 2012. p. 392–407.
20. Spriet LL. New insights into the interaction of carbohydrate and fat metabolism during exercise. Sports Med. 2014;44(Suppl 1):87–96.
21. Hannukainen JC, Nuutila P, Ronald B, et al. Increased physical activity decreases hepatic free fatty acid uptake: a study in human monozygotic twins. J Physiol. 2007;578:347–58.
22. Mul JD, Stanford KI, Hirshman MF, et al. Exercise and regulation of carbohydrate metabolism. Prog Mol Biol Transl Sci. 2015;135:17–37.
23. Jeppesen J, Kiens B. Regulation and limitations to fatty acid oxidation during exercise. J Physiol. 2012;590:1059–68.
24. Karlic H, Lohninger A. Supplementation of L-carnitine in athletes: does it make sense? Nutrition. 2004;20:709–915.
25. Violante S, Ijlst L, Te Brinke H, et al. Peroxisomes contribute to the acylcarnitine production when the carnitine shuttle is deficient. Biochim Biophys Acta. 1831;2013:1467–74.
26. Seim I, Ma S, Gladyshev VN. Gene expression signatures of human cell and tissue longevity. NPJ Aging Mech Dis. 2016;2:16014. https://doi.org/10.1038/npjamd.2016.14.. eCollection 2016
27. Conley KE, Lindstedt SL. Energy-saving mechanisms in muscle: the minimization strategy. J Exp Biol. 2002;205:2175–81.
28. Baker JS, McCormick MC, Robergs RA. Interaction among skeletal muscle metabolic energy systems during intense exercise. J Nutr Metab. 2010;2010:905612. https://doi.org/10.1155/2010/905612.
29. Spriet LL. Anaerobic metabolism in human skeletal muscle during short-term, intense activity. Can J Physiol Pharmacol. 1992;70:157–65.
30. Weiss K, Schar M, Panjrath GS et al. Fatigability, exercise intolerance, and abnormal skeletal muscle energetics in heart failure. Circ Heart Fail. 2017;10(7). pii: e004129. https://doi.org/10.1161/CIRCHEARTFAILURE.117.004129
31. Kushnarevae Y. Bioenergetics and cell death. Ann N Y Acad Sci. 2010;1201:50–7.
32. Beis I, Newsholme EA. The contents of adenine nucleotides, phosphagens and some glycolytic intermediates in resting muscles from vertebrates and invertebrates. Biochem J. 1975;152:23–32.
33. Scott CB. Energy expenditure of heavy to severe exercise and recovery. J Theor Biol. 2000;207:293–7.
34. Nakazawa MH, Ando H, Nishimoto A, et al. Anaerobic respiration coupled with mitochondrial fatty acid synthesis in wax ester fermentation by Euglena gracilize. FEBS Lett. 2018;592:4020–7.
35. Bogdanis GC, Nevill ME, Boobis LH, et al. Contribution of phosphocreatine and aerobic metabolism to energy supply during repeated sprint exercise. J Appl Physiol (1985). 1996;80:876–84.

36. Gray SR, Söderlund K, Ferguson RA. ATP and phosphocreatine utilization in single human muscle fibers during the development of maximal power output at elevated muscle temperatures. J Sports Sci. 2008;26:701–7.
37. Noy T, Vergnolle O, Hartman TE, et al. Central role of pyruvate kinase in carbon co-catabolism of Mycobacterium tuberculosis. J Biol Chem. 2016;291:7060–9.
38. Greenhaff PL, Nevill ME, Soderlund K, et al. The metabolic responses of human type I and II muscle fibers during maximal treadmill sprinting. J Physiol. 478(Pt 1):149–55.
39. Melzer K. Carbohydrate and fat utilization during rest and physical activity. e-SPEN, Eur e-J Clin Nutr Metabol. 2011;6:e45–52.
40. Hoppel CL, Tandler B, Fujioka H, et al. Dynamic organization of mitochondria in human heart and in myocardial disease. Int J Biochem Cell Biol. 2009;41:1949–56.
41. Rui L. Energy metabolism in the liver. Compr Physiol. 2014;4:177–97.
42. Lemasters JJ, Holmuhamedov H. Voltage-dependent anion channel (VDAC) as mitochondrial governator – thinking outside the box. Biochim Biophys Acta. 1762;2006:181–90.
43. Margolis LM, Pasiakos SM. optimizing intramuscular adaptations to aerobic exercise: effects of carbohydrate restriction and protein supplementation on mitochondrial biogenesis. Adv Nutr. 2013;4:657–64.
44. Menshikova EV, Ritov VB, Dube JJ, et al. Calorie restriction-induced weight loss and exercise have differential effects on skeletal muscle mitochondria despite similar effects on insulin sensitivity. J Gerontol A Biol Sci Med Sci. 2018;73:81–7.
45. Harvey RA. Biochemistry. 5th ed. Baltimore, MD: Wolters Kluwer; 2011.
46. Ball D. Metabolic and endocrine response toexercise: sympathoadrenal integration with skeletal muscle. J Endocrinol. 2015;224:R79–95.
47. Lodish H, Berk A, Kaiser CA, et al. Molecular cell biology. 6th ed. New York: W.H. Freeman and Company; 2008.
48. Murray RK, Granner DK, Mayes PA, et al. Harper's illustrated biochemistry. New York, NY: Lange Medical Books/MgGraw Hill; 2003. p. 96.
49. Alberts B, Johnson A, Lewis J, et al. Molecular biology of the cell. 4th ed. New York: Garland Science; 2002.
50. Rich PR. The molecular machinery of Keilin's respiratory chain. Biochem Soc Trans. 2003;31(Pt 6):1095–105.
51. Calvani M, Reda E, Arrigoni-Martelli E. Regulation by carnitine of myocardial fatty acid and carbohydrate metabolism under normal and pathological conditions. Basic Res Cardiol. 2000;95:75–83.
52. Voet D, Voet JG, Pratt CW. Fundamentals of biochemistry: life at the molecular level. 4th ed. Hoboken, NJ: Wiley; 2013.
53. Zechner R, Strauss JG, Haemmerle G, et al. Lipolysis: pathway under construction. Curr Opin Lipidol. 2005;16:333–40.
54. Martin WH. Effects of acute and chronic exercise on fat metabolism. Exerc Sport Sci Rev. 1996;24:203–31.
55. Mittendorfer B, Horowitz JF, Klein S. Effect of gender on lipid kinetics during endurance exercise of moderate intensity in untrained subjects. Am J Physiol Endocrinol Metab. 2002;283:E58–65.
56. Kushwaha P, Wolfgang MJ, Riddle RC. Fatty acid metabolism by the osteoblast. Bone. 2018;115:8–14.
57. Romijn J, Coyle EF, Sidossis LS, et al. Relationship between fatty acid delivery and fatty acid oxidation during strenuous exercise. J Appl Physiol. 1996;79:1939–45.
58. Cheneviere X, Borrani F, Ebenegger V, et al. Effect of a 1-hour single bout of moderate-intensity exercise on fat oxidation kinetics. Metabolism. 2009;58:1778–86.
59. Jocken JW, Goossens GH, van Hees AM, et al. Effect of beta-adrenergic stimulation on whole-body and abdominal subcutaneous adipose tissue lipolysis in lean and obese men. Diabetologia. 2008;51:320–7.
60. Gonzalez-Haro C, Galilea PA, Gonzalez-de-Suso JM, et al. Maximal lipidic power in high competitive level triathletes and cyclists. Br J Sports Med. 2007;41:23–8.

61. Braun B, Horton T. Endocrine regulation of exercise substrate utilization in women compared to men. Exerc Sport Sci Rev. 2001;29:149–54.
62. Anderson CM, Stahl A. SLC27 fatty acid transport proteins. Mol Aspects Med. 2013;34:516–28.
63. Stryer L. Biochemistry. 4th ed.. New York: W.H. Freeman and Co; 1995. pp. 510–515, 581–613, 775–778.
64. Houten SM, Wanders RJ. A general introduction to the biochemistry of mitochondrial fatty acid β-oxidation. J Inherit Metab Dis. 2010;33:469–77.
65. Sewell DA, Watkins P, Griffin M. Sport and exercise science, an introduction. 2nd ed. London and New York: Routladge, Taylor and Francis Group; 2012.
66. LaForgia J, Withers RT, Gore CJ. Effects of exercise intensity and duration on the excess post-exercise oxygen consumption. J Sports Sci. 2006;24:1247–64.
67. Speakman JR, Selman C. Physical activity and resting metabolic rate. Proc Nutr Soc. 2003;62:621–34.
68. Donahoo WT, Levine JA, Melanson EL. Variability in energy expenditure and its components. Curr Opin Clin Nutr Metab Care. 2004;7:599–605.
69. McArdle WD. Exercise physiology. 8th ed. Philadelphia: Lea & Febigier; 2014.
70. Roberts SB, Dallal GE. Energy requirements and aging. Public Health Nutr. 2005;8:1028–36.
71. Speakman JR, Krol E, Johnson MS. The functional significance of individual variation in basal metabolic rate. Physiol Biochem Zool. 2004;77:900–15.
72. Pasco JA, Nicholson GC, Brennan SL, et al. Prevalence of obesity and the relationship between the body mass index and body fat: cross-sectional, population-based data. PLoS One. 2012;7(1):e29580. https://doi.org/10.1371/journal.pone.0029580.
73. Stephens FB. Does skeletal muscle carnitine availability influence fuel selection during exercise? Proc Nutr Soc. 2018;77:11–9.
74. Melanson EL, MacLean PS, Jo H. Exercise improves fat metabolism in muscle but does not increase 24-h fat oxidation. Exerc Sport Sci Rev. 2009;37:93–101.
75. Melanson EL, Sharp TA, Seagle HM, et al. Resistance and aerobic exercise have similar effects on 24-h nutrient oxidation. Med Sci Sports Exerc. 2002;34:1793–800.
76. Miller T, Mull S, Aragon AA, et al. Resistance training combined with diet decreases body fat while preserving lean mass independent of resting metabolic rate: a randomized trial. Int J Sport Nutr Exerc Metab. 2018;28:46–54.
77. Alberts B, Johnson A, Lewis J, et al. molecular biology of the cell. 4th ed. New York: Garland Science; 2002.
78. Weil AS, Hernández GP, Suslow T, et al. Implicit affect and autonomous nervous system reactions: a review of research using the implicit positive and negative affect test. Front Psychol. 2019;10:1634. https://doi.org/10.3389/fpsyg.2019.01634.. eCollection 2019
79. Mizuno K, Ueno Y. Autonomic nervous system and the liver. Hepat Res. https://doi.org/10.1111/hepr.12760.
80. Mullur R, Liu YY, G Brent GA. Thyroid hormone regulation of metabolism. Physiol Rev. 2014;94:355–82.
81. Molina PE. Endocrine physiology: McGraw-Hill Education; 2018.
82. Bartuzi D, Kaczor AA, Matosiuk D. Signaling within allosteric machines: signal transmission pathways inside G protein-coupled receptors. Molecules. 2017;22(7). pii: E1188. https://doi.org/10.3390/molecules22071188
83. Shuster M. Biology for a changing world, with physiology. 2nd ed. New York, NY: Scientific American Biology for changing world; 2014.
84. Neave N. Hormones and behaviour: a psychological approach. Cambridge: Cambridge University Press; 2008.
85. Ruhs S, Nolze A, Hbschmann R, et al. 30 years of the mineralocorticoid receptor: nongenomic effects via the mineralocorticoid receptor. J Endocrinol. 2017;234:T107–24.
86. Lall S. Clearopathy: Partridge Publishing India; 2013.
87. Pia RV, Bingbing W, Yixian L, et al. Pancreatic regulation of glucose homeostasis. Exp Mol Med. 2016;48(3):e219. https://doi.org/10.1038/emm.2016.6.

88. Hauraix H, Goislard DE Monsabert B, et al. Force-length relationship modeling of wrist and finger flexor muscles. Med Sci Sports Exerc. 2018;50:2311–21.
89. di Prampero PE, Boutellier U, Marguerat A. Efficiency of work performance and contraction velocity in isotonic tetani of frog sartorius. Pflugers Arch. 1988;412:455–61.
90. Willoughby D, Cooper DM. organization and Ca2+ regulation of adenylyl cyclases in cAMP microdomains. Physiol Rev. 2007;87:965–1010.
91. Tresguerres M, Levin LR, Buck J. Intracellular cAMP signaling by soluble adenylyl cyclase. Kidney Int. 2011;79:1277–88.
92. Zhang F, Zhang L, Qi Y, et al. Mitochondrial cAMP signaling. Cell Mol Life Sci. 2016;73:4577–90.
93. Batty NJ, Fenrich KK, Fouad K. The role of cAMP and its downstream targets in neurite growth in the adult nervous system. Neurosci Lett. 2017;652:56–63.
94. Khurana (2008). Essentials of medical physiology. Elsevier India. p. 460.
95. Thompson BR, Lobo S, Bernlohr DA. Fatty acid flux in adipocytes; the in's and out's of fat cell lipid trafficking. Mol Cell Endocrinol. 2010;318(1–2):24–33.
96. Voet D, Voet JG. Biochemistry. 4th ed. New York: Wiley; 2011.
97. Dimitriadis G, Mitrou P, Lambadiari V, et al. Insulin effects in muscle and adipose tissue. Diabetes Res Clin Pract. 2011;93(Suppl 1):S52–9.
98. Stryer L. Biochemistry. 4th ed. New York: W.H. Freeman and Company; 1995. p. 773–4.
99. Sonksen P, Sonksen J. Insulin: understanding its action in health and disease. Br J Anaesth. 2000;85:69–79.
100. Lundsgaard AM, Fritzen AM, Kiens B. Molecular regulation of fatty acid oxidation in skeletal muscle during aerobic exercise. Trends Endocrinol Metab. 2018;29:18–30.
101. Karlic H, Lohninger A. Supplementation of L-carnitine in athletes: does it make sense? Nutrition. 2004;20:709–15.
102. Violante S, Ijlst L, Te Brinke H, et al. Peroxisomes contribute to the acylcarnitine production when the carnitine shuttle is deficient. Biochim Biophys Acta. 2013;183:1467–74.
103. Kersten S. Mechanisms of nutritional and hormonal regulation of lipogenesis. EMBO Rep. 2001;2:282–6.
104. Assimacopoulos-Jeannet F, Brichard S, Rencurel F, et al. In vivo effects of hyperinsulinemia on lipogenic enzymes and glucose transporter expression in rat liver and adipose tissues. Metab Clin Exp. 1995;44:228–33.
105. Reece J, Campbell N. Biology. San Francisco: Benjamin Cummings; 2002.
106. Hoffman JR, Falvo MJ. Protein-which is the best? J Sports Sci Med. 2004;3:118–30.
107. Ranabir S, Reetu K. Stress and hormones. Ind J Endocrinol Metab. 2011;15:18–22.
108. Davari D, Karimzadeh I, Khalili H. The potential effects of anabolic-androgenic steroids and growth hormone as commonly used sport supplements on the kidney: a systematic review. BMC Nephrol. 2019;20(1):198. https://doi.org/10.1186/s12882-019-1384-0.
109. Marliss EB, Vranic M. Intense exercise has unique effects on both Insulin release and Iits roles in glucoregulation. implications for diabetes. Diabetes. 2002;51:S271–83.
110. Shambrook P, Kingsley M, Taylor N, et al. Accumulated or continuous exercise for glycemic regulation and control: a systematic review with meta-analysis. BMJ Open Sport Exerc Med. 2018;4(1):e000470. https://doi.org/10.1136/bmjsem-2018-000470.. eCollection 2018
111. Textbook of diabetes mellitus. 2nd ed. New York: Wiley; 1997. Pp. 315–336.
112. Adams OP. The impact of brief high-intensity exercise on blood glucose levels. Diabetes Metab Syndr Obes. 2013;6:113–22.
113. Salgueiro RB, Gerlinger-Romero F, Guimaraes-Ferreira L, et al. Exercise training reverses the negative effects of chronic L-arginine supplementation on insulin sensitivity. Life Sci. 2017;191:17–23.
114. Barth E, Albuszies G, Baumgart K, et al. Glucose metabolism and catecholamines. Crit Care Med. 2007;35:S508–18.
115. Bell DR. Medical physiology: principles for clinical medicine. 3rd ed. Philadelphia: Lippincott Williams & Wilkins; 2009. p. 312.

116. Saxton SN, Withers SB, Heagerty AM. Emerging roles of sympathetic nerves and inflammation in perivascular adipose tissue. Cardiovasc Drugs Ther. 2019;33:245–59.
117. Umpierre D, Ribeiro PA, Kramer CK, et al. Physical activity advice only or structured exercise training and association with HbA1c levels in type 2 diabetes: a systematic review and meta-analysis. JAMA. 2011;305:1790–9.
118. Colberg SR, Sigal RJ, Fernhall B, et al. Exercise and type 2 diabetes: the American College of Sports Medicine and the American Diabetes Association: joint position statement executive summary. Diabetes Care. 2010;33:2692–6.
119. Iqbal A, Heller S. Managing hypoglycaemia. Best Pract Res Clin Endocrinol Metab. 2016;30:413–30.
120. Lafontan M, Sengenes C, Galitzky J, et al. Recent developments on lipolysis regulation in humans and discovery of a new lipolytic pathway. Int J Obes Relat Metab Disord. 2000;24:S47–52.
121. Bircher S, Knechtle B. Relationship between fat oxidation and lactate threshold in athletes and obese women and men. J Sports Sci Med. 2004;3:174–81.
122. Romijn JA, Coyle EF, Sidossis LS, et al. Regulation of endogenous fat and carbohydrate metabolism in relation to exercise intensity and duration. Am J Phys. 1993;265:E380–91.
123. Romijn JA, Coyle EF, Sidossis LS, et al. Relationship between fatty acid delivery and fatty acid oxidation during strenuos exercise. J Appl Physiol. 1995;79:1939–45.
124. Vranic M, Kawamori R, Wrenshall GA. The role of insulin and glucagon in regulating glucose turnover in dogs during exercise. Med Sci Sports. 1975;7:27–33.
125. Adlanmerini M, Solinhac R, Abot A, et al. Mutation of the palmitoylation site of estrogen receptor α in vivo reveals tissue-specific roles for membrane versus nuclear actions. Proc Natl Acad Sci U S A. 2014;111:E283–90.
126. Figueira FR, Umpierre D, Bock PM, et al. Effect of exercise on glucose variability in healthy subjects: randomized crossover trial. Biol Sport. 2019;36:141–8.
127. Wilk M, Petr M, Krzysztofik M, et al. Endocrine response to high intensity barbell squats performed with constant movement tempo and variable training volume. Neuro Endocrinol Lett. 2018;39:342–8.
128. Mangine GT, Hoffman JR, Gonzalez AM, et al. The effect of training volume and intensity on improvements in muscular strength and size in resistance-trained men. Physiol Rep. 2015;3(8):e12472. https://doi.org/10.14814/phy2.12472.
129. Frystyk J. Exercise and the growth hormone-insulin-like growth factor axis. Med Sci Sports Exerc. 2010;42:58–66.
130. French DN, Kraemer WJ, Volek JS, et al. Anticipatory responses of catecholamines on muscle force production. J Appl Physiol. 2007;102:94–102.
131. West DW, Phillips SM. Associations of exercise-induced hormone profiles and gains in strength and hypertrophy in a large cohort after weight training. Eur J Appl Physiol. 2012;112:2693–702.
132. Kim E, Gregg LD, Kim LD, et al. Hormone responses to an acute bout of low intensity blood flow restricted resistance exercise in college-aged females. J Sports Sci Med. 2014;13:91–6.
133. Kraemer WJ, Ratamess NA. Hormonal responses and adaptations to resistance exercise and training. Sports Med. 2005;35:339–61.
134. Garber CE, Blissmer B, Deschenes MR, et al. American College of Sports Medicine position stand. Quantity and quality of exercise for developing and maintaining cardiorespiratory, musculoskeletal, and neuromotor fitness in apparently healthy adults: guidance for prescribing exercise. Med Sci Sports Exerc. 2011;43:1334–59.
135. Coll AP, Yeo GS. The hypothalamus and metabolism: integrating signals to control energy and glucose homeostasis. Curr Opin Pharmacol. 2013;13:970–6.
136. Le Tissier P, Campos P, Lafont C, et al. An updated view of hypothalamic-vascular-pituitary unit function and plasticity. Nat Rev Endocrinol. 2017;13:257–67.
137. Lopez M, Nogueiras R, Tena-Sempere M, et al. Hypothalamic AMPK: a canonical regulator of whole-body energy balance. Nat Rev Endo. 2016;12:421–32.

138. Purkayastha S, Cai D. Neuroinflammatory basis of metabolic syndrome. Mol Metab. 2013;2:356–63.
139. Morena M, Patel S, Bains JS, et al. Neurobiological interactions between stress and the endocannabinoid system. Neuropsychopharmacology. 2016;41:80–102.
140. Yoshida M, Murakami M, Ueda H, et al. An unusual case of hypopituitarism and transient thyrotoxicosis following asymptomatic pituitary apoplexy. Neuro Endocrinol Lett. 2014;35:342–6.
141. Kai D, Khor S. Hypothalamic microinflammation paradigm in aging and metabolic diseases. Cell Metab. 2019;30:19–35.
142. Gray S, Kim JK. New insights into insulin resistance in the diabetic heart. Trends Endocrinol Metab. 2011;22:394–403.
143. Weiss RG, Maslov M. Normal myocardial metabolism: fueling cardiac contraction. Adv Stud Med. 2004;4:S457–63.
144. Doenst T, Nguyen TD, Abel ED. Cardiac metabolism in heart failure – implications beyond ATP production. Circ Res. 2013;113:709–24.
145. Rosano GM, Fini M, Caminiti G, et al. Cardiac metabolism in myocardial ischemia. Curr Pharm Des. 2008;14:2551–62.
146. Jogl G, Tong L. Crystal structure of carnitine acetyltransferase and implications for the catalytic mechanism and fatty acid transport. Cell. 2003;112:113–22.
147. Boehme J, Le Moan N, Kameny RJ, et al. Preservation of myocardial contractility during acute hypoxia with OMX-CV, a novel oxygen delivery biotherapeutic. PLoS Biol. 2018;16(10):e2005924. https://doi.org/10.1371/journal.pbio.2005924.
148. Shao D, Tian R. Glucose transporters in cardiac metabolism and hypertrophy. Compr Physiol. 2015;6:331–51.
149. Zhou MS, Schulman IH, Zeng Q. Link between the renin-angiotensin system and insulin resistance: implications for cardiovascular disease. Vasc Med. 2012;17:330–41.
150. Ames MK, Atkins CE, Pitt B. The renin-angiotensin-aldosteron systen and its suppression. J Vet Intern Med. 2019;33:363–82.
151. Kopelman PG. Obesity as a medical problem. Nature. 2000;404:635–43.
152. Lee L, Sanders RA. Metabolic syndrome. Pediatr Rev. 2012;33:459–68.
153. Brown AE, Walker M. Genetics of insulin resistance and the metabolic syndrome. Curr Cardiol Rep. 2016;18(8):75. https://doi.org/10.1007/s11886-016-0755-4.
154. Kohjima M, Enjoji M, Higuchi N, et al. Re-evaluation of fatty acid metabolism-related gene expression in nonalcoholic fatty liver disease. Int J Mol Med. 2007;20:351–8.
155. Avramoglu RK, Basciano H, Adeli K. Lipid and lipoprotein dysregulation in insulin resistant states. Clin Chim Acta. 2006;368:1–19.
156. Burks DJ, White MF. IRS proteins and beta-cell function. Diabetes 2001;50:140–5.
157. Wilcox G. Insulin and insulin resistance. Clin Biochem Rev. 2005;26:19–39.
158. Schmitz O, Rungby J, Edge L, et al. On high-frequency insulin oscillations. Ageing Res Rev. 2008;7:301–5.
159. Fu Z, Gilbert ER, Liu D. Regulation of insulin synthesis and secretion and pancreatic beta-cell dysfunction in diabetes. Curr Diabetes Rev. 2013;9:25–53.
160. Mauvais-Jarvis F. Epidemiology of gender differences in diabetes and obesity. Adv Exp Med Biol. 2017;1043:3–8. https://doi.org/10.1007/978-3-319-70178-3_1.
161. Ryu TY, Park J, Scherer PE. Hyperglycemia as a risk factor for cancer progression. Diabetes Metab J. 2014;38:330–6.
162. Olokoba AB, Obateru OA, Olokoba LB. Type 2 diabetes mellitus: a review of current trends. Oman Med J. 2012;27:269–73.
163. Martín-Timon I, Sevillano-Collantes C, Segura-Galindo A, et al. Type 2 diabetes and cardiovascular disease: have all risk factors the same strength? World J Diabetes. 2014;5:444–70.
164. Feldstein CA, Akopian M, Renauld A, et al. Insulin resistance and hypertension in postmenopausal women. J Hum Hypertens. 2002;16:S145–50.

165. Xu G, Liu B, Sun Y, et al. Prevalence of diagnosed type 1 and type 2 diabetes among US adults in 2016 and 2017: population-based study. BMJ. 2018;362:k1497.
166. Asif M. The prevention and control the type-2 diabetes by changing lifestyle and dietary pattern. J Educ Health Promot. 2014;3:1. https://doi.org/10.4103/2277-9531.127541.
167. Freeman J. Management of hypoglycemia in older adults with type 2 diabetes. Postgrad Med. 2019;131:241–50.
168. About diabetes. World Health Organization. Archived from the original on 31 March 2014.
169. Chiang JL, Kirkman MS, Laffel LM, et al. Type 1 diabetes through the life span: a position statement of the American Diabetes Association. Diabetes Care. 2014;37:2034–254.
170. Gong Z, Muzumdar RH. Pancreatic function, type 2 diabetes, and metabolism in aging. Int J Endocrinol. 2012;2012:320482. https://doi.org/10.1155/2012/320482.
171. Zemel MB. Insulin resistance vs. hyperinsulinemia in hypertension: insulin regulation of Ca2+ transport and Ca(2+)-regulation of insulin sensitivity. J Nutr. 1995 Jun;125(6 Suppl):S1738–43.
172. Little RR, Rohlfing CL. The long and winding road to optimal HbA1c measurement. Clin Chim Acta. 2013;418:63–71.
173. Andersen H. Motor dysfunction in diabetes. Diabetes Metab Res Rev. 2012;28:S89–92.
174. Obanda DN, Cefalu WT. Modulation of cellular insulin signaling and PTP1B effects by lipid metabolites in skeletal muscle cells. J Nutr Biochem. 2013;24:1529–37.
175. Morigny P, Houssier M, Mouisel E, et al. Adipocyte lipolysis and insulin resistance. Biochimie. 2016;125:259–66.
176. Triggle CR, Ding H. A review of endothelial dysfunction in diabetes: a focus on the contribution of a dysfunctional eNOS. J Am Soc Hyperten. 2010;4:102–15.
177. Wu G, Meininger CJ. Nitric oxide and vascular insulin resistance. BioFactors (Oxford, England). 2009;35:21–7.
178. Wang G. Raison detre of insulin resistance: the adjustable threshold hypothesis. J R Soc Interface. 2014; https://doi.org/10.1098/rsif.2014.0892.
179. Haffner SM, Mykkanen L, Festa A, et al. Insulin-Resistant prediabetic subjects have more atherogenic risk factors than insulin-sensitive prediabetic subject's implications for preventing coronary heart disease during the prediabetic state. Circulation. 2000;101:975–80.
180. Schwartsburd P. Insulin resistance is a two-sided mechanism acting under opposite catabolic and anabolic conditions. Med Hypotheses. 2016;89:8–10.
181. Bellinger DA, Merricks PE, Nichols TC. Swine models of type 2 diabetes mellitus: insulin resistance, glucose tolerance, and cardiovascular complications. ILAR J. 2006;47:243–58.
182. Hojlund K. Metabolism and insulin signaling in common metabolic disorders and inherited insulin resistance. Dan Med J. 2014;61(7):B4890.
183. Kahn BB, Flier JS. Obesity and insulin resistance. J Clin Invest. 2000;106:473–81.
184. Boutens L, Stienstra R. Adipose tissue macrophages: going off track during obesity. Diabetologia. 2016;59:879–94.
185. World Health Organization (WHO). Use of glycated hemoglobin (HbA1c) in the diagnosis of diabetes mellitus abbreviated report of a WHO consultation. Geneva: WHO; 2011.
186. Sherwani SI, Khan HA, Ekhzaimy A, et al. Significance of HbA1c test in diagnosis and prognosis of diabetic patients. Biomark Insights. 2016;11:95–104.
187. Cooke DW, Plotnick L. Type 1 diabetes mellitus in pediatrics. Pediatr Rev. 2008;29:374–84.
188. Maffi P, Secchi A. The burden of diabetes: emerging Data. Dev Ophthalmol. 2017;60:1–5.
189. Hadaegh F, Shafiee G, Ghasemi A, et al. Impact of metabolic syndrome, diabetes and prediabetes on cardiovascular events: Tehran lipid and glucose study. Diabetes Res Clin Pract. 2010;87:342–7.
190. Saleh J. Glycated hemoglobin and its spinoffs: cardiovascular disease markers or risk factors? World J Cardiol. 2015;7:449–53.
191. Misra S, Oliver NS. Diabetic ketoacidosis in adults. BMJ. 2015;351:h5660. https://doi.org/10.1136/bmj.h5660.
192. Kitabchi AE, Umpierrez GE, Miles JM, et al. Hyperglycemic crises in adult patients with diabetes. Diabetes Care. 2009;32:1335–43.

193. Sivanandan S, Sinha A, Jain V, et al. Management of diabetic ketoacidosis. Indian J Pediatr. 2011;78:576–84.
194. Casqueiro J, Casqueiro J, Alves C. Infections in patients with diabetes mellitus: a review of pathogenesis. Indian J Endocrinol Metab. 2012;16:S27–36.
195. Ahmad I, Davis L. Management of diabetic patients with low blood pH. J Pak Med Assoc. 1977;27:311–2.
196. Gershuni VM, Yan SL, Medici V. Nutritional ketosis for weight management and reversal of metabolic syndrome. Curr Nutr Rep. 2018;7:97–106.
197. Laffel L. Ketone bodies: a review of physiology, pathophysiology and application of monitoring to diabetes. Diabetes Metab. https://doi.org/10.1002/(SICI)1520-7560(199911/12)15:6 <412::AID-DMRR72>3.0.CO;2-8.
198. Lynch CJ, Adams SH. Branched-chain amino acids in metabolic signalling and insulin resistance. Nat Rev Endocrinol. 2014;10:723–36.
199. Karslioglu E, Donihi AC, Korytkowski MT. Diabetic ketoacidosis and hyperosmolar hyperglycemic syndrome: review of acute decompensated diabetes in adult patients. BMJ 2019; 365. https://doi.org/10.1136/bmj.l1114
200. Reddy MA, Zhang E, Natarajan R. Epigenetic mechanisms in diabetic complications and metabolic memory. Diabetologia. 2015;58:443–55.
201. Wu Y, Yanping D, Tanaka Y, et al. Risk factors contributing to type 2 diabetes and recent advances in the treatment and prevention. Int J Med Sci. 2014;11:1185–200.
202. McNicholas E, Abdelhafiz AH. Diabetes management in older people: a focus on cardiovascular risk reduction. Postgrad Med. 2017;129:169–77.
203. Zheng J, Cheng J, Zhang Q et al. Novel insights into DNA methylation and its critical implications in diabetic vascular complications. Biosci Rep. 2017;37(2). pii: BSR20160611. https://doi.org/10.1042/BSR20160611
204. Marín-Penalver JJ, Martín-Timon I, Sevillano-Collantes C, et al. Update on the treatment of type 2 diabetes mellitus. World J Diabetes. 2016;7:354–95.
205. Florez H, Pan Q, Ackermann RT, et al. Impact of lifestyle intervention and metformin on health-related quality of life: the diabetes prevention program randomized trial. J Gen Intern Med. 2012;27:1594–601.
206. Li J, Zeng Z, Zhao Y, et al. Effects of low-intensity pulsed electromagnetic fields on bone microarchitecture, mechanical strength and bone turnover in type 2 diabetic db/db mice. Sci Rep. 2017;7(1):10834. https://doi.org/10.1038/s41598-017-11090-7.
207. Dokken BB. The pathophysiology of cardiovascular disease and diabetes: beyond blood pressure and lipids. Diabetes Spectrum. 2008;21:160–5.
208. Amara F, Hafez S, Orabi A, et al. Review of diabetic polyneuropathy: pathogenesis, diagnosis and management according to the consensus of Egyptian experts. Curr Diabetes Rev. 2019;15:340–5.
209. Voulgari C, Papadogiannis D, Tentolouris N. Diabetic cardiomyopathy: from the pathophysiology of the cardiac myocytes to current diagnosis and management strategies. Vasc Health Risk Manag. 2010;6:883–903.
210. Luscher TF, Creager MA, Beckman JA, et al. Diabetes and vascular disease. Pathophysiology, clinical consequences, and medical therapy: part II. Circulation. 2003;108:1655–61.
211. Yan ST, Jia JH, Lv XF, et al. Glycemic control and comprehensive metabolic risk factors control in older adults with type 2 diabetes. Exp Gerontol. 2019;28:110713. https://doi.org/10.1016/j.exger.2019.110713.
212. Ozder A. Lipid profile abnormalities seen in T2DM patients in primary healthcare in Turkey: a cross-sectional study. Lipids Health Dis. 2014;13:183. https://doi.org/10.1186/1476-511X-13-183.
213. Bernard M, Cheung Y, Li C. Diabetes and hypertension: is there a common metabolic pathway? Curr Atheroscler Rep. 2012;14:160–6.
214. Steinberger J, Daniels SR. Obesity, insulin resistance, diabetes, and cardiovascular risk in children. Circulation. 2003;107:1448–53.
215. Ames MK, Atkins CE,Lantis AC, Brunnen JZ. Evaluation of subacute change in RAAS activity Ias indicated by urinary aldosterone: criatinin, after pharnacologoc provoca-

tion) and the responce of aCE inhibition. J Aldosteron Renin Syst. 2016. https://doi.org/10.1177/1470320316633897.
216. Weir MR. Effects of renin-angiotensin system inhibition on end-organ protection: can we do better? Clin Ther. 2007;29:1803–24.
217. Takatori S. Function of perivascular nerves in insulin resistance-induce hypertension. [Article in Japanese]. Yakugaku Zasshi. 2017;137:1367–71.
218. Castro-Chaves P, Leite-Moreira AF. Renin-angiotensin system and its role in cardiovascular physiopathology and therapy. [Article in Portuguese]. Rev Port Cardiol. 2004;23(Suppl 2):II61–1177.
219. Shrimanker I, Bhattarai S. Electrolytes. StatPearls [Internet]. Treasure Island (FL): StatPearls Publishing; 2019 Jan–2019 May 3.
220. Hall JE, do Carmo JM, da Silva AA, et al. Obesity, kidney dysfunction and hypertension: mechanistic links. Nat Rev Nephrol. 2019;15:367–85.
221. Dooley R, Harvey BJ, Thomas W. The regulation of cell growth and survival by aldosterone. Front Biosci (Landmark Ed). 2011;16:440–57.
222. Singh R, Kishore L, Kaur N. Diabetic peripheral neuropathy: current perspective and future directions. Pharmacol Res. 2014;80:21–35.
223. Singleton JR, Smith AG, Bromberg MB. Increased prevalence of impaired glucose tolerance in patients with painful sensory neuropathy. Diabetes Care. 2001;24:1448–53.
224. James K. Effect of an intensive glucose management protocol on the mortality of critically Ill adult patients. Mayo Clinic Proc. 2004;79(8):992–1000.
225. Gleason CE, Gonzalez M, Harmon JS, et al. Determinants of glucose toxicity and its reversibility in pancreatic islet beta-cell line, HIT-T15. Am J Physiol Endocrinol Metab. 2000;279:E997–E1002.
226. Pradhan L, Nabzdyk C, Andersen ND, et al. Inflammation and neuropeptides: the connection in diabetic wound healing. Expert Rev Mol Med. 2009;11:e2. https://doi.org/10.1017/S1462399409000945.
227. Malik RA. Pathology of human diabetic neuropathy. Handb Clin Neurol. 2014;126:249–59. https://doi.org/10.1016/B978-0-444-53480-4.00016-3.
228. Yuan H, Reddy MA, Deshpande S, et al. Epigenetic histone modifications involved in profibrotic gene regulation by 12/15-lipoxygenase and Its oxidized lipid products in diabetic nephropathy. Antioxid Redox Signal. 2016;24:361–75.
229. Villeneuve LM, Natarajan R. The role of epigenetics in the pathology of diabetic complications. Am J Physiol Renal Physiol. 2010;299:F14–25.
230. Fox CS, Golden SH, Anderson C, et al. Update on prevention of cardiovascular disease in adults with type 2 diabetes mellitus in light of recent evidence: a scientific statement from the American heart association and the American diabetes association. Diabetes Care. 2015;38:1777–803.
231. Foster-Schubert KE, Alfano CM, Duggan CR, Xiao L, et al. Effect of diet and exercise, alone or combined, on weight and body composition in overweight-to-obese post-menopausal women. Obesity (Silver Spring). 2012;20:1628–38.
232. Stewart KJ. Role of exercise training on cardiovascular disease in persons who have type 2 diabetes and hypertension. Cardiol Clin. 2004;22:569–86.
233. Patel H, Alkhawam H, Madanieh R, et al. Aerobic *vs* anaerobic exercise training effects on the cardiovascular system. World J Cardiol. 2017;9:134–8.
234. Jensen J, Rustad PI, Kolnes AJ, et al. The role of skeletal muscle glycogen breakdown for regulation of insulin sensitivity by exercise. Front Physiol. 2011;2:112. https://doi.org/10.3389/fphys.2011.00112.
235. Hawley J. The fuels for exercise. Aust J Nutr Diet. 2001;58:S19–22.
236. Knowler WC, Barrett-Connor E, et al. Reduction in the incidence of type 2 diabetes with lifestyle intervention or metformin. N Engl J Med. 2002;346:393–403.
237. Della Valle E, Grimaldi R, Farinaro E. Importance of physical activity for prevention of chronic diseases. [Article in Italian]. Ann Ig. 2008;20:485–93.
238. EACPR, Corra U, Piepoli MF, et al. Secondary prevention through cardiac rehabilitation: physical activity counselling and exercise training: key components of the position paper

from the Cardiac Rehabilitation Section of the European Association of Cardiovascular Prevention and Rehabilitation. Eur Heart. 2010;31:1967–1974.
239. Lee JH, Lee R, Hwang MH, et al. The effects of exercise on vascular endothelial function in type 2 diabetes: a systematic review and meta-analysis. Diabetol Metab Syndr. 2018;10:15. https://doi.org/10.1186/s13098-018-0316-7.. eCollection 2018
240. Hollekim-Strand SM, Hoydahl SF, Follestad T, et al. Exercise training normalizes timing of left ventricular untwist rate, but not peak untwist rate, in Individuals with Type 2 diabetes and diastolic dysfunction: a pilot study. J Am Soc Echocardiogr. 2016;29(5):421–430.e2. https://doi.org/10.1016/j.echo.2016.01.005.
241. Elizabeth M, Robert SH. A dictionary of biology. 6th ed: Oxford University Press; 2008. p. 315–6.
242. Vandermosten L, Pham TT, Knoops S et al. Adrenal hormones mediate disease tolerance in malaria. Nat Commun 2018; 9: Article number 4525.
243. Rooijackers HMM, Wiegers EC, Tack CJ, et al. Brain glucose metabolism during hypoglycemia in type 1 diabetes: insights from functional and metabolic neuroimaging studies. Cell Mol Life Sci. 2016;73:705–22.
244. Kalra S, Gupta Y. The insulin:glucagon ratio and the choice of glucose-lowering drugs. Diabetes Ther. 2016;7:1–9.
245. Tesfaye N, Seaquist ER. Neuroendocrine responses to hypoglycemia. Ann N Y Acad Sci. 2010;1212:12–28.
246. Madsen SM, Thorup AC, Overgaard K, et al. High intensity interval training improves glycaemic control and pancreatic β cell function of type 2 diabetes patients. PLoS One. 2015;10(8):e0133286. https://doi.org/10.1371/journal.pone.0133286.
247. Winnick JJ, Sherman WM, Habash DL, et al. Short-term aerobic exercise training in obese humans with type 2 diabetes mellitus improves whole-body insulin sensitivity through gains in peripheral, not hepatic insulin sensitivity. J Clin Endocrinol Metab. 2008;93: 771–8.
248. Zanuso S, Jimenez A, Pugliese G, et al. Exercise for the management of type 2 diabetes: a review of the evidence. Acta Diabetol. 2010;47:15–22.
249. Hamasaki H. Daily physical activity and type 2 diabetes: a review. World J Diabetes. 2016;7:243–51.
250. American Diabetes Association. Standards of medical care in diabetes. Diabetes Care. 2013;36:S11–66.
251. Brazeau AS, Rabasa-Lhoret R, Strychar I, et al. Barriers to physical activity among patients with type 1 diabetes. Diabetes Care. 2008;31:2108–9.
252. Younk LM, Mikeladze M, Tate D, et al. Exercise-related hypoglycemia in diabetes mellitus. Expert Rev Endocrinol Metab. 2011;6:93–108.
253. González M, Rojas S, Avila P, et al. Insulin reverses D-glucose-increased nitric oxide and reactive oxygen species generation in human umbilical vein endothelial cells. PLoS One. 2015;10(4):e0122398. https://doi.org/10.1371/journal.pone.0122398.
254. Brownlee M. The pathobiology of diabetic complications: a unifying mechanism. Diabetes. 2005;54:1615–25.
255. Mozos I, Malainer C, Horbańczuk J, et al. Inflammatory markers for arterial stiffness in cardiovascular diseases. Front Immunol. 2017;8:1058. https://doi.org/10.3389/fimmu.2017.01058.
256. Shirwany AN, Zou NH. Arterial stiffness: a brief review. Acta Pharmacol Sin. 2010;31:1267–76.
257. Finkel T, Holbrook NJ. Oxidants, oxidative stress and the biology of ageing. Nature. 2000;408:239–47.
258. Frazziano G, Champion HC, Pagano PJ. NADPH oxidase-derived ROS and the regulation of pulmonary vessel tone. Am J Physiol Heart Circ Physiol. 2012;302:H2166–77.
259. Sahoo S, Meijles ND, Pagano JP. NADPH oxidases: key modulators in aging and age-related cardiovascular diseases? Clin Sci (Lond). 2016;130:317–35.
260. McArdle A, Pattwell D, Vasilaki A, et al. Contractile activity-induced oxidative stress: cellular origin and adaptive responses. Am J Physiol Cell Physiol. 2001;280:C621–7.

261. McArdle A, Pollock N, Staunton CA, et al. Aberrant redox signaling and stress response in age-related muscle decline: role in inter-intra-cellular signaling. Free Radic Biol Med. 2019;132:50–7. https://doi.org/10.1016/j.freeradbiomed.2018.11.038.
262. Powers KS, Nelson WB, Hudson BM. Exercise-induced oxidative stress in humans: cause and consequences. Free Radic Biol Med. 2011;51:942–50.
263. Carracedo J, Buendia P, Merino A, et al. Klotho modulates the stress response in human senescent endothelial cells. Mech Ageing Dev. 2012;133:647–54.
264. Palmer RM, Ferrige AG, Moncada S. Nitric oxide release accounts for the biological activity of endothelium-derived relaxing factor. Nature. 1987;327:524–6.
265. Engin A. Endothelial dysfunction in obesity. Adv Exp Med Biol. 2017;960:345–79.
266. Brandes RP, Weissmann N, Schroder K. Redox-mediated signal transduction by cardiovascular Nox NADPH oxidases. J Mol Cell Cardiol. 2014;73:70–9.
267. Batchuluun B, Inoguchi T, Sonoda N, et al. Metformin and liraglutide ameliorate high glucose-induced oxidative stress via inhibition of PKC-NAD(P)H oxidase pathway in human aortic endothelial cells. Atherosclerosis. 2014;232:156–64.
268. Massart J, Sjogren RJO, Lundell LS, et al. Altered miR-29 expression in type 2 diabetes influences glucose and lipid metabolism in skeletal muscle. Diabetes. 2017;66:1897–18.
269. Park DR, Park KH, Kim BJ, et al. Exercise ameliorates insulin resistance via Ca2+ signals distinct from those of insulin for GLUT4 translocation in skeletal muscles. Diabetes. 2015;64:1224–34.
270. Mul JD, Stanford KI, Hirshman MF, et al. Exercise and regulation of carbohydrate metabolism. Prog Mol Biol Transl Sci. 2015;135:17–37.
271. Swulius MT, Waxham MN. Ca(2+)/calmodulin-dependent protein kinases. Cell Mol Life Sci. 2008;65:2637–57.
272. Deshmukh AS, Hawley JA, Zierath JR. Exercise-induced phospho-proteins in skeletal muscle. Int J Obes. 2008;32(Suppl 4):S18–23.
273. Fang P, He B, Shi M, et al. Crosstalk between exercise and galanin system alleviates insulin resistance. Neurosci Biobehav Rev. 2015;59:141–6.
274. Way KL, Hackett DA, Baker MK, et al. The effect of regular exercise on insulin sensitivity in type 2 diabetes mellitus: a systematic review and meta-analysis. Diabetes Metab J. 2016;40:253–71.
275. Dos Santos JM, Moreli ML, Tewari S, et al. The effect of exercise on skeletal muscle glucose uptake in type 2 diabetes: an epigenetic perspective. Metabolism. 2015;64:1619–28.
276. Popov DV. Adaptation of skeletal muscles to contractile activity of varying duration and intensity: the role of PGC-1α. Biochemistry (Mosc). 2018;83:613–28.
277. Rose AJ, Frosig C, Kiens B, et al. Effect of endurance exercise training on Ca2+ calmodulin-dependent protein kinase II expression and signalling in skeletal muscle of humans. J Physiol. 2007;583(Pt 2):785–95.
278. Cartee GD. Roles of TBC1D1 and TBC1D4 in insulin- and exercise-stimulated glucose transport of skeletal muscle. Diabetologia. 2015;58:19–30.
279. Borghouts LB, Keizer HA. Exercise and insulin sensitivity: a review. Int J Sports Med. 2000;21:1–12.
280. Kelley GA, Kelley KS. Effects of aerobic exercise on lipids and lipoproteins in adults with type 2 diabetes: a meta-analysis of randomized-controlled trials. Public Health. 2007;121:643–55.
281. Way KL, Hackett DA, Baker MK, et al. The effect of regular exercise on insulin sensitivity in type 2 diabetes mellitus: a systematic review and meta-analysis. Diabetes Metab J. 2016;40:253–71.
282. Hawley JA, Lessard SJ. Exercise training-induced improvements in insulin action. Acta Physiol (Oxf). 2008;192:127–35.
283. Shambrook P, Kingsley M, Taylor N, et al. Accumulated or continuous exercise for glycaemic regulation and control: a systematic review with meta-analysis. BMJ Open Sport Exerc Med. 2018;4(1):e000470. https://doi.org/10.1136/bmjsem-2018-000470.
284. Madden KM. Evidence for the benefit of exercise therapy in patients with type 2 diabetes. Diabetes Metab Syndr Obes. 2013;6:233–9.

285. Braith RW, Stewart KJ. Resistance exercise training Its role in the prevention of cardiovascular disease. Circulation. 2006;113:2642–50.
286. Pesta DH, Goncalves RLS, Madiraju AK et al. Resistance training to improve type 2 diabetes: working toward a prescription for the future. Nutr Metab 2017;14:Article number: 24.
287. Ryan AS, Hurlbut DE, Lott ME, et al. Insulin action after resistance training in insulin resistant older men and women. J Am Geriatr Soc. 2001;49:247–53.
288. American Diabetes Association. Diabetes mellitus and exercise (Position Statement). Diabetes Care. 2000;23(Suppl. 1):S50–4.
289. Zorzano A, Palacín M, Guma A. Mechanisms regulating GLUT4 glucose transporter expression and glucose transport in skeletal muscle. Acta Physiol Scand. 2005;183:43–58.
290. Wyne KL. Preventing cardiovascular disease and diabetes: a call to action from the ADA and AHA. J Cardiometab Syndr. 2006;1:220–1.
291. Bowden Davies KA, Sprung VS, Norman JA, et al. Short-term decreased physical activity with increased sedentary behavior causes metabolic derangements and altered body composition: effects in individuals with and without a first-degree relative with type 2 diabetes. Diabetologia. 2018;61:1282–94.
292. Garber CE, Blissmer B, Deschenes MR, et al. American College of Sports Medicine position stand. Quantity and quality of exercise for developing and maintaining cardiorespiratory, musculoskeletal, and neuromotor fitness in apparently healthy adults: guidance for prescribing exercise. Med Sci Sports Exerc. 2011;43:1334–59.
293. Pereira LO, Lancha AH Jr. Effect of insulin and contraction up on glucose transport in skeletal muscle. Prog Biophys Mol Biol. 2004;84:1–27.
294. Hall KE, McDonald MW, Grise KN, et al. The role of resistance and aerobic exercise training on insulin sensitivity measures in STZ-induced Type 1 diabetic rodents. Metabolism. 2013;62:1485–94.
295. Chen WS, Xu PZ, Gottlob K, et al. Growth retardation and increased apoptosis in mice with homozygous disruption of the Akt1 gene. Genes Develop. 2001;15:2203–8.
296. Holten MK, Zacho M, Gaster M, et al. Strength training increases insulin-mediated glucose uptake, GLUT4 content, and insulin signaling in skeletal muscle in patients with type 2 diabetes. Diabetes. 2004;53:294–305.
297. Krisan AD, Collins DE, Crain AM, et al. Resistance training enhances components of the insulin signaling cascade in normal and high-fat-fed skeletal muscle. J Appl Physiol. 2004;96:1691–700.
298. Sigal R, Wasserman D. Physical activity/exercise and type 2 diabetes. Diabetes Care. 2004;27:2518–39.
299. Newton RL Jr, Johnson WD, Larrivee S, et al. A randomized community-based exercise training trial in African American men: ARTIIS. Med Sci Sports Exerc. 2019;52(2):408–16. https://doi.org/10.1249/MSS.0000000000002149.
300. Pette D, Staron RS. Cellular and molecular diversities of mammalian skeletal muscle fibers. Rev Physiol Biochem Pharmacol. 1990;116:1–76.
301. Hussey SE, McGee SL, Garnham A, et al. Exercise increases skeletal muscle GLUT4 gene expression in patients with type 2 diabetes. Diabetes Obesity Metab. 2012;14:768–71.
302. Lira VA, Soltow QA, Long JH, et al. Nitric oxide increases GLUT4 expression and regulates AMPK signaling in skeletal muscle. Am J Physiol Endocrinol Metab. 2007;293:E1062–8.
303. Li Q, Zhu X, Ishikura S, Zhang D, et al. Ca^{2+} signals promote GLUT4 exocytosis and reduce its endocytosis in muscle cells. Am J Physiol Endocrinol Metab. 2014;307:E209–24.
304. Kraniou GN, Cameron-Smith D, Hargreaves M. Acute exercise and glut4 expression in human skeletal muscle: influence of exercise intensity. J Appl Physiol. 2006;101:934–7.
305. Gurley JM, Ilkayeva O, Jackson RM, et al. Enhanced GLUT4-dependent glucose transport relieves nutrient stress in obese mice through changes in lipid and amino acid metabolism. Diabetes. 2016;65:3585–97.

306. Lehnen AM, Angelis KD, Markoski MM, et al. Changes in the GLUT4 expression by acute exercise, exercise training and detraining in experimental models. J Diabetes Metab. 2012;S10:002. https://doi.org/10.4172/2155-6156.S10-002.
307. Kjobsted R, Hingst JR, Fentz J, et al. AMPK in skeletal muscle function and metabolism. FASEB J. 2018;32:1741–77.
308. Klip A, McGraw TE, James DE. Thirty sweet years of GLUT4. J Biol Chem. 2019;294:11369–81.
309. Yu H, Fujii NL, Toyoda T, et al. Contraction stimulates muscle glucose uptake independent of atypical PKC. Physiol Rep. 2015;3(11):e12565. https://doi.org/10.14814/phy2.12565.
310. Richter EA, Hargreaves M. Exercise, GLUT4, and skeletal muscle glucose uptake. Physiol Rev. 2013;93:993–1017.
311. Balducci S, Zanuso S, Cardelli P, et al. Effect of high- versus low-intensity supervised aerobic and resistance training on modifiable cardiovascular risk factors in type 2 diabetes; the Italian Diabetes and Exercise Study (IDES). PLoS One. 2012;7(11):e49297. https://doi.org/10.1371/journal.pone.0049297.
312. Colberg SR, Albright AL, Blissmer BJ, et al. Exercise and type 2 diabetes: American College of Sports Medicine and the American Diabetes Association: joint position statement. Exercise and type 2 diabetes. Med Sci Sports Exerc. 2010;42:2282–303.
313. Pan B, Ge L, Xun YQ, et al. Exercise training modalities in patients with type 2 diabetes mellitus: a systematic review and network meta-analysis. Int J Behav Nutr Phys Act. 2018;15(1):72. https://doi.org/10.1186/s12966-018-0703-3.
314. Kennedy JW, Hirshman MF, Gervino EV, et al. Acute exercise induces GLUT4 translocation in skeletal muscle of normal human subjects and subjects with type 2 diabetes. Diabetes. 1999;48:1192–7.
315. Colberg SR, Sigal RJ, Yardley JE, et al. Physical activity/exercise and diabetes: a position statement of the American Diabetes Association. Diabetes Care. 2016;39:2065–79.
316. Hurt RT, Frazier TH, McClave SA, et al. Obesity epidemic: overview, pathophysiology, and the intensive care unit conundrum. JPEN J Parenter Enteral Nutr. 2011;35(5 Suppl):S4–13.
317. Lopez RP, Hynes HP. Obesity, physical activity, and the urban environment: public health research needs. Environ Health. 2006;5:25. https://doi.org/10.1186/1476-069X-5-25.
318. Kones R. Primary prevention of coronary heart disease: integration of new data, evolving views, revised goals, and role of rosuvastatin in management. A comprehensive survey. Drug Des Devel Ther. 2011;5:325–80.
319. Antonopoulos AS, Tousoulis D. The molecular mechanisms of obesity paradox. Cardiovasc Res. 2017;113:1074–86.
320. Stralfors P, Jonsson C, Castor Batista AP, et al. Insulin and β-adrenergic receptors mediate lipolytic and anti-lipolytic signalling that is not altered by type 2 diabetes in human adipocytes. Biochem J. 2019;476(19):2883–908. https://doi.org/10.1042/BCJ20190594.
321. Lennon H, Sperrin M, Badrick E, et al. The obesity paradox in cancer: a review. Curr Oncol Rep. 2016;18(9):56. https://doi.org/10.1007/s11912-016-0539-4.
322. Schafer MJ, White TA, Evans G, et al. Exercise prevents diet-induced cellular senescence in adipose tissue. Diabetes. 2016;65:1606–15.
323. U.S. Department of Health & Human Services, Content source: Division of Nutrition, Physical Activity, and Obesity, National Center for Chronic Disease Prevention and Health Promotion.
324. Alpert MA, Omran J, Bostick BP. Effects of obesity on cardiovascular hemodynamics, cardiac morphology, and ventricular function. Curr Obes Rep. 2016;5:424–34.
325. Lavie CJ, Ventura HO, Messerli FH. Left ventricular hypertrophy. Its relationship to obesity and hypertension. Postgrad Med. 1992;91:131–43.
326. Lavie CJ, Milani RV. Effects of cardiac rehabilitation, exercise training, and weight reduction on exercise capacity, coronary risk factors, behavioral characteristics, and quality of life in obese coronary patients. Am J Cardiol. 1997;79:397–401.

327. Bader DS, Maguire TE, Spahn CM, et al. Clinical profile and outcomes of obese patients in cardiac rehabilitation stratified according to National Heart, Lung, and Blood Institute criteria. J Cardpulm Rehabil. 2001;21:210–7.
328. Meires J, Christie C. Contemporary approaches to adult Obesity treatment. Nurse Pract. 2011;36:37–46.
329. Hu G, Tuomilehto J, Silventoinen K, et al. Joint effects of physical activity, body mass index, waist circumference and waist-to-hip ratio with the risk of cardiovascular disease among middle-aged Finnish men and women. Eur Heart J. 2004;25:2212–9.
330. Giannopoulou I, Ploutz-Snyder LL, Carhart R, et al. Exercise is required for visceral fat loss in postmenopausal women with type 2 diabetes. J Clin Endocrinol Metab. 2005;90:1511–8.
331. Tilg H, Moschen AR. Adipocytokines: mediators linking adipose tissue, inflammation and immunity. Nature. 2006;6:772–83.
332. Del Campo JA, Gallego P, Grande L. Role of inflammatory response in liver diseases: therapeutic strategies. World J Hepatol. 2018;10:1–7.
333. Manduteanu I, Simionescu M. Inflammation in atherosclerosis: a cause or a result of vascular disorders? J Cell Mol Med. 2012;16:1978–90.
334. Tzotzas T, Evangelou P, Kiortsis DN. Obesity, weight loss and conditional cardiovascular risk factors. Obesity Rev. 2011;12:e282–9.
335. Xu Y, Wang Y, Yan S et al. Regulation of endothelial intracellular adenosine via adenosine kinase epigenetically modulates vascular inflammation. Nat Commun 2017;8:Article number: 943
336. Antonov A, Kolodgie FD, Munn DH, et al. Regulation of macrophage foam cell formation by αVβ3 integrin: potential role in human atherosclerosis. Am J Pathol. 2004;165:247–58.
337. Bournat JC, Brown CW. Mitochondrial dysfunction in obesity. Curr Opin Endocrinol Diabetes Obes. 2010; 17(5): 446–452
338. Mansouri A, Gattolliat CH, Asselah T. Mitochondrial dysfunction and signaling in chronic liver diseases. Gastroenterology. 2018;155:629–47.
339. The British Medical Association (BMA) (2002). Illustrated medical dictionary. A Dorling Kindersley Book, p. 199.
340. Hojlund K, Mogensen M, Sahlin K, et al. Mitochondrial dysfunction in type 2 diabetes and obesity. Endocrino Metabolism Clin North Am. 2008;37:713–31.
341. Bogacka I, Xie H, Bray GA, et al. Pioglitazone induces mitochondrial biogenesis in human subcutaneous adipose tissue in vivo. Diabetes. 2005;54:1392–9.
342. Gao CL, Zhu C, Zhao YP, et al. Mitochondrial dysfunction is induced by high levels of glucose and free fatty acids in 3T3-L1 adipocytes. Mol Cell Endocrinol. 2010;320:25–33.
343. Koh EH, Park JY, Park HS, et al. Essential role of mitochondrial function in adiponectin synthesis in adipocytes. Diabetes. 2007;56:2973–81.
344. Shutherland LN, Capozzi LC, Turchinsky NJ, et al. Time course of high-fat diet-induced reductions in adipose tissue mitochondrial proteins: potential mechanisms and the relationship to glucose intolerance. Am J Physiol Endocrinol Metab. 2008;295:E1076–83.
345. Wan Z, Root-McCaig J, Castellani L, et al. Evidence for the role of AMPK in regulating PGC-1 alpha expression and mitochondrial proteins in mouse epididymal adipose tissue. Obesity (Silver Spring). 2014;22:730–8.
346. Aatsinki SM, Buler M, Hi HS, et al. Metformin induces PGC-1α expression and selectively affects hepatic PGC-1α functions. Br J Pharmacol. 2014;171:2351–63.
347. Islam H, Edgett BA, Gurd BJ. Coordination of mitochondrial biogenesis by PGC-1α in human skeletal muscle: a re-evaluation. Metabolism. 2018;79:42–51.
348. Scarpulla RC. Transcriptional paradigms in mammalian mitochondrial biogenesis and function. Physiol Rev. 2008;88:611–38.
349. Kappel VD, Pereira DF, Cazarolli LH, et al. Involvement of GLUT-4 in the stimulatory effect of routine on glucose uptake in rat soleus muscle. J Pharm Pharmacol. 2013;65(8):1179–86. https://doi.org/10.1111/jphp.12066.

350. Amy C, Maher AC, Akhtar M, et al. Women have higher protein content of β-oxidation enzymes in skeletal muscle than men. PLoS One. 2010;5(8):e12025. https://doi.org/10.1371/journal.pone.0012025.
351. Sagiv SM. Exercise cardiopulmonary function in cardiac patients. London: Springer; 2012.
352. Redman LM, Heilbronn LK, Martin CK, et al. Effect of calorie restriction with or without exercise on body composition and fat distribution. J Clin Endocrinol Metab. 2007;92:865–72.
353. Vuori IM. Dose-response of physical activity and low back pain, osteoarthritis, and osteoporosis. Med Sci Sports Exerc. 2001;33(6 suppl):S551–86.
354. Hansen D, Dendale P, Berger J, et al. The effects of exercise training on fat-mass loss in obese patients during energy intake restriction. Sports Med. 2007;37:31–46.
355. Donnelly JE, Blair SN, Jakicic JM, et al. American College of Sports Medicine. American College of Sports Medicine position stand: appropriate physical activity intervention strategies for weight loss and prevention of weight regain for adults. Med Sci Sports Exerc. 2009;4:459–71.
356. Achten J, Gleeson M, Jeukendrup AE. Determination of the exercise intensity that elicits maximal fat oxidation. Med Sci Sports Exerc. 2002;34:92–7.
357. Ibrahimi A, Bonen A, Blinn WE, et al. Muscle-specific overexpression of FAT/CD36 enhances fatty acid oxidation by contracting muscle, reduces plasma triglycerides and fatty acids, and increases plasma glucose and insulin. J Biol Chem. 1999;274:26761–6.
358. Tunstall RJ, Mehan KA, Wadley GD, et al. Exercise training increases lipid metabolism gene expression in human skeletal muscle. Am J Physiol Endocrinol Metab. 2002;283:E66–72.
359. Hirabara SM, Silveira LR, Abdulkader F, et al. Time-dependent effects of fatty acids on skeletal muscle metabolism. J Cell Physiol. 2007;210:7–15.
360. Frayn KN. Fat as a fuel: emerging understanding of the adipose tissue–skeletal muscle axis. Acta Physiol. 2010;199:509–18.
361. Windecker S, Allemann Y, Billinger M, et al. Effect of endurance training on coronary artery size and function in healthy men: an invasive follow up study. Am J Physiol Heart Circ Physiol. 2002;282:H2216–23.
362. van Hall G. The physiological regulation of skeletal muscle fatty acid supply and oxidation during moderate-intensity exercise. Sports Med. 2015;45(Suppl 1):23–32.
363. Horowitz JF, Klein S. Lipid metabolism during endurance exercise. Am J Clin Nutr. 2000;72:S558–63.
364. Purdom T, Kravitz L, Dokladny K, et al. Understanding the factors that affect maximal fat oxidation. J Int Soc Sports Nutr. 2018;15:3. https://doi.org/10.1186/s12970-018-0207-1.
365. Yeo WK, Carey AL, Burke L, et al. Fat adaptation in well-trained athletes: effects on cell metabolism. Appl Physiol Nutr Metab. 2011;36:12–22.
366. Helge JW, Wulff B, Kiens B. Impact of a fat-rich diet on endurance in man: role of the dietary period. Med Sci Sports Exerc. 1998;30:456–61.
367. Kiens B, Kristiansen S, Jensen P, et al. Membrane associated fatty acid binding protein (FABPpm) in human skeletal muscle is increased by endurance training. Biochem Biophys Res Commun. 1997;231:463–5.
368. Bonen A, Dyck DJ, Ibrahimi A, et al. Muscle contractile activity increases fatty acid metabolism and transport and FAT/CD36. Am J Physiol Endocrinol Metab. 1999;276:E642–9.
369. Berthon PM, Howlett RA, Heigenhauser GJF, et al. Human skeletal muscle carnitine palmitoyltransferase I activity determined in isolated intact mitochondria. J Appl Physiol. 1998;85:148–53.
370. Spina RJ, Chi MM-Y, Hopkins MG, et al. Mitochondrial enzymes increase in muscle in response to 7-10 days of cycle exercise. J Appl Physiol. 1996;80:2250–4.
371. Kerksick CM, Arent S, Schoenfeld BJ, et al. International society of sports nutrition position stand: nutrient timing. J Int Soc Sports Nutr. 2017;14:33. https://doi.org/10.1186/s12970-017-0189-4.
372. Bangsbo J, Graham TE, Kiens B, et al. Elevated muscle glycogen and anaerobic energy production during exhaustive exercise in man. J Physiol. 1992;451:205–27.

373. Pollock ML, Franklin BA, Franklin BA. Resistance exercise in individuals with and without cardiovascular disease: benefits, rationale, safety, and prescription: an advisory from the Committee on Exercise, Rehabilitation, and Prevention, Council on Clinical Cardiology, American Heart Association. Circulation. 2000;101:828–33.
374. Pescatello LS, Franklin BA, Fagard R, et al. American College of Sports Medicine position stand. Exercise and hypertension. Med Sci Sports Exerc. 2004;36:533–53.
375. Sigal RJ, Kenny GP, Wasserman DH, et al. Physical activity/exercise and type 2 diabetes. Diabetes Care. 2004;27:2518–39.
376. Bateman LA, Slentz CA, Willis LH, et al. Comparison of aerobic versus resistance exercise training effects on metabolic syndrome (from the Studies of a Targeted Risk Reduction Intervention Through Defined Exercise – STRRIDE-AT/RT). Am J Cardiol. 2011;108:838–44.
377. Stone MH, Fleck SJ, Triplett NT, et al. Health- and performance-related potential of resistance training. Sports Med. 1991;11:210–31.
378. Thomson RL, Buckley JD, Moran LJ, et al. Comparison of aerobic exercise capacity and muscle strength in overweight women with and without polycystic ovary syndrome. BJOG. 2009;116:1242–50.
379. Byrne NM, Wood RE, Schutz Y, et al. Does metabolic compensation explain the majority of less-than-expected weight loss in obese adults during a short-term severe diet and exercise intervention? Int J Obes (Lond). 2012;36:1472–8.
380. Kumar S, Maiya GA, Shastry BA, et al. Correlation between basal metabolic rate, visceral fat and insulin resistance among type 2 diabetes mellitus with peripheral neuropathy. Diabetes Metab Syndr. 2019;13(1):344–8. https://doi.org/10.1016/j.dsx.2018.10.005.
381. Trommelen J, Holwerda AM, Kouw IW, et al. Resistance exercise augments postprandial overnight muscle protein synthesis rates. Med Sci Sports Exerc. 2016;48:2517–25.
382. Kerksick CM, Wismann-Bunn J, Fogt D, et al. Changes in weight loss, body composition and cardiovascular disease risk after altering macronutrient distributions during a regular exercise program in obese women. Nutr J. 2010;9:59. https://doi.org/10.1186/1475-2891-9-59.
383. Fry AC. The role of resistance exercise intensity on muscle fibre adaptations. Sports Med. 2004;34:663–79.
384. Kraemer WJ, Volek SJ, Clark KL, et al. Physiological adaptations to a weight-loss dietary regimen and exercise programs in women. Physiology. 1997;83:270–9.
385. Ingelsson E, McCarthy MI. Human genetics of obesity and type 2 diabetes mellitus: past, present, and future. Circ Genom Precis Med. 2018;11(6):e002090. https://doi.org/10.1161/CIRCGEN.118.002090.
386. Cheung WW, Mao P. Recent advances in obesity: genetics and beyond. ISRN Endocrinol. 2012;2012:536905. https://doi.org/10.5402/2012/536905.
387. Tank A, Wong D. Peripheral and central effects of circulating catecholamines. Compr Physol. 2015;5:1–15.
388. Kreisman SH, Mew NA, Arsenault M, et al. Epinephrine infusion during moderate intensity exercise increases glucose production and uptake. Am J Physiol Endocrinol Metab. 2000;278(5):E949–57. https://doi.org/10.1152/ajpendo.2000.278.5.E949.
389. Pelley JW. Epinephrine receptors. Membranes and intracellular signal transduction. Elsevier's integrated review biochemistry. https://doi.org/10.1016/B978-0-323-03410-4.50011-0.
390. Pelley JW. Integration of carbohydrate, fat, and amino acid Metabolism. Elsevier's integrated review biochemistry. https://doi.org/10.1016/B978-0-323-03410-4.50019-5
391. Poirier P, Després JP. Exercise in weight management of obesity. Cardiol Clin. 2001;19:459–70.
392. Farooqi IS, O'Rahilly S. Mutations in ligands and receptors of the leptin-melanocortin pathway that lead to obesity. Nat Clin Pract Endocrinol Metab. 2008;4:569–77.
393. Speakman JR. The Fat Mass and Obesity Related (FTO) gene: mechanisms of impact on obesity and energy balance. Curr Obes Rep. 2015;4:73–91. https://doi.org/10.1007/s13679-015-0143-1.
394. Silventoinen K, Jelenkovic A, Sund R, et al. Differences in genetic and environmental variation in adult BMI by sex, age, time period, and region: an individual-based pooled analysis of 40 twin cohorts. Am J Clin Nutr. 2017;106:457–66.

395. Mazzone R, Zwergel C, Artico M, et al. The emerging role of epigenetics in humanautoimmune disorders. Clin Epigenetics. 2019;11:34. https://doi.org/10.1186/s13148-019-0632-2.
396. Tina R, Ling C. The impact of exercise on DNA methylation of genes associated with type 2 diabetes and obesity in human adipose tissue. US Endocrinology. 2014;10:64–6.
397. McGee SL, Hargreaves M. Epigenetics and exercise. Tren Endocrinol Metab. 2019;30:636–45.
398. Greene NP, Martin SE, Crouse SF. Acute exercise and training alter blood lipid and lipoprotein profiles differently in overweight and obese men and women. Obesity (Silver Spring). 2012;20:1618–27.
399. Arnarson A, Ramel A, Geirsdottir OG, et al. Changes in body composition and use of blood cholesterol lowering drugs predict changes in blood lipids during 12 weeks of resistance exercise training in old adults. Aging Clin Exp Res. 2014;26:287–92.
400. Mogelvang R, Pedersen SH, Flyvbjerg A, et al. Comparison of osteoprotegerin to traditional atherosclerotic risk factors and high-sensitivity C-reactive protein for diagnosis of atherosclerosis. Am J Cardiol. 2012;109:515–20.
401. Slentz CA, Houmard JA, Johnson JL, et al. Inactivity, exercise training and detraining, and plasma lipoproteins. STRRIDE: a randomized, controlled study of exercise intensity and amount. J Appl Physiol (1985). 2007;103:432–42.
402. Silverman MG, Ference BA, Im K, et al. Association between lowering LDL-C and cardiovascular risk reduction among different therapeutic interventions: a systematic review and meta-analysis. JAMA. 2016;316:1289–97.
403. Parto P, Lavie CJ, Swift D, et al. The role of cardiorespiratory fitness on plasma lipid levels. Expert Rev Cardiovasc Ther. 2015;13:1177–83.
404. Phillips WT, Kiratli BJ, Sarkarati M, et al. Effect of spinal cord injury on the heart and cardiovascular fitness. Curr Probl Cardiol. 1998;23:641–716.
405. Karvinen S, Jergenson MJ, Hyvarinen M, et al. Menopausal status and physical activity are independently associated with cardiovascular risk factors of healthy middle-aged women: cross-sectional and longitudinal evidence. Front Endocrinol (Lausanne). 2019;10:589. https://doi.org/10.3389/fendo.2019.00589.. eCollection 2019
406. Rotkis HB, Cote R, Coly E, et al. Relationship between high density lipoprotein cholesterol and weekly running mileage. J Cardiac Rehab. 1982;2:109–12.
407. Black K, Slater J, Brown RC, et al. Low energy availability, plasma lipids, and hormonal profiles of recreational athletes. J Strength Cond Res. 2018;32:2816–24.
408. Hernan Jimenez O, Ramírez-Velez R. Strength training improves insulin sensitivity and plasma lipid levels without altering body composition in overweight and obese subjects. Endocrinol Nutr. 2011;58:169–74.
409. Tucker LA, Silvester LJ. Strength training and hypercholesterolemia: an epidemiologic study of 8499 employed men. Am J Health Promot. 1996;11:35–41.
410. Kohl HW III, Gordon NF, Scott CB, et al. Musculoskeletal strength and serum lipid levels in men and women. Med Sci Sports Exerc. 1992;24:1080–7.
411. Boyden TW, Pamenter RW, Going SB, et al. Resistance exercise training is associated with decreases in serum low-density lipoprotein cholesterol levels in premenopausal women. Arch Intern Med. 1993;153:97–100.
412. Halverstadt A, Phares DA, Wilund KR, et al. Endurance exercise training raises high-density lipoprotein cholesterol and lowers small low-density lipoprotein and very low-density lipoprotein independent of body fat phenotypes in older men and women. Metabolism. 2007;56:444–50.
413. Sagiv M, Goldbourt U. Influence of physical work on high density lipoprotein cholesterol: implications for the risk of coronary heart disease. Int J Sports Med. 1994;15:261–6.
414. Thompson PD, Tsongalis GJ, Seip RL, et al. Apolipoprotein E genotype and changes in serum lipids and maximal oxygen uptake with exercise training. Metabolism. 2004;53:193–202.
415. Angelopoulos TJ, Sivo SA, Kyriazis GA. Do age and baseline LDL cholesterol levels determine the effect of regular exercise on plasma lipoprotein cholesterol and apolipoprotein B levels? Eur J Appl Physiol. 2007;101:621–8.

416. Kelley GA, Kelley KS, Franklin B. Aerobic exercise and lipids and lipoproteins in patients with cardiovascular disease: a meta-analysis of randomized controlled trials. J Cardiopulm Rehabil. 2006;26:131–44.
417. Kodama S, Tanaka S, Saito K, et al. Effect of aerobic exercise training on serum levels of high-density lipoprotein cholesterol: A meta-analysis. Arch Intern Med. 2007;167:999–1008.
418. Eapen DJ, Kalra GL, Rifai L, et al. Raising HDL cholesterol in women. Int J Women's Health. 2010;1:181–91.
419. Joseph LJ, Davey SL, Evans WJ, et al. Differential effect of resistance training on the body composition and lipoprotein-lipid profile in older men and women. Metabolism. 1999;48:1474–80.
420. Goldberg AP, Busby-Whitehead MJ, Katzel LI, et al. Cardiovascular fitness, body composition, and lipoprotein lipid metabolism in older men. J Gerontol A Biol Sci Med Sci. 2000;55:M342–9.
421. Corvilain B. Lipoprotein metabolism. Rev Med Brux. 1997;18:3–9.
422. Ben Ounis O, Elloumi M, Makni E, et al. Exercise improves the ApoB/ApoA-I ratio, a marker of the metabolic syndrome in obese children. Acta Paediatr. 2010;99:1679–85.
423. Morris R, Digenio A, Padayachee GN, et al. The effect of a 6-month cardiac rehabilitation programme on serum lipoproteins and apoproteins A1 and B and lipoprotein a. S Afr Med J. 1993;83:315–8.
424. Morris R, Digenio A, Padayachee GN, et al. The effect of 5-months cardiac rehabilitation programme on serum lipoprotein and apoprotein A1 and B and lipoprotein a. S Afr Med J. 1993;83:315–8.

Chapter 3
Oxygen Uptake and Anaerobic Performances

Introduction

During aerobic exercise, oxygen supply is essential to be continuously delivered for the muscle to contract aerobically, since oxygen during metabolism is of importance to the cell [1]. VO_2 is the transportation and the body's ability to utilize oxygen, and at times it is referred to as aerobic power. The appropriateness of tissue oxygenation rests on the rate of oxygen delivered to the tissues and the rate of oxygen utilized by the tissues. The abbreviation of VO_2 is derived from V = volume and O_2 = oxygen. It is the amount of oxygen consumed at rest and during exercise at any given time and effort.

Oxygen delivery is the product of cardiac output (Q) and arterial oxygen content; the amount of oxygen reaching the mitochondria is defined as oxygen uptake. When the cells receive satisfactory oxygen amount, metabolism is through the aerobic pathway. However, when oxygen delivery to the tissues is not adequate, anaerobic metabolism occurs. Oxygen has two components: supply and demand; therefore, factors affecting VO_2 are also divided into supply and demand [2]. Compared to resting conditions, during exercise, increased oxygen demand by the contracting muscles is an illustration of an increase in 'oxygen demand by the tissue' [3]. A fundamental requirement for many daily living activities is the ability to perform predominantly aerobic work. Since blood flow and oxygen supply are unvarying, it involves the combined efforts of the cardiopulmonary system for transporting oxygen to the working muscles and remove VCO_2 [4]. Therefore, assessment of functional capacity or exercise lasting time or VO_{2max} provides important information in health and disease. In addition, numerous clinical trials, especially those in patients with coronary heart disease, have used exercise lasting time or VO_{2max} as a primary or secondary endpoint.

Most common expression of VO_2 is as milliliters (mL) of oxygen uptake per kg body weight per min ($mLO_2\ kg^{-1}\ min^{-1}$), expression for whole body is ($LO_2\ min^{-1}$). The best way to describe VO_2 is by the Fick equation, developed by Adolf Eugen

M. S. Saghiv, M. S. Sagiv, *Basic Exercise Physiology*,
https://doi.org/10.1007/978-3-030-48806-2_3

Fick, which has been applied to the measurement of Q. The Fick equation states that the amount of VO_2 is equal to Q multiplied by the $(a\text{-}v)O_2$ [5]. More simply, VO_2 depends on the quantity of blood distributed by the cardiovascular system as well as the working muscle's ability to extract oxygen from the blood.

The Fick equation defines VO_2 as follows:

$$VO_2 = Q \cdot (a - v)O_2,$$

where Q = cardiac output, $(a\text{-}v)O_2$ = arteriovenous oxygen content difference.

Or,

$$Q = VO_2 / (a - v)O_2$$

To calculate arterial and venous oxygen blood content, there is a need to know: hemoglobin % and hemoglobin saturation % in the blood. One gram of hemoglobin can carry 1.34 mL of O_2.

Oxygen content of the blood can be calculated as follows:

$$\text{Oxygen blood content for 1 dL} = \text{Hb content in grams} \cdot 1.34\, O_2$$

when Hb = 14 g, oxygen content in 1 dL will be as follows:

$$O_2 \text{ blood content for } 1\,\text{dL} = 14\,\text{g Hb} \cdot 1.34\,\text{mL}O_2 = 18.76\,\text{mL}O_2\,\text{dL}^{-1}$$

$$\text{Oxygen Content of blood} = [\text{Hb}](\text{g/dl}) \cdot 1.34(\text{ml}\,O_2/\text{g of Hb}) \cdot O_2 \text{ saturation fraction} + 0.0032 \cdot PO_2(\text{torr})$$

If saturation of the blood is 95%, then oxygen content of arterial blood will be:

$$\text{Arterial oxygen content for dL} = 18.78 \cdot 0.95 = 17.82\, O_2\,\text{dL}^{-1}$$

$$\text{Oxygen Content of blood} = [\text{Hb}](\text{g/dl}) \cdot 1.34(\text{ml}\,O_2/\text{g of Hb}) \cdot O_2 \text{ saturation fraction} + 0.0032 \cdot PO_2(\text{torr})$$

Mixed venous blood saturation in healthy individual is around 75% at rest. This figure decreases significantly during exercise, depending on the intensity of effort.

Oxygen Diffusing Capacity in the Lung

Oxygen diffusing capacity is a measure of the rate at which oxygen can diffuse from the alveoli into the blood. It measures the variables at work, concerning oxygen passage from the alveoli to the hemoglobin. Functional capacity depends on the rate of oxygen diffusion from the lung into the bloodstream and has an importance in health and disease. In a healthy individual, lung diffusing capacity is above what it is needed to transfer oxygen from the lung into the blood. However, exercise hypoxia and disease may limit and interfere with lung diffusing capacity. Such oxygen limitation can lead to anaerobic metabolism and thus, further body complication. Therefore, the lung diffusing capacity is a vital factor in oxygen delivery [6]. Lung diffusing capacity is known to vary with hemoglobin concentration and alveolar lung volume. For the determination of lung diffusing capacity, carbon monoxide (CO) is used since it is a gas tracer with high affinity for binding to the hemoglobin [7]. An increase in diffusing capacity is observed in a state of maximal exercise intensity. At rest, blood flow through many of the pulmonary capillaries is very low since the pulmonary blood circulation is with a low pressure [8]. Therefore, large capillary numbers are without blood flow through them. Diffusing capacity in normal subjects can be doubled as Q increases from rest to peak exercise. Increased blood flow through the lungs causes all pulmonary capillaries to be perfused at their maximal level [9]. This provides a greater surface area through which oxygen can diffuse into the pulmonary capillary blood. Capillaries' recruitment is critical for maintaining normal arterial oxygen saturation as VO_2 increases. Trained individuals, who require greater amounts of oxygen per minute, have been found to have higher diffusing capacities, but the exact reason for that is not yet known. The effectiveness of the capillary's recruitment can be described by how the diffusing capacity/Q ratio changes as exercise load increases; when diffusing capacity/Q ratio declines below a critical level, arterial oxygen saturation falls [10]. Although one would expect arterial oxygen partial pressure (PaO_2), to decrease during strenuous exercise and CO_2 pressure of venous blood to increase far above normal, this is not the case. Both of these values remain close to normal. In addition, recently it has been demonstrated that genetic variation of the α-subunit of ENaC, at amino acid 663, influences lung diffusion at peak exercise in healthy humans, suggesting differences in alveolar Na+ and, therefore, fluid handling [11]. These findings may be important in determining who may be susceptible to pulmonary edema in response to various clinical or environmental conditions [11].

Oxygen Deficit and Excess Postexercise Oxygen Consumption (EPOC)

An exercising individual that begins to exercise needs immediately sufficient ATP, in order to perform the exercise bout. However, the aerobic energy system is increasing slow oxygen supply and thus the muscle is unable to regenerate aerobically the necessary ATP for several minutes (2–3 min) into exercise [12]. At the onset of an aerobic exercise, three energy pathways are the sources for ATP regeneration: phosphagen system, glycolysis and aerobic (Fig. 3.1).

When exercise bout commences, ATP is supplied solely from Creatine phosphate (CP) and ATP stored in the working muscles. Thus, the anaerobic systems contribute ATP to the muscles, but their contribution does not reflect VO_2, and is defined as oxygen debt (Fig. 3.2).

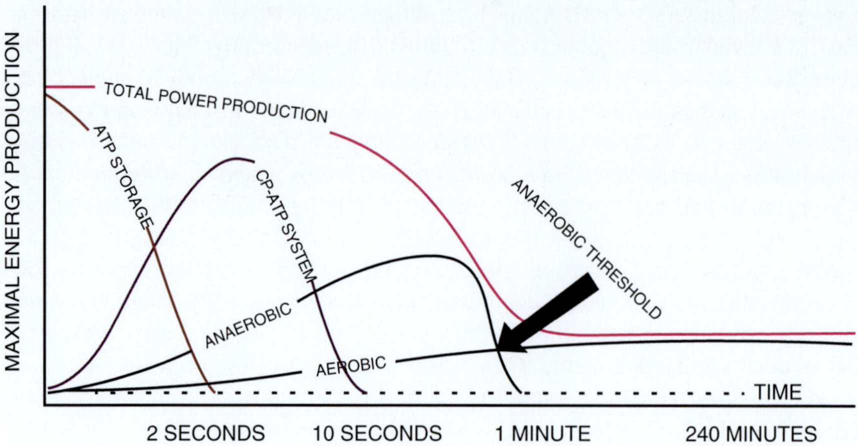

Fig. 3.1 Relative contribution of each energy pathway along time. At the anaerobic threshold, the total energy sources for ATP regeneration are 50% by the aerobic pathway and 50% anaerobically

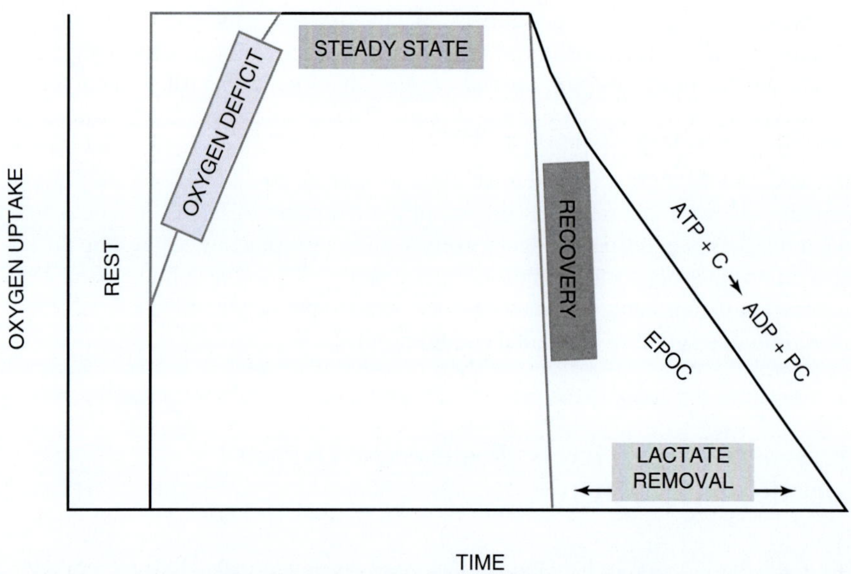

Fig. 3.2 Oxygen deficit and excess postexercise oxygen consumption (EPOC)

During exercise, it is difficult to deliberate the metabolic pathways in unqualified relationships; both pathways, aerobic and anaerobic, work together at the same time, whereas, one of them dominates. Yet, the aerobic pathway and the glycolytic pathway are autonomous as physiological systems, with close association with ATP production.

It reflects ATP regeneration in the muscles by non-aerobic means. This a-lactic process is oxygen-independent and lasts up to about 10 s without lactate production. Due to imbalance between oxygen demand–supply, the following stage is glycolysis or oxidative phosphorylation that, at this phase glycolysis, is the major energy source to resynthesize ATP along with the production of lactate. The a-lactic (CP) and lactic phases (glycolysis) contribute to regenerate ATP in muscles without utilizing oxygen. Therefore, this lasts a short period of time. Thus, oxygen has to be utilized due to an increase in ATP demand [13]. However, although CP and glycolysis do not use oxygen, after CP donated energy to ADP to become ATP, the re-phosphorylated C to CP requires ATP that occurs with the conclusion of the exercise, at exercise recovery time. With extra oxygen required during recovery time, lactate is converted back to glycogen. However, most of the lactate is converted back to pyruvate during exercise [14]. With the presence of oxygen, during exercise, lactate converts to pyruvate utilized later as an energy source, in the slow twist fibers of skeletal muscles and myocardium. Pyruvate enters the Krebs cycle to produce ATP.

During the first minute of postexercise recovery, with the excessive oxygen supply, lactate is removed by being metabolized, partially converted to glucose by the liver. Moreover, these two energy processes occur during aerobic exercise utilizing oxygen. Thus, the anaerobic and aerobic pathways ultimately pay back the deficit following the exercise bout. With the onset of a submaximal exercise, VO_2 increases gradually, as shown above (Fig. 3.2), reflecting aerobic contribution for regenerating ATP. Within 3–5 min, VO_2 reaches a steady state indicating that the aerobic system is supplying all of the energy required by the muscles [15]. Even though VO_2 has reached a steady state, where the aerobic system is providing all of the ATP necessary, anaerobic glycolysis is still operative. This is due to the fact that fat cannot be consumed without the breakdown of glucose or glycogen in the Krebs cycle at the same time. During steady state, lactate, a metabolic product of anaerobic glycolysis, is mostly converted into pyruvate in slow twist muscle fibers which enter the Krebs cycle. Furthermore, during the recovery period, VO_2 does not immediately return to resting levels, rather, VO_2 makes a gradual decrease even though energy demands are only at resting levels. VO_2 during the exercise recovery is above what is needed to maintain the resting metabolic rate. The term excess postexercise VO_2 (EPOC) is used to describe the elevated VO_2 during exercise recovery [16]. Comparing areas between oxygen deficit and EPOC (Fig. 3.2) reveals that EPOC is usually greater than that of oxygen deficit. This excessive oxygen supply during the recovery period is to cover the oxygen debt and phosphagens' renewal. The decrease in VO_2 is not linear and it is composed of two phases (a) rapid one, lasting 2–3 min and, (b) slow phase, lasting up to 15 min. Causes of the EPOC are numerous. The rapid EPOC phase reflects the energy needed to resynthesize depleted ATP and CP stores. Muscle myoglobin, a mobile carrier of oxygen, is replenished back with oxygen. With increased oxygen demand during exercise, myoglobin that exists in red skeletal muscle and heart cells facilitates oxygen transportation from blood hemoglobin through the sarcolemma to the mitochondria. In addition, myoglobin serves as a site to store oxygen in small amounts. Mechanisms of the slow EPOC phase are more complex. Some of the components that are responsible for the slow phase include elevated temperature, catecholamine, and replenishment of muscle glycogen. A different explanation is that mitochondria function repossesses causing interference to generate contraction by the Ca++ released, otherwise, disconnection of ATP production via the oxidation phosphorylation process. The flux of oxygen facilitated by myoglobin is balanced by an identical efflux of unloaded myoglobin. This decreases the efficiency of the aerobic system to deliver more oxygen required to produce ATP. All of these mechanisms further increase the recovery period.

Maximal Oxygen Uptake

VO_2 relates linearly to workload and as exercise intensity increases, VO_2 increases proportionally until VO_{2max} is defined (Fig. 3.3).

VO_{2max} suggests maximal rate of aerobic energy expenditure of an individual, which relates to achievement in endurance sports. VO_{2max} is the maximum amount of oxygen an individual can consume at peak strenuous exercise [17] and describes individual's fitness level. It is the maximum capacity of an individual's body to transport and use oxygen during incremental exercise, which reflects the physical fitness of the individual. VO_{2max} is widely accepted as the single best measure of cardiovascular fitness and maximal aerobic power. In an attempt to compare the cardiopulmonary system in different populations, VO_2 is usually expressed relative to bodyweight, measured as milliliters of oxygen used in 1 min per kg of body weight ($mLO_2\ kg^{-1}\ min^{-1}$) because VO_2 and total energy demands are determined

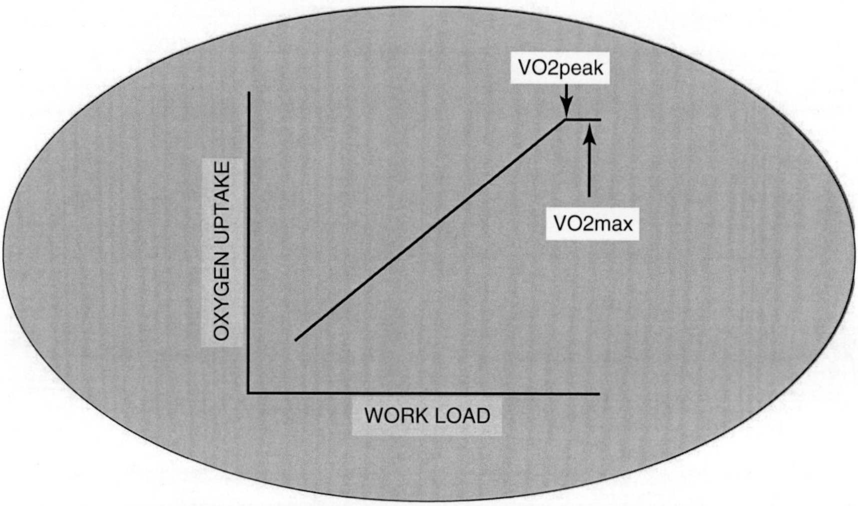

Fig. 3.3 Two different reference points determining VO_{2peak} and VO_{2max}. Determining VO_{2peak} differs from that of VO_{2max}. VO_{2peak} is defined when the subject withdraws voluntarily from the test, while VO_{2max} is defined whenever VO_2 does not increase further with increased intensity

by body size. Another possibility is to normalize VO_2 to body surface area, which can be more accurate when comparing children and gender. VO_{2max} is achieved under maximal aerobic metabolism. This rate is found to increase only 10% with the effect of training. VO_2 and aerobic capacity are all terms used interchangeably with VO_{2max}. One factor that can determine an individual's capacity to perform sustained exercise and is linked to aerobic endurance is VO_{2max}.

In marathon runners, VO_{2max} is 45% greater than that of a moderately trained person. This is believed to be partly genetically determined: stronger respiratory muscles, larger chest size in relation to body size, and partly due to long-term training. The fastest rate at which the body can utilize oxygen during heavy exercise is defined as the maximum rate of VO_2, which is an index of maximal cardiopulmonary function [18]. Therefore, it might be hypothesized that VO_{2peak} or VO_{2max} is dependent on oxygen delivery and by the tissues' ability for $(a\text{-}v)O_2$ [19, 20]. While pulmonary capacity, maximal Q, and total hemoglobin content are in uncertainty considering central conditioning factors for VO_2, while the total hemoglobin content and oxygen release capacity of the blood are defined by the $(a\text{-}v)O_2$ [20]. VO_{2max} represents the maximal amount of ATP that the aerobic system can produce. This is believed to be partly genetically determined: stronger respiratory muscles, larger chest size in relation to body size, and partly due to long-term training. In addition, VO_{2max} reflects the aerobic physical fitness of the individual and is an important determinant of their cardiopulmonary and endurance capacity during prolonged submaximal exercise. Thus, VO_2 max is an assessment of mitochondrial ability and efficiency, and, therefore, mitochondria loss due to aging results in reduced energy production capacity, less rapidity, and less stamina [21].

VO_{2max} assesses the maximal ability of the body to deliver and utilize oxygen and relate to the ability to perform a prolonged exercise. The kinetics of VO_2 in humans depends on the diffusion gradient between the lung and red blood cell and from blood cell to the tissue's cell. Therefore, measuring VO_{2max} provides a good indication to the physician of how adequate the oxygen-extraction is functioning and how healthy the individual is.

VO_2 is of vital importance for both health and sports performance [22]. Although VO_{2max} is generally considered the best indicator of cardiopulmonary function and endurance, it is more useful as an indicator of a person's aerobic potential or upper limit than as a predictor of success in endurance events. While a high VO_{2max} may be a prerequisite for performance in endurance events at the highest level, other markers such as lactate threshold are more predictive of performance [23]. The speed at which lactate threshold is crossed is more significant than the actual value itself. Yet, VO_{2max} varies greatly between individuals as does Q between elite athletes that compete in the same sport. Therefore, VO_{2max} should not be confused with the lactate threshold (or anaerobic threshold). Lactate threshold, during exhaustive, all-out exercise refers to the point at which lactate builds up in the muscles. With proper training, individuals are often able to substantially increase their anaerobic threshold and thus exercise lasts longer at a higher intensity. With a sudden increase in exercise workload, VO_2 increases very rapidly with a half-time of around 30 s. This may occur either as a result of oxygen delivery to the tissues, by the increased *Q*, or of an increase in the $(a\text{-}v)O_2$, or a combination of both factors. One of the fundamental beliefs in exercise physiology is that there is a lag between the onset of an increase and the arrival of the hypoxic blood from the exercising muscle at the lungs. This traditional belief, however, was questioned by Casaburi et al. [24], who found that pulmonary artery desaturation occurred in humans within 4 s after a transition between rest and 150 W cycle exercise, i.e. earlier than the predicted circulation time of around 12–17 s [25]. A high VO_{2max} is related with a high lean body mass and, therefore, low body fat. This results in lower oxygen demand relative to body mass. Increased adipose tissue adds a dead weight to be carried, without the ability of the adipose tissue to utilize oxygen. Therefore, it increases significantly the energy cost to move of a heavier body during exercise and thus it reduces performance [26]. Theoretically, the more oxygen the individual can use during high-level exercise, the more energy he/she can produce. This is often the case with elite endurance athletes who typically have very high VO_{2max} values. Absolute values of VO_{2max} are typically significantly far higher in men than in women and are considered a standard of maximal aerobic power. However, many elite female athletes have higher VO_{2max} values than most trained men. VO_{2max} results vary greatly; the average for a sedentary individual is close to 35 $mLO_2\ kg^{-1}\ min^{-1}$ while elite endurance athletes often average 70 $mLO_2\ kg^{-1}\ min^{-1}$. One of the highest recorded VO_{2max} results (94 $mLO_2\ kg^{-1}\ min^{-1}$) was that of a cross-country skier.

Postexercise Recovery

Following exercise stress, recovery time is an important part of the body adaptation to training [27]. During this period of recovery, the body, mainly the muscles, replaces depleted energy stores such as muscle ATP, phosphagens, glycogen, water loss and mending damaged tissues. Indications of insufficient recovery time include a feeling of general dissatisfaction, reduced physical performance ability, and increased risk of muscular damage. There are two classes of recovery which are important for best sports performance: (a) recovery from intense training session or sports affair and (b) the enduring recovery needs between training sessions and sports events.

Recovery from intense training is divided into two subperiods: active recovery and without motion recovery or, cool-down recovery. Active recovery occurs by a decrement of the exercise intensity to a low intensity in order to allow the cardiopulmonary and skeletal muscles to cool down along with oxygen supply while [28] without motion recovery last hours and days after the active recovery [29]. During the second recovery period, muscle protein synthesis increases to avoid muscle degradation thus resulting in muscular hypertrophy. This time is important for muscles, tendons, ligament restoration, and the elimination of waste chemicals accumulated as a result of cell reactions and metabolism during exercise.

VO_{2max} Determination

Measuring precisely VO_{2max} involves an aerobic exercise with adequate duration and intensity to completely stress the aerobic energy pathway. Usually, VO_{2max} testing, in clinical setting or for athletic performance, typically includes a graded exercise test, where workload is gradually increased. Physiological variables that are measured include minute ventilation (VE), oxygen, and CO_2 concentration of the air inhaled and exhaled.

VO_{2max} can be determined through a number of physical evaluations. These tests can be direct or indirect. Direct testing requires the utilization of a metabolic cart to quantify volume and gas concentrations in breathed air and in exhaled air. There are many protocols used on treadmills, cycle ergometers, and other exercise equipment to measure VO_{2max} directly. Precisely measuring VO_{2max} requires an all-out effort usually on a treadmill or bicycle, intensity sufficient in duration that fully stresses the aerobic energy system. It should be performed under a strict protocol in an exercise physiology laboratory. The subject is typically connected to a metabolic cart to measure VO_2, and intensity is increased over a limited period of time. These protocols for clinical and athletic testing measure the volume and oxygen concentration in breathed air and in exhaled air. This determines how much oxygen the individual is using.

The metabolic cart is an Open Circuit Spirometry that measures at the same time an online percentage of oxygen and CO_2, multiply by VE in order to have VO_2 and volume of CO_2 production. An individual's VO_2 increases in a linear relationship with exercise intensity up to a point in which VO_2 plateaus even if the exercise intensity increases [17]. This plateau marks the VO_{2max}. At this point, the individual moves from aerobic metabolism to full anaerobic metabolism. From here, it is not long before muscle fatigue forces the individual to withdrawal exercising due to the large lactate production. VO_{2max} varies greatly between individuals and even between elite athletes that compete in the same sport [30]. The test usually takes between 6 and 8 min. Tests that last longer than that may force the individual to withdraw from the test due to fatigue. In addition, normal length of the test requires from the individual to be completely rested and motivated to endure long enough to define the true VO_{2max}. Defined VO_{2max} at peak exercise depends on the mode of exercise and the specific training of the individual. An incremental aerobic stress test on a treadmill compared to cycle ergometry has higher VO_{2max} values; however, world-class cyclists can achieve a VO_{2max} value on cycle ergometry similar to that on a treadmill [31]. Anyhow, due to larger muscle mass involved, in most cases VO_{2max} values measured on a treadmill will be higher by 14% than that of the cycle ergometry [31]. Calculating VO_2 is somehow a simple process: in addition to the Fick equation, by subtracting the volume of oxygen expired from the volume of oxygen inspired, it enables expressing the volume of oxygen consumed, therefore:

$$VO_2 = VE\left(FIO_2 - FEO_2\right)$$

where VO_2 = oxygen uptake; VE = minute ventilation; FEO_2 = fractional(%) of oxygen in inspired air; FIO_2 = fractional(%) of oxygen in expired air.

Peak Oxygen Uptake

Compared to VO_2, with the increased workload, peak oxygen uptake (VO_{2peak}) do not reach a plateau as VO_{2max}. Namely, VO_{2peak} is defined as the point that the subject withdraws voluntarily from the test. VO_{2peak} is the highest value of VO_2 achieved on a specific exercise test, while VO_{2max} is the maximum rate of VO_2 that is considered possible by a subject. In spite of this, differences between VO_{2peak} and VO_{2max} are often mistakenly used interchangeably. VO_{2peak} does not designate the level of intensity that the individual is able to attain. Therefore, VO_{2max} is the top indicator of an individual's cardiopulmonary aerobic capacity.

Myocardial functional changes in cardiovascular disease patients include decline in maximum HR, SV, and left ventricular contractility, and an increase in peripheral vascular resistance. The changes in structure and function of the cardiopulmonary system result in VO_{2peak} decline, which is the best single indicator of declined physical working capacity. During the past two decades, VO_{2peak} compared to VO_{2max} has been consistently confirmed as an independent predictor of mortality in patients with cardiovascular disease [32].

Because maximal exertion is not without risk in elderly and cardiac patients, VO_{2peak} is used for maximal exercise testing since the attempt to reach the plateau to define VO_{2max} may put the patient in a very dangerous situation. An early study suggested that a specific VO_{2peak} cut-point was able to distinguish patients with low and high probability of survival [33]. High-intensity aerobic interval exercise is superior to moderate exercise for increasing VO_{2peak} in stable CAD patients. As VO_{2peak} seems to reflect a continuum between health and cardiovascular disease, the present data may be useful in designing effective training programs for improving health status. Several limitations have been attributed to VO_{2peak} as an indicator of the severity of cardiac performance. It can be affected by age, sex, motivation, body composition, and muscle deconditioning [34]. It has also been claimed that the exercise capacity for a given level of cardiac dysfunction may vary with the etiology of disease. Thus, these data suggest that VO_{2peak} is only an indirect indicator of the cardiac response to exercise. In addition, it may be unfavorably affected by alterations in skeletal muscle characteristics. Also, one has to keep in mind that muscle mass plays an important role in the definition of VO_{2peak}. Cardiovascular disease patients exhibited in general considerably lower total skeletal muscle mass and VO_{2peak} compared to healthy controls. Other VO_2 exercise-derived parameters have emerged as predictors of outcomes in cardiovascular disease patients [35]. They mainly include percent achieved of predicted VO_{2peak} and VO_2 kinetics during and after exercise [36]. Patients who are unable to reach 50% of predicted VO_{2peak} have increased mortality, but the prognostic value of VO_2 percent over VO_{2peak} alone has not been confirmed in a number of studies [36]. Yet, it is still a matter of debate whether VO_2 variables are superior to VO_{2peak} in the prognostic evaluation of cardiovascular disease [37].

In healthy subjects, the VO_{2peak} or VO_2 is limited by maximal capacity of oxygen delivery to active tissues and/or oxygen release from hemoglobin. Since metabolic demands during incremental exercise are increased over time, diverse populations may maintain the energy supply in an imbalance between oxygen delivery and extraction and hence different VO_2 rate of reaction. In the mechanisms of exercise intolerance in cardiovascular disease patients in comparison with rest indices of cardiac performance, measures of pulmonary function and peripheral vasodilator capacity were more closely related to exercise VO_{2peak}. Furthermore, the associations were independent of each other and together accounted for 50% of the variance in VO_{2peak} [38]. This suggests that pulmonary and peripheral vascular adaptations may be important determinants of exercise intolerance in cardiovascular disease. In fact, it has been suggested that under elevated pressure loads as during isometric bouts, a failure in the performance of the left ventricle in susceptible subjects may compromise left ventricular contractility and SV, with the consequent lower oxygen delivery to the working muscles [39].

Energy Sources at Peak Anaerobic Exercise

During heavy all-out exercise that engenders a significant lactic acidosis, VO_2 does not achieve steady state but continues to rise until exercise is terminated, or exhaustion ensues [40]. The nature of the contractile and metabolic properties of skeletal muscle suggests that the speed of shortening influences energy turnover during contractions. Indeed, it has been observed in vitro using isolated muscles and single muscle fibers that energy varies with shortening velocity [41]. Metabolic responses during incremental exercise differ between men and women despite a similar fitness [42]. In healthy subjects, it has been found that during aerobic exercises, VO_2 of men exceeds that of women due to gender-related variance in heart size [43].

Anaerobic all-out exercise such as the Wingate Anaerobic Test exposes the subjects to a very high degree of sudden strenuous exercise relying primarily on the glycolysis pathway [44]. Accordingly, performance time can be limited by lactate levels in the blood and active muscles [45]. With aging, two main factors influence the physical performance: (a) the decrease in cardiopulmonary function and (b) a progressive decline in muscle mass and strength that occurs with aging [46]. Consequently, the VO_{2max} and power ability decrease, irrespective of lifestyle because of genetic factors, thereby severely limiting body function [47]. However, structured endurance training after 40 years of age is powerful enough to induce beneficial cardiovascular adaptations in later life [48]. Although data are available on metabolic cost during maximal anaerobic dynamic exercise testing [49], little data are available on the physiological variables that maintain the source of metabolic costs with aging.

Graded exercise up to VO_{2max} is usually maintained because of a balance between oxygen delivery and $(a\text{-}v)O_2$. At peak all-out anaerobic bout, there are no significant differences between young well-trained athletes and master athletes in absolute VO_2 values, at peak strenuous exercise due to aging. However, one has to keep in mind the significant higher workload and peak power output that the young group performed. VO_2 in masters is compensated for the very lower blood delivery to the working muscles by the significant increase in $(a\text{-}v)VO_2$.

Both groups achieves similar absolute VO_2 values attained at peak anaerobic exercise, which accounts for 20% of the total energy expenditure in both groups. All energy systems are active at a given time and the primary energy source utilized depends on the intensity and duration. The maximal (a-v)O_2 at peak graded exercise decreases from perhaps 140–150 mL dL^{-1} in a young adult to 120–130 mL dL^{-1} in elderly, thus, in untrained elderly, (a-v)VO_2 is limited [50].

With the reduced cardiac delivery in the masters, compared to the young once, the similar VO_2 reached at peak Wingate Anaerobic Test is due to different muscle fiber recruitment patterns or different efficiencies of fibers that were recruited during exercise. Previous findings indicated that in untrained normal population, young have larger muscle fibers size than aged subjects; however, training increases entirely muscle fiber hypertrophy in particular type II muscle fiber [51]. These differences may influence metabolism and the adaptive response to exercise [52]. In addition, it seems that the lower contraction velocity in the masters, compared to the young, enabled the masters to utilize relatively more slower twitch fibers and hence, increasing the ability to extract oxygen. From in vitro studies, it has been observed that the efficiency of different fiber types is closely related to contraction velocity [53]. At low contraction velocities as in the masters, the efficiency of slow-twitch fibers is higher than that of fast-twitch fibers and the reverse is observed at high speeds [53].

Values for Q during the strenuous exercise are low in young and masters, compared to what one would have expected, during maximal graded aerobic exercise. The lower Q in the young and to the greater extent in masters during strenuous exercise might be related to the short duration of the bout, which in turn, does not bring about the appropriate adjustment of the circulation, due to the relatively high TPR which is not low as seen during aerobic exercise.

Echocardiographic indices and BP response during strenuous exercise are similar to those seen during isometric maneuver, indicating that autoregulation is not involved while performing sudden strenuous all-out exercise. This assumed lack of autoregulation mechanism indicates that the reduced SV in the young and masters, during strenuous exercise, is due to the relatively high-pressure overload opposing left ventricular ejection. Cardiac adaptations in the elderly following aerobic exercise training improve SV and left ventricular systolic function, at peak aerobic exercise [52].

Although in the elderly left ventricular systolic function is preserved, SV reduction at peak exercise is significantly larger in the masters compared to young individuals due to the lesser increased left ventricular contractility, attributed to functional changes with age in the myocardium [54].

Gender Differences During Anaerobic Bout

During strenuous exercise that engenders a significant lactic acidosis, VO_2 does not achieve steady state but continues to rise until exercise is terminated or exhaustion ensues. The nature of the contractile and metabolic properties of skeletal muscle suggests that the speed of shortening influences energy turnover during contractions. Indeed, it has been observed in vitro using isolated muscles and single muscle fibers that energy varies with shortening velocity [52].

Gender differences in metabolic and left ventricular to strenuous exercise have been well documented. In addition, metabolic and left ventricular responses during incremental exercise also differ between men and women, despite a relatively similar fitness [55]. In healthy subjects, it has been found that during aerobic exercises, VO_2 of men exceeds that of women due to gender-related variance in heart size [56]. Since metabolic demand during sudden strenuous all-out exercise is increased over time, oxygen delivery to the working muscle may be reduced. Under these conditions, females may maintain the energy supply due to the balance between oxygen delivery and extraction. In addition, sudden strenuous all-out exercise alters left ventricular contractility and function in young healthy adults [57]. Thus, oxygen delivery to the working muscle may be reduced. The gender difference may bring about a failure in the performance of the female's myocardium, such as reduced SV under elevated pressure loads [58].

Significant differences in VO_{2max} values of the working muscles at peak sudden strenuous all-out exercise exist due to gender. This is in spite of the significant differences in peak power output, body weight and lean body mass. However, VO_2 in women is compensated for the lower Q by a significant increase in $(a\text{-}v)O_2$ accompanied with unchanged SV.

Compared to men, in the women, oxygen donates at peak sudden strenuous all-out exercise, about 12% of the total energy expenditure, while men have 10%. The higher VO_2 at the peak sudden strenuous all-out exercise in the women may be due to different muscle fiber recruitment patterns or different efficiencies of fibers that are recruited during exercise. Men have larger muscle fibers than women, but similar fiber composition [59]. However, gender-related differences exist in the size of individual skeletal fibers which may influence metabolism and the adaptive response to exercise [52]. In addition, it seems that the lower velocity of contraction in the women compared to men enables women to utilize, relatively to men, additional slower twitch fibers and hence increasing the ability to extract oxygen. From in vitro studies, it has been observed that the efficiency of different fiber types is closely related to the velocity of contraction [53]. At low contraction velocities as in women, the efficiency of slow-twitch fibers is higher than that of fast-twitch fibers and the opposite is observed at high speeds [53].

Rating of Perceived Exertion

The simplest and most controversial measurement traditionally used as a VO_{2max} criterion was developed by Borg [60], known as Borg's rating of perceived exertion. This rating of perceived exertion is not a physiological response measurement; it is the individual's remarks at each increased intensity during the increased exercise test. This personal perceived exertion is very important for the general correctness of the test [61]. In addition, it has been found that 94% of the individuals defined the anticipated rating of perceived exertion, despite the fact that only 60% of them established a true VO_{2max} plateau [62]. Generally, the presumptions of using a rating of perceived exertion standard rest on the individual's careful thought of the measure and related verbal identifications, and ability to distinguish between uneasiness drive and physiological response.

Factors Affecting VO_{2max}

Factors that affect VO_{2max} include genetics, aging, body size, gender, fitness and training, changes in altitude, and action of the ventilatory muscles [19].

Genetic factors and training (epigenetic) regulate the various physiological factors that contribute to the body's ability to transport oxygen. Genetics determines the upper value of VO_{2max}, once reached [63]. If an individual has been training for a prolonged time, in high intensities and has been including the correct quantity of interval and endurance work, VO_{2max} will not increase any further. Consequently, due to genetic limitations, not everyone can become a competitive athlete. It is well known that different individuals have different DNA sequences. In general, DNA sequence differences occurring in more than 1% of the population and are termed polymorphisms. Polymorphisms may account for some of the differences in performance capacity between individuals, including VO_{2max}. Genetics plays a major role in the variability of VO_2 and Q, or in the responsiveness to training, it is suggested that heredity can account for up to 25–50% of the variance seen between individuals [64]. Mitochondrial DNA is of particular interest because it contains the genes for several enzymes involved in VO_2. Mitochondrial DNA contains genes for 13 proteins of the electron transport chain as well as 22 transfer RNAs and two ribosomal RNAs required for their intra-mitochondrial synthesis [65]. Variation in the DNA sequence of these genes could therefore contribute to the differences in endurance performance between individuals. Research shows that although VO_{2max} has a genetic component, it can also be increased through training. Research in this area has focused on mitochondrial DNA sequences, genes for creatine kinase, and genes for angiotensin-converting enzyme [66]. The two methods for increasing VO_{2max} include increases in both training volume and intensity. Research also indicates that the more sedentary an individual is, the further he/she can increase its VO_{2max} through training. In fact, novice exercisers have been able to increase VO_{2max} by 20% through proper training. Well-trained athletes have a harder time increasing their VO_{2max}, most likely because they are already so near to their genetic potential. Results of a genomic scan based on 289 polymorphic markers covering all 22 pairs of autosomes performed on the Caucasian families of the HERITAGE Family Study indicated that in response to a 20 weeks standardized endurance training program, results showed linkages between VO_{2max} and markers on 4q, 8q, 11p, and 14q for VO_{2max} before training and with markers on 1p, 2p, 4q, 6p, and 11p for the change in VO_{2max}. These chromosomal regions harbor many genes that may qualify as candidate genes for these quantitative traits [67].

Accurately Measuring VO_{2max}

It involves an aerobic effort sufficient in duration and intensity to maximal stress on the aerobic energy pathway. The testing protocol's duration to reach VO_{2max} lasts between 8 and 10 min for healthy, moderately to highly trained subjects. For coronary heart disease patients, testing time is lengthened by modifying it to the patient's health status and physiological abilities to reach VO_{2peak}. In general, clinical and athletic testing involves a graded exercise test, either on a treadmill or on a cycle ergometer, in which exercise intensity is elevated progressively while measuring ventilation, VO_2, VCO_2 volumes, and concentrations of the inhaled and exhaled air [68]. Measuring VO_{2max} involves an aerobic effort sufficient in duration and intensity to maximal stress the aerobic energy pathways. The VO_{2peak} attained on a maximum-effort incremental test is likely to be a valid index of VO_{2max}, despite no evidence of a plateau in the data themselves. However, without additional tests, one cannot be certain [69]. The VO_{2peak} is usually determined in patients with cardiopulmonary complications and the elderly.

Important note: Tests measuring VO_{2max} can be dangerous for individuals who are considered not healthy, due to cardiopulmonary dysfunction. Thus, the Balke protocol and Bruce protocol for estimating VO_{2max} were developed for those whom a traditional VO_{2max} test would be too risky [68]. These generally are similar to a VO_{2max} test, but do not reach the maximum of the cardiopulmonary system and are called submaximal tests reaching VO_{2peak}.

Muscle Mass, VO_2, and Cardiopulmonary Function

The normal cardiovascular response to aerobic exercises is its ability to deliver oxygen, as an answer to peripheral oxygen demand, during different intensities, regardless of age or state of fitness [70]. In healthy individuals, the relation between Q and VO_2 depends on the muscle mass involved in the effort. Consequently, HR is faster and SV lesser at any given level of VO_2 for small compared to large contracting muscle mass [71]. The lesser SV during exercise utilizing small muscle mass is due to lower myocardial filling pressures and relatively higher afterload. A larger difference occurs between BP responses to numerous methods of exercise. BP is excessively larger at a given level of VO_2 for small muscle mass involved, probably due to a lesser reduction in TPR [72].

Aging Effect

In western countries, individuals 65 years and above are almost 20% of the entire population [73]. Involvement in a chronic exercise regime is an effective intervention to prevent a number of functional declines associated with aging [74]. In this regard, current guidelines point out the importance of aerobic exercise in the maintenance and improvement of various aspects of cardiovascular function, health, and strength in the attenuation of muscle strength declines associated with normal aging [75]. Two main factors influence the physical performance with aging: (a) the decrease in cardiopulmonary function and (b) a progressive decline in muscle mass and strength that occurs with aging [46] Consequently, the VO_{2max} and power ability decrease, irrespective of lifestyle because of genetic factors, thereby severely limiting body function, quality of life, and longevity [47] However, structured endurance training after 40 years of age is powerful enough to induce beneficial cardiovascular adaptations in later life [48]. The average untrained healthy male will have a VO_{2max} of approximately 37–42 and female 32–38 $mLO_2\ kg^{-1}\ min^{-1}$. These scores can improve with training and decrease with age, though the degree of trainability also varies widely. No significant decline in the VO_{2max} over a 10-year period is observed in a group of competitive master athletes, who maintained their training intensity and continued to compete. Aging process decrease blood flow to the working muscles, thus reducing oxygen delivery, which causes mitochondrial dysfunction. This reduced mitochondrial function compared to young adult muscle is the result of a significant decline in oxygen delivery. VO_{2max} declines from a peak at about age 20-year to almost half of that at age 80-years, mostly due to declines in Q and lung capacity.

The annual decrease rate in VO_{2max} for the general population is estimated to be 0.4–0.5 $mLO_2\ kg^{-1}\ min^{-1}\ year^{-1}$, or about 1% per year after the age of 25 years. However, constant exercise training can reduce the rate of VO_{2max} decline to $5\% \cdot 10\ years^{-1}$ during aging [76]. Constant exercise allows biological systems continuously to make immediate alterations for the best and effective physiological function, regardless of variations in internal and external conditions [77]. It is generally recognized that aging is related to an advanced deterioration in the work capacity as well as with a decline in lean body mass. Muscle mass is related with VO_{2max} decline; however, it seems that VO_{2max} in aged individuals have a lower reduced quantity, influenced by muscle mass compared to young adults [78]. Main reason for this decline is the reduced oxygen delivery to the working muscles and thus lower VO_{2max}. A decline in muscle oxygen supply mainly is due to reduced HRmax and myocardial contractility which decrease SV consequently, Q is significantly reduced. The aging process also brings about a decline in skeletal muscle oxidative capacity, partially due to mitochondrial dysfunction [79].

Delivery–Demand Ratio

Delivery is the transport of oxygen from the lungs to the mitochondria including lungs, Q, oxygen-carrying capacity, and capillary density of the skeletal muscle, while demand is the rate at which the mitochondria can extract oxygen, in the process of oxidative phosphorylation [3]. Of these, in healthy individuals, during large muscle mass exercise, such as cycling, running, and swimming, VO_{2max} is limited by blood flow (Q), delivering oxygen to the contracting muscles and not by the

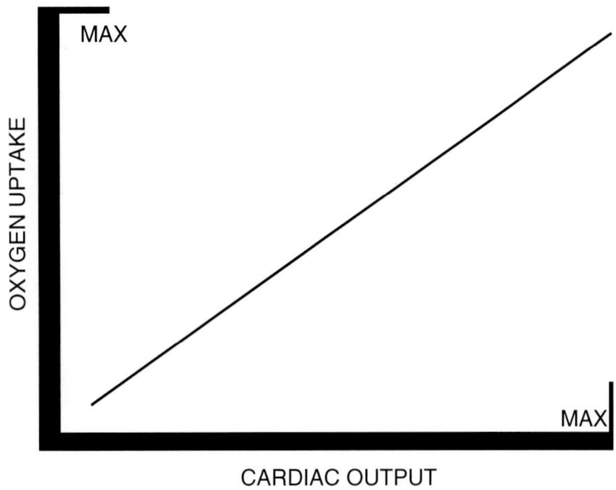

Fig. 3.4 Leaner relationship exists between Q_{max} and VO_{2max}

muscle's (a-v)O_2 ability [19]. Thus, VO_{2max} seems to associate strongly with Q_{max} (Fig. 3.4). However, it has also been argued that while trained individuals probably are delivery limited, untrained individuals can indeed have a muscle (a-v)O_2 limitation, in addition to demand limitation [80].

The extent of the rise in VO_{2max} following endurance training rests on numerous factors, particularly the original fitness level of the individual, the period length of the training program, intensity, length, and rate of the individual training sessions. VO_{2max} is a relatively poor predictor of performance in runners due to variations in running economy and fatigue resistance during prolonged exercise. Pulmonary diffusion capacity, Q, oxygen-carrying capacity, and other peripheral limitations like muscle diffusion capacity, mitochondrial enzymes, and capillary density are all examples of VO_{2max} determinants [19]. If one of these factors is not fully functioning, then the whole system loses its capacity to function properly. In addition, VO_{2max} performance relates to several physiological variables, such as the anaerobic threshold and long-lasting aerobic exercise training.

Gender has a key role in the determinization of the aerobic capacity, owed to body gender differences regardless of age [81]. The VO_{2max} average gender difference has been reported earlier [82]. Average female value of VO_{2max} is about 30–25% less of male in both absolute terms and relative to body mass [83]. During strenuous exercise that genders a significant lactic acidosis, VO_2 endures to increase until exercise ends or fatigue follows [84]. VO_2 values, observed in sedentary trained men and women at the same age, vary primarily due to body composition and differences in maximal SV between subjects, due to variance in heart size while insignificant differences exist in HR_{max} and systemic (a-v)O_2 [85]. The highest ever recorded VO_{2max} is 94 $mLO_2\ kg^{-1}\ min^{-1}$ in men and 77 $mLO_2\ kg^{-1}\ min^{-1}$ in women. Both were cross-country skiers [86]. However, when comparing VO_2 max between genders, relative to lean body mass, differences are reduced by half. Untrained girls and women typically have a VO_{2max} 20–25% lower than untrained boys and men. However, when comparing elite athletes, the gap tends to be reduced to about 10%. Taking it further, if VO_2 is adjusted to account for lean body mass in trained and untrained males and females, the differences disappear in some studies [87], suggesting that sex-specific essential fat stores account for the majority of metabolic differences in running between men and women. On comparing gender responses during a 1-h treadmill run, similar gender physiologic and perceptual responses were generally found [88]. In addition, it has been suggested that normal exercise capacity as determined by VO_{2peak} is related to the indexed SV in males and left atrial pressure in females. These native normal differences between genders may explain the known vulnerability of women to endurance exercise compared to men [88].

Comparing cardiac function during endurance exercise of young women and men suggest that compared to women, men had a higher blood volume and hemoglobin. At HRmax, men had a higher Q (31.0 ± 0.6 vs. 24.1 ± 0.9 L min^{-1}) and SV than women. When expressed relative to body mass (kg), there were no gender differences in maximal SV index [88]. Factors linked mainly to oxygen-carrying capacity and cardiac size have been shown to contribute significantly to this gender-related difference [89].

VO_{2max} Following Training

The extent by which VO_{2max} can change with training also depends on the starting point. Considerable training is required to reach the upper limit for VO_{2max}. However, much less is required to maintain it. Previously, sedentary individuals trained at 60–75% of VO_{2max}, for 30 min, 3 times a week over 6 months increased VO_{2max} on average 15–20% [90]. However, this is an average and there are large individual variations, ranging from 4 to 93% increase in VO_2 [90]. In healthy young adults, when volume of exercise is controlled, higher intensities of exercise are more effective for improving VO_{2max} than lower intensities [91].

There are three primary ways to improve VO_{2max}: (a) since VO_{2max} is based on weight, the individual could just simply lose weight. Then, the individual's VO_{2max} will instantly go up. However, this is not advised as a method of improving VO_{2max} for most runners since losing weight compromises sometimes health. (b) The individual can improve VO_{2max} through running, by increasing volume and workout sessions per week. Going from being sedentary, to run 40 km per week can improve VO_{2max} by up to 16%. Moving from 40 to the 80 km per week range can improve it up to another 16%. And moving from 80 to 110 km per week improves VO_{2max} by additional 3%. Beyond 110 km per week there is a minimal effect on VO_{2max}. One has to keep in mind that with each increase in kilometer, there is an enormous increase in injury incidences. There also is a genetic upper limit beyond which further increases in either intensity or volume have no effect on VO_2 [90]. Athlete's peak VO_{2max}, is reached early in their career and therefore the potential of improving VO_2 is low. From the standpoint of athletic performance, sometimes running faster is a better mode of exercise to improve VO_{2max}. However, if training is to improve health without participating in competition, it is preferably to increase the volume of exercise above the anaerobic threshold than increasing speed. Once a plateau in VO_{2max} has been reached, further improvements in performance are still seen with training. Two major reasons for this are improvements in anaerobic threshold and running economy.

Intense resistive training and anaerobic training have a small effect on VO_{2max} [92]. Resistance training alone does not increase VO_{2max} [92], even when short rest intervals are used between sets and exercises [93]. However, others found a small but significant increase in VO_{2max}. A study undertaken to determine if heavy resistance training results in an increase in VO_{2max} found a significant 4% increase in VO_{2max} during bicycle exercise following resistive training, but no significant differences were observed when expressed in ($mLO_2\ kg^{-1}\ min^{-1}$). Strength training had no effect on VO_{2max} when measured during tested on treadmill exercise. The data suggest that the small increase in leg cycle VO_{2max} may be due to adaptations in oxidative capacity and increased mass of the strength-trained muscles [94]. In a recent study, significant improvements in aerobic capacity and treadmill time to exhaustion can be obtained in older adults as a consequence of either high- or low-intensity resistance exercise. These findings suggest that increased strength, as a consequence of resistance exercise training, may allow older adults to reach and/or improve their aerobic capacity [95]. The discrepancy in the results following resistive training may be due to different cohorts and protocols used during the different studies.

VO_{2max} Response to Insufficient Training Stimulus

Detraining can be defined as the partial or complete loss of physiological adaptations, gained following long-lasting training, due to an insufficient training stimulus [96]. Strength loss is gradually when muscles start to show atrophy signs, after 2–3 weeks, and it can be preserved without training up to 5 weeks, then it is slowly vanished thereafter. But, it is possible to regain it back, fast with retraining. VO_{2max} drops 6–20% in well-trained athletes at around 4 weeks of detraining. Beginners can maintain endurance performance for at least 2 weeks without training, though, recent VO_{2max} gains can be reversed after 4 weeks. Although VO_{2max} drops significantly, still it stays above those untrained individual. This is because of the reduced blood volume [97], SV, and thus, Q and pulmonary function, in spite of increased HR [98]. Consequently, aerobic exercise performance is decreased. Furthermore, lactate threshold is lowered, although it remains above untrained values in the highly trained. At the muscle level, following 1 week of detraining, capillarization, (a-v)O_2, and mitochondrial oxidative capacity, in trained individuals, are entirely inverted.

VO_{2max} Prediction

Test for VO_{2max}, as an indicator of fitness, is direct and accurate measurement of maximal work capacity and serves as a gold standard of exercise testing.

Yet, compared to submaximal tests, this enhanced precision is expensive, time-consuming, and effort applied up to maximal HR. In addition, the ability to test a large group of individuals to VO_{2max} is limited since tests are performed only at the laboratory; therefore, it is out-of-the-way to many active people. Bringing individuals to reach a true VO_{2max} needs expensive laboratory equipment and to stress the individuals to volitional fatigue. In addition, VO_{2max} test is sometimes dangerous for populations such as heart disease patients and elderly. Thus, submaximal exercises predicting VO_{2max} tests have been developed especially for the layman [99]. VO_{2max} test may require from the individuals a few minutes continual maximal all-out exercise to the point where the individual will no longer be able to keep exercise pace. Reasons for terminating the test besides the physiological factor can be also a motivational factor that may add strain. VO_{2max} tests are typically performed continuously, without rest between workload increments. Therefore, submaximal exercises predicting VO_{2max} have been suggested and validated to estimate VO_{2max}. Submaximal exercises predicting VO_{2max} are methods for improving fitness, without exposing the individual to dangerous events [69]. Using precise submaximal procedures to evaluate VO_{2max} is an appropriate substitute to maximal exercise testing. Submaximal testing is applied since it offers an economical method and allows a large number of individuals to be tested in a short time.

VO_{2max} is the standard measure of physical fitness and an unbiased measure of the individual's physical work capacity and performance. Several submaximal exercises predicting VO_{2max} tests have been developed and uniformed to predict VO_{2max}. The predictability of the VO_{2max} is calculated using progressive and steady-state protocols on different ergometers [100]. Alternatively, estimated ways using submaximal exercise tests usually utilize the HR response and its linear relationship with VO_2, between 40% and 85% of VO_{2max} [101]. HR is often utilizing to precisely predict VO_{2max}, by submaximal exercise intensities, and estimate maximal aerobic work capacity [101]. Alternatively, it is possible to prescribe an exercise training intensity, based on the individual's percentage of its predicted VO_{2max}. Moreover, since HR relates linearly to VO_2 for aerobic activities with large muscle mass, it can facilitate the estimation of energy cost during exercise [102]. However, one has to keep in mind that the utilization of the HR response to estimate VO_2 is based on the norms that they are not at all times correct, norms such as an unvarying maximal HR during the life span, and the linear relations between HR and VO_2 [103].

Performance based on these submaximal tests should be independent of strength, workload, body size, and skill.

Field-Tests for VO_{2max} Prediction

While laboratory-testing using metabolic cart (indirect calorimetry) is the most accurate method to determine maximal aerobic capacity, the procedure is expensive and time-consuming. Evolving field tests are faster and easily based on their correlation with laboratory data [104]. The most usually used submaximal exercise tests, applying testing tools such as stairs, cycle ergometer, or treadmill, are Harvard test, Cooper 12 min walk, PWC170 and the Astrand-Rhyming test.

Cooper's 12-min and 1.5-mile runs are two of the most broadly recognized and utilized field tests. But, these tests also need a highly motivated individual exercising to voluntary fatigue, in order to maximize their prediction capacity. Not all individuals have the motivation and ability to perform a maximal test and some contraindications prevent maximal testing of a diseased individual. Consequently, a test base on HR response was developed to estimate VO_{2max} grounded on submaximal workloads. These methods usually are able to rapidly test large groups of individuals applying step climbing [105], cycle ergometry [106], and walking/running protocols [104]. Some of the well-known prediction tests include the Harvard-Step-Test and the Åstrand-Rhyming nomogram.

The Harvard Step Test

The test is a kind of cardiac stress test for detecting and diagnosing cardiovascular disease. Its advantages are in measuring the fitness level and recovery rate capability, following an energetic exercise. During recovery, shorter time for HR to return to resting values, suggests better fitness level [107]. The test is given for noticing and identifying the existence of cardiovascular disease. It is also a good measurement of fitness and a person's ability to recover, after an all-out exercise. The Harvard-Step-Test characteristic is an endurance fitness test. The test calculates the ability to exercise nonstop for lengthy intervals of time without exhausting.

The Harvard-Step-Test is conducted as follows:

The individual taking the test steps up and down 150 times on a platform in a cycle of one-step per 2 s for 5 min.

- The platform is at a height of about 50 cm or 20 inches.
- Subjects must hold rate of 30 steps per minute up for 5 min or until exhaustion.
- To ensure the right speed, a metronome is used.
- Exhaustion is the point at which the subject cannot maintain the stepping rate for 15 s.
- With the test completion, the subject immediately sits down and measures its heartbeats postexercise, between 1 to 1.5 min (pulse 1), 2 to 2.5 min (pulse 2), and 3 to 3.5 min (pulse 3).
- Results are written down and plotted as a simple fitness index equation:

$$(\mathrm{te} \cdot 100) / (\mathrm{HR} \cdot 2)$$

where te = time until exhaustion in seconds; HR = total heartbeats counted. Table 3.1 grades the score by subjects.

Table 3.1 The outcome of the equation is rated as follows

Rating	Fitness level
Excellent	96 and above
Good	83–95
Average	68–82
Low average	54–67
Poor	53 and less

Example
Following 5 min (300 s), HRs measured were as follows: Pulse 1 = 50 beats, Pulse 2 = 55 beats, Pulse 3 = 60 beats. Total heart beats = 50 + 55 + 60 = 165

$$\text{Fitness index} = (300 \cdot 100)/(165 \cdot 2) = 91.$$

The Cooper 12 min Walk/Run Test

The Cooper 12 min walk, run fitness test is a field-test method for measuring aerobic capacity and to provide an estimated VO_2 max. The point of the test is to run as far as possible within 12 min. It is supposed to be running at a steady pace instead of sprints and fast running. There is a very high correlation between the distance that the individual can run (or walk) within 12 min and VO_{2max} [108]. This test is still one of the basic fitness tests used by the military, and by many professional trainers to determine endurance capacity over time. Cooper 12 min walk, run test likewise allows us to compare interindividual differences in endurance fitness levels, in the life cycle in males and females.

Performing the Test

The Cooper 12-min run test requires the person being tested to run or walk as far as possible in a 12 min time. The purpose of the 12-min run/walk test is to cover a maximum distance during the 12 min period (Table 3.2). The test is usually executed in a stadium, on the running track, where the distance of each lap is known. It is an easy test to perform on large groups. For athletes, the length of a good run is considered to be everything above 3 km.

Equipment Needed
A stopwatch is needed to ensure that the individual runs for the correct amount of time.

- Safety First. This field test is a stress fitness test and therefore it is recommended to have a medical backup while performing the test.

Table 3.2 Comparing individual results for group's norms by age and gender

Age males	Excellent	Above average	Average	Below average	Poor
20–29	2800	2400 = 2800	2200–2399	1600–2199	1600
30–39	2700	2300–2700	1900 = 2299	1500–1999	1500
40–49	2500	2100–2500	1700–2099	1400–1699	1400
50 + >	2400	200–2400	1600–1999	1300–1599	1300
Age females	Excellent	Above average	Average	Below average	Poor
20–29	>2700	2200–2700	1800–2199	1500–1799	<1500
30–39	>2500	2000–2500	1700–1999	400–1699	<1400
40–49	>2300	1900–2300	1500–1899	1200–1499	<1200
50 + >	>2200	1700–2200	1400–1699	1100–1399	<1100

- Location. This test takes place on a stadium track with clearly marked distance. In addition, it is possible to perform the test on a treadmill, with an inclination of one degree. Speed can be changed as the individual feels to simulate outdoor running.
- Before starting the test, a short warm-up for 10–15 min is advised.
- Following the warm-up, the individual runs or walks as far as he/she can within 12 min.
- The total number of miles or kilometers covered in 12 min should be recorded.

Results can be transferred to VO_{2max} as follows:

$$\text{In meters}: VO_2 = (d12 - 504.9)/44.73$$

where d12 = distance covered (in meters) in 12 min.
Alternatively,

$$\text{In miles}: VO_{2max} = (35.97 \cdot \text{miles}) - 11.29.$$

Calculating Test Results
The above equations allow us to compare results between different individuals. The following address supply's calculator online: https://exrx.net/Calculators/MinuteRun

The PWC170 Test

Purpose: This test measures aerobic fitness.

PWC stands for physical work capacity. PWC170 estimates the working capacity at an HR of 170 beats min^{-1}, while PWC75% estimates the working capacity at 75% of maximum HR [109]. The procedures are very similar for the two tests;

below the method for the PWC170 test is described. For the PWC75%, HR aim for each stage may be different.

Required Resources

- Bicycle Ergometer
- Heart rate monitor
- Stopwatch
- Staff

How to Conduct the Test

This test requires the athletes to cycle 10 min for warm-up while maintaining their HR between 110 and 120 beats min^{-1} followed by a 10-min rest.

Test procedure: athletes perform three consecutive workloads on a cycle ergometer. Check subjects' exercise history to determine the starting workload (maybe an HR between 100 and 115 beats min^{-1}). At rest, measure HR each minute for 3 min. The subject sits on the bicycle following sitting height adjustment (knee slightly bent at bottom of the cycle). Put HR monitor on the subject. Start the test, each exercise stage will last 3–4 min in order to reach steady HR. Test workloads will be elevated two additional stages, set to achieve an HR between 120 and 145 beats min^{-1}.

Assessment

Scoring: Each steady-state HR and workload should be graphed, with the line of best fit for the three points extrapolated to estimate the workload that would elicit an HR of 170 beats min^{-1} (or 75% of maximum HR for the PWC75% test). This workload value can then be compared to norms.

Advantages

- Having the athletes seated during this test makes it appropriate for ECG monitoring.
- Minimal equipment required.
- Simple to set up and conduct.

Disadvantages

- As the PWC170 test is performed on a cycle ergometer, the test favors cyclists.
- Determining the line of best fit and extrapolating to 170 beats min^{-1} can be subjective and open to error.
- Specialist equipment is required.
- Assistant required administering the test.

Test versions: As well as the PWC170 and 75%, other levels may also be used, such as PWC140 for measuring elderly, cardiac patients or other populations in which one do not wish to excessively elevate HR (Fig. 3.5).

Analysis

Analysis of the recent test's results is done by comparing with subject's previous results. This may help to compare training between each test effect.

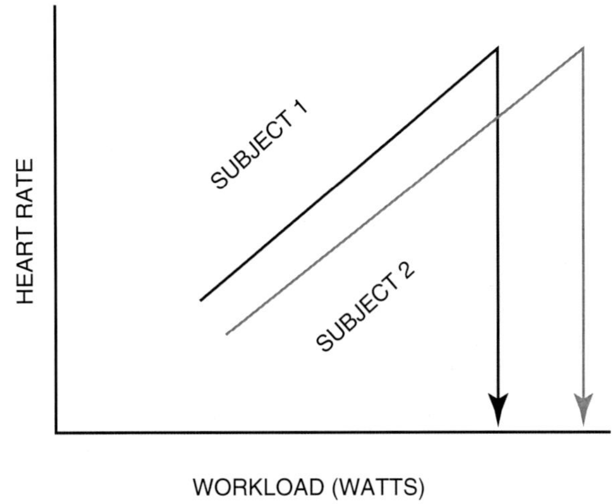

Fig. 3.5 Workload (watts) or oxygen uptake predicted from the three different workloads in two different subjects

Estimation of VO_{2max} from Submaximal Exercise HR

Estimation of VO_{2max} from submaximal exercise HR is based on two well-established physiological principles:

(a) HR and VO_2 are linearly related to one another and both variables are linearly related to exercise workload, and (b) maximal HR and VO_{2max} are produced at the same exercise workload [110].

Thus, if a given subject's maximal HR is known and his/her HR can be measured at two or more submaximal VO_2 levels, VO_{2max} can be estimated by extrapolating the HR−VO_2 curve to the maximal HR point.

Anaerobic Threshold

The anaerobic threshold is defined as the highest sustained exercise intensity for which measurement of VO_2 can account for the entire energy requirement. It is the exercise intensity at which the blood concentration of lactate begins to increase exponentially [111]. Blood lactate level is the balance between lactate production and removal. Lactate threshold normally is expressed as a percentage of an individual VO_{2max}. For example, an individual with VO_{2max} that occurs at 20 km h^{-1} on a treadmill test will experience a sharp rise in blood lactate concentration. Lactate levels are expected in the above individual, at 10 km h^{-1}, meaning that the lactate threshold is about 50% of VO_{2max} [80]. At intensities lower than the anaerobic threshold, the majority of the energy needed to perform the exercise's load applied comes from the aerobic energy pathway. When exercising below the anaerobic threshold work intensity, any lactate produced by the muscles is removed by the body without its buildup. The anaerobic threshold is a useful measure that determines exercise intensity for training and racing in endurance sports such as long-distance running, cycling, rowing, swimming, and cross-country skiing. At intensities above the anaerobic threshold, anaerobic energy production is needed to supplement aerobic energy production. At the anaerobic threshold, the rate at which lactate appears in the blood will be equal to the rate of its disappearance. Following training, the decreases in blood lactate level at any assumed workload results from a decrease in lactate rate production and an increase in lactate declining [112]. Yet, endurance training affects only lactate disappearance rather than buildup [113]. The intensity at the lactate threshold represents the maximal intensity at which steady-state exercise can be maintained. In sedentary individuals, the anaerobic threshold occurs at approximately 50–60% of their VO_{2max}. The amount of lactate in the blood at the end of a maximal exercise test reflects: (a) the intensity of the exercise for the subject and answer the question if the individual gave its maximal effort? (b) the degree to which the individual is needed to supplement aerobic energy production, with anaerobic energy production, and (c) the individual's lactate tolerance. It should be emphasised that the amount of blood lactate at any given time depend on what is the amount entering the circulation (produced) and how much is being removed from the circulation (removed).

At rest, the lactate blood level does not differ significantly between trained and untrained individuals. However, trained individuals have a significant advantage in blood lactate removal, during active recovery [114].

Lactate and lactic acid differ and are not the same molecule. Because lactic acid is a strong acid, it appears usually in the blood, as lactate [115]. Formation of lactate occurs once lactic acid loses hydrogen ions. Since blood samples are taken from the blood, only lactate can be measured.

Differences Between Lactic Acid and Lactate

Lactic acid is a carboxylic acid and also an alpha hydroxyl acid. In addition, in lactic acid, an intramolecular hydrogen bonding cannot hold its proton decisively. Therefore, lactic acid's proton tends to be removed easily [116]. Lactic acid is produced from pyruvate by fermentation during glycolysis [117]. Regularly, at rest, fermentation does not take place in cells, nonetheless, during exercise, lactic acid can be formed in great quantities in muscles; it is the anion produced from lactic acid. When lactic acid is liquified in water, it tends to detach and forms a lactate ion and a proton. Consequently, the majority of lactic acid in the body is detached and exists as lactate. Lactate can be seen as a leftover product for one cell; on the other hand, a valuable molecule for another, lactate has a very high reutilizing rates in transitional metabolism. To summarize the differences between lactic acid and lactate, Table 3.3 is presented.

Table 3.3 Differences between lactic acid and lactate

Variable	Lactic acid	Lactate
Produced	Anaerobic	Deprotonation
Give proton	Yes	No
In solution	No	Yes
Anion	Yes	No
Intra_molecule hydrogen	Yes	No
Pass membraNE	Yes	No

Competitive Sports

When increasing aerobic workload, there is a point at which lactate begins to accumulate. This point is vital as lactate can interfere with muscle contraction and energy production causing pain and a discomfort sensation. The anaerobic threshold is a very sensitive indicator measurement and has a high value to define the fitness level. For example, following an effective training regime, the anaerobic threshold will be reached in a longer time and at a higher workload. Therefore, lactate threshold is the utmost reliable interpreter of performance in endurance actions. It has been frequently found a high association between endurance ability (running, cycling, etc.), and the highest steady-state intensity at the lactate threshold [118].

Effective aerobic training will move the onset of blood lactate accumulation (OBLA) to a higher level [119], up to near the maximal work capacity (Fig. 3.6). One reason why blood lactate is an important variable to be assessed is that the hydrogen ions resulting from lactic acid production in the cell is known to cause muscle fatigue by several different mechanisms [120]. An athlete who can perform at high intensities with minimal lactate accumulation in the blood circulation can better avoid fatigue. On the other hand, during high-intensity exercise, the ability to use anaerobic energy systems is very important. Thus, it is common to find very high maximal blood lactate values in athletes competing at high-intensity events. Resting blood lactate is typically close to 1 mmol L^{-1}, the anaerobic threshold is said to be around 4 mmol L^{-1}, one criterion for a valid VO_{2max} test is a value exceeding 4 mmol L^{-1}, elite athletes competing in anaerobic events lasting 2–6 min have been reported to have values exceeding 25 mmol L^{-1}. Well-trained athletes reach lactate threshold at high workload reaching 90–92% of their VO_{2max}, while untrained individuals reach it at 55–60% of their VO_{2max}. Females have a significantly lower anaerobic threshold compare to males [121].

In two individuals with the same VO_{2max}, the one with the higher lactate threshold (Fig. 3.7) will perform better during long-lasting endurance events.

Thus, increased anaerobic threshold is a good treasure for participants in competitive events and physiologically efficient, when exercising or competing. Therefore, anaerobic threshold is one of the most important measurements that can be obtained during recovery of an endurance athlete [122].

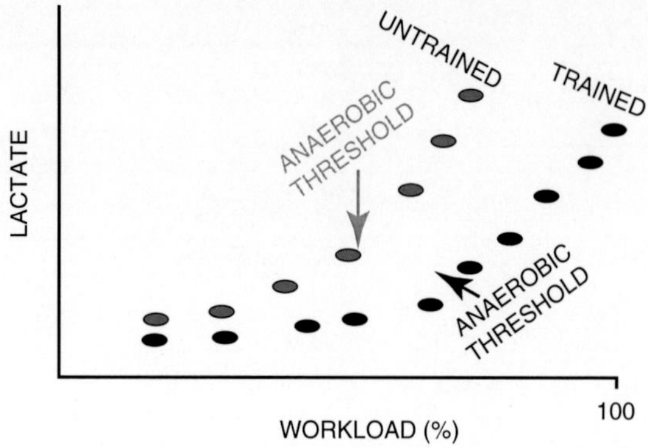

Fig. 3.6 Effect of training on the anaerobic threshold

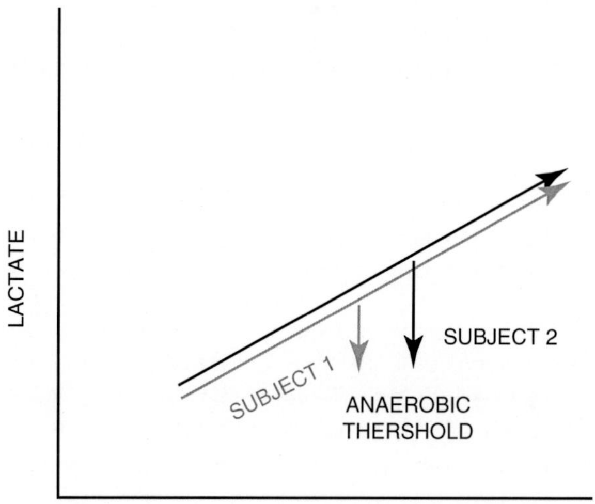

Fig. 3.7 Although both subject 1 and subject 1 reached VO_{2max} at a similar workload, subject 2 has a higher anaerobic threshold than subject 2. Theoretically, subject 2 can maintain a higher pace compared to subject 2

The Lactate Anaerobic Threshold

The test is most precise and consistent, by drawing blood samples from the fingertip or ear's lobe, at rest and several times during the graded exercise test at different workloads. The test is performed during maximal graded aerobic exercise on a bicycle or treadmill. Lactate threshold is defined at a workload intensity that lactate values start to increase. Lactate threshold values are sometimes also called anaerobic threshold, OBLA, HR deflection point, ventilatory anaerobic threshold, and maximum lactate steady state. The onset of OBLA is often confused with the lactate threshold. OBLA happens when the concentration of blood lactate is about 4 mmol l L^{-1}. Since it is almost impossible to determine the differences between the OBLA and that of the lactate threshold, scientists often prefer the OBLA measurement. In addition, high levels of lactate accumulation along with depletion of muscle fuels may lead to an early muscular fatigue. Higher workload at the defined anaerobic threshold means a higher aerobic performance's ability. The anaerobic

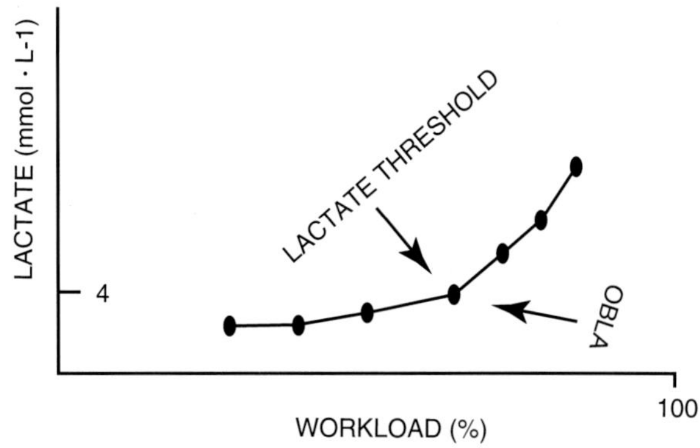

Fig. 3.8 Lactate threshold and OBLA determination

threshold represents a point where the accumulation of lactate reaches the OBLA point (Fig. 3.8).

Methods Defining Anaerobic Threshold

Several methods [123] are utilized to determine the anaerobic threshold: blood lactate levels [124], pulmonary gas exchange [125], and the Conconi test [126].

Pulmonary Gas Exchange Test

Additional method to define lactic acidosis (anaerobic) threshold is the pulmonary gas exchange. The pattern of alterations in arterial HCO_3 and pulmonary gas exchange supports this threshold concept. Anaerobic threshold achieved during physical exercise can be detected from the relationship between VCO_2 (CO_2) and VO_2 (Fig. 3.9). At the beginning of the exercise, CO_2 and VO_2 increase linearly with workload. However, with an increase in aerobic exercise intensity, at a certain point, the utilization of glucose (glycolysis) increases muscle lactic acidosis that in turn increases VCO_2 production. In addition, the acidification of the blood at strenuous exercise intensities is due to the muscle releasing H^+ ions that are transported from the muscle to the bloodstream through the mono-carboxylate transporter owing to the high rates of ATP hydrolysis. However, glycolysis becomes limited by the fact that it cannot shuttle the NADH + H^+ (from Step 6 of glycolysis) thus, NADH + H^+ forms lactic acid in the cytoplasm, which begins to accumulate. Secondly, bicarbonate in the blood begins to react. During this energy demand, mitochondrial respiration is unable to keep up with the pace of the oxygen demand. Consequently, the VCO_2 curve increases steeply, while the VO_2 curve is linear.

The result is that the elevated VCO_2 is higher than that of VO_2. At this point, the anaerobic threshold is defined (Fig. 3.8). Beyond the anaerobic threshold, larger glycolysis increases muscle lactic acidosis, buffered mostly by bicarbonate, thus a

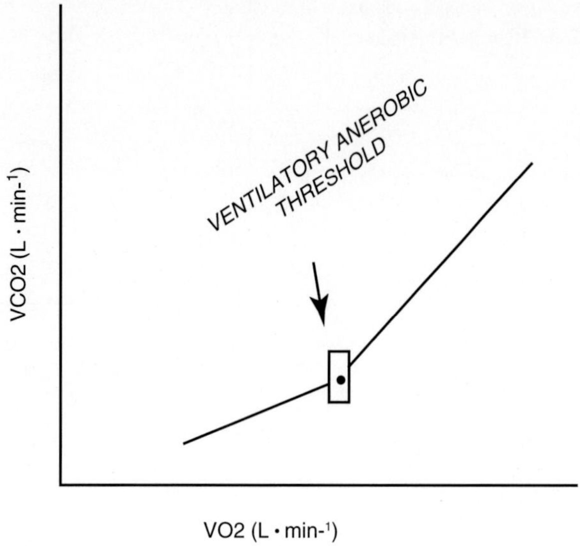

Fig. 3.9 Relationships between VCO_2 and VO_2 which define the anaerobic threshold

higher increase in VCO_2 output relative to VO_2 is a sign that lactate begins to accumulate. Plotting graph from minute ventilation against VO_2 L min^{-1} values can barely estimate the lactate threshold.

The Conconi Test

The third method is the Conconi test, which is a relatively simple way for cyclists to define their anaerobic threshold. The test does not require drawing blood samples to calculate anaerobic threshold.

This means that this test can be performed outdoors by bicycle and any cyclist can perform this test to define his anaerobic threshold (Fig. 3.10). Using the HR deflection approach enables to estimate when lactate begins to accumulate. Assessing the validity of the Conconi method, with the pulmonary gas exchange revealed, that the Conconi test is a worthy mean for measuring anaerobic threshold. Yet, for athletes participating in high demanding sports competitions, validity of the HR deflection test is questionable among physiologists regarding its precision for high aerobic sports events.

The Conconi anaerobic threshold is defined from the relationship between workload applied and HR. This is an out of laboratory test performed in the field. The lactate threshold as reflected from the HR deflection test, for most young males, values are between 160 and 175 beats min^{-1}; in females, it beings slightly higher at about 170–180 beats min^{-1}. Of all measurements obtained during testing, the workload and HR at which the lactate threshold is obtained are probably the most important values to use when planning training.

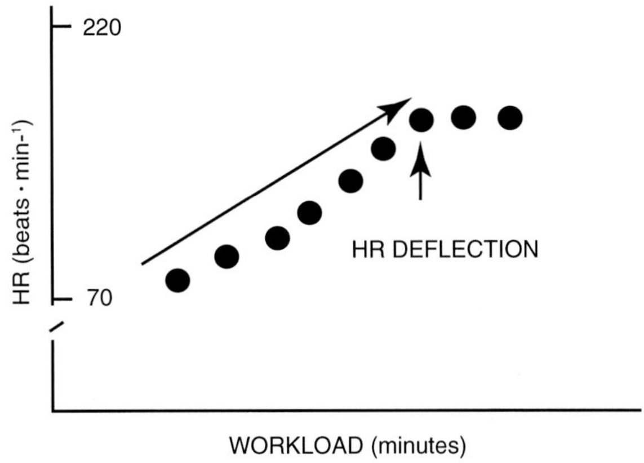

Fig. 3.10 Heart rate (HR) deflection point during the Conconi test

Training Means to Improve Aerobic Capacity

The cardiopulmonary system is the prime limiting factor of VO_{2max} by the ability to deliver oxygen to the contracting muscles, and not skeletal muscle $(a\text{-}v)O_2$. Thus, training at the anaerobic threshold intensity or directly above enhances both the aerobic capacity (VO_{2max}) and anaerobic threshold level. In order to have a higher aerobic performance, it is necessary to train at high workloads above the anaerobic threshold. Training regimes that are a mixture of high volume, maximal steady-state in high volume, and interval workouts have the most marked result on lactate threshold enhancement [127]. "Maximal steady-state" is a substitute expression for training steady-state at the lactate threshold. During vigorous all-out exercise bout in very well-trained athletes, lasting 30 up to 120 s, lactate values may reach 15–25 mM, recorded 3–8 min postexercise. Measurements during progressive, incremental exercise, lactate values rise increasingly at first, and then, when exercise turns out to be more intense, lactate values increase sharply. This training approach at high lactate levels improves significantly anaerobic threshold toward VO_{2max}. World-class athletes reach their anaerobic threshold around 90–92% of their VO_{2max}. Endurance training above the anaerobic threshold result in an increase in mitochondrial size and enzyme activities [128], which enhances performance by reducing lactate production due to the improved fat oxidation at any given intensity [129].

The anaerobic threshold is influenced by training and genetics [130]. Endurance performance is a multifaceted phenotype influenced by a few factors such as training intensities, mode of exercise, and nutrition. In addition, genetics determined by the genotype correlates highly with endurance capacity. Interval training and long-lasting aerobic training are suggested as means to increase anaerobic threshold and to improve performance. One has to consider the workload applied at which the anaerobic threshold is achieved. Interval training (Fartlek training) is a type of training that includes a sequence of high-intensity bouts classically, at or near to anaerobic exercise, combined with rest periods [131]. Fartlek training and interval training are similar, the main difference being the structure of the exercise. Interval training can take the form of many different types of exercise and should closely duplicate the movements found in the sport of interest.

When the training purpose is to increase the lactate threshold, it will depend on the kind of competition that the athlete has to participate. For example, endurance-trained individuals tend to have lower blood lactate values at any given aerobic workload, or intensity, compared to untrained individuals; this difference is due in part to the utilization of aerobic metabolism pathway and to the better removal ability of lactate from the blood circulation, in the trained individuals. In theory, an individual could exercise at any intensity up to their anaerobic threshold, for an indefinite period. Still, when exercising above the anaerobic threshold, as exercise intensity approaches VO_{2max}, a sharp increase in blood lactate accumulation occurs; subsequently, fatigue comes about. Having a high anaerobic threshold is more important for long-distance runners (10,000 m or marathon) and having a high VO_{2max} is probably more important for mid-distance runners (e.g. 1500–5000 m). A long-distance runner runs at velocity corresponding to the lactate threshold which correlated to the runner's performance [132]. By running at the lactate threshold, muscular performance becomes more effectual by lactate removal, from the muscle and its recycling to glucose [133]. The body adapts to stress overload placed upon it by becoming more efficient. Stressing the body to cause an increase in efficiency, the individual needs to increase the lactate above that which individual can be tolerate. This is attained by exercising at a high intensity for several short periods and recovering periods between hard bouts. The recovering periods allow excess lactate to be clear from the muscle. A proper combination of hard efforts with short recovery, and some medium-hard efforts with longer recovery, will raise both the lactate threshold and the VO_{2max}. Aerobic training is known to increase oxygen delivery to the working muscles and mitochondria size, number, and density, and myoglobin in active muscles. Consequently, it increases muscle's oxygen supply-extraction ability thus, increasing the anaerobic threshold and VO_{2max}. In addition, while training the aerobic energy system, the individual will also be training the phosphagen and glycolytic system. Other ways for the elimination of lactate levels during aerobic exercise are the utilization by the myocardium, which is reached with mitochondria and is an aerobic organ, and recycling lactate through the Cori Cycle that converts lactate to glycogen in the liver.

Anaerobic Power Measurements

The force–velocity relationship characterizes the opposite association between force and velocity. The interface between force and velocity and their impact on exercise choice is important for strength and conditioning athletes [134]. The most commonly used anaerobic power tests that utilize force–velocity tests are stairs, Vertical Jump Test, and the Wingate Anaerobic Test. Analysis of the athlete's progress is an important component in the design of training regimes and in the analysis of exercise training efficiency. It is important for any testing procedure that the conditions remain consistent throughout the assessment session. All tests should be conducted in an environment in which the floor surface and weather conditions are consistent, as these could directly affect the results, particularly for a Vertical Jump Test. Without a consistent environment to test, the results may not be reliable or valid and would therefore have to be disregarded. Two methods of anaerobic tests exist; one measures anaerobic power and includes tests such as cycle ergometer, vertical jump, force-velocity tests, and staircase. The second testing method is the anaerobic capacity measurement by calculating maximal oxygen debt. However, the maximal oxygen debt is not effective. Though it is sufficient to be utilized to measure anaerobic capacity. Anaerobic power reflects the ability of the ATP and CP (ATP-CP) energy pathways to produce energy for muscle contraction. This system depletes quickly and is used for short bursts of intense power output. Sprint or track cyclists, sprint runners, hockey players, and other athletes that use short, high-intensity efforts benefit from training the ATP and CP system. For example, a cyclist, who has a high peak and average power, is a strong sprinter. How long-last the bout, the individual would depend on his anaerobic fatigue delay.

The Wingate Anaerobic Test requires from the individual to pedal a mechanically braked bicycle ergometer for 30 s, at an "all-out" pace. A vertical jump power test may be a more specific test, for ballplayers and the ability to perform this test on the court may make it more appealing than the Wingate Anaerobic Test. However, technical aspects (requirements of a force platform) and administrative concerns (only a single individual can be tested at any time) may also prevent this test from being widely accepted. The Line Drill is a field test that has been proposed as a measure of anaerobic capacity in athletes. It has an advantage for testing basketball players: the test is also commonly used for basketball, conditioning practice, and can be used to test several athletes at the same time.

The Vertical Jump Test

Two laboratory fields are used to assess anaerobic power, vertical jump test and the line drill. The Vertical Jump Test or the Sargent Jump, Vertical Leap Test, is a simple and quick to perform and use, inexpensive and portable device that can measure an athlete's vertical jump height, measuring tape, or marked wall. In addition, the vertical jump test can use the special device, Vertec. This test has many variations and methods to measure lower-limbs explosive power, such as direct and indirect vertical jump. Vertical Jump Test has been shown to correlate with maximal strength and sprint performance and is therefore of interest to many strength and conditioning professionals. It is also possible to use this test to compare pre-training to post-training results for each individual. The Vertical Jump Test requires the individual to displace the slats at the peak of their jump, and therefore, requires at least a single-arm swing. Due to the vertical force vectors, this test is effective and used for team sports. Other method to measure height jump is the mat and timing mat by the hip's displacement. In order to measure precisely, the individual is obligated that when landing back on the mat, legs must be maximally extended.

Vertical jumps measure the lower-limbs explosive power, by jumping to a maximal height; vertical force vectors provide data about the individual's anaerobic capability. This Vertical Jump Test may therefore be appropriate for assessing performance in a variety of athletics such as team sports, weightlifting, American football, and soccer. However, the way that the athlete jumps can affect the results namely. jumps backwards, forwards, sideways, and other considerations such as fatigue, motivation, displacement from the take-off point, single or double-arm swing, placement of the non-dominant hand if a single-arm swing is used and flexing of the hips, knees, or ankles.

The test procedures include the jump as high as the individual can to hit the vanes of the Vertec device and land at the same spot that he started his jump. The individual has three jump trials. The average or the best jump out of three can be considered. Jump height is determined by the difference between the individual's standing reach height and the highest vane displaced. The results of the test are reported in centimeters or inches.

The Line Drill Test

The line drill is a field test common to ballplayers and is a highly intense bout of exercise to measure the anaerobic capacity of the individual. It consists of a continuous 143.4 m sprint with several changes of direction. The line drill is performed on a regulation-size basketball court. Specifically, the subject begins from a standing position, at the baseline and run, at maximal workload to four separate cones placed, at the near free-throw line (5.8 m), half-court line (14.3 m), far free-throw line (22.9 m), and far baseline (28.7 m). As the athlete arrives to each cone, he runs back as fast as he can to the original baseline and proceed as quickly as possible to the next line. The line drill is performed three times (T1, T2, and T3) with a 2-min passive recovery between each sprint.

A fatigue index is developed by dividing the slowest recorded time of the three sprints by the fastest recorded time of the three sprints.

The Wingate Anaerobic Test

Because the effort lasts only 30-s, its easiness and time-efficiency, the Wingate Anaerobic Test is considered the most common test of anaerobic fitness. The Wingate Anaerobic test, was developed to measure anaerobic power and anaerobic capacity. It is mainly used to measure an individual's anaerobic capacity and anaerobic power outputs. The Wingate Anaerobic Test is the most important and effective tool defining anaerobic capacity and anaerobic power. These data have vital aspects for competitive athletes that request short-lasting strenuous bouts. As a result, this specific test can serve as valuation means for sportspersons competing in sports of a comparable nature.

The accessibility of the Wingate Anaerobic Test to most kind of ballplayers has most likely impeded the widespread acceptance and use of this test for most basketball teams. The Wingate Anaerobic Test is a laboratory-based cycle ergometer test, using a Bodyguard or Monark cycle ergometer. In addition, it can also be performed on an arm crank ergometer.

The Wingate Anaerobic Test is designed to measure peak power (PP), mean power (mP), relative peak power (RPP), anaerobic capacity (AC), and percent of fatigue.

Before the tests start, it is important to ensure the following items:

- Cycle ergometer
- Computer connected to the bicycle
- Weighing scales
- Performance recording sheet
- Stopwatch

Following warm-up, two subjects will perform the 30 s all-out Wingate Anaerobic Test, utilizing a weight-adjusted Monark cycle-ergometer (model 864). The subject seats on the ergometer with his feet fastened to the pedals by means of racing-type toe-clips, after seat height was adjusted. The Wingate Anaerobic Test consisted of 30 s supramaximal pedaling against a resistance determined relative to the subject's body mass at 80 g kg^{-1} for the young group, and a lower resistance of 40 g kg^{-1} bodyweight for elderly. This load is used for men and women. However, the original Wingate test load is still using the original test weight of 7.5% of body mass.

Subjects commenced cranking as fast as they could against the ergometer's inertial resistance only. The full, predetermined resistance load is applied within 3–4 s once the inertial resistance had been overcome. Pedal revolution count started at that instant by means of an electromechanical counter and subjects maintained an all-out effort throughout the test. Strong verbal encouragement is given to ensure maximal effort. HR and electrocardiogram are monitored continuously using 12-lead ECG recorder system and oscilloscope. Five-second recordings are obtained at rest and at peak exercise. BP is taken using a standard sphygmomanometer cuff and mercury manometer mounted at eye level, in the sitting position at rest and at peak exercise.

Calculations

Peak power output (P) in L min^{-1} for 30 s was calculated as follows:

$$P = \left[\left(D \cdot \text{kg} \cdot 2\,\text{mLO}_2 \right) + \text{VO}_{2\text{rest}} \right] / 2,$$

where D = distance (rpm = 6m); kg = pedaling resistance; 2 mLO_2 = energy cost per pedal revolution; VO_{2rest} = oxygen uptake at rest.

Fatigue index is calculated as the absolute difference between the highest and the lowest paddling rate divided by the highest paddling rate, expressed as a percent of the highest paddling rate.

Mean power (mP) is determined as the average power throughout the 30 s of testing.

Relative peak power (RPP) is determined simply by dividing peak power by body mass and is expressed as W kg^{-1}.

Anaerobic capacity (AC) is the total amount of work accomplished over a 30-s bout.

Field Test: Vertical Jump Test

Before jumping, two subjects stand on a rubber-coated contact platform; individuals are in a semi-squat position. The platform is connected to a digital timer to record the airlift time sum of all jumps. The timer is triggered by the release of the subject's feet from the platform and stopped on touchdown. Subjects' hands remained on hips throughout all jumps. Each subject performed three jumps. The highest vertical jump achieved is recorded. An anaerobic power jump test is also measured by using a 15-s vertical jump test on the Ergo-jump. Anaerobic power is calculated from the sum of the flight time and the number of jumps executed during the exercise lasting time.

Physical Inactivity

Regular and chronic exercise training primarily prevents, or delays, chronic diseases, and maintains physiological functional capacity. If for any reason, functional capacity homeostasis is disrupted, due to an external stress, it may impose maladaptive response that leads to metabolic dysfunction [135], and a life-threatening situation. Functional capacity is not kept without muscular contraction; it decreases quickly following long-lasting physical inactivity and during aging although more slowly. On the other hand, regular exercise increases significantly functional capacity. Physical inactivity is a major risk factor for causing numerous chronic diseases. An aerobic functional capacity in individuals with work capacity of 4 METs, which is very low work capacity, is a sign for cardiac complication or long-term medical risk [136].

The body adapts to inadequate physical activity, which results in a substantial decrease in total bodywork capacity. A main prognosticator of functional capacity is VO_{2max}, which suggests cardiopulmonary fitness, muscle function, and other physiologic machineries [137]. Physical inactivity intensifies the absolute risk of CAD, stroke, hypertension, and osteoporosis [135]. Physical inactivity impairs insulin sensitivity, which is worsened with aging. Insulin act increases skeletal muscle blood flow glucose removal while reduced daily physical activity weakens glycemic control and insulin sensitivity [138]. In young healthy individuals, the femoral artery becomes vulnerable to acute hyperglycemia following 5 days of physical inactivity, which results in a decreased insulin sensitivity. This response is not present in healthy active individuals [139].

Gender Variances in (a-v)O_2

In general, the exercise-related measurements established for women follow the same general principles as those established for men, except for the quantitative differences caused by differences in body size, body composition, and levels of testosterone. In healthy subjects, it has been found that during aerobic exercise, VO_2 of male exceeds that of female due to gender-related variance in heart size [140]. Mean values for VO_2 in all ages are consistently greater in males than females (Fig. 3.11), whether VO_2 is expressed in absolute terms or relative to body mass.

Factors linked mainly to oxygen-carrying capacity and cardiac size have been shown to contribute significantly to this gender-related difference [89]. The mechanism that facilitates (a-v)O_2 by males is due to the different skeletal muscle fiber recruitment patterns [89]. Males have larger skeletal muscle fibers than females but similar fiber composition, influencing metabolism and the adaptive response to exercise [141]. The mechanism that enables the females to increase their (a-v)O_2 during exercise is the lower velocity of contraction by skeletal muscles [142] . From in vitro studies, it has been observed that at low contraction velocities, the efficiency of slow-twitch fibers is higher than of fast-twitch fibers, and the reverse is observed at high speeds [143]. Cardiopulmonary responses (delivery) and metabolic responses during incremental exercise differ between men, women, adolescents, and elderly despite a relatively similar fitness [144]. During the period of adolescence, many structural, hormonal, biochemical, and physiological changes take place interfere with the oxygen delivery and extraction relationship [145].

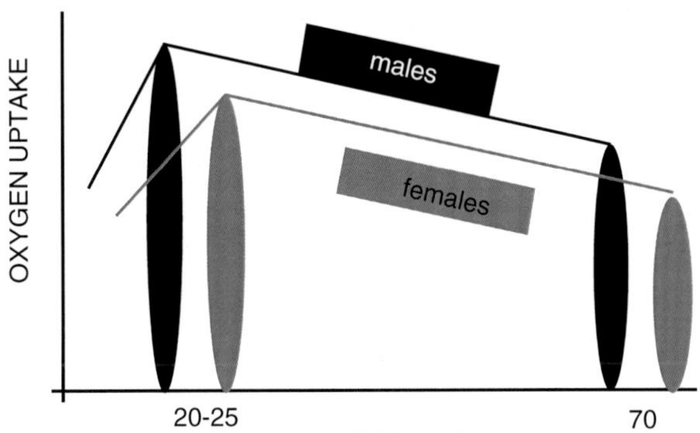

Fig. 3.11 Maximal VO_2 is achieved between the ages of 20–25 years and decreases afterward. Maximal oxygen uptake in children, adolescents, and adults are consistently greater in males than females

Overtraining Syndrome

Effective training must involve high exercise intensities or overload, nonetheless, need to keep adequate rest periods to regain physiological functional capacity. This training approach can have great benefits that ultimately lead to an enhancement in performance [146]. Thus, as part of training routine, professional athletes are stressed to the limits of their physical capabilities. However, insufficient time to recover from the exercise bouts result in lasting-fatigue and loss to perform physical tasks at high intensities, in career-ending damage [147]. Overtraining signs and symptoms include persistent fatigue even after satisfactory rest, raised resting HR, even following night sleep, chronic muscle and/or joint pain, reoccurrence musculoskeletal damages, the feeling of working harder than normal once exercising, recurrent infections, increased BP, sharp weight loss, and the need to eat is lost.

Overtraining syndrome can be seen as a prolonged abnormal adaptation reaction to disproportional exercise training without satisfactory rest. This results in distresses of biological, neurochemical, and hormonal regulation mechanisms, and immunologic related [148] with psychological changes [149].

An additional complication is medial tibial stress syndrome or shin splints, which is the greatest common in athletes, exercising at high intensities such as long-distance runners or military personnel. Medial tibial stress syndrome is painful in the shins. It is the annoyance of muscles, tendons, and other tissue of the lower limbs. Medial tibial stress syndrome is most of the times triggered by an unexpected increase in the workload or rate of exercise. The muscles and tendons over the shin become maddened and swollen. Both the inflammation and the pressure it creates cause discomfort and pain that may be throbbing, aching, or sharp. Without treatment and rest, medial tibial stress syndrome symptoms can get worse over time following activity. Early treatment and rest immediately with the first symptom's appearance may reduce the symptoms. However, without treatment, as medial tibial stress syndrome develops to be more severe, time to be cured will be very long even with rest. Medial tibial stress syndrome complications may include: (a) shin pain at a very specific point, (b) while running, pain becomes more severe, (c) pain when standing or walking due to the body's weight, and (d) any touch over the sensitive area will result in pain.

Treatment is the main part, and then, concentrate on inflammation reduction by medications and ease pain, all of the above, to be used during recovery. Sometime, in seldom cases, surgery may be needed although it is not clear if surgery is an effective treatment for medial tibial stress syndrome. Muscular overtraining may cause muscle fibers tear and if there is not enough time for the muscle to heal, early return to physical activity may tear the fibers again. Consequently, muscles will not reach the desired extent or strength that actually get weaker or smaller.

Exercising every day may be a routine training regime, however, the right way is to alternate each day, between different modes of exercise. As an example of a combined aerobic workout will be one-day cycling exercise, and a day of walk/run intermissions.

High-intensity exercise may cause tissue damage, production of stress hormones, increased lactate levels, and alterations in the function and quantity of various immune cells [150]. To improve sprinter's anaerobic performance, athletes devote large portions of their training program to speed and anaerobic work [151]. Thus, increases in blood lactate concentration, is associated with improved maximal work capacity [152] which in turn may alter lymphocyte subsets [153].

The immune response to anaerobic exercise has received little attention compared to aerobic exercise. An acute bout of aerobic exercise induces mobilization of all lymphocyte subsets to the blood [154]. During intense exercise, neutrophil, and during the recovery period, lymphocyte concentrations increase and fall below values recorded at rest [155].

In athletes subjected to acute strenuous exercise such as sprints, transmural pressure created by muscle contraction reduces blood flow, causing relative hypoxia and as a result leukocytosis [156].

Acute strenuous exercise evoked immediate responses in the number and activity of natural killer lymphocytes [157]. This is explained by the critical plasma epinephrine necessary before lymphocyte subsets rise above resting level after which any additional increases in catecholamine concentration do not entail a further increase in lymphocytes subsets [158]. The close relationship between catecholamine and lactate turnover is demonstrated previously by several investigators [158] in which a critical plasma epinephrine level is necessary before blood lactate rises above resting values. In addition, the breakpoint in plasma catecholamine moves in an identical manner and occurs simultaneously with that of lactate during both running and cycling in trained subjects [31]. Therefore, it has been suggested that the stress-induced changes from rest to exercise in the immune system are controlled by the beta-adrenergic mechanisms [159]. Exercise-induced catecholamine modulates the expression of adhesion molecules on immune cells, resulting in increased neutrophil count concomitant with the recruitment of natural killer cells and T cells into the circulation as reflected by the elevation of the total lymphocyte count [160].

It seems that the effects of aerobic exercise outlines on circulating lymphocyte subsets have been likely mediated by interactions between the central nervous and endocrine systems. The pathways by which the central nervous system, in particular the sympathetic division, may communicate with the periphery including neuroendocrine outflow such as growth hormone, catecholamine, and glucocorticoids via the hypothalamic-pituitary-adrenal axis and the autonomic nervous system through direct nerve fiber connections with cells or the organs of the immune system [161].

Natural killer cells are a distinct but heterogeneous subset of lymphocytes that play a role in inflammatory diseases. It is capable of recognizing and killing virally infected cells, certain tumor cells, and some microorganisms without prior exposure. Natural killer cells are stimulated by exercise contribute to host defense, growth, and repair. Resting natural killer cell number and percentage of lymphocytes appear to be normal in athletes compared with untrained subjects [162]. Changes in natural killer cells are also in relation to the changes in lactate levels. The magnitude of changes in natural killer cell number is the function of exercise intensity [163]. The pattern of peripheral blood mononuclear cells gene expression suggests that the initiation of an immunologic "danger" signal is associated with a sudden change in energy expenditure [164]. Training program for speed and anaerobic work increases blood lactate concentration, in direct association with an increased maximum performance [153]. Changes in plasma lactate concentration [165] correlates with changes in natural killer cell activity and number in peripheral blood mononuclear cells. Lactate activates natural killer cells through lymphotactin to form chemokine XCLia. Human lymphotactin was previously expressed by the lactate bacterium Lactococcus lactis. Biological activity of lymphotactin is confirmed by chemo-attraction of human T-cells by chemotaxis [165].

These suggest that a critical level of lactate should be reached before lymphocyte subsets rise above resting levels. Any additional increase in lactate concentration does not entail a further increase in lymphocyte subsets. This is not the case with the natural killer cells; it appears that during exercise, the recruitment of natural killer cells into circulation from the margined pool depends on the levels of lactate, i.e., higher levels of acetate increase natural killer cell activity and number.

Exertional Rhabdomyolysis

Exertional rhabdomyolysis is a potentially lethal syndrome, with clinical and biochemical indications. Exercise such as military basic training, intense weight lifting, and marathon running in hot and humid environments, inappropriate hydration, insufficient recovery time between training sessions of exercise, and very high intense exercise bouts may cause exertional rhabdomyolysis [166]. Dehydration is one of the main features that influence instantly the clinical significance of acute kidney injury [167]. Following an extreme strenuous exercise, the muscle tissue is crushed with nearby tissue and bone causing exertional rhabdomyolysis, which leads to necrosis of skeletal muscle cells. Intracellular muscle degeneration from rhabdomyolysis extinguishes the myosin and actin filaments in the affected tissue, their components, leaks into the bloodstream and extracellular space, while non-physical causes interfere with muscle cell metabolism [168]. Consequently, serum creatine kinase and myoglobin increase their activity. Yet, its precise occurrence and frequency are not clear. Circulatory creatine kinase and myoglobin values define the severity of the muscle damage. When myoglobin considerably accumulates in the blood and precipitates in the kidney, that lead to an acute renal failure [169].

The results are devastation of Z line architecture, intrusion in excitation-contraction coupling, reduced ability to stain desmin and dystrophin, and elevation of pro-inflammatory cytokines [170]. These cytokines have pro-inflammatory and anti-inflammatory responses followed by muscle damage. In addition, they inhibit inflammatory responses by reducing the actions of pro-inflammatory cytokines. On the other hand, skeletal muscle damage positively induces muscle hypertrophy by stimulating the production of growth factors, involved in myogenic satellite cell regulation.

The range of muscle damage and the increased creatine kinase levels following an extreme strenuous exercise shows extensive inconsistency.

Recurrent rhabdomyolysis may result from intrinsic muscle enzyme deficiencies [171], and thus, rhabdomyolysis cannot be predicted by age, race, body composition, and the individual's level of physical activity or inactivity. It is thus not possible to determine with certainty which factors are responsible for the variability in the phenotypic expression of exertional rhabdomyolysis. Nevertheless, recent data suggest that genetic factors may contribute to the development and progression of exercise-induced rhabdomyolysis [172]. Positive relationship was established between exertional rhabdomyolysis and polymorphisms, suggesting that genetic etiologies should be considered [173].

Muscle recovery relies on satellite cells, behavior which is tightly regulated by the surrounding microenvironment through multiple signals derived from local cells and systemic factors. Among epigenetic mechanisms, histone deacetylation has been shown to affect muscle regeneration. As a matter of fact, pan-histone deacetylase inhibitors improved muscle regeneration [174].

Insulin-like growth factors (IGF) are proteins involved in satellite cell differentiation and proliferation [175]. In animal models, IGFs play a key role in cell growth, and their level increases in response to cell degeneration and regeneration, thereby influencing satellite cell activation [176]. In human males, IGF-II genes polymorphism were significantly related to muscle damage indicators. The main associations of IGF-II SNPs with indicators of muscle damage could be due to the role of IGFs in muscle repair.

Lactic Acidosis

Lactic acidosis is characteristically the consequence of a primary severe or long-lasting medical condition, medication, or carbon monoxide poisoning. The symptoms are usually due to the above-mentioned reasons but may include lung's hyperventilation in response to acidosis (Kushal breathing), widespread paleness, and vomiting. The prediction of lactic acidosis rests on the fundamental source; in some circumstances, it shows a bigger risk of death [177]. Lactic acidosis or metabolic acidosis happens when lactate and protons gather in tissues and fluids in a higher rate and quantities, while their disposal is not sufficient. In addition, it occurs due to reduce in oxygen delivery, consequently, muscles cannot utilize glucose and glycogen. Subsequently, acidity levels in the blood increase sharply above resting production, which reduce pH level to 7.35 and below [178], defined as lactic acidosis. This reduced pH affects negatively two alkaline digestive glands, the liver, and the pancreas function [179], and kidney disposal ability [180]. Daily production of lactate is 1500 mmol which equals to the rate of removal by the liver and kidneys, while blood plasma level is in range of 0.5–1.5 mmol l L^{-1}.

Lactate, the anion that results from the detachment of lactic acid, is the end product of anaerobic glycolysis, where pyruvate is converted to lactate by the enzyme lactate dehydrogenase. Tissue hypoxia is a frequent cause of several critical sickness; consequently, measuring lactate is used to screen tissue oxygenation.

Nature lactate exists in two isoforms: L-lactate and D-lactate [181], owed to the existence of an uneven carbon atom. Both isomers alternate plane-polarized light in contrary directions: L-lactate rotates light clockwise and D-lactate at the opposite direction. In humans, most forms of lactic acidosis are caused by too much L-lactate. Both are results of the enzyme lactate dehydrogenase (LDH), acting on pyruvate. However, each lactate needs it specific LDH; L-LDH for L-lactate and D-LDH for D-lactate. The quantity of D-lactate usually existing in healthy adults is very low 5–20 μmol l L^{-1}, compared to 1000 μmol l L^{-1} for L-lactate. When D-LDH is not present, the body can process D-lactate to pyruvate by the act of the mitochondrial enzyme D-2-hydroxyacid-dehydroganse [182]. Lactic acidosis has many causes and can often be treated. But if left untreated, it may be life-threatening.

Running Injuries

The knee is the greatest commonly injured joint in runners at all distances. Injuries usually occur when the runner pushes himself too firm or overuse, excess body weight, and distance run per week are high-risk factors for injuries; these factors cause medical problems in the runners' knees. The effects of other factors such as shoes, stretching, and biomechanics are not well defined. Runner has numerous diverse causes that regularly occurs when the kneecap is not supported by the alignment. In addition to the fact that running is related to health improvement, it also bears a high risk of injuries. Almost half of the long-lasting runners experience more than one injury per year. Some injuries are a result of overtraining [183]. Factors that affect injuries include patellofemoral pain syndrome, stress fractures, Shin splint (tibia), Achilles tendinopathy, and teared muscle.

Patellofemoral pain syndrome is the most frequent diagnoses and is the likely reason for anterior knee pain, without any structural changes or significant chondral damage [184]. It includes (a) runner's knee, an overuse injury in runners. The exact etiology of it is not well understood [185], and (b) plantar fasciitis, a foot pain, precisely plantar heel pain [186].

Stress fractures are a variety of injuries extending from periostitis, due to inflammation of the periosteum, to a whole stress breakage [187]. Over time, the cartilage on the kneecap can be degraded down, with significant pain, around the kneecap. This may happen predominantly when running stairs, squatting, and following long sitting with the knee bent. Stress fracture is most common in runners and other athletes and typically occur in the lower extremities. This is a small crack in a bone that causes pain and discomfort. It typically affects runners in the shin and feet. It is often due to working too hard before the body gets used to a new activity. Fractures are the result of irregular, recurring loading on regular healthy, bone resulting in local cortical absorption into the circulation of cells or tissue and fracture [188]. Pain becomes worse with exercise that can result in a further injury and recovers with rest.

Shin splint (tibia) is a leg pain triggered by intense exercise [189], the pain is typically more spread out lengthwise in the bone. This happens in the front or inside of the lower leg along the shin bone. Shin splints are common with increased running distances or increasing without enough rest between sessions. The pain is usually more spread out along the bone. Shin splints may also develop in individuals with flat feet. People with flat feet are more likely to develop shin splints. Treatment may include rest, stretching exercises, and slow return to physical activity after some weeks of medication.

Achilles tendinopathy, regularly described, is stiffness that worsens by means of long rest and pain that is intensified by exercise [190]. Formerly called tendinitis, this is inflammation of the Achilles tendon. That is the large tendon that attaches the calf to the back of the heel. It causes pain and stiffness in the area of the tendon, especially during the morning and physical activity. It is usually caused by repetitive stress to the tendon due to increasing running distance. Treatment includes rest, frosting the area and calf stretches.

Teared muscle is associated with dysfunction, pain, and muscular weakness. This is a small tear in the muscle, also called a muscle strain. It is often caused by overstretching a muscle. Treatment include few months of exercise therapy for individuals with muscle tears [191].

Laboratory Assignments

The tool for measuring VO_2 will be the Open Circuit Spirometry that measures simultaneously, online in real time. Fractional percentage of oxygen and CO_2 in the expired air is multiplied by VE volume, in order to calculate VO_2 and volume of CO_2 production. Calculating VO_2 is somehow a simple process: in addition to the Fick equation discussed previously, by subtracting the volume of oxygen expired from the volume of oxygen inspired enables to express VO_2, therefore:

$$VO_2 = VE(FIO_2 - FEO_2)$$

where VO_2 = oxygen uptake; VE = minute ventilation; FEO_2 = fractional of oxygen inspired (%); FIO_2 = fractional of oxygen expired (%).

Equipment

- Metabolic cart.
- Treadmill or stationary bike as an Ergometer
- Trials are control by a stopwatch.

Testing procedures for VO_{2max}

1. Subject is connected to a pulse meter and to the metabolic cart.

2. The subject sits for 5 min in full relaxation to be measured during the last 3 min: RMR, heart rate, resting VO_2, VCO_2, and RQ.
3. Before the actual test, the subject walks (on a treadmill) for 5 min at a pace of 4.5 km h^{-1} (2.3 mph), without elevation after that run for 5 min at a pace of 8–8.5 km h^{-1}. If the ergometer is a bicycle, the subject will pedal at a resistance of 100 W. This is in order to be familiar with the ergometer utilized. Following the running or cycling, the subject sits until HR and VO_2 return to near (±) resting values.
4. The treadmill workload is set to a previously determined 8–10 km h^{-1} with 0% elevation (5.5–6.5 mph). Thereafter, workload is increased by elevation of 1.5% each minute up to 6 min. On the bike ergometer it will be based on the individual's physical ability. As an example, initial workload of 75–100 W, thereafter workload is increased by 25–50 W.
5. The test ends when VO_2 remains at steady state despite an increase in workload.
6. Immediately following the test, the individual is asked to sit for the next 10 min for recovery measurements.

VO_2 test report

Name _____________ Height_____________

Weight _____________ Date_____________

Workload $km \cdot h^{-1}$	Time (min)	HR $(b \cdot min^{-1})$	VO2) $(mLO_2 \cdot kg^{-1} \cdot min^{-1})$	RQ (index)	METs
REST					
RUNNING 8-12	1				
	2				
	3				
	4				
	5				
	6				
ECOVERY	7				
	8				
	9				
	10				
	11				
	13				
	13				
	14				

Laboratory on estimating VO_{2max}:

For WPC170

Required Resources

- Bicycle ergometer
- HR monitor

- Stopwatch
- Rating of Perceived Exertion (RPE) table
- Inflatable cuff and a mercury

Procedures

Select two subjects and record body weight.

(a) Adjust the bicycle ergometer to an appropriate height for the subjects (knee straight on downstroke).
(b) If available, attach an HR monitor to the subject's ear lobe.
(c) Instruct the subjects to begin pedaling at 50 revolutions per minute (RPM), adjust the resistance on the bicycle to 1.0 kg (300-kg m min^{-1}) if sedentary or to 1.5 kg (450-kg m min^{-1}) if active. Immediately start the stopwatch.
(d) Subject will work for 3 min in each stage, at 50 RPM and should not stop pedaling until the test ends. HR should be recorded at the end of each minute. At the beginning of the third minute of each stage, the subject's BP should be measured using the BP cuff and stethoscope. Also, ask for the Rating of Perceived Exertion (RPE). Record the HR, BP, and RPE on the form provided.
(e) At the end of 3 min stage, increase the workload according to the HR response using the chart below:

HR measurement at minute 3, for sedentary subjects

- Start workload with 0.5 kg resistance at 50 RPM (25 W)
- Increase workload to 1 kg resistance at 50 RPM (50 W)
- Increase workload to 1.5 kg resistance at 50 RPM (75 W)
- Increase workload to 2.0 kg resistance at 50 RPM (100 W)
- Increase the workload to 2.5 kg resistance at 50 RPM (125 W)
- Increase workload to 3 kg resistance at 50 RPM (150 W)

HR measurement at minute 3, for active subjects

- Start workload of 1.0 kg resistance at 50 RPM (50 W)
- Increase workload to 1.5 kg resistance at 50 RPM (75 W)
- Increase workload to 2.0 kg resistance at 50 RPM (100 W)
- Increase workload to 2.5 kg resistance at 50 RPM (125 W)
- Increase the workload to 3.0 kg resistance at 50 RPM (150 W)
- Increase workload to 3.5 kg resistance at 50 RPM (175 W)

Again, monitor HR, RPM, BP during each minute, RPE and BP at the start of the sixth minute of this workload. At the end of the test, keep paddling for an additional 3 min to cool down at a low workload or no resistance.

- Present a graph of heart rate versus workload for at least three workloads time to find values of VO_2 at heart rate 170

Questions: What are five different ways to estimate one's maximum oxygen uptake?

(a) What are the two physiologic assumptions that must be made in order to predict one's maximum oxygen uptake from a submaximal stress test?

(b) What are the different factors that can affect one's maximum oxygen uptake?
(c) What is the difference between an absolute and relative measure of oxygen uptake?
(d) Briefly explain the effects of increasing age on one's VO_{2max}.

Report form.

Subject's Name: ___________ Age: ________
Date: ________ Weight: _______ lbs. _______ kg.
Maximum Predicted HR: ________ 85% Predicted HR: _________
Fitness Status: _____ active _____ sedentary
Precautions: ____
____

Rest HR__________ BP________/________ RPE_________
1st Workload______ HR 1__ HR 2__ HR3__ BP_____ RPE _____
2nd Workload______ HR 1__ HR 2__ HR3__ BP_____ RPE _____
3rd Workload______ HR 1__ HR 2__ HR3__ BP_____ RPE _____
Estimated maximum oxygen uptake_______________
Fitness classification____________________________
Comments:______ ____

Laboratory on anaerobic threshold

During this laboratory, two subjects will be tested.
Pretest instructions for subjects include:

(a) Subjects should minimize activity at the day of the test.
(b) Subjects must fast (including all food and beverages, except water) for 10 h prior to the test.

Test procedures:

(a) Obtain subject's data and environmental conditions (temperature, barometric pressure, and humidity). Record it on the laboratory sheet.
(b) Secure the HR monitor and prepare receiver to record HR.
(c) Calibrate the metabolic cart.
(d) Put subject's personal data into computer.
(e) Secure the mouthpiece and headgear.
(f) Instruct the subject to stand quietly on the treadmill for 3 min. Record pre-exercise metabolic data.
(g) At the end of the 3 min, record resting HR, VO_2, and VCO_2.
(h) Begin exercise. The subject will walk on the treadmill at a speed of 4.6 km h^{-1} (3.5 mph) and a grade of 0%. Thereafter, elevation will increase by 2% each 3 min.
(i) Record lactate values, HR, VO_2, and VCO_2 at the end of each 2 min.

The Conconi test

It includes the following steps:

- Ten–fifteen minutes of warm-up at a comfortable pace, avoid intensive activity that will cause fatigue.

- Start by pedaling 80–100 rpm on a stationary bicycle and 90–110 rpm for road race. Write pace and speed.
- Each minute increase speed by 1 mill h^{-1} or 1.6 km h^{-1}.
- Write down HR at rest and at each minute into exercise, before raising speed.
- Keep increasing speed until subject, is no longer able to continue the test.
- With the test conclusion, cool down for 10–15 min or more to decrease HR to resting value.

The anaerobic threshold is defined at the HR deflection point.

Data Analysis:

(a) Measure blood lactate levels at rest and each 2 min throughout the test.
(b) Calculate the lactate anaerobic threshold at rest and during exercise. Use the final minute of rest for resting calculations.
(c) Enter the descriptive data.
(d) Based on data from laboratory 3, calculate ventilatory gas exchange anaerobic threshold.
(e) Create the following two graphs, one from blood lactate collected from ventilatory gas exchange anaerobic threshold
(f) Variables (data from laboratory 3) for two: trained and untrained subjects.
(g) Be sure graphs include a title and descriptive caption.
(h) Attach data collection tables and sample calculations to the end of your laboratory report.

Laboratory report instructions

(a) You should write up a report with introduction, methods, results, and discussion including the information requested in the data analysis section above.
(b) What were the differences between the three testing methods for the determination of the anaerobic threshold? Explain.
(c) How did lactate levels change from rest to exercise?
(d) Discuss the physiological factors that may explain inter-subject differences in the anaerobic threshold.
(e) Be sure to reference material properly if you used it to obtain answers to laboratory report questions.

Collection Sheet

Note: You must calculate RER.
Name ___________ Height _______ Weight _______
Date_______

Time (min)	VO_2 L min^{-1})	VCO_2 (L min^{-1})	Lactate (mmol L^{-1})	HR (beats min^{-1})
Rest				
3				
Exercise				
1–2				
3–4				

Time (min)	VO_2 L min^{-1})	VCO_2 (L min^{-1})	Lactate (mmol L^{-1})	HR (beats min^{-1})
5–6				
7–8				
9–10				
11–12				
13–14				

Laboratory on the Wingate Anaerobic Test

Calculate the Wingate Anaerobic Test results, for two subjects, female (F) or male (m): _____

1. F or M body weight ___ (kg) age ___ (years) resistance ___ (kg)
2. F or m body weight ___ (kg) age ___ (years) resistance ___ (kg)

Time (s)	Revolutions	Watts	kpm min^{-1}	kcal min^{-1}
0–5				
5–10				
10–15				
15–20				
20–25				
25–30				

(a) Plot test's results.

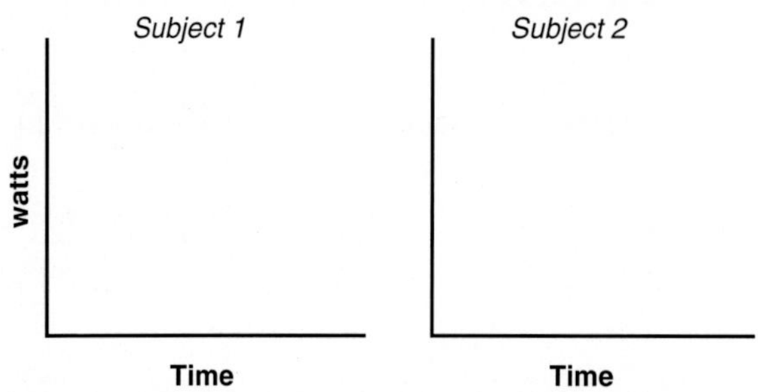

(b) Subject's peak power?

Subject 1. ______Watts ______Watts kg-1______kcal min^{-1}
Subject 2. ______Watts ______Watts kg-1______kcal min^{-1}

(c) Subject's mean power?

Subject 1. ______Watts ______Watts kg-1______kcal min−1
Subject 2. ______Watts ______Watts kg^{-1}______kcal min^{-1}

(d) Percentage power decline (FI) from peak power for 1______ 2______ (Subtract minimum observed power in 5 s period from the maximum observed 5 s power output; that is the power drop. Then calculate the percentage drop)

Vertical jump anaerobic test

Subject 1 body wt. ________kg _________lbs.
Subject 2 body wt. ________kg _________lbs.
Sum of riser heights ________cm _________inches
Subject 1
Times trial 1_______s Trial 2________s Trial 3________s
Subject 2
Times trial 1_______secs Trial 2________s Trial 3________s
Subject 1
Work output_______kgm s^{-1}_________ft-lbs s^{-1}?
Subject 2
Work output_______kgm s^{-1}_________ft-lbs s^{-1}?
v5. Convert these values to Watts. Are they the same as the values for the Wingate Test? Explain why they are the same or why they differ.

References

1. Garcia-Ochoa F, Gomez E, Santos VE, et al. Oxygen uptake rate in microbial processes: an overview. Biochem Eng J. 2010;49:289–307.
2. Pittman RN. Regulation of tissue oxygenation. San Rafael, CA: Morgan and Claypool Life Sciences; 2011.
3. Joyner MJ, Casey DP. Regulation of increased blood flow (hyperemia) to muscles during exercise: a hierarchy of competing physiological needs. Physiol Rev. 2015;95:549–601.
4. Cano I, Roca J, Wagner PD. Effects of lung ventilation-perfusion and muscle metabolism-perfusion heterogeneities on maximal O2 transport and utilization. J Physiol. 2015;593:1841–56.
5. Dorfman KD, Yariv E. Assessing corrections to the Fick-Jacobs equation. J Chem Phys. 2014;141(4):044118. https://doi.org/10.1063/1.4890740.
6. Roy TK, Secomb TW. Theoretical analysis of the determinants of lung oxygen diffusing capacity. J Theor Biol. 2014;351:1–8.
7. Mehra R, Strohl KP. Sleep disorders medicine. In: Chokroverty S, editor. Basic science, technical considerations, and clinical aspects. 3rd ed. Amsterdam: Elsevier; 2009. p. 188–97. https://doi.org/10.1016/B978-0-7506-7584-0.00014-8.
8. Naeije R, Chesler N. Pulmonary circulation at exercise. Compr Physiol. 2012;2:711–41.
9. Taylor BJ, Johnson BD. The pulmonary circulation and exercise responses in the elderly. Semin Respir Crit Care Med. 2010;31:528–38.
10. Hsia CCW. Recruitment of lung diffusing capacity. Chest. 2002;122:1774–83.
11. Baker SE, Wheatley CM, Cassuto NA, et al. Genetic variation of αENaC influences lung diffusion during exercise in humans. Respir Physiol Neurobiol. 2011;179:212–8.
12. Scott CB, Kemp RB. Direct and indirect calorimetry of lactate oxidation: implications for whole-body energy expenditure. J Sports Sci. 2005;23:15–9.
13. Sedlock DA, Lee MG, Flynn MG, et al. Excess postexercise oxygen consumption after aerobic exercise training. Int J Sport Nutr Exerc Metab. 2010;20:336–49.
14. Fairchild TJ, Armstrong AA, Rao A, et al. Glycogen synthesis in muscle fibers during active recovery from intense exercise. Med Sci Sports Exerc. 2003;35:595–602.

15. Melzer K. Carbohydrate and fat utilization during rest and physical activity. Eur J Clin Nutr Metab. 2011;6:e45–52.
16. Borsheim E, Bahr R. Effect of exercise intensity, duration and mode on post-exercise oxygen consumption. Sports Med. 2003;33:1037–60.
17. Dlugosz EM, Chappell MA, Meek TH, et al. Phylogenetic analysis of mammalian maximal oxygen consumption during exercise. J Exp Biol. 2016;216:4712–21.
18. Gonzalez-Alonso J, Calbet JA. Reductions in systemic and skeletal muscle blood flow and oxygen delivery limit maximal aerobic capacity in humans. Circulation. 2003;107:824–30.
19. Bassett DR Jr, Howley ET. Limiting factors for maximum oxygen uptake and determinants of endurance performance. Med Sci Sports Exerc. 2000;32:70–84.
20. Tolle J, Waxman A, Systrom D. Impaired systemic oxygen extraction at maximum exercise in pulmonary hypertension. Med Sci Sports Exec. 2008;40:3–8.
21. Sun N, Youle RJ, Finkel T. The mitochondrial basis of aging. Molecular Cell. 2016;16:654–66.
22. Sheel AW, MacNutt MJ, Querido JS. The pulmonary system during exercise in hypoxia and the cold. Exp Physiol. 2010;95:422–30.
23. Van de Veirea NR, Van Laethemb C, Philippéc J, De Winterd O, et al. VE/VCO2 slope and oxygen uptake efficiency slope in patients with coronary artery disease and intermediate peakVO2. Eur J Cardiovasc Prev Rehabil. 2006;13:916–23.
24. Casaburi R, Patessio A, Ioli F, et al. Reductions in exercise lactic acidosis and ventilation as a result of exercise training in patients with obstructive lung disease. Am Rev Respir Dis. 1991;143:9–18.
25. Koga S, Poole DC, Shiojiri T, et al. Comparison of oxygen uptake kinetics during knee extension and cycle exercise. Am J Physiol Regul Integr Comp Physiol. 2005;288:R212–20.
26. Sagiv M, Sagiv M, Amir R, et al. Left ventricular systolic function during treadmill walking with load carriage in adolescents. J Sports Sci Med. 2006;5:202–7.
27. Sands WA, Apostolopoulos NC, Kavanaugh AA, et al. Recovery-adaptation. Strength Cond J. 2016;38:10–26.
28. Wiewelhove T, Schneider C, Schmidt A, et al. Active recovery after high-intensity interval-training does not attenuate training adaptation. Front Physiol. 2018;9:415. https://doi.org/10.3389/fphys.2018.00415.
29. Heiss R, Lutter C, Freiwald J, et al. Advances in delayed-onset muscle soreness (DoMS) – Part II: Treatment and prevention. Sportverletz Sportschaden. 2019;33:21–9.
30. Midgley AW, Carroll S, Marchant D, et al. Evaluation of true maximal oxygen uptake based on a novel set of standardized criteria. Appl Physiol Nutr Metab. 2009;34:115–23.
31. Millet GP, Vleck VE, Bentley DJ. Physiological differences between cycling and running lessons from triathletes. Sports Med. 2009;39:179–206.
32. O'Neill JO, Young JB, Pothier CE, et al. Peak oxygen consumption as a predictor of death in patients with heart failure receiving beta-blockers. Circulation. 2005;111:2313–8.
33. Woods PR, Bailey KR, Wood CM, et al. Submaximal exercise gas exchange is an important prognostic tool to predict adverse outcomes in heart failure. Eur J Heart Fail. 2011;13:303–10.
34. Robbins M, Francis G, Pashkow FJ, et al. Ventilatory and heart rate responses to exercise: better predictors of heart failure mortality than peak oxygen consumption. Circulation. 1999;100:2411–7.
35. Lang CC, Karlin P, Haythe J, et al. Peak cardiac power output, measured noninvasively, is a powerful predictor of outcome in chronic heart failure. Circ Heart Fail. 2009;2:33–8.
36. Dufour DM, Prud'homme D, Boulay P. Time-of-day variation in cardiovascular response to maximal exercise testing in coronary heart disease patients taking a beta-blocker. Appl Physiol Nutr Metab. 2007;32:664–9.
37. Sagiv M, Amir O, Goldhammer E, Ben-Sira D, et al. Left ventricular contractility in response to upright isometric exercise in heart transplant recipients and healthy men. J Cardiopulm Rehabil Prev. 2008;28:17–23.
38. Gabbett TJ, Johns J, Riemann M. Performance changes following training in junior rugby league players. J Strength Cond Res. 2008;22:910–7.

39. Kemps HM, Schep G, Zonderland ML, et al. Are oxygen uptake kinetics in chronic heart failure limited by oxygen delivery or oxygen utilization? Int J Cardiol. 2010;142:138–44.
40. Sousa A, Rodríguez FA, Machado L, et al. Exercise modality effect on oxygen uptake off-transient kinetics at maximal oxygen uptake intensity. J Expl Physiol. 2015;100:719–29.
41. Widrick JJ, Stelzer JE, Shoepe TC, et al. Functional properties of human muscle fibers after short-term resistance exercise training. Am J Physiol Reg Integrat Comp Physiol. 2002;283:R408–16.
42. Scott JM, Esch BT, Haykowsky MJ, et al. Sex differences in left ventricular function and beta-receptor responsiveness following prolonged strenuous exercise. J Appl Physiol (1985). 2007;102:681–7.
43. Sagiv M, Amir O, Ben-Sira D, et al. Oxygen delivery of men and women at peak wingate test. Braz J Biomotricity. 2007;1:65–75.
44. Sahlin K. Muscle energetics during explosive activities and potential effects of nutrition and training. J Sports Med. 2014;44:S167–73.
45. Girard O, Mendez-Villanueva A, Bishop D. Repeated-sprint ability – part I: factors contributing to fatigue. J Sports Med. 2011;41:673–94.
46. Gregorich ZR, Peng Y, Cai W, et al. Top-down targeted proteomics reveals decrease in myosin regulatory light-chain phosphorylation that contributes to sarcopenic muscle dysfunction. J Prot Res. 2016;15:2706–16.
47. Baumann CW, Kwak D, Liu HM, et al. Age-induced oxidative stress: how does it influence skeletal muscle quantity and quality? J Appl Physiol. 1985. 2016;121:1047–52.
48. Matelot D, Schnell F, Kervio G, et al. Cardiovascular benefits of endurance training in seniors: 40 is not too late to start. Int J Sports Med. 2016;37:625–32.
49. Hill DW, Smith JC. Gender difference in anaerobic capacity: role of aerobic contribution. Br J Sports Med. 1993;27:45–8.
50. Shephard RJ. Health and aerobic fitness. Champaign, IL: Human Kinetics; 1993.
51. Nilwik R, Snijders T, Leenders M, et al. The decline in skeletal muscle mass with aging is mainly attributed to a reduction in type II muscle fiber size. J Exp Gerontol. 2013;48:492–8.
52. Fox J, Garber P, Hoffman M, et al. Morphological characteristics of skeletal muscles in relation to gender aging. J Clin Exp Res. 2003;15:264–9.
53. He Z-H, Bottinelli R, Pellegrino MA, et al. ATP consumption and efficiency of human single muscle fibres with different myosin isoform composition. Biophys J. 2000;79:945–61.
54. Carrick-Ranson G, Hastings JL, Bhella PS, et al. The effect of lifelong exercise dose on cardiovascular function during exercise. J Appl Physiol. 2014;116:736–45.
55. Kang J, Hoffman JR, Chaloupka EC, Ratamess NA, et al. Gender differences in the progression of metabolic responses during incremental exercise. J Sports Med Phys Fitness. 2006;46:71–8.
56. Sagiv M, Metrany R, Fisher N, et al. Comparison of hemodynamic and left ventricular responses to increased after-load in healthy males and females. Int. J. Sports Med. 1991;12:41–5.
57. Sagiv M, Ben-Sira D, Goldhammer E, et al. Left ventricular contractility and function at peak aerobic and anaerobic exercises. Med Sci Sports Exerc. 2000;32:1197–201.
58. Sagiv M, Ben-Sira D, Goldhammer E. Direct vs. indirect blood pressure measurement at peak anaerobic exercise. Int J Sports Med. 1999;20:275–8.
59. Toft I, Lindal S, Bonaa KH, et al. Quantitative measurements of muscle fiber composition in a normal population. Muscle Nerve. 2003;28:101–8.
60. Borg GAV. Psychophysical bases of perceived exertion. Med Sci Sports Exerc. 1982;14:377–81.
61. Coquart JB, Garcin M, Parfitt G, et al. Prediction of maximal or peak oxygen uptake from ratings of perceived exertion. Sports Med. 2014;44:563–78.
62. Magnan RE, Kwan BM, Ciccolo JT, et al. Aerobic capacity testing with inactive individuals: the role of subjective experience. J Phys Act Health. 2013;10:271–9.
63. Bouchard C, Sarzynski MA, Rice TK, et al. Genomic predictors of the maximal O2 uptake response to standardized exercise training programs. J Appl Physiol (1985). 2011;110:1160–70.
64. Lippi G, Longo UG, Maffulli N. Genetics and sports. Br Med Bull. 2010;93:27–47.

65. Herbers E, Kekalainen NJ, Hangas A, et al. Tissue specific differences in mitochondrial DNA maintenance and expression. Mitochondrion. 2019;44:85–92.
66. Hagberg JM, Moore GE, Ferrell RE. Specific genetic markers of endurance performance and VO2max. Exerc Sport Sci Rev. 2001;29:1519.
67. Bouchard C, Rankinen T, Chagnon YC, et al. Genomic scan for maximal oxygen uptake and its response to training in the HERITAGE family study. J Appl Physiol. 2000;88:551–9.
68. Beltz NM, Gibson AL, Janot JM, et al. Graded exercise testing protocols for the determination of VO2max: historical perspectives, progress, and future considerations. J Sports Med. 2016. https://doi.org/10.1155/2016/3968393.
69. Loe H, Nes BM, Wisloff U. Predicting VO2peak from submaximal- and peak exercise models: the HUNT 3 fitness study, Norway. PLoS One. 2016;11(1):e0144873. https://doi.org/10.1371/journal.pone.0144873.
70. Wolsk E, Bakkestrom R, Thomsen JH, et al. The influence of age on hemodynamic parameters during rest and exercise in healthy individuals. JACC Heart Fail. 2017;5:337–46.
71. Schneider DA, Wing AN, Morris NR. Oxygen uptake and heart rate kinetics during heavy exercise: a comparison between arm cranking and leg cycling. Eur J Appl Physiol. 2002;88:100–1061.
72. Sugie M, Harada K, Takahashi T, et al. Relationship between skeletal muscle mass and cardiac function during exercise in community-dwelling older adults. ESC Heart Fail. 2017;4:409–16.
73. WHO, World Health Organization. Physical activity. 2007. http://www.euro.who.int/hepa.
74. Henwood TR, Taaffe DR. Short-term resistance training and the older adult: the effect of varied programmes for the enhancement of muscle strength and functional performance. Clin Physiol Funct Imag. 2006;26:305–13.
75. Chodzko-zajko WJ, Proctor DN, Fiatarone Singh MA, et al. American College of Sports Medicine position stand. Exercise and physical activity for older adults. Med Sci Sports Exerc. 2009;41:1510–30.
76. Betik AC, Hepple RT. Determinants of VO2 max decline with aging: an integrated perspective. Appl Physiol Nutr Metab. 2008;33:130–40.
77. Pomatto LCD, Davies KJA. The role of declining adaptive homeostasis in ageing. J Physiol. 2017;595:7275–309.
78. Kim CH, Wheatley CM, Behnia M, et al. The effect of aging on relationships between lean body mass and VO2max in rowers. PLoS One. 2016;11:e0160275. https://doi.org/10.1371/journal.pone.0160275.
79. van der Zwaard S, de Ruiter CJ, Noordhof DA, et al. Maximal oxygen uptake is proportional to muscle fiber oxidative capacity, from chronic heart failure patients to professional cyclists. J Appl Physiol (1985). 2016;121:636–45.
80. Wilmore JH, Costill DL. Physiology of sport and exercise. 3rd ed. Champaign, IL: Human Kinetics; 2005.
81. Stoggl T, Welde B, Supej M, et al. Impact of incline, sex and level of performance on kinematics during a distance race in classical cross-country skiing. J Sports Sci Med. 2018;17:124–33.
82. Nalcakan GR, Songsorn P, Fitzpatrick BL, et al. Decreasing sprint duration from 20 to 10 s during reduced-exertion high-intensity interval training (REHIT) attenuates the increase in maximal aerobic capacity but has no effect on affective and perceptual responses. Appl Physiol Nutr Metab. 2018;43:338–44.
83. Kenney WL, Wilmore JH, Costill DL. Physiology of sport and exercise. 5th ed. Champaign, IL: Human Kinetics; 2012.
84. Endo MY, Kobayakawa M, Kinugasa R, et al. Thigh muscle activation distribution and pulmonary VO2 kinetics during moderate, heavy, and very heavy intensity cycling exercise in humans. Am J Physiol Regul Integr Comp Physiol. 2007;293:R812–20.
85. Ben-Sira D, Sagiv M. The effect of gender on left ventricular systolic function at Peak Wingate Anaerobic Test. Eur J Appl Physiol. 1977;75:549–53.
86. Astrand P-O, Rodahl K, Dahl HA, Stromme SB. Textbook of work physiology: physiological bases of exercise. 4th ed. Champaign, IL: Human Kinetics; 2003.

87. Pardaens K, Van Cleemput J, Vanhaecke J, et al. Peak oxygen uptake better predicts outcome than submaximal respiratory data in heart transplant candidates. Circulation. 2000;101:1152–7.
88. Lindqvist P, Morner S, Henein MY. Cardiac mechanisms underlying normal exercise tolerance: gender impact. Eur J Appl Physiol. 2011. [Epub ahead of print].
89. Batterham A, George K, Mullineaux D. Allometric scaling of left ventricular mass by body dimensions in males and females. Med Sci Sports Exerc. 1997;29:181–6.
90. Costill DL. Inside running: basics of sports physiology. Indianapolis: Benchmark Press; 1986.
91. Gormley SE, Swain DP, High R, et al. Effect of intensity of aerobic training on VO2max. Med Sci Sports Exerc. 2008;40:1336–43.
92. Gergley JC. Comparison of two lower-body modes of endurance training on lower-body strength development while concurrently training. J Strength Cond Res. 2009;23:979–87.
93. Kraemer WJ, Patton JF, Gordon SE, et al. Compatibility of high-intensity strength and endurance training on hormonal and skeletal muscle adaptations. J Appl Physiol. 1995;78:976–89.
94. Van Zant RS, Bouillon LE. Strength cycle training: effects on muscular strength and aerobic conditioning. J Strength Cond Res. 2007;21:178–82.
95. Vincent KR, Braith RW, Feldman RA, et al. Improved cardiorespiratory endurance following 6 months of resistance exercise in elderly men and women. Arch Intern Med. 2002;162:673–8.
96. Mujika I, Padilla S. Cardiorespiratory and metabolic characteristics of detraining in humans. Med Sci Sports Exerc. 2001;33:413–21.
97. Eastwood A, Bourdon PC, Snowden KR, et al. Detraining decreases Hbmass of triathletes. Int J Sports Med. 2012;33:253–7.
98. Mujika I, Padilla S. Detraining: loss of training-induced physiological and performance adaptations. Part I: short term insufficient training stimulus. Sports Med. 2000;30:79–87.
99. Storniolo JL, Pavei G, Minetti AE. A “wearable” test for maximum aerobic power: real-time analysis of a 60-m sprint performance and heart rate off-kinetics. Front Physiol. 2017;8:868. https://doi.org/10.3389/fphys.2017.00868.
100. Zhang R, Zhan L, Sun S, et al. Validity of a newly-designed rectilinear stepping ergometer submaximal exercise test to assess cardiorespiratory fitness. J Sports Sci Med. 2017;16:357–64.
101. Mann T, Lamberts RP, Lambert MI. Methods of prescribing relative exercise intensity: physiological and practical considerations. Sports Med. 2013;43:613–25.
102. Colosio AL, Pedrinolla A, Da Lozzo G, et al. Heart rate-index estimates oxygen uptake, energy expenditure and aerobic fitness in rugby players. Sports Sci Med. 2018;17:633–9.
103. Murias JM, Pogliaghi S, Paterson DH. Measurement of a true [formula: see text]O2max during a ramp incremental test Is not confirmed by a verification phase. Front Physiol. 2018;9:143. https://doi.org/10.3389/fphys.2018.00143.
104. Batista MB, Possamai Romanzini CL, Castro-Pinero J, et al. Validity of field tests to estimate cardiorespiratory fitness in children and adolescents: a systemic review. Rev Paul Pediatr. 35:222–33. https://doi.org/10.1590/1984-0462/;2017;35;2;00002.
105. Bennett H, Parfitt G, Davison K, et al. Validity of submaximal step tests to estimate maximal oxygen uptake in healthy adults. Sports Med. 2016;46:737–50.
106. Bjorkman F, Ekblom-Bak E, Ekblom O, et al. Validity of the revised Ekblom Bak cycle ergometer test in adults. Eur J Appl Physiol. 2016;116:1627–163.
107. Chiu-Ping Y, Hsien-Cheng H, Yaju C, et al. Test to predict cardiopulmonary fitness in healthy older men. Biomed Res Int. 2018;2018:4863454. https://doi.org/10.1155/2018/4863454.
108. Leone M, Duverge S, Kalinova E, et al. Comparison of bioenergetics of walking during a multistage incremental shuttle walk test and a 6-min walk test in active older adults. Aging Clin Exp Res. 2017;29:239–46.
109. Klusiewicz A, Borokowski L, Sitkowski D, et al. Indirect methods of assessing maximal oxygen uptake in rowers: practical implications for evaluating physical fitness in a training cycle. J Hum Kinet. 2016;50:187–94.
110. Uth N, Sorensen H, Overgaard K, et al. Estimation of VO2max from the ratio between HRmax and HRrest – the heart rate ratio method. Eur J Appl Physiol. 2004;91:111–5.

111. Wahl P, Manunzio C, Vogt F, et al. Accuracy of a modified lactate minimum test and reverse lactate threshold test to determine maximal lactate steady state. J Strength Cond Res. 2017;31:3489–96.
112. Rosset R, Lecoultre V, Egli L, et al. Endurance training with or without glucose-fructose ingestion: effect on lactate metabolism assessed in a randomized clinical trial on sedentary men. Nutrients. 2017;9(4). pii: E411. https://doi.org/10.3390/nu9040411.
113. Hall MM, Rajasekaran S, Thomsen TW, et al. Lactate: friend or foe. PM R. 2016;8:S8–S15.
114. Gmada N, Bouhlel E, Mrizak I, et al. Effect of combined active recovery from supramaximal exercise on blood lactate disappearance in trained and untrained man. Int J Sports Med. 2005;26:874–9.
115. Plowman SA, Smith DL. Anaerobic metabolism during exercise. In: Donatelli, editor. Sports-specific rehabilitation. London: Churchill Livingstone. https://doi.org/10.1016/B978-044306642-9.50006-X.
116. Su F, Li C, Li R, et al. Effects of chemical composition on the in vitro degradation of micelles prepared from poly(D,L-lactide-co-glycolide)-poly(ethylene glycol) block copolymers. Polym Degrad Stab. 2018;158:202–11.
117. de Araujo Cavalcante W, Leitao RC, Gehring TA, et al. Anaerobic fermentation for n-caproic acid production: a review. Process Biochem. 2017;54:106–19.
118. McArdle WD, Katch FI, Katch VL. Exercise physiology: energy, nutrition, and human performance. Baltimore, MD: Williams & Wilkins; 2014.
119. Santos-Concejero J, Granados C, Irazusta J, et al. Time to exhaustion at the onset of blood lactate accumulation in runners with different athletic ability. Rev Bras Med Esporte. 19:297–302. https://doi.org/10.1590/S1517-86922013000400014.
120. Allen DG. Skeletal muscle function: role of ionic changes in fatigue, damage and disease. Clin Exp Pharmacol Physiol. 2004;31:485–93.
121. Gibson AR, Ojiambo R, Konstabel K, et al. Aerobic capacity, activity levels and daily energy expenditure in male and female adolescents of the Kenyan Nandi sub-group. PLoS One. 2013;8(6):e66552. https://doi.org/10.1371/journal.pone.0066552.
122. Radak Z. Fundamentals of endurance training. The physiology of physical training. ScienceDirect. 2018. p. 81–109.
123. Abreu P, Mendes SVD, Leal-Cardoso JH, et al. Anaerobic threshold employed on exercise training prescription and performance assessment for laboratory rodents: a short review. Life Sci. 2016;151:1–6.
124. Goodwin ML, Harris JE, et al. Blood lactate measurements and analysis during exercise: a guide for clinicians. J Diabetes Sci Technol. 2007;1:558–69. https://doi.org/10.1177/193229680700100414.
125. Nikooie R, Gharakhanlo R, Rajabi H, et al. Noninvasive determination of anaerobic threshold by monitoring the %SpO2 changes and respiratory gas exchange. J Strength Cond Res. 2009;23:2107–13.
126. Cabo JV, Martinez-Camblor P, del Valle M. Validity of the modified Conconi Test for determining ventilatory threshold during on-water rowing. Sports Sci Med. 2011;10:616–23.
127. Beneke R, Leithauser RM, Ochentel O. Blood lactate diagnostics in exercise testing and training. Int J Sports Physiol Perform. 2011;6:8–24.
128. Menshikova EV, Ritov VB, Fairfull L, et al. Effects of exercise on mitochondrial content and function in aging human skeletal muscle. J Gerontol A Biol Sci Med Sci. 2006;61:534–40.
129. Sloth M, Sloth D, Overgaard K, et al. Effects of sprint interval training on VO2max and aerobic exercise performance: a systematic review and meta-analysis. Scand J Med Sci Sports. 2013;23:e341–52.
130. Guth LM, Roth SM. Genetic influence on athletic performance. J Gerontol A Biol Sci Med Sci. 2006;61:534–40.
131. MacInnis MJ, Gibala MJ. Physiological adaptations to interval training and the role of exercise intensity. J Physiol. 2016;595:2915–30.

132. Zinner C, Sperlich B, editors. Marathon running: physiology, psychology, nutrition and training aspects. Cham: Springer; 2016.
133. Garcia-Tabar I, Gorostiaga EM. A blood relationship between the overlooked minimum lactate equivalent and maximal lactate steady state in trained runners. Back to the old days? Front Physiol. 2018;9:1034. https://doi.org/10.3389/fphys.2018.01034.
134. Jaafar H, Attiogbe E, Rouis M, et al. Reliability of force-velocity tests in cycling and cranking exercises in men and women. Biomed Res Int. 2015;2015:954780. https://doi.org/10.1155/2015/954780.
135. Booth FW, Lees SJ. Fundamental questions about genes, inactivity, and chronic diseases. Physiol Genomics. 2007;28:146–57.
136. Eagle KA, Berger PB, Calkins H, et al. ACC/AHA guideline update for perioperative cardiovascular evaluation for noncardiac surgery – executive summary a report of the American College of Cardiology/American Heart Association Task Force on Practice Guidelines (Committee to Update the 1996 Guidelines on Perioperative Cardiovascular Evaluation for Noncardiac Surgery). Circulation. 2002;105:1257–67.
137. Kunutsor SK, Kurl S, Khan H, et al. Oxygen uptake at aerobic threshold is inversely associated with fatal cardiovascular and all-cause mortality events. Ann Med. 2017;49:698–709.
138. Reynolds LJ, Credeur DP, Holwerda SW, et al. Acute inactivity impairs glycemic control but not blood flow to glucose ingestion. Med Sci Sports Exerc. 2015;47:1087–94.
139. Credeur DP, Reynolds LJ, Holwerda SW, et al. Influence of physical inactivity on arterial compliance during a glucose challenge. Exp Physiol. 2018;103:483–94.
140. Ben-Sira D, Sagiv M. The effect of gender on left ventricular systolic function at Peak Wingate Anaerobic Test. Eur J Appl Physiol. 1997;75:549–53.
141. Tossavainen M, Nummela A, Paavolainen L, et al. Comparison of two maximal anaerobic cycling tests. Int J Sports Med. 1996;17(Suppl 2):S120–4.
142. Fox J, Garber P, Hoffman M, et al. Morphological characteristics of skeletal muscles in relation to gender. Aging Clin Exp Res. 2003;15:264–9.
143. Gurd BJ, Peters SJ, Heigenhauser GJ, et al. Prior heavy exercise elevates pyruvate dehydrogenase activity and muscle oxygenation and speeds O2 uptake kinetics during moderate exercise in older adults. Am J Physiol Regul Integr Comp Physiol. 2009;297:R877–84.
144. Hoffman JR, Chaloupka EC, Ratamess NA, Weiser PC. Gender differences in the progression of metabolic responses during incremental exercise. J Sports Med Phys Fitness. 2006;46:71–8.
145. Jansen RK, Nassas G. Growth of segment principal movements of interia between four and twenty years. Med Sci Sports Exerc. 1988;20:594–604.
146. Meeusen R, Duclos M, Foster C, et al. Treatment of the overtraining syndrome: joint consensus statement of the European College of Sport Science and the American College of Sports Medicine. Med Sci Sports Exerc. 2013;45:186–205.
147. Carfagno DG, Hendrix JC 3rd. Overtraining syndrome in the athlete: current clinical practice. Curr Sports Med Rep. 2014;13:45–51.
148. Kreher JB, Schwartz JB. Overtraining syndrome: a practical guide. Sports Health. 2012:128–38.
149. Brenner JS; Council on Sports Medicine and Fitness. Sports specialization and intensive training in young athletes. Pediatrics. 2016;138(3). pii:e20162148. https://doi.org/10.1542/peds.2016-2148.
150. Kanda K, Sugama K, Sakuma J, et al. Evaluation of serum leaking enzymes and investigation into new biomarkers for exercise-induced muscle damage. Exerc Immunol Rev. 2014;20:39–54.
151. Friedmann B, Frese F, Menold E, et al. Effects of acute moderate hypoxia on anaerobic capacity in endurance-trained runners. Eur J Appl Physiol. 2007;101:67–73.
152. Minahan C, Chia M, Inbar O. Does power indicate capacity? 30-s Wingate anaerobic test vs. maximal accumulated O2 deficit. Int J Sports Med. 2007;28:836–43.

153. Illi SK, Held U, Frank I, et al. Effect of respiratory muscle training on exercise performance in healthy individuals: a systematic review and meta-analysis. Sports Med. 2012;42:707–24.
154. Walsh NP, Gleeson M, Shephard RJ, et al. Position statement. Part one: Immune function and exercise. Exerc Immunol Rev. 2011;17:6–63.
155. Pedersen BK, Toft AD. Effects of exercise on lymphocytes and cytokines. Br. J. Sports Med. 2000;34:246–51.
156. Kruger K, Frost S, Most E, et al. Exercise affects tissue lymphocyte apoptosis via redox-sensitive and Fas-dependent signaling pathways. Am J Physiol Regul Integr Comp Physiol. 2009;296:R1518–27.
157. Kakanis MW, Peake J, Brenu EW, et al. The open window of susceptibility to infection after acute exercise in healthy young male elite athletes. Exerc Immunol Rev. 2010;16:119–37.
158. Hopker JG, Jobson SA, Pandit JJ. Controversies in the physiological basis of the 'anaerobic threshold' and their implications for clinical cardiopulmonary exercise testing. Anaesthesia. 2011;66:111–23.
159. Kanemi O, Zhang X, Sakamoto Y, et al. Acute stress reduces intraparenchymal lung natural killer cells via beta-adrenergic stimulation. Clin Exp Immunol. 2005;139:25–34.
160. García JJ, Bote E, Hinchado MD, et al. A single session of intense exercise improves the inflammatory response in healthy sedentary women. J Physiol Biochem. 2011;67:87–94.
161. Jonsdottir IH. Exercise immunology: neuroendocrine regulation of NK-cells. In. J Sports Med. 2000;21:S20–3.
162. Watt MJ, Hargreaves M. Effect of epinephrine on glucose disposal during exercise in humans: role of muscle glycogen. Am J Physiol Endocrinol Metab. 2002;283:E578–83.
163. Li TL, Gleeson M. The effects of carbohydrate supplementation during the second of two prolonged cycling bouts on immunoendocrine responses. Eur J Appl Physiol. 2005;95:391–9.
164. Radom-Aizik S, Zaldivar F Jr, Leu SY, et al. Brief bout of exercise alters gene expression in peripheral blood mononuclear cells of early- and late-pubertal males. Pediatr. Res. 2009;65:447–52.
165. Zavala-Flores LM, Villatoro-Hernandez J, Gamez-Escobedo A, et al. Production of biologically active human lymphotactin (XCL1) by Lactococcus lactis. Biotechnol. Lett. 2009;31:215–20.
166. Khan FY. Rhabdomyolysis: a review of the literature. Neth J Med. 2009;67:272–83.
167. Hodgson LE, Walter E, Venn RM, et al. Acute kidney injury associated with endurance events-is it a cause for concern? A systematic review. BMJ Open Sport Exerc Med. 2017;3(1):e000093. https://doi.org/10.1136/bmjsem-2015-000093.
168. Sauret JM, Marinides G, Wang GK. Rhabdomyolysis. Am Fam Physician. 2002;65:907–12.
169. Rojas-Valverde D, Sanchez-Ureña B, Pino-Ortega J, et al. External workload indicators of muscle and kidney mechanical injury in endurance trail running. Int J Environ Res Public Health. 2019;16(20). pii: E3909. https://doi.org/10.3390/ijerph16203909.
170. Chavez LO, Leon M, Einav S, Varon J. Beyond muscle destruction: a systematic review of rhabdomyolysis for clinical practice. Crit Care (London, England). 2016;20:135. https://doi.org/10.1186/s13054-016-1314-5.
171. Bosch X, Poch E, Grau JM. Rhabdomyolysis and a cute kidney injury. N Engl J Med. 2009;361:62–72.
172. Yamin C, Meckel Y, Oliveira J, et al. Genetic aspects of exercise and rhabdomyolysis. Pediatr Endocrinol Rev. 2014;11:400–8.
173. Chatzizisis YS, Misirli G, Hatzitolios AI, et al. The syndrome of rhabdomyolysis: complications and treatment. Eur J Intern Med. 2008;19:568–74.
174. Renzini A, Marroncelli N, Noviello C, et al. HDAC4 regulates skeletal muscle regeneration via soluble factors. Front Physiol. 2018;9:1387. https://doi.org/10.3389/fphys.2018.01387.
175. Bella P, Farini A, Banfi S, et al. Blockade of IGF2R improves muscle regeneration and ameliorates Duchenne muscular dystrophy. EMBO Mol Med. 2020;12(1):e11019. https://doi.org/10.15252/emmm.201911019.

176. Zebrowska A, Waskiewicz Z, Zając A, et al. IGF-1 response to arm exercise with eccentric and concentric muscle contractions in resistance-trained athletes with left ventricular hypertrophy. Int J Sports Med. 2013;34:116–22.
177. Booth FW, Roberts CK, Laye MJ. Lack of exercise is a major cause of chronic diseases. Compr Physiol. 2012;2:1143–211.
178. Gutierrez G, Williams JD. The riddle of hyperlactatemia. Crit Care. 2009;13(4):176. https://doi.org/10.1186/cc7982.
179. Melamed P, Melamed F. Chronic metabolic acidosis destroys pancreas. JOP. 2014;15:552–60.
180. Kraut JA, Madias NE. Adverse effects of the metabolic acidosis of chronic kidney disease. AKCD. https://doi.org/10.1053/j.ackd.2017.06.005.
181. Seheult J, Fitzpatrick G, Boran G. Lactic acidosis: an update. Clin Chem Lab Med. 2017;55:322–33.
182. Kato O, Youn JW, Stansen KC, et al. Quinone-dependent D-lactate dehydrogenase Dld (Cg1027) is essential for growth of Corynebacterium glutamicum on D-lactate. BMC Microbiol. 2010;10:321. https://doi.org/10.1186/1471-2180-10-321.
183. Tschopp M, Brunner F. Diseases and overuse injuries of the lower extremities in long distance runners. (Article in German). Z Rheumatol. 2017;76:443–50.
184. Petersen W, Ellermann A, Gosele-Koppenburg A, et al. Patellofemoral pain syndrome. Knee Surg Sports Traumatol Arthrosc. 2014;22:2264–74.
185. Foch E, Reinbolt JA, Zhang S, et al. Associations between iliotibial band injury status and running biomechanics in women. Gait Posture. 2015;41:706–10.
186. Huffer D, Hing W, Newton R, et al. Strength training for plantar fasciitis and the intrinsic foot musculature: a systematic review. Phys Ther Sport. 2017;24:44–52.
187. Tsakotos GA, Tokis AV, Paganias CG. Tension band plating of an anterior tibial stress fracture nonunion in an elite athlete, initially treated with intramedullary nailing: a case report. J Med Case Rep. 2018;12(1):183. https://doi.org/10.1186/s13256-018-1718-8.
188. Matcuk GR Jr, Mahanty SR, Skalski MR, et al. Stress fractures: pathophysiology, clinical presentation, imaging features, and treatment options. Emerg Radiol. 2016;23:365–75.
189. Saxena A, Fullem B, Gerdesmeyer L. Treatment of medial tibial stress syndrome with radial soundwave therapy in elite athletes: current evidence, report on two cases, and proposed treatment regimen. J Foot Ankle Surg. 2017;56:985–9.
190. Chimenti RL, Cychosz CC, Hall MM, et al. Current concepts review update: insertional Achilles tendinopathy. Foot Ankle Int. 2017;38:1160–9.
191. Christensen BH, Andersen KS, Rasmussen S, et al. Enhanced function and quality of life following 5 months of exercise therapy for patients with irreparable rotator cuff tears – an intervention study. BMC Musculoskelet Disord. 2016;17:252. https://doi.org/10.1186/s12891-016-1116-6.

Chapter 4
Pulmonary Function

Introduction

The pulmonary system contains air passages, pulmonary vessels, the lungs and breathing muscles. The prime structures of the respiratory system are the lungs, which carry out the exchange of gases as one inhales and exhales. The pulmonary system includes airways for air passage and filtration and lungs where gas exchange takes place in the alveolus. The total inside surface of the lungs is very large owing it to the assembly of the alveolus structure. The pulmonary system is composed of two main parts: (a) where air passes through bronchioles without any gas exchange, which includes the trachea, the bronchi and bronchiole and (b) the act of respiration taking place in this part of the system, namely, gas exchange in the alveoli [1]. The trachea is a tube that connects the throat which divides into two bronchi. One bronchus emerges to the left lung, the other bronchus goes to the right lung (Fig. 4.1). Within the lungs, the bronchi are further divided into smaller bronchioles that connect to the pulmonary alveolus. Pulmonary alveoli are tiny air sacs defined by a single-layer membrane surrounded with blood capillaries. The organs of the lower respiratory tract are located in the chest cavity. They are outlined and protected by the ribcage, the sternum and the diaphragm and the muscles between the ribs [2].

One function of the pulmonary system is to deliver oxygen for essential metabolic processes in the body cells and to eliminate CO_2 production from the tissues [3]. Humans are able to extract oxygen from the atmosphere and transport it to their cells where it is utilized for essential metabolic processes. The body cells need a continuous supply of oxygen for the metabolic processes that are necessary to maintain life. The pathway for oxygen from the atmosphere to the muscle contains a series of steps that include mechanical work, pressure gradient and chemical reactions. Previously, the thought was that breathing is the only activity of the pulmonary system. However, the lungs have large activities: adding to breathing the lungs

M. S. Saghiv, M. S. Sagiv, *Basic Exercise Physiology*,
https://doi.org/10.1007/978-3-030-48806-2_4

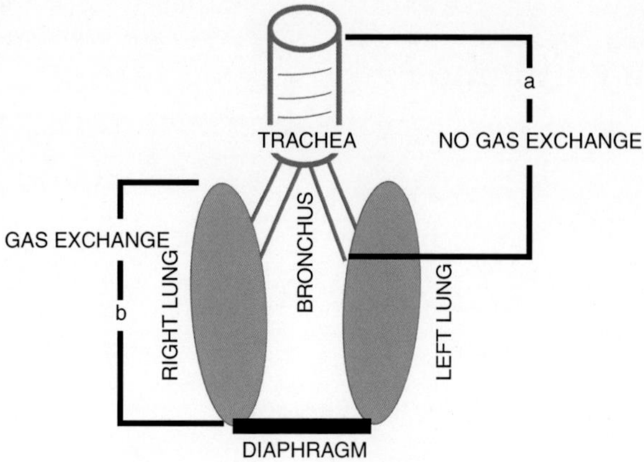

Fig. 4.1 The two main parts: (**a**) air passes through without any gas exchange and (**b**) gases exchange in the alveoli

provide oxygen to the tissues and remove waste metabolites and CO_2. In addition, the respiratory system also helps to the following:

(a) Regulate acid–base balance (pH balance) of the blood.
(b) Filtering small blood clots, removing small air embolisms.
(c) Acts as a buffer for the heart in a certain kind of impact.
(d) Certain membranes within the lungs secrete protective immunoglobulin A against infections.
(e) Mucus traps dust particles and bacteria. Cilia (hairs) move these particles upward and by coughing out of the body.
(f) The lungs have blood reservoirs containing 500–1000 mL of blood, which can be useful, during exercise.
(g) Regulate and eliminate hormones.

At the trachea, the air has been warmed and humidified since the vapor pressure in the trachea is 47 mmHg, the oxygen partial pressure (PO_2) in the trachea when breathing will be 150 as follows:

$$PO_2 = \left[(760 - 47\,\text{mmHg})(21/100\%)\right] = 150\,\text{mmHg}.$$

When oxygen has reached the alveoli, the oxygen partial pressure has fallen to about 100 mmHg. This is because the oxygen partial pressure of the gas in the alveoli (PAO_2) is the balance between two processes: the removal of oxygen by the pulmonary capillaries and its continual supply by alveolar ventilation (VA)

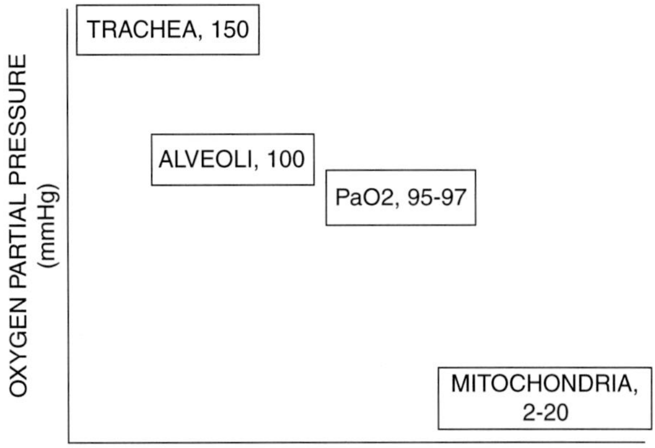

Fig. 4.2 The oxygen cascade

(breathing). Oxygen moves down the pressure or, concentration gradient from a relatively high level in air (760 mmHg), to the levels in the respiratory tract and then alveolar gas (100 mmHg) by means of diffusion, where the oxygen partial pressure reaches the lowest level (4–20 mmHg) in the mitochondria (Fig. 4.2). This decrease in oxygen partial pressure from air to the mitochondrion is known as the oxygen cascade and the size of any one step in the cascade may be increased under pathological circumstances and may result in hypoxia.

Substrate exchange occurs in every living cell. Due to that exchange, oxygen is consumed and CO_2 produced. The oxygen utilization reduces its partial pressure in the cell, and since the oxygen partial pressure is higher in the bloodstream, oxygen enters into the cell by diffusion due to the differences in partial pressures. The opposite is right for CO_2 in which its partial pressure in the cell is higher than that of the blood capillary, and therefore, CO_2 by diffusion leaves the cell and moves into the blood (Fig. 4.3).

The pulmonary system exchanges gases between the atmospheric air and circulating blood, between the blood and the alveoli. The alveolar tubes are accountable for real gas exchange [4]. The primary function of the respiratory system is to permit oxygen of atmospheric air to be used by the body, while it allow us to exhale CO_2 to the atmosphere (Fig. 4.4). An additional function of the respiratory system is to keep appropriate blood pH. High CO_2 levels result in acidosis; on the contrary, when CO_2 levels are low, the pH increases because of a decrease in the concentration of H^+ ions, resulting in alkalosis [5].

The oxygen pathway from the atmosphere to mitochondria is a cascade of resistances in chain [4], each one being overwhelmed by a specific pressure gradient. However, there are several elements in the oxygen transport pathway from mouth to mitochondria that have the individual potential to limit oxygen supply during exercise and, therefore, VO_{2max} [6]. Inhaled oxygen reaches the lungs and then to the alveoli. The layers of cells coating the alveoli and the nearby capillaries are each one cell dense and in close contact between them. These alveoli microscopic air sacs are surrounded by a large amount of blood and with close contact [7]. The

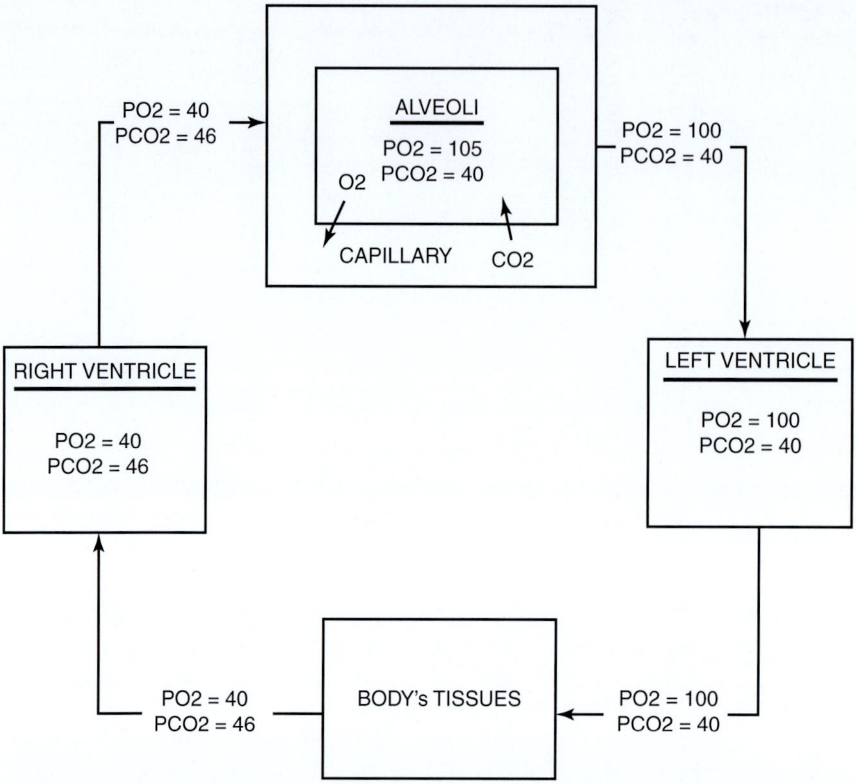

Fig. 4.3 Oxygen and CO_2 partial pressure in the blood following gas exchange in the lungs and tissues

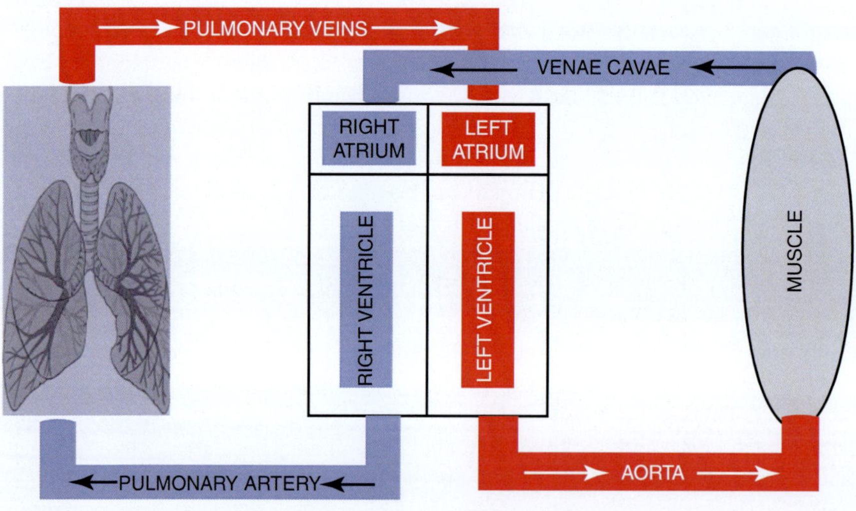

Fig. 4.4 Exchanges of gases between the atmosphere air and circulating blood, between the blood and the alveoli, to the mitochondria and carbon dioxide from the cell to the atmosphere

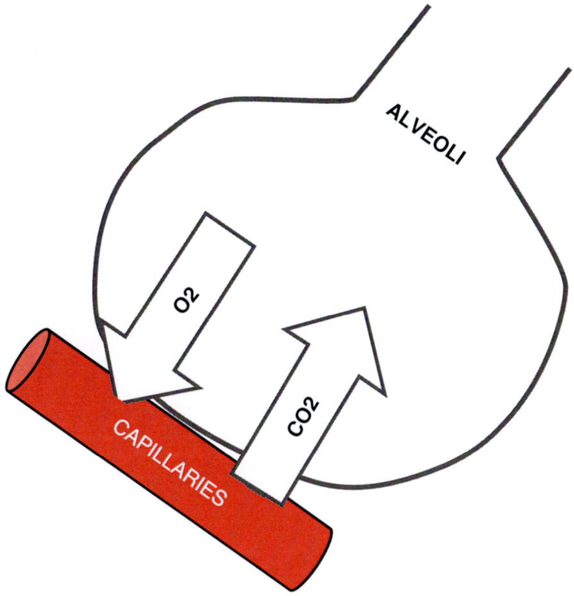

Fig. 4.5 Gas exchange between alveolar and space capillaries

distance that the oxygen and CO_2 molecules have passed in order to diffuse is about 1 micron (1/10,000 cm). Gases such as oxygen diffuse rapidly over the barrier into the capillaries and CO_2 from the blood into the alveoli (Fig. 4.5). Changes of VO_2 could be induced by altering the blood oxygen carrying capacity, peripheral function, or both. When oxygen diffuses from the alveolus to the capillary, it goes on and diffuses into the red blood cell and bound to hemoglobin. Oxygen is very low soluble in liquids, and only 1.5% of the total oxygen is transported by soluble liquids.

The widely held oxygen is carried from the lungs to the tissues by red blood cells, containing hemoglobin, in which oxygen binds to the heme, which is a portion of hemoglobin that contains iron, where oxygen binds chemically to become oxyhemoglobin ($Hb–O_2$). Therefore, one hemoglobin molecule is capable of delivering up to four molecules of oxygen [8]. The first oxygen molecule binding makes a formation change of hemoglobin, which rises the affinity for the other three oxygen molecules.

Pulmonary Gas Exchange

The minute pulmonary ventilation is in action, and the lungs are filled with air, the next phase of pulmonary gas exchange comes about, in the millions of small air sacs called alveoli. The process that gases exchange is passively by diffusion across a very tiny membrane of the alveoli. These microscopic alveoli are with very large blood supply, and consequently, oxygen and the blood are very close. Diffusion rate, which is a transfer factor, measures the transfer of oxygen from air in the alveoli to the red blood cells in capillary blood. This is defined by Fick's low [9], essentially states that the diffusion rate that gas passes through a membrane is in linear relation with surface area (A) and depends on diffusion coefficient (D) and the partial pressure difference of the gas $P1 - P2$ between both sides of the membrane and in reversed relation to the membrane thickness (T). Therefore, the diffusion (V) equation is as follows:

$$V = (P1 - P2)D(A/T)$$

where $V = T$ = membrane thickness, A = diffusion area, D = diffusion coefficient, $P1 - P2$ = partial pressure differences.

Hence, the diffusion rate is increased when the surface area is larger and has bigger partial pressure differences. The lungs are ideally built for gas diffusion with a large surface and a very thin membrane. This lung structure is important during strenuous exercise where VO_2 and CO_2 production is increased 20–25-fold from resting values.

The diffusion process is affected by a number of factors that are

(a) Contact surface, ideal diffusion is linear to the contact area of air with blood and opposite to the surface thickness. In addition, the surface resistance for gas exchange has to be minimal.
(b) The inspired air has to be with water and warm at the same tissue temperature, as temperature increases the average kinetic energy of particles increases.
(c) The inhaled air must arrive at the bottom of the lungs where the blood passes near the alveolus due to gravity.
(d) Density of the diffusing substance. Density is defined as the amount of material that exists within a given volume.
(e) Medium of diffusion. Diffusion also depends upon the medium in which it takes place.
(f) Concentration gradient.

The major functions of the pulmonary system are

Supply oxygen required for energy production, remove CO_2 produced in the cells and control H_2 to preserve acid–base equilibrium.

In the pulmonary system, oxygen is constantly consumed and CO_2 produced by cellular and metabolic reactions, so it is an efficient system for gas exchange between the interior of the tissue and the exterior environment. This happens in the alveoli, which is the inner surface of the lung.

Breathing is a sequence of events that affects the exchange of oxygen and carbon dioxide between the atmosphere and the body cells. Nerve impulses stimulate breathing to transfer air through a series of passages from the atmosphere to the alveoli and out back to the atmosphere. During breathing, inhalation (amount inspired air) happens when atmospheric air enters the lungs by creating negative pressures in the thoracic cavity (passive phase), while exhalation (amount expired air) is the movement of air from the lungs by the pulmonary muscle contraction to the atmosphere (active phase). This is accomplished by the skeletal muscles of inspiration (diaphragm and inter-costal) and expiration (inter-costal and abdominals), which change the volume within the thoracic cavity the greater the change in volume or tidal volume (VT). Catabolism results in exhaling CO_2 following the oxygen entry to the lungs as the final electron acceptor in the Krebs cycle. Following the gas entry to the lungs, external respiration of gases between the lungs and the blood occurs. At the tissue level, the utilization of oxygen and the elimination of CO_2 are the internal respiration. Finally, the cells utilize the oxygen for their specific activities: this is called cellular metabolism or cellular respiration. External and internal respiration activities establish breathing.

Minute Ventilation Regulation During Exercise

At rest, in normal healthy individual, VE is constant and expected. VT and *f*, which are constant , do not need effort to control respiratory functions. Regulation of ventilation is complex and under the influence of several mechanisms. However, under most conditions, the partial pressure of CO_2 (PCO_2), or concentration of CO_2, controls the respiratory rate. The control of the pulmonary function includes chemical and mechanical sensors. The medulla regulates breathing and the respiratory muscle activity (Fig. 4.6).

The peripheral chemoreceptors are branches of the peripheral nervous system into blood circulation that detect chemically, changes in blood oxygen and CO_2 levels [10]. The peripheral chemoreceptors are located in the arterial aortic bodies and the carotid bodies. Information from the peripheral chemoreceptors is conveyed along nerves to the respiratory groups of the respiratory center. Peripheral chemoreceptors are activated by changes in the partial pressure of oxygen and activate to maintaining a normal partial pressure level [11]. Central chemoreceptors are an anatomical collection of neuronal chemoreceptors located just under the ventral surface of the brainstem's medulla oblongata near to the brainstem respiratory centers. Central chemoreceptors are primarily sensitive to changes in blood pH since central chemoreceptors are crucial sensors in maintaining

Fig. 4.6 Scheme of the respiratory parts controlling inspiration

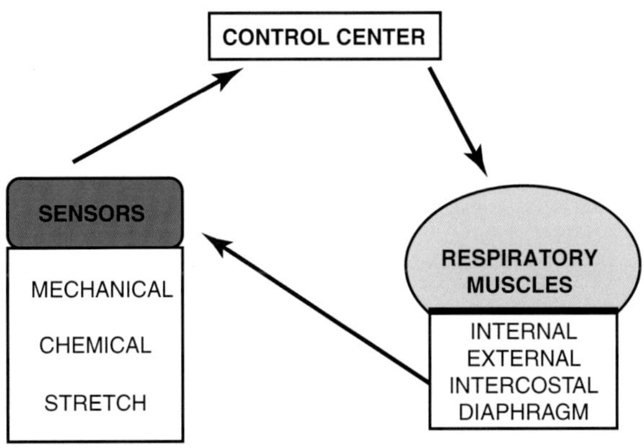

comparatively constant levels of arterial CO_2 [12]. Chemoreceptors sense pH (linked to CO_2) levels in the blood, send signs to the respiratory centers in the brain to alter *f*, to balance acid–base by increasing or reducing the CO_2 elimination. Blood's pH is regulated by the central chemoreceptors located on the medulla oblongata near to the medullar respiratory groups of the respiratory center are primarily sensitive to changes in the blood pH. Information from the peripheral chemoreceptors is carried along nerves to the respiratory groups of the respiratory center. From the center, the diaphragm, internal and external intercostal nerves are recruited to act. There are four respiratory groups, two in the medulla and two in the pons [13]. From the respiratory center, the muscles of respiration, in particular the diaphragm, are activated to cause air to move in and out of the lungs [14]. Although the processes of respiratory muscular contraction are not well understood, the motor cortex is the center for voluntary respiratory neural control and thus initiates contraction by transferring impulses to the spinal cord, and from the spinal cord on to the diaphragm and internal and external intercostal muscles involved in voluntary contraction and relaxation. This path is known as the ascending respiratory pathway. Different parts of the cerebral cortex control different forms of voluntary respiration. Initiation of the voluntary contraction and relaxation of the internal and external intercostal muscles takes place in the superior portion of the primary motor cortex, while the inferior part of the main motor cortex may be involved in controlled exhalation. The VE characterizes the mixture of a neural output and a mechanical response. At rest and during submaximal and maximal exercise intensities, arterial oxygen and CO_2 partial pressure stays stable because of the tight and efficient ventilation control. Mechanical sensors are located in the joints and lungs. The joint sensors increase VE when exercise commence while the mechanical sensors are sensitive to the lung and chest recoil pressure, thus keeping VT within the mechanical lung work efficiency. Figure 4.7, describes the stages of the VE increase during exercise. Before exercise begins

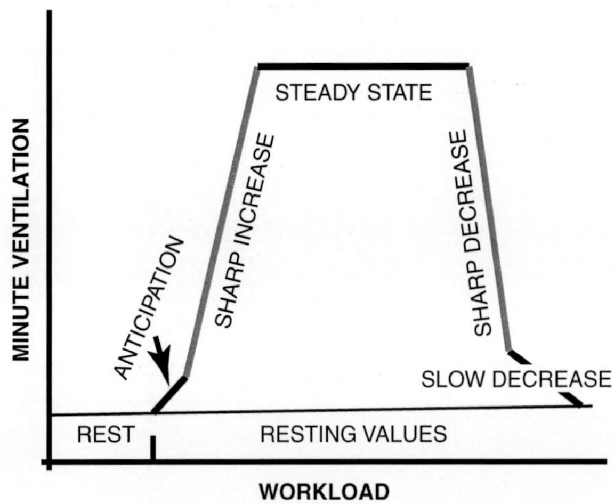

Fig. 4.7 Pattern of the VE increment during steady state aerobic exercise

due to the bout anticipation, the cerebral cortex sends signals to increase VE. When exercise commences, rapid and a very sharp increase in VE occurs within a very short time, seconds.

This probably relates to joints' receptors due to movements, which thereafter, the nervous system stimulates the respiratory system to increase VE [15].

This sharp rise in VE lasts short period (few seconds) which after it comes a slower rise phase. During submaximal aerobic exercise, the slower rise phase reaches steady state. At all-out exercise bout, VE continues to increase until the individual reaches exhaustion. This stage is chemically controlled namely; CO_2 level stimulation produced during exercise. During recovery, at first, there is a rapid decease in VE because there is no receptor stimulation from the muscles and joints [4]. Following the rapid recovery phase , there is a slower VE decrease to resting values. The length of the slow recovery phase for VE to return to resting values depends on few factors; work intensity, individual's fitness level, core body temperature and workout time.

Minute Ventilation at Rest and During Exercise

Minute ventilation (VE) is the volume of gas inhaled or exhaled from lungs in 1 min. It is a vital variable in respiratory medicine owed to its association with blood CO_2 levels. The oxygen path from the atmosphere to the mitochondria can be seen as chain resistance of airflow, which is overwhelmed by a pressure incline (Fig. 4.8).

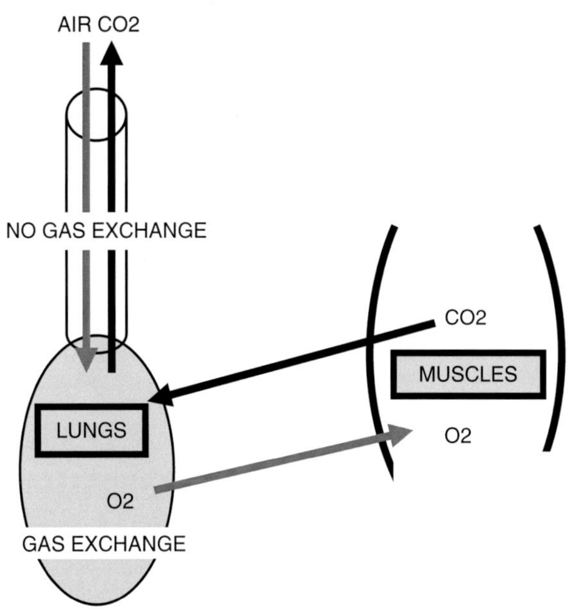

Fig. 4.8 Oxygen pathway from the atmosphere to the lungs and the elimination of CO_2 to the atmosphere minute ventilation

During heavy exercise, a large amount of oxygen is transported through by the cardiopulmonary system to the mitochondria to synthesize ATP. Nevertheless, the body keeps homeostasis by balancing diffusion of oxygen and carbon by regulating minute ventilation (VE). At sea level, VE is sensitive to VCO_2 produced than to the oxygen demand [16]. An individual that hyperventilates (over-breathing) will lower blood CO_2 levels. Minor increases in blood CO_2 partial pressure result in a proportional increase in VE.

At high altitudes (above 2000 m), with hypoxic conditions, when oxygen delivery is lessen and CO_2 production increases at the cell level, the chemoreceptors are stimulated to increase VE (Fig. 4.9). VCO_2 and oxygen changes stimulating breathing are not the only mechanism to increase VE, the voluntary control mechanism also controls breathing. Although hypoxia is frequently an extreme disorder, differences in arterial oxygen concentrations are normal physiologically, such as during all-out strenuous exercise.

Arterial partial pressure of CO_2 ($PaCO_2$) levels generally varies inversely with VE. For example, a person with increased VE due to hyperventilation should demonstrate a lower arterial blood CO_2 level. Hyperventilation is the term for having a VE higher than physiologically need. Hypoventilation and hyperventilation define VE quantity, which is physiologically inappropriate [17]. The healthy human body alters VE in an effort to preserve physiological homeostasis. A normal VE in humans at rest is about 5–8 L min^{-1} and increases with exercise. For example, during light aerobic activities VE may be around 12 L min^{-1}. Paddling a bicycle increases VE by

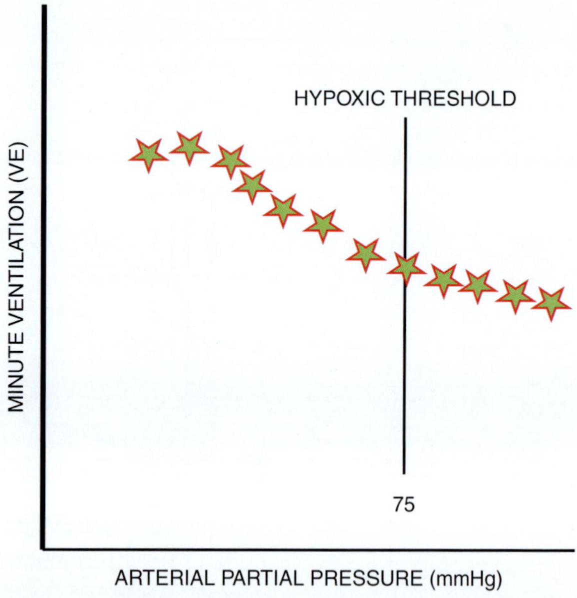

Fig. 4.9 At high altitudes with hypoxic conditions, CO_2 production increases and thus increases VE

a factor of 2–4 depending on the exercise intensity involved. During moderate exercise, VE may be between 40 and 60 L min^{-1} [18]. Note, although VE implies that it is a volume, VE is actually a flow, it represents a volume change over time, expressed in liters. VE is vital for gas exchange between the atmosphere and cell's metabolism. VE values are smaller in women compared to those of men and vary with body mass. Tidal volume (VT) and respiratory frequency (*f*) vary even more than VE. The VE amount can be determined by the VT and the *f*, namely, VE equals to the product of *f* multiplied by VT. At rest, VT values are usually 500 mL and vary between 400 and 600 mL, while *f* ranges from 10 to 25 breaths min^{-1}. The ventilation control mechanism determines the interplay between *f* and VT. Thus, VE can be calculated by multiplying VT by (*f*)

Hence,

$$\mathrm{VE} = \mathrm{VT}\cdot f$$

Therefore,

$$\mathrm{VE} = 500 \cdot 10 = 5000\,\mathrm{mL\,min^{-1}} \text{ or } 5\,\mathrm{L\,min^{-1}}\left(\mathrm{BTPS}\right)$$

Thus, increasing *f* or VT will increase VE (Fig. 4.10). However, only air delivered to the terminal bronchioles and alveoli is available for gas exchange.

This is because of the anatomical dead space (VD) that equals to 150 mL, it is the volume of air that does not take part in the gas exchange either because it (a) remains in the conducting airways and (b) reaches alveoli that are not perfused or poorly perfused [19]. VD volume contains air in the mouth, trachea, and all but the smallest bronchioles. In order to quantify the ratio of physiological dead space to the total tidal volume, the Bohr equation [20] is used, which gives a hint of the amount of unused ventilation.

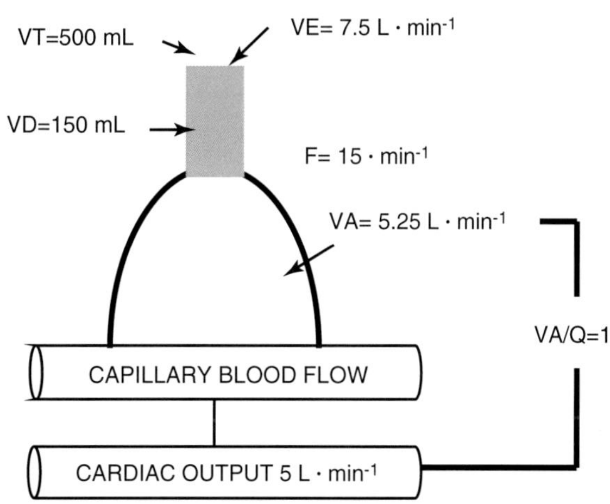

Fig. 4.10 Characteristic values for lungs' volumes and cardiac output

Consequently, minute VA is less than VE. Thus, minute VA equals to the product of *f* and tidal volume minus the VD or

$$VA = (VT - VD) \cdot f$$

or

$$VA = (500\,mL - 150\,mL) \cdot (10\,breaths\,min^{-1}) = 3500\,mL\,min^{-1}$$

At rest, in normal young males, weighting 70 kg its VE equals 6–7.5 L min^{-1} from 12 to 15 breathes min^{-1} and a VT of 0.5 L. At peak, all-out exercise intensities, VO_2, CO_2 formation, total pulmonary and alveolar ventilation (VA) increase by approximately 20-fold. At relatively low exercise intensities , VT and *f* increase proportionally. However, at higher relative intensities, VT reaches a plateau and further increases in VE depend exclusively upon increasing breathing rate. During maximal exercise, VE in normal sedentary subjects may reach 120–150 L min^{-1}. The increase in pulmonary ventilation is attributable to a combination of increases in VT and *f* and closely matches the increase in VO_2 and CO_2 output. At the same time, hemoglobin continues to be fully saturated with oxygen, throughout exercise in people with normal respiratory function.

At rest, in children weighing 30–35 kg, *f* will be around 20 times min^{-1} and VE 4.3 L min^{-1}. The increase in *f* reduces the alveoli ventilation since during each inhalation air lost because of the dead space. Well trained individual, breath slower and deeper compared to untrained individual. The slower breathing rate benefits the athlete by reducing air loss through the physiological dead space. Physiological dead space is a result of VD expansion during exercise. Therefore, minute VA needs to exceed that of VD. In addition, increasing VT volume rises VA more effectively than does increasing respiratory rate since; with each breath, minute VD is increased. During exercise, VE increases because of the rise in VCO_2 production and the oxygen supply amount needed. The increase in VE levels generally is linearly between workload applied, and CO_2 produced by the working muscles (Fig. 4.11). During

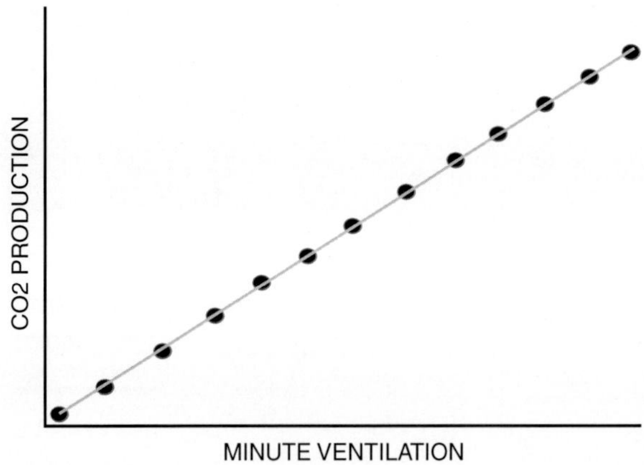

Fig. 4.11 Relationship between minute ventilation and carbon dioxide production (CO_2)

high intensity workloads, VE levels increase significantly due to a larger VCO_2 production and respiratory compensation for metabolic acidosis. Thus, the respiratory system regulates acid–base balance. At rest, for most individual's VE range from 5 to 10 L min^{-1} at maximal exercise maximal minute ventilation (VEmax), values depend on age and gender, ranging from 100 to 170 L min^{-1} (BTPS).

This means that about 20–36-fold increase over resting values. During exercise, in sedentary individuals VT is 3.0–3.5 L, while *f* is 40–55 min^{-1}. Top athletes and their VE at peak exercise may reach VT of 4.0 L and *f* of 60–70 min^{-1} (Fig. 4.12). Elite male athletes can reach VEmax values up to 200 L min^{-1} (BTPS), while elite female athletes reach 180–190 L min^{-1} (Fig. 4.5).

Elite athletes increase VE by increasing both; VT and *f* that closely matches the increase in VO_2 and CO_2 output. Untrained relies by increasing mainly *f* since VT, which by that they lose more air to VD. Thus, the respiratory system does not limit oxygen delivery to the working muscles. Untrained males and females, compared to trained individuals, have lower, VE max and breathing efficiency. Thus, untrained males and females, compared to trained ones, have higher VE values at all relatively given efforts. VE differs not only at any given workload but also during recovery. An oxygen increase in a linear fashion with workload up to about 60% of VO_{2max} (Fig. 4.13).

Beyond this point VE, increases at a higher rate due to the CO_2 control. This shows that during maximal exercise VE is regulated mainly by CO_2 removal than to oxygen demand. This suggests that the increase in VE does not normally limit the capacity to deliver oxygen through the cardiorespiratory system. In addition, during exercise at a given workload trained individuals, compared to untrained, have a lower VE and VCO_2 production. This response is probably the differences in oxygen availability and fuel consumed.

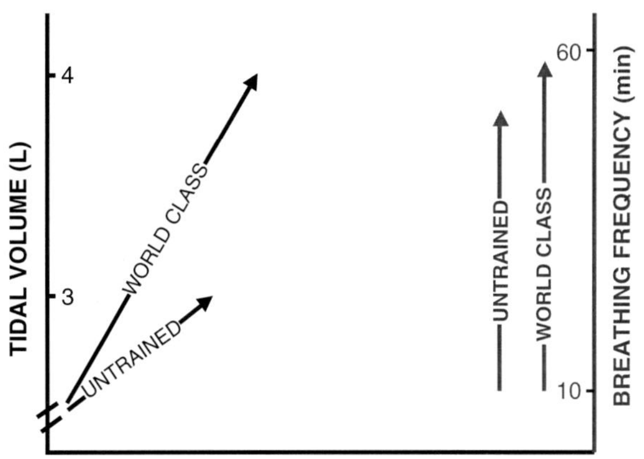

Fig. 4.12 Relationships between breathing frequency and tidal volume, in untrained and world class athlete. Untrained subject is limited in tidal volume and, consequently, increases mainly breathing frequency, while world class athlete increases both: world class athlete and tidal volume

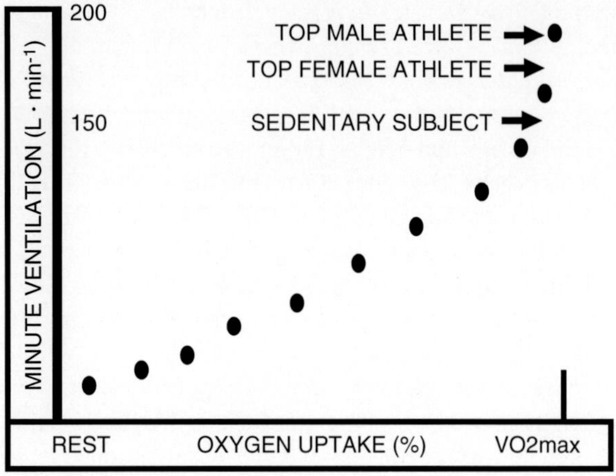

Fig. 4.13 Comparison between untrained, well trained, top female athlete and top male athlete in VE at all-out exercise intensity

Alveolar Ventilation/Cardiac Output Ratio

Alveoli ventilation/perfusion ratio (VA/*Q*) that equals 1, it is the ratio of air quantity reaching the alveoli each minute over, the total blood reaching the alveoli every minute [21], a VA/*Q* mismatch may be a source for respiratory failure. VA/*Q* is used to evaluate the effectiveness if these variables match. Therefore, VA/*Q* ratio may be defined as the ratio of volumetric flow rates. These two variables, VA and *Q*, establish the key factors of the blood oxygen and CO_2 levels.

At rest, if oxygen provided by VA saturates 1 L of capillaries' blood fully, then it will carry about 200 mL of oxygen. Hence, the best ventilation-perfusion ratio would be close to 1.

During maximal aerobic exercise, the VA/*Q* ratio is increased by four- to fivefold since VA reaches 140–150 L min^{-1} and *Q* equals 25 L min^{-1}. This increased ratio is a positive response that secures the red blood cell to be engulfed in the alveoli by high oxygen quantities since the red blood cell, time in the lungs is short due to the significant increase in HR [16].

Maximal Voluntary Ventilation

Maximal voluntary ventilation (MVV) is most widely used for pulmonary function test, it is defined as the maximum minute volume of ventilation that the individual can maintain for 12–15 s. The subject simply breathes in and out of a spirometer without the need for removal of carbon dioxide; although simple, the test is exhausting to perform and is now seldom used. This test clearly demonstrates all the mechanical factors of breathing, becomes abnormal and provides information regarding an increase in airway resistance, reduced compliances or respiratory muscle force [22]. The average fit young male adult at the end of exercise, maximal minute ventilation (VE_{max}) as a fraction of predicted MVV (VE_{max}/MVV) for all subjects is 0.61 ± 0.14 [23], about 300 L min − 1 or MVV is approximately 40% greater than the actual pulmonary VE_{max}. However, normal values depend on body size, age and gender.

Respiratory Muscle Work

In healthy individuals, at sea level, the respiratory system has the capacity to deliver oxygen up to VO_{2max} and to eliminate CO_2. This, however, may change at peak strenuous exercise. But, in the case of respiratory muscle fatigue, the major consequence is an increased sympathetic vasoconstrictor outflow to working skeletal muscle through a respiratory muscle metabolic reflex, thereby reducing working muscle blood flow and increasing the severity of exercise-induced anaerobiosis and working muscle fatigue [24]. The energy expenditure required by the respiratory muscles during exercise is a function of their work rate, cost of breathing and efficiency. During exercise, ventilatory requirements increase further, worsen the potential imbalance between inspiratory muscle load and capacity. Up to 75% of VO_{2max}, respiratory mechanical loading and unloading affect total VO_2 and *Q*. However, it does not change the needed amount of blood flow to working skeletal muscles. Exercise intensities, above 75% of VO_{2max} work of breathing, required during maximal exercise compromises blood flow to working skeletal muscles and reduces exercise performance because of sympathetically mediated vasoconstriction. Cardiovascular disease patients and sedentary subjects have a reduced cardiac function and increased work of breathing due to the low VT and high *f* and thus result in pulmonary insufficiency. This, in turn, increases the demand for oxygen by the diaphragm and expiratory abdominal muscles.

This results in increased skeletal working muscle blood flow demand and the increased demand for oxygen by the diaphragm and expiratory abdominal muscles. This also results in a competition between the diaphragm, expiratory abdominal muscles and skeletal working muscle vascular beds. Thus, peripheral fatigue of working skeletal muscles resulting from high-intensity sustained exercise is, in part, due to the increased demand for oxygen by the diaphragm and expiratory abdominal muscles.

The respiratory muscle work normally expends during maximal exercise [25]. The diaphragm has multiple sources of blood supply and a distinctive high aerobic enzymatic capacity that make it the most fatigue-resistant of all the skeletal muscles. In addition, a unique resistance to vasoconstrictor influences on vascular diameter and therefore a substantial fatigue resistance [26]. In most of the cases, the diaphragm has no reduction in force output or velocity of shortening. This has two significant effects on the cardiovascular system: first, the total blood flow to the skeletal muscles 14–16% of the *Q* is directed to the respiratory muscles (almost 3 L min^{-1}) and second, local reflex vasoconstriction significantly compromises blood flow to working muscles [27].

High level of exercise intensity in conjunction with contracting respiratory muscles is the reason for respiratory muscle fatigue in healthy subjects. Fatigue appears to be due to elevated levels of respiratory muscle work combined with an increased competition for blood flow with skeletal working muscles. These suggest that respiratory muscle fatigue may be involved in limiting exercise tolerance or that other factors including alterations in the sensation of dyspnea or mechanical load may be important. This places a large demand on the myocardium muscle even in well-trained subjects because of the ventilation required by strenuous sustained exercise. This is a significant limitation to exercise performance by reducing blood flow to skeletal working muscle.

Lung's Mechanical Work Efficiency

Gas exchange cost energy because of the active exhaling of the respiratory muscles to affect VA. There are mechanisms which control *f* and VT accordingly, and VA might be attained with a negligible energy cost. At rest, the mechanism of ventilation during inspiration depends on the diaphragm which contracts and moves toward the lung's apex, causing a reduction in the chest cavity pressure that, in turn, increases lung's volume and reduces lung's pressure; thus, ambient air with higher pressure enters the lungs due to pressure gradient. This phase is active and costs energy. Expiration is a passive process due to the diaphragm relaxation, lung's elasticity and greater lung's pressure than ambient pressure; thus, air is exhaled. During aerobic exercise, the breathing work has applied importance as a possible limiting factor during strenuous exercise. During exercise respiratory muscles, work is significantly increased due to the increased physiological dead space. Consequently, during exercise, inspiration and expiration are active and energy costly. Lung's mechanical efficiency can be compared to a balloon inflation, which has three stages (Fig. 4.14):

(a) At the first stage, the inflator blows hard (cost energy) for the first opening of the balloon, with little change in volume.
(b) At the second stage, the blow is easier with a large increase in the balloon's volume.
(c) At the third stage, inflator blows again, hard (cost energy) with a small change in balloon's volume.

Therefore, it will be best to inflate the lungs in stage (b) with low energy expenditure.

The first stage in the lung is pararenal to residual volume (RV), and thus when arriving at the maximal volume of stage (b) (500 mL of VT), instead of trying to reach stage (c), it would be better to increase *f*. During aerobic exercise in a world-class athlete, the VT value can reach 4 L, and therefore, it takes full advantage of stage (b) without crossing over to stage (c). This interplay between VT and *f* is controlled very tight by the mechanical sensors, which reduces energy expenditure.

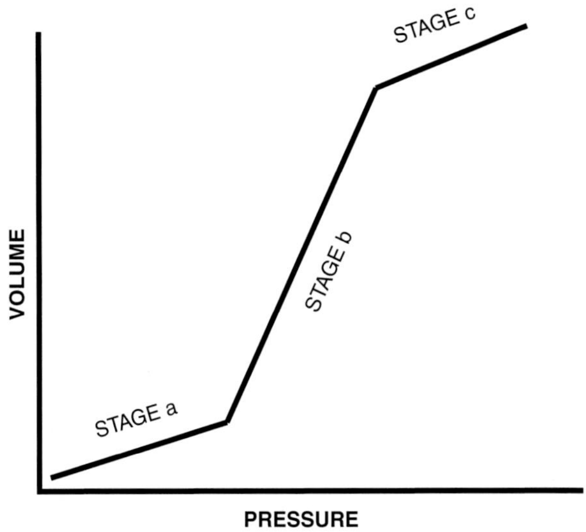

Fig. 4.14 Lung's mechanical work efficiency and pulmonary function

However, as mentioned earlier, during exercise, inhalation and exhalation are active due to respiratory muscles and diaphragm work [28].

Physiological Dead Space

The response to hypoxia is enhanced by hypercarbia, acidosis and exercise. During maximal exercise, highly trained individuals often reach the mechanical limits of the lung and respiratory muscle for producing large VA. This level of ventilation is achieved at a considerable metabolic cost but with a mechanically optimal pattern of breathing and respiratory muscle recruitment and without the sacrifice of significant alveolar hyperventilation [29]. It is important to realize that the potential of gas exchange between the lung and blood is the amount of gas reaching the alveolus. At rest, the VA fraction is 70% of the VT, while 30% is the anatomical dead space (VD), while at peak, exercise VA increases up to 85–90% (10–15% is the physiological dead space).

Physiological dead space is the part of the tidal volume, which does not participate in gas exchange. The difference between anatomical dead space and physiological dead space is alveolar dead space. Physiological dead space can be calculated using Bohr's method [30] as follows:

$$\text{Physiological dead space} = [CO_2]A \cdot (VT - VD) = [CO_2]\text{expired VT}$$

where $(CO_2)A$ = alveolar carbon dioxide concentration, VT = tidal volume, VD = dead apace, (CO_2) expired VT = tidal volume and the expired concentration of CO_2.

In other words, the product of alveolar CO_2 concentration and the amount of inspired air reaching the alveoli equals the expired CO_2 concentration times the VT. It is easy to measure the tidal volume and the expired concentration of CO_2, so all which needed to determine physiological dead space is the alveolar CO_2 concentration. In most cases, it can simply substitute in the CO_2 arterial partial pressure, as the two are generally very similar.

The Oxygen Dissociation Curve

Hemoglobin is the main carrier for conveying oxygen in the blood. Oxygen is also carried dissolved in the blood's plasma, in a low amount. The amounts of hemoglobin and hematocrit are important factors for the oxygen association. Both hemoglobin and hematocrit concentrations increase during aerobic exercise due to a change in plasma volume. During exercise, part of the plasma migrates to other compartments, where there is a need to keep water balance. The movement of cardiovascular water to other compartments is due to the increase in BP. Other important factor during exercise is the percentage of oxygen saturation in the arterial blood. Saturation is the ratio of the amount of oxygen bound to the hemoglobin to the oxygen-carrying capacity of the hemoglobin. The quantity of oxygen bound to the hemoglobin at any time is linked, in large part, to the oxygen partial pressure (PO_2) to which the hemoglobin is exposed. Resting saturation levels are stable up to strenuous effort, suggesting that saturation percentage is not a limiting factor.

The oxygen–hemoglobin dissociation curve, also called the oxyhemoglobin dissociation curve, explain how oxygen is loaded, carried and releases in the arterial blood. Each gram of hemoglobin has the potential to carry 1.34 mLO_2. At low partial pressures in the cells, oxyhemoglobin releases the oxygen by diffusion to the cell form hemoglobin. The oxygen–hemoglobin dissociation curve is a curve that plots the proportion of hemoglobin in its saturated form on the vertical axis against the prevailing oxygen tension on the horizontal axis (). The oxygen dissociation curve (Fig. 4.15) expresses the amount of hemoglobin saturated with oxygen to

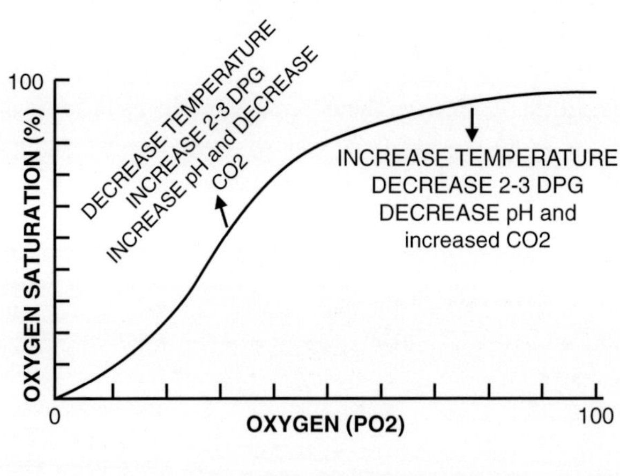

Fig. 4.15 The oxygen–hemoglobin dissociation curve

form oxyhemoglobin once the blood is entirely saturated with oxygen. Factors that affect the standard dissociation curve shift or reform the oxyhemoglobin dissociation curve. The strength with which oxygen binds to hemoglobin is affected by several factors.

These factors shift or reshape the oxyhemoglobin dissociation curve. A rightward shift indicates that the hemoglobin has a decreased affinity for oxygen. This requires a higher PO_2 to achieve the same oxygen saturation; however, oxygen unloading at the tissue is easier, such as during exercise. If the curve shifts to the left it suggests that the hemoglobin has an increased affinity for oxygen so hemoglobin loads oxygen without difficulty, However, at the tissue level, oxygen release is more difficult. In short, left shift: higher O_2 affinity and right shift: lower O_2 affinity.

Four factors help to dissociate oxygen from hemoglobin:

(a) When cell's metabolism is increased, oxygen utilization increases; thus, cell's oxygen partial pressure (PO_2) is reduced. Those conditions increase pressure gradient between arterial blood PO_2 and cell's PO_2. Consequently, it facilitates oxygen dissociation from the arterial blood into the cell.

$$HbO_2 \rightarrow Hb + O_2$$

(b) Concentration of 2–3 diphosphoglycerate (2–3 DPG), a product of metabolize red blood cell and ATP. Increased 2–3 DPG concentration moves the oxygen–hemoglobin dissociation curve to the right. High 2–3 DPG levels can be seen at high altitude (low PO_2), following all-out exercise bout and during interval training.

(c) The Bohr effect states that hemoglobin's oxygen binding affinity is inversely related to both acidity and the concentration of CO_2 [31]. It means that during strenuous bouts, lactate levels rise above the anaerobic threshold, and consequently, the oxygen–hemoglobin dissociation curve moves to the right due to increased PCO_2, which decreases pH [32]. Table 4.1, describes the oxygen saturation percentage under different PO_2.

Table 4.1 PO_2 effects on oxygen saturation

PO_2 (mmHg)	Hb saturation (%)
10	13.5
20	35
30	57
40	75
50	83.5
60	89
70	92.5
80	94.5
90	96.5
100	97.5

Table 4.2 Summarizes factor that affects the oxygen-hemoglobin curve shift direction

Variable	Alteration	Shift's direction
Carbon dioxide Partial pressure	Reduced Increased	Move to the left Moved to the right
2,3-BPG	Reduced Increased	Move to the left Moved to the right
pH	Acidity Alkaline	Move to the left Moved to the right
Temperature	Increased Reduced	Moved to the right Moved to the left

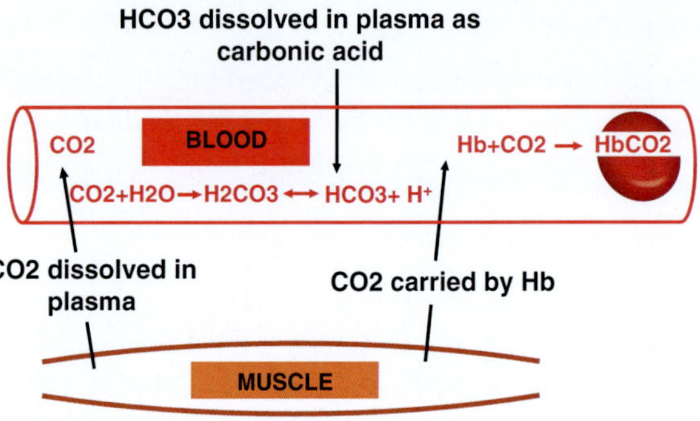

Fig. 4.16 CO_2 is transported in the blood in three ways: $HbCO_2$, HCO_3 and dissolved CO_2 in plasma

When blood temperature is increased, the oxygen–hemoglobin dissociation curve moves also to the right. Table 4.2 summarizes the different effects of these factors on the shift direction.

Carbon dioxide Transportation

Tissue CO_2 production from metabolism leavings the cell to the bloodstream where it become an important factor in the regulation of the pulmonary system [33]. The normal CO_2 value in the arterial blood is 40 mmHg. Pulmonary ventilation is also an important factor that determines CO_2 in the arterial blood. Hyperventilation decreases lung's $PaCO_2$ and thus $PaCO_2$ decreases in the arterial blood, while hypoventilation affects the opposite [34]. CO_2 is transported (Fig. 4.16) in the blood in three ways [35]:

Some of the CO_2 is transported by being dissolved in the plasma. This dissolvement equals about 25–27 mL of CO_2 in one blood liter when arterial blood CO_2 partial pressure is 40 mmHg. CO_2 is 20 times more soluble than oxygen; it obeys Henry's law. In physical chemistry, Henry's law is a gas law stating that the amount of dissolved gas in a liquid is proportional to its partial pressure above the liquid. In other words, the number of molecules in solution is proportional to the partial pressure at the liquid surface, solubility increases as the temperature falls [36].

Q of 5 L min^{-1} delivers 150 mL of dissolved CO_2 to the pulmonary, and only, 25 mL is eliminated. For the high solubility and diffusing capacity, PCO_2 at alveolar–pulmonary capillary blood is nearly 40 mmHg for both.

(a) Buffering, CO_2 associations with water to form carbonic acid, a very weak acid, a reaction accelerated by the enzymatic action of carbonic anhydrase. The carbonic acid then freely dissociates. As blood flows through the tissues, CO_2 diffuses into red blood cells, where it is converted into bicarbonate $HCO_3 + H^+$ and carbonic acid as follows:

$$CO_2 + H_2O \rightarrow H_2CO_3 \leftrightarrow HCO_3 + H^+$$

some CO_2 is transported as carbaminohemoglobin ($HbCO_2$).

(b) About 75% of CO_2 is carried by bounding to hemoglobin and 25% in the plasma. The moderately minor quantity in plasma is due to a deficiency of carbonic anhydrase in plasma subsequently, connection with water is measured; plasma has a minor role in buffering and blend with plasma proteins is deprived.

Difference exists among the amount of the total CO_2 delivered in each way and the quantity exhaled from them. As an instance, 5% of the total CO_2 is delivered in the form of solution; nonetheless, 10% of CO_2 breathe out is 10% driven from solution.

Acid–Base Balance

Acid–base balance is the homeostatic control of the pH of the body's extracellular fluid [37]. The appropriate equilibrium between the acids and bases, in the extracellular fluid has a key role in the body's normal physiological function, and cellular metabolism [37]. Acid–base imbalance happens once a substantial malfunction occurs; blood pH then shifts out of the 7.32–7.42 range. An abnormally low pH in the extracellular fluid is called an acidemia (acidosis), and an abnormally high pH is called an alkalemia (alkalosis). When a strong base is added, the acid present in the buffer neutralizes the hydroxide ions (OH), when a strong acid is added, the base present in the buffer neutralizes the hydronium ions (H^+).

There is a normal pH value in each body compartment: extracellular fluid, plasma and intracellular fluid. Intracellular pH is problematic to quantify due to its variation in unlike kinds of body cells and in variation cell's body. The intracellular compartment is the space within the organism's cells; it is separated from the extracellular compartment by cell membranes. Extracellular fluid means all body fluid external to the cells, containing one-third of total body water [38]. The pH of the intracellular fluid and the extracellular fluid needs to be maintained at a constant level [39]. Correct extracellular pH must be maintained in its very narrow fluctuation limits due to high cell's plasma proteins sensitivity to this variation, which affects their three-dimensional structures. If not within the adequate variety of pH, protein's 3-D structure is dislocated; consequently, functional failure of antibiotic resistance, enzymes and ion channels is noted.

Blood is usually somewhat basic ranging from 7.35 to 7.45. Typically, humans keep the pH of blood close to 7.40. To evaluate pH, the relationship between HCO_3^- (a base) and CO_2 (an acid) ratio in the blood is used. pH of a solvent is a logarithm based on 10 with relationship with H^+. Acid strength depends on the concentration of H^+ and the amount of acid separated in water. Water dissolves to H^+ ion and OH ions as follows:

$$H_2O \rightarrow H^+ + OH$$

The equilibrium of this reaction in room temperature is

$$H^+ \cdot OH = 1014$$

Water produced an equal amount of H^+ and OH; therefore, water's pH is 7.0.

Since acid has higher concentration of H^+ compared to water, it results in lower pH (below 7.0), while base has low H^+ concentration and thus is higher than 7.0. The pH in humans' blood is 7.4 meaning that the blood is base. During intense aerobic exercise, pH decreases due to increased CO_2 and the inability to buffer lactate production quantities, following the utilization of the anaerobic pathway to produce ATP. The acid/base ratio is the balance between HCO_3^- and CO_2, where the kidney regulates HCO_3^- blood's levels and the lungs regulate the CO_2 values. Pulmonary role in regulating acid/base balance is more important than that of the kidney. Three mechanisms [40] help to keep pH at the level of 7.4 in the external fluid compartment [41]:

Chemical buffering of strong acids and bases produced by the body or those entering the body is the first reaction in order to keep pH levels within the normal range. The body has four main systems that buffer chemicals and helping to keep constant pH at 7.4; phosphate buffer system, protein buffer system, and hemoglobin buffering and bicarbonate buffer system, all of which, lessen pH variations out of the normal range. Buffers serve to decrease the degree of the variation and do not correct pH changes. Chemical buffer is a special solution that stops massive changes in pH levels. Every buffer that is made has a certain buffer capacity. Buffer can be defined as the quantity of strong acid or base that must be added to change the pH of 1 L of solution by 1 pH unit.

Phosphorus buffering relies on phosphates found in the blood in two forms: sodium dihydrogen phosphate ($Na_2H_2PO_4^-$), which is a weak acid, and sodium mono-hydrogen phosphate ($Na_2HPO_4^{2-}$), which is a weak base. When $Na_2HPO_4^{2-}$ interacts with a strong acid, the base takes a second hydrogen ion to form the weak acid $Na_2H_2PO_4^-$. When $Na_2HPO_4^{2-}$, the weak acid, comes to interact with a strong base, the weak acid returns to be weak base and produces water. It is mainly intracellular and is an important process for red blood cells and other cells, especially in kidney's tubules cells that exert H^+ ions. The phosphate buffer system operates in the internal fluids of all cells. The concentration of phosphate is low in the extracellular fluid, but the phosphate buffer system is an important urinary buffer. It's made up of a weak base. The pH of a phosphate buffer is usually maintained at a physiological pH of 7.4 [37].

Protein buffering is active mainly in the cell and plasma. Plasma proteins play a limited role in extracellular buffering, whereas intracellular proteins play an important role in the total buffer response of the body. The buffer effect of proteins is the result of their dissociable side groups. Hemoglobin is responsible for more than 80% of the non-bicarbonate buffering capacity of whole blood, whereas plasma proteins contribute 20%. Of the plasma proteins, albumin is much more important than are the globulins [42].

Hemoglobin buffering of red blood cells. The buffering action of hemoglobin is to catch the extra H^+ and CO_2. If the quantities of H^+ and CO_2 are above the hemoglobin capacity, carbonic acid equilibrium is affected and thus lowers blood pH, initiating acidosis [43].

The carbonic acid–bicarbonate buffer system, the largest system in the body, has a very significant part in preserving blood pH balance. The carbonic acid–bicarbonate buffer system converts strong bases to a weak base (bicarbonate ion) and strong acids to a weak acid (carbonic acid). The normal ratio between the HCO_3^- and H_2CO_3 in the blood is 20:1.

The carbonic acid–bicarbonate buffer system acts in the external fluid compartment and is unique in the sense that it results in H_2O and CO_2. A bicarbonate ion, which is a weak base, and carbonic acid, a weak acid, keep chemical equilibrium between them to stand any sudden changes in blood pH levels. For instance, when blood pH becomes more alkaline, the equilibrium shifts to the right. Consequently, a right shift forces more carbonic acid to detach; thus, pH is reduced. The opposite happens, when the blood pH becomes more acidic the equilibrium shifts to the left. A left shift justifies more bicarbonate to bind with protons [44]. Carbonic acid (H_2CO_3) detaches in a reversible reaction and releases bicarbonate ions (HCO_3^-) and protons (H^+) as follows:

If pH is increased, then H^+ proton donor

$$H_2CO_3 \rightarrow\rightarrow HCO_3^- + H^+$$

If pH is decrease, then H^+ proton acceptor

$$HCO_3^- + H^+ \rightarrow\rightarrow H_2CO_3$$

Adding both equations shows the reversibly of each reaction

$$CO_2 + H_2O \leftrightarrow H_2CO_3 \leftrightarrow HCO_3^- + H^+$$

Henderson–Hasselbalch Equation

An adapted form of the Henderson–Hasselbalch equation is utilized to express buffer calculations associated with the blood pH [45]. The Henderson–Hasselbalch calculation is valuable for approximating the pH of a buffer solution and to discover the equilibrium pH in an acid–base reaction. In addition, the equation can determine the quantity of acid, joined together with base that is needed to make a buffer solution of certain pH. The pH of a buffer solution hangs exclusively on the relation of the molar concentrations of the frail acid to the weak base. High concentration of the frail acid in the solution decrease the solution's pH. Similarly, if the weak base dominates the higher the consequential pH [45]. This opinion is for the pH regulation of the extracellular fluids. Molar ratio between weak acid to frail base is 1:20; therefore, pH will be 7.4, and the opposite, when the pH of the extracellular fluids is 7.4, then the ratio of carbonic acid to bicarbonate ions in that fluid is 1:20 [46]. The relationship is defined precisely by the Henderson–Hasselbalch calculated pH is as follows:

$$\mathrm{PH} = \mathrm{pKa}\,\mathrm{H_2CO_3} + \log_{10}\left(\mathrm{HCO_3}^- / \mathrm{H_2CO_3}\right)$$

where pH is the negative logarithm of molar concentration of hydrogen ions in the extracellular fluids. It indicates the acidity in the extracellular fluids in an inverse manner: the lower the pH, the greater the acidity of the solution; pKa H_2CO_3 is the cologarithm of the acid separation constant of carbonic acid. It is equal to 6.1; HCO_3^- is the molar concentration of bicarbonate in the blood plasma; H_2CO_3 is the molar concentration of carbonic acid in the extracellular fluids.

The normal value of H_2CO_3/HCO_3^- is 20, and pKa equals to 6.1, therefore,

$$\mathrm{pH} = 6.1 + \log \mathrm{HCO_3}^- / \mathrm{H_2CO_3}$$

The amount of H_2CO_3 in a pH of 6.1 equals to the amount of HCO_3^- under normal conditions; HCO_3^- value in the external fluid compartment is 27 mEq L^{-1} and H_2CO_3 is 1.35 mEq L^{-1}; thus, pH can be calculated as follows:

$$\mathrm{pH} = 6.1 + \log\left(27 / 1.35\right) = 20$$

This equation can be changed to

$$\mathrm{pH} = 6.1 + \log 20$$

Since log 20 equals 1.3, then

$$\mathrm{pH} = 6.1 + 1.3 = 7.4$$

Since H_2CO_3 (mEq L^{-1}) and the dissolved CO_2 (mmHg) are in a linear relationship with CO_2; therefore, it can be said that

$$\mathrm{pH} = 6.1 + \log\left(\mathrm{HCO_3}^- / \mathrm{CO_2}\right)$$

There are two main, equilibrium negative feedback systems regulating plasma pH: pulmonary regulation of blood CO_2 partial pressure, related to the carbonic acid concentration in the plasma [44]. Pulmonary regulation mechanisms for compensation come to action within a few minutes. It helps to regulate and eliminate H_2CO_3. The second equilibrium negative feedback system is the kidney, which helps to eliminate acids and bases. The metabolism of body's cells produces CO_2, which is quickly changed to H^+ and HCO_3^- through the action of carbonic anhydrase. When the extracellular fluids become acidic, the kidney secretes the H^+ ions and at the same time, HCO_3^- ions are secreted into the blood plasma, which increases the bicarbonate ion level in the plasma; consequently, the carbonic acid/bicarbonate ion ratio decreases, and thus plasma pH increases [39].

Some disruption effects of the normal 7.35–7.45 pH occur in the arterial blood. When arterial blood becomes more acidic (lower than 7.35 pH), it is acidosis, and when arterial blood becomes more basic (higher than 7.45 pH), it is alkalosis. Because of the inability of the chemical buffers to preserve the right acid/base balance, the interplay between the lungs and kidneys does it. Basically, there are four forms of acid/base disruption due to different causes. Two are from the inappropriate lung's dysfunction: Respiratory alkalosis occurs when the breathing rate is high; therefore, CO_2 is reduced to a very low level. On the other hand, when the blood turns to be too acidic (pH decrease), respiratory acidosis occurs, when acute and chronic renal failure and other types of renal damages occur, metabolic acidosis or metabolic alkalosis happens. Respiratory acidosis is the result of reduced pulmonary ventilation that retains and increases CO_2 levels (above 40 mmHg) in the blood. Kidney compensation is increased in H^+ elimination that increases HCO_3^- levels (above 27 mEq L^{-1}) in the blood. Initial signs of acute respiratory acidosis include headache, anxiety, blurred vision, restlessness and confusion. Respiratory acidosis is usually caused by pulmonary dysfunction due to disease or condition that CO_2 is not properly, eliminated. Some conditions that cause chronic respiratory acidosis are asthma, chronic obstructive pulmonary disease (COPD), acute pulmonary edema, severe inability to expand lungs, neuromuscular disorders such as sclerosis or muscular dystrophy and scoliosis. Kidney compensation is done by eliminating H^+ that increases HCO_3^- levels (above 27 mEq L^{-1}) in the blood. Kidney response to respiratory acidosis may help to ameliorate the effect on the plasma pH; it is insufficient to completely normalize pH, but compensation brings pH levels up to 7.39–7.38.

Respiratory alkalosis is the result of increased pulmonary ventilation (hyperventilation) that eliminates significantly, CO_2 levels (below 40 mmHg) in the blood. Panic attacks and anxiety are the most usual reasons for hyperventilation. However, there are more possible causes that include heart attack, pain, drug use, asthma, fever, chronic obstructive pulmonary disease, infection, pulmonary embolism and pregnancy. Someone who is hyperventilating breathes very profoundly or quickly. Marker for the development of respiratory alkalosis is the increased breathing rate that significantly reduces blood CO_2. However, the low blood CO_2 concentration may have physical discomforts such as dizziness, feeling lightheaded, hands and feet coldness or muscle spasms, distress feeling in the chest, confusion, dry mouth, itchy in the arms, heart palpitations and *f* short of breath. Kidney compensation is done by retaining H^+ that reduces HCO_3^- levels (below 27 mEq L^{-1}) in the blood. Kidney response to a respiratory acidosis may help to ameliorate the effect on the plasma pH; it is insufficient to completely normalize pH, but compensation brings pH levels down to 7.41–7.42.

Metabolic acidosis is characterized by the inability of the kidney to eliminate HCO_3^-; thus, blood levels are increased above 27 mEq L^{-1}. As a compensation, respiratory hypoventilation retains CO_2 above 40 mmHg; consequently, blood CO_2 levels are increased. Metabolic acidosis is the most common acid–base disorder. It is a syndrome that happens when the body forms, extreme quantities of acid, such as ketoacids or lactic acid. The kidneys are incapable of taking away sufficient acid formed from usual metabolism or the body loses too much HCO_3^- ions [47]. The bicarbonate buffer equation is shifted to the left in metabolic acidosis. The H^+ cation cannot be added without an anion. Consequently for each HCO_3^- expended, a negative charge of some other type is added to body fluids [48]. Metabolic acidosis can lead to acidemia, which is arterial blood pH lower than 7.37 due to increased production of H^+ ions and, consequently, produce bicarbonate (HCO_3^-) in the kidney or gastrointestinal tract [47]. Metabolic acidosis grounds varied, and its significances can be serious, including coma and death. Together with respiratory acidosis, it is one of the two general causes of acidemia fluids [48]. Indications are not precise, and analysis can be problematic except the individual have strong signs from arterial blood gas samples. Pulmonary response to metabolic acidosis may help to ameliorate the effect on the plasma pH; it is insufficient to completely normalize pH, but compensation brings pH levels up to 7.39–7.38.

Metabolic alkalosis develops when the body loses too much acid or gains too much base. This can be due to excess vomiting, which causes electrolyte loss, overuse of diuretics, adrenal disease, a large loss of potassium or sodium in a short amount of time, antacids, accidental ingestion of bicarbonate, which can be found in baking soda, laxatives and alcohol abuse [49]. During the early stages of the condition, an individual may have symptoms, such as nausea, numbness, prolonged muscle spasms, muscle twitching and hand tremors, and symptoms of alkalosis can vary. If alkalosis is not treated immediately, severe indicators may develop. These indicators might lead to shock or coma, dizziness, difficulty breathing, confusion, state of senselessness and shock or coma. Metabolic alkalosis is characterized by the inability of the kidney's retention of HCO_3^-; thus, its levels are decreased (below 27 mEq L^{-1}) in the blood. Respiratory compensation is done by the high CO_2 elimination (below 40 mmHg); consequently, its levels are reduced in the blood. Pulmonary response to metabolic acidosis may help to ameliorate the effect on the plasma pH; it is insufficient to completely normalize pH, but compensation brings pH levels down to 7.41–7.42. Table 4.3 summarizes acid–base compensation mechanisms for pulmonary acidosis and alkalosis and for metabolic acidosis and alkalosis.

Table 4.3 Acid–base imbalance and compensation

Condition	HCO_3^- (mEq L^{-1})	CO_2 (mmHg)
Respiratory acidosis	Compensation: increase above 27	Problem: increase above 40
Respiratory alkalosis	Compensation: decrease below 27	Problem: decrease below 40
Metabolic acidosis	Problem: decrease below 27	Compensation: decrease below 40
Metabolic alkalosis	Problem: increase above 27	Compensation: increase above 40

Training Effects on the Pulmonary System

Long-lasting aerobic exercise improves pulmonary functions at rest and during exercise. However, it does not improve the diffusion capacity or the lung's tissue surface. Following chronic training at a given, submaximal aerobic exercise, VE is reduced owing to the increase in the heart work efficiency and the increased (a-v)O_2 by the working muscles. As an example, an untrained individual with a maximal work capacity of 12 km h^{-1} runs at a submaximal speed of 9 km h^{-1} meaning his workload is 75% of his maximal work capacity. Following long-lasting aerobic exercise, his work capacity increases to14 km h^{-1}, therefore running at a speed of 9 km h^{-1} means that his workload is 65% of his maximal work capacity. Although there are no significant differences in lung's function between trained and untrained individuals, except for VO_2 that is significantly higher in the trained individual. The reason for the luck in pulmonary function differences between trained and untrained is that the pulmonary has a very wide reserve capacity (MVV of about 300 L min^{-1}) that is larger than the maximal VA achieved during all-out maximal aerobic exercise. The improvement in cardiovascular function and muscle oxygen extraction reduces pulmonary *f* along with VT increment due to an increase in sensitivity of the mechanic and the pulmonary regulatory center. Following these pulmonary improvements, respiratory muscle energy cost decreased.

About the Gases

Gas exchange acts in the pulmonary system, so it is vital to know the aspects that affect gases:

Gas volume changes under conditions such as temperature, pressure and water vapor.

Temperature: As the temperature of an amount of gas increases, so does its volume (Charles's law),

$$V = K\,T \text{ or } V1/T1 = V2/T2$$

where V = volume, T = temperature (in kelvin), K = constant (in $m^3\,K^{-1}$).

At STP, one mole (6.02 · 1023 representative particles) of any gas occupies a volume of 22.4 L. A mole of any gas occupies 22.4 L at standard temperature and pressure (0 °C and 1 atm). The actual value for CO_2 and N_2O is about 22.2 L.

Pressure (*P*)–volume of a gas inversely relates to the pressure exerted on it (Boyle's law):

$$P \cdot V = K$$

where V = volume, P = pressure, K = constant

Water vapor—The more the gas is dry, the larger is its volume.

Gas expressions:

- STPD = it means that the gas under standard conditions, in which ambient temperature corrected to 0 °C, barometric pressure of 760 mmHg at sea level and dry air. Expressing gas in STPD units enables us to compare gas volumes measured under different temperatures, ambient pressure, and percentage of water in the air.
- BTPS units relate to a gas volume expressed under conditions of, core body temperature, a given barometric pressure and saturated with water vapor of 47 mmHg.
- ATPS units relate to a gas volume expressed under conditions of ambient temperature, a given barometric pressure and saturated with water vapor of the ambient.
- ATPD units relate to a gas volume expressed under conditions of ambient pressure and temperature and dry.

In general, gases move from an area of high pressure to areas of low pressure. If there are mixtures of gases in a container, each gas behaves as if it were on its own: it exerts a partial pressure that is independent of that exerted by other gases in the mixture (Dalton's law). At sea level, the atmospheric pressure equals 760 mmHg (1 atmosphere of pressure = 760 mmHg = 101 kPa = 15 l bs/sq. in). The air composition is 20.9% oxygen, 78% nitrogen and small quantities of CO_2, argon and helium. The pressure exerted by the main two gases individually, when added together, equals the total surrounding pressure or atmospheric pressure. The oxygen partial pressure of dry air at sea level is, therefore,

$$(20.9/100\%) \cdot 760\,\text{mmHg} = 159\,\text{mmHg}.$$

However, at the trachea the air is warm and humidified since the vapor pressure in the trachea is 47 mmHg, the oxygen partial pressure (PO_2) in the trachea when breathing is

$$PO_2 = [(760 - 47\,\text{mmHg})(21/100\%)] = 150\,\text{mmHg}.$$

When oxygen has reached the alveoli, the oxygen partial pressure has fallen to about 100 mmHg. This is because the oxygen partial pressure of the gas in the alveoli (PAO_2) is a balance between two processes: the removal of oxygen by the

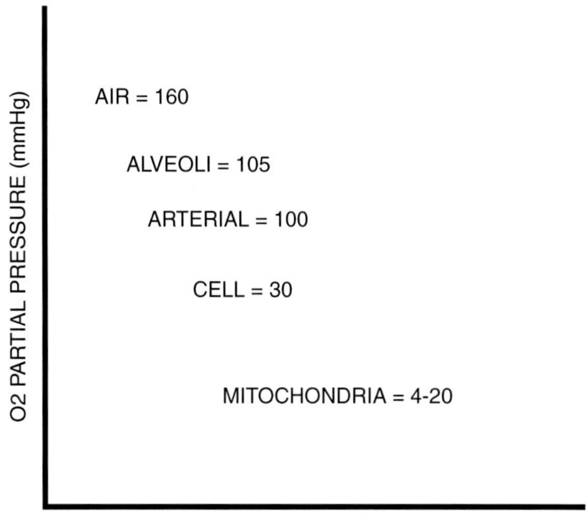

Fig. 4.17 The oxygen partial pressure cascade from the atmosphere to the tissue

pulmonary capillaries and its continual supply by alveolar ventilation (breathing). Oxygen moves down the pressure or concentration gradient from a relatively high level in air (760 mmHg) to the levels in the respiratory tract and then alveolar gas (100 mmHg) by means of diffusion, the arterial blood (95–97 mmHg), capillaries and finally the cell. The oxygen partial pressure reaches the lowest level (4–20 mmHg) in the mitochondria (Fig. 4.17).

This decrease in oxygen partial pressure from air to the mitochondrion is the oxygen cascade and the value of any one-stage in the cascade may be increased under pathological circumstances and may result in hypoxia.

Lung's Volumes

Lung volumes and lung capacities are the volume of air associated with different phases of the respiratory cycle. The average total lung capacity of an adult male is about 6 L of air. VT is one of the measurements made with metabolic cart at rest and during exercise. Ventilation can be divided into distinct volumes and capacities, as shown in Fig. 4.18.

Pulmonary function testing VT, vital capacity, inspiratory capacity and expiratory reserve volume can be measured directly with a spirometer. These are the basic elements of a ventilatory pulmonary function test. Pulmonary capacity is the sum of two or more primary lung volumes.

Inspiratory Capacity (IC): 3.5 L for males and 2.4 L for females, it is the sum of IRV and TV.

$$IC = TV + IRV$$

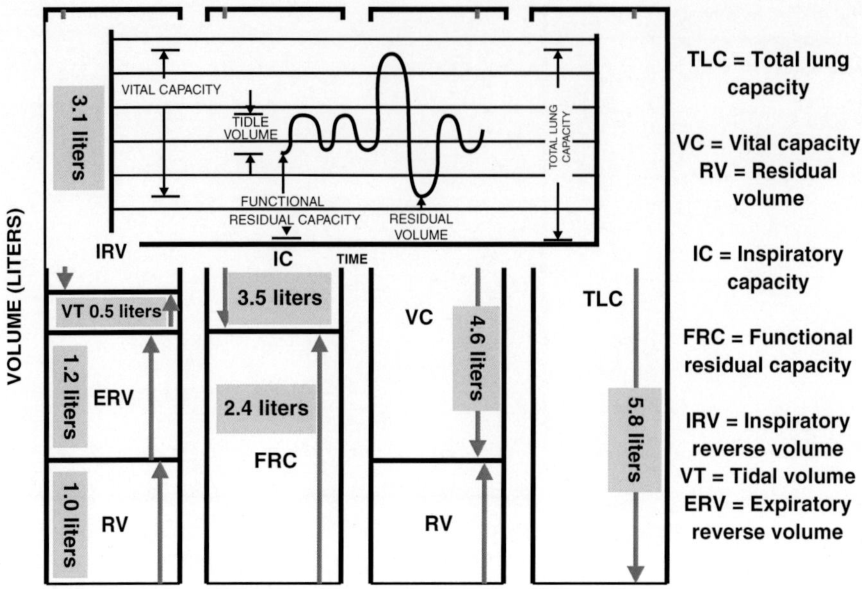

Fig. 4.18 Lung volumes and capacities

Functional Residual Capacity (FRC): about 2.4 L and 1.8 L for females, it is the volume in the lungs at the end-expiratory position

$$FRC = ERV + RV$$

Vital Capacity (VC): about 4.6 L for males and 3.2 L for females, the volume of air breathed out after the deepest inhalation.

$$VC = IRV + TV + ERV$$

Total Lung Capacity (TLC): about 5.8 L for males and 4.4 L for females, the volume in the lungs at maximal inflation.

$$TLC = IRV + TV + ERV + RV$$

IVC = Inspiratory vital capacity: the maximum volume of air inhaled from the point of maximum expiration.

Residual volume as percent of TLC:

RV/TLC

TV = tidal volume: about 500 mL, volume of air moved into or out of the lungs during quiet breathing (VT indicates a subdivision of the lung; when VT is precisely measured, as in gas exchange calculation, the symbol TV or VT is used.

IRV = Inspiratory reserve volume: about 3.1 L, the maximal volume inhaled from the end-inspiratory level.

ERV = expiratory reserve volume: about 1.2 L, the maximal volume of air exhaled from the end-expiratory position.

RV = residual volume: about 1.0 L, the volume of air remaining in the lungs after a maximal exhalation.

Actual volume of the lung (VL) including the volume of the conducting airway.

Pulmonary volumes and capacities are a measure, usually, when it is suspected, of the existence of a respiratory disease because volumes and capacities change with respiratory diseases.

A pulmonary function test (PFT) can measure the amount (volume) and workload (flow rate) of inspiration and expiration. PFTs can be used to identify the health and capacity of the pulmonary system. Most PFTs can measure VT, as well as the following:

- Forced vital capacity (FVC)— maximum volume forcibly expired after maximum inspiration (in liters).
- Forced expiratory volume (FEV1)—Volume of air exhaled in the first second after maximal inhalation; used as a diagnostic tool for limitations in flow rates (expressed as L s^{-1}).
- FEV1.0/FVC ratio—this ratio is a common measure of pulmonary disease. In healthy individuals, about 70–85%, however, the ratio is reduced in COPD patients but may be normal in restrictive diseases.

- Peak expiratory flow (PEF): The maximum instantaneous flow achieved during the FVC maneuver. The maximum expiratory flow during a forced expiration from TLC; expressed in L min^{-1} or L s^{-1}, used to provide a measure of airway diameter and airflow (yet, dependent not only on airway caliber but also on lung elastic recoil, subject effort, and cooperation; generally, less specific than FEV1 as a diagnostic measure.
- FEF: Forced expiratory flow relates to some portion of the FVC curve; modifiers refer to amount of FVC already exhaled.
- FIF: Forced inspiratory flow: (Specific measurement of the forced inspiratory curve is denoted by nomenclature analogous to that for the forced expiratory curve. For example, the maximum inspiratory flow is denoted as FIFmax. Unless otherwise specified, volume qualifiers indicate the volume inspired from RV at the point of measurement.)

Maximum voluntary ventilation (MVV); maximal amount of air expired in 1 min (L min^{-1}).

Lung volumes are largely a function of age, height and sex. Exercise training does not significantly affect lung size per se, but it may improve the measurement of flow rates.

Expressions for PFT volume results are usually in terms of BTPS (body temperature and pressure, saturated). Volumes of gases depend on temperature, barometric pressure and level of humidity or water saturation. As a result, expression of gas volumes needs to have a systematic unit of expression. Since the air in the lungs is at body temperature, ambient pressure and nearly 100% humidity, the units of BTPS are common for PFT results. However, when exhaling air and it cools to the temperature of the room and loses humidity or condenses, its volume changes. Thus, it may be necessary to convert from ambient (or room) temperature, ambient pressure, saturated (ATPS). The equation for converting ATPS to BTPS is as follows:

$$\text{VBTPS} = \text{VATPS} \cdot \left[310/(273+\text{TA})\right] \times \left[(\text{PB}-\text{PH2O})/(\text{PB}-47)\right]$$

where VATPS = volume of gas at ATPS, TA = temperature in the laboratory, PB = barometric pressure in the laboratory and PH_2O = water vapor pressure at the TA.

Important Gas Measurements

VCO_2 (CO_2) is a by-product of cellular metabolic processes. Most of the CO_2 given off by the body comes from this cellular respiration. However, during high-intensity exercise some of the CO_2 that the subject is blowing off comes from buffering of the blood (in order to maintain proper pH).

FIO_2 is the fraction of expired air that is oxygen percentage.

FEO_2 is the percent of expired air that is oxygen. The air we breathe in is 20.93% oxygen and we typically extract at rest 3% and during exercise, it rises up to 6%. Thus, the air that we exhale usually contains 15–18% O_2. Low values for FEO_2 suggest that the subject is extracting O_2 well and that gas exchange in the alveoli is good.

$FECO_2$ is the percent of expired air that is oxygen. The air that we breathe has very little amount of VCO_2 (0.03%). As stated previously CO_2 results from cellular metabolism, and thus most of the CO_2 exhaled comes from metabolic processes.

Pulmonary Diseases

Lung's problem can affect life-quality since inhaled air over the dead space directly into the lungs. This course exposes the lungs to numerous pollutions and diseases. Avoiding precautions of taking care in time may have unpleasant consequences and at times death. Symptoms of the lung diseases in general, rest on its kind and differ from one individual to another. Main disease features include cough, shortness of breath, wheezing, tiredness and increased heartbeat. Lung diseases can be physiologically categorized into three types of lung diseases:

Obstructive lung disease caused by the obstruction of the airways. Asthma, emphysema and bronchitis are considered as obstructive lung diseases.

- Restrictive lung disease results from the incapacity of the lungs to keep the air in the alveoli. In restrictive lung disease, there is a reduction in lung's elasticity or the loss of chest wall expansion ability.

- Reduced diffusion capacity diseases are associated with the failure of the alveoli to supply oxygen or to move it to the blood. This causes a reduction in the delivery of oxygen by the blood to the body organs.

Obstructive Lung Diseases

Pulmonary function test outcomes are an indicative standard of respiratory diseases, which are considered into two dissimilar kinds: obstructive and restrictive. Describing a disruptive (obstructive) lung disease of an acute or chronic obstruction in the bronchi leads to the alveoli. Frequently, these obstructions are denoted as chronic obstructive pulmonary diseases (COPDs). COPD occurs most often; in older adults, it takes several years for COPD to develop. It happens mostly in individuals over the age of 40, when symptoms of COPD first appear. It does not appear in younger adults since it takes several years for COPD to develop. Males and females are affected equally. Individuals with these diseases typically have normal lung volumes but restricted flow rates. The respiratory test is performed recurrently three times. With the Good method, reproducibility can be expected if the standards attained for maximal expiratory flow cannot be augmented additionally by powerful maneuver. By testing FEV1, FEV1/FVC ratio, PEF and other flow rate measures allow diagnosing COPDs. Examples of COPDs include asthma and emphysema.

Between all pulmonary diseases, COPD is with the maximal death rate [50], it accounts for the highest quantity of time being disabled [51]. COPD is a chronic inflammatory lung disease that causes obstructed airflow from the lungs, categorized by long-term breathing problems and poor airflow [52]. COPD is a disruptive disease of the respiratory system that includes main articles: chronic obstructive bronchitis and pulmonary emphysema. The pathogenesis of airway obstruction in COPD results from variations in the airways and in the pulmonary functional tissue (parenchyma). Bronchitis/bronchiolitis and emphysema both contribute to the limitation of expiratory flow found in PFT [53]. Under normal conditions, airways are open during expiration; however, COPD patient due to reduced airways elasticity, in addition to inflammation, that narrows the airways, results in an increase in small airways resistance [54]. Therefore, individuals with COPD are very limited to expire entirely their vital capacity volume, detected by the FEV1/FVC ratio. COPD strictness is found on FEV1 from capacities found during maximal forced expiratory maneuvers [55].

Chronic bronchitis and emphysema are used for different types of COPD [56]. Symptoms include shortness of breath, cough, mucus (sputum) production and wheezing [52]. Individuals may put a stop to COPD by reducing exposure to risk factors [57], such as the rate of smoking and refining surrounding environments air. Although treating COPD may reduce the rate of the disease progression, yet it is a poorly reversible disease [58]. COPD is a progressive disease, which deteriorates over time, resulting in narrowing of the small airways and deterioration of lung tissue [59]. Some individuals can profit after lasting oxygen therapy treatment or lung transplantation [60]. It is caused most often from cigarette smoke [61] and factors such as air pollution and genetics. Epidemiologic studies suggested that individuals that never smoked had a chance of about 5% to have COPD [62]. Reasons for having COPD in nonsmokers consist of asthma, aging, exposure to toxic agents and infections during childhood. People with COPD are at increased risk of developing heart disease, lung cancer and a variety of other conditions. Ultimately, physical activities such as walking or carrying light weights turn out to be problematic. Chronic bronchitis patients are at a greater risk of developing COPD [63]. Emphysema is a chronic, progressive disease that results in the destruction of the alveoli in the lungs, leading to the abnormal presence of air or other gas within tissues. This results in a greatly reduced surface area for gas exchange in the lungs, which typically leads to hypoxia [64]. COPD is treatable. With proper management, most people with COPD can achieve good symptom control and quality of life, as well as reduced risk of other associated conditions. The disease diagnosis is based on deprived airflow as measured by lung function tests [65]. Pulmonary function test (PFT) is a complete evaluation of the respiratory system by testing the pulmonary function. The key drive of pulmonary function testing is to categorize the severity of pulmonary impairment [66]. Pulmonary function testing has diagnostic and therapeutic roles and helps clinicians answer some general questions about patients with lung disease. Different from asthma, the airflow reduction does not improve much with the use of a bronchodilator [67]. Pulmonary function testing is a diagnostic and management tool used for a variety of reasons, such as chronic shortness of breath, asthma, chronic obstructive pulmonary disease, restrictive lung disease, preoperative testing, impairment or disability and early morning wheezing.

Asthma

Asthma differs from chronic obstructive bronchitis and pulmonary emphysema that it can be temporary and reversible. Exercise-induced asthma is a temporary inflammatory response during exercise that obstructs ventilatory flow rates. Exercise-induced asthmatics may have normal flow rates at rest but produce a positive test: 15% decrease from pre-exercise measured FEV1, following 6–8 min of vigorous exercise at a target intensity of 85–90% of HRmax.

Asthma is a common chronic inflammatory disease of the lung's airways, it persists for a long period of time in its effects. It is characterized by variable and recurring symptoms, reversible airflow obstruction spontaneously or with treatment, airway hyperresponsiveness and easily triggered bronchospasms [68]. Asthma triggers respiratory infections, such as the common cold, exercise-induced asthma in cold air, air pollutants and irritants, such as smoke. Asthma is a condition in which airways narrow, swell and produce extra mucus. Pathophysiologic factors include bronchiolar inflammation with airway constriction and resistance that shows episodes of coughing, shortness of breath and wheezing. In addition, asthma can affect the trachea, bronchi and bronchioles. This can lead to bronchoconstriction with bronchospasm of bronchial smooth muscle; consequently, edema from microvascular leakage narrows the airways. For some individuals, asthma is a minor annoyance; for others, it can be a major problem that interferes with daily activities and may lead to a life-threatening while having an asthma attack. Asthma cannot be cured, but its symptoms can be controlled.

Asthma symptoms vary between individuals. Asthma episodes may be infrequent at certain times—such as when exercising or having it all the time. Asthma signs and symptoms include shortness of breath, chest tightness or pain, sleeping problems due to breathing shortness, coughing or wheezing, A whistling or wheezing sound when exhaling (especially in children), coughing or wheezing attacks that are worsened by a respiratory virus, such as a cold or the flu. Signs that your asthma is probably worsening include increasing difficulty in breathing, the need to use a quick-relief inhaler more often, exercise-induced asthma, which may be worse when the air is cold and dry, occupational asthma, triggered by workplace irritants such as chemical fumes, gases or dust, allergy-induced asthma, triggered by airborne substances, such as pollen, mold spores, cockroach waste or particles of skin and dried saliva shed by pets (pet dander).

Asthma is triggered by a mixture of ecological and genetic factors playing a smaller role [69]. Environmental factors include exposure to air pollution and allergens, in addition, medications such as aspirin and beta blockers. Diagnosis leans typically on the nature of symptom response to treatment over-time and spirometry lung function testing [70]. Asthma is classified according to the frequency of symptoms, FEV1, and peak expiratory flow rate [71]. Unfortunately, asthma is not curative. However, taking care of factors that trigger asthma mainly by inhaled corticosteroids that reduce airway inflammation or the allergen effect to alleviate the stress [72]. In addition to inhaled corticosteroids, long-acting beta agonists or antileukotriene agents may be used if asthma symptoms persist to be unrestrained [73].

Restrictive Lung Diseases

Obstructive lung diseases may be caused by an intrinsic disorder, usually pulmonary fibrosis [74] or extrinsic [75]. Restrictive diseases can be diagnosed by measuring FVC or a slow VC. Examples of restrictive diseases include pulmonary fibrosis, scar tissue and tumors. Obstructive lung diseases differ from COPD by having reduced lung volumes, compared to restrictive disease, which is related to a decreased TLC. Obstructive lung diseases are a group of pulmonary diseases such as pleural effusion, extrapulmonary tuberculosis and parenchymal all, which restrict lung enlargement. Consequently, they are categorized by reduced lung volumes, a rise in breathing energy cost and insufficient VE and/or blood oxygenation. Although individuals may have normal flow rates, restrictive lung diseases reduce VC volumes either because of changes in lung parenchyma or due to a disease of the pleura, chest wall, or neuromuscular dysfunction. Expiratory airflow measurements are conserved, and airway resistance to flow is normal, while the FEV1/FVC ratio is increased. If obstructive lung diseases are caused by parenchymal lung disease, restrictive lung disorders are accompanied by reduced gas diffusion capacity; thus, arterial blood is significantly desaturated following exercise bout [76].

Pulmonary Exercise

After cautious patients screening, it is advisable to be careful when prescribing exercising intensities for COPD patients; nonetheless, shortness of breath does not always mean that the patient has to stop every time. In addition, consulting the physician is a must about when the patient has to stop exercising and rest. Chronic aerobic exercise training is part of healthy individuals and COPD patients. Beneficial physiological alterations are noted subsequent to a planned exercise training program in patients with COPD, without lung function alterations at rest [77]. Aerobic exercise is the key non-pharmacological action, which is tolerated by COPD patients [78]. COPD patients benefit from long-lasting exercise training by chronic steady state aerobic exercise utilizing large muscle mass [79]. Dyspnea, reduced exercise tolerance is a common criticism in patients with COPD [80]. Bronchodilators may expand modestly, pulmonary function in COPD [81] but does not essentially have a strong effect on exercise capacity [82]. Aerobic exercise and resistive training are a very important part of a wide-ranging pulmonary restoration program in COPD patients. This has many advantages such as reductions in dyspnea and fatigue [83], improve myocardium and cardiovascular function and oxygen delivery and improve COPD symptoms, VO_{2max} and $(a\text{-}v)O_2$. Therefore, increase patient's anaerobic threshold where patients become short of breath due to increased endurance. In addition, high-intensity endurance training may induce better, exercise tolerance, functional ability and work capacity improvements compared to lower-intensity exercise [84].

Exercise Training in Asthmatic Patients

Asthma treatment aim is to preserve a normal activity, where important part of it is aerobic exercise and resistive training. This is in addition to taking medications in order to avoid triggers. This along of inspection symptoms to verify how well the lungs are working. Exercising sometimes must be avoided because of symptoms, therefore exercising under these conditions, a discussion with a physician is needed to provide asthma relief and safety during exercise [85]. Regarding the occurrence of Spartan asthma in children, the prevalence of severe asthma is 0.5% in all 10-year old, and 4.5% among current asthmatics [86].

Exercise is a stressor calculable that can be altered and can be considered as a model of stress [87]. The endocrine effect [87], the nervous system [88] and the immune systems [89] are significantly affected by exercise. Exercise-induced asthma (EIA) and the clinical exercise-induced bronchoconstriction (EIB) are frequently used interchangeably to define symptoms of asthma such as cough, wheeze or dyspnea, triggered by forceful exercise. Mechanisms behind EIA and EIB is due to airway drying and thermal hypothesis [90]. EIA and EIB were well-defined, designating bronchoconstriction happening in relation to exercise in vulnerable non-asthmatic and asthmatic individuals. Some evidence suggested that the pathogenesis of this condition is connected to airway damage owed to lengthy hyperventilation and destructive environmental factors. If the reliable diagnostic measurements are undistinguishable for the diagnosis of EIA and EIB, the analyses are based on the high sensitivity of aggravation by exercise or the eucapnia voluntary hyperventilation aggravation test [91]. Activities of team sports, which include brief spurts of exertion, are recommended, adding, single or group relaxation activities such as walking, biking and hiking. Swimming is a good decision since breathing is done, in sufficiently warm, humid air. Swimming forces the upper-body muscle to work hard for ventilation, which turns out to be an inordinate way to build upper-body strength. Before any exercise session, the individual has to use asthma medicine (inhaled bronchodilators or cromolyn). The following step is warm-up, with low-intensity exercises. In addition, upon stopping the training session, one must have a significant cool-down period. In cold weather, pollen counts or air pollution counts are high, exercise training will take place indoors, especially if the individual has allergic asthma. If exercising outdoors, it is recommended to wear a mask or scarf over the nose and mouth. EIA and bronchial hyper-responsiveness are often seen, in elite athletes competing outdoor toughness (mainly aerobic) winter sports, mostly in cross-country and biathlon skiers [92]. Therefore, physical activity in the cold-weather such as ice hockey, cross-country skiing, and ice skating is not recommended for asthmatic patients [93]. Also, activities that involve long-lasting bouts of effort, such as soccer, distance running, basketball, and field hockey can impose a heavy burden on the patient. In spite of the above-mentioned recommendations, some asthmatic patients are able to participate in the above activities.

Exercise for Restrictive Lung Disease Patients

Patients with restrictive lung diseases, frequently are agonized from dyspnea, fatigue and reduced muscle force generation and exercise intolerance. This is owed to a difficult condition of reduced ventilation and gas diffusion in addition to variations skeletal muscles and myocardial function [94]. Therefore, exercise training is advantageous for patients with restrictive lung disease. Yet, it has a tendency to be impossible for those with severe disease [95]. Patients with restrictive lung disease [96] improved their 6-min walk distance, and in health-related quality of life. No significant improvements were seen in peak exercise work capacity [97]. Similar results were reported by Salhi and colleagues [98]. Compared to the baseline values, patients following 12 weeks of aerobic exercise training (walking), enhanced, their 6-min walk distance performance by, 79 m, and 102 m at the end of the 24th week. In addition, patient's symptoms, dyspnea and VO_{2max} were improved. The increment in both; peak work capacity and VO_{2max}, indicated of the physiological training achievement.

Resistance Exercise for Lung Disease Patients

Exercising against various loads, build and improve individual's skeletal muscle function in health and in lung disease patients. Physical inactivity in those patients results in disuse atrophy, where muscle proteins are broken down, which, naturally there is a loss of muscle mass and strength, which results in further decreased work capacity and exercise. Thus, this condition progresses to a point where the patient is unable to perform, basic daily tasks. This condition has a viscous cycle, where physical capacity drops, and affecting, quality of life. Resistance exercise can break downhill this viscous cycle. Lightweight loads and more repetitions are preferred than heavyweight loads with lower repetitions. It is recommended to train under the supervision of a professional staff and a personal trainer, at least to begin with. Resistance training is good involvement for COPD and asthmatic patients suffering from skeletal muscle weakness and marked indications of dyspnea during exercise [99].

Resistance training is an important addition to aerobic exercise and an operative involvement for the acute increase in the severity of COPD [100]. COPD patients use more energy to breath than other individuals do. Resistive exercise improves significantly upper respiratory muscles and bone strength. Due to the increased total muscles' strength, breathing effort becomes easier, which reduces the perceived intensity of breathlessness [100]. It leads to significant improvements in dyspnea and functional exercise capacity. When added to a program of aerobic exercise, strength training confers additional benefits in muscle force, but not in overall exercise capacity or health status.

Laboratory on Pulmonary Exercise Test

This test measures VE, VT and *f* when exercising on a treadmill or on a stationary bike. The test may pinpoint where the breathing problems begin if any, which may begin in the lungs, the heart or both.

Before the Testing Session

Subject has to check in at least 30 min before testing. Subject will be asked to do the following:

(a) Have a light breakfast no less than 2 h before the test.
(b) Stop eating and drinking fluids for 2 h before the test.
(c) Stop smoking for at least 8–12 h before the test, better at all.
(d) Stop taking certain medications before the test.
(e) Do not exercise in the morning before the test.

(f) Wear comfortable exercise clothing and walking shoes, such as tennis shoes.
(g) If you have dentures, wear them.

Testing Session

The method for measuring pulmonary exercise stress test at rest and throughout the exercise, utilizing metabolic cart that measures at once and on-line oxygen uptake, CO_2 production, VE, VT and *f*.

Equipment:
Use treadmill or stationary bike as an Ergometer to measure physiological parameters and to control exercise intensity.

The subject will wear a mask with a breathing tube connected to a metabolic cart to measure the ventilator function. Trials are timed with a stopwatch.

Testing procedures for pulmonary exercise stress test:
Small electrodes placed on the subject's upper body for monitoring HR. A device clipped to the finger will measure the oxygen saturation in the blood. The subject will also wear nose clips and breathe through a mouthpiece.

The subject is connected to a pulse meter and to the metabolic cart,

(a) Before starting to exercise, a spirometry test may be done. This will measure the amount of air inhaled and exhaled while at rest.
(b) The subject sits for 5 min in full relaxation to measure during the last 3 min: RMR, HR, resting VO_2, VCO_2, VE, VT, *f* and RQ.
(c) While exercising, HR, breathing and BP are monitored. When the test is over, post-exercise measurements will be taken.
(d) Following rest, before the actual test, the subject walks (on a treadmill) for 5 min at a pace of 4.5 km h^{-1} (2.3 mph), without elevation after that run for 5 min at a pace of 8–8.5 km h^{-1}. If cycling, subject will pedal at a resistance of 100 W. This is done to accustom the subject with the ergometer utilized. Following the running or cycling, the subject sits until HR and VO_2 return to near (±) resting values.
(e) The treadmill workload is set to at 8–10 km h^{-1}, 0% elevation (5.5–6.5 mph). Thereafter, the workload is increased by an elevation of 1.5% each minute up to 6 min. The initial workload on the bike ergometer will be based on the subject's physical ability. As an example: the initial workload is 75–100 W, and thereafter the workload is increased by 25–50 W.
(f) While exercising, HR, *f*, VT, VE, VO_2, CO_2 and BP are monitored. When the test is over, mouthpiece stay's in place until told to remove it.
(g) The test ends when VO_2 does not increase more, despite an increase in workload.
(h) Immediately following the test, the subject will be asked to sit for the next 10 min for recovery measurements.

References

1. Hsia CC, Hyde DM, Weibel ER. Lung structure and the intrinsic challenges of gas exchange. Comp Physiol. 2016;6:827–95.
2. Gilroy AM, MacPherson BR, Ross LM. Atlas of anatomy. Stuttgart: Thieme; 2008. p. 108–11.
3. Smith SL. Postoperative perfusion deficits. ScienceDirect. 1990;(2):567–78.
4. Sagiv M. Exercise cardiopulmonary function in cardiac patients. London: Springer; 2012.
5. Patel S, Majmundar SH. Physiology, carbon dioxide retention. StatPearls Publishing; 2019. https://www.ncbi.nlm.nih.gov. NBK482456.
6. Wagner PD, Simonson TS, Wei G, et al. Sea-level haemoglobin concentration is associated with greater exercise capacity in Tibetan males at 4200 m. Exp Physiol. 2015;100:1256–62.
7. Knudsen L, Ochs M. The micromechanics of lung alveoli: structure and function of surfactant and tissue components. Histochem Cell Biol. 2018;150:661–76.
8. Pittman RN, Rafael S. Regulation of tissue oxygenation. Morgan & Claypool Life Sciences, CA; 2011. https://www.ncbi.nlm.nih.gov. NBK54104.
9. Lane SJ, Moran AL, Shishido CM, et al. Cuticular gas exchange by Antarctic sea spiders. J Exp Biol. 2018;221(Pt 8). pii: jeb177568. https://doi.org/10.1242/jeb.177568.
10. Donnelly DF. Voltage-gated Na(+) channels in chemoreceptor afferent neurons—potential roles and changes with development. Respir Physiol Neurobiol. 2013;185:67–74.
11. Kumar P, Prabhakar NR. Peripheral chemoreceptors: function and plasticity of the carotid body. Compr Physiol. 2012;2:141–219.
12. Johnson JO. pharmacology and physiology for anesthesia. ScienceDirect. 2013:208–17. https://doi.org/10.1016/B978-1-4377-1679-5.00012-0.
13. Tortora G. Principles of anatomy and physiology. 12th ed. Hoboken, NJ: Wiley; 2008. p. 905–9.
14. Tortora GJ, Derrickson BH. Principles of anatomy and physiology – maintenance and continuity of the human body. 12th ed. Danvers: Wiley; 2009.
15. Tipton NJ, Harper A, Paton JFR, et al. The human ventilatory response to stress: rate or depth? J Physiol. 2017;595:5729–52.
16. Lumb AB. Nunn's applied respiratory physiology. 8th ed. 2017. p. 227–34. https://doi.org/10.1016/B978-0-7020-6294-0.00013-7.
17. Diest IV, Tegen K, Van de Woestijne KP, et al. Hyperventilation and attention: effects of hypocapnia on performance in a stroop task. Biol Psychol. 2000;53:233–52.
18. Gastinger S, Sorel A, Nicolas G, et al. A comparison between ventilation and heart rate as indicator of oxygen uptake during different intensities of exercise. J Sports Sci Med. 2010;9:110–8.
19. Quinn M, Rizzo A. Anatomy, anatomic dead space. StatPearls [Internet]. Treasure Island, FL: StatPearls Publishing; 2019–2018.
20. West JB, Luks MD, Andrew M. West's respiratory physiology: the essentials. LWW. 10th ed. 2015.
21. Powers KA, Dhamoon AS. Physiology, pulmonary, ventilation and perfusion. StatPearls [Internet]. Treasure Island (FL): StatPearls Publishing; 2019.
22. Verma S, Goni M, Kudyar RP, et al. Assessment of respiratory muscle endurance in type 1 & 2 diabetes mellitus. JK Sci. 2012;4:168–71.
23. Blackie SP, Fairbarn MS, McElvaney NG, et al. Normal values and ranges for ventilation and breathing pattern at maximal exercise. Chest. 1991;100:136–42.
24. Romer LM, Polkey MI. Exercise-induced respiratory muscle fatigue: implications for performance. J Appl Physiol. 2008;104:879–88.
25. Olson TP, Joyner MJ, Dietz MM, et al. Effects of respiratory muscle work on blood flow distribution during exercise in heart failure. J Physiol. 2010;588:2487–501.
26. Dempsey JA, Romer L, Rodman J, et al. Consequences of exercise-induced respiratory muscle work. Respir Physiol Neurobiol. 2006;151:242–50.
27. Harms CA, Wetter TJ, McClaran SR, et al. Effects of respiratory muscle work on cardiac output and its distribution during maximal exercise. J Appl Physiol. 1998;85:609–18.

28. Gravier G, Delliaux S, Delpierre S, et al. Inter-individual differences in breathing pattern at high levels of incremental cycling exercise in healthy subjects. Respir Physiol Neurobiol. 2013;189:59–66.
29. Guyton AC, Hall JE. Textbook of medical physiology. 13th ed. Philadelphia, PA: WB Saunders; 2015.
30. Klocke R. Dead space: simplicity to complexity. J Appl Physiol. 2006;100:1–2.
31. Hall JE. Guyton and hall textbook of medical physiology. 12th ed. Saunders/Elsevier: Philadelphia, PA; 2010. p. 502.
32. Voet D, Voet JG, Pratt CW. Fundamentals of biochemistry: life at the molecular level. 4th ed. Hoboken, NJ: Wiley; 2013. p. 189.
33. Ganong WF. Review of medical physiology. 21st ed. New York: Lange Medical Books; 2003.
34. Fischer K, Yamaji K, Luescher S, et al. Feasibility of cardiovascular magnetic resonance to detect oxygenation deficits in patients with multi-vessel coronary artery disease triggered by breathing maneuvers. J Cardiovasc Magn Reson. 2018;20(1):31. https://doi.org/10.1186/s12968-018-0446-y.
35. Arthurs GJ, Sudhakar M. Carbon dioxide transport. Contin Edu Anaesth Crit Care Pain. 2005;5:207–10.
36. Sander R. Compilation of Henry's law constants (version 4.0) for water as solvent. Atmos Chem Phys. 2015;15:4399–981.
37. Hamm LL, Nakhoul N, Hering-Smith KS. Acid-base homeostasis. Clin J Am Soc o Nephrol. 2015;10:2232–42.
38. Rhoades RA, Bell DR. Medical physiology: principles for clinical medicine. Baltimore, MD: Lippincott Williams & Wilkins; 2012. p. 5–6.
39. Tortora G, Derrickson B. Principles of anatomy & physiology. 13th ed. Hoboken, NJ: Wiley; 2012. p. 42–3.
40. Unglaub SD. Human physiology. An integrated approach. 7th, Global ed. Harlow: Pearson; 2016. p. 607–608, 666–673.
41. Adrogue HE, Adrogue HJ. Acid-base physiology. Respirat Care. 2001;46:328–41.
42. DiBartola SP. Introduction to acid-base disorders, fluid, electrolyte, and acid-base disorders in small animal practice (4th, ed.). ScienceDirect; 2012:237–52. https://doi.org/10.1016/B978-1-4377-0654-3.00016-0.
43. Swietach P, Tiffert T, Mauritz JM, et al. Hydrogen ion dynamics in human red blood cells. J Physiol. 2010;588:4995–5014.
44. Silverthorn DU. Human physiology. An integrated approach. 7th, Global ed. Harlow: Pearson; 2016. p. 607–608, 666–673.
45. Bhagavan NV. Water, acids bases and buffers, in Medical biochemistry (4th ed). ScienceDirect. 2002:1–16. https://doi.org/10.1016/B978-012095440-7/50003-2.
46. Johnson LR, editor. Essential medical physiology. 3rd ed. Amsterdam: Elsevier Academic; 2003.
47. Costanzo L. Physiology. Philadelphia, PA: Elsevier; 2010.
48. Engelking LR. Metabolic acidosis. Textbook of veterinary physiological chemistry (3rd ed.). SienceDirect. 2015;561–7. https://doi.org/10.1016/B978-0-12-391909-0.50087-6.
49. Yee AH, Rabinstein AA. Neurologic presentations of acid-base imbalance, electrolyte abnormalities, and endocrine emergencies. Neurol Clin. 2010;28:1–16.
50. Vos T, Flaxman AD, Naghavi M, Lozano R, et al. Years lived with disability (YLDs) for 1160 sequelae of 289 diseases and injuries 1990-2010: a systematic analysis for the Global Burden of Disease Study 2010. Lancet. 2012;380:2163–96.
51. Lozano R, Naghavi M, Foreman K, et al. Global and regional mortality from 235 causes of death for 20 age groups in 1990 and 2010: a systematic analysis for the Global Burden of Disease Study 2010. Lancet. 2012;380:2095–128.
52. Vogelmeier CF, Criner GJ, Martinez FJ, Anzueto A, et al. Global strategy for the diagnosis, management and prevention of chronic obstructive lung disease 2017 report: GOLD executive summary. Am J Respirol Crit Care Med. 2017;22:575–601.
53. Hogg JC, Chu F, Utokaparch S, et al. The nature of small-airway obstruction in chronic obstructive pulmonary disease. N Engl J Med. 2004;350:2645–53.

54. Gibson GJ. Clinical tests of respiratory function. 3rd ed. London: Hodder Arnold; 2009.
55. Johns DP, Walters JA, Walters EH. Diagnosis and early detection of COPD using spirometry. J Thorac Dis. 2014;6:1557–69.
56. Craig JA. Ferri's Netter patient advisor. 2nd ed. Philadelphia: Saunders; 2012. p. 913.
57. Pirozzi C, Scholand MB. Smoking cessation and environmental hygiene. Med Clin North Am. 2012;96:849–67.
58. Devine JF. Chronic obstructive pulmonary disease: an overview. Am Health Drug Benefits. 2008;1:34–42.
59. Abdool-Gaffar MS, Ambaram A, Ainslie GM, et al. Guideline for the management of chronic obstructive pulmonary disease—2011 update. S Afr Med J. 2011;101:63–73.
60. Rabe KF, Hurd S, Anzueto A, et al. Global strategy for the diagnosis, management, and prevention of chronic obstructive pulmonary disease: GOLD executive summary. Am J Respir Crit Care Med. 2007;176:532–55.
61. Vestbo J. Definition and overview. Global strategy for the diagnosis, management, and prevention of chronic obstructive pulmonary disease. Am J Respir Crit Care Med. 2013;187:1–7.
62. Leung JM, Sin DD. COPD in never smokers: prognosis unveiled. Lancet Respir Med. 2013;1:502–4.
63. Restrepo MI, Sibila O, Anzueto A. Pneumonia in patients with chronic obstructive pulmonary disease. Tuberc Respir Dis (Seoul). 2018;81:187–97.
64. Taylor J. Bailliere's dictionary e-Book: for nurses and health care workers. Edinburgh: Elsevier Health Sciences; 2019. p. 128.
65. Nathell L, Nathell M, Malmberg P, et al. COPD diagnosis related to different guidelines and spirometry techniques. Respir Res. 2007;8:89.
66. Collins EG, Bauldoff G, Carlin B, et al. Clinical competency guidelines for pulmonary rehabilitation professionals: position statement of the American Association of Cardiovascular and Pulmonary Rehabilitation. J Cardiopulm Rehabil Prev. 2014;34:291–302.
67. Lee TS, Goyal P, Li C, et al. Computational fluid dynamics to evaluate the effectiveness of inferior turbinate reduction techniques to improve nasal airflow. JAMA Facial Plast Surg. 2018;20:263–70.
68. Global Initiative for Asthma. Archived from the original (PDF) on 2012-11-20. 2011.
69. Decramer M, Janssens W, Miravitlles M. Chronic obstructive pulmonary disease. Lancet. 2012;379:1341–51.
70. Lemanske RF, Busse WW. Asthma: clinical expression and molecular mechanisms. J Allergy Clin Immunol. 2010;125:S95–102.
71. Yawn BP. Factors accounting for asthma variability: achieving optimal symptom control for individual patients. Prim Care Resp J. 2008;17:138–47.
72. Ernst P. Inhaled corticosteroids moderate lung function decline in adults with asthma. Thorax. 2006;61:93–4.
73. Scott JP, Peters-Golden M. Antileukotriene agents for the treatment of lung disease. Am J Respir Crit Care Med. 2013;188:538–44.
74. Vuga LJ, Tedrow JR, Pandit KV, et al. C-X-C motif chemokine 13 (CXCL13) is a prognostic biomarker of idiopathic pulmonary fibrosis. Am J Respir Crit Care Med. 2014;189:966–74.
75. Hurrass J, Heinzow B, Aurbach U, et al. Medical diagnostics for indoor mold exposure. Int J Hyg Environ Health. 2017;220:305–28.
76. Neves J, Leitz D, Kraut S, et al. Disruption of the hepcidin/ferroportin regulatory system causes pulmonary iron overload and restrictive lung disease. EBiomedicine. 2017;20:230–9.
77. Vogiatzis I, Terzis G, Nanas S, et al. Skeletal muscle adaptations to interval training in patients with advanced COPD. Vogiatzis I. Chest. 2005;128:3838–45.
78. Martín-Valero R, Cuesta-Vargas AI, Labajos-Manzanares MT. Effectiveness of the physical activity promotion programme on the quality of life and the cardiopulmonary function for inactive people: randomized controlled trial. BMC Public Health. 2013;13:127. https://doi.org/10.1186/1471-2458-13-127.

79. Garvey C, Bayles MP, Hamm LF, et al. Pulmonary rehabilitation exercise prescription in chronic obstructive pulmonary disease: review of selected guidelines: an official statement from the American association of cardiovascular and pulmonary rehabilitation. J Cardiopulm Rehabil Prev. 2016;36:75–83.
80. Paneroni M, Simonelli C, Vitacca M, et al. Aerobic exercise training in very severe chronic obstructive pulmonary disease: a systematic review and meta-analysis. Am J Phys Med Rehabil. 2017;96:541–8.
81. Cazzola M, Matera MG. Bronchodilators: current and future. Clin Chest Med. 2014;35:191–201.
82. Voduc N, Alvarez GG, Amjadi K, et al. Effect of theophylline on exercise capacity in COPD patients treated with combination long-acting bronchodilator therapy: a pilot study. Int J Chron Obstruct Pulmon Dis. 2012;7:245–52.
83. Maltais F, LeBlanc P, Jobin J, et al. Intensity of training and physiologic adaptation in patients with chronic obstructive pulmonary disease. Am J Respir Crit Care Med. 1997;155:555–61.
84. Morris NR, Walsh J, Adams L, et al. Exercise training in COPD: what is it about intensity? Respirology. 2016;21:1185–92.
85. Del Giacco SR, Firinu D, Bjermer L, et al. Exercise and asthma: an overview. Eur Clin Respir J. 2015;2:27984. https://doi.org/10.3402/ecrj.v2.27984.
86. Lang A, Carlsen KH, Haaland G, et al. Severe asthma in childhood: assessed in 10 year old in a birth cohort study. Allergy. 2008;63:1054–60.
87. Eliakim A. Endocrine response to exercise and training-closing the gaps. Pediatr Exerc Sci. 2016;28:226–32.
88. Fu Q, Levine BD. Exercise and the autonomic nervous system. Handb Clin Neurol. 2013;117:147–60.
89. Vallet H, Fali T, Sauce D. Aging of the immune system: from fundamental to clinical data. (Article in French). Rev Med Interne. 2019;40:105–11.
90. Del Giacco SR, DFirinu D, Bjermer L, Carlsen KH. Exercise and asthma: an overview. Eur Clin Respir J. 2015;2:27984. https://doi.org/10.3402/ecrj.v2.27984.
91. Poussel M, Chenuel B. Exercise-induced bronchoconstriction in non-asthmatic athletes. (Article in French). Rev Mal Respir. 2010;27:898–906.
92. Carlsen KH. Sports in extreme conditions: the impact of exercise in cold temperatures on asthma and bronchial hyper-responsiveness in athletes. Br J Sports Med. 2012;46:796–9.
93. Fitch KD, Sue-Chu M, Anderson SD, et al. Asthma and the elite athlete: summary of the International Olympic Committee's consensus conference, Lausanne, Switzerland, January 22-24, 2008. J Allergy Clin Immunol. 2008;122:254–60.
94. COPD Working Group. Pulmonary rehabilitation for patients with chronic pulmonary disease (COPD): an evidence-based analysis. Ont Health Technol Assess Ser. 2012;12:1–75.
95. Ho SC, Lin HC, Kuo HP, et al. Exercise training with negative pressure ventilation improves exercise capacity in patients with severe restrictive lung disease: a prospective controlled study. Respir Res. 2013;14:22. https://doi.org/10.1186/1465-9921-14-22.
96. Nishiyama O, Kondoh Y, Kimura T, et al. Effects of pulmonary rehabilitation in patients with idiopathic pulmonary fibrosis. Respirology. 2008;13:394–9.
97. Varadi RG, Goldstein RS, Stanbrook MB. Outcomes of pulmonary rehabilitation (PR) in idiopathic pulmonary fibrosis (IPF) [abstract]. Am J Respir Crit Care Med. 2008;177:A792.
98. Salhi B, Troosters T, Behaeqel M, et al. Effects of pulmonary rehabilitation in patients with restrictive lung diseases. Chest. 2010;137:273–9.
99. O'Shea SD, Taylor NF, Paratz J. Progressive resistance exercise improves muscle strength and may improve elements of performance of daily activities for people with COPD: a systematic review. Chest. 2009;36:1269–83.
100. Gloeckl R, Marinov B, Pitta F. Practical recommendations for exercise training in patients with COPD. Eur Resp Rev. 2013;22:178–86.

Chapter 5
Blood Pressure

Introduction

Blood pressure is a vital sign, in addition to respiratory rate, HR, oxygen saturation, and body temperature. Blood pressure (BP) is the pressure of circulating blood on the walls of blood vessels. BP usually refers to the pressure in the arteries; however, measuring invasively venous and pulmonary vessels is an important variable in medical care. Pressures in the vessels of the venous and pulmonary are not as arterial pressure; left atrium pressure is 6–12 mmHg, while the right atrium pressure is 12 mmHg.

Pressure is the force applied vertical to the surface of a body per unit area over which that force is dispersed [1]. The myocardium and the circulatory system are the engine behind the blood flow and pressure throughout the circulatory system (Fig. 5.1).

Blood flows through the cardiovascular system owing to pressure gradient between one part of the system to the following part. BP drops sequentially throughout the cardiovascular system, thus, blood at one point will flow to the next where the pressure is lower. BP varies from minute to minute with exercise, emotional reactions, sleep, digestion, circadian rhythm, age, and gender. Q is eventually responsible for increases or decreases in BP.

The highest values recorded were in the early morning and evenings while at night is the lowest. The heart's ejection elevates BP at the aorta to create pressure gradient in the system. In addition, aortic elastic walls expand following left ventricular ejection and compress back to ensure continuous blood flow throughout the cardiac cycle, at the left ventricular diastole phase, when ventricular BP has dropped to near zero. The aortic expansion is due to the force of the blood surging from the left ventricular ejection. BP values are expressed in millimeters of mercury (mmHg).

BP is the pressure applied on the walls of the large arteries of the systemic circulation. BP usually has two values: SBP and DBP. Although BP is measured in the arms, SBP and DBP represent the pressures in the aorta. Blood pressure is affected

M. S. Saghiv, M. S. Sagiv, *Basic Exercise Physiology*,
https://doi.org/10.1007/978-3-030-48806-2_5

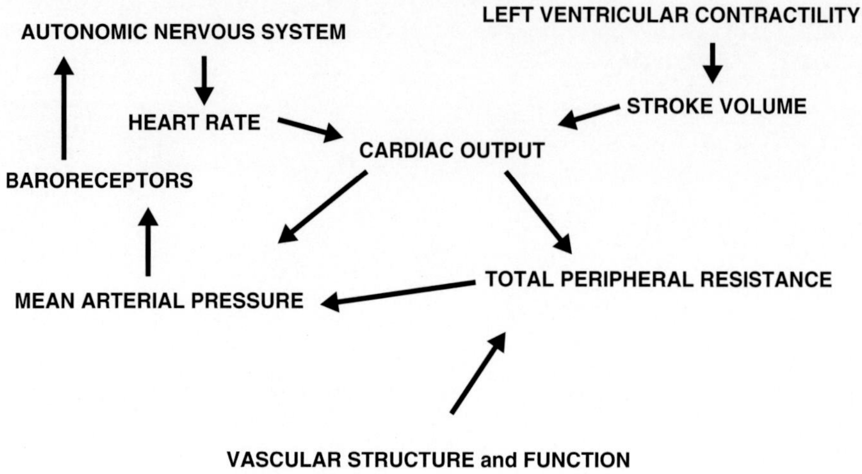

Fig. 5.1 Factors affecting blood pressure

by Q, TPR, and arterial stiffness. In addition, it fluctuates depending on the state, exercise, health status, and emotional stress. Measurements of arm BP differ slightly between both arms. SBP is the pressure exerted, on the aortic walls, immediately, following the left ventricular ejection. While DBP is the pressure exerted, on the aortic walls, before the left ventricular ejection. Resting BP values for an adult male are approximately 120 mmHg (16 kPa) for SBP, over 80 mmHg (11 kPa) for DBP; female's values are significantly lower for SBP 100 mmHg and DBP 60–70 mmHg [2]. In healthy children, height is a factor that influences BP response which is lower than the adult's values. Aging process has the tendency to increase SBP, while DBP has a trend to drop. Consequently, SBP has higher values compared to young normal adults due to arteries stiffening with aging. BP with low values is determined as hypotension (SBP < 90 and DBP < 60 mmHg), while BP that is steadily high is hypertension (SBP 140–160 and DBP 90–100 mmHg). Long-term hypertension is more common than long-term hypotension, which often goes undetected because of rare monitoring and the nonappearance of indications.

Venous Blood Pressure

The veins deliver deoxygenated blood to the right atrium and then to the lungs. Veins are basically tubings that collapse when their lumens are not filled with blood. Compared with arteries, the tunica media of veins, which contains smooth muscle or elastic fibers allowing for contraction, is much thinner, resulting in less ability to deliver pressure. Two actions help to generate venous pressure, one is the skeletal-muscle pump that squeezes the blood toward the heart and the thoracic pump of breathing during respiration that sucks the blood also toward the heart direction.

Venous pressure, in a vein or the atria of the heart, is significantly lower than that of the arterial pressure, with common values of 5 mmHg in the right atrium and 8 mmHg in the left atrium. These low pressures aid the blood to return to the atria. BP usually discusses the arterial pressure in the systemic circulation, yet it is important for intensive care medicine to measure pressures in the venous system. In addition, venous return to the atria is a very important physiological variable since it is a key role in the determination of stroke volume and it is a section in the entire closed circulatory system. Variants of venous pressure include: (a) central venous pressure, a good approximation of right atrial pressure, which is a major determinant of right ventricular end diastolic volume; (b) jugular venous pressure, the indirectly observed pressure over the venous system. It can be useful in differentiating different forms of heart and lung disease, and (c) portal venous pressure or the blood pressure in the portal vein. It is normally 5–10 mmHg.

Mean Arterial Pressure

Another derived measurement is the mean arterial pressure (MABP), which is the average BP in the cardiovascular arteries throughout the cardiac sequence. MABP is an important diagnostic measurement in identifying chronic hypertension. MABP enables to ignore the continual fluctuations of BP. It is the calculated average arterial BP during a single cardiac cycle. There are three significant factors that affect MABP: *Q*, TPR, and blood volume; thus, MABP increment is a result of increases of these three factors. TPR is the total resistance to flow of blood in the circulatory system. Therefore, when resistance increases, BP also increases. Blood volume is also directly related to BP since the circulatory system is a closed system loop, increases fluid in a closed system, and rises pressure to higher values. In normotensive individuals, MABP increases from about 93 mmHg at rest to about 120 mmHg at maximal aerobic exercise, determined by *Q*, TPR, and central venous pressure (usually ignored). MABP can be calculated as follows:

$$\text{MABP}(\text{mmHg}) = 1/3(\text{SBP} - \text{DBP}) + \text{DBP}$$

Pulse Pressure

Pulse pressure characterizes the physical arterial palpation of the pulsation which is easily reachable. The difference in pressure between SBP and DBP is the pulse pressure, which is a useful diagnostic measure for cardiovascular function [3]. The distinction of 40 mmHg between the systolic (120 mmHg) and diastolic (80 mmHg) is also considered to be a normal pulse pressure. For individuals who are 60 years and above, a pulse pressure higher than 60 mmHg serves as a good predictor of myocardial infarction, or other problems in the cardiovascular system. In some instances, a low pulse pressure less than 40 mmHg is indicative of poor myocardial contractility, a lower pulse pressure of 25 mmHg is regarded to be very low. Following vigorous exercise bout, of healthy individuals, low pulse pressure may be noted, though the pulse pressure returns to normal values after a short period of time. Low pulse pressure may also be caused by a drop in the volume of the left ventricular SV, blood loss, shock, congestive heart failure, and aortic valve stenosis. Abnormal pulse pressure values do not always designate health problems; sometimes the difference between systolic and diastolic pressures is not a good cardiovascular complication predictor. It is vital to keep in mind that low pulse pressure is not the only factor taken to define any fundamental medical problem. As an example, at rest, a healthy woman with a BP of 105/60 mmHg has the same pulse pressure (45 mmHg) as a patient woman with a BP of 180/135 mmHg, although the latter patient is clearly at higher risk for adverse events. Aging that brings about the loss of arterial compliance may explain the high pulse pressures found in elder individuals [4].

Total Peripheral Resistance

Changes in SBP, DBP, and MABP influence total peripheral resistance (TPR) or afterload, which is the resistance created by all the factors opposing ventricular ejection. Therefore, low TPR will facilitate left ventricular ejection and supports decrement of left ventricular end systolic volume, compared to resting values and hence increase in SV. TPR increases due to binding of norepinephrine and epinephrine to the $\alpha 1$ receptor on vascular smooth muscles, causing vasoconstriction, thus decreasing the radius of the vessels in the periphery. Vasculature throughout the entire body can be thought of as two separate circuits—one is the systemic circulation, while the other is the pulmonary circulation. TPR is the sum of the resistance of all peripheral vasculature in the systemic circulation. This should not be confused with pulmonary vascular resistance, which is the resistance in the pulmonary circulation. Total peripheral resistance is represented mathematically by the following equation:

$$\mathrm{TPR} = \left(\mathrm{MABP}\cdot 80\right) / Q$$

where TPR = Total Peripheral Resistance; MABP = Mean Arterial Blood Pressure; 80 is a constant number; Q = Cardiac output.

In young and elderly healthy individuals and coronary artery disease patients (without hypertension), TPR value at rest is approximately 1400 dynes s^{-1} cm^{-5}. During aerobic exercise, values drop to around 400 dynes s^{-1} cm^{-5} [5]. This change occurs because of the tremendous increase in blood flow to working skeletal muscles created by a fivefold increase in Q. Even though SBP increases, TPR may fall to one-third of the resting values during exercise. Vasoconstriction in non-exercising tissue is not enough to compensate for the vasodilatation in active muscles. The reduced total peripheral resistance during exercise helps to decrease forces opposing left ventricular ejection and end systolic volume compared to resting values and hence, increase in stroke volume. During dynamic anaerobic bout, TPR values are reduced to 700 dynes s^{-1} cm^{-5} [5]. However, this reduction is not similar to the level observed at peak dynamic aerobic exercise but still, during anaerobic exercise, values are lower than that obtained during isometric exercise (1600 dynes s^{-1} cm^{-5}) [6]. Young and older hypertensive patients when cycling do not reduce significantly their TPR from resting value 2049 dynes s^{-1} cm^{-5} to about 2046 dynes s^{-1} cm^{-5} during exercise [7].

Blood Pressure Regulation

An individual's BP is regulated on a second-to-second basis by a mechanism known as the arterial baroreceptors or baroreflex. Individual's BP varies on a heartbeat to heartbeat controlled mainly by the baroreceptors or baroreflex. Baroreceptors are groups of neurons located in the carotid arteries, at the arch of aorta, and the right atrium, that at rest and during exercise sense changes in the blood oxygen level. When the blood oxygen level decreases, parasympathetic activity decreases and sympathetic activity increases, in addition to changes in BP. An increase in BP during exercise results in a decrease in parasympathetic activity, when the sympathetic activity overrides the parasympathetic activity. Baroreceptors and chemoreceptor supply information to the cardiovascular control areas in the brain.

In hypertensive patients, abnormal variations are in the baroreflex arch. The important change that is caused by the baroreflex is the elevation of BP levels by altering the set point. When this point is set, baroreflex sensitivity for BP changes decreases [8]. Therefore, when BP changes, it triggers the baroreflex to rise or decrease BP. Baroreceptors, controlled by the brain, regulate BP for the short term through the nervous system that influences the endocrine systems. BP can be changed based on activity levels and on body position. For example, when a person is standing, blood will tend to be drawn into the extremities (particularly the legs) with the force of gravity. The baroreflex regulates BP levels, mostly by HR, which is a measurable physiological variable which expresses baroreflex sensitivity and activity. Thus, the heart will need to pump harder in order to recover blood and to deliver it to the brain against gravity; consequently, BP will become elevated. In contrast, if a person is reclining, blood tends to pool in the abdomen and thorax, and the effects of gravity are lessening, thus the heart does not need to pump blood as rigorously to ensure adequate circulation and thus BP will tend to become lower.

The baroreceptor reflex, parasympathetic and sympathetic nervous systems, and the hormonal action regulate BP by reducing and increasing BP, respectively, in order to keep normal values (5.2). Baroreceptors are stretch-sensitive fibers for high-pressure receptors [9], sensitive to blood pressures above 60 mmHg, and transmit the information to the cardio-regulatory center in the medulla oblongata, in the brain. Pressure fluctuates with each beat of the heart, which is one cardiac cycle.

Cardiac cycle has two phases: diastole is the left ventricle filling without contraction, and systole, which is the left ventricle ejection phase (Fig. 5.2).

Therefore, BP is continuously observed and regulated within the normal values range. This is achieved by the baroreceptors, located in the aortic arch and carotid sinus and within the walls of the blood vessels. Baroreceptors are receptors that perceive any minor variations in BP. From the baroreceptors, an impulse is transmitted to the central nervous system.

If the blood pressure within the aorta walls or the carotid sinus increases, it increases baroreceptors' activity and response. Following the information from the baroreceptors, the action increases in vagal neural activity that reduces heartbeats and causes general vasodilation. The cardio-regulatory center of the medulla initiates mechanisms that decrease the blood pressure to a normal level [10].

Lowering BP depends on the sympathetic and parasympathetic input interplay, which increases or decreases BP: the sympathetic nervous system causes an increase in sympathetic nerve activity to the myocardial SA node, that increases HR and SV and thus, Q, while the parasympathetic nervous system does exactly the opposite. Additionally, when baroreceptors sense that BP is elevated, vasodilation results from the increase in action potentials sent to the cardio-regulatory center which, in turn, reduces sympathetic input on the blood vessels. This also lowers TPR values along with BP. The contrary occurs once the baroreceptors of the aorta or carotid sinus sense a reduction in BP. Additional mechanism is the increase in sympathetic stimulation to the blood vessels, resulting in the contraction of arterioles smooth muscle, producing vasoconstriction, which increases TPR and BP. Long-term BP adjustment is achieved by the renin-angiotensin system, by letting the kidneys through angiotensin II vasoconstrictor to keep higher blood volume and thus, increasing arterial BP. Angiotensin II acts to release a steroid hormone Aldosterone released from the adrenal cortex. Sodium retention and potassium excretion are

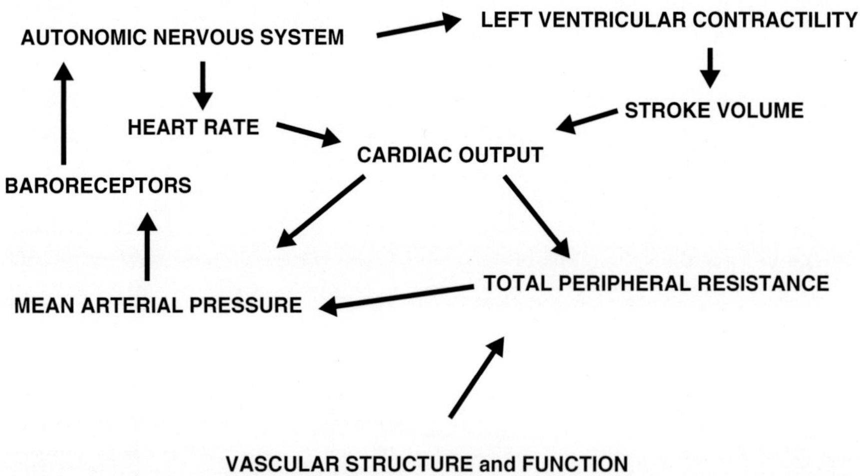

Fig. 5.2 Role of baroreceptors in increasing/decreasing body BP

affected by Aldosterone in the kidneys. Sodium is characterized by its osmotic ability, which increases fluid retention and thus, BP. When BP reaches low levels, the baroreceptors act to secrete antidiuretic hormone that in turn increase BP and *Q*.

In addition to the above, there are four main factor determinants of BP; they are blood volume, *Q*, TPR, and compliance. Increased blood volume increases pressure in the cardiovascular system. Increase in *Q* will increase also arterial pressure. TPR depends on the overall compliance. When TPR decreases, the overall compliance decreases and thus, the arterial BP is reduced (DBP) as during aerobic exercise. The arterials' elastic characteristics enable to control BP by recoiling and expanding and by that compliance changes, consequently, blood pressure varies. When the arterials' expansion occurs, BP decreases and when expending BP increases.

Measuring BP

Measuring BP is a basic skill at rest and during exercise since it provides a good indication of the opposing forces to left ventricular ejection. The indirect measurement of blood pressure monitors the sounds of blood flow in the brachial artery (Korotkoff phases, IV or V) that are audible through the stethoscope due to turbulent blood flow following the arm occlusion by the cuff making a variety of sounds. Choosing the right or left arm to measure blood pressure does not make a big difference since differences are negligible. Different-size cuffs are available for children, women, and adults It should be kept in mind that also body size affects the cuff used. For accurate resting measurements, individuals should avoid vigorous physical activity of any kind for 6-h before the measurement takes place.

Usually, BP is measured noninvasively using auscultation with a mercury-tube sphygmomanometer [11]. BP value of 120/80 mmHg (SBP/DBP) is normal response for males and it is even lower in woman [2]. BP measurement is subject to mistakes, by measuring an individual due to bias, consequently, an error occurs. Training of the measuring individual may decrease but not eliminate totally these causes of mistake [12]. Because accuracy of measurement is particularly desirable in clinicals, efforts have been made to devise devices that would minimize or abolish observer error.

Auscultatory (noninvasive) readings in clinical settings are considered to be the best method for the accurate determination of BP values [13]. BP, like all biological variables, varies widely in different individuals and, in the same individual, at different times of the day. Auscultatory measurements of BP, in apparently healthy individuals, show values that can be 20 or even 40 mmHg higher or lower than the average values [14]. This and the fact that BP varies considerably during the day make it very difficult to decide on the basis of a single auscultatory measurement, whether a patient suffers from hypertension [15]. Perhaps the four most common errors with the measurement of BP by means of auscultation are: (a) using the wrong size cuff, (b) applying the cuff incorrectly, (c) not positioning the stethoscope directly above the artery as incorrect placement will result in too low of a systolic and too high of a diastolic pressure measurement, and (d) incorrect interpretation of the sounds heard. There are many more ways to obtain inaccurate results such as trying to obtain a measurement in a noisy environment, not supporting the arm at the correct level, wrong definition of systole or diastole, not pumping the cuff high enough, releasing the pressure too rapidly, too much (or even too little) finger pressure on the stethoscope, manometer not at the eye level, and hand not palm up. The point is that it is easy to obtain an inaccurate measurement without realizing it, and a correct measurement is only obtained with a great deal of attention being paid to every detail. This attention is seldom present in a clinical environment [16]. Measuring BP in hypertensives in a clinical environment, accuracy may be of high importance. Even a few mmHg can make the difference between being prescribed medication or just having the BP monitored. In general, if the measurement is erroneously low, for example, the patient may be denied the most valuable drug treatment to prevent future stroke and heart attack, whereas if, on the other hand, the measurement is incorrectly high, the individual may be commenced on lifelong blood pressure-lowering drugs unnecessarily, with side effects. Unfortunately, some of the least accurate measurements (auscultatory) take place in the clinical environment where accuracy is most important [17]. Therefore, accurate measurement of arterial pressure is necessary for the diagnosis of hypertension and for assessment of its therapy. It has been also shown to be valuable for defining the effects of antihypertensive drugs therapy. Results of such studies are crucial for advancing antihypertensive management [18]. On the other hand, Invasive measurement is the only direct measurement of BP, which in theory, is a method that could be highly accurate [14].

Blood Pressure Responses to Aerobic Exercise

At rest, Q distribution reveals that only about 20%, namely, 1 L min^{-1} out of 5 L min^{-1}, goes to the skeletal muscles. During exercise, in sedentary young adults Q increases up to 25 L min^{-1}, working muscles receives almost 90% of it, and 10% goes to the liver, kidneys, digestive track, bones, and brain. In world class athletes, Q may reach values of 35–40 L min^{-1}. Due to an increase in right ventricle activity during aerobic exercise, blood flow to the lungs is also increased. When submaximal aerobic exercise commences, SBP rises fast and then levels off at the steady-state phase. This increase in pressure is the resultant increased Q, due to the autoregulation mechanism and reduced TPR. Autoregulation is a local mechanism at the arterioles smooth muscles and it is a myogenic response. During aerobic exercise, the autoregulation mechanism regulates blood flow to the working muscles, which, correlates with work intensity. This mechanism's action preserves DBP similar to resting values and allows SBP to increase.

Two factors may determine BP response at rest and during exercise namely, *Q* and TPR. When arteries narrow, TPR rises. When they widen, TPR drops. More *Q* and greater TPR result in higher blood pressure. During submaximal and maximal dynamic exercise, baroreflex remains active in controlling levels of BP. However, during exercise, the required pressure is rearranged (pressure set-point) to much higher values compared to rest.

The level at which BP is retuned during aerobic exercise depends on work intensity and to the extended influence of the autoregulation mechanism. Compared to rest, blood flow changes remarkably to exercising muscle, which, at maximal effort, can increase up to 35-fold. This significant increment of blood flow may cause a considerable decrease in pressure. This obliges the myocardium to increase significantly *Q* for preserving suitable pressure and oxygen supply. This may explain the linear relationship between HR and VO_2 during exercise. DBP changes little during dynamic aerobic exercise in normal individuals. Typically, there is either no change or a slight decrease during exercise.

There is also a small decrease during recovery of less than postexercise hypotension. A significant increase in DBP (above 15 or above 120 mmHg) is associated with a greater prevalence of coronary artery disease. It seems that the autoregulation mechanism is responsible for keeping DBP values similar and sometimes lower than those values at rest.

SBP during submaximal exercise range from 140 to 160 mmHg, it rises steadily during exercise. It increases from about 120 mmHg at rest to 180–220 mmHg or more during maximal exercise (Fig. 5.3). However, DBP changes very little, if at all. In a young, healthy individual, DBP typically rises by no more than a few mmHg, even during a strenuous effort while some times it decreases.

Following strenuous exercise, SBP drops quickly. This fast drop (postexercise hypotension) is a consequence of blood pooling in the arteries that were expanded during the heavy exercise. As a result of the blood volume pooling in the venous, there is a reduction in venous return to the myocardium and thus, decreased SV and thus, *Q*.

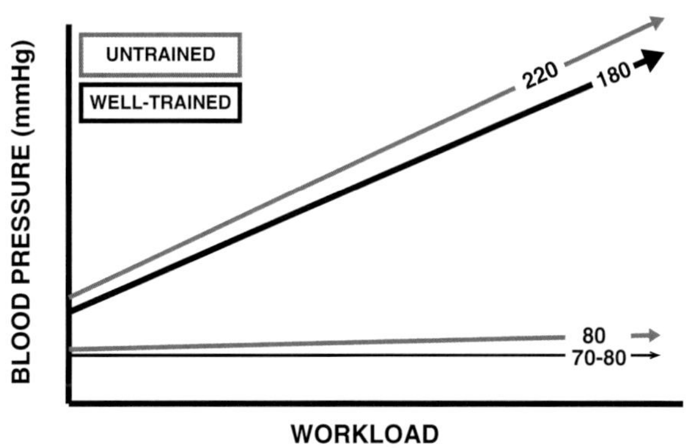

Fig. 5.3 Blood pressure differences between well-trained and sedentary from rest to peak exercise

Gender Effects on Blood Pressure

Studies using the technique of 24-h ambulatory BP monitoring have shown that BP is higher in men compared to women at similar ages [19]. Previous work has shown a gender difference in BP response to exercise measured by auscultatory means. Men had significantly higher absolute SBP responses at rest, 50%, 75%, and 100% peak HR on all modalities [20]. Men have a higher systolic BP in spite of the fact that they had similar sympathetic nervous system response as indicated by urinary norepinephrine excretion [21].

Although the mechanisms responsible for the gender differences in BP control are not clear, there is significant evidence that testosterone plays an important role in gender-associated differences in BP regulation [22]. Comparing direct vs indirect BP measurement method of DBP measurement has shown that in men at peak treadmill exercise and bicycle, the indirect method tends to bias low compared with the direct one. Therefore, the indirect method is not valid for the assessment of DBP [23]. Instead, invasive measurement is the only measurement that is highly accurate. This method reveals that the mean values of SBP and DBP are higher in the males compared to that of the females.

During exercise, there is a significantly higher rise in SBP in males than females. Studies [24] show that the rise in SBP during exercise is mainly due to increase in *Q*, reflecting the level of sympathetic and parasympathetic drive, with men demonstrating greater sympathetic activity and higher baroreflex responsiveness compared to females [25].

The average adult male HR is higher from that of adult women differences largely accounted for by the smaller heart in females than males. Thus, female pumps less blood with each beat, and therefore has to beat at a faster rate to match the larger male heart's output. Aerobic exercise is associated with an increase in sympathetic tone and skeletal blood flow but a decrease in TPR [26]. With exercise, a rise in SBP is expected, whereas the change in DBP can vary 10-mmHg increase or decrease in TPR [26].

Blood Pressure Response During Resistance Exercise

Resistance exercise is associated with a resetting of the operating range and an increase in the reflex gain of the arterial baroreflex control of muscle sympathetic nerve activity [27]. In isometric muscle activity, large increases in SBP and DBP are noted resulting in increased MABP, which is seemingly inappropriate to the metabolic demand (Fig. 5.4).

During isometric muscle activity, a sharp and rapid increase in SBP (300–360 mmHg) and DBP (120–140 mmHg) and thus, MABP (140 mmHg and above) can be seen. This increase in BP response is seemingly inappropriate to the metabolic demand (Fig. 5.5). This phenomenon, known as pressor response, is

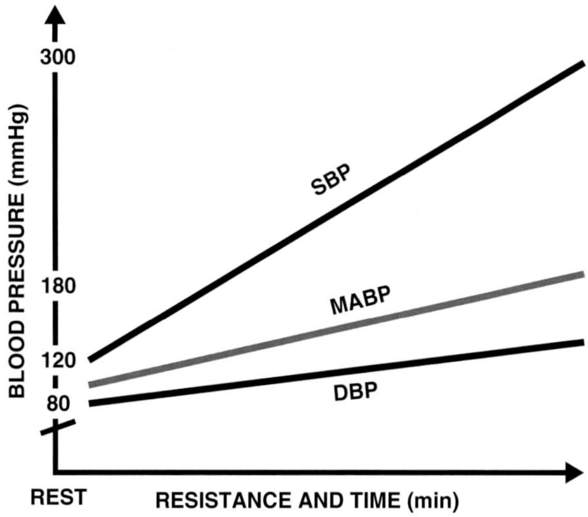

Fig. 5.4 Systolic, diastolic, and mean arterial blood pressures response to isometric bout percentage of maximal voluntary contraction

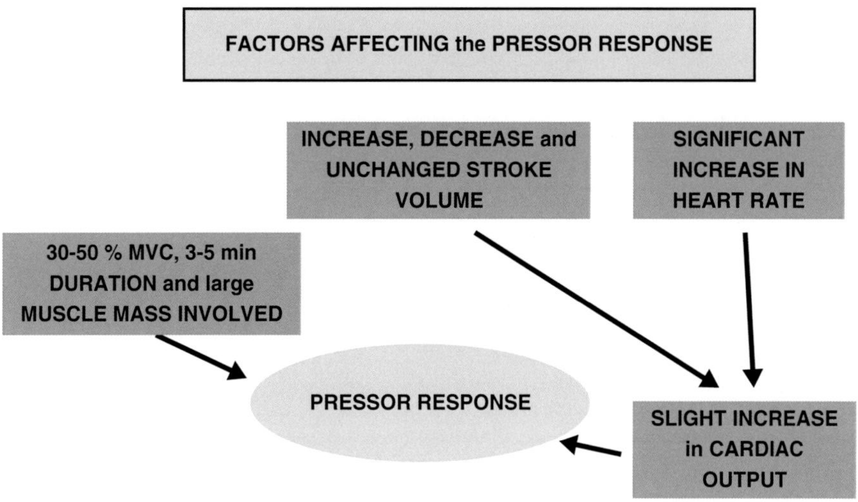

Fig. 5.5 Relations between physiological factors that influence BP response during isometric exercise: muscle mass involved, exercise duration, and MVC. *MVC* Maximal voluntary contraction

characterized by a sustained rise in SBP and DBP to higher levels than during dynamic exercise at an equal level of oxygen uptake and *Q* demand [28]. The abrupt and sharp increase in BP occurs in contractions above 20–30% of maximal voluntary contraction (MVC).

The pressor response is a result of the small increase in *Q* to values of 8–9 L min^{-1}, due to the moderate increase in HR and TPR, and with possible vasoconstriction, in non-active muscles, due in part to the reflex mechanism originating in active muscles. It has been found that the pressor response during isometric work increases proportionally to the contraction intensity, duration, and muscle mass along the integrity of the sensor pathway from active muscles [29].

Figure 5.5 summarizes the effects and relationships between cardiac output, muscle mass, percentage of maximal voluntary contraction, and exercise duration on the pressor response during isometric exercise.

During a resistance exercise bout, in patients with cardiovascular disease, or known risk factors, SBP and DBP may show significant increases above that of normotensives, which advise to be cautioned when performing resistance bouts. The extent of the increase in blood pressure is dependent on the time the contraction is held, the intensity of the contraction, and the amount of muscle mass involved in the contraction [30].

More dynamic forms of resistance training, such as circuit training, that involve moderate resistance and high repetitions with short rests are related with reductions in BP, others have shown decreases in DBP [31], or no change in blood pressure, and decreases in systolic blood pressure. The effects of resistance training on blood pressure are varied due largely to differences in study design, which suggests that more research is necessary to clearly understand the role of resistance training in blood pressure management. Thus, even though those studies have shown that resistance training improves blood pressure in people with borderline hypertension [32], many are still concerned that blood pressure may rise to dangerous levels during the workout itself. A study that followed 10 experimental and 16 control subjects through 9 weeks of resistance training found a marked increase in strength, but more importantly, the conclusion was that circuit weight training did not exacerbate resting or exercise blood pressure [31].

BP Response During Anaerobic Exercise

During all-out physical activity (anaerobic bout), it is almost impossible to determine blood pressure values utilizing the conventional indirect sphygmomanometer. Such activity is the Wingate Anaerobic Test which is characterized by exposing the subjects to an all-out strenuous exercise on a cycle ergometer [33]. Measurements comparing auscultation and intra-arterial values reveal that SBP can be measured indirectly at peak anaerobic exercise with satisfactory results. However, at peak all-out anaerobic exercise, the indirect method is not valid for the assessment of DBP [33].

While anaerobic exercise is dynamic in nature, BP responses are similar to those seen during isometric exercise, i.e., SBP and DBP are increased and thus MABP. Figure 5.6 summarizes the effects and relationships between blood volume, HR, SV, TPR, and viscosity on BP at rest and during dynamic aerobic and anaerobic exercises. Viscosity or thickness of the fluid gets thicker; the result is an increase in blood pressure. Certain medical conditions can change the viscosity of the blood.

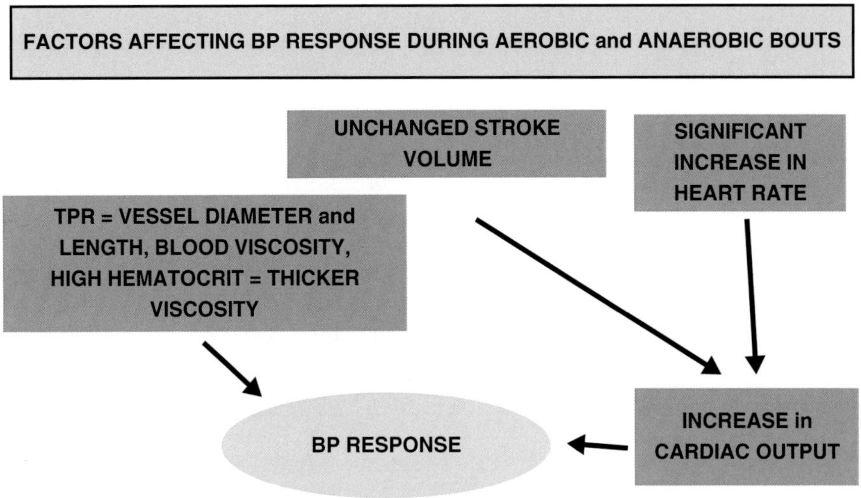

Fig. 5.6 Relations between physiological factors that influence BP response at rest and during aerobic and anaerobic exercises. *TPR* Total peripheral resistance

For instance, low red blood cell concentration, anemia, reduces viscosity, whereas increased red blood cell concentration increases viscosity.

In practice, each individual's autonomic nervous system responds to and regulates all these interacting factors so that, although the above issues are important, the actual BP response of a given individual varies widely because of both split-second and slow-moving responses of the nervous system and end organs. These responses are very effective in changing the variables and resulting in blood pressure from moment to moment.

With the cessation of the dynamic aerobic exercise, there are rapid and large changes in the return of blood to the heart. This is because once the muscles have stopped contracting, there is no muscle pump action to massage blood back to the heart (venous return) and therefore blood remains in the active limbs. If the removal of the muscle pump is combined with an upright posture, the influence of metabolites such as lactic acid and elevated body core temperature, it will be quite possible that the individual will experience hypotension and symptoms of fainting. Values of post-exercise hypotension decline between 5 and 20 mmHg compared to pre-exercise values.

Hypertension

BP results by both the amount of blood that the myocardium ejects (Q) and the total of resistance to blood flow in the arteries. The more blood the heart pumps and the narrower the arteries, the higher BP. Maintaining constant adequate pressure in the circulatory system is compulsory for blood flow. Significant low blood pressure reduces perfusion of organs, while unusually high blood pressure may result in myocardial disease, vascular disease, and stroke. Lack of exercise, type 2 diabetes, and obesity raises the risk of BP disorders, which include hypertension. The disturbing long-term results of high BP include: cardiac hypertrophy, heart failure, stroke, kidney disease, and atherosclerosis. Hypertension is called also the silent killer [34] since, most individuals with hypertension do not have any signs or indications. Sometimes, few individuals with high BP may experience headaches, shortness of breath, or nosebleeds, however, these symptoms occur also in other diseases or medical complications. The precise sources for high BP are not clear enough, nonetheless numerous effects might be part of the cause, including: smoking, physical inactivity, alcohol consumption, aging, genetics, kidney dysfunction, and obesity. In 95% of high BP of individuals, the primary reason cannot be detected [35]. These individuals have primary essential high BP due to general lifestyle and genetic factors [35]. General lifestyle includes: surplus salt in the diet, obesity, smoking, and alcohol use. This kind of increased BP is called essential hypertension or primary essential high BP [36]. High BP is defined if the resting BP is persistently at or above 130/80 or 140/90 mmHg [37]. Essential hypertension has been related to certain risk factors: family history, gender, likely to affect men more than women, aging, and race. It has been found that African Americans, compared to whites suffer twice from high BP. This race difference begins to be reduced around age 44 years. Still, black women have the uppermost occurrence of high BP after the age of 65 years. Essential hypertension is significantly influenced and related to daily salt amount intake. Though individuals that consume low daily salt demonstrate nearly no signs of essential hypertension. Most individuals with high BP are sensitive to salt intake, and any amount above the body needs results in an increased BP. Additional effectors that can increase essential hypertension include physical inactivity, obesity, diabetes, chronic alcohol consumption stress, magnesium, potassium, and calcium imbalance.

Once the factor affecting increased BP is recognized, the disorder is defined as secondary hypertension caused mainly by kidney disease. Hypertension can be developed by the secretion of large quantities of the hormones by the adrenal glands located on top of the kidneys, hormones that increase BP.

Inflammation and lipids narrow the arterioles' diameter by the creation of plaque in the arteries that cause a significant increase in BP [38]. Plaque buildup in the walls of the arteries from hypercholesterolemia, fatty deposits, other substances, and inflammatory mechanisms combine dyslipidemia to atheroma development (Fig. 5.7) to cause hypertension [39]. The additional arterial tension and subsequent damage due to hypertension cause the coronary arteries to become narrowed from accumulation of fat, cholesterol, and other molecules that collectively is the slow process of chronic inflammatory disease [40].

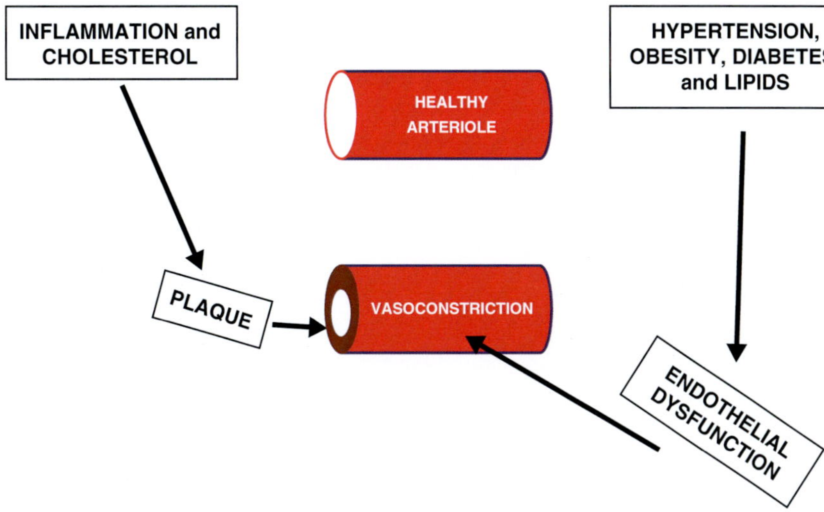

Fig. 5.7 Endothelium dysfunction resulting in an arteriole vasoconstriction

Aging

Aging is accompanied by cardiovascular changes that include a decrease in elasticity and an increase in stiffness of arteries, which increases the pressure opposing (afterload) left ventricle ejection, resulting in systolic hypertension. At the age 55 and above, the risk of developing hypertension in normotensive women is very high [2], for men the rate is 60%.

Role of Kidneys in Hypertension

Kidney genetics has a key role in the progression of hypertension [41]. Previously, it has been suggested that kidney recipients, from a genetically hypertensive donor, develop higher BPs [41]. Kidney mechanisms that regulate arterial pressure for the long term are related to the blood volume, sodium balance, and extracellular fluid volume control. Sodium intake is a function of diet and environment, while excretion is defined by the kidneys. Sodium creates the main osmolarity activity of the extracellular fluid [42], while the regulation of plasma osmolality is done mainly by vasopressin hormone and the water drinking quantity. Consequently, net sodium equilibrium regulates the level of extracellular fluid volume. Balancing sodium is mainly done by means of the kidneys, by adjusting sodium excretion rate, as a response to large variations in sodium intake. In addition, the effects of increases in salt intake on extracellular fluid volume is the actually increase in plasma sodium level which influences vascular reactivity, and thus, the hypertensive progression [43].

Increased blood volume expands the vessels' ability which in turn, increases MABP which reflects the actual blood volume. Through its direct effects on venous return and Q, the MABP influences arterial pressure that alters renal sodium and water excretion balance. Thus, at any given balanced condition, a direct association among renal perfusion pressure and sodium excretion exists. Hypertension pathophysiology is a complicated problem since there are numerous connections between neural, hormonal, and cell–cell communication (paracrine) mechanisms regulating cardiovascular and renal function. Systemic BP can be influenced by various factors, such as Q, TPR, and cardiovascular expandable capacity [44]. Mechanisms regulating sodium excretion keep sodium balance by suitably controlling the relations, pressure-disproportionate loss of sodium in the urine, resulting in a normal arterial pressure.

The Renin-Angiotensin System

Renin-angiotensin system (RAS) is an influential inflector of blood pressure and the opposite for RAS resulting in hypertension [45]. Blocking RAS with renin inhibitors, angiotensin-converting enzyme (ACE) inhibitors, lowers significantly BP in the majority of hypertensive patients [46], suggesting RAS activation as a source of human hypertension. Similarly, genetic manipulation on rodent models revealed that when the RAS gene was deleted, BP decreased, and vice versa [47]. The influential RAS facilitates its significant effects on BP by thru effects on the kidney. Moreover, there is classified regulation of the RAS laterally to the nephron causing origination and development of hypertension.

Obesity Effect on Hypertension

Obesity is the main risk for essential hypertension, type 2 diabetes mellitus, and development of kidney disease [48]. Obesity effect on the kidney results in impaired pressure natriuresis that causes volume expansion. The relationship between obesity and hypertension is well known [49] in its association with fat dysfunction, characterized by adipocytes growth, increased penetration by macrophages, and significant variations in the excretion of adipokines and free fatty acids. However, still there is no one cause to explain obesity effect on hypertension [50]. This causes chronic vascular inflammation, oxidative stress, activation of the renin-angiotensin-aldosterone system, and sympathetic overdrive, ultimately resulting in hypertension [49]. The sympathetic nervous system, sodium retention, and the RAS are factors affecting the pathogenesis of obesity-related hypertension [51]. Several alterations in renal structure and function have been associated with obesity. Activation of the sympathetic nervous system to the kidney [52], and the RAS as well as increases in plasma aldosterone levels can cause abnormal sodium retention and raise arterial pressure [51]. RAS controls the balance of sodium and water retention in the body. Obesity interferes these two factors that regulate body BP leading to hypertension. In addition, the adipose tissue deposition can affect kidney function and thus, altering BP. Almost 60% of the individuals with obesity are at risk of developing hypertension. So, obesity should be seen as a complex and chronic medical disease with its mechanisms leading to hypertension. To control obesity-associated hypertension, weight loss and exercise are an operative way for handling the disease. Weight loss and exercise result in a significant reduction in BP.

Atherosclerosis

Atherosclerosis is the very significant leading cause of death, has a key role in the development of arterial diseases, and it is responsible for myocardial infractions and strokes. Although the exact cause of atherosclerosis is unknown, it is believed that atherosclerosis begins with an injury in the endothelium. When cholesterol-laden plaques build up in the arteries, it results in atherosclerosis, known as a chronic inflammatory disorder. This inflammatory condition results from the connections between modified lipids and monocyte-derived macrophages, T-lymphocytes, and a variety of cytokines secreted by these and other cells in the artery wall [53]. These factors excluding lipids are various components of the immune system. With time, plaques enlarge progress, the arteries narrow, causing blood flow impairment. Many times, individuals with atherosclerosis do not have any symptoms until an artery is narrowed by 40%. Many factors contribute to the development of atherosclerosis, but the oxidation of the protein and lipid components of low-density lipoprotein particles is believed to be among the most important initiating events [54]. Plaque causes arteries to become stiff and inflexible. Plaque becomes thick and hard reducing blood delivery through the arteries. Risk factors for atherosclerosis include being male (if female, being past menopause), hypertension, high low-density lipoprotein cholesterol or triglycerides circulating in the blood, diabetes, obesity, smoking, a family history of heart disease, and physical inactivity.

Inflammation Effect on Hypertension

Arterial endothelium assists vasomotor tone and function by producing and discharging nitric oxide [55]. Arterial endothelium is characterized by cells in the inner layer of all blood arteries and lymphatic system. The inner layer of arteries and veins is the tunica Intima. In arteries, this layer is composed of an elastic membrane lining and smooth endothelium which is an unusual type of epithelial tissue that is covered by elastic tissues. The small thickness of endothelium at the capillaries level permits molecules' movements actively and passively as well as ions among blood and lymph by means of the tissues [56]. Following the endothelial layer, elastic fibers' thickness varies, which is in a straight line associated with its ability to change vessels' blood volume, pressure, and flow velocity [57]. Arterial endothelium contributes to the regulation of blood coagulation, platelet and leukocyte action, vessel's tone, and inflammatory responses [58]. Inflammatory response is a defensive answer to damage or infection. The endothelium plays an important role in blood vessel functions by producing various signaling molecules [59]. It is a vital regulator of blood flow and blood pressure in the circulatory system. The endothelium is the internal coating of cells that have several critical functions, it controls the correct vessels' radius namely, constriction or dilation, thus changing rapidly blood's amount delivered to the various organs. The endothelium also protects the tissues from various toxic substances by (a) coagulation or formation of a fibrin clot, which is a mechanism that blood changes from a liquid to a gel, it includes activation, bond, and accumulation of platelets along with deposition and maturation of fibrin; (b) controls the fluid; (c) regulates electrolytes' levels; and (d) exchange of numerous substances between the blood and the tissues and vice versa; and (e) regulates inflammation damages in the tissues. Consequently, endothelium function is critical for the regular function of the arteries, tissues, and organs [60].

Existence of hypertension and dyslipidemia are directly associated with endothelial dysfunction resulting in inflammatory (Fig. 5.8) and arterial damage [61]. The coronary artery response to acetylcholine rests on the capacity of the endothelium and the endothelial nitric oxide pathway [62]. Arteriosclerosis affects the major elastic and muscular arteries in the extracellular intermediate of arteries' elasticity resulting in arterial stiffness [63].

Arterial stiffness results after a deteriorating course affecting mostly the extracellular medium of elastic arteries. It expresses the arterial wall viscoelastic capacity which affects arterial smooth muscle and therefore, the regulation of blood flow,

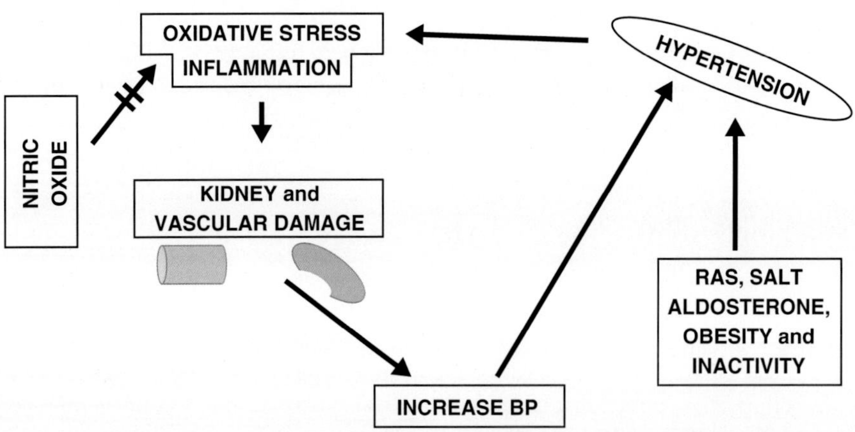

Fig. 5.8 Factors causing hypertension that result in inflammation, kidney, and vascular damage

blood pressure, arterial pulse, permeability, and inflammation [64, 65]. Changes in extracellular medium proteins and mechanical properties of the arterial wall associated to vessel stiffening can trigger other mechanisms implicated in the progress of atherosclerosis. With time, arterial stiffness increases progressively, exposing the individuals to a greater pressure variation related to increased risk of stroke and renal impairment [66]. Changes in arterial pulsatile lead to an increase in the slow frictional force of blood flow counter to the vessel wall [67]. However, increased wall shear stress reduces nitric oxide production by the endothelial that in turn, increases the rate of atheromatous (artery inflammatory) development [68].

Dysregulation of the nitric oxide normal smooth muscle contraction properties response is usually characterized by endothelial dysfunction, following advanced atherosclerosis and increased oxidative stress [69] that induces endothelial dysfunction and atherosclerosis progression by reducing nitric oxide availability [70]. Endothelial dysfunction is a significant moderator in the progress of atherosclerosis and exists long ahead of the creation of atherosclerotic plaques [71]. Endothelial dysfunction caused by inflammation is a possible mechanism causing hypertension. Endothelial dysfunction is mostly triggered by decreased production or action of relaxing mediators. At this point, it is clear that reduced nitric oxide levels cause endothelial dysfunction in blood arterial walls which leads to vasoconstriction of the artery [72] and thus increases blood pressure to high values (hypertension). In addition, it also activates platelets leading to blood clotting that increase the stimulation for artery walls' inflammation, consequently, arterial walls' penetrability to destructive lipoprotein, oxidized free radical species, and various toxins are increased.

Physical Inactivity Effect on Hypertension

Humans genetically are planned to exercise for physiological benefits, while physical inactivity is a major risk factor for many diseases and reason for untimely deaths [73]. Although physical inactivity has a significant role in the pathogenesis of many chronic diseases, however, exercise training reduces their occurrence [74]. Physical inactivity has emerged as a risk factor for hypertension (Fig. 5.9) and numerous diseases [75] while regular exercise decreases the risk of coronary vascular disease. Regular exercise has the capacity to defend against the development of coronary artery disease and to diminish the risk of repeated cardiac episodes [76], while physical inactivity is a substantial threat for coronary vascular disease itself.

Physical inactivity causes less fluid to enter the cells which interfere with the cell's metabolic process resulting in disuse syndrome thus, less ATP is produced. Due to this relationship, metabolic syndrome is diagnosed by BP. Hypertension is an important essential of the metabolic syndrome [77], yet the causal mechanisms for triggering hypertension by metabolic syndrome are yet to be clarified. Physical inactivity has been suggested to be a probable factor, to cause hypertension, by affecting metabolic syndrome which in turn, may affect; obesity, insulin resistance,

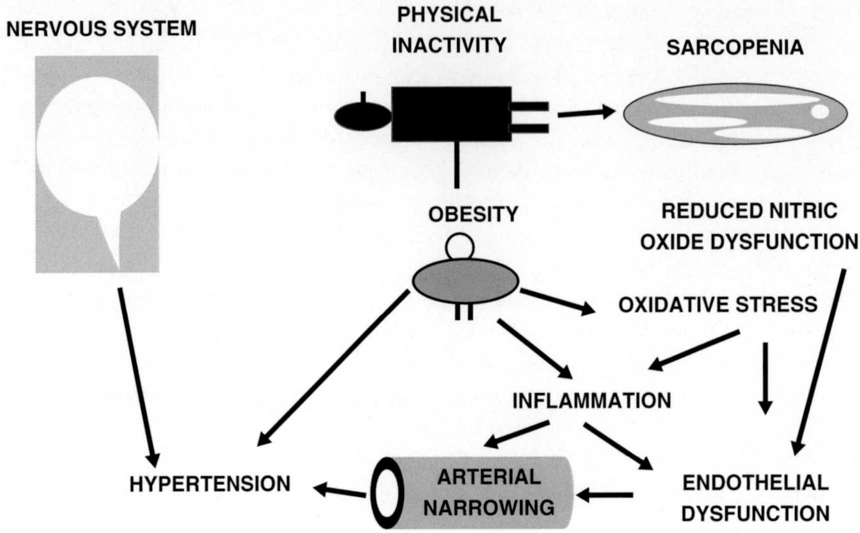

Fig. 5.9 Factors affecting endothelial dysfunction, which narrow the arterials resulting in hypertension

sympathetic overactivity, oxidative stress, endothelial dysfunction, activated RAS, and increased inflammatory mediators have been suggested to be possible factors to develop hypertension in the metabolic syndrome.

Arterial Stiffening

Arterial stiffening is a term that designates the decreased elastic vessel wall properties [78]; it occurs as a result of biological aging process and arteriosclerosis, thus, it turns to be a key donor to arteries stiffening [79], and is related with an increased risk of cardiovascular events such as myocardial infarction and stroke [80]. Usually, the elasticity and distensibility of arteries preserve a moderately stable BP, in spite of the pulsating characteristic of the blood flow by every heart ejection [81]. Increased arterial stiffness decreases the capacity of arterial capacitance along with an increase in pulse pressure, causing increased afterload [82]. As a consequence, increased afterload force left ventricular to augment contractility in order to eject SV; this increases myocardial demand for the demand of oxygen and may develop left ventricular hypertrophy and times, heart failure [83]. Inadequate arterial capacitance also spreads the augmented pulsatile pressure to the peripheral arteries where it causes injuries in the microvasculature of the kidneys, brain, and others [83].

Endothelial dysfunction is characterized by reduced vascular nitric oxide levels that cause some irregularities in blood artery function [84]. These functional irregularities result from atherosclerosis that can cause vasoconstriction of the small arteries and is a major factor in producing hypertension and possibly, left ventricle diastolic dysfunction. Nitric oxide is a key controller of vascular stability, acting through different means to control vascular tone. Nitric oxide formed within the cytosol, of endothelial cells, has a significant role in arterial smooth muscle relaxation. In addition, nitric oxide preserves arterial blood flow by modifying platelet-endothelial relations [85]. Usually, nitric oxide employs defense to the artery by numerous ways, as well as the decrease of endothelial cell apoptosis and initiation, through nuclear factor-kappa factor (NF-κB) reserve and inflammatory gene expression [86].

Hypertension is a serious health problem that increases the risk of aortic stiffness. These increases in arterial stiffness are thought to contribute to age-related increases in the incidence of cardiovascular disease [87]. Endothelial dysfunction in hypertension and diabetes mellitus appears to be due, in part, to an imbalance in reactive oxygen species (ROS) generation and a decrease in nitric oxide accessibility [84]. In rats, a decrease in nitric oxide obtainability donates to endothelial dysfunction, causing essential hypertension and diabetes mellitus [88]. Endothelial NOS plays a prominent role in arterial stiffness, which may predict coronary heart disease beyond classic risk factors. In a longitudinal study, the predictive value of arterial stiffness on coronary heart disease in patients with essential hypertension and without known clinical cardiovascular disease has been assessed [89]. It was found that measurements of arterial stiffness correlate significantly with those of endothelial function, mainly influenced by the relationship of angiotensin-converting enzyme (a key component in the RAS) and bradykinin. Data show that different molecular mechanisms are responsible for hypertension, and aortic stiffness is seen in elderly and in cardiac patients. Moreover, there is growing evidence for a genetic contribution to the pathophysiology of hypertension and aortic stiffness. Genetic studies may help to understand the mechanisms underlying the involvement of the RAS in arterial regulation. Hypertension may produce arterial stiffening by both functional and structural mechanisms. Therefore, the distending or MABP is an important confounder of measurements of arterial stiffness [90].

Arteries serve a dual role of conducting blood to the peripheral tissues and buffering the pressure pulsations that are a necessary accompaniment of intermittent ventricular pumping. Isolated systolic hypertension may increase arterial stiffness, especially in older subjects, but not peripheral artery stiffness, although the underlying mechanisms are somewhat unclear [91].

Aortic Stiffening

Aortic elasticity is an important factor of left ventricular performance and coronary blood flow. As a consequence of arterial stiffness, left ventricular workload is increased, and myocardial oxygen demand, leading at times to left ventricular hypertrophy, [92] impairs ventricular relaxation [93] and causes subendocardial ischemia in the presence or absence of coronary artery stenosis [94]. It is well appreciated that the aorta does not only provide a conduit function but plays important roles in regulating left ventricular performance, coronary blood flow, and normal arterial function, throughout the entire cardiovascular system. Therefore, loss of vascular distensibility reduces the capacitance function of the aorta revealed as elevated pulse pressure, which adds load on the myocardium and likely injures large and small vessels as well. Several studies have shown that disease states, aging, and pharmacological agents may alter the elastic properties of the aorta and that stiffening of the aorta may be associated with an increased incidence of cardiovascular events [95]. These effects may increase the susceptibility of the aged heart to ischemia and ventricular dysfunction. An increase in the stiffness of large conduit vessel may represent either a cause or a consequence of endothelial dysfunction [96].

Aerobic Exercise Effect on Hypertension

Aerobic exercise has preventive effects and anti-atherogenic properties [97], as described in different animal models.

A previous study demonstrates that long-lasting aerobic exercise training reduces cardiovascular risk factors including hypertension [98]. Following long-lasting aerobic exercise, results of arterial stiffness are meaningfully lesser in aerobically well-trained individuals [99], and in subjects who were on an aerobic exercise program [100].

Long-lasting aerobic exercise training increases the preventive ability of the vasculature to oxidative stress, atherosclerosis, inflammation, and reductions in arterial stiffness [100].

Nitric oxide has inhibitory effects on platelet and leukocytes as well as induces proliferation of arterial smooth fibers [101]. The nitric oxide synthesis rate seems to be parallelly related to an increase in amino acid arginine availability [102]. The antithrombotic effect of aerobic exercise decreases serum levels and activity of inflammatory factors, such as interleukin-6, C-reactive protein, and tumor necrosis factor-α, pointing out about the antithrombotic effect of aerobic exercise [103]. Following the aerobic bout, BP decreases to lesser levels than those recorded at rest, identified as postexercise hypotension [104]. This autoregulation mechanism response of aerobic exercise coupled with the nitric oxide vasodilator affects positively the endothelium function and thus blood flow by decreasing TPR [105].

Exercise has a key role in the therapeutic process for the prime prevention and treatment of hypertension [106]. Epidemiological studies suggested that an inverse relationship between BP and exercise exists. Meta-analysis study suggested that aerobic exercise training decreases SBP and DBP of hypertensive patients [107]. In addition, it was found that training in healthy individuals resulted in a significant reduction in BP. This course of reduction was more marked in hypertensive patients. Similarly, the outcomes of 15 meta-analysis studies, on aerobic training effects on ambulatory blood pressure in hypertensive individuals, demonstrated a substantial decrement of SBP and DBP [107].

Aerobic exercise training reduces BP through the autoregulation mechanism, which attenuates TPR, following neurohormonal and arterioles local responses, along with decreases in sympathetic nerve activity and an increase in arterial lumen diameters [108]. Post-exercise hypotension is a response following aerobic exercise, it remains almost 24 h [104]. The most marked effects are found in those with higher baseline blood pressure [109]. Long-lasting aerobic exercise response is further continued with decreases in BP [110]. Other studies proved exercise had appreciated effects on hypertension by decreasing both systolic and diastolic BP by 5–7 mmHg in hypertensive patients [111]. On the other hand, anti-hypertensive response to exercise is very inconstant, differences in exercise routines, ecological factors, and genetic factors are accountable for substantial inconsistency in the BP response to physical efforts [112]. Almost up to 25% of hypertensive patients are non-responders without decrement in BP following aerobic exercise training [113]. Following resistance training, revealed also a reduction in BP.

Aerobic Effect on Arterial and Aortic Stiffness

Regular exercise is related to a reduced risk of cardiovascular disease [114]. In the Baltimore Longitudinal Study of Aging, it was found that aortic arterial pulse wave velocity increased progressively and that older adult males who performed endurance exercise on a regular basis demonstrated lower levels of aortic arterial pulse wave velocity than their sedentary peers did. Previous studies have reported that in endurance athletes, arterial stiffness is decreased by prolonged endurance training [115], and aortic distensibility is increased, particularly in those with the ACE II genotype. In addition, even a moderately intense exercise may be as brief as 10 min in duration is needed in order to elicit a decrease in resting blood pressure and may have potential benefits as a non-pharmacological aid to hypertension [116]. These effects represent an extracardiac adaptation to chronic prolonged training in athletes [117].

Arterial compliance is an important therapeutic target in individuals in whom stiffening of the proximal aortic circulation is thought to underlie systolic hypertension and increased cardiac work. This indicates a positive association between systemic arterial compliance and fitness level in healthy older people and an inverse association between systemic arterial compliance and systolic blood pressure. These findings are consistent with either acquisition of a more compliant circulation and lower blood pressure due to enhanced physical activity or that a more compliant arterial circulation and lower blood pressure permits greater athletic performance

[118]. For healthy subjects between the ages of 25 and 65 years, there is an interactive effect between age and gender and an independent effect of physical training on peripheral vascular function [119]. It was suggested that interventions to improve aerobic capacity alleviates the stiffening of the arterial tree that accompanies normative aging [120].

Resistance Exercise Effect on Hypertension

Meta-analysis that analyzed 64 different studies, on a total of 2344 hypertensive patients, suggested that resistance exercise is an operative stand-alone antihypertensive treatment [121]. Resistance training lowers SBP, DBP, and MABP. The scale of effect is greater than that before described during aerobic exercise or resistance training. It has been suggested that resistance training has the possibility to reduce BP [122]. In studies that hypertensive patients performed above eight resistance bouts each session, demonstrated greater decreases in SBP compared to studies using a smaller amount than eight exercises bouts per session. DBP was reduced also, following resistance exercise, however, greater reductions were seen in untreated hypertension individuals, performing more than three resistance exercise bouts sessions per week.

Genetic alteration has been revealed in family and twin studies, but the exact nature of the postulated genetic defect remains largely unknown [110]. So, evolving data propose genetic associations to the BP reductions related to acute and chronic exercise.

Combined Resistance and Aerobic Exercise

Combined resistance and aerobic exercise training is an approach suggested by the American College of Sports Medicine in hypertensive patients [109]. It is recommended to train 30 min of aerobic exercise at 50–60% VO_2 reserve 4–5 days a week, accompanied by moderate resistance exercise 60%–80% 1 RM, 8–12 repetitions, 2–3 days per week [109]. A recent meta-analysis of 68 trials on hypertensive patients supports the efficacy of combined resistance and aerobic exercise training as antihypertensive therapy [123]. Together, the studies demonstrated in patients with hypertension a decrease in SBP and DBP [123]. A previous meta-analysis, informed of an extensive difference in the BP decreases in SBP and DBP, also demonstrated that studies with a developed procedural quality gained greater BP drops related to lesser methodological quality trials [124].

In a recent controlled trail, hypertensive patients, performed an aerobic bout and isometric handgrip maneuvert [125], results suggested to further investigate, if resistance exercise can reduce BP.

Hypotension During Exercise

Usually, BP response to aerobic exercise is an increase in SBP, while the diastolic values stay the same or may even decrease slightly. SBP levels that reduce during aerobic exercise reveal of hypotension. Arterial pressure is the greatest controlled circulatory component, thus far, there is a continued drop in arterial pressure (post-exercise hypotension), following aerobic exercise of moderate duration and intensity [126]. Post-exercise hypotension is seen subsequent to an acute bout of either resistance or submaximal dynamic exercise [127]. However, both, aerobic and resistance exercises do not result with the same BP response, although in both modes of exercise, there is a reduction in post-exercise pressure [128]. A decrease in SBP, which is not extensive, does not yield any symptoms at all. Following aerobic exercise reductions in BP values, are well known, this hypotensive response holds for short periods of time, particularly if an individual exercise at high intensities. With hypotension following aerobic exercise, SBP values decrease, sometimes reaching lower values compared to those, taken at rest, before the exercise bout performance. Continual post-exercise hypotension is facilitated by both central neural mechanisms and local vasodilatory mechanisms. The arterial baroreflex decrease sympathetic activity, related to HR beats recovering from the effort, due to lessening sympathetic outflow [129]. In addition to the sympathetic activity removal, during recovery HR is mainly determined by parasympathetic reactivation [130].

The mechanism underlining vasodilation or autoregulation mechanism is a manifestation of local blood flow regulation, during aerobic exercise and during post-exercise hypotension occurs mainly within the vascular of the active skeletal muscle [131], and is which is independent of nitric oxide [132]. The mechanism underlining vasodilation or autoregulation mechanism is a manifestation of local blood flow regulation during exercise and during post-exercise hypotension occurs mainly within the vascular of the active skeletal muscle [131]. In general, vasodilation in active skeletal muscles results from local autoregulation, namely, CO_2 and pH along with arterial baroreflex retuning. Autoregulatory mechanism is intrinsic, happens when neural and hormonal effects.

It is defined as the fundamental capacity of a tissue to preserve continuous blood flow in spite of variations in perfusion pressure. Redistribution of Q, during exercise, does not increase blood flow to splanchnic, cutaneous, and cerebral; due to the sympathetic nervous system activity, vascular conductance is unchanged from rest [133]. Autoregulation mechanism is defined as the fundamental capacity of a tissue to preserve continuous blood flow in spite of variations in perfusion pressure. Redistribution of Q, during exercise, does not increase blood flow to splanchnic, cutaneous, and cerebral; due to the sympathetic nervous system activity, vascular conductance is unchanged from rest [133]. Continual post-exercise hypotension is facilitated by both central neural mechanisms and local vasodilatory mechanisms. The arterial baroreflex decreases sympathetic activity related to HR beats recovering from the effort, due to lessening sympathetic outflow [129]. In addition to the sympathetic activity removal, during recovery, HR is mainly determined by parasympathetic reactivation [130].

Vasodilation which is independent of nitric oxide [132] results from local substances, namely, CO_2 and pH along with arterial baroreflex retuning. Autoregulation mechamism is intrinsic affected mainly by CO_2 and pH.

When the subject suffers from hypotension, any stress test performed has to be with the personal presence of medical professionals.

Post-exercise hypotension is a phenomenon of a prolonged decrease in resting BP in the minutes and up to 7–12 h following acute exercise. In the occasion that BP heights last longer than a few hours, medical care is required.

However, its occurrence in normotensive humans is inconsistent and can be detected in normotensive individuals [98]. Still, it was found in normotensives to be much less steady and of smaller magnitude than in hypertensive individuals. This may be due to other compensatory mechanisms, such as the baroreflex that are activated in normotensive subjects and prevent the degree of post-exercise hypotension from affecting orthostatic tolerance.

Several other mechanisms have been proposed to account for this post-exercise hypotension including the release of vasodilator metabolites from the exercising muscle, resetting of the baroreflex control of sympathetic tone, and inhibition of muscle sympathetic nervous activity. A further possibility relates to the need to dissipate heat accumulated during the exercise period [134]. The amount of pressure reduction appears to be dependent on the intensity of the exercise performed, the mode of exercise, duration, and the recovery posture of the individual. The greater the exercise intensity performed, the larger the post-exercise hypotension value following exercise. As exercise begins, the increase in BP forces water to leave the cardiovascular compartment to the extracellular compartment, which in turn increases albuminous concentration of the blood. During exercise recovery period, water returns to the cardiovascular system due to osmotic pressure. Although there are reports of gender differences in BP and sympathetic nerve activity, post-exercise hypotension appears to be unaffected by gender, since gender-specific and mixed-gender studies have found similar degrees of hypotension. It also occurs independent of age, having been observed in young middle-aged and older adults [135].

Hypotension Following Resistive Exercise

Resistive exercise elicits more pronounced increments in both SBP and DBP, without the autoregulation mechanism. Mechanical compression of the vessels causes SBP and DBP to rise rapidly to extremely high values during resistive contractions, reaching mean pressures of 320/250 mmHg, with pressures in one subject exceeding 480/350 mmHg. A portion of the observed increase in BP was caused by a Valsalva maneuver. Thus, the mechanical compression of blood vessels during weightlifting bouts, combined with a potent pressor response and a Valsalva response produces extreme elevations in blood pressure. Hypotension following resistive exercise may be mainly the result of mechanical compression withdrawal that abolishes vasoconstriction. Pressure then declines, often to below resting values at the completion of the lift, and then increases again during the lowering phase of the exercise [136]. The restoration of baseline BP is similar to that of endurance exercise, although the transient pressure is often more pronounced following heavy resistance exercise [137]. During anaerobic bouts, systolic, diastolic, and mean arterial BP values are higher than that of aerobic exercise and lower than during weightlifting [33].

Laboratory on Blood Pressure

Equipment required:
Sphygmomanometers and stethoscope.

Measurement of resting BP

- Have the subject rest quietly for at least 5 min with the elbow slightly flexed.
- Wrap the cuff firmly around the left upper arm at heart level; align the cuff with the brachial artery.
- Place the stethoscope bell or diaphragm below the antecubital space over the brachial artery and below the lower cuff edge.
- Quickly inflate the cuff pressure to 180 mmHg in females and 200 mmHg for males, respectively. Slowly release pressure at a rate equal to 2–3 mmHg s^{-1}, noting first Korotkoff sound.
- Continue releasing pressure, noting when sound becomes muffled (fourth phase DBP) and when sound disappears (fifth phase DBP).
- Record SBP value, fourth and fifth DBP.
- Repeat this procedure in the right arm.

Follow these procedures when using a stethoscope:
Insert the ear tips of the stethoscope directly down each ear canal so that the ear tips of the stethoscope point forward. If you fail to position the ear tips properly, it will be difficult to hear the Korotkoff sounds.

Gently tap the diaphragm of the stethoscope to be sure you can pick up sound through the stethoscope.

Position the stethoscope over the brachial artery in the antecubital fossa.

Hold the diaphragm of the stethoscope firmly against the skin. Placing the stethoscope over clothing increases the chances of interference in hearing Korotkoff sounds.

Sample for laboratory report

Name	Supine BP	Sitting BP	Standing BP	Exercise BP

References

1. Giancoli DG. Physics: principles with applications. Upper Saddle River, NJ: Pearson Education; 2004.
2. NCD Risk Factor Collaboration (NCD-RisC). Worldwide trends in blood pressure from 1975 to 2015: a pooled analysis of 1479 population-based measurement studies with 19·1 million participants. Lancet. 2017;389:37–55.
3. Khattar RS, Swales JD, Banfield A, et al. Prediction of coronary and cerebrovascular morbidity and mortality by direct continuous ambulatory blood pressure monitoring in essential hypertension. Circulation. 1999;100:1071–6.
4. Tan CH, Low KA, Kong T, et al. Mapping cerebral pulse pressure and arterial compliance over the adult lifespan with optical imaging. PLoS One. 2017;12:e0171305. https://doi.org/10.1371/journal.pone.0171305.eCollection2017.
5. Sagiv M, Ben-Sira D, Sagiv M, et al. Left ventricular function at peak all-out anaerobic exercise in older men. Gerontology. 2005;51:122–5.
6. Sagiv M, Ben-Sira D. Weight lifting training and left ventricular function in adolescents subjects. J Sports Med Phys Fitness. 2007;47:329–34.
7. Sagiv M, Sagiv Beevers G, Lip GYH, O'Brien E. Blood pressure measurement. Part II-Conventional sphygmomanometry: technique of auscultatory blood pressure measurement. BMJ. 2001;322:1043–7.
8. Ben-Sira D, et al. The effect of hypertension and aging on left ventricular function during isometric exercise. J Aging Phys Activ. 1996;4:69–79.
9. Kougias P, Weakley SM, Yao O, et al. Arterial baroreceptors in the management of systemic hypertension. Med Sci Monit. 2010;16:RA1–8.
10. Kougias P, Weakley SM, Yao O, Lin PH, Chen C. Arterial baroreceptors in the management of systemic hypertension. Med Sci Monit. 2010;16:RA1–8.
11. Hart EC, Wallin BG, Barnes JN, et al. Sympathetic nerve activity and peripheral vasodilator capacity in young and older men. Am J Physiol Heart Circ Physiol. 2014;306:H904–9.

12. Beevers G, Lip GYH, O'Brien E. Blood pressure measurement. Part II-Conventional sphygmomanometry: technique of auscultatory blood pressure measurement. BMJ. 2001;322:1043–7.
13. Grim CE, Grim CM. Auscultatory BP: still the gold standard. J Am Soc Hyperten. 2016;10:191–3.
14. Saghiv MS, Goldhammer E, Ben-Sira D, et al. Gender effects on direct vs indirect blood pressure measurements at rest and during biking. Int J Clin Cardiol Res. 2018;2:17–21.
15. Papaioannou TG, Protogerou AD, Stamatelopoulos KS, et al. Non-invasive methods and techniques for central blood pressure estimation: procedures, validation, reproducibility and limitations. Curr Pharm Des. 2009;15:245–53.
16. Clark JA, Lieh-Lai MW, Sarnaik A, et al. Discrepancies between direct and indirect blood pressure measurements using various recommendations for arm cuff selection. Pediatrics. 2002;110:920–3.
17. O'Brien E, Atkins N, Stergiou G, et al. On behalf of the Working Group on Blood Pressure Monitoring of the European Society of Hypertension. European Society of hypertension international protocol revision 2010 for the validation of blood pressure measuring devices in adults. Blood Press Monit. 2010;15:23–38.
18. Krakoff LR. Ambulatory blood pressure improves prediction of cardiovascular risk: implications for better antihypertensive management. Curr Atheroscler Rep. 2013;15:317.
19. Reckelhoff JF. Gender differences in the regulation of blood pressure. Hypertension. 2001;37:1199–208.
20. Joyner MJ, Charkoudian N, Wallin BG. Sympathetic nervous system and blood pressure in humans individualized patterns of regulation and their implications. Hypertension. 2010;56:10–6.
21. Itoh H, Ajisaka R, Koike A, et al. Heart rate and blood pressure response to ramp exercise and exercise capacity in relation to age, gender, and mode of exercise in a healthy population. J Cardiol. 2013;61:71–8. https://doi.org/10.1016/j.jjcc.2012.09.010.
22. Fischer M, Baessler A, Schunkert H. Renin angiotensin system and gender differences in the cardiovascular system. Cardiovac Res. 2002;53:672–7. https://doi.org/10.1016/S0008-6363(01)00479-5672-677.
23. Saghiv M, Goldhammer E, Sagiv M, et al. Comparison of direct vs indirect blood pressure measurements on treadmill and bicycle in hypertensive responders. Clin Exp Pharmacol. 2016;6:4.
24. Ganong WF. Review of medical physiology. 22nd ed. Singapore: International Edition; 2005. p. 571.
25. Kaye DM, Esler MD. Autonomic control of the aging heart. Neuromolecular Med. 2008;10:179–86.
26. Kohli P, Gulati M. Exercise stress testing in women: going back to the basics. Circulation. 2010;14(122):2570–80.
27. Kamiya A, Michikami D, Fu Q, et al. Static handgrip exercise modifies arterial baroreflex control of vascular sympathetic outflow in humans. Am J Physiol Regul Integr Comp Physiol. 2001;281:R1134–9.
28. Sagiv M, Amir O, Goldhammer E, et al. Left ventricular contractility in response to upright isometric exercise in heart transplant recipients and healthy men. J Cardiopulm Rehabil Prev. 2008;28:17–23.
29. Sagiv M, Sagiv M, Ben-Sira D. Weight lifting training and left ventricular function in adolescent subjects. J Sports Med Phys Fitness. 2007;47:329–34.
30. Sagiv M, Hanson P, Besozzi M, et al. Left ventricular responses to upright isometric handgrip and deadlift in men with coronary artery disease. Am J Cardiol. 1985;55:1298–302.
31. Harris KA, Holly RG. Physiological response to circuit weight training in borderline hypertensive subjects. Med Sci Sports Exerc. 1987;19:246–52.
32. Kelley GA, Kelley KS. Progressive resistance exercise and resting blood pressure: a meta-analysis of randomized controlled trials. Hypertension. 2000;35:838–43.

33. Sagiv M, Ben-Sira D, Goldhammer E. Direct vs. indirect blood pressure measurement at peak anaerobic exercise. Int J Sports Med. 1999;20:275–8.
34. Kellici TF, Liapakis G, Tzakos AG, et al. Pharmaceutical compositions for antihypertensive treatments: a patent review. Expert Opin Ther Pat. 2015;25:1305–17.
35. Poulter NR, Prabhakaran D, Caulfield M. Hypertension. Lancet. 2015;386:801–12.
36. Monticone S, D'Ascenzo F, Moretti C, et al. Cardiovascular events and target organ damage in primary aldosteronism compared with essential hypertension: a systematic review and meta-analysis. Lancet Diabetes Endocrinol. 2018;6:41–50.
37. Whelton PK, Carey RM, Aronow WS, et al. Guideline for the prevention, detection, evaluation, and management of high blood pressure in adults: A report of the American College of Cardiology/American Heart Association Task Force on Clinical Practice Guidelines. Hypertension. 2018;71:e13–e115.
38. Hotamisligil GS. Endoplasmic reticulum stress and atherosclerosis. Nat Med. 2010;16:396–9.
39. Frostegard J. Low level natural antibodies against phosphorylcholine: a novel risk marker and potential mechanism in atherosclerosis and cardiovascular disease. Clin Immunol. 2010;134:47–54.
40. Aziz M, Yadav K. Pathogenesis of atherosclerosis. Med Clin Rev. 2016;2:22. https://doi.org/10.21767/2471-299X.1000031.
41. Navar LG. The Role of the Kidneys in Hypertension. J Clin Hypert. 2007; https://doi.org/10.1111/j.1524-6175.2005.04130.x.
42. Carroll RG. Elsevier's integrated physiology 3 – body fluid distribution. ScienceDirect. 2007:19–26. https://doi.org/10.1016/B978-0-323-04318-2.50009-1.
43. Walker HK, Hall WD, Hurst JW, editors. Serum sodium in clinical methods: the history, physical, and laboratory examinations. 3rd ed. Boston: Butterworths; 1990.
44. Alsharari R, Lip GYH, Shantsila A. Assessment of arterial stiffness in patients with resistant hypertension: additional insights into the pathophysiology of this condition? Am J Hypertens. 2019; https://doi.org/10.1093/ajh/hpz169.
45. Crowley SD, Coffman TM. The inextricable role of the kidney in hypertension. J Clin Invest. 2014;124:2341–7.
46. Messerli FH, Bangalore S, Bavishi C, et al. Angiotensin-converting enzyme inhibitors in hypertension: to use or not to use? J Am Coll Cardiol. 2018;71:1474–82.
47. Le TH, Coffman TM. Targeting genes in the renin-angiotensin system. Curr Opin Nephrol Hypertens. 2008;17:57–63.
48. Jiang SZ, Lu W, Zong XF, et al. Obesity and hypertension. Exp Ther Med. 2016;12:2395–9.
49. Dorresteijn JA, Visseren FL, Spiering W. Mechanisms linking obesity to hypertension. Obes Rev. 2012;13:17–26.
50. Portela DS, Vieira TO, Matos SM, et al. Maternal obesity, environmental factors, cesarean delivery and breastfeeding as determinants of overweight and obesity in children: results from a cohort. BMC Pregnancy Childbirth. 2015;15:94.
51. Re RN. Obesity-related hypertension. Ochsner J. 2009;9:133–6.
52. Vaz M, Jennings G, Turner A, et al. Regional sympathetic nervous activity and oxygen consumption in obese normotensive human subjects. Circulation. 1997;96:3423–9.
53. Ancion A, Tridetti J, Nguyen Trung ML, Oury C, et al. A review of the role of Bradykinin and nitric oxide in the cardioprotective action of angiotensin-converting enzyme inhibitors: Focus on Perindopril. Cardiol Ther. 2019. https://doi.org/10.1007/s40119-019-00150-w.
54. Rafieian-Kopaei M, Setorki M, et al. Atherosclerosis: process, indicators, risk factors and new hopes. Int J Prev Med. 2014;5:927–46.
55. Gartner LP, Hiatt JL. Color atlas and text of histology. 6th ed. Baltimore, SA: Lippincott. Wiliams & Wilkins Ed; 2013.
56. Fullstone G, Wood J, Holcombe M, et al. Modelling the transport of nitric oxide particles under blood flow using an agent-based approach. Sci Rep. 2015;5:10649.
57. Higashi Y, Ueda K, Yoshizumi M, et al. Endothelial function and oxidative stress in cardiovascular diseases. Circ J. 2009;73:411–8.

58. Godo S, Shimokawa H. Divergent roles of endothelial nitric oxide synthases system in maintaining cardiovascular homeostasis. Free Radic Biol Med. 2017;109:4–10.
59. Wang I, Dong X, Gao J, et al. Targeting inflammatory vasculature by extracellular vesicles. AAPS J. 2018;20:37.
60. De Rosa S, Cirillo P, Paglia A, et al. Reactive oxygen species and antioxidants in the pathophysiology of cardiovascular disease: does the actual knowledge justify a clinical approach? Curr Vasc Pharmacol. 2010;8:259–75.
61. Zeiher AM, Drexler H, Saurbier B, et al. Endothelium-mediated coronary blood flow modulation in humans: effects of age, atherosclerosis, hypercholesterolemia, and hypertension. J Clin Invest. 1993;92:652–62.
62. Palombo C, Kozakova M. Arterial stiffness, atherosclerosis and cardiovascular risk: pathophysiologic mechanisms and emerging clinical indications. Vascul Pharmacol. 2016;77:1–7.
63. Palatini P, Casiglia E, Gąsowski J, et al. Arterial stiffness, central hemodynamics, and cardiovascular risk in hypertension. Vasc Health Risk Manag. 2011;7:725–39.
64. Huveneers S, Daemen MJ, Hordijk PL. Between Rho(k) and a hard place: the relation between vessel wall stiffness, endothelial contractility, and cardiovascular disease. Circ Res. 2015;116:895–908.
65. Thorin-Trescases N, Thorin E. Lifelong cyclic mechanical strain promotes large elastic artery stiffening: increased pulse pressure and old age-related organ failure. Can J Cardiol. 2016;32:624–33.
66. Zhao SZ, Ariff B, Long Q, et al. Inter-individual variations in wall shear stress and mechanical stress distributions at the carotid artery bifurcation of healthy humans. J Biomech. 2002;35:1367–77.
67. Dietz J. Arterial stiffness and extracellular matrix. Adv Cardiol. 2007;44:76–95.
68. Kleinstreuer C, Hyun S, Buchanan JR Jr, et al. Hemodynamic parameters and early intimal thickening in branching blood vessels. Crit Rev Biomed Eng. 2017;45:319–82.
69. Vanhoutte PM, Shimokawa H, Feletou M, et al. Endothelial dysfunction and vascular disease – a 30th anniversary update. Acta Physiol (Oxf). 2017;219:22–96.
70. Barthelmes J, Matthias P, Nagele PM, et al. Endothelial dysfunction in cardiovascular disease and Flammer syndrome-similarities and differences. EPMA J. 2017;8:99–109.
71. Avin KG, Coen PM, Huang W, et al. Skeletal muscle as a regulator of the longevity protein, Klotho. Front Physiol. 2014;5:189.
72. Lim K, Groen A, Molostvov G, et al. α-Klotho expression in human tissues. J Clin Endocrinol Metab. 2015;100:E1308–18.
73. Lees SJ, Booth FW. Sedentary death syndrome. Can J Appl Physiol. 2004;29:447–60.
74. Stewart KJ. Physical activity and aging. Ann NY Acad Sci. 2005;1055:193–206.
75. Venkat NKM, Mohammed KA, Jeffrey PK. Global non-communicable diseases – where worlds meet. New Engl J Med. 2010;363:1196–8.
76. Stear S. Health and fitness series—1. The importance of physical activity for health. J Fam Health Care. 2003;13:10–3.
77. Yanai H, Tomono Y, Ito K, Furutani N, et al. The underlying mechanisms for development of hypertension in the metabolic syndrome. Nutr J. 2008; https://doi.org/10.1186/1475-2891-7-10.
78. Cameron JD, Asmar R, Struijker-Boudier H, et al. Current and future initiatives for vascular health management in clinical practice. Vasc Health Risk Manag. 2013;9:255–64.
79. Mozos I, Malainer C, Horbańczuk J, et al. Inflammatory markers for arterial stiffness in cardiovascular diseases. Front Immunol. 2017; https://doi.org/10.3389/fimmu.2017.01058.
80. Adel M, ELSheikh A, Sameer S, et al. Arterial stiffness in metabolic syndrome. J Saudi Heart Assoc. 2016;28:249–56.
81. Shirwany NA, Zou M. Arterial stiffness: a brief review. Acta Pharmacol Sin. 2010;31:1267–76.
82. Avolio A, Butlin M, Liu YY, et al. Regulation of arterial stiffness: cellular, molecular and neurogenic mechanisms. Artery Res. 2011;5:122–7.
83. Hughes TM, Craft S, Lopez OL. Review of the potential role of arterial stiffness in the pathogenesis of Alzheimer's disease. Neurodegener Dis Manag. 2015;5:121–35.

84. Jin RC, Loscalzo J. Vascular nitric oxide: formation and function. J Blood Med. 2010;1:147–62.
85. Bellien J, Thuillez C, Joannides R. Contribution of endothelium-derived hyperpolarizing factors to the regulation of vascular tone in humans. Fundam Clin Pharmacol. 2008;22:363–77.
86. Malekmohammad K, Sewell RDE, Rafieian-Kopaei M. Antioxidants and aherosclerosis: mechanistic aspects. Biomolecules. 2019; https://doi.org/10.3390/biom9080301.
87. Ray CA, Carrasco DI. Isometric handgrip training reduces arterial pressure at rest without changes in sympathetic nerve activity. Am J Physiol Heart Circ Physiol. 2000;279:H245–9.
88. Leo MD, Zhai X, Yin W, et al. Impaired trafficking of β1 subunits inhibits BK channels in cerebral arteries of hypertensive rats. Hypertension. 2018;72:765–75.
89. Boutouyrie P, Tropeano AI, Asmar R, et al. Hypertensive patients: a longitudinal study or of primary coronary events in hypertensive patients: a longitudinal study. Hypertension. 2002;39:10–5.
90. McEniery CM, Wilkinson IB, Avolio AP. Age, hypertension and arterial function. Clin Exp Pharmacol Physiol. 2007;34:665–71.
91. Benetos A, Waeber B, Izzo J, et al. Influence of age, risk factors, and cardiovascular and renal disease on arterial stiffness: clinical applications. Am J Hypertens. 2002;15:1101–8.
92. Stefanadis C, Dernellis J, Tsiamis E, et al. Aortic stiffness as a risk factor for recurrent acute coronary events in patients with ischaemic heart disease. Eur Heart J. 2000;21:390–6.
93. Kohno F, Kumada T, Kamabayashi M, et al. Change in aortic end-systolic pressure by alterations in loading sequence and its relation to left ventricular isovolumic relaxation. Circulation. 1996;93:2080–7.
94. Laurent S, Katsahian S, Fassot C, et al. Aortic stiffness is an independent predictor of fatal stroke in essential hypertension. Stroke. 2003;34:1203–6.
95. Kontopoulos AG, Athyros VG, Pehlivanidis AN, et al. Long-term treatment effect of atorvastatin on aortic stiffness in hypercholesterolaemic patients. Curr Med Res Opin. 2003;19:22–7.
96. Nigam A, Mitchell GF, Lambert J, et al. Relation between conduit vessel stiffness (assessed by tonometry) and endothelial function (assessed by flow-mediated dilatation) in patients with and without coronary heart disease. Am J Cardiol. 2003;92:395–9.
97. Okabe T-A, Kishimoto C, Murayama T, et al. Effects of exercise on the development of atherosclerosis in apolipoprotein E-deficient mice. Exp Clin Cardiol. 2006;11:276–9.
98. Duncker DJ, Bache RJ. Regulation of coronary blood flow during exercise. Physiol Rev. 2008;88:1009–86.
99. Santos-Parker JR, LaRocca TJ, et al. Aerobic exercise and other healthy lifestyle factors that influence vascular aging. Adv Physiol Educ. 2014;38:296–307.
100. Ashor AW, Lara J, Siervo M, et al. Effects of exercise modalities on arterial stiffness and wave reflection: a systematic review and meta-analysis of randomized controlled trials. PLoS One. 2014;9:e110034.
101. Rush JWE, Ford RJ. Nitric oxide, oxidative stress and vascular endothelium in health and hypertension. Clin Hemorh Microc. 2007;37:185–92.
102. Liu TH, Wu CL, Chiang CW, et al. Nitric oxide effect of short-term arginine supplementation on nitric oxide production, metabolism and performance in intermittent exercise in athletes. J Nutr Biochem. 2009;20:462–8.
103. Petersen BK. Exercise-induced myokines and their role in chronic diseases. Brain Behav Immun. 2011;25:811–6.
104. Brito LC, Fecchio RY, Peçanha T, et al. Postexercise hypotension as a clinical tool: a "single brick" in the wall. J Am Soc Hypertens. 2018;12:e59–64.
105. Romero SA, Ely MR, Sieck DC, et al. Effect of antioxidants on histamine receptor activation and sustained postexercise vasodilatation in humans. Exp Physiol. 2015;100:435–49.
106. Lopes S, Mesquita-Bastos J, Alves AJ, et al. Exercise as a tool for hypertension and resistant hypertension management: current insights. Integr Blood Press Control. 2018;11:65–71.
107. Cornelissen VA, Smart NA. Exercise training for blood pressure: a systematic review and meta-analysis. J Am Heart Assoc. 2013;2(1):e004473.

108. Hamer M. The anti-hypertensive effects of exercise: integrating acute and chronic mechanisms. Sports Med Auckl NZ. 2006;36:109–16.
109. Pescatello LS, Franklin BA, Fagard R, et al. Exercise and hypertension. Med Sci Sports Exerc. 2004;36:533–53.
110. Pescatello LS. Exercise and hypertension: recent advances in exercise prescription. Curr Hypertens Rep. 2005;7:281–6.
111. Diaz KM, Shimbo D. Physical activity and the prevention of hypertension. Curr Hypertens Rep. 2013;15:659–68.
112. Ash GI, Eicher JD, Pescatello LS. The promises and challenges of the use of genomics in the prescription of exercise for hypertension: the 2013 update. Curr Hypertens Rev. 2013;9:130–47.
113. Pescatello LS, Blanchard BE, Tsongalis GJ, et al. A comparison of the genetic and clinical profile of men that respond and do not respond to the immediate antihypertensive effects of aerobic exercise. Appl Clin Genet. 2008;1:7–17.
114. Blair SN, Kohl HW 3rd, Barlow CE, et al. Changes in physical fitness and all-cause mortality. JAMA. 1995;273:1093–8.
115. Kakiyama T, Sugawara J, Murakami H, et al. Effects of short-term endurance training on aortic distensibility in young males. Med Sci Sports Exerc. 2005;37:267–71.
116. MacDonald JR, MacDougall JD, Hogben CD. The effects of exercise duration on post-exercise hypotension. J Hum Hypertens. 2004;14:125–9.
117. Tanriverdi H, Evrengul H, Kaftan A, et al. Effects of angiotensin-converting enzyme polymorphism on aortic elastic parameters in athletes. Cardiology. 2005;104:113–9.
118. Cameron JD, Rajkumar C, Kingwell BA, et al. Higher systemic arterial compliance is associated with greater exercise time and lower blood pressure in a young older population. J Am Geriatr Soc. 1999;47:653–6.
119. Martin WH 3rd, Ogawa T, Kohrt WM, et al. Effects of aging, gender, and physical training on peripheral vascular function. Circulation. 1991;84:654–64.
120. Vaitkevicius PV, Fleg JL, Engel JH, et al. Effects of age and aerobic capacity on arterial stiffness in healthy adults. Circulation. 1993;88:1456–62.
121. De Simone G. Left ventricular concentric geometry is associated with impaired relaxation in hypertension: the HyperGEN study. Eur Heart J. 2005;26:1039–45.
122. Carlson DJ, Dieberg G, Hess NC, et al. Isometric exercise training for blood pressure management: a systematic review and meta-analysis. Mayo Clin Proc. 2014;89:327–34.
123. Corso LM, Macdonald HV, Johnson BT, et al. Is concurrent training efficacious antihypertensive therapy? A meta-analysis. Med Sci Sports Exerc. 2016;48:2398–406.
124. Va C, Buys R, Na S. Endurance exercise beneficially affects ambulatory blood pressure: a systematic review and meta-analysis. J Hypertens. 2013;31:639–48.
125. Ash GI, Taylor BA, Thompson PD, et al. The antihypertensive effects of aerobic versus isometric handgrip resistance exercise. J Hypertens. 2017;35:291–9.
126. Halliwill JR, Buck TM, Lacewell AN, et al. Postexercise hypotension and sustained postexercise vasodilatation: what happens after we exercise? Exp Physiol. 2013;98:7–18.
127. MacDonald JR, MacDougall JD, Interisano SA, et al. Hypotension following mild bouts of resistance exercise and submaximal dynamic exercise. Eur J Appl Physiol Occup Physiol. 1999;79:148–54.
128. Romero SA, Minson CT, Halliwill JR. The cardiovascular system after exercise. J Appl Physiol (1985). 2017;122:925–93.
129. Pecanha T, Bartels R, Brito LC, et al. Methods of assessment of the post-exercise cardiac autonomic recovery: a methodological review. Int J Cardiol. 2017;227:795–802.
130. Seiler S, Haugen O, Kuffel E. Autonomic recovery after exercise in trained athletes: intensity and duration effects. Med Sci Sports Exerc. 2007;39:1366–73.
131. Halliwill JR, Taylor JA, Eckberg DL. Impaired sympathetic vascular regulation in humans after acute dynamic exercise. J Physiol. 1996;495:279–88.

132. Halliwill JR, Minson CT, Joyner MJ. Effect of systemic nitric oxide synthase inhibition on postexercise hypotension in humans. J Appl Physiol (1985). 2000;89:1830–6.
133. Pricher MP, Holowatz LA, Williams JT, et al. Regional hemodynamics during postexercise hypotension. I. Splanchnic and renal circulations. J Appl Physiol (1985). 2004;97:2065–70.
134. Franklin PJ, Green DJ, Cable NT. The in fluence of thermoregulation mechanisms on post exercise hypotension in humans. J Physiol. 1993;470:231–41.
135. MacDonald JR. Potential causes, mechanisms, and implications exercise hypotension. J Hum Hypertens. 2002;16:225–36.
136. Lentini AC, McKelvie RS, McCartney N, et al. Left ventricular response in healthy young men during heavy-intensity weightlifting exercise. J Appl Physiol. 1993;75:2703–10.
137. Sagiv SM. Exercise cardiopulmonary function in cardiac patients. London: Springer; 2012.

Chapter 6
Cardiovascular Function

Introduction

The cardiovascular system connects a pump to blood vessels. It is comprised of the heart, arteries, capillaries, veins, and about 4.7–5.7 L of blood, which is up to 8% an individual's body weight, in an average adult. Blood contains plasma, red and white blood cells, and platelets. All these components are organized to circulate blood to each of the body cells. Left ventricular contraction is the driving force that propels the blood and oxygen through the arteries, arterioles, and capillaries and then through the venous back to the right atrium of the heart. In the capillaries at the boundary of the arterial and venous, exchange of oxygen and substances between the blood and body tissues occurs, while waste product, CO_2, and blood poor oxygenated pass to the venous system. The cardiovascular system has numerous main functions namely: oxygen and CO_2 transport, nutrient and waste product transport, disease protection and healing, hormone delivery, and body temperature regulation.

Cardiovascular abilities are significant issues in sport, health, and exercise. Oxygen delivery is the most vital function of the cardiovascular system. About 98.5% of the oxygen is delivered by hemoglobin molecules and 1.5% dissolves in the blood liquids, without a link to hemoglobin. The oxygen that enters the lungs is absorbed into the bloodstream, which is pumped by the heart into the arterial circulation. In the capillaries, oxygen diffuses into the cells, while CO_2 is diffused into the venous blood and transported to the lungs and on to the atmosphere. In addition to CO_2, the circulatory system picks up metabolic waste products and toxins and transports them to the liver, kidneys, and lungs for eventual elimination from the body.

It delivers nutrients that include sugars, fats, vitamins, minerals, and protein-building blocks called amino acids [1]. Each of these nutrients is vital to healthy body function. For example, glucose is the body's primary fuel for the generation of ATP energy. As oxygen, nutrients diffuse from the bloodstream into body cells through the capillaries.

M. S. Saghiv, M. S. Sagiv, *Basic Exercise Physiology*,
https://doi.org/10.1007/978-3-030-48806-2_6

The circulatory system serves as a delivery tool for the immune system, where immune white blood cells eliminate invading germs. In case that infection occurs, immune white blood cells are transported to the site of the infection. Hormones produced by endocrine glands affect distant organs. The cardiovascular system connects the endocrine glands to the tissues controlled by hormones. Body temperature regulation is an important role of the cardiovascular system. Temperature regulation happens within a comparatively slight temperature range, and therefore is closely controlled. Dilation and vasoconstriction of blood vessels permit the body to eliminate surplus heat or preserve body heat through the skin.

System's Uniqueness

It is a closed loop system, meaning that the blood does not leave the blood system.

The cardiovascular system is filled with fluid with positive pressure without connection to the pump contraction.

Most of the time the atria are filed 80% passively, with minor energy expenditure, while the other 20% are active by the atria contraction.

System's regulation is governed by peripheral factors and not by heart variables, due to passive filling.

The system is flexible.

Physical exercise is the stronger stimulation that increases Q to its maximum. Other stimulations such as emotional thrill, gravitational changes, ambient and body core temperature variations, sexual activity, and pregnancy are less strong than exercise.

Additional part of the circulatory system is the lymphatic system, which is a network of lymphatic vessels and lymph capillaries, lymph nodes, and organs, and lymphatic tissues and circulating lymph. One of its major functions is to carry the lymph, draining and returning interstitial fluid back toward the heart for return to the cardiovascular system, by emptying into the lymphatic ducts. Its other main function is in the adaptive immune system [2].

In the cardiovascular system, there are two circulations that originate from the heart: (a) the pulmonary circulation and (b) the systemic circulation (Fig. 6.1).

The pulmonary circulation is a short circle from the heart to the lungs and back again. It is a part of the cardiovascular system that carries deoxygenated blood from the right ventricle to the lungs, and delivers oxygenated blood to the left atrium and left ventricle. The pulmonary circulation blood vessels are the pulmonary arteries and pulmonary veins. The systemic circulation conveys blood from the heart to the body's organs and back again. The systemic circulation is the main part of the cardiovascular system that transports oxygenated blood and nutrients from the left atrium and ventricle through the aorta to the rest of the body and returns oxygen-depleted blood to the right atrium.

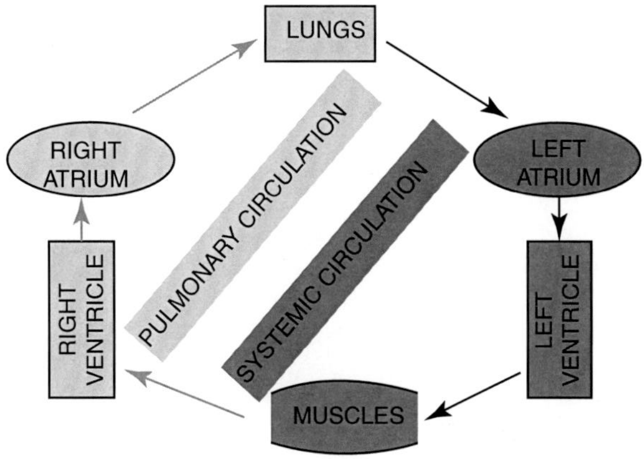

Fig. 6.1 The cardiovascular system

The Myocardium

The heart lies left to the middle of the chest and positioned between the two lungs. The heart is enveloped by the pericardium, a fibrous covering the entire heart. It grasps the heart in place however, still flexible to move while beating. The heart is divided into four chambers; two atria and two ventricles. The heart can be seen also as having two sides, left and right. The left side includes the left atrium and left ventricle, while the right side includes the right atrium and right ventricle. The heart functions as the body's main pump in the blood circulation; oxygenated blood is propelled from the left ventricular through series of arterials to the body's organs, and oxygen-depleted blood returns to the heart through sequences of vines. The heart is an involuntary muscle with contraction abilities as any other skeletal muscle, meaning it reacts to an electrical impulse. Its size is approximately as the hand fist and is roughly cone-shaped. The myocardium through a cycle of contraction and relaxation pumps blood through the cardiovascular system. When HR is 70 beats·min^{-1}, the cardiac cycle lasts 0.8 s with relaxation time between heartbeats of 0.4–0.5 s. Numerous factors besides size can affect the HR; motion plays an important key role in just how fast a heartbeat, the HR rises when the body position of an individual changes; if standing rather than lying down, HR can increase due to physical activity, body and surrounding temperatures variations, and emotional stress.

During the initial portion of the the cardiac cycle generates an electrical signal by pacemaker cells through an electrical conduction system and then it is spread through the myocardium. Electrical impulses from the heart cause it to contract. The electrical signal begins in the sinoatrial (SA) node (heart's pacemaker), located at the top of the right atrium. Electrical impulse originating in the pacemaker causes the atrium to contract. That impulse then travels through the right atrium to the atrioventricular (AV) node. From the AV node the impulse travels via the bundle of His and Purkinje fibers to cause contraction of the heart muscle. This signal stimulates contraction, first of the right and left atrium, and then the right and left ventricles.

In response to electrical stimulation, the atria contracts first and then the ventricles (systole), followed by sequential relaxation (diastole) of the two ventricles, with a fraction of a second delay between the ventricles. The left atrium has oxygenated blood arriving from the lungs, while the right atrium receives oxygen-depleted blood from the body's organs. The ventricles have a significant thicker wall compared to atria enabling them to perform hard work. The role of the right ventricle is to pump through the pulmonary system, which is a low-pressure system, oxygen-depleted blood to the lungs, through the pulmonary trunk and pulmonary arteries. The role of the left ventricle is to pump newly oxygenated blood to the body, through the aorta. This cycle of compressing blood in the ventricles during systole followed by the filling of the ventricles during diastole induces pressure changes in the ventricles that cause one-way valves in the heart to close, audibly at different intervals of the cardiac cycle. The result of the blood ejection into the arteries by the ventricles undergoing systole is the generation of body's BP, the primary driving force for blood flow through the body.

The Aorta

The ventricles eject blood into the aorta and pulmonary artery in an alternation manner, yet, blood flow in the entire cardiovascular system is continuously due to the aorta feature. The aorta receives the ejected SV amount and expends its radius, then aorta elasticity brings about the radius to its initial stage, meanwhile blood flows continuously through the system. The aorta is the major vessel in the systemic circuit, ascending from the left ventricle. It is usually said to have three regions: the ascending aorta, the arch of the aorta, and the descending aorta; the latter may be further subdivided into the thoracic and the abdominal aorta. Originating from the ascending portion of the aorta are the right and left coronary arteries, which supply the heart with oxygenated blood.

Electrical Stimulation

Electrical impulses (Fig. 6.2) from the myocardium causes heart to contract. Action potentials formed spontaneously at regular intervals in specialized (pacemaker cells) enables impulse to originate action potential.

The impulse travels through four pivot points found in the electrical conducting system. The sinoatrial node (SA node), located in the right atrial wall near the junction for the superior vena cava, undergoes spontaneous depolarizations at a higher rate than any of the other pacemaker cells in the heart. As a result, the SA node sets the basic rhythm of contraction. The electrical signal begins in the SA node (heart's pacemaker); action potentials originating in the SA spread rapidly through both atriums through tracts of pacemaker cells node, causing the atrium to contract. That

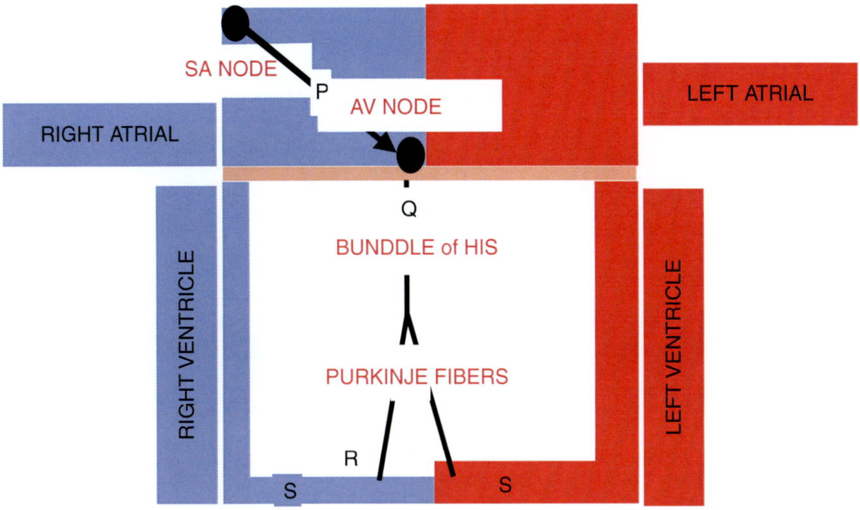

Fig. 6.2 The myocardium's electrical impulse path SA node to AV node, bundle of His, and to the Purkinje fibers

impulse then travels through the right atrium to the atrioventricular (AV) node. The AV node's cells are with very low transmission rates and delay ensures that atrial systole is complete and increases ventricles filling time. From the AV node along the bundle of His and through the Purkinje fibers, the impulse travels to cause myocardium contractions. This signal stimulates contraction first of the right and left atrium, and then the right and left ventricles. Electrocardiogram (ECG) records the electrical activity from the surface of the body to detect any myocardial pathology. Ten electrodes are adhered with adhesive pads to the skin, one to each limb and six to the chest surface, to record the electrical impulse progression through the myocardium. The skin is able to register the ECG as the final electrical impulse receiver carried through the body displayed graphically. Twelve-lead ECG can compare normal ECG activity to many normal rhythms in the healthy heart. The Twelve-lead ECG measurement is an important medical tool; however, in order to have beats' counts, for reviewing HR variations at rest and during exercise, a 3-electrode configuration satisfies. With each heartbeat, depolarization of the myocardium activates its contraction. A normal ECG graph has some components, representing different waveforms with the letters P, Q, R, S, and T (Fig. 6.3). The P wave generated when the atria depolarize as the impulse spreads out from the SA node. The P-R interval is the time between the first deflection of the P wave and the first deflection of the QRS complex. The QRS complex is triggered by the depolarization of the AV node to bundle of His, which conducts an impulse through the Purkinje fibers toward the ventricles' apex. Once the signal reaches the apex of the myocardium, it is steered up the lateral walls of the ventricles. The slow transmission velocity of the AV node causes a delay evident by the P-R interval of the impulse originated in the atrial. This delay ensures to have enough ventricular filling time which is passive, and that

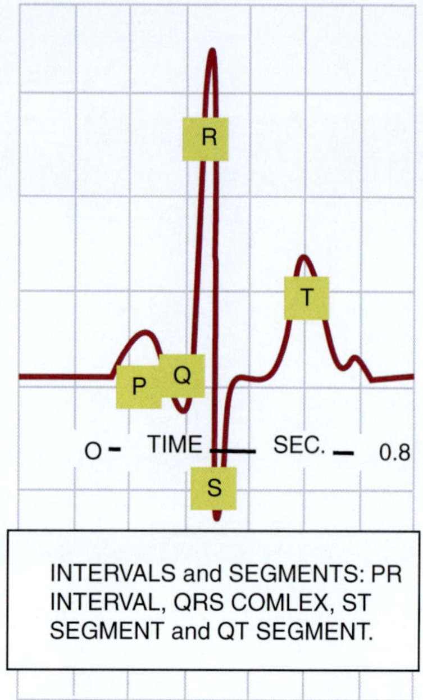

Fig. 6.3 An electrocardiogram for a single cardiac cycle. Note the three distinctive waveforms: the P wave, QRS waves, and the T wave. Four important diagnostic segments (P-R, QRS complex, S-T, and QT) are also noted

atrial systole is complete at the onset of ventricular systole. The great QRS complex shads the atrial repolarization signal. The three waves of the QRS complex represent ventricular depolarization. The small Q and R waves deflaction reflects depolarization of the main mass of the ventricles, followed by large Swave. The S wave is the final depolarization of the ventricles at the base of the heart. The S-T segment is the time between the end of the QRS complex and the start of the T wave. It reflects the period of zero potential between ventricular depolarization and repolarization. T waves represent ventricular repolarization.

Electrocardiograph (ECG) Leads

Electrodes form leads, which are electrical view (can view as an electrical camera) recordings of the heart from different directions and from any one combination of the ten electrodes, altogether, 12 views or leads (Fig. 6.4).

The 12 leads are 3 bipolar, 3 unipolar, and 6 unipolar precordial obtained from ten electrodes attached to the left arm, the right arm, left leg, and chest. The electrode located in the right leg serves as an earth electrode.

Lead I is the electrical link from the right arm electrode to the left arm electrode, lead II is the electrical link from the right arm electrode to the left leg electrode, and lead III is the electrical link from the left leg to the left arm electrode.

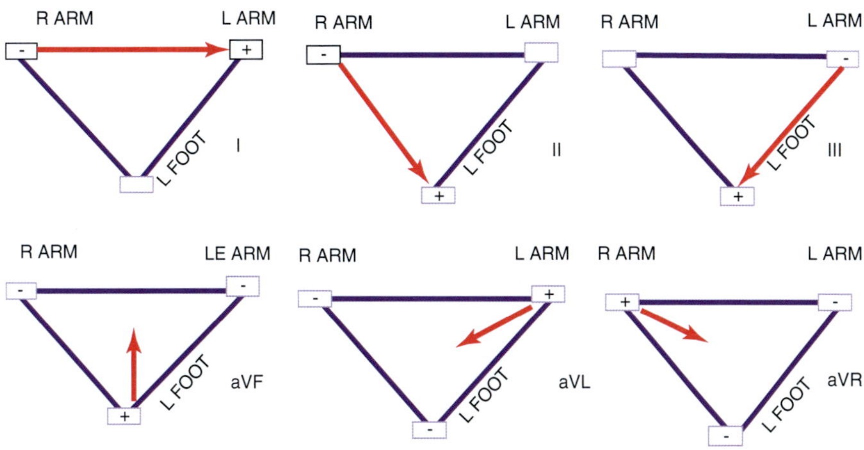

Fig. 6.4 The limb leads looking at the heart in a vertical plane

Fig. 6.5 Placement of the precordial electrodes V1–V6

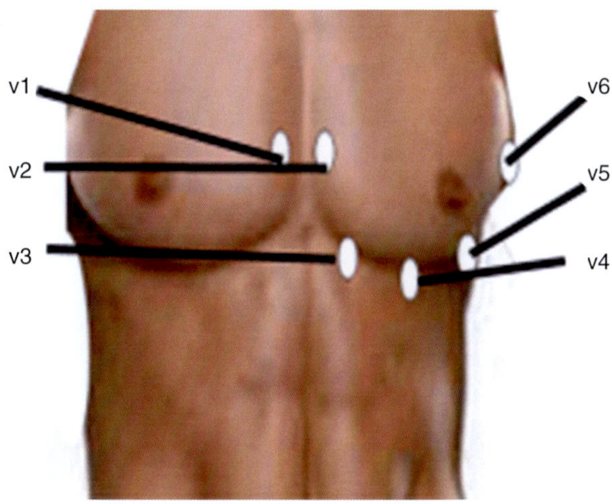

The bipolar limb leads are: The potential difference between one of the three limb electrodes of the unipolar leads duplicate and estimate of zero potential derived from the remaining two limb electrodes. The unipolar limb leads are aVR lead located in the right arm, aVL lead located in the left arm, and aVF lead located in the left leg.

Further, six unipolar precordial leads are on the chest wall, V1–V6 look at the heart in a vertical plane (Fig. 6.5), whereas the chest leads look at the heart in a horizontal plane.

Finally, the ST interval is an important diagnostic interval, it may become raised because of myocardial infarction, and in individuals with ischemic heart ST interval it is depressed.

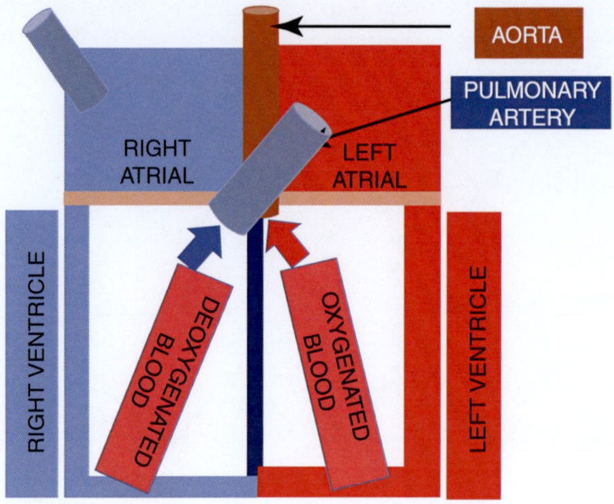

Fig. 6.6 Ventricular systole to the pulmonary artery (blue) and aorta (red)

The Cardiac Cycle

Cardiac cycle has few stages, the first phase is early diastole, when the pulmonary valve and the aortic valve are close, the mitral valve and the tricuspid valve are open, and the whole myocardium is relaxed. The second stage is ventricles' active filling when the atrial contracts through the tricuspid and aortic valves.

The third stage is isovolumic (unchanged volumes) contraction seconds that last 0.4–0.6 as soon as the ventricles start the isometric contraction, the pulmonary artery valve and the aortic valve are closed. Following the increased pressure within the ventricles, the ventricles eject blood into the pulmonary artery and aorta, the semilunar valves are open (Fig. 6.6).

The fourth stage is the isovolumic relaxation time, blood does not enter the ventricles, pressure falls, the ventricles do not contract and are in relaxation, and the semilunar valves are close owing to the pressure of blood in the aorta. The cardiac cycle is synchronized by electrical conducting system that is produced by specialized pacemaker cells found within the SA node and the AV node. The myocardium is composed of myocytes which initiate their own contraction without the help of external nerves (with the exception of modifying HR due to metabolic demand). Under normal circumstances, each cycle takes 0.8 s.

Cardiovascular Function

In a normal individual, Q at rest ranges between 4 and 6 L·min$^{-1,}$ but during all-out exercise, it may reach 25 L·min^{-1}. Q is determined and regulated by two main ways: (a) intrinsic cardiac regulation of blood volume and (b) the autonomic nervous

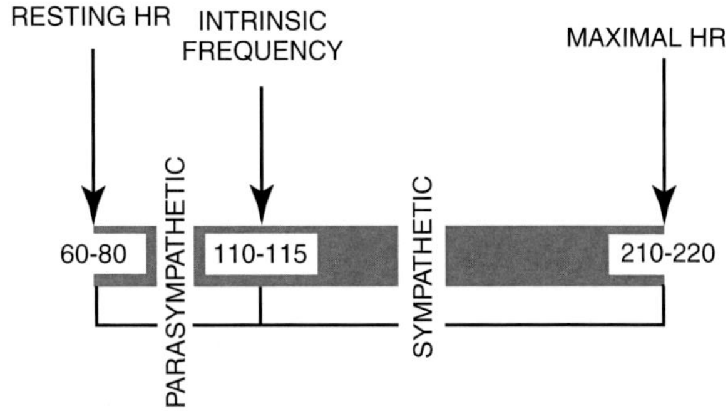

Fig. 6.7 Parasympathetic and sympathetic effect on the natural myocardial intrinsic frequency

system that normally is determined by the pacemaker activity of the sinoatrial (SA) node, accordingly, controls HR and left ventricular contractility [3].

The Frank-Starling mechanism is the intrinsic capacity of the myocardium that changes SV. This response depends on venous return; the more the myocardium is extended blood volume increases, consequently, left ventricular contractility increase, and, thus, SV ejected to the aorta is also increased. The fundamental base for this condition is associated with sarcomeres lengths and the functional subunits of actin and myosin to form crossbridges. Sympathetic stimulation has a positive chronotropic effect, resulting in HR increment, while an increase in vagal tone has a negative chronotropic effect on HR. Sympathetic and parasympathetic interplay regulates HR (Fig. 6.7). The heart exhibits a natural frequency of 100–115 beats·min^{-1}, this intrinsic frequency decreases with age. Frequency is decreased at rest to 60–80 beats·min^{-1}, slower than in a totally denervated heart, due to the activation of the vague nerve that innervates the SA node, and inhibits adrenergic activity [4]. Normally, at rest, there is a significant vagal tone on the SA node. During exercise, sympathetic stimulation overcomes parasympathetic stimulation (vagal) and therefore maximal HR reaches 220 beats·min^{-1}. Maximal HR with age decreases by 1 beat·min^{-1}·year^{-1}, consequently, in an individual 55 years old, maximal HR is 220–55 = 165 beats·min^{-1}.

Blood chemical composition controls the cardiovascular function through the baroreceptors, located in the walls of some arteries and veins, by releasing numerous hormones [5]. More specifically, baroreceptors located in the carotid sinus exist to regulate BP in the brain, while those located in ascending arch of the aorta control the systemic BP. Chemoreceptors, located close to the baroreceptors of the carotid sinus and arch of the aorta, react to the chemical configuration of blood. The chemoreceptors notice variations in blood levels of O_2 (hypoxia), CO_2 (hypercapnia), and H^+ (acidosis), and excites the chemoreceptors that send signals to the cardiovascular control centers in the brain. In response to this, increase signaling causes an increased sympathetic stimulation to arterioles and veins, producing vasoconstriction and a subsequent increase in BP. At the same time, the chemoreceptors send neural signal to the respiratory control centers in the brain, which in turn, the pulmonary increases O_2 supply and eliminates CO_2 levels.

Autonomic Nervous System

The autonomic nervous system responds to pressure sensors in the cardiovascular system via the baroreceptors [6], which technically are mechanoreceptors. These major stretch receptors are located in the proximal aorta and in the area of the carotid sinus. Additional stretch receptors are found in the great veins, atria, and pulmonary artery and in other areas of the cardiovascular system.

At rest and during exercise, the autonomic nervous system also responds to levels of CO_2 sensors through chemoreceptors. A portion of the changes related to the heart performance during exercise are similar to those that result from stimulation of the sympathetic nervous pathways to the myocardium. Thus, it can be seen that an increase in sympathetic nervous activity occurs during exercise and is an important factor in HR response [6]. There is considerable evidence that parameters thought to reflect the complex interaction between the autonomic nervous system and the cardiovascular system during exercise can provide significant prognostic information. Specific variables of great importance include HR response to exercise (reserve), HR recovery after exercise, and multiple components of HR variability both at rest and during exercise. Poor HR response to exercise has been strongly associated with sudden cardiac death and HR recovery from a standard exercise test has been shown to be predictive of mortality. In addition, there are limited studies evaluating the components of HR variability at rest, during exercise, and their prognostic significance [7].

Parasympathetic may influence the high-frequency-peak shift during strenuous exercise. HR variability and the usual indices of sympathetic activity do not accurately reflect changes in autonomic modulation during exhaustive exercise [8]. From the control center located in the ventro-lateral medulla, the two channels of the autonomic nervous system originate the sympathetic and parasympathetic components. The sympathetic components increase HR by releasing the neural hormone catecholamines—epinephrine and norepinephrine. These hormones are cardio accelerators. The parasympathetic nervous system located in the brain stem and upper or sacral portion of the spinal cord slows HR. The parasympathetic releases the neurohormone acetylcholine, which inhibits HR. During aerobic exercise, norepinephrine levels increase threefold. It seemed that exercise induces a significant response of the sympathetic nervous system [9]. The cardiac muscle cells of the ventricular myocardium are richly enervated by sympathetic nerve fibers. Release of norepinephrine by these fibers during aerobic exercise causes an increase in myocardial contractility, thus increasing SV. Norepinephrine is thought to increase the intracellular concentration of calcium in myocardial cells, thus facilitating faster actin/myosin cross bridging interaction. Also, a general sympathetic response by the body will induce the release of epinephrine from the adrenal medulla. Epinephrine, like norepinephrine, will stimulate an increase in the strength of myocardial contraction and thus increase again stroke volume [10].

The Blood

The blood is a mixture of two components: liquid as plasma and solids as cells and cell fragments. Blood contents include in addition to plasma, solved ions and different organic molecules. The amount of blood in an adult individual weighing about 70 kg is around 5.5 L. Blood elements include 55% plasma in the form of liquid portion of the blood, it contains salts, proteins, and blood cells, 40% red blood cells (erythrocytes) greatest abundant cell type in the blood, 1% white blood cells, and 4% platelets (thrombocytes). Almost 90% of plasma is water which acts as a solvent, suspension of the blood components, molecules transportation and thermoregulation, by controlling body thermal energy.

In the plasma, there are hormones and proteins such as albumin, globulin, (alfa, beta, and gamma), and fibrinogen. Additional solutes in plasma include electrolytes, nutrients, gases; O_2, large amounts of CO_2 and N_2; regulatory substances (enzymes and hormones); and waste products (urea, uric acid, creatine, creatinine, bilirubin, and ammonia).

The elements that constitute the blood are erythrocytes that contain blood platelets, hemoglobin, and leucocytes. The blood component of the cardiovascular system is a kind of connective tissue in the form of liquid plasma matrix. Two main purposes of the blood consist of plasma dissolved substances such as electrolytes, nutrients and vitamins, hormones, clotting factors, and proteins such as albumin and immunoglobulins (antibodies), providing immunity and defense against infections.

Stem cells formed within the spongy material inside the bones mature into red blood cells, live about 120 days, white blood cells, circulate from hours up to some days and platelets, last 6–9 days. The cellular portion of the blood is produced in the bone marrow and comprises red blood cells, white blood cells, and platelets. Red blood cells' surface (disc shape) bends inward like the interior of a sphere (biconcave shape). This flexible disc shape increases the surface area/volume ratio. Red blood cells contain millions of hemoglobin molecules, which bind oxygen molecules. In addition, CO_2 is transported to the lungs. Red blood cells (erythrocytes) mainly carry oxygen from the lungs and it drifts in the plasma. In a healthy individual, red blood cells constitute almost 99% of cells which play a key role in oxygen and CO_2 transportation and exchange. Hemoglobin, which binds oxygen, permits to transport 40–50 times the quantity of oxygen compared to the plasma capacity. White Blood Cells (leukocytes) are essential for immune processes; contain monocytes, lymphocytes, neutrophils, basophils, and eosinophils; and have a key role in the immune system and lymphatic system. Platelets (thrombocytes) play a primary role in blood clotting; when the minute blood vessel is injured, it clusters together to block the injured site. In a normal functioning cardiovascular system, the continuous motion of blood aids in keeping cells to be detached all over the plasma of the greater width vessels. At high altitude, where air oxygen content is low, bone marrow responds by increasing significantly red blood cells production, and once infection is present in the body, extra white blood cells are formed.

The Microcirculation Webs

Microcirculation means blood flow through the smallest vessels, within organ tissues and is the link between blood and single cell. In addition, the microcirculation delivers oxygen and nutrients and removes carbon dioxide. The microcirculation is the largest component in the cardiovascular system and is composed of four different sections: arterioles, capillaries, sphincters, and venules [11]. Each component has sole assembly and purpose. The blood flow through the microvasculature is in vessels with diameter less than 120 μm, within organ tissues. It is composed of terminal arterioles contracted by sympathetic nerves with a diameter of 75 μm [12]. The arteriole regulates blood flow and thus, influencing blood pressure and consequently, total peripheral resistance [13]. The sphincters before the capillaries still with a small amount of smooth muscle respond to chemical compounds as the autoregulation mechanism [14]. Blood flow continues through the capillary bed with a diameter of 7 μm almost as the red blood cell size, and intima tone with a thin wall that facilitates the exchange of gasses by diffusion. Then, blood flows to the venules with a diameter of 120 μm. The major site of exchange of materials occurs in the capillaries and in the junction of capillary-venules [15]. The coronary arteries' microcirculation consists the myocardium and large coronary arteries. When the arteries are occluded, myocardial infarction might happen. The coronary microcirculation has an important role to maintain oxygen demand–supply balance. To achieve this, it will be essential that the myocardium's coronaries resistance will integrate the involvement of inherent and extrinsic signals. The myocardium almost is solely dependent on the ability of aerobic metabolism [16], with maximal oxygen extraction at rest. On the other hand, the anaerobic pathway is inadequate to produce energy. With increase left ventricular contractility and heartbeats demand for ATP and therefore, oxygen is matched by myocardial aerobic metabolism due, to large and exceptional amounts of myocardial mitochondria. Since at rest oxygen extraction by the myocardium is maximal, the only way at maximal oxygen uptake to balance the increased myocardial oxygen demand is by linear increase in coronary blood flow by almost sixfold [17].

The oxidative production of ATP is mainly achieved from fatty acids in the Kreb's cycle [16]. However, in normal subjects, increase in myocardial oxygen demand to maintain viability, oxygen delivery is usually balanced and met by augmentation. Coronary blood flow is normally regulated by resistance in the arterioles of the coronary bed. Flow resistance in the coronary circulation has mixture signals coming from small arteries and arterioles. Resistance in the arterioles of the coronary is influenced mainly by myogenic activity, metabolic activity influencing the autoregulation mechanism, neurological and humoral factors, and by the endothelium. The greatest important and leading mechanism is the metabolic one. The oxygen requirements of the myocardium control the coronaries diameter and thus coronary artery diameter is increased by the vasodilation mechanism [17]. Normal arteries are capable to maintain the coronary blood flow at levels appropriate to the needs of the myocardium by the autoregulation mechanism. Myogenic and endothelial regulatory mechanisms effect on coronary diameters by stretch. Myogenic coronary arterioles narrow and then dilate, while shear stress induced dilation [18]. The vascular myogenic response refers to the acute reaction of a blood vessel to a change in transmural pressure. This response is critically important for the development of resting vascular tone [18]. So, in healthy individuals, coronary blood flow does not limit aerobic metabolism of the myocardium during maximal aerobic exercise [19].

Red Blood Cell Mechanical Properties

Red blood cell is involved through a number of mechanisms in order to increase oxygen delivery as needed by the working muscles during aerobic exercise. The increase in skeletal muscle demand is met by increasing oxygen delivery namely, *Q*, increases in muscle's (a-v)O_2, and increasing microcirculation blood flow [20]. Associated with this there is an increased release into the circulation of catecholamine. The large fraction of the blood volume is contained in the microcirculation webs and is affected as well. These alterations of blood rheology are related to muscular activity and intensity [21]. These may give rise in the traffic of Thallium 201 while testing vascularity changes for myocardial perfusion.

The mechanical properties and filterability of circulating red blood cells play a central role in the function of microcirculation [22].

Cell membrane mechanical fluctuation of red blood cells plays a vital role in health and disease. Circulatory resistance depends mainly on red cell concentration, plasma viscosity, and on red cell deformability [23]. The ability of red blood cells to deform is crucial to allow their circulation, especially when passing in narrow capillaries, this, in turn, determines vessel's resistance to flow [22]. Catecholamine appears to play an important role in membrane fluctuations and filterability of red blood cells due to changes in lymphocyte levels which are dependent on β-adrenergic receptor expression on the respective cell types and also on cyclic adenosine monophosphate (cAMP) which activates protein kinas A. It was shown that following the administration of prostaglandin E1 and adrenaline, cell membrane fluctuations and filterability of red blood cells were elevated [24] through a cAMP-dependent pathway.

Effect of Aerobic Exercise Training

It is a type of dynamic exercise of low to high intensity that relies largely on the Krebs aerobic pathway to produce ATP [25]. It means that forming ATP is mainly done with the presence of oxygen [26]. Aerobic exercise is characterized by sustained and prolonged activity that can last from few minutes up to hours [25] whereas vasculature plays a critical role in the physiological adaptations of the myocardium in response to physical exercise. The practical measurement of the cardiopulmonary system is the aerobic capacity measured as VO_{2max}.

Based on evidence that chronic repeated aerobic exercise changes structural and functional adaptations within the cardiovascular system [17], it: (a) increases myocardial capillary density; (b) increases vascularity diameters [27] and therefore, reduces resistance of coronary blood flow that is greatest in the coronary arterioles smooth muscle; (c) changes red blood cell mechanical properties [28] namely, change in cell shape and cytoskeletal rearrangement.

Following aerobic exercise, training adaptations in the coronary microcirculation increase the capacity of the coronaries blood flow [29]. Therefore, the microcirculation network stimulates erythropoiesis and consequently, augments blood hemoglobin levels and red blood cells [30].

Aerobic exercise and training regimens have large and lasting effects on hemoglobin by altering plasma levels and augment its levels [21, 31]. This, in turn, increases the ability of the blood to carry a larger amount of oxygen to the working muscles and myocardium [32].

Acute exercise disrupts hemorheological in a high exercise intensity, however, more important is hemoconcentration. The rheological properties of blood may control, in part, the alteration rate of oxygen delivery–supply at the onset of exercise [33].

Following long-lasting high aerobic exercise intensity training, in well-trained subjects' erythrocyte deformability was unchanged following test to maximal oxygen uptake [34]. It has been suggested that the reduced oxygen content in the arterials due to strenuous exercise [35] in well-trained subjects is a positive adaptive mechanism that facilitates the muscle's contractions. However, a significant increase in erythrocyte deformability response is found in well-trained athletes following four submaximal running sessions, for 30 min each, at 70% VO_{2peak} [36]. Therefore, it can be suggested that the effect of submaximal aerobic exercise bouts at 70% of VO_{2max} influence red blood cell deformability [37]. On the other hand, red blood cell deformability does occur under strenuous aerobic exercise in well-trained athletes, following only one session.

Blood Flow

Metabolic demand define blood flow amount to the various body organs, nonetheless blood flow is relative unresponsive to vessel pressure. The requirement is that blood flows nearly unceasingly over all the capillaries. In brief, blood flows through vessels depends on lengths (L) and internal radii (r), as a result of pressure difference (ΔP) between the vessel ends. In addition, when blood flows through the vessel, friction (vascular resistance) develops between the blood and the vessels walls [38]. The relationship between vascular flow, pressure difference, and resistance can be calculated as follows:

$$\text{Flow} = \text{pressure difference} / \text{resistance}$$

Or

$$Q = \Delta P / R$$

where flow rate (volume·time^{-1}), ΔP = pressure difference (mmHg), and R = resistance to flow (mmHg (time·volume^{-1}).

It is known that the resistance to flow through a cylindrical tube or vessel depends on several factors (described by Poiseuille) including: (a) radius, (b) length, (c) viscosity of the fluid, and (d) inherent resistance to flow, as follows [39]:

$$R = 8 l \eta / \pi r4$$

where r = inside radius of the vessel, L = vessel length, and η = blood viscosity.

Any small change in r has a significant impact (4th power) on its resistance to flow. As an example, a reduction in vessel's r by 50% rises its resistance to flow by almost 16-fold. As resistance decreases, blood flow increases despite the presence of reduced perfusion pressure.

It is possible to combine the above two equations into one, Poiseuille equation which can be utilized to estimate factors influencing flow through a tubular vessel:

$$Q = \left(\Delta P \pi r4\right) / 8 l \eta$$

Because of the pressure difference between the vessel's ends, arterial BP regulation is the greatest controlled physiological variable in the cardiovascular system. Therefore, BP flow moment-to-moment regulation is mainly controlled by regulating the radii of vessels. Vessel length and blood viscosity are not factoring that regulated BP flow moment-to-moment [40]. Regardless, the primary function of the heart is to keep pressure within arteries higher than those in veins, hence to create a pressure gradient to induce flow. Under normal conditions, MABP in systemic arteries is around 93 mmHg, and it is near 0 mmHg in the superior and inferior vena cava.

Blood Flow During Aerobic Exercise

During aerobic exercise, blood vessels in skeletal muscles dilate allowing to increase significantly blood flow and oxygenated blood to the working muscle. The mechanism that dilates the vessel is the autoregulation mechanism, produced by adenosine, hydrogen ions, and CO_2.

When exercising, Q redistribution occurs; a majority of the Q goes to the contracting muscles and less to the internal organs. This is possible through the interplay between the autonomic nervous system that causes vasoconstriction and autoregulation mechanism that causes vasodilation.

Metabolic demand define blood flow amount to the various body organs, nonetheless blood flow is relative unresponsive to vessel pressure. Therefore, muscle

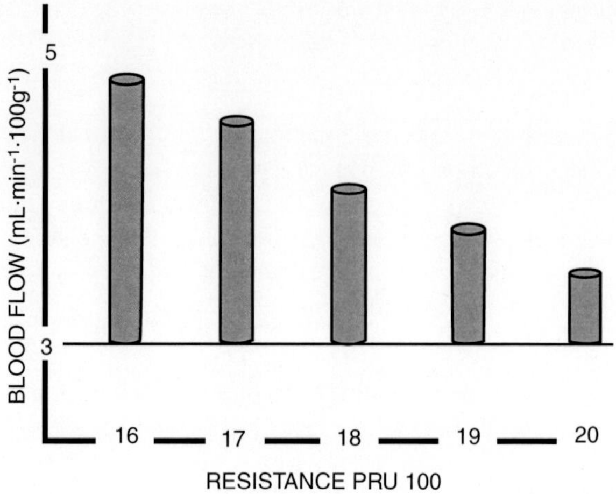

Fig. 6.8 Relationship between resistance and blood flow. When resistance value is high as 20 PRU100, blood flow is 3.3 $mL \cdot min^{-1} \cdot dL^{-1}$, while when resistance equals to 16 PRU100, blood flow is 4.7 $mL \cdot min^{-1} \cdot dL^{-1}$

blood flow is strictly corresponding to the metabolic demands of contraction [41]. Local control of blood flow ensures matching of perfusion to spatially varying oxygen demand, thereby achieving efficient oxygen delivery [42]. Blood flow regulation during exercise must guarantee blood and oxygen delivery, needed in tissues. Blood flow to tissues depends on the relationship between BP and the vessel. During exercise, fiber oxidative capacity among fiber types defines blood flow to skeletal muscles [43]. Regular endurance training has shown an adaptation of increased mitochondrial volume densities in skeletal muscle, slow twist and fast twist fibers [44]. Local blood flow is regulated by chemical factors, metabolites, active hyperemia, paracrine, and surrounding temperatures (warm or cold). A paracrine regulator is a hormone formed by a tissue to regulate its own activity. Therefore, paracrine regulators differ from endocrine hormones; they are not secreted into the bloodstream. The paracrine regulation is mainly regulated by nitric oxide, histamine release, and prostacyclin. Nitric oxide diffuses to smooth muscle and causes vasodilation by reducing calcium entry into smooth muscle. Muscular autoregulation occurs almost entirely in response to decreased oxygen concentrations. With increased demand for oxygen by the working muscles, it is crucial to increase blood flow to the working muscles [45]. Therefore, during exercise, vascular smooth muscle tone plays a fundamental role in regulating BP, blood flow, microcirculation, and other cardiovascular functions. As resistance decreases, blood flow increases despite the presence of reduced perfusion pressure (Fig. 6.8).

Coronary Arteries

Although all the blood enters the myocardium itself, it is impossible for nutrients to diffuse due to the myocardial layers. Thus, the coronary circulation delivers blood to the myocardium itself. The heart functions as an aerobic organ, with very slight anaerobic metabolism capacity, to form ATP. The myocardium is rich with mitochondria and therefore, at rest 70–80% of the oxygen delivered is extracted by the myocardium. Consequently, the myocardium is limited to increase oxygen by extraction, thus, the only way for the myocardium to have more oxygen supply is by increasing coronaries blood flow. In the case that oxygen deliver/demand is imbalanced, myocardial ischemia results, which may result in anginal pain, ECG changes, and left ventricular dysfunction.

In microcirculatory bed, the highest resistance to coronary blood flow occurs in the arterioles. As noted earlier, blood flow in the arterioles varies with the fourth power of these vessels' radii. Like systemic vascular beds, coronary arteriolar smooth muscle tone is normally regulated by few mechanisms that include neural, hormonal, and autoregulation, which are the most mechanisms [46]. Coronary effective perfusion pressure of the coronary bed is regulated by the pressure aortic pressure and the coronary vascular bed; however, coronary blood flow varies in a way that matches the aortic pressure, since coronary venous pressure is near zero. Similarly, during systole, the contracting myocardium produces a high level of intramyocardial pressure that compresses the coronary arteries and the microvasculature, resulting in almost zero flow [47]. On the other hand, during diastole, intraventricular pressures conveyed into the left ventricular wall apply a minor compressive force on the intramural vasculature [17], generating pressure differences at the arterioles and the venules, yet, blood flow compensation is relatively high during diastole [48]. Under normal and resting conditions, MABP of 60–140 mmHg helps coronary blood flow to be continual and constant, beyond these limits; flow turnouts to be pressure-dependent. Possible mechanisms include the myogenic response to rapid intraluminal pressure vagaries and slow metabolic control. Autoregulation range depends on myocardial oxygen tension, vasoconstrictors, or vasodilators.

Note: autoregulation of the coronaries and arterial leading blood to skeletal muscles is the same mechanism. During exercise, aortic pressure increases very slightly and therefore, actual perfusion pressure increases about 30%. So, during aerobic, coronary blood flow increases four- to sixfold, achieved mainly by the autoregulation mechanism and a decrease in coronary vascular resistance.

During aerobic exercise, compressive forces on the arteries is increased; however, in normal coronary circulation, this is not a physiological problem, since coronary vasodilator reserve capacity keep it up even during VO_{2max} [17]. However, during aerobic exercise under anemia, hypoxia, and decreased vascular diameter (atherosclerosis), coronaries compress, reduce oxygen delivery to the myocardium, which may yield a cardiac dysfunction.

Arterial Autoregulation Mechanism

The cardiovascular system is a closed-loop system in which billions of vessels interact with each other, and it enables the control of the systemic arterial pressure and varying organ flow through autoregulatory mechanisms [49]. At rest, skeletal muscles are considerably vasoconstricted, attributed to high sympathetic stimulation of muscle's vasomotor. In brief, vasomotor tone is almost exclusively determined by local metabolic oxygen demand. Hypoxia causes coronary vasodilatation directly but also releases adenosine and opens ATP-sensitive potassium channels. Pre-capillary sphincters are relaxed and more capillaries are recruited.

Thus, the autoregulation mechanism is significant in keeping coronary arteries and arterioles blood flow. It is an indicator of local blood flow regulation defined as the intrinsic ability of an organ to maintain constant blood flow despite changes in perfusion pressure, or, the inherent capacity of an organ to preserve constant blood flow in spite of changes in perfusion pressure. Arterioles are the smallest arteries and are the most important in regulating blood flow into the various capillary beds through smooth muscle bands in the arterioles and at the entrance to capillaries. The autoregulation mechanism is characterized by its vasodilatation effect on the small arteries and arterioles, despite the domination of the sympathetic nervous system. Since the circulatory system is a closed-loop system, it facilitates the control of the systemic arterial pressure and other organs' flow by autoregulatory mechanisms. The autoregulation mechanism is a local blood flow regulation that expands (vasodilatation) small arteries and arterioles, even though the sympathetic nervous system is in control of the arterial vasomotor tone. Alpha receptors, beta receptors, muscarinic receptors, angiotensin receptors, and endothelin receptors are factors affecting coronary vasomotor tone. The change in blood vessels' diameter brought by the autoregulation mechanism is the result of an imbalance between muscle oxygen demand–supply. During exercise, accumulation of local metabolites appears to mediate increased blood flow and thus, oxygen delivery. Such reactive hyperemia may be impaired in disease states, for example in cardiac failure. Arteriolar diameter is controlled by contraction or relaxation of vascular smooth muscle. The level of tone generated by a vascular smooth muscle cell depends on several factors, including its length and tension, local levels of metabolites and signaling molecules, communications with other cells via gap junctions, and neural inputs.

Autoregulation mechanism is a process that diverts oxygenated blood where it is greatly required and is active during aerobic exercises and not during anaerobic bouts. The autoregulatory response occurs in the absence of neural and hormonal influences and therefore is intrinsic to the organ [50, 51].

The autoregulation mechanism is utmost evidently observed in skeletal muscle, kidney, myocardial coronaries, and the brain. Given that arterioles are the main effectors of blood flow regulation, a crucial aspect of the metabolic response is the need for arterioles to receive information about the metabolic status of the tissue that they supply [52]. It has been previously suggested that the term autoregulation should also include a variation of blood flow in proportion to metabolic rate and the compensatory changes in blood flow which occur in the face of varying arterial oxygen content [53]. Currently, two theories are leading for the control of the autoregulation: the metabolic mechanism and myogenic mechanism. Both contribute to the autoregulation responses in arteries and arterioles to maintain blood flow in the face of increased oxygen demand by the working muscles (Fig. 6.9).

Fig. 6.9 Blood flow in vasoconstriction with the large increase in perfusion pressure, compared to the autoregulation effect, in spite of changes in perfusion pressure. Following 4 min of exercise, the autoregulation mechanism increases blood flow from 3.3 to 4.7 mL·min^{-1}·100 g^{-1}

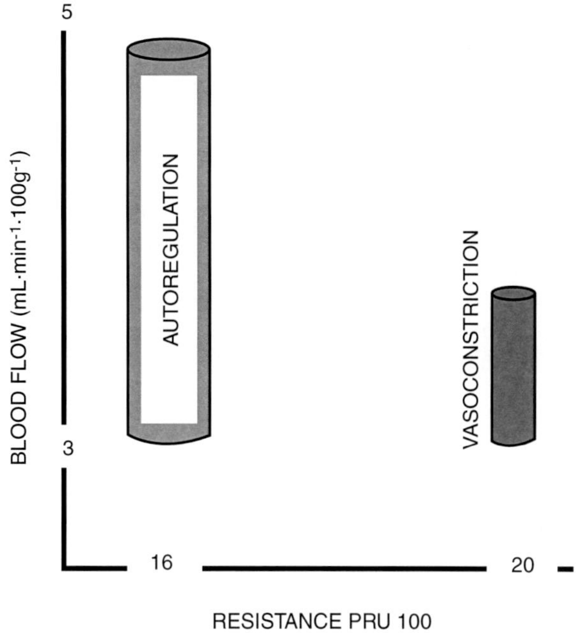

In the skeletal muscle, VO_2 is low at rest averaging 3.5 mLO_2·kg^{-1}·min^{-1} in addition to lower blood flow than other organs, such as the heart, kidneys, liver, and brain. The low rate of blood flow in inactive skeletal muscles results from a comparatively partial arteries and arterioles constriction due to the sympathetic nerves activity innervating them. Given that arterioles are the main effectors of blood flow regulation, a crucial aspect of the metabolic response is the need for arterioles to receive information about the metabolic status of the tissue that they supply [52]. It has been previously suggested that the term autoregulation should also include a variation of blood flow in proportion to metabolic rate and the compensatory changes in blood flow which occur in the face of varying arterial oxygen content [53].

When aerobic exercise commences, the sympathetic nervous system signals the arteries and arterioles muscles to contract reducing vessels radius. However, on the one hand, oxygen demand is increased, while, on the other hand, blood flow and oxygen delivery to the working muscles are reduced. Under these conditions, there is an accumulation of vasodilator metabolites in the tissue, increasing CO_2 which reduces pH, and ADP due to hypoxia. In general, ADP and the reduced pH have a strong effect on blood vessels' diameter in spite of the sympathetic activity. Increase or decrease of BP activates the myogenic mechanism to keep blood flow suitable to the oxygen demand by changing vessel's diameter. Myogenic mechanisms in the working muscles are part of the autoregulation mechanism which maintains a constant muscle blood flow at varying arterial pressure.

During maximal aerobic exercise, the significant increase in skeletal muscle blood flow is essential to deliver oxygen demanded (70–80 $mLO_2 \cdot kg^{-1} \cdot min^{-1}$) and substrates required to improve the metabolic activity of the working muscles. The autoregulation plays a key role in the *Q* redistribution of blood flow to the working muscles during exercise, and similarly it contributes greatly to the efficient delivery of the demanded oxygen by the tissues.

The autoregulation mechanism is not activated in working muscles during anaerobic and isometric exercises, likely because of greater concentrations of vasoactive substances owing to tissue hypoxia and acidosis [54]. Therefore, during anaerobic exercise, diastolic blood pressure is elevated and total peripheral resistance is modestly reduced compared with aerobic-type effort [55]. For example, during aerobic effort, if for any reason blood flow to an organ is reduced, the autoregulation mechanism then acts to restore normal blood flow.

Blood flow to the exercising muscles will be regulated through the arterioles leading blood in proportion to its needs and is controlled intrinsically by modifying the diameter of local arterioles. During exercise, the autoregulation mechanism keeps, on the one hand, DBP values similar or lower than that at rest, while, on the other hand, the elevated SBP increases blood flow to the working muscles [56].

Following long-term exercise training, autoregulation develops over weeks or months, an increase in the size of existing blood vessels and an increase in the number of vessels in a specific area, a process called angiogenesis. Shear stress-induced intussusceptive angiogenesis gives rise to the longitudinal splitting of blood capillaries. The cells of the vessel wall can regulate angiogenesis by producing growth factors, proteolytic enzymes, extracellular matrix components, cell adhesion molecules, and vasoactive factors. This property enables preexisting blood vessels to generate new vessels in the absence of exogenous angiogenic stimuli [57].

Heart Rate Control

The control of HR has two branches of the autonomic nervous system. The sympathetic nervous system and the parasympathetic nervous system. The sympathetic nervous system releases the hormones catecholamines—epinephrine and norepinephrine—to accelerate the HR. The parasympathetic nervous system (PNS) releases the hormone acetylcholine to slow the HR. Exercise increases HR which remains elevated throughout the exercise. At the beginning of exercise, body removes the parasympathetic stimulation, which enables the HR to gradually increase. During an increase in exercise intensity, the sympathetic system increases its stimulation that raises HR further. Long-lasting aerobic exercise training decreases resting HR due to increasing myocardial cavities, left ventricular contractility and filling tome. The reduced HR results from the increased influence of the parasympathetic nervous system, which restrains sympathetic nervous system activity.

There are two different factors involved in HR regulation: intrinsic and extrinsic controls. Intrinsic regulation of HR is a unique nature phenomenon of the heart; it maintains its own pace, while extrinsic controls involve both hormonal responses as well as nervous system regulation. Extrinsic regulation can change HR swiftly since blood chemicals have direct action of nerves that innervates the myocardium. Therefore, the HR depends on the balance between sympathetic and parasympathetic. Exercise is the most potent factor that increases HR to its maximal. This increase in HR is due to the increased metabolism and thus, oxygen demand by the contracting muscles. Other factor is the interplay between SV and HR.

Both the vagal and the sympathetic nerves are constantly stimulating the SA node. Long-lasting aerobic exercise reduces sympathetic activity and/or increases parasympathetic stimulation. In healthy middle-aged, older men and sedentary individuals, strenuous and prolonged endurance training elicited reductions in HR at rest, due to an increase in cardiac vagal tone [58].

Alterations in the autonomic nervous system have been correlated with a reduction in HR and BP, either in normal subjects or hypertensive subjects. However, the underlying mechanisms by which physical exercise produce bradycardia and reduces BP has not been fully understood. It seems that the role of the receptor and transduction signaling pathways on the heart and blood vessels influence the heart's response to exercise training [59].

HR increases at the start and sometimes in anticipation for exercise. The rapid onset of tachycardia is probably due to impulses from the cerebral cortex relayed via the hypothalamus to the cardio-inhibitory and vasomotor centers. Increased activity in the cardiac sympathetic nerves as well as a decrease in vagal tone appears to be involved. Each individual has an exercise intensity that will elicit his maximal Q. Once this limit is achieved, further increases in work intensity will result in no further increase in heart rate and anaerobic metabolism. By definition, this, then, is maximal HR, which in humans varies from individual to individual and decreases with age. Therefore, the only way to know precisely what a specific person's maximal HR is to conduct a maximal exercise stress test. Without such precise knowledge, professionals often use the Karvonen formula (220 minus ages), to approximate maximum HR [60]. This formula gives general results within plus or minus 10 beats·min^{-1}.

True maximum HR may not be achieved in some forms of exercise that do not employ a large muscle mass, or if the person is unfamiliar with the mode of exercise employed. For example, one person may have a true maximal HR of 195 beats·min^{-1} in uphill running, but only 191 beats·min^{-1} in a cycling test, and 187 beats·min^{-1} while swimming. These latter HRs are termed HR_{peak} and should be used as a basis for determining training intensity for a specific exercise mode. The important thing to remember is that maximal HR does not increase following chronic aerobic training. It stays sometimes, it even decreases slightly.

Factors Affecting Stroke Volume

Average aortic blood pressure, left ventricular contractility, and EDV are factors that control SV. At rest and during exercise, SV size depends mainly on venous return, which affects the force of ventricular contraction. As a result, SV increases as suggested by the Frank-Starling law that characterizes the association between SV and EDV: "Energy of contraction is proportional to the initial length of cardiac muscle fiber" [61], i.e., the force of cardiac contraction increases in proportion to the degree of diastolic stretch of its muscle fibers (EDV). The Frank-Starling mechanism happens due to the length–tension relationship detected in skeletal muscle [62] and myocardium [63]. In healthy subjects, EDV is often referred to as the preload. If the EDV increases, SV also increases. With an increased EDV, a slight stretching of the cardiac muscle fibers occurs, which increases the force of contraction. Thus, any increase in contractility will augment ejection fraction (EF) and SV. The sliding filament hypothesis has been used to explain the fundamental property of the cardiac muscle. Active tension is built when muscle fiber is stretched, produced by shifting the connection of thick and thin filaments. It has been found that the maximal contraction force is created when the sarcomere is stretched at the end of diastole between 2 and 2.2 nm [64].

Sympathetic stimulation constricts the skeletal muscle veins. This causes greater blood to flow to the arterials. The muscle pump is the rhythmic contraction and relaxation of skeletal muscle compresses the veins and thus, drains the skeletal muscle [65]. This causes greater blood flow back to the heart, i.e., venous return. The muscle pump is very important during both resting and aerobic exercise conditions. During exercise, breathing is one of the mechanisms that facilitates venous return. The pressure within the chest decreases and abdominal pressure increases with inhalation (pulmonary pump), thus, easing blood flow back to the heart. This pressure change results in reduced pulmonary blood vessels diameter, pushing more blood through the pulmonary veins into the left atrium. In addition, pressure in the right atrial decreases, due to its stretching when the ventricle contract that leads to an increase in venous return, and vice versa. Those increases in venous return, in turn increases left ventricular EDV and hence augments SV, as in the Frank-Starling law. At the same time, the increase in sympathetic discharge increases inotropic strength of the myocardium, which increases ventricular contractility (Fig. 6.10). Strength of ventricular contractility is increased by the secretion of epinephrine and norepinephrine which increases calcium concentration within the cardiac muscle fiber. Epinephrine and norepinephrine allow for greater calcium entry through the calcium channels in cardiac muscle fiber membranes. This allows for greater myosin and actin interaction and an increase in left ventricular contractility. Inotropy is the term applied to myocardial contraction, independent of alterations in preload and afterload. Inotropy varies with a variety of factors: increase rate of contraction, effect of post-extra-systolic potentiation, catecholamine, glucagon, and inotropic drugs. Inotropic state decreases with myocardial ischemia and heart failure.

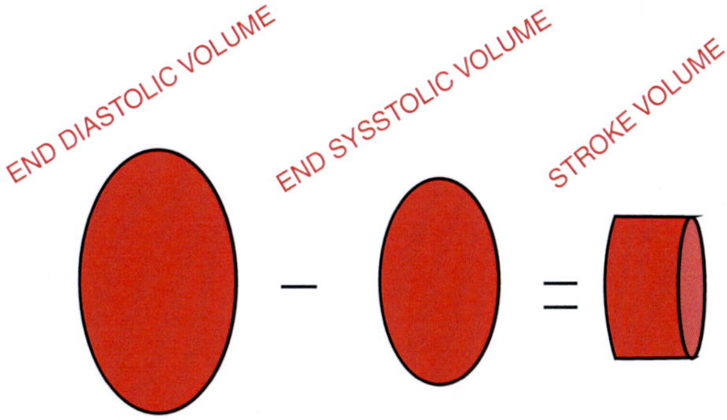

Fig. 6.10 Left ventricular volumes affecting stroke volume namely, left ventricular end-diastolic volume and left ventricular end-systolic volume

Blood ejected to the aorta is the result of differences in pressure developed in the left ventricle and that existing in the aorta often referred to as afterload. The SV is inversely proportional to the aortic BP. During aerobic exercise, the afterload is reduced, which allows for an increase in SV. Following the ejection of SV, about 30–50 mL of blood is left in each ventricle at the end of systole, end-systolic volume (ESV), which is dependent of the afterload and state of left ventricular contractility. Increase in afterload will reduce SV unless there is sufficient augmentation in left ventricular contractility to match the increase in afterload.

Stroke Volume During Aerobic Exercise

In sedentary individuals, SV during dynamic aerobic exercise in the supine position is almost 160 compared to 130 mL during erect dynamic aerobic exercise. The differences in SV during dynamic exercise between erect and supine can be attributed to the decreased resting SV normally found in the erect position [66]. This orthostatic effect is compensated for by increase in the muscle pump and HR.

In response to endurance exercise, the heart adapts by increasing SV at rest, during submaximal exercise, and during maximal exercise. There is some debate, as to whether SV increases because HR decreases which is due to increasing diastolic filling time, or, because of an increase in left ventricular volume due to eccentric hypertrophy of the heart muscle. Based on available data, both factors probably contribute. Both changes also rapidly revert toward normal with the cessation of training (Fig. 6.11). One other important change that takes place is increased blood volume, which helps to take advantage of the increased filling capacity of the heart and facilitates stroke volume increase. This adaptation occurs fairly rapidly with training but is also the first adaptation lost if training is stopped for a week or two.

Understanding the Fick equation is of principal importance for appreciating the utility of the stroke volume prediction. Since (a-v)O_2 do not change dramatically during steady-state exercise, the final equation for the estimation of stroke volume was suggested by Whipp et al. [67] as follows:

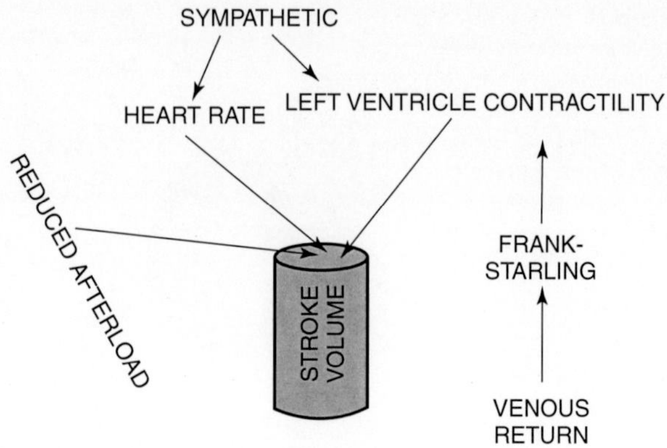

Fig. 6.11 Factors controlling stroke volume at rest and during exercise such as sympathetic nervous system, Frank-Starling law, venous return, heart rate, left ventricular contractility, and afterload

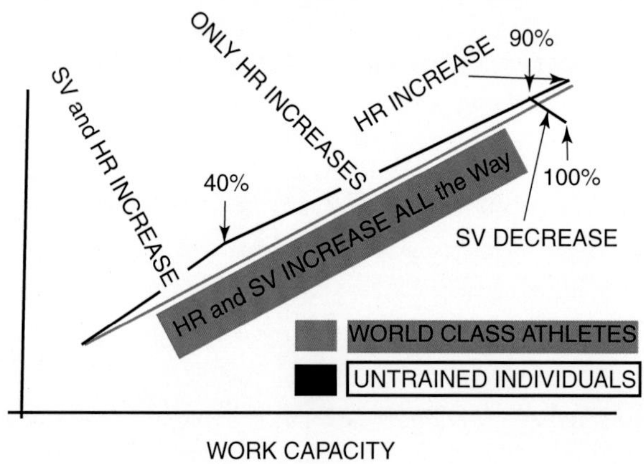

Fig. 6.12 The inverse relationships between heart rate (HR) and stroke volume (SV) in three different workloads. In untrained individuals and world-class athletes

$$\text{Exercise SV}(\%) = 5(\Delta VO_2 / \Delta HR) \bullet 100$$

where VO_2 = oxygen uptake, HR = heart rate, Δ = delta, and 5 = a constant number.

Thus, during exercise, stroke volume is significantly related to oxygen pulse ($p < .05$) but not to (a-v)O_2 [68]. However, this relationship is not valid in patients with lung disease, anemia, or hyperemia; normal subjects at high altitude or subjects who desaturate during exercise, such as some highly trained athletes [67].

The Interplay Between HR and SV

In sedentary individuals, contribution of SV and HR to Q varies at three different intensities of aerobic exercise: (a) from rest to 40% of VO_{2max}, (b) between 40 and 90% of VO_{2max}, and (c) from 90% to VO_{2max} (Fig. 6.12).

During the first intensity from rest to 40% of VO_{2max}, Q is increased mainly by augmented SV along with a moderate increase in HR, in untrained individuals.

When 40% of VO_{2max} is reached in sedentary subjects, SV attains its maximal value. During the second intensity, of 40–85% of VO_{2max}, any further increase in *Q* is met only by increased HR since SV decline. Figure 6.13 describes the linear relations between work intensity and HR between 120 and 180 beats·min^{-1}. In world-class athletes and trained individuals, SV and HR increase with the increment of workload up to VO_{2max} [69], which point on their capacity to utilize the Frank-Starling effect throughout the incremental exercise.

The Frank-Starling mechanism describes the ability of the heart to change its force of contraction and hence SV in response to changes in venous return. In other words, if the end-diastolic volume (EDV) increases, there is a corresponding increase in SV (Fig. 6.14). The Frank-Starling mechanism can be explained on the basis of preload. As the heart fills with more blood than usual, there is an increase in the load experienced by each muscle fiber. Maximal force is produced with an initial sarcomere length of 2.2 μm [70]. This stretching of the muscle sarcomere increases the affinity of troponin C to Ca^{2+} ions causing a greater number of cross-bridges to interact with actin and thus, increases contractility.

Evidence indicates that endurance-trained athletes are able to increase their SV throughout incremental upright exercise, probably due to a progressively greater effect of the Frank-Starling mechanism. This is contrary to the widely held belief that stroke volume reaches a plateau at a submaximal HR irrespective of fitness level, owing to a limitation in the time for diastolic filling [66].

During the third load, between 85% and VO_{2max} (100%), in sedentary subjects, *Q* increases as a result of a sharp increase in HR due to slightly decreased SV, owing to the shorter left ventricular filling time and thus, smaller left ventricular EDV (Fig. 6.15).

Therefore, in untrained individual, during dynamic exercise at the third load, working muscles will rely more and more on the anaerobic pathway to produce

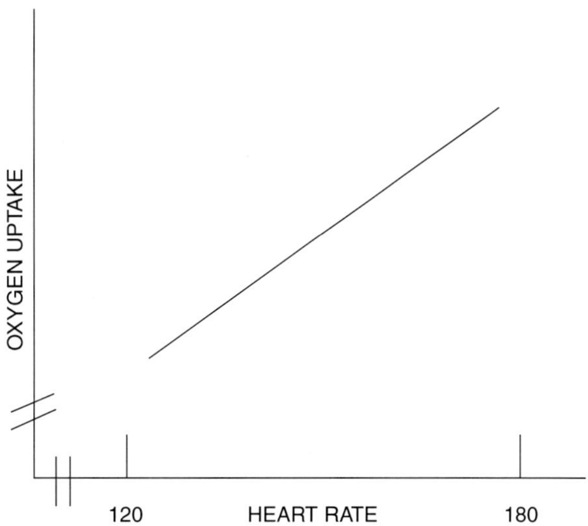

Fig. 6.13 In untrained individuals, a linear relation exists between heart rate and workload between 120 and 180 beats·min^{-1}

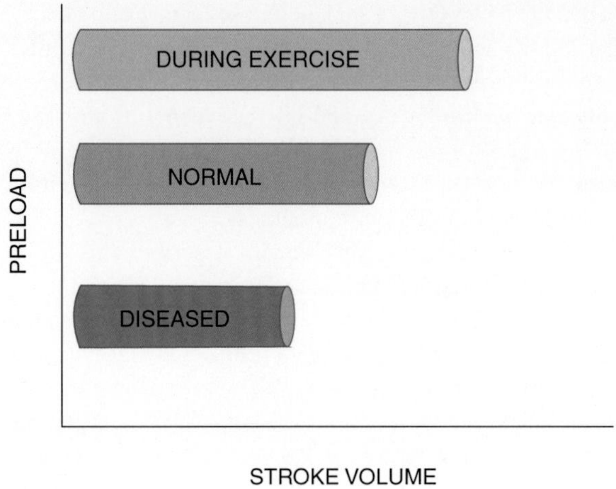

Fig. 6.14 Figure illustrates the Frank-Starling law effect on stroke volume: increased venous return increases left ventricular end-diastolic volume, which in turn, left ventricular contractility increases ejected blood amount (stroke volume)

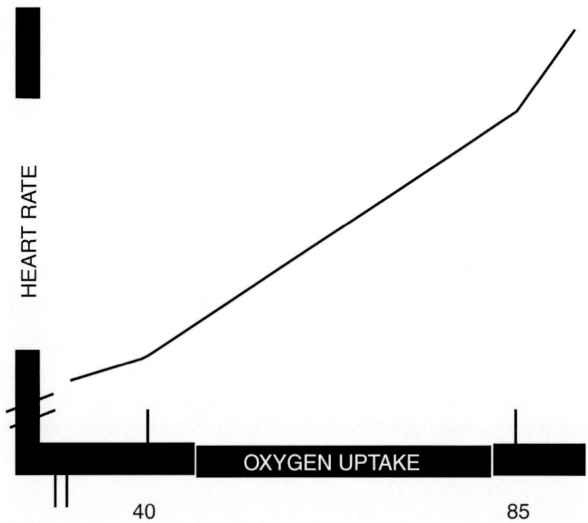

Fig. 6.15 Heart rate responses during dynamic aerobic exercise in relation to different workloads

ATP, with elevated levels of lactic acid. This in turn will increase fatigue in untrained individual's SV at rest averages 80 mL·beat^{-1} and it increases up to 130 mL·beat^{-1} during intense aerobic exercise. In elite athletes, resting SV averages 110 mL·beat^{-1} and in Olympic cross-country skiers, SV can increase at peak exercise up to 200 mL·beat^{-1} [71].

During aerobic exercise at a given *Q*, the interplay between HR and SV is inversely related. If SV increases markedly, HR will increase moderately. Under the same given *Q*, SV will increase only slightly if HR increases significantly.

The relationship between HR and stroke volume is best defined at the end of the filling stage, which is expressed as left ventricular EDV. Ventricular volume depends on filling time, which is dependent on beats·min^{-1}. Increased heart rate reduces diastolic period and thus, filling time as in untrained individuals. Therefore, for *Q* to increase during vigorous exercise, it is reasonable to argue that there is a progressive utilization of myocardial contractility, tachycardia, and the Frank-Starling mechanism. Very little information exists regarding the cardiovascular responses to changes in postural position during incremental to maximal exercise in endurance-trained athletes. Previously it has been suggested that when comparing erect vs. supine exercises, peak diastolic filling rate, end-diastolic volume, and SV are consistently lower throughout incremental exercise in the upright position, reflecting differences in the effective filling volume of the myocardium. Investigators have shown that there are minor changes in SV during incremental exercise in the supine position in healthy individuals [72], indicating that the myocardium may approach its limits for ventricular filling during supine exercise. At rest, SV and resting HR of the average person is approximately 70 mL·beat^{-1} and 70 beats·min^{-1}, respectively. This gives us about 5 L·min^{-1} resting *Q*, determined by oxygen demand at rest which is approximately 250 mLO_2·min^{-1}, and also by the need for high blood flow to the kidneys for filtration purposes. It does not change appreciably with endurance training. However, the manner in which the heart delivers this resting demand does change following aerobic training.

Cardiac Output in Normal Individuals

Total amount of blood flow circulating through the heart, lungs, and organs represents *Q*, defined as the blood amount ejected from one ventricle during 1 min. Cardiac function is the ability of the myocardium to deliver oxygen to the muscles' metabolic demands. *Q* is an important indicator of the heart work efficiency in oxygen delivery.

The adjustment of the oxygenated blood flow through the cardiovascular system is by the tissue's oxygen demand. Tissue's oxygen request increases when metabolic needs increase as during exercise. To meet the increased metabolic demands, *Q* increases linearly (Fig. 6.16) with VO_2, by the rise in HR, SV, and $(a\text{-}v)O_2$ by the contracting muscles.

In addition, Alveoli ventilation increases, which increases oxygen partial pressure and hence better oxygen diffusion. The stage of respiration with intrathoracic pressure changes influencing diastolic heart filling and therefore *Q*. Inhaling reduces intrathoracic pressure, filling the heart and increasing *Q*, while exhaling increases intrathoracic pressure, reduces heart filling, and *Q* [73]. Most individual tissues determine their own flow in proportion to their metabolic rate. The skin is a notable exception where the priority is thermal rather than metabolic. Brain, heart, skeletal muscle, and the splanchnic area all vary their blood flows according to local tissue metabolic rate. Summation of peripheral blood flows constitutes venous return and hence *Q* [53].

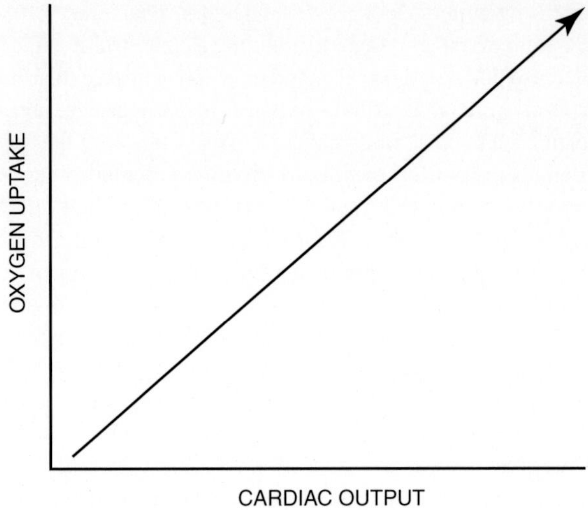

Fig. 6.16 The relationship between oxygen demand and cardiac output

Venous return (preload) forces opposing left ventricular ejection (afterload), myocardial contractility, and filling volume of the atrial affect Q function. Q is the product of the HR, and the volume of blood propelled by the ventricle for each beat namely, SV.

Thus,

$$Q = \mathrm{HR} \times \mathrm{SV}.$$

Values for Q are usually signified as $\mathrm{L{\cdot}min^{-1}}$. The total volume of blood in the circulatory system of an average person is about 5 L (5000 ml). Since at rest Q equals also 5 $\mathrm{L{\cdot}min^{-1}}$, it means that the entire volume of blood within the circulatory system is pumped by the heart each minute. At rest, for a healthy individual weighing 70 kg, Q norms are about 5–5.5 $\mathrm{L{\cdot}min^{-1}}$ presuming a HR of 70–75 $\mathrm{beats{\cdot}min^{-1}}$, and SV of 70–80 mL. Maximal HR of young sedentary persons is approximately 220 $\mathrm{beats{\cdot}min^{-1}}$ at maximal work, in well trained it is slightly lower 215–220 $\mathrm{beats{\cdot}min^{-1}}$, and world-class athlete it may be as low as 205 $\mathrm{beats{\cdot}min^{-1}}$. Figure 6.17 describes Q in sedentary, well-trained, and world-class individuals during vigorous aerobic exercise can increase from resting values up to eightfold or 25–40 $\mathrm{L{\cdot}min^{-1}}$ [74].

Factors Affecting Cardiac Output

Numerous factors affect HR and SV and thus, Q (Fig. 6.18).

There are two different factors involved in HR control: intrinsic and extrinsic controls [75]. Intrinsic regulation of HR is by the unique nature of cardiac tissue through the natural pacemaker (SA node). However, the heart does not regulate its

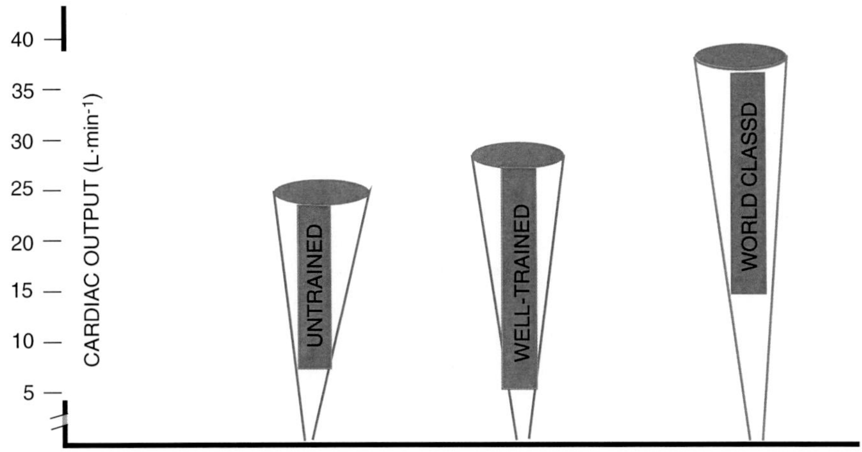

Fig. 6.17 Cardiac output values for different fitness levels: untrained, well-trained, and world-class athletes

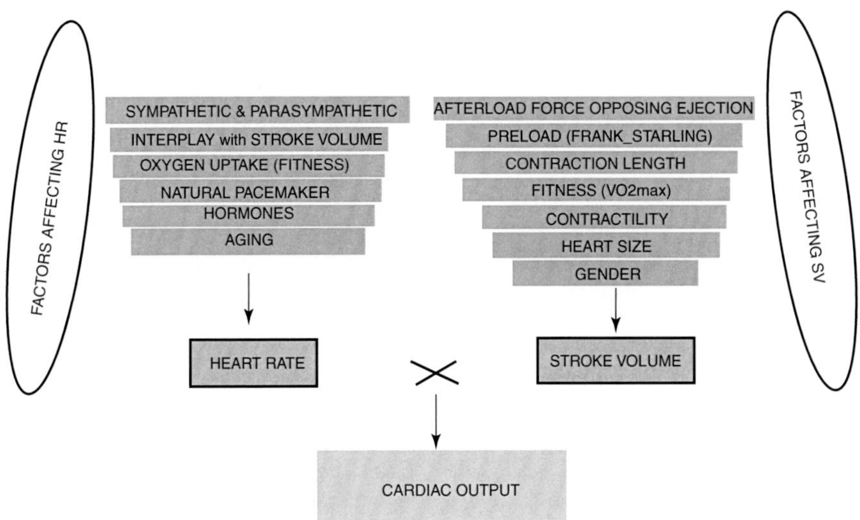

Fig. 6.18 Factors affecting cardiac output

own output. Extrinsic controls are the effect of hormonal responses through the central and autonomic nervous systems. Extrinsic regulation is done by chemical influence such as catecholamine (adrenalin) that increase HR rapidly due to its direct action by the nerve innervating the myocardium [76]. The second but, low additional HR increase occurs, due to the circulating catecholamine in the blood. The vagal tone (parasympathetic system) has the opposite effect on HR by reducing it [77]. Increase in SV reduces HR due to the interplay between those variables. In addition, an increase in fitness levels increase the vagal tone and thus, HR is reduced at rest and during submaximal aerobic exercise. Aging is also an additional factor that decreases maximal HR about, 1 beat per year.

Factors affecting SV include gender differences, HR, left ventricular contraction length, left ventricular contractility, preload, afterload, fitness level, and aging. Gender differences exist in cardiac structure and function, men have bigger hearts than women; therefore, in females, SV is smaller and HR higher to compensate for the smaller SV. In addition, it seems probable that gender hormonal differences play also an important role. Changes in HR influence the interplay with SV, increased HR reduces ventricle filling time consequently, SV decreases. In healthy individuals, left ventricular contraction is short, which in turn, increases diastolic phase meaning, an increase in ventricles' filling time.

During aerobic exercise, left ventricular contractility is an intrinsic stretch of the myocardium independent of preload along reduced TPR so, SV increases. An echocardiogram or radionuclide methods are employed to calculate end-diastolic volume (EDV), end-systolic volume (ESV), and SV, as follows:

$$SV = EDV - ESV.$$

EDV is the blood volume in the ventricle before ejection namely, preload. It is the initial stretching of the myocardium myocytes prior to systole, related to ventricular filling and sarcomere length. Preload increases due to increase in venous return depend (Fig. 6.19) on venous tone, volume of circulating blood, increased sympathetic tone, body posture position, pressure gradient between arterial to atrium, respiratory pump, and skeletal muscle pump.

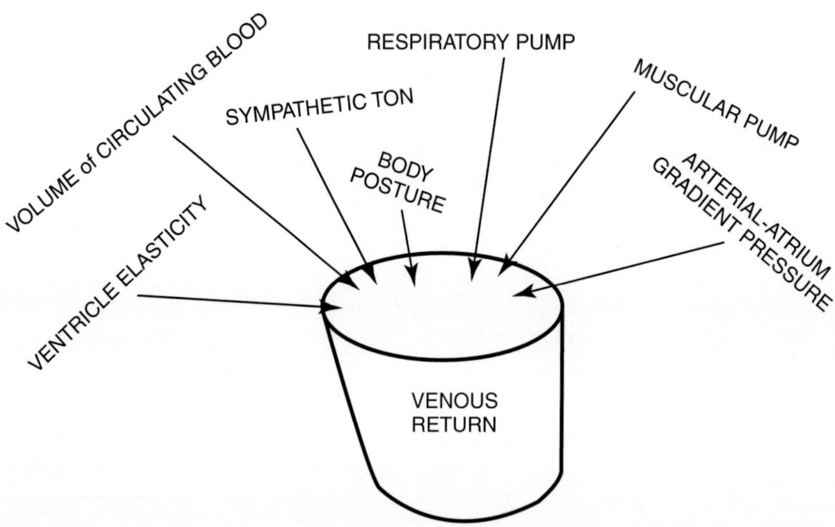

Fig. 6.19 Factors affecting venous return to the heart

Increase in EDV increases myocardium's fibers stretching and consequently, ventricle contractility increase (Starling's law or Frank–Starling mechanism). This law represents the relationship between SV and EDV. Afterload is the force opposing ejection that the ventricle must overcome to open the aortic valve. At rest, afterload depends on the sympathetic charge, vasculature elasticity, and blood viscosity.

Q responses to exercise at the sea level depend mainly on the mode of exercise performed, metabolic demand, and fitness level. Modes of exercise include dynamic aerobic, dynamic anaerobic, and isometric. During aerobic exercise such as running, biking, walking, and swimming, muscular metabolic rate increases significantly and thus, oxygen demand. In order to deliver the needed oxygen, *Q* increases along with its redistribution to the different organs. *Q* increases linearly with workload expressed as percentage of VO_{2max}. The increase in *Q* is achieved by increases in HR and SV. The augmented SV results from the increase in venous return that increases EDV and thus, left ventricular contractility. Untrained reaches its maximal SV at 40% VO_2, while in world athletes, SV increases from resting to maximal aerobic exercise (Fig. 6.20).

When exercising in the supine position, SV reaches at 40% of maximal work capacity its maximal volume, which thereafter, it does not change, consequently, *Q* is increased mainly by HR (Fig. 6.21). In well-trained athletes, SV can reach values of 200 mL during aerobic supine exercise.

During aerobic supine exercise HR increases linearly with VO_2 from the beginning to maximal effort (Fig. 6.22).

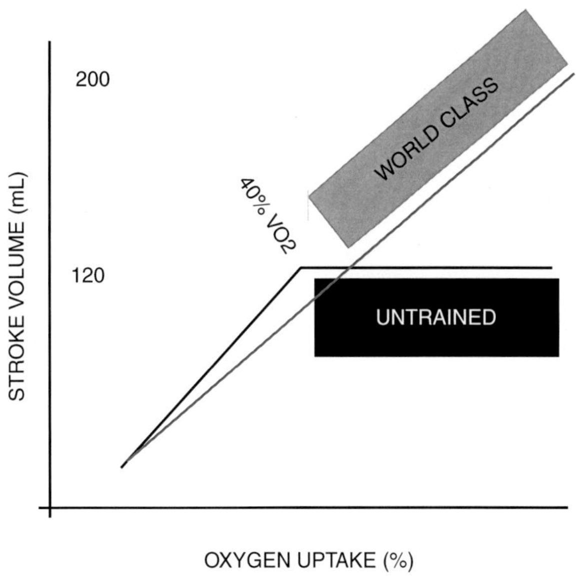

Fig. 6.20 Difference in stroke volume increment during aerobic exercise in untrained individuals and world-class athletes

Fig. 6.21 Stroke volume response during supine aerobic exercise

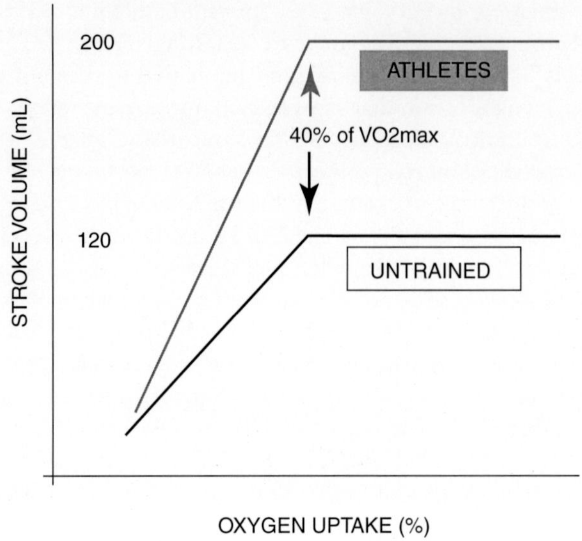

Fig. 6.22 Relationships between heart rate and workload during supine exercise

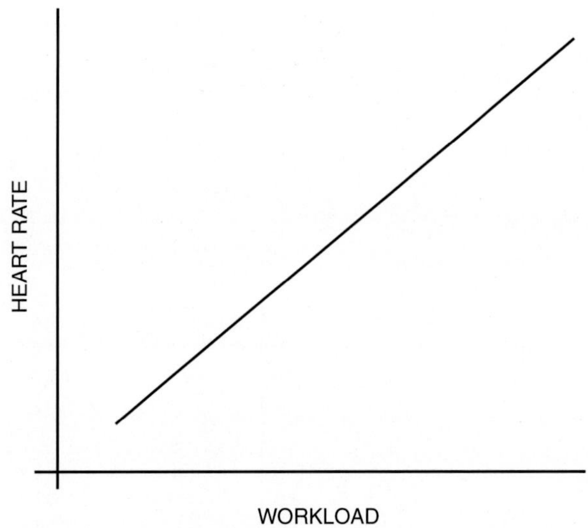

However, during aerobic exercise, HR is linear only between 40 and 85% of VO_2 (Fig. 6.23), while SV when exercise commences, increases.

Q increases only to its peak at 85% VO_2. Well-trained individual increases its SV all the way to peak exercise (Fig. 6.24).

During isometric bout such as deadlift utilizing large muscle mass, Q increases modestly (11–13 $L{\cdot}min^{-1}$), SV decreases, increases, or remain unchanged, while HR increased moderately (120–130 $beats{\cdot}min^{-1}$).

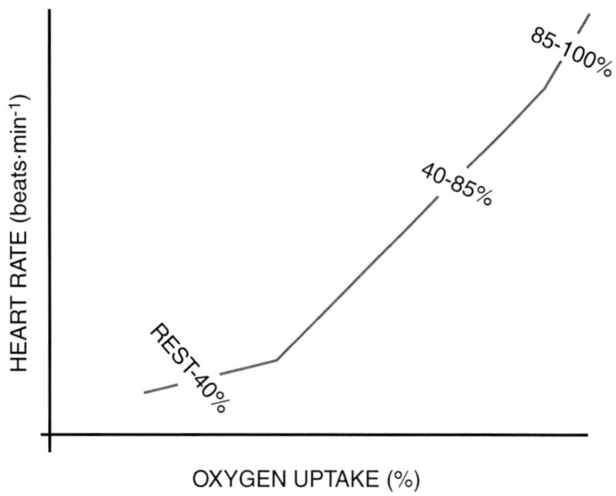

Fig. 6.23 Heart rate response during aerobic exercise

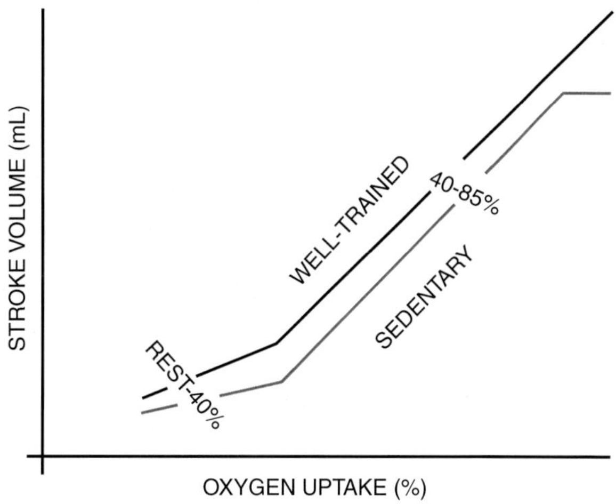

Fig. 6.24 Stroke volume response during aerobic exercise in well-trained and sedentary

Cardiac Output in Adolescents

Previous investigations have indicated that cardiac output values for adolescents at VO_{2peak} are low compared to young adults during maximal aerobic exercise [78]. The explanation for the lower Q in both male's and female's adolescent at VO_{2peak} might be related to the smaller ventricular cavities, increased afterload, and lower VO_{2max} [79]. Cardiac adaptive changes to aerobic exercise training in adolescent are infrequent. The changes in left ventricular volumes during exercise may be due to reduced contractility, diastolic and systolic left ventricular functions, and loading conditions [79]. The extent that enhanced ventricular function contributes to superior aerobic fitness of trained athletes compared to sedentary adolescents is unclear. Recent study [78] revealed no differences between trained adolescents and untrained adolescents in *Q* responses to progressive exercise, implying that greater aerobic fitness in these athletes reflected volume expansion of the cardiovascular system without the contribution of enhanced systolic or diastolic ventricular function. Such findings should be considered limited to the context of trained adolescents with limited duration of athletic training. No significant differences are attained for SV, *Q*, and left ventricular dimensions when they are scaled to body surface area at rest and whatever the exercise intensity. The patterns of SV, as well as its underlying mechanisms, are not age-related during an upright maximal aerobic exercise test [80].

Cardiac Output During Aerobic Exercise

The significant increase in VO_2 during dynamic aerobic exercise forces the heart to increase *Q* and to dilate the arterioles (autoregulation). During dynamic aerobic exercise such as walking, running, swimming, and cycling, oxygen demand by the working muscles increases proportionally to the intensity and by diverting blood from the liver, kidneys, and digestive tract. During submaximal dynamic aerobic exercise at the same absolute load, *Q* will be similar in trained and untrained subjects. However, the way to achieve that *Q* differs significantly between these two subjects. In untrained subjects, the increase in *Q* is achieved mainly by a significant increase in HR and a moderate increase in SV, while in trained subject, the increase is due to a significant increase in SV and HR.

Arterial-Venous Oxygen Difference

(a-v)O_2 reflects the capacity of active skeletal muscles and the respiratory muscles to extract and consume oxygen from the blood for ATP production during exercise. Relative to young adults, maximal (a-v)O_2 of older adults is reduced consistent with the marked reductions in capillary density and mitochondrial enzyme activities [81]. During exercise, the increased demand for oxygen is met by increasing muscle blood flow [20] and by improved oxygen unloading from hemoglobin achieved by decreasing hemoglobin to oxygen affinity [82], Thus, the (a-v)O_2 could be explained by the ability of hemoglobin to release oxygen or instead by the tissue capacity to extract it (peripheral factor) from hemoglobin [83].

A review on the age-related decline in VO_{2max} suggests that there is only a minor age-related decline in maximal *Q*, and therefore peripheral factors such as a decrease in (a-v)O_2 may better account for the age-related decline in VO_{2max} [84]. One possible mechanism for the inability to increase (a-v)O_2 could be mild arterial hypoxemia at high-intensity endurance exercise. This is related to a misdistribution of *Q*, i.e., overperfusion of organs other than those in which perfusion and diffusion limitations are rapidly compromised [85].

Other explanation for low (a-v)O_2 in the elderly may be due to physical inactivity or sarcopenia. As a consequence, the result is mitochondrial partial dysfunction, due to unmatched or perhaps intrinsic mitochondrial function and regulation [86]. The outcome of abnormal mitochondrial function is the accumulation of reactive oxygen species generated by the addition of a single electron to the oxygen molecule [87], resulting in a reduction in muscle VO_2 at rest and during exercise. Therefore, defects in the biogenesis of this organelle can impact the ability of muscle to oxidize substrates and can have serious consequences on the action of insulin on glucose uptake [88]. The molecular defect in subjects with mitochondrial myopathies commonly involves the mitochondrial genome, with either single large-scale deletions or point mutations detected [89]. This may potentially result in an inequality between tissue perfusion and metabolic demand in the active muscle tissue, ultimately compromising the ability to extract oxygen. Thus, reducing oxidative capacity, due in part to age-related mitochondrial dysfunction, which contributes to the decline in aerobic performance in aging skeletal muscles [90].

Changes during aging are partly due to oxidative stress and inflammation resulting from mitochondrial malfunction and abnormal mitochondrial-nuclear signaling. The aging process is characterized by an imbalance between an increase in the production of reactive oxygen species in the organism and the antioxidant defenses as a whole. However, it has been suggested that as the level of oxidative stress in skeletal muscle increases with age, the production of some antioxidant enzymes increases adaptively to compensate in part [87]. Mitochondrial maintenance failure may result from trade-offs between mitochondrial turnover versus growth and reproduction and nuclear mutations and loss of epigenetic regulation.

Several interventions increase longevity of a mitochondrial unfolded protein response (UPRmt), observed during normal aging, indicating hormesis. The UPRmt may increase longevity by stimulating mitochondrial turnover through autophagy, and/or by inhibiting the production of hormones and toxic metabolites [91]. It has been suggested that reactive oxygen species production is responsible for the higher (a-v)O_2, while mitochondrial loss of function results from the increased oxidative damage to proteins, lipids, and DNA of this organelle [92]. The metabolic signatures of exercise-trained skeletal muscle reflected reprogramming of mitochondrial function and intermediary metabolism and correlated with changes in cardiometabolic fitness. Mitochondrial background has been demonstrated to influence VO_{2max}, but this genetic influence can be compensated for by regular exercise. A positive correlation among electron transport chain coupling, ATP, and reactive oxygen species production has been established, and mitochondrial variants have been reported to show differences in their electron transport chain performance [93]. The latter is most likely associated with poor muscle microcirculatory network and capillary-myofiber dissociation [94], due to higher arterial perfusion pressures at rest and at any submaximal dynamic exercise than their younger counterparts. Thus, attenuate leg blood flow during doderate intensity while paddling, suggesting a heightened level of vasoconstriction in the legs that could result from structural and/or functional alterations within the arterial vasculature [95]. The reduced blood flow during exercise could be related to an elevation in tonic sympathetic nerve activity to the α-adrenergic receptor within the vasculature of the active muscles [96]. In fact, in the young, α-adrenergic vasoconstriction is blunted during leg exercise in an intensity-dependent manner [97]. An additional explanation suggests that due to the observed differences in leg α-adrenergic vasoconstriction, sensitivity could be attributed to the fact that older subjects compared to young subjects usually do not reach absolute work rates high enough to provoke full attenuation of this receptor [98].

The endothelin-1 vasoconstrictor pathway contributes to age-related elevations in resting peripheral vascular tone primarily through activation of the endothelin subtype α-adrenergic vasoconstriction receptor. Findings have identified an age-related increase in endothelin subtype α-adrenergic vasoconstriction—mediated vasoconstrictor activity that persists during exercise, suggesting an important role of this pathway in the regulation of exercising skeletal muscle blood flow and maintenance of arterial blood pressure in the elderly [99]. Other data challenged the concept of reduced sympatholytic in the elderly, suggesting instead that the inhibition of α-adrenergic vasoconstriction in the exercising leg is associated with work performed and not attributable to vascular aging, therefore, more closely related to the rate of oxidative metabolism than to age per se [98].

In healthy subjects, during maximal aerobic exercise, a large amount of the available oxygen is extracted from blood perfusing the working muscles. At rest, for a subject with approximately 15 $g \cdot dL^{-1}$ hemoglobin 98% saturated, oxygen content of arterial blood is approximately 200 $mLO_2 \cdot L^{-1}$. In the exercising legs, blood flow and oxygen delivery leveled off at 73–88% of VO_{2peak} accompanying rise in systemic oxygen extraction to 84 ± 2%, particularly in well-trained athletes. This brings about a fall in oxygen content to about 20–30 $mLO_2 \cdot L^{-1}$ in the venous blood [100]. However, these high percentages of oxygen extraction are not applied to sedentary individuals, whose maximal percentage of oxygen extraction is around 70–75% of blood oxygen content [101]. Consequently, particularly in athletes there is little oxygen left to be extracted out of the blood during heavy exercise. Hence, the dominant mechanism for the increase in VO_2 with training must be an increase in Q and oxygen delivery.

Anaerobic metabolism becomes increasingly important during periods of ischemia resulting from an inadequate blood flow that fails to meet the energy demands of tissues. If tissues are subject to ischemia, they try to compensate by extracting more oxygen from the blood. Tissue groups such as muscle typically use at rest only a third of the oxygen available to them, and therefore are able to increase oxygen extraction up to maximal exercise. Conversely, the myocardium extracts most of the available oxygen from the blood even during periods when the body is at rest. The myocardium, then, has a very little physiological reserve to respond to episodes of high energy demand. Rather, the myocardium depends almost entirely on increased coronary blood flow to satisfy high energy demand. Low Q can cause cardiac ischemia—perhaps more so for the myocardium than other organs because of the heart's already high rate of oxygen extraction. A vicious cycle ensues. Cardiac ischemia forces a shift toward anaerobic metabolism from the much more efficient aerobic metabolism. With less energy available and increased intercellular acidity, the left ventricular contractility weakens, causing a further reduction in SV and *Q*.

Under physiological conditions, VO_2 is independent of oxygen delivery. In order to maintain a constant VO_2, the body can respond by increasing $(a\text{-}v)O_2$ corresponding to a reduction of venous oxygen content and/or by increasing cardiac output.

Cardiac Output Response to Supine Exercise

Postural position has a significant effect on the relative contribution of heart rate, myocardial contractility, and the Frank-Starling mechanism to the increase in cardiac output during exercise conditions [66]. During supine bicycle exercise in sedentary individuals, *Q* increases linearly with the increase in VO_2. During the supine position, end-diastolic volume, stroke volume, and peak diastolic filling rate are increased reflecting an increase in preload. Thus, with an increase in exercise workload, stroke volume has a small contribution and therefore it relies mainly on the increase of HR [102]. It is believed that the supine position prevents blood from pooling in the lower extremities enhancing venous return. Linear regression analysis suggested by Warburton et al., revealed that in top athletes, SV increases in a linear fashion throughout incremental exercise in both the supine and upright positions. The highest level for SV was reached during maximal exercise in both the supine and upright positions. In absolute terms, stroke volume was significantly larger in the supine than the upright position [66].

Maximal HR is higher in the erect position due to the smaller SV compared to the lower maximal HR in the erect condition (Fig. 6.25).

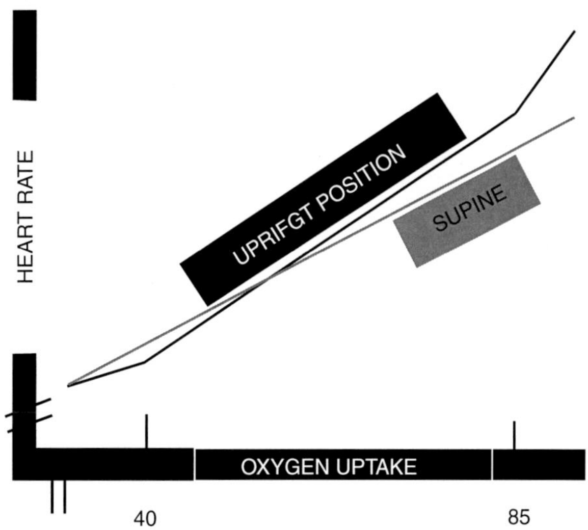

Fig. 6.25 Heart rate increment up to maximal effort: a comparison between upright erect exercise and supine exercise

Supine exercise results in a significantly larger cardiac output than upright exercise at rest and throughout incremental exercise. Cardiac output increased throughout incremental exercise in both exercise positions.

The contribution of SV and HR to *Q* during supine dynamic exercise varies during two different stages of exercise intensities: (a) from rest to 90% of VO_{2max} and (b) from 90% to VO_{2max}. During the first intensity, from rest to 90% VO_{2max}, *Q* is increased mainly by increased HR along with a small increase in SV. Interestingly, the highest SV attainable in supine exercise is only slightly greater than the resting value. During supine exercise, such as swimming, SV also increases by only about 20–40% from resting values, not nearly as much as in an upright position such as in cross-country skiing athletes. Such a difference depends on body position; when the body is in the supine position, blood does not pool in the lower extremities. Because of this, venous returns more easily to the heart, which means that at rest, SV values are much higher in the supine position than in the upright position and hence, lower HR.

During the second intensity, between 90% to VO_{2max}, *Q* increases as a result of a sharp increase in HR due to a slight decrease in SV, owing to smaller left ventricular EDV, as an outcome of short ventricular filling time [102]. Therefore, during supine dynamic exercise at intensities above 90% VO_{2max}, working muscles will rely more and more on the anaerobic pathway to produce ATP, with elevated levels of lactate. This in turn will increase fatigue.

Left Ventricle Function and Volumes

The complex interplay of the different physiological responses to exercise with genetic polymorphism can be found in the origins of left ventricular (LV) volumes and ejection fraction (EF) in healthy and in coronary artery disease (CAD) patients. It has been assumed that resistive exercise can be very demanding due to the pressor response [103, 104], which might have a significant effect on LV volumes [105]. However, in healthy subjects, the effects of successful adaptations to pressure overload training are still controversial, as pressure per se does not always produce intrinsic depression of the myocardial inotropic state. Though, this is not the case in patients with heart disease and hypertension in which the adaptation to pressure overload is pathological.

Although hypertrophy is supposed to compensate for hemodynamic stress, echocardiographic-documented myocardial hypertrophy is an independent risk factor for cardiovascular morbidity and mortality [106]. Pressure overload leads to a marked increase in LV mass of the myocardium, which is probably an adaptive reaction aimed at normalizing the increased wall stress, thus, reduces LV volumes and EF [107].

On the other hand, endurance training improves maximal work capacity as evidenced by higher maximal oxygen uptake, stroke volume (SV), LV end-diastolic volume (EDV), EF, cardiac output and by lower end-systolic volume (ESV), heart rate, and blood pressure [108].

Aerobic exercise, such as running, walking, cycling, and swimming, involves the movement of large muscle groups. In the marked arteries' dilatation, i.e., volume overload that produces LV eccentric hypertrophy by increasing venous return and consequently, LV EDV is increased and LV ESV is reduced and hence, SV and EF are augmented [109]. This physiological hypertrophy is characterized by chamber enlargement and a proportional change in wall thickness. In contrast, isometric, static exercise or resistive training involves a sharp increase in afterload rather than volume overload resulting in LV concentric hypertrophy [110]. Physiological hypertrophy in athletes is an unusual adaptation by which the myocardium compensates for an increase in volume overload. At some point, this adaptation can be defined as an athletic heart in healthy subjects or as pathological in cardiac patients. An important problem is determining when LV hypertrophic ceases to be physiological and begins to be pathological. However, animal studies have demonstrated that physiological and pathological hypertrophy have distinct structural and molecular bases [111]. These LV Hypertrophies change volumes, mass, and function in relation to the mode of exercise emphasizing the role of exercise in epigenetic.

LV Volumes and EF

LV systolic function is generally assessed by measuring the extent and velocity of fiber shortening, EDV, ESV, EF, and velocity of circumferential fiber shortening [112]. LV EF is a measurement of the fraction or ratio of blood ejected from EDV leaving the ventricles with each contraction [113]. LV EF represents the volumetric fraction of blood pumped out of the ventricle at every heartbeat. It is calculated by subtracting ESV from EDV divided by EDV and is an essential parameter for the evaluation of LV function [114]. In addition, the relationship between LV EDV and its SV refers to the capacity of the LV to increase contractility state at systole, i.e., ejection indices, which can be obtained by echocardiography and radionuclide methods [114]. Higher EF values mean an increase in LV contractility.

LVSV is an absolute measurement as Q where LVEF is a relative quantity, and it is the amount of blood pumped past the aortic valve into the aorta and then into the systemic circulation, while right ventricular SV is driven through the pulmonary valve into the pulmonary circulation.

At rest and at peak aerobic exercise, left ventricular contractility and volume measurements can be obtained in a clinical setting by noninvasive methods [56]. LVEF is expressed as a percentage measured by several methods which include: (a) Echocardiogram, (b) Magnetic resonance imaging, and (c) Multiple gated acquisition radio nuclear scan, probably more reliable than echo with a smaller variation [115].

If the LV EF is reduced below normal values, it can be a sign of myocardial dysfunction. Ranges of LV EF to defined health status are as follows: (a) 50–75% is a normal range, (b) 40–49% slightly below normal, and (c) 39% and below, mild-to-moderate-to-severe heart failure. This is with the understanding that mitral regurgitation and segmental wall motion abnormalities are not present. The accuracy in the determination of LV volumes which is a function of chamber size and shape is very much dependent on the experience of the technologist or cardiologist and thus influences the determination of LV EF. A change in LV EF of 5–10% is likely to represent a genuine fall in LV EF. Anything <5% is likely to be within the limits of test–retest reproducibility for the measurement of LVEF >10% is highly likely to represent a real decline. LVEF is a routine clinical standard to assess cardiac function. It is valuable to determine LV function and contractility in health and disease. It is often useful in determining contractility in the basal state of chronic heart disease [116]. LVEF is a complex function of LV EDV, ESV, and SV which are preload and afterload dependent [116]. Thus, it is understandable why LVEF would not detect subtle changes in myocardial contractile state.

LVEF and Volumes Responses During Aerobic Exercise

Aerobic exercise alters the loading conditions of the myocardium. The main changes are in heart rate, LV EDV, ESV, and SV which are the components of cardiac output. Total peripheral resistance declines with a slight-to-moderate rise in systolic blood pressure, while diastolic blood pressure decreases or remains unchanged. Therefore, the contracting LV works against volume overload. Following physical activity or inactivity, LV function, EDV, ESV, SV, and EF may adjust subsequent to the changes in muscle mass. [117]. In healthy young individuals, LV function indices measured have been shown to increase with aerobic exercise and training [118]. In healthy subjects during aerobic exercise, the male's SV exceeded that of the female due to gender differences in LV geometry and variance in heart size. During maximal aerobic exercise, in healthy young males and healthy young women, oxygen uptake values and SV are significantly higher in men, while LV EF increases similarly in both groups from rest to maximal effort [119]. At peak exercise, men and women increase EDV and decrease ESV, thus increases in SV and EF are noted [120]. Such responses may reflect both an increase in myocardial inotropic state and a significant reduction in total peripheral resistance [121]. Therefore, the increase in cardiac output during aerobic exercise was due to an increase in heart rate and SV in both men and women [122].

EF during dynamic aerobic exercise measured at maximal oxygen uptake may reach 80% and 90% in top aerobic athletes. In normal active subjects, contractility and SV are increased along with EF reaching 70–80% [56]. During submaximal exercise, however, the decrease in ESV is less in women than in men.

Healthy aging appears to have no effect on LV EF, EDV, ESV, and SV at rest. However, during exercise rather or not EF increases, decreases, or remains unchanged depends on the individual's physical fitness. Aging reduces LV function due to environmental influence, genetics, and changes attributed to normal physiological aging [123]. Prior to training of elderly subjects, LV EF increases only modestly during exercise from resting values. Following submaximal endurance exercise training in the healthy elderly, LV EF increases significantly at peak exercise compared to that observed before the initial training program [124]. The increase in LV contractility and the reduction in total peripheral resistance following training bringing about a significant reduction in ESV and thus, increase in SV which correlates with an increase in venous return [125].

Cardiovascular system, morphological, and physiological changes have been identified in the CAD patients. An important factor that influences the physical performance of the cardiac patient is aging associated with a shift in the mechanism by which cardiac output is maintained during submaximal exercise. Myocardial functional changes include declines in maximum heart rate, SV, LV function, and an increase in total peripheral resistance. In spite of having lower early diastolic filling rate, LV response to submaximal exercise in CAD patients with normal resting LV EF showed an increase in LV EF and LV ejection rate during aerobic exercise. The LV EF response of CAD patients to aerobic exercise is a significant increase in values from rest to exercise. In addition, during submaximal exercise, EDV is maintained and even increased. As a result, CAD patients appear to have a greater reliance on the Frank-Starling mechanism for the maintenance of cardiac output during submaximal exercise [126].

Exercise-based cardiac rehabilitation increases peak oxygen uptake and improves work capacity in cardiac patients. Following 6 months, exercising cardiac patients compared with controls improve SV, EDV, and EF with reduced total peripheral resistance at rest [127].

Marked changes are notable in the CAD patients during maximal effort in which EF, SV, cardiac output, contractility, and oxygen uptake are increased, while total peripheral resistance, systolic, and diastolic blood pressure are decreased, thus, lowering afterload which in turn facilitates LV systolic and diastolic function [128]. Furthermore, a decrease in resting and during submaximal exercise in heart rate, systolic and diastolic blood pressure are observed.

LV Volumes and EF During Resistance Exercise

During static exercise, force is developed without or with minimal muscle shortening. Cardiac output is increased slightly from 5 to 10–12 $L \cdot min^{-1}$, mainly by the increase of heart rate. In addition, sharp increases in systolic, diastolic, and mean arterial blood pressures are seen. These blood pressure elevations bring about a pressure overload on LV. This pressor response may be important in maintaining perfusion of the active muscle groups during sustained contraction However, it opposes LV ejection.

Resistive training known as power and strength training improves work capacity in sport and daily tasks. Also, it may attenuate the aging effects on muscle strength and mass. Resistive training of small muscle mass, such as the handgrip maneuver, affects LV function in a variable manner both in healthy young men and women. EDV, ESV, SV, and EF have been found [129] to increase, decrease, or remain unchanged. By contrast, the corresponding effects of resistive training with active large muscle mass is related to the size of the activated muscle mass, intensity, and duration of contraction [130, 131]. No changes were noted in males and females in SV and EDV. However, EF rises by exercise only in the males due to decrease in ESV, while it remained unchanged in the females.

During resistive bouts in elderly men, in the face of the pressor response, EF remains unchanged, while it increases significantly in the young subjects. In elderly and young individuals, a significant increase in EDV and ESV is observed from rest to moderate resistive exercise. However, during heavy isometric exercise such as deadlift, ESV is significantly reduced from resting values in the young group, while in the elderly, ESV is increased [129]. The effects of long-lasting resistive training on power athletes show slight or minor changes in LV end-diastolic and systolic dimensions resulting in a slight decrease or unchanged LV EF [132].

LV function is normally maintained or improves in normal subjects and cardiac patients with mild impairment of LV contractility. CAD patients with poor left ventricular function may show deterioration during isometric exercise, although this pattern of response is difficult to predict from resting studies. Many CAD patients who show ischemic ST depression or angina during dynamic exercise may have a reduced ischemic response during isometric or combined isometric and dynamic exercise [133]. Patients with CAD have an impaired health status and reduced exercise work tolerance due to LV dysfunction and reduced peripheral adjustment ability. Therefore, isometric exercise is usually discouraged for patients with CAD because of the possible adverse effects of increased blood pressure on LV function. In addition, it presents several constraints in the assessment of cardiac function in CAD patients. Furthermore, even if resting LV volumes and LV EF could give a complete picture of LV systolic function during exercise, it would not simultaneously quantify LV diastolic function in heart failure [133, 134]. In patients documented CAD or myocardial infarction which were in a supervised aerobic exercise program, had a significantly lower LV EF at rest than normal subjects and did not have significant changes in LV volumes and EF during handgrip or deadlift. These suggest that LV responses are similar in normal and exercise-trained CAD patients during upright submaximal isometric exercise utilizing small or large muscle mass [103].

LV EF and Volumes Responses During Anaerobic Exercise

Anaerobic effort is characterized by exposing the subject to a very high degree of sudden strenuous all-out exercise and therefore, it is not recommended for CAD patients. It consists of short, intense bouts of physical activity, such as weightlifting and sprints, where oxygen demand exceeds oxygen supply. While aerobic exercise energy supply relies on oxygen, anaerobic exercise utilizes through glycolysis.

Although one would expect to obtain increases in LV function with an increase in the adrenergic state produced by anaerobic exercise intervention, echocardiographic indices and blood pressure response resembles those seen during isometric exercise [104], indicating that autoregulation is not involved while performing a sudden strenuous all-out exercise. During all-out anaerobic, effort in healthy young men and women values for *Q* is low compared to what one would expect during maximal dynamic aerobic exercise; 14.4 L·min^{-1} for men and 12.3 L·min^{-1} for women [135]. The explanation for the lower cardiac output might be related to the short duration of the tests; in turn, it did not bring about the appropriate adjustment of the LV volumes and LV EF and due to the relatively high total peripheral resistance which was not as low as seen during aerobic exercise [135].

The assumed luck of the autoregulation mechanism may indicate that the reduced SV and LV EF in men are due to the relatively high-pressure opposing LV ejection although they increased LV contractility. On the other hand, women keep their SV and LVEF values at peak all-out anaerobic exercise as at rest probably due to the lesser increase in mean arterial blood pressure along with increased contractility. Previous report suggested that during strenuous exercise, well-trained healthy young adult subjects did not increase SV and LVEF and sometimes SV was reduced [136].

Following the Wingate Anaerobic Test, in older subjects, few elderlies may experience ECG abnormalities and thus, all-out anaerobic exercise may be dangerous for the older population, due to hypoxia and inappropriate blood pressure response [137]. During exercise in the elderly, compared to resting values, LV EF and ESV are unchanged, while EDV and SV are lower. However, it is very important to emphasis that these elderlies were with a work capacity of 12.5 METs. These LV function responses in the elderly are most likely due to the relatively higher total peripheral resistance opposing the LV ejection [138]. Data suggest that during all-out anaerobic exercise, forces opposing ejection were not reduced enough to avoid reduced left ventricular function in the older subjects [139]. This is attributed to functional changes with age in the myocardium and associated blood vessels, leading to impairment of left ventricular function and blunted inotropic and chronotropic responses to catecholamine. Therefore, an all-out anaerobic-type effort should not be given to an older subject due to the great hazardous potential.

Gender Effect on the Cardiovascular During Anaerobic Bout

Values for *Q* during the sudden strenuous all-out exercise are low in men and women compared to what one would have expected during maximal dynamic exercise; 14.4 L·min^{-1} for the men and 12.3 L·min^{-1}, respectively. The explanation for the lower *Q* in men and to the greater extend in women, during the sudden strenuous all-out exercise, might be related to the short duration of the exercise bout, which in turn, does not bring about the appropriate adjustment of the circulation, due to the relatively high TPR, which was not as low as seen during aerobic exercise.

Echocardiographic indices and BP response are similar to those seen during isometric exercise, indicating that autoregulation is not involved while performing sudden strenuous all-out exercise. This assumed lack of the autoregulation mechanism may indicate that the reduced SV in the men during the sudden strenuous all-out exercise is due to the relatively high pressure opposing left ventricular ejection. Women, on the other hand, keep their SV values at peak exercise as at rest values, probably due to the lesser increase in MABP and increased contractility [140, 141].

Training Effects on the Cardiovascular System

In brief, long-lasting aerobic, work capacity increases vagal tone, circulatory blood volume, left ventricular contractility and SV while reduces TPR and HR at rest and during submaximal exercise. At peak aerobic exercise, *Q* increases due to an increase in both components, HR and SV [142]. Although the contribution of each is different, it depends on HR and SV increased levels during exercise and on individual's VO_{2max}. It results in an increase in myocardial myofibrils, accordingly, ventricles' contractility is increased during ventricle contraction and hence, increased SV. Consequently, HR does not rise as much as during submaximal aerobic exercise to produce similar *Q*. This enables the individual who exercises regularly to maintain a similar level of exercise lasting a longer time and to recover faster. Responses of *Q* during exercise at the sea level depend mainly on the mode of the exercise bout, metabolic demand, and fitness level if an individual exercises chronically.

In healthy sedentary young individuals, during aerobic exercise, HR may increase from 70 at rest to 220 beats·min^{-1} at maximal aerobic exercise. SV increases from 70 to 75 mL to approximately 120–130 mL due to increased ventricular contractility. This would increase *Q* to 24–25 L·min^{-1}.World-class athletes increase their SV to 200 mL and *Q* to 40 L·min^{-1} and maximal HR of 200 beats·min^{-1} (Fig. 6.26).

When exercise commences, the load on the cardiovascular increases to provide the needs of the cardiorespiratory and skeletal muscles [143]. Therefore, the cardiopulmonary system increases HR, SV, *Q*, respiration *f*, and VT. During aerobic exercise, *Q* is redistributed, reducing blood flow to the internal organs and increases blood flow availability to the contracting muscles (exercise hyperemia) [142]. The role of the autoregulation has a key role in cardiac redistribution of *Q*, while the sympathetic nervous system reduces blood flow to the kidneys, liver, and gastrointestinal tract [41]. Skin blood flow depends mainly on the surrounding temperatures and exercise intensity. Under hot conditions, blood flow to the skin increases on the account of the contracting muscles. The alterations in the cardiovascular system during exercise are needed to increase oxygen delivery and nutrients, mainly to the working muscles [144].

During exercise, $(a\text{-}v)O_2$ from the blood delivered to the working muscles is increased, which in turn reduces oxygen content in blood venous. This in turn, facilitates oxygen diffusion from the lungs to the blood.

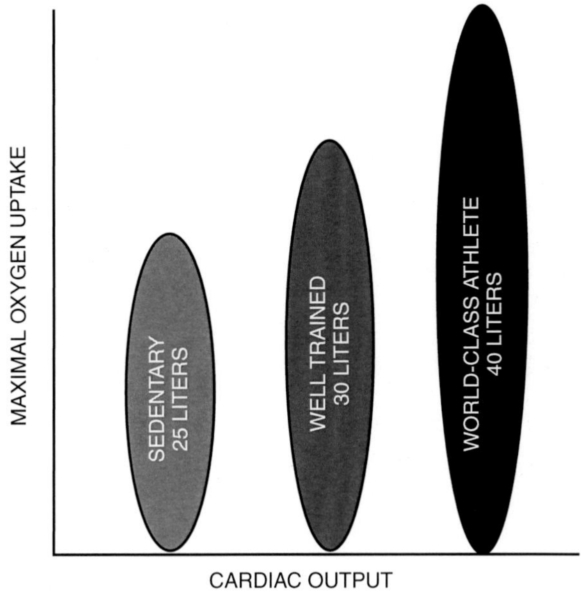

Fig. 6.26 Cardiac output at maximal aerobic exercise in sedentary, well-trained, and world-class athletes

During vigorous exercise, almost up to 80% of the arterial oxygen content is extracted, suggesting that the 20% arterial oxygen content leftover can be used as a reservoir when exercise commences. This gives time to the cardiopulmonary system to adjust to the high metabolic demands. Unloading of oxygen from hemoglobin to the cell, facilitated by lower cell's PO_2 [145], changes in the oxygen dissociation curve due to increased blood hydrogen ion and CO_2 levels [146]. This facilitates unloading of oxygen from hemoglobin in the contracting skeletal muscles.

Following vigorous long-lasting aerobic exercise training, the VO_{2max} increases very significantly, in young adults as well as in elderly mainly due to the adaptive increases in myocardial blood delivery and to the lesser extent to the $(a\text{-}v)O_2$ [147]. The increase in maximal Q during maximal aerobic exercise in trained elderly and young adults is the result of a large increase in stroke volume increased due to a significant increase in left ventricular end-diastolic volume, while left ventricular end-systolic volume decreased [148]. These significant changes in left ventricular volumes at VO_{2max} exercise seen result from an integrated mechanism that involves changes in contractility and afterload. Longitudinal studies have shown that the training-induced increase in VO_{2max} results primarily from an increase in maximal Q, rather than a widening of the systemic $(a\text{-}v)O_2$. Indeed, it is estimated that 70–85% of the training-induced increase in VO_2 is explained by the increase of maximal Q [149]. Capelli et al. [149] examined VO_2 in sedentary individuals after 20 days of bed rest and 50 days of training. The difference in VO_2 between the deconditioned and trained states resulted mostly from a significant differences in Q. Earlier [150], it has been suggested that following 16 weeks of aerobic training, VO_2 was increased from 3.15 to 3.68 $L{\cdot}min^{-1}$, due to significant Q increment, and increase in $(a\text{-}v)O_2$ (from 138 to 143 $mL{\cdot}L^{-1}$). In sedentary individuals, after 6 months of endurance training, resting HR decreased to 55 beats·min^{-1}, in athletes, it decreases in the 30s and low 40s, while resting SV increase to about 90 mL, in athletes it reaches 200 mL. These improvements in HR and SV do not change Q at rest following training since oxygen demand at rest did not change. These cardiac improvements are carried on during exercise, where individuals work at similar intensities with lower HRs and less myocardial energy demand. Following dynamic aerobic exercise, HR is decreased at rest and during submaximal exercise, and sometimes at maximal effort. At peak exercise, maximal heart rate does not increase after training. It stays the same or might even decrease slightly. In untrained individuals, maximal heart rate is achieved at peak exercise. It might seem that untrained individuals have an advantage over trained individuals in their ability to attain higher maximal HR trained individual. However, the opposite is true, the trained individuals and world-class athletes have a significant increase in left ventricular filling time during submaximal and maximal exercise. Consequently, at peak exercise, the slightly reduced HR attenuates increase EDV and thus, SV. Following training, during submaximal dynamic aerobic exercise at the same absolute workload, Q is similar in trained and untrained subjects. However, the way to achieve that Q differs significantly between these two subjects.

Individuals that exercise long-lasting aerobic exercises, resting HR decreases to 40–60 beats·min^{-1}, with normal ECG, which is a significant reduction from the normal resting HR of 70 beats·min^{-1}. In addition, SV increases significantly up to 100–110 mL compared to the normal sedentary individuals with values of 70–80 mL. If training regimes are vigorous long-lasting aerobic exercise, expansion of the normal myocardial structure occurs, a phenomenon called "athletic heart." Athletic heart itself is related to numerous structural variations that overlay those seen in individuals with cardiomyopathies [151]. However, in addition to transformation in myocardial structure, functional capacity increases significantly almost in any physiological parameter [152].

Individuals that take part in long-lasting aerobic exercise such as long-distance running, the eccentric left ventricle hypertrophy, in which the thickness of ventricle is not large whereas the left ventricular wall is a relatively increased, will be produced [153].

Cardiovascular Responses During Isometric and Anaerobic Exercises

When utilizing large muscle mass, isometric exercise differs significantly from those during dynamic exercise. Isometric maneuver causes a mechanical compression of the blood vessels, by the contracting muscles, and therefore, a reduction in muscles' blood flow. As a result, TPR increases significantly almost four times of the resting values usually 1100–1200 dynes·s^{-1}·cm^{-5} reaching 1600–1800 dynes·s^{-1}·cm^{-5} and four times of that during dynamic exercise (400 dynes·s^{-1}·cm^{-5}). In addition to the mechanical compression, activation of the sympathetic system further increases SBP and DBP, while the increase in HR and Q is far from that noted at peak aerobic exercise. The increase in SBP, DBP, and MABP increases forces opposing left ventricular ejection. Therefore, left ventricular works against pressure overload, while during aerobic exercise it works against volume overload due to the autoregulation mechanism. Resistance training therefore does not increase VO_{2max} and have no influence on the pulmonary function. Progressive resistance training leads to increases in maximal strength of the mechanism responsible for strength increment is not clear yet. Although the neural mechanisms responsible for the strength increase, previous reports suggested that the effect of resistance training is the increase in muscle mass (hypertrophy) due to increase in muscle cross-sectional area.

Echocardiographic measurements reveal that the myocardium goes through a positive morphologic alteration in response to long-lasting aerobic exercise. These alterations may be defined as athletic heart. Athletic heart is characterized by eccentric myocardial hypertrophy, which increases in left ventricular cavities and volumes. Characteristic weightlifters have concentric myocardial hypertrophy and thus, reduce left ventricular cavities due to the pressure overload training.

During isometric and anaerobic exercises, the increase in VO_2 is not significant. Compared to aerobic exercise, Q is significantly lower, during isometric exercise and anaerobic exercises with the lowest values during isometric exercise (Fig. 6.27).

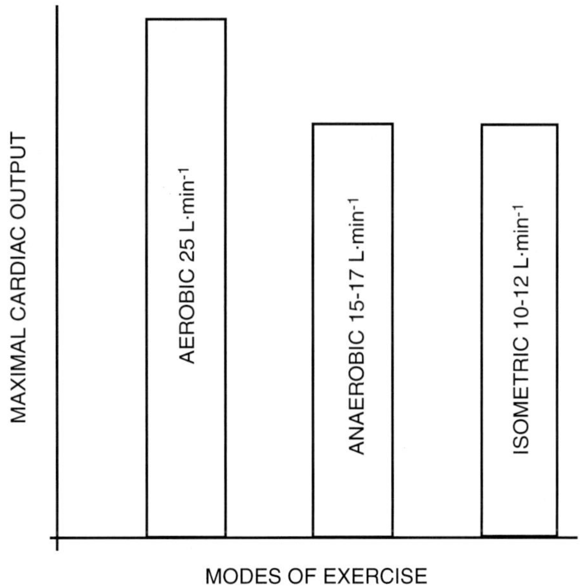

Fig. 6.27 Maximal cardiac output in different modes of exercise

Note that during isometric bouts there is no autoregulation mechanism, due to mechanical compression of vessels undergoing muscle contraction. The same is true during anaerobic exercise regarding the autoregulation mechanism. Because anaerobic effort is characterized by exposing the subjects to a very high degree of sudden strenuous all-out exercise [154]. Therefore, there is a need to consider such effects in individuals with impaired cardiovascular function [58].

Cardiac Output Redistribution

Several mechanisms affect redistribute Q at rest and during exercise, mainly, vasodilatation reduced TPR and increased vagal tone during aerobic exercise. Vagal tone refers to the action of the vagus nerve, an important element of the parasympathetic nervous system. At rest, skeletal muscle arterioles and other body organs are under strong sympathetic tone influence. However, during aerobic exercise, the autoregulation mechanism overrides the sympathetic input, and consequently, blood flow to the working muscles increases significantly while the internal organs have a reduced blood supply, due to lack of autoregulation in these organs. By that, the body diverts blood and thus, oxygen is supplied where it is most needed. Q redistribution during exercise is relatively related to the metabolic activity of the organ. Therefore, during aerobic exercise, the majority of Q diverts to the working muscles. At rest, blood flow differs greatly among tissues, as shown in Fig. 6.28.

Sedentary young adults weighing 70 kg, have at rest, Q of approximately 5 L·min^{-1}, while during dynamic endurance exercise, Q increases to 25 L·min^{-1}. Figure 6.28 discloses that skeletal muscles at rest receive 1.4 L·min^{-1} from a total of

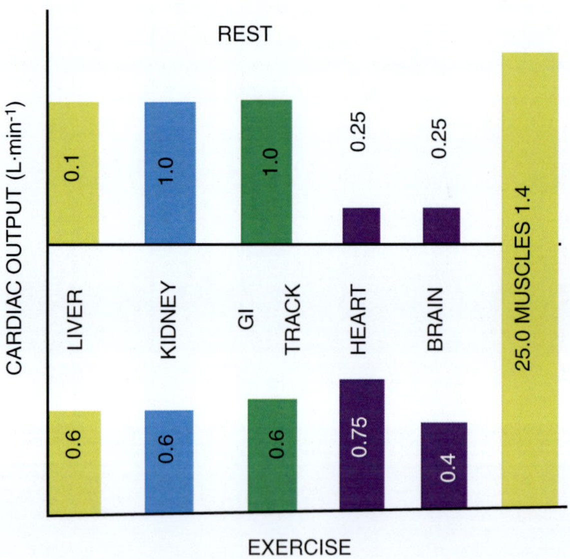

Fig. 6.28 Cardiac distribution at rest (5 L·min^{-1}) and during exercise (25 L·min^{-1})

5 L·min^{-1}, while the kidneys, digestive tract, liver, heart, and brain receive 3.6 L·min^{-1}. During aerobic exercise, working muscles blood flow increases to 25 L·min^{-1}, the brain and myocardium increase blood flow to 0.75 L·min^{-1} each, while flows to the liver, kidneys, and digestive track have approximately 2 L·min^{-1} combined. Thus, during endurance exercise, blood flow to the internal organs is reduced relatively and absolutely. Values for blood flow to the skeletal muscles also include values of skin blood flow at rest and during exercise.

Cardiac Output During Isometric Exercise

It is well accepted that a resistive training program results in an increase in muscular strength [155]. Isometric exercise is an activity in which muscles undergo no change in length with contraction. As a result, no physical work is done despite an increase in the metabolic rate. Isometric exercise is a long-lasting muscle contraction in which intramuscular pressure is increased, thereby resulting in compression of the blood vessels and mechanical occlusion of blood flow. It has been observed that contractions at 20–30% of maximal voluntary contraction can be maintained for 3–5 min. In addition, it has been assumed that in normal subjects, resistive training could be very demanding due to the dramatic increase in SBP, DBP, and MABP known as the pressor response, mediated by a slight increase in *Q*, but not to the level observed during dynamic aerobic and anaerobic exercises. No change from rest in TPR and sometimes it is increased; this might have a significant effect on *Q*, especially on SV [156].

Cardiovascular response during isometric exercise in young, elderly, adolescents, and coronary artery disease patients' males and females are in general, similar [131, 157, 158]. In normal subjects, isometric exercise causes a small increase in *Q* (10–12 L·min^{-1}) resulting almost entirely from an increase in HR, which increases to approximately 140 beats·min^{-1}, while SV increases slightly and at times remains unchanged or decreases.

Cardiac Output During Anaerobic Exercise

Dynamic anaerobic exercise is characterized by exposing the subject to a very high degree of sudden strenuous all-out exercise. It places special demands on the cardiovascular system and skeletal muscles. In order to improve athlete's physical performance, the athlete devotes a large portion of the training program to speed and strenuous exercise [159]. This mode of effort brings about oxygen debt, which increases blood lactate concentration directly associated with an increase in work intensity [160]. It has been assumed that this type of activity is potentially dangerous for elderly and cardiac patients, due to hypoxia and seemingly inappropriate BP

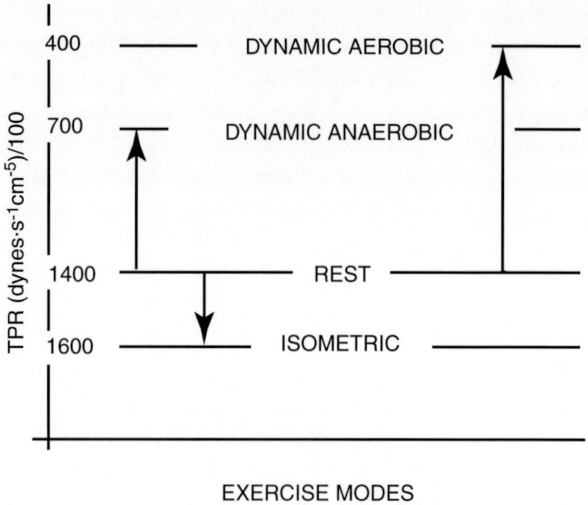

Fig. 6.29 Different values for total peripheral resistance (TPR) during different modes of exercise. The maximal reduction in total peripheral resistance is obtained for aerobic exercise, while the lowest reduction is found for isometric exercise

response. This has the effect of placing a large load on the left ventricular affecting systolic function and hence, on *Q* [137]. During anaerobic exercise, SV in young adult subject remains unchanged compared to the resting values, while it decreases in the elderly [161].

Compared to responses observed during aerobic-type effort, strenuous anaerobic efforts do not produce the expected increases in the left ventricular volumes, EF, SV, and *Q*. During anaerobic exercise, maximal *Q* achieved is approximately 16–17 $L{\cdot}min^{-1}$ [56]. This may suggest that the increase in left ventricular inotropic state during the strenuous bouts are not adequate to increase left ventricular volumes and hence, EF and SV [148]. In addition, during all-out exercise, total TPR and afterload values are higher for isometric exercise than at rest, while it decreased at peak aerobic exercise (Fig. 6.29). Such inadequate reduction in TPR and afterload during anaerobic bout could be responsible for the relatively small augmentation in left ventricular function, seen during all-out anaerobic exercise [56].

The end result is a lower *Q* than that expected at VO_{2max}. An additional explanation for the relatively lower *Q* during anaerobic exercise bout may be related to the short duration of the exercise bout. Thus, insufficient time to allow an appropriate adjustment of the circulatory system such as impeded venous return. This, coupled with lower values of HR, compared to the maximal HRs that healthy subjects can reach during dynamic exercise when achieving VO_{2max} result in a lower *Q*.

Cardiac Output During Isodynamic Exercise

Isodynamic exercise such as backpack carriage or load carriage is a form of exercise which combines isometric modes or combines volume and pressure loads on the left ventricular. The effect of the isometric stress during graded aerobic exercise test in normal subjects revealed that at the same absolute VO_2, the isometric component augmented significantly afterload and HR, compared to dynamic exercise alone [162].

Previously, it has been shown that prolonged isodynamic exercise at constant speed and grade with different backpack carriage loads did not change the steady state of left ventricular systolic function and hemodynamic throughout 240 min of efforts [154].

During isodynamic exercise, changes in grade of walking are the most important factor determining *Q* response, i.e., with an increase in grade VO_2 increases, without any change in speed or load [162]. Blood flow to the working muscles during isodynamic exercise is sufficient to avoid heavy reliance on anaerobic glycolysis as a source of energy.

Isodynamic and Cardiac Function in CAD Patients

Coronary blood flow is closely related to myocardial VO_2, which can be estimated by the double product of HR and SBP, during dynamic exercise in normal subjects and in patients with CAD [163]. It has been assumed that isometric maneuver can be very demanding, due to the pressor response [156], and it might have a significant effect on left ventricular global performance. Therefore, patients with CAD have traditionally been advised to avoid activities that require isometric muscle contraction due to the pressor response. However, ischemic responses indicated by ECG, 0.1 mV ST depression, were absent in CAD patients during isometric or combined isometric and dynamic exercise (isodynamic), in contrast to dynamic exercise alone at the similar rate-pressure product [164].

During isodynamic exercise, left ventricular function and contractility do not decrease compared to aerobic alone. In addition, an attenuation of the ST-segment is noted in the face of increased HR, SBP, DBP, and MABP [165]. This response may be due to relatively low HR and the increase in MABP, during isometric or isodynamic exercises. This, in turn, causes better perfusion of the myocardium and improves the oxygen demand–supply balance [162].

Left ventricular performance does not differ significantly during aerobic and isodynamic exercises namely, left ventricular EF, SV and left ventricular ESV and EDV are similar. Although significant differences are observed in rate pressure product between the two exercises modes, however, in well-trained CAD patients, additional load carriage does not alter left ventricular performance. These physiological responses are different from what one would expect to find with an increased adrenergic state during isodynamic exercise due to the isometric component [166].

These results are explained by the interaction of opposite influences that dynamic exercise and isometric exercise have on BP. It seems that the autoregulation mechanism, keeps afterload during isodynamic exercise at a relatively low level, with nearly the same values as seen during the dynamic exercise, while also enabling to increase blood flow through the coronary arteries, which in turn helps to maintain appropriate left ventricular contractility and attenuate ST depression.

All Causes Cardiovascular Mortality

Exercise is a nonpharmacological treatment tool in the prevention of and treatment in cardiovascular disease. It reduces the general risk factors of different cardiac diseases [167]. Risk factors for cardiovascular diseases that involve the heart or blood vessels included cigarette smoking, hypertension, blood lipid levels, diabetes, and overweight, in association with mortal myocardial infarction [168]. Previously, it has been proven that a contrary relationship exists between energy expenditure and mortal myocardial infarction [169]. Chronic exercise training that increases significantly VO_{2max} is related to cardiovascular risk factor reduction. On the other hand, physical inactivity is apparently linked with risk factors and cardiac abnormalities, and reduced morbidity [170]. Higher VO_{2max} has a more noticeable positive effect on cardiovascular risk factors and metabolic risk [171]. Another study suggested that the relation between high VO_{2max} and the reversal of all-cause and cardiovascular mortality is close-fitting [172].

Myocardial Blood Flow

Aerobic exercise is the most physiological stressor stimulus for increased myocardial oxygen demand and thus, blood flow in the coronaries. This needs an increase in Q that results in increases in HR and myocardial contractility [17]. The coronary circulation is autoregulated so the heart receives a sufficient supply of oxygen [17]. Therefore, flow through the coronary circulation is maintained in most circumstances and over a range of pressures. Blood flow to the myocardium is profoundly disturbed in the presence of obstructive coronary atherosclerosis.

Myocardial oxygen demands during vigorous aerobic exercise are met by about fivefold increase in coronary blood flow. Since the myocardium has a very limited anaerobic capacity, it is therefore dependent on a continuous supply of oxygen through the coronary circulation.

At rest, the myocardium extracts almost 75% from oxygen supplied, compared to 25% extraction by skeletal muscles. Accordingly, myocardial oxygen extraction reserve is very limited during exercise. Therefore, the relationship between myocardial VO_2, coronary blood flow, and the $(a\text{-}v)O_2$ from the blood is unique. This requires tight control of coronary blood flow, because an increase in myocardial oxygen demand, as occurs during exercise, cannot be obtained by a significant further increase in oxygen extraction [173]. Nevertheless, the myocardium is ultimately dependent on oxygen delivery and therfore, it extract all of the delivered oxygen.

The human heart weight is on average 300 g in men and 200 g in females [174]. Resting coronary blood flow averages 250 $mLO_2 \cdot min^{-1} \cdot 300 \cdot g^{-1}$ (0.8 $mLO_2 \cdot g^{-1} \cdot min^{-1}$ of heart muscle), and has one of the highest energy demands per tissue mass of all human organs [175]. This relationship is an application of the Fick Principle and expressed as:

$$MVO_2 = CBF \cdot (a\text{-}v)O_2$$

where MVO_2 = myocardial oxygen uptake; CBF = coronary blood flow; (a-v) O_2 = arterial-venous oxygen content difference.

Due to sensitive oxygen demand–supply balance, ATP produced via glycolysis is around 5% and the rest 95% are primarily dependent on oxidative phosphorylation [176]. As mentioned earlier, the normal oxygen concentration in arterial blood is about 200 $mLO_2 \cdot L^{-1}$, normal function of the at rest heart requires about 60–80 $mLO_2 \cdot L^{-1} \cdot min^{-1}$, during exercise, it increase by fivefold [177]. High capillary density facilitates myocardial $(a\text{-}v)O_2$ [178]. In rats, long-lasting aerobic exercise training increases significantly myocardial capillary density [179]. This is different from skeletal muscles which during exercise are able to increase both (a-v) O_2 and blood flow [180].

Exercise training also alters local control of coronary resistance vessels. Thus, arterioles exhibit increased myogenic tone, likely due to a calcium-dependent protein kinase C signaling-mediated alteration in voltage-gated calcium channel activity in response to stretch [17].

Hemodynamic alterations that accompany exercise can augment the ability to increase coronary blood flow during exercise. Failure of coronary blood flow to increase adequately during exercise could limit myocardial oxygen availability, thereby impairing cardiac performance (Fig. 6.30).

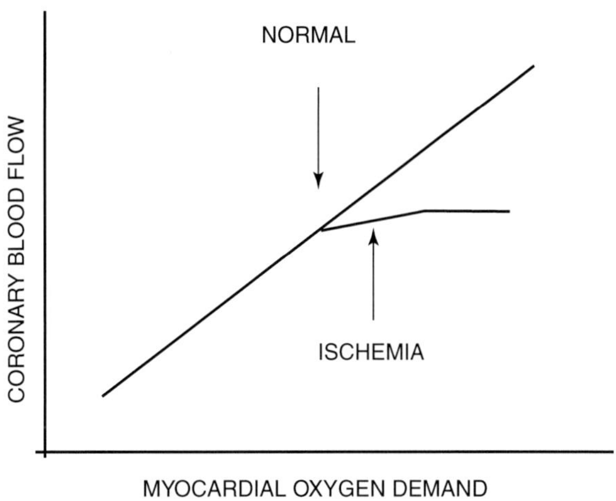

Fig. 6.30 In the event that blood flow does not keep up with oxygen demand as in coronary artery disease patients, ischemia is reached (red line)

Coronary Artery Disease

The CAD is a long-lasting course that starts during adolescence and gradually advances all over life. Coronary artery disease (CAD) or ischemic heart disease is the single most common of the cardiovascular diseases and cause of death globally [181]. Physiologically, CAD means reduced blood supply to the myocardium. CAD happens almost always due to atheromatous, narrowing and subsequent occlusion of the vessel, consequently, blood flow and oxygen to the myocardium are significantly reduced [182]. The heritability of coronary artery disease has been estimated between 40% and 60% [183]. Genome-wide association studies have identified around 60 genetic susceptibility loci for coronary artery disease [184]. Risk factors include high lipid levels, high BP, diabetes, cigarette smoking, family history, and unhealthy diet. These risk factors increase the rate or alter a multifaceted and enduring inflammatory course that eventually is established as fibrous atherosclerotic plaque. The streak consists of cholesterol, white blood cells, and other cellular matter. The fatty streak by itself does not cause symptoms of heart disease but can develop into a more advanced form of atherosclerosis.

The coronary arteries deliver oxygenated blood to the myocardium, thus, any reduction in coronaries blood flow reduces oxygen supply to the myocardium. Cardiac veins drain away the blood once it has been deoxygenated. Different from skeletal muscles that have anastomosis, an interconnection of vessels is present to form a sort of a web that can let blood to circulate to an organ, in the event of partial arterial blockage in another branch. This is not the case with the heart, there are no anastomoses in between the coronaries. Therefore, partial blockage of coronary results in ischemic condition and times even [185].

CAD include stable angina, unstable angina, myocardial infarction, and sudden cardiac death [186]. Signs and symptoms of CAD include chest pain (angina), which is the most common symptom; heart burn; shortness of breath; discomfort which may travel into the shoulder, arm, back, neck, or jaw; and myocardial infarction.

Myocardial Oxygen Uptake

Myocardial oxygen demand is determined by HR, the intramyocardial tension, contractile state, preload, and afterload [187]. Increase in HR increases metabolic activity, which adds myocardial oxygen demand. This HR increase raises myocardial wall stress, which in turn, left ventricular contractility needs to be more forceful and hence, increase in myocardial oxygen. Consequently, myocardial work, Q, coronary blood flow, and HR all increase in a parallel manner and peak together.

Intramyocardial tension or the pressure developed by the ventricular wall is a meaningful determinant of myocardial VO_2. Left ventricular wall tension is defined by Laplace's law which states "that the wall tension is equal to the pressure within the cylinder times the radius of the curvature of the wall". The tension is also related to the thickness of the wall and the radius of the ventricle. Therefore, at a fixed pressure and wall thickness, an increase in venous return will increase tension and thus, myocardial VO_2.

Left ventricular contractile state of the left ventricular is achieved by an increase in the metabolism of myocardial cells, influenced by the sympathetic nervous system, via catecholamine that innervates the heart. Preload and afterload also influence contractility state, while contractility state determines the velocity of contraction.

The coronary arteries deliver oxygenated blood to the myocardium, thus, any reduction in coronaries blood flow reduces oxygen supply to the myocardium [187]. Cardiac veins drain away the blood once it has been deoxygenated. Different from skeletal muscles that have anastomosis, an interconnection of vessels is present to form a sort of a web that can let blood to circulate to an organ, in the event of partial arterial blockage in another branch. This is not the case with the heart, there are no anastomoses in between the coronaries. Therefore, partial blockage of a coronary artery results in ischemic condition due to the imbalance of oxygen delivery–demand [185]. Most ischemic heart disease is caused by atherosclerosis, usually present even when the artery lumens appear normal by angiography. Coronary blood flow is a dynamic process modulated via multiple parameters. Cardiac metabolism is affected by several factors. Under normal physiologic conditions, there is a balance between oxygen supply and demand. In fact, there is a significant reserve on the supply side. Under certain pathologic states such as coronary artery disease, the supply of oxygen may be exhausted and an imbalance between supply and demand occurs which is translated into ischemia. The area of myocardium most susceptible to ischemia is the subendocardium due to mechanical and metabolic forces. In therapy of coronary artery disease, attention should be directed to directional changes in factors influencing supply and demand to improve blood flow to the most susceptible area. Figure 6.31 summarizes the relation between myocardial oxygen demand and blood flow, and the oxygen imbalance of delivery–supply. However, with the reduction of blood flow in coronary artery disease patients, oxygen supply is reduced significantly, bringing about an imbalance in oxygen supply–demand.

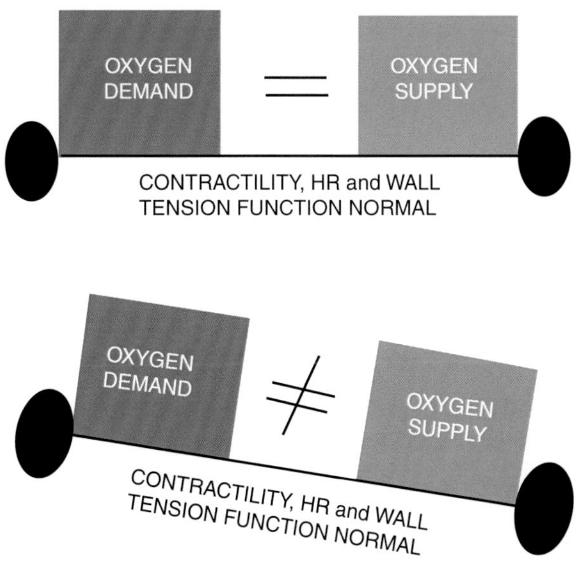

Fig. 6.31 Variables affecting left ventricular oxygen demand as balanced by coronary blood flow, normal myocardial function when oxygen supply–demand is balanced and abnormal function when it is imbalanced

Estimation of myocardial VO_2 is calculated by multiplying HR and SBP, called double product, or the rate-pressure product (RPP) enables the estimation of myocardial VO_2, see equation below:

$$(\mathrm{HR} \cdot \mathrm{SBP})/100$$

where HR = heart rate, SBP = systolic blood pressure, 100 = 100%

This equation correlates very well in a wide range of exercise intensities. The double product is also a good tool to estimate the angina threshold, i.e., the point in which the imbalance between oxygen supply–demand occurs. This is a very important variable for the clinician and cardiac patients.

The RPP has been shown to correlate well with myocardial VO_2 during both isometric effort and dynamic exercise. During aerobic exercise, the increase in HR and SBP increase requires more oxygen. Since long-lasting aerobic exercise reduces HR at rest and during submaximal intensities, well-trained healthy individual and cardiac patient are better able to stand myocardial oxygen demands [188].

Cardiac Output in Coronary Artery Disease Patients

Compared to healthy subjects, values for Q during aerobic exercise at VO_{2peak} are low in coronary artery disease patients. In CAD patients while determining their VO_2 and hemodynamic values at rest and during maximal treadmill exercise, maximal Q is reduced significantly approximately by 4–5 $L \cdot min^{-1}$ compared to matched healthy sedentary subjects. The lower Q in CAD patients at peak aerobic exercise is related to the medications, limited maximal HR, SV, left ventricular contractility, and to the inappropriate adjustment of the circulation [189].

The determinants of Q during aerobic exercise in patients with CAD are multifactorial and incompletely understood [190]. Exercise performance and its hemodynamic differ in CAD patients with symptomatic and asymptomatic left ventricular dysfunction. Asymptomatic CAD patients have greater increases in HR, systolic blood pressure, left ventricular EF, and Q than symptomatic CAD patients.

The reduction in maximal HR is a greater factor in the reduction of Q than SV. As compared with healthy controls, Q is more frequently preserved in young than older CAD patients with short disease duration, although the left ventricular response to exercise is already reduced [191].

Sudden Cardiac Death

Pathophysiology of Sudden Cardiac Death

Sudden cardiac death is the major cause of natural death among the general population and is accountable for nearly 50% of all heart disease deaths. It is an unanticipated death, triggered by alterations in heart rhythm, i.e., sudden cardiac arrest [192].

Sudden cardiac arrest differs from myocardial infarction; however, it can happen during a myocardial infarction. The main risk factor causing sudden cardiac death is the existence of CAD [193]. Ischemic condition due to arterial obstruction results in myocardial infarction, damaging the myocardium. In contrast, sudden cardiac arrest arises due to electrical system failures and unexpectedly develops arrhythmias, causing an abnormal HR increase to a hazardous level. Although during most of the sudden cardiac arrest events occurrence are without prior symptoms, sometimes, the individual may experience symptoms, such as dizziness, forceful and fast heartbeats. Consequently, the ventricles may develop fibrillation, the most frequent fatal arrhythmia, resulting in a drastic reduction in blood delivery first, to the brain and subsequently, consciousness loss. This condition, if left untouched, can cause death within a few minutes, except emergency treatment is given immediately.

Most sudden cardiac deaths are caused by abnormal heart rhythms called arrhythmias, secondary to intrinsic abnormalities in the myocytes or can occur due to dynamic outflow tract obstruction during exercise. The most common life-threatening arrhythmia is ventricular fibrillation, which is an erratic, disorganized firing of impulses from the ventricles (the heart's lower chambers).

Numerous physical conditions may be accountable for sudden cardiac death during exercise. Three main conditions causing sudden death include myocardial ischemia, structural abnormalities, and arrhythmias. Myocardial ischemia includes atherosclerotic coronary artery disease, coronary artery spasm, de novo coronary artery thrombus, intramyocardial bridging, hypoplastic coronary artery, anomalous coronary arteries, and CAD.

Structural abnormalities include hypertrophic cardiomyopathy, mitral valve prolapse, valvular heart disease, lipomatous infiltration of the right ventricle, Marfan's syndrome, right ventricular dysplasia, aortic stenosis, and sickle cell trait. Arrhythmias includes the following: Wolfe-Parkinson-White, Lown-Ganong-Levine, long QT syndrome, ventricular arrhythmias and Medial hyperplasia, and intimal proliferation of the main sinus node artery. Repolarization is due to potassium channel downregulation and it may also be explained by the presence of factors such as athlete's heart, hypertrophic cardiomyopathy, increased sympathetic tone, and genetic faults. These factors together may increase repolarization homogeneity which sometimes leads to the risk of arrhythmias and sudden cardiac death [194]. Sudden death is probably a consequence of an electrically unstable and unpredictable myocardial substrate with reentrant ventricular tachyarrhythmias, as evidenced by the presence of histopathological markers of disorganized myocardial architecture and replacement scarring (the consequence of microvascular abnormalities and myocardial ischemia) [195]. Factors involved in the generation of ventricular tachycardia or ventricular fibrillation are an imbalance of autonomic tone, tissue reentry, wave break, action potential duration alteration, cellular triggered activity, subcellular abnormal activation, or deactivation of ion channels (automaticity level). At present, the exact mechanism causing exercise-related sudden death in individuals with established disease, especially of the CAD, is not known. Whereas plaque rupture or thrombosis is present in up to 95% of sudden cardiac deaths in the general population, the incidence seems to be lower in exercise-related deaths. Thus, exercise-induced ischemia or coronary spasm may be involved in exercise-related deaths.

The most common electrophysiologic mechanisms leading to sudden cardiac death are tachyarrhythmias such as ventricular fibrillation or ventricular tachycardia [196]. At the subcellular level, especially in ischemia, altered intracellular Ca^{2+} currents alter intracellular K^+ currents or mutations resulting in dysfunction of a sodium channel (Na^+ channelopathy) which may increase the likelihood of ventricular tachycardia and ventricular dysfunction. Other causes include inherited arrhythmia syndromes (e.g., commotio cordis, long QT syndrome, catecholaminergic polymorphic ventricular tachycardia, and Brugada syndrome). Some young athletes die of aortic aneurysm rupture (in Marfan's syndrome).

The cardiac renin-angiotensin system has been suggested to be involved in the development of left ventricular hypertrophy. Some of the components of the renin-angiotensin system have been reported to be synthesized in the ventricle of the heart [197]. In humans, a strong correlation has been found between plasma angiotensin I-converting enzyme activity and the insertion/deletion (I/D) polymorphism of the angiotensin-converting enzyme gene, which has been reported to be associated with myocardial infarction, ischemic and idiopathic dilated cardiomyopathy, sudden death in hypertrophic cardiomyopathy, and restenosis after percutaneous transluminal coronary angioplasty [198]. In humans, the plasma level of angiotensin-converting enzyme is determined genetically.

Hypertrophic Cardiomyopathy

Hypertrophic cardiomyopathy (Fig. 6.32) is the most-common mono-genetically inherited form of heart disease, related to the thickening of the left ventricular wall, contractility dysfunction, and potentially fatal arrhythmias [199]. If asymmetric, the hypertrophy is usually greatest in the ventricular septum, but variations occur in which the hypertrophy may be maximal at the mid-ventricular level, at the apex, or rarely, in the free wall of the left ventricle. Although rare in the general population, this is the most common cause of sudden cardiac death in young athletes. Hypertrophic cardiomyopathy is a diverse clinical and pathophysiologic disorder of unknown cause that principally involves the left ventricle and is manifested as asymmetric or concentric hypertrophy. Hypertrophic cardiomyopathy is a disease of the myocardium in which a portion of the myocardium is thickened without any obvious cause [200]. It is characterized by abnormal hypertrophy, primary of the left ventricular. It is perhaps most well known as a leading cause of sudden cardiac death in young athletes [201]. It is inherited and 60% of individuals with this have an affected first-degree relative. The prevalence of hypertrophic cardiomyopathy is about 0.2–0.5% of the general population. Hypertrophic cardiomyopathy characterized by an asymmetrically hypertrophied and non-dilated left ventricle is a relatively common genetic cardiac disease (with an incidence of 1 in 500 individuals in the general population), with heterogeneous clinical, morphologic, and genetic expression [195].

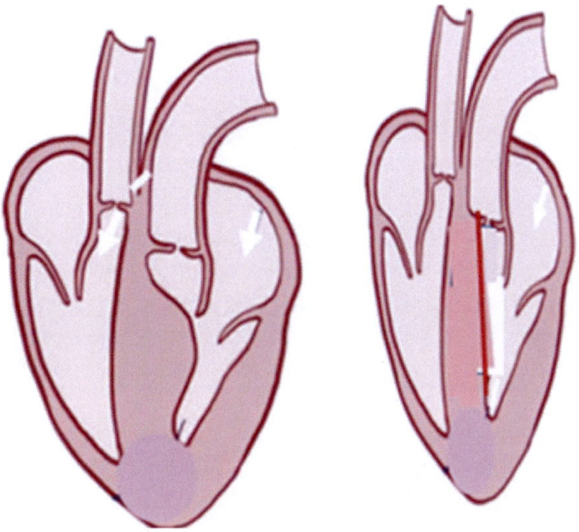

Fig. 6.32 View of left ventricular hypertrophic cardiomyopathy walls and those of normal left ventricular

Individuals with hypertrophic cardiomyopathy have some degree of left ventricular hypertrophy. Right ventricular involvement is usually less evident. Usually, this is asymmetric hypertrophy, involving the interventricular septum, and is known as asymmetric septal hypertrophy. This is in contrast to the concentric hypertrophy seen in aortic stenosis or hypertension.

Diastolic dysfunction is common to most hypertrophic cardiomyopathy: patients, irrespective of the presence or absence of outflow obstruction [202]. About two-thirds of individuals with hypertrophic cardiomyopathy have asymmetric septal hypertrophy.

With hypertrophic cardiomyopathy, histological features of myocyte hypertrophy, myofibril disarray, and interstitial fibrosis occur. In hypertrophic cardiomyopathy—myofiber disarray—not all fibers are pulling in the same direction. Thus, the contraction is ineffective. However, the cardiac conduction system might cause arrhythmias and sudden death. Also, the normal configuration of muscle cells is disrupted, a phenomenon known as myocardial disarray. Hypertrophic cardiomyopathy is inherited as an autosomal dominant trait and is attributed to mutations in one of a number of genes that encode for one of the sarcomere proteins. Thus, hypertrophic cardiomyopathy is most commonly due to a mutation in one of nine sarcomere genes that involve genes encoding cardiac sarcomere protein [199]. About 50–60% of patients with a high index of clinical suspicion for hypertrophic cardiomyopathy will have a mutation identified in at least one of nine sarcomere genes. Approximately 45% of these mutations occur in the β myosin heavy chain gene on chromosome 14 q11.2-3, while approximately 35% involve the cardiac myosin binding protein C gene. Interestingly, most of these genes encode sarcomere proteins, such as myosin-7 (also known as cardiac muscle β-myosin heavy chain; MYH7), cardiac myosin-binding protein C (MYBPC3), and cardiac muscle troponin T (TNNT2). However, the molecular events that ultimately lead to the clinical phenotype of HCM are still unclear [199].

About 25% of individuals with hypertrophic cardiomyopathy demonstrate an obstruction to the outflow of blood from the left ventricle during rest. In others, obstruction only occurs under certain conditions. The principal abnormality in systole is the obstruction to left ventricular outflow caused by systolic anterior motion of the anterior or posterior mitral leaflet(s) with mitral leaflet–septal contact [202]. This is known as dynamic outflow obstruction because the degree of obstruction is variable and is dependent on the amount of blood in the ventricle immediately before ventricle systole. Mitral regurgitation invariably accompanies the obstruction to outflow.

Another, nonobstructive variant of hypertrophic cardiomyopathy is apical hypertrophic cardiomyopathy [203], also called Yamaguchi Syndrome or Yamaguchi Hypertrophy, first described in individuals of Japanese descent.

Sudden Cardiac Death in Athletes

Unexpected sudden cardiac death deterrence in athletes is a difficult medical task, usually happening during exercise bout or cardiac arrest within 1 h post-exercise [204]. Most reasons are related to hereditary or developed cardiovascular disease without symptoms noted before the deadly event. The single-most important predictor is fainting or near-fainting during exercise, which should require detailed explanation and investigation [205]. The sudden cardiac death of a young athlete is a catastrophic event, particularly in the absence of prior notice symptoms. Coronary artery anomalies are a well-described cause of cardiac symptoms and sudden cardiac death, nonetheless, the diagnosis is usually missed by utilizing usual methods (noninvasive) to detect myocardial ischemia. It is difficult to distinguish anomalous coronary arteries by methods of electrocardiograph since resting electrocardiographs are typically normal and exercise stress testing may be negative. Although some cases may be suspected on echocardiography, a definitive diagnosis is made by coronary angiography, computed tomography, or magnetic resonance imaging. Sudden cardiac death is preventable by correction of the anomaly [206].

A variety of inherited coronary artery abnormalities combine to represent the second leading cause of sudden death in young athletes, most frequent cardiovascular cause of sudden death in young athletes in the United States [201]. Only about one-third of affected individuals are thought to be symptomatic (experiencing angina, syncope, or shortness of breath with exercise) before sudden death. It is a congenital coronary-artery anomaly in which the artery arises from the wrong aortic sinus most commonly. The most common coronary anatomy associated with risk of sudden death is the left main coronary arising from the right sinus with an intra-arterial course between the aorta and pulmonary artery [207]. The mechanism of sudden death in all athletic cases is thought to be an arrhythmia triggered by heart tissue ischemia or infarction. These anomalies, which are not usually related to coronary-artery atherosclerosis, may be more common than previously thought. Their diagnosis requires a high index of suspicion and is particularly important since surgical correction is feasible. Echocardiography is the method to be utilized in order to detect the existence of anomalous coronary arteries.

Anemia

Anemia is a disorder that occurs when production and amount of the healthy red blood cells are below normal; therefore, hemoglobin content in the blood is significantly low. Consequently, oxygen delivery to body tissues and organ will not be enough. Women, young children, and individuals with long-term diseases are more likely to have anemia. Women are at risk of iron-deficiency due to blood loss during pregnancy. There are various forms of anemia, with some types posing greater health problems than others. Therefore, because of the different reasons, medical treatments differ also. Anemia comes in various forms, with some types posing greater health problems. One of the primary symptoms of anemia is fatigue. Anemia is passed on to the next generations through the genes, it may appear in infants from birth. Aged individuals have a greater risk of anemia arising from other medical diseases, or due to wrong and insufficient diet. Anemia is a frequent problem of many diseases; however, mechanisms linking reduced blood oxygen content to the long-term consequences of anemia are not understood yet.

Aerobic exercise may easily bring upon the subject fatigue and short of breath. Thus, a drop in athletic performance or a general feeling of fatigue may indicate the existence of anemia. Because the blood is iron deficient and carries less oxygen to working muscles, when exercising moderately, the individual may be relatively exercising significantly more strenuous. The maintenance of oxygen supply to the tissues during anemia involves complex cardiovascular adaptations, including an increase in cardiac output, reduced peripheral resistance, and increased oxygen extraction from hemoglobin [208]. Reduced arterial oxygen content by the lower hemoglobin concentration would result in a higher blood flow, as observed with a low arterial partial pressure, and maintenance of oxygen delivery. A low arterial oxygen partial pressure induces an elevation in submaximal muscle blood flow [209]. In addition, hypoxia-inducible factors are associated with the transcriptional activation of genes involved in adaptive mechanisms that increase oxygen delivery and provide alternative metabolic pathways. The normal hemodynamic responses to chronic anemia include increased cardiac output, decreased systemic vascular resistance, widened arteriovenous oxygen gradient, and increased plasma volume [210].

Cardiac Rehabilitation

Cardiac rehabilitation includes providing complete long-term supervised environment connecting the following core components: medical evaluation, prescribed exercise, coronary risk factor reduction, and secondary prevention. Rapid evolution in the management of CAD and physical activity decreases mortality rates. The aim behind the cardiac rehabilitation is to restore physical work capacity by advancing fitness, changes in lifestyle and health of CAD patients [211]. This evolution changed the demographics of cardiac patients who can be advised to participate in a rehabilitation training program. Cardiac rehabilitation programs have been consistently shown to improve objective measures of exercise tolerance and psychosocial well-being without increasing the risk of significant complications.

Partaking in a cardiac rehabilitation program can reduce risk factors and myocardial reinfarction and even return to normal life. Cardiac rehabilitation program involves aerobic and resistance exercises, risk factors alteration, and psychological stress. Exercise is the most significant element of a cardiac rehabilitation program, since chronic exercise recovers cardiovascular system, weight control, increases work capacity and response to stress. In addition, those who participate in exercise rehabilitation achieve a significantly lower risk of mortality, and of having repeated myocardial infarction.

CAD, postmyocardial infarction, bypass, congestive heart failure, and other cardiovascular diseases benefit from exercise. Exercise may reduce recovery time period, improve myocardial, and sometimes using less medications. The application of cardiac rehabilitation programs in cardiac patients has demonstrated their efficacy in increasing functional capacity, controlling coronary risk factors, reducing symptoms, and countering psychological deterioration. They have also been shown to improve cost-efficiency. Given that ischemic heart disease is a progressive, chronic illness and that low-risk patients are included in cardiac rehabilitation programs, possible beneficial effects of these therapeutic programs on morbidity and mortality require long-term evaluation. The results of different studies support the hypothesis that multifactor cardiac rehabilitation programs with secondary prevention measures maintained in the long-term favorably influence prognosis in cardiac patients. In general, the benefits of regular exercise for the general population exceed the possible danger associated with it. While all data are not yet accumulated to assure us of longer life through exercise, it does appear this may be the case. Aerobic exercise training is associated with health benefits and specifically with decreased cardiovascular mortality in a few large observational studies. Recently, it was suggested that following training the mechanism most likely to be involved is a change in the cardiac autonomic balance producing an increase, or a relative dominance, of the vagal component. Physical and physiological improvements turn out to be a safe integral part of cardiac rehabilitation programs for work capacity restoration [212]. In good cardiac rehabilitation program as primary prevention modality, physical activity is the cornerstone. It applies to patients with coronary artery disease, ischemic heart disease, heart failure, and post-heart surgery. Cardiac rehabilitation programs demonstrated their efficacy in mortality rate reduction, reducing symptoms, improved exercise tolerance and functional capacity, coronary risk factors modification, and countering psychological deterioration [213]. Yet, beneficial effects of these therapeutic programs require long-term evaluation of their impacts on morbidity and mortality rates. The results of different studies support the hypothesis that multifactor cardiac rehabilitation programs with secondary prevention measures maintained in the long term produce favorable influences prognosis in cardiac patients. In general, the benefits of regular exercise for the overall population exceed the possible danger associated with it. While all data are not yet accumulated to assure us of longer life through exercise, it does appear this may be the case. Exercise training was associated with health benefits and specifically with decreased cardiovascular mortality in a few large observational studies. Recently, it was suggested that following training the mechanism most likely to be involved is a change in the cardiac autonomic balance producing an increase, or a relative dominance, of the vagal component. It has been long known that exercise training reduces resting and submaximal heart rate. Alterations in left ventricular structure and function are a well-described and accepted component of the response to physical conditioning. This is true if cardiac patients or healthy subjects are engaged in similar endurance training. Both groups demonstrated decreases in heart rate, ventricular end-systolic volume and wall stresses, and increases in ventricular end diastolic volume and passive active phase's ratio during diastole, following exercise training, effect on cardiopulmonary function during submaximal exercise of a fixed absolute work rate is similar for younger and coronary artery disease patients. The degree of the VO_{2max} improvement with training in the coronary artery disease patients expressed in relative terms appears to be comparable to that demonstrated by younger subjects. The increased VO_{2max} in the active subjects may be due to an increase in mitochondrial respiratory capacity. Endurance exercise training in coronary artery disease patients decreases resting and submaximal exercise heart rate, systolic and diastolic blood pressure, while stroke volume increased. Marked changes are notable in the coronary artery disease during peak effort in which stroke volume, cardiac output, contractility, and oxygen uptake are increased, while total peripheral resistance, systolic and diastolic blood pressure decreased, thus lowering afterload which in turn facilitates left ventricular systolic and diastolic function. Endurance training in CAD patients induces also changes in pulmonary function and skeletal muscle. These changes include a reduction in lactic acid levels and increased maximal ventilation. The most important adaptive response of skeletal muscle to endurance exercise is an augmentation of respiratory capacity with increases in the ability to oxidize pyruvate, fatty acids, and ketones. As a result of increases in the levels of the enzymes of the malate aspartate shuttle, there is also an enhancement of the capability for mitochondrial oxidation of the reducing equivalents generated in the cytoplasm during glycolysis. The alterations in mitochondrial composition enhance the extraction of oxygen by the muscles and therefore, reduce the need for high Q and HR.

When previously sedentary cardiac patients are retested at the same absolute submaximal work rate after adapting to endurance exercise, their endurance is found to be markedly increased. However, metabolic factors do not appear to determine the magnitude of VO_{2max}. The changes in muscle oxidative potential may play a major role in the patients' capacity to perform submaximal work.

Physical inactivity is a risk factor for CAD, while regular aerobic exercise endures health benefits. Myocardial functional changes with cardiac complications in sedentary adult include a decline in maximum HR, SV, and left ventricular contractility, and an increase in RVS. The changes in the structure and function of the cardiovascular system result in VO_{2max} decline, which is the best single indicator of physical working capacity. Whether maximal Q actually decreases depends on many interactive factors. First, the primary aging process, which has a genetic component, occurs in the absence of disease and independent of lifestyle. If oxygen transport is indeed limited by maximal Q, this may be calculated as the product of maximal HR, (a-v)o2, and maximal SV.

The second possible cause that may reduce VO_{2max} with cardiac disease is the reduced $(a\text{-}v)O_2$ at maximal effort. Cardiac patients are not generally anemic and the red blood cell content is usually well maintained, although it may decrease in subjects with maximum oxygen intake. Physical inactivity is directly related to cardiovascular mortality. Aerobic exercise training is the basis of a well-meaning cardiac rehabilitation program. Aerobic exercise training through moderate-intensity activity (55–70% of VO_2) of about 1000–2000 $\text{kcal} \cdot \text{week}^{-1}$ of energy expenditure increases work capacity and plays an important part in the prevention of CAD. Aerobic training increases cardiovascular functional capacity and improves myocardial oxygen delivery–demand balance in health and disease. The decreased myocardial oxygen demand, at a similar workload, is very beneficial to patients with CAD, indicating the delay of ischemic appearance. Partial control of key risk factors may be affected by aerobic exercise training. The mainstream data on the beneficial effects of aerobic exercise were focused on studies in men; however, similar results are in women. Cardiac rehabilitation program often involves exercise training and education about lifestyle changes. The aim is to have the patient return to an active life, and to reduce further heart problems. Cardiac rehabilitation can improve myocardial performance and thus, increase quality of life; prevent future heart problems, for patients suffering from, heart attack, CAD, following myocardial infarction, nonobstructive cardiomyopathy, angioplasty and stents or coronary artery bypass, heart valve repair or replacement, angina, and heart failure. The cardiac rehabilitation workforce will design a program to meet the individual's needs. Enrolling in such a program requires a battery of test, ECG at rest, and an ECG stress test on an ergometer or/and cardiac imaging tests. During cardiac rehabilitation, subjects learn to exercise at a target HR, calculated following the given stress test. This makes sure that the patient exercises safely in an intensity that increases its work capacity or VO_2. Participants in the cardiac rehabilitation program can expect to regain strength, avoiding from worsening the disease and reduce future complications and death. However, cardiac rehabilitation is not suitable for every patient. Seldom, patients may suffer from tense muscles due to overtraining or high work intensities. In addition, there is likewise a minor danger of cardiovascular difficulties. Cardiac rehabilitation is a long-term physical program with medical supervision during that period, which means continuing the behaviors change. After 2–3 months of exercising, exercise routine develops.

A cardiac rehabilitation program combines aerobic exercise training and emotive actions intended to help patients recover from a cardiovascular event and to prevent future events. Therefore, the cardiac rehabilitation program contains medical and well-being programs. Training at prescribed target HR for aerobic training reduces incidences of CAD and myocardial infarctions. CAD is the result of major blood arteries damaged or sickly so, plaque in the arteries and inflammation are usually the cause for CAD. Due to the plaque narrowed coronaries developed, consequently, blood flow and oxygen supply to the myocardium are reduced. Ultimately, an imbalance between oxygen delivery–demand may cause angina and chest pain. During stress testing, myocardial response is controlled by ECG and often-imaging studies during an induced episode of increased cardiac demand so that ischemic areas potentially at risk of infarction can be identified. Exercise stress testing is a validated diagnostic test for CAD in symptomatic patients and is used in the evaluation of patients with known cardiac disease.

It has been long known that aerobic exercise training reduces resting and submaximal HR; alterations in left ventricular structure and function are a well-described and accepted component of the response to physical conditioning. This is true if cardiac patients or healthy individuals engaged in similar endurance training. Both groups demonstrated cardiac changes previously documented to occur with aerobic exercise training, including lower HR, ventricular ESV and wall stresses, higher ventricular EDV, and passive/active phase ratio during diastole. The training effect on cardiopulmonary function during submaximal exercise of a fixed absolute work rate is similar for younger and CAD patients. The degree of the VO_{2max} improvement with training in CAD patients expressed in relative terms appears to be comparable to that demonstrated by younger subjects. The increased VO_{2max} in the active subjects may be due to an increase in mitochondrial respiratory capacity. Endurance exercise training in CAD patients decreases resting and submaximal exercise HR, SBP, and DBP, while SV increases. Marked changes are notable in the CAD during peak effort in which SV, Q, contractility, and VO_{2peak} are increased, while RVS decreases, thus lowering afterload which in turn facilitates left ventricular systolic and diastolic function. Endurance training in CAD patients induces also changes in pulmonary function and skeletal muscle. These changes include a reduction in blood lactate levels and increased maximal VE. The most important adaptive response of skeletal muscle to endurance exercise is an augmentation of cell's respiratory capacity with increases in the ability to oxidize pyruvate, fatty acids, and ketones. As a result of increases in the malate-aspartate shuttle enzymatic levels, there is also an enhancement of the capability for mitochondrial oxidation of the reducing equivalents generated in the cytoplasm during glycolysis. The alterations in mitochondrial composition enhance $(a\text{-}v)O_2$ by the muscles and therefore, reduce the need for high HR and thus, Q.

In summary, these data indicate that the CAD patients' skeletal muscle, cardiovascular system, and pulmonary function retain a high degree of trainability, with much of the improvement occurring peripherally, just as in younger individuals.

Figure 6.33 summarizes some of the changes associated with coronary artery disease and training on cardiovascular and hemodynamic variables, while Fig. 6.34 reveals that the efficiency of breathing is improved in coronary artery disease patients following an endurance training program. These changes include a reduction in blood lactate levels and increased maximal ventilation.

Table 6.1 reviews cardiovascular and hemodynamic variables change in CAD patients, following training on at rest and during submaximal aerobic exercise, while Table 6.2 summarizes pulmonary changes in CAD patients, following training on at rest and during submaximal aerobic exercise.

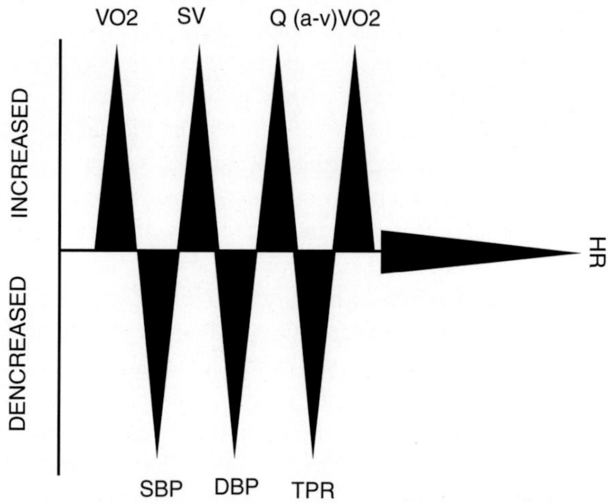

Fig. 6.33 Endurance training-induced changes in cardiovascular and hemodynamic variables in CAD patients. *VO_2* oxygen uptake, *SBP* systolic blood pressure, *DBP* diastolic blood pressure, *SV* stroke volume, *HR* heart rate, *Q* cardiac output, *$(a\text{-}v)O_2$* arterio-venous oxygen differences

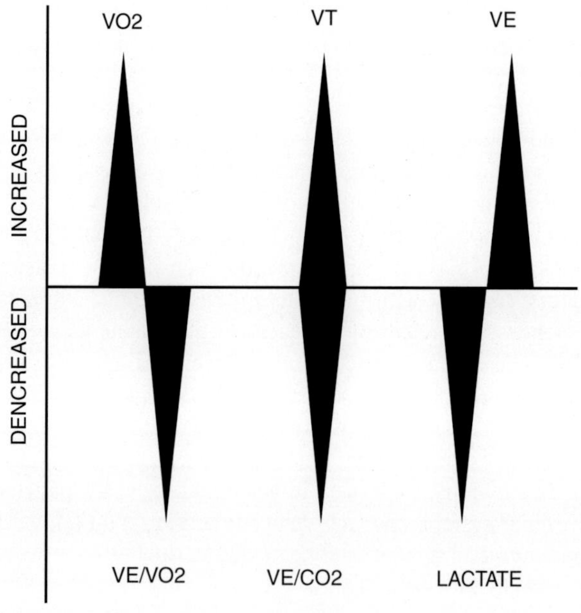

Fig. 6.34 Endurance training-induced changes in pulmonary variables in coronary artery disease patients. *VO_2* oxygen uptake, *VT* tidal volume, *VE* minute ventilation

Table 6.1 Endurance training-induced changes in cardiovascular and hemodynamic variables in coronary artery disease patients

Variable	Rest	Submaximal exe
VO_2	Unchanged	Unchanged
SBP	Decreases	Decreases
DBP	Decreases	Decreases
SV	Decreases	Increases
HR	Decreases	Decreases
Q	Unchanged	Unchanged
Contract	Unchanged	Unchanged
TPR	Unchanged	Unchanged
$(a\text{-}v)O_2$	Unchanged	Unchanged

VO_2 oxygen uptake, *SBP* systolic blood pressure, *DBP* diastolic blood pressure, *SV* stroke volume, *HR* heart rate, *Q* cardiac output, *$(a\text{-}v)O_2$* arterio-venous oxygen differences

Table 6.2 Endurance training-induced changes in pulmonary function in coronary artery disease patients

Variable	Rest	Submaximal exe
MVV	Unchanged	
VO_2	Unchanged	Unchanged
VE	Unchanged	Decreases
VE/VO_2	Unchanged	Increases
VE/VCO_2	Unchanged	Unchanged
Lactate	Unchanged	Decreases

MVV maximal voluntary ventilation, *VO_2* maximal oxygen uptake, *VE* minute ventilation, *VCO_2* carbon dioxide volume

Table 6.2 reveals that efficiency of breathing is improved in the coronary artery disease patients following an endurance training program. These changes include a reduction in blood lactate levels and increased maximal ventilation.

In addition to the above-mentioned benefits, the multiple benefits of regular aerobic exercise are enumerated as follows:

Increased physiological capacity:

- Increased physical performance capacity.
- Lower rate of breathing required.
- Better air passage and lung function.
- Greater efficiency and less perceived effort.
- More appropriate peripheral blood distribution and return.
- Enhanced blood volume and red blood cell mass.
- Less tendency for hemo-concentration.
- Less tendency for thrombophlebitis and pulmonary emboli.
- Stimulus for increased arterial size.
- Stimulus for collateral vessels to bypass obstruction.
- Less rapid loss of bone minerals due to aging.
- Stimulus for better cartilage protection of joints.

- Improved function and appearance of the skin.
- Reduced allergic reactions (asthma, hay fever).

Expanded diet permits a greater intake of vitamins, minerals, and other desirable dietary elements:

- Reduction in low-density lipoprotein cholesterol.
- Increase in high-density lipoprotein cholesterol.
- Reduced total serum cholesterol.
- Reduced serum triglycerides.
- Increased electiveness of insulin.
- Reduced platelet adhesiveness.
- Increased fibrinolysis (early dissolution).
- Decreased inappropriate catecholamine (adrenaline) responses to stress.
- Decreased vulnerability or susceptibility to cardiac dysrhythmia.
- Enhanced feelings of energy, enthusiasm, and well-being with more stamina.
- Decreased depression.
- Enhanced self-image.
- Better relaxation and sleep.
- More satisfactory sexual responses.
- Less craving for stimulants and tranquilizers.

Socioeconomic benefits:

- More optimism, enthusiasm, creativity.
- Less illness, absenteeism.
- Shorter illness course and fewer complications.
- Reduced medical expenses.

Resistance Training in CAD Patients

Previously, patients with CAD were advised to avoid re-resistance bouts for the possible adverse effects of increased pressure opposing left ventricular ejection. However, cardiovascular responses are alike in aerobically trained CAD patients and healthy subjects during submaximal isometric exercise utilizing small or large muscle mass. Resistance training is safe and beneficial and currently recommended for its effects on strength's maintenance, increased flexibility, increased muscle mass, and bone mineral density. Patients with CAD are advised to workout with both arms and legs resistance training at bouts of moderate intensity.

At a voluntary contraction equal or greater than 20–30% of maximal voluntary contraction, the intramuscular pressure exceeds the intravascular pressure in the contracting muscle and significantly reduces localized blood flow, causing muscle ischemia and hypoxia [214]. Therefore, isometric exercise is usually discouraged in CAD patients as of the potential adverse effects of increased afterload and associated ischemia on left ventricular function [215]. During resistive exercise, CAD patients show the same pressor response as normal subjects.

Increase in HR is the dominant mechanism for the slightly increased Q [158]. Despite the increased Q, blood flow to the non-contracting muscles does not significantly increase, probably because of reflex vasoconstriction. The increase in Q, however, is modest 8–10 L·min^{-1} [158], compared to 20 L·min^{-1} during aerobic exercise [161].

The increase in HR in CAD patients is probably explained by the limited ability of the patient to increase SV. It remains largely unchanged except at high levels of tension (>50% maximal voluntary contraction), in which it may decrease. Whereas normal subjects can increase both SV mildly and HR in order to increase Q. Therefore, in order to compensate for the low SV, an increase in HR is greater in patients with limited SV [216]. Increased ventricular performance induced by handgrip maneuver appears to be mediated by increased contractility, but in addition by varying degrees of reliance upon the Frank-Starling mechanism, depending upon the adequacy of inotropic to increase contractility reserve to meet the stress imposed by isometric exercise.

Mild-to-moderate-intensity isometric exercise, in CAD patients, results in a low rate-pressure product, compared to maximum treadmill aerobic exercise. However, while gaining strength, no alterations in HR, SV and thus in Q are noted [214].

Exercise Stress Testing

An exercise stress test most often performed for the following reasons evaluate patient's complaints and determine if CAD worsens; irregular heart rhythm occurs during physical work and myocardial performance following bypass. The following training HR (THR) equation is employed following the stress test, calculated by the Karvonen's equation:

$$\mathrm{THR} = \left[\left(\mathrm{HR}_{\mathrm{max}} - \mathrm{HR}_{\mathrm{resting}}\right) \cdot \%\mathrm{intensity}\right] + \mathrm{HR}_{\mathrm{rest}}$$

Example,

A cardiac patient aged 60 years with resting HR 70 beats·min^{-1}, maximal HR of 165 beats·min^{-1}, intensity assigned 65% of its work capacity thus,

$$\mathrm{THR} = \left[\left(165\,\mathrm{beats} \cdot \mathrm{min}^{-1} - 70\,\mathrm{beats} \cdot \mathrm{min}^{-1}\right)65\%\right] + 70 = 132\,\mathrm{beats} \cdot \mathrm{min}^{-1}$$

Exercise stress test is performed up to 85% of individual's estimated maximal HR by age or reached at the secession of the test or until ECG or angina changes develop. Risks of stress testing include infarction and sudden death, which occur in about 1 out of 5000 tests.

A number of tests may help with diagnoses including ECG, cardiac stress testing, coronary computed tomographic angiography, and coronary angiogram, among others.

Exercise work capacity testing is a significant indicative and predictive means for evaluating healthy subjects and patients with alleged or known myocardial disease [217]. Exercise stress test is done on an ergometer, preferably suitable to the sport or typical exercise means of the individual. The exercise intensities are designated to gradually progress in increases from modest to utmost intensity. The exercise laboratory needs a special fitting; exercise testing should be performed in a well-ventilated room with a comfortable temperature (20–22 °C) with humidity around 50%. Stress tests are performed on a treadmill and bicycle to assess symptoms occurring during or following the exercise stress test subsequent to patient's complaints about chest pain due to coronary artery disease and arrhythmia aggravation. Stress tests were developed as a clinical test to evaluate patients with suspected coronary heart disease, though it can also be used to estimate cardiovascular athletic capacity [218]. Since these stress tests serve as maximal test performed commonly on people with heart problems, the test should be performed under the supervision of appropriately trained medical staff [219].

An exercise stress test with sensitivity of 78% and a specificity of 70% for detecting coronary artery disease usually involves walking on a treadmill or riding a stationary bike, while the HR, BP, ECG, BP, and breathing are monitored [220]. Three protocols are used for maximal treadmill stress tests; Balke (Table 6.3), Bruce (Table 6.4), and Ellestad (Table 6.5), each test varied in the workload incremental namely, speed or grade. Measuring VO_{2max}, maximal HR, SBP, and DBP values reveal no significant differences between the protocol tests [221]. Furthermore, the

Table 6.3 Balke protocol test

Time (min^{-1})	Speed ($km \cdot h^{-1}$)	Grade (%)	$mLO_2 \cdot kg^{-1} \cdot min^{-1}$
1	5.3	0	9.6
2	5.3	1	11.6
3	5.3	2	17.2
4	5.3	3	18.6
5	5.3	4	20.3
6	5.3	5	22.6
7	5.3	6	23.5
8	5.3	7	25.2
9	5.3	8	26.6
10	5.3	9	28.4
11	5.3	10	29.8
12	5.3	11	31.5
13	5.3	12	32.9
14	5.3	13	34.7
15	5.3	14	36.4
16	5.3	15	37.8

Table 6.4 Bruce protocol test

Stage	Time (min)	Grade (%)	Speed ($km \cdot h^{-1}$)	mph	$O_2 mL \cdot kg^{-1} \cdot min^{-1}$
1	3	10	2.7	1.7	16.1
2	6	12	4.0	2.5	21.1
3	9	14	5.4	3.4	31.7
4	12	16	6.7	4.2	43.4
5	15	18	8.0	5.0	56.5
6	18	20	8.8	5.5	67.4
7	21	22	9.6	6.0	79.4

Table 6.5 Ellestad stress protocol

Stage	Time	Grade (%)	Speed ($km \cdot h^{-1}$)	Mph	$O_2 mL \cdot kg^{-1} \cdot min^{-1}$
1	3	10	2.74	1.7	16.1
2	2	10	4.83	3.0	20.1
3	2	10	6.44	4.0	33.5
4	2	10	8.05	5.0	41.1
5	2	15	8.05	5.0	53.1
6	2	15	9.66	6.0	63.1
7	2	15	11.27	7.0	73.0
8	2	5	4.83	3.0	18.5

energy cost of each stage increased during the Balke protocol test is 2 $mLO_2 \cdot kg^{-1} \cdot min^{-1}$, while the energy expenditure for the Bruce protocol and Ellestad protocol tests are increased by 3.5–5.5 $mLO_2 \cdot kg^{-1} \cdot min^{-1}$ [222]. The clinical test of Balke protocol is used to achieve peak VO_2 in cardiac patients; however, it estimates also cardiovascular work capacity in sportspersons. However, compared to the other protocols, the test lasts a long time. The Balke protocol keeps a constant speed but increases grade by 1% each minute. The Bruce protocol increases speed and grade every 3 min (Table 6.3). Table 6.5 summarizes the Ellestad protocol that increases speed each stage until the 10th minute upon the introduction of a single increase in grade (to 5%) followed by subsequent increases in speed [221].

Bruce and Balke protocols are the most commonly used on treadmill enabling the estimation of cardiopulmonary function and aerobically work capacity by measuring VO_{2max} or VO_{2peak}. Usually, all three tests yield similar results. During the Balke protocol test, subjects can undergo the test at a slow pace, making it suitable for cardiac patients as well as healthy elderly people. Both protocol tests need treadmill, timer, and monitor (ECG or pulse meter) to track HR and myocardial function during the test.

All tests start with a warm-up on the treadmill or another exercise device in order to accustom the subject and increase metabolic efficiency by increasing core body temperature.

The Balke protocol is administered with a constant speed of 5.3 $km \cdot h^{-1}$ (3.3 mph) for men and 4.8 $km \cdot h^{-1}$ (3.0 mph) for woman. The test begins with the treadmill grade at 0% for 1 min. Thereafter, it is increased by 1% every minute. The idea is to keep up the pace and increase the incline until subjects are exhausted and cannot continue. It is important to record all physiological available variables. The Bruce and Ellestad graded stress tests are significantly shorter than the Balke graded stress test. Compared to Balke, Bruce and Ellestad protocol tests start at the beginning with higher workloads and thereafter it becomes harder due to the increment of speed and grade [222].

Borg's Rating of Perceived Exertion Scale

The most common method used to express intensity relatively to individual's capacity is by the rating of perceived exertion scale [223]. The scale is used to document the individual's exertion as he feels during a stress test. In sports, the scale is used during the training sessions to evaluate the intensity of training workloads. The original scale was on a scale of 6–20 (Table 6.6), later it was modified to have a scale of 10, utilized in the clinical diagnosis of breathlessness and dyspnea, chest pain, angina, and skeletal muscles with pain. The use of the Borg scale turned to be very helpful in determining patient's effort.

When previously sedentary cardiac patients were retested at the same absolute submaximal work rate after adapting to endurance exercise, their endurance capacity was found to be markedly increased, by the Borg scale. However, metabolic factors do not appear to determine the magnitude of VO_{2max}. The changes in muscle oxidative potential may play a major role in the patients' capacity to perform submaximal work.

Table 6.6 The Borg's rating of perceived exertion levels

Perceived exertion	Intensity level
6	No exertion at all
7	Extremely light
8	
9	Very light
10	
11	Light
12	
13	Somewhat hard
14	
15	Hard (heavy)
16	
17	Very hard
18	
19	Extremely
20	Maximal exertion

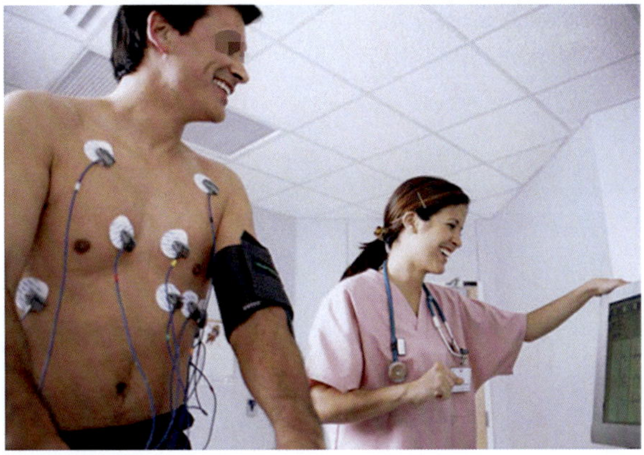

Fig. 6.35 Exercise stress test with 12 leads ECG

Exercise stress testing:
Several imaging tests can detect ischemia after exercise:

- ECG
- Radionuclide perfusion imaging
- Echocardiography

ECG and BP recordings are used during exercise stress testing (Fig. 6.35) to diagnose CAD and help define prognosis. Exercise stress testing is the recording of the heart's electrical activity and blood pressure, while the myocardium is under physical stress. ECG is most useful in patients with a probability of having CAD based on age, gender, and a normal ECG at rest. In men, sensitivity averages 67%; and specificity 72%, while in women sensitivity and specificity are lower.

BP Response

In cardiac patients, SBP rises with the progressive increase in ergometer workload, whereas DBP usually remains unchanged and might increase slightly. A rising DBP usually signs the presence of CAD. However, sometimes it is a hypertension response, which may develop to CAD. When exercise is stopped at once, about 10% of the individuals will experiance hypotension. to avoid hypotension, individuals should keep working with light workloads, on the ergometer.

Generally, the earlier these changes occur during the test, the more severe the CHD. Not all patients who test positive have CHD. Similarly, not all patients who test negative do not have CHD.

Laboratory on ECG Testing

Experiment 1. Electrocardiogram (ECG).

- Carrying out an ECG
- Ask the subject to undress down to the waist and lie down
- Remove excess hair of the chest, where necessary
- Attach limb leads (anywhere on the limb)
- Attach the chest leads as follows:
  - V1 and V2: either side of the sternum on the fourth rib (count down from the sternal angle, the second rib insertion)
  - V4: on the apex of the heart (feel for it)
  - V3: halfway between V2 and V4
  - V5 and V6: horizontally laterally from V4 (not up toward the axilla)

In all preexcitation syndromes, at least one more conductive pathway is present. Physiologically, the normal electrical depolarization wave is delayed at the atrioventricular node to allow the atria to contract before the ventricles. However, there is no such delay in the abnormal pathway, so the electrical stimulus passes to the ventricle by this tract faster than via normal atrioventricular/bundle of His system, so the ventricles are depolarized (excited) before (pre-) normal conduction system.

The student will be recording two different ECG tracings. First, record the ECG of a subject and follow the instructions below. Then, unsnap the cable leads from the adhesive electrodes. Have the subject exercise (run around in the hallway or outside, up/down staircases, do jumping jacks, etc.) to get their HR elevated. Then, quickly reconnect the cable leads to the electrodes and record a second tracing.

For the Subject

1. Remove any metal from wrists and left ankle.
2. Apply an adhesive disk electrode to the skin on the inside of each wrist and to the left ankle.
3. Attach the ECG electrode cables to the electrode disks in the following manner: o White (−)—left arm o Red (+)—right arm o Black (ground)—left ankle the disks snap on to the cables.
4. Attach the clip on the ECG cable to a belt loop of the edge of your pocket to help keep the cables from moving.
5. Attach the pulse plethysmograph transducer to the tip of your index or middle finger so that the white top of the electrode is in contact with the pad of the finger. It should be secured firmly (but not so hard that it cuts off circulation) with the Velcro strap.
6. Sit down on a stool and rest your arms on top of your legs. Be sure that the ECG cables are slack, and that you are relaxed. Sit quietly during the recording, and avoid moving, talking, etc.—electrical signals from your skeletal muscles will distort the recording.

For the Operator

1. Once the subject is ready, click on start of the screen.
2. You should see the ECG being complex on the screen and the BP being recorded. If you do not see a distinctive ECG pattern or a regular pulse pattern, ask the instructor for help.
3. Record for 5–8 s, then mark on stop in the upper right corner of the strip, and then, explain the strip by data. Analysis

Determine the following:

1. Duration of a single cardiac cycle, measured from the peak of one R wave of the QRS complex to the peak of the next R wave. Do this for 4–5 intervals and note any differences that may occur in the duration of these ECG cycles. Using this value, calculate the HR (beats·min^{-1}) by dividing 60 over the cardiac cycle duration.
2. Duration of the P-R interval, measured from the peak of the P wave to the peak of the R wave in the QRS complex. This is an indication of the time it takes for an electrical signal to travel from the SA node to the ventricles.
3. Duration of the R-T interval, measured from the peak of the R wave to the peak of the T wave in the same ECG cycle. This is a measure of how long the ventricles are depolarized when they undergo an action potential, and is roughly the same duration as ventricular systole.
4. Duration of the T-R interval, measured from the peak of the T wave of one cardiac cycle to the peak of the R wave in the same ECG cycle. This is a measure of how long the ventricles remain repolarized after undergoing an action potential, and is roughly the same duration as ventricular diastole.
5. Duration of a single heartbeat measures the amount of time that elapses between one heartbeat on the pulse pressure recording and the next. You can use either the start of the heartbeat or the peak or the heartbeat as your reference point. Compare this value with that of the duration of a single cardiac cycle.
6. Blood conduction time. Measure the duration between the R wave of a cardiac cycle and the peak of the corresponding pressure wave in the pulse recording. This is roughly how long it takes for blood to travel from your heart to the tip of your finger through arteries. If you assume that the distance from your heart to your finger is ~1 m, calculate the velocity that blood is traveling by dividing 1 m by the blood conduction time.

Experiment 2: changes in R with exercise

1. Have the subject step up and down on a stair, for five times, allowing 3 s total time for each repetition.
2. Immediately after completion of the fifth repetition, measure the subject's HR for 15 s and multiply the number by 4 to have beats·min^{-1}, record this value.
3. Repeat the 15-s HR values at 30, 60, 90, and 120 s post-exercise. Record the time it takes HR to return to resting values.
4. Score the subject based upon (a) the difference between HR immediately post-exercise and normal standing HR and (b) the time needed for HR to return to normal standing rate after cessation of exercise.

References

1. Guyton A, Hall J. Guyton textbook of medical physiology. 10th ed. 2000.
2. Alberts B, Johnson A, Lewis J, et al. Molecular biology of the cell. 4th ed. New York: Garland Science; 2002.
3. Merz CNB, Elboudwarej O, Mehta P. The autonomic nervous system and cardiovascular health and disease a complex balancing act. JACC Heart Fail. 2015; https://doi.org/10.1016/j.jchf.2015.01.008.
4. Doughty RN, White HD. Carvedilol: use in chronic heart failure. Expert Rev Cardiovasc Ther. 2007;5:21–31.
5. O'Connell MN, Falchier A, McGinnis T, et al. Dual mechanism of neuronal ensemble inhibition in primary auditory cortex. Neuron. 2011;69:805–17.
6. Raven PB, Fadel PJ, Ogoh S. Arterial baroreflex resetting during exercise: a current perspective. Exp Physiol. 2006;91:37–49.
7. Freeman JV, Dewey FE, Hadley DM, et al. Autonomic nervous system interaction with the cardiovascular system during exercise. Prog Cardiovasc Dis. 2006;48:342–62.
8. Pichon AP, de Bisschop C, Roulaud M, et al. Spectral analysis of heart rate variability during exercise in trained subjects. Med Sci Sports Exerc. 2004;36:1702–8.
9. Zuckerman-Levin N, Zinder O, Greenberg A, et al. Physiological and catecholamine response to sympathetic stimulation in turner syndrome. Clin Endocrinol (Oxf). 2006;64:410–5.
10. Zanesco A, Antunes E. Effects of exercise training on the cardiovascular system: pharmacological approaches. Pharmacol Ther. 2007;114:307–17.
11. Xu H, Mao C, Liu JM, et al. Microanatomic study of the vascular and duct system of the submandibular gland. J Oral Maxillofac Surg. 2011;69:1103–7.
12. Schubert R, Lidington D, Bolz SS. The emerging role of Ca2+ sensitivity regulation in promoting myogenic vasoconstriction. Cardiovasc Res. 2008;77:8–18.
13. Hong K, Li M, Nourian Z, Meininger GA, et al. Angiotensin II type 1 receptor mechanoactivation involves RGS5 (regulator of G protein signaling 5) in skeletal muscle arteries: impaired trafficking of RGS5 in hypertension. Hypertension. 2017;70:1264–72.
14. Hill MA, Meininger GA, Davis MJ, et al. Therapeutic potential of pharmacologically targeting arteriolar myogenic tone. Trends Pharmacol Sci. 2009;30:363–74.
15. Piagnerelli M, Ince C, Dubin A. Microcirculation. Crit Care Res Pract. 2012;2012:867176.
16. Martín-Fernández B, Gredilla R. Mitochondria and oxidative stress in heart aging. Age (Dordr). 2016;38:225–38.
17. Duncker DJ, Bache RJ. Regulation of coronary blood flow during exercise. Physiol Rev. 2008;88:1009–86.
18. Roman RJ, Van Dokkum RP. Commentary on the special issue on the impact of myogenic tone in health and disease. Curr Vasc Pharmacol. 2014;12:779.
19. Laughlin MH. Physical activity in prevention and treatment of coronary disease: the battle line is in exercise vascular cell biology. Med Sci Sports Exerc. 2004;36:352–62.
20. Laughlin MH, Davis MJ, Secher NH, et al. Peripheral circulation. Compr Physiol. 2012;2:321–447.
21. Brun JF, Varlet-Marie E, Connes P, et al. Hemorheological alterations related to training and overtraining. Biorheology. 2010;47:95–115.
22. Pretorius E. Erythrocyte deformability and eryptosis during inflammation, and impaired blood rheology. Clin Hemorheol Microcirc. 2018;69:545–50.
23. Stuart J, Nash GB. Red cell deformability and hematological disorders. Blood Rev. 1990;4:141–7.
24. Muravyov AV, Tikhomirova I. Role molecular signaling pathways in changes of red blood cell deformability. Clin Hemorheol Microcirc. 2013;53:45–59.
25. Plowman SA, Smith DL. Exercise physiology for health, fitness, and performance. Lippincott Williams & Wilkins; 2007. p. 61

26. McArdle WD, Katch FI, Katch VL. Essentials of exercise physiology. Lippincott Williams & Wilkins; 2006. p. 204
27. Leung FP, Yung LM, Laher I, et al. Exercise, vascular wall and cardiovascular diseases. Sports Med. 2008;38:1009–24.
28. Kilic-Toprak E, Kilic-Erkek O, Abban-Mete G, et al. Contribution of heme oxygenase 2 to blood pressure regulation in response to swimming exercise and detraining in spontaneously hypertensive rats. Med Sci Monit. 2018;24:5851–9.
29. Laughlin MH, Bowles DK, Duncker DJ. The coronary circulation in exercise training. J Physiol Heart Circ Physiol. 2012;302:H10–23.
30. Hu M, Lin W. Effects of exercise training on red blood cell production: implications for anemia. Acta Haematol. 2012;127:156–64.
31. Mairbauri H. Red blood cells in sports: effects of exercise and training on oxygen supply by red blood cells. Front Physiol. 2013;4:1–13.
32. Montero D, Cathomen A, Jacobs RA, et al. Haematological rather than skeletal muscle adaptations contribute to the increase in peak oxygen uptake induced by moderate endurance training. J Physiol. 2015;593:4677–88.
33. Simmonds MJ, Tripette J, Sabapathy S, et al. Cardiovascular dynamics during exercise are related to blood rheology. Clin Hemorheol Microcirc. 2011;49:231–41.
34. Alis R, Sanchis-Gomar F, Ferioli D, et al. Exercise effects on erythrocyte deformability in exercise-induced arterial hypoxemia. Int J Sports Med. 2015;36:286–91.
35. Harms CA, McClaran SR, Nickele GA, et al. Exercise-induced arterial hypoxaemia in healthy young women. J Physiol. 1998;507:619–28.
36. Wahl P, Bloch W, Mester J, et al. Effects of different levels of compression during submaximal and high-intensity exercise on erythrocyte deformability. Eur J Appl Physiol. 2012;112:2163–9.
37. Simmonds MJ, Connes P, Sabapathy S. Exercise-induced blood lactate increase does not change red blood cell deformability in cyclists. PLoS One. 2013;8(8):e71219.
38. Jadeja RN, Thounaojam MC, Khurana S. Characterization of pressure-mediated vascular tone in resistance arteries from bile duct-ligated rats. Oncotarget. 2017;8(19):30706–22. https://doi.org/10.18632/oncotarget.15409.
39. Rand PW, Lacombe E, Hunt HE, et al. Viscosity of normal human blood under normothermic and hypothermic conditions. J Appl Physiol. 1964;19:117–22.
40. Fieldman JS, Phong DH, Saint-Aubin Y, et al. Rheology. Biology and mechanics of blood flows, Part II: Mechanics and medical aspects. Springer; 2007. pp. 119-123
41. Joyner MJ, Casey DP. Regulation of increased blood flow (hyperemia) to muscles during Exercise: a hierarchy of competing physiological needs. Physiol Rev. 2015;95:549–601.
42. Lu DY, Chen EY, Wong DJ, et al. Vein graft adaptation and fistula maturation in the arterial environment. Surg Res. 2014;188:162–73.
43. Ferguson SK, Holdsworth CT, Wright JL, et al. Microvascular oxygen pressures in muscles comprised of different fiber types: impact of dietary nitrate supplementation. Nitric Oxide. 2015;48:38–43.
44. Michel RN, Chin ER, Chakkalakal JV, et al. Ca2+/calmodulin-based signalling in the regulation of the muscle fibre phenotype and its therapeutic potential via modulation of utrophin A and myostatin expression. Appl Physiol Nutr Metab. 2007;32:921–9.
45. Harms CA. Effect of skeletal muscle demand on cardiovascular function. Med Sci Sports Exerc. 2000;32:94–9.
46. Tortora GJ, Derrickson B. The cardiovascular system: the blood. Principles of anatomy and physiology. 13th ed. Wiley; 2012. pp. 729–732.
47. Westerhof N, Boer C, Lamberts RR, et al. Cross-talk between cardiac muscle and coronary vasculature. Physiol Rev. 2006;86:1263–308.
48. Versluis JP, Heslinga JW, Sipkema P, et al. Microvascular pressure measurement reveals a coronary vascular waterfall in arterioles larger than 110 microm. Am J Physiol Heart Circ Physiol. 2001;281:H1913–8.

49. Kim HJ, Jansen KE, Taylor CA. Incorporating autoregulatory mechanisms of the cardiovascular system in three-dimensional finite element models of arterial blood flow. Ann Biomed Eng. 2010;38:2314–30.
50. Sun D, Huang A, Koller A, et al. Flow-dependent dilation and myogenic constriction interact to establish the resistance of skeletal muscle arterioles. Microcirculation. 1995;2:289–95.
51. Murrant CL, Sarelius IH. Local and remote arteriolar dilations initiated by skeletal muscle contraction. Am J Physiol Heart Circ Physiol. 2000;279:H2285–94.
52. Timothy W. Theoretical models for regulation of blood flow. Microcirculation. 2008;15:765–75.
53. Wolff CB. Normal cardiac output, oxygen delivery and oxygen extraction. Adv Exp Med Biol. 2007;599:169–82.
54. Sagiv M, Sagiv M, Goldhammer E, et al. Adolescents' left ventricular response at peak Wingate Anaerobic Test. 2006;17:22–9.
55. Sagiv M, Ben-Sira D, Goldhammer E. Direct vs. indirect blood pressure measurement at peak anaerobic exercise. Int J Sports Med. 1999;20:275–8.
56. Sagiv M, Ben-Sira D, Goldhammer E, et al. Left ventricular contractility and function at peak aerobic and anaerobic exercises. Med Sci Sports Exerc. 2000;32:1197–201.
57. Nicosia RF, Villaschi S. Auto-regulation of angiogenesis by cells of the vessel wall. Int Rev Cytol. 1999;185:1–43.
58. Wilson LB, Andrew D, Craig AD. Activation of spinobulbar lamina I neurons by static muscle contraction. J Neurophysiol. 2002;87:1641–5.
59. Sandercock GR, Hardy-Shepherd D, Nunan D, et al. The relationships between self-assessed habitual physical activity and non-invasive measures of cardiac autonomic modulation in young healthy volunteers. J Sports Sci. 2008;26:1171–7.
60. Camarda SR, Tebexreni AS, Pafaro CN, et al. Comparison of maximal heart rate using the prediction equations proposed by Karvonen and Tanaka. Arq Bras Cardiol. 2008;91:311–4.
61. Widmaier EP, Hershel R, Strang KT. Vander's Human physiology: the mechanisms of body function. 14th ed. New York, NY: McGraw-Hill Education; 2016.
62. West JM, Humphris DC, Stephenson DG. Differences in maximal activation properties of skinned short- and long-sarcomere muscle fibers from the claw of the freshwater crustacean Cherax destructor. J Muscle Res Cell Motility. 1992;13:668–84.
63. Katz M. Ernest Henry Starling, his predecessors, and the "Law of the Heart". Circulation. 2002;106:2986–92.
64. Margulies KB, Houser SR. Chapter 3 – Cellular basis for heart failure. In: Heart failure: a companion to Braunwald's heart disease. 2nd ed. ScienceDirect; 2011. pp. 32–47. https://doi.org/10.1016/B978-1-4160-5895-3.10003-8
65. Cassey D, Hunt E. Cardiovascular function in humans during exercise: role of the muscle pump. J Physiol. 2008;586:5045–6.
66. Warburton DE, Haykowsky MJ, Quinney HA, et al. Myocardial response to incremental exercise in endurance-trained athletes: influence of heart rate, contractility and the Frank-Starling effect. Exp Physiol. 2002;87:613–22.
67. Whipp BJ, Higgenbotham MB, Cobb FC. Estimating exercise stroke volume from asymptotic oxygen pulse in humans. J Appl Physiol. 1996;81:2674–9.
68. Bhambhani Y, Norris S, Bell G. Prediction of stroke volume from oxygen pulse measurements in untrained and trained men. Can J Appl Physiol. 1994;19:49–59.
69. Zhou B, Conlee RK, Jensen R, et al. Stroke volume does not plateau during graded exercise in elite male distance runners. Med Sci Sports Exerc. 2001;33:1849–54.
70. Moss RL, Fitzsimons DP. Frank-Starling relationship. Circulation Res. 2002;90:11–3.
71. Warburton DER, Gledhill N. Counterpoint: stroke volume does not decline during exercise at maximal effort in healthy individuals. J Physiol Appl. 2008;104:274–6.
72. Kanstrup IL, Marving J, Gagsboll N, et al. Left ventricle haemodynamics and vaso-active hormones during graded supine exercise in healthy male subjects. Eur J Appl Physiol. 1995;72:86–94.
73. Guyton AC, John E. Textbook of medical physiology. 11th ed. Philadelphia: Elsevier; 2006.

74. Wilmore JH, Costill DL. Physiology of sport and exercise. 3rd ed. Champaign, IL: Human Rate of Reaction; 2005.
75. Daniłowicz-Szymanowicz L, Figura-Chmielewska M, Raczak A, et al. The assessment of influence of long-term exercise training on autonomic nervous system activity in young athletes preparing for competitions. (Article in Polish). Pol Merkur Lekarski. 2011;30:19–25.
76. Razanskaite-Virbickiene D, Danyte E, Mockeviciene G, et al. Can coefficient of variation of time-domain analysis be valuable for detecting cardiovascular autonomic neuropathy in young patients with type 1 diabetes: a case control study. BMC Cardiovasc Disord. 2017;17(1):34. https://doi.org/10.1186/s12872-016-0467-0.
77. Laborde S, Mosley E, Thayer JF. Heart rate variability and cardiac vagal tone in psychophysiological research – recommendations for experiment planning, data analysis, and data reporting. Front Psychol. 2017;8:213. https://doi.org/10.3389/fpsyg.2017.00213.
78. Rowland TW, Garrard M, Marwood S, et al. Myocardial performance during progressive exercise in athletic adolescent males. Med Sci Sports Exerc. 2009;41:1721–8.
79. Obert P, Mandigouts S, Nottin S, et al. Cardiovascular responses to endurance training in children: effect of gender. Eur J Clin Invest. 2003;33:199–208.
80. Nottin S, Vinet A, Stecken F, et al. Central and peripheral cardiovascular adaptations during a maximal cycle exercise in boys and men. Med Sci Sports Exerc. 2002;34:456–63.
81. Tanaka H, Seals DR. Endurance exercise performance in Masters athletes: age-associated changes and underlying physiological. J Physiol. 2008;586:55–63.
82. Mairbaurl H, Weber RE. Oxygen transport by hemoglobin. Compr Physiol. 2012;2:1463–89.
83. Jensen FB. Red blood cell pH, the Bohr Effect, and other oxygenation linked phenomena in blood O2 and CO2 transport. Acta Physiol Scand. 2008;182:215–27.
84. DeLorey DS, Paterson DH, Kowalchuk JM. Effects of ageing on muscle O2 utilization and muscle oxygenation during the transition to moderate-intensity exercise. Appl Physiol Nutr Metab. 2007;32:1251–62.
85. Taylor RW, Turnbull DM. Laboratory diagnosis of mitochondrial disease. In: Applegarth DA, Dimmick JE, Hall JG, editors. Organelle diseases clinical features, diagnosis, pathogenesis and management. New York: Chapman & Hall; 1997. p. 341–50.
86. McGuire DK, Levine BD, Williamson JW, et al. A 30-year follow-up of the Dallas Bedrest and Training Study: II. Effect of age on cardiovascular adaptation to exercise training. Circulation. 2001;104:1358–66.
87. Rossi P, Marzani B, Giardina S, et al. Human skeletal muscle aging and the oxidative system: cellular events. Curr Aging Sci. 2008;1:182–91.
88. Joseph AM, Hood DA. Relationships between exercise, mitochondrial biogenesis and type 2 diabetes. Med Sport Sci. 2014;60:48–61.
89. Metivier F, Marchais SJ, Guerin AP, et al. Pathophysiology of anaemia: focus on the heart and blood vessels. Nephrol Dial Transplant. 2000;15(Suppl 3):S14–8.
90. Hagen JL, Krause DJ, Baker DJ, et al. Skeletal muscle aging in F344BN F1-hybrid rats: I. Mitochondrial dysfunction contributes to the age-associated reduction in VO2max. Gerontol A Biol Sci Med Sci. 2004;59:1099–110.
91. Tower J. Mitochondrial maintenance failure in aging and role of sexual dimorphism. Arch Biochem Biophys. 2014. pii: S0003-9861(14)00375-0.
92. Figueiredo PA, Mota MP, Appell HJ, et al. The role of mitochondria in aging of skeletal muscle. Biogerontology. 2008;9:67–84.
93. Martínez-Redondo D, Marcuello A, Casajus JA, et al. Human mitochondrial haplogroup H: the highest VO2max consumer–is it a paradox? Mitochondrion. 2010;10:102–7.
94. Wilkerson DP, Rittweger J, Berger NJ, et al. Influence of recombinant human erythropoietin treatment on pulmonary O2 uptake kinetics during exercise in humans. J Physiol. 2005;568:639–52.
95. Proctor DN, Parker BA. Vasodilation and vascular control in contracting muscle of the aging human. Microcirculation. 2006;13:315–27.
96. Nishiyama SK, Wray DW, Richardson RS. Aging affects vascular structure and function in a limb-specific manner. J Appl Physiol. 2008;105:1661–70.

97. Brothers RM, Haslund ML, Wray DW, et al. Exercise-induced inhibition of angiotensin II vasoconstriction in human thigh muscle. J Physiol. 2006;577:727–37.
98. Wray DW, Fadel PJ, Smith ML, et al. Inhibition of alpha-adrenergic vasoconstriction in exercising human thigh muscles. J Physiol. 2004;555:545–63.
99. Barrett-O'Keefe Z, Ives SJ, Trinity JD, et al. Endothelin-A-mediated vasoconstriction during exercise with advancing age. J Gerontol A Biol Sci Med Sci. 2015;70:554–65.
100. Mortensen SP, Dawson EA, Yoshiga CC, et al. Limitations to systemic and locomotor limb muscle oxygen delivery and uptake during maximal exercise in humans. J Physiol. 2005;566:273–85.
101. Ogawa T, Spina RJ, Martin WH III, et al. Effects of aging, sex, and physical training on cardiovascular responses to exercise. Circulation. 1992;86:494–503.
102. Sampath S, Derbyshire JA, Ledesma-Carbayo MJ, et al. Imaging left ventricular tissue mechanics and hemodynamics during supine bicycle exercise using a combined tagging and phase-contrast MRI pulse sequence. Magn Reson Med. 2011;65:51–9.
103. Sagiv M, Hanson P, Besozzi M, et al. LV responses to upright isometric and deadlift in men with coronary artery disease. Am J Cardiol. 1985;55:1298–304.
104. Sagiv M, Sagiv M, Ben-Sira D. Weight lifting training and LV function in adolescent subjects. J Sports Med Fitn. 2007;47:329–34.
105. Utomi V, Oxborough D, Whyte GP, et al. Systematic review and meta-analysis of training mode, imaging modality and body size influences on the morphology and function of the male athlete's heart. Heart. 2013;99:1727–33.
106. McKinsey TA, Olson EN. Cardiac hypertrophy: sorting out the circuitry. Curr Opin Genet Dev. 1999;9:267–74.
107. Rodrigues JC, Rohan S, Dastidar AG, et al. The relationship between LV wall thickness, myocardial shortening, and ejection fraction in hypertensive heart disease: insights from cardiac magnetic resonance imaging. J Clin Hypertens (Greenwich). 2016;18:1119–27.
108. Goodman JM, Liu PP, Green HJ. LV adaptations following short-term endurance training. J Appl Physiol (1985). 2005;98:454–60.
109. Barbier J, Ville N, Kervio G, et al. Sports-specific features of athlete's heart and their relation to echocardiographic parameters. Herz. 2006;31:531–43.
110. Rossi MA, Carillo SV. Cardiac hypertrophy due to pressure and volume overload: distinctly different biological phenomena? Int J Cardiol. 1991;31:133–41.
111. Fernandes T, Soci UP, Oliveira EM. Eccentric and concentric cardiac hypertrophy induced by exercise training: microRNAs and molecular determinants. Braz J Med Biol Res. 2011;44:836–47.
112. Gong FF, Campbell DJ, Prior DL. Noninvasive cardiac imaging and the prediction of heart failure progression in preclinical stage A/B subjects. JACC Cardiovasc Imaging. 2017;10:1504–19.
113. Armstrong WF, Ryan T, Feigenbaum H. Feigenbaum's echocardiography. Lippincott Williams & Wilkins; 2010.
114. Byrd BF III, Abraham TP, Buxton DB, et al. A summary of the American Society of Echocardiography Foundation value-based healthcare: Summit 2014: the role of cardiovascular ultrasound in the new paradigm. Am Soc Echocardiogr. 2015;28:755–69.
115. Chan J, Jenkins C, Khafagi F, et al. What is the optimal clinical technique for measurement of left ventricular volume after myocardial infarction? A comparative study of 3-dimensional echocardiography, single photon emission computed tomography, and cardiac magnetic resonance imaging. J Am Soc Echocardiogr. 2006;9:192–201.
116. Moulton MJ, Secomb TW. A low-order model for left ventricle dynamics throughout the cardiac cycle. J Mathem Med Biol. 2013;30:45–63.
117. Tanaka H, Seals DR. Endurance exercise performance in Masters Athletes: age-associated changes and underlying physiological mechanisms. J Physiol. 2008;586:55–63.
118. Foster C, Gal RA, Port SC, et al. Left ventricular ejection fraction during incremental and steady state exercise. Med Sci Sports Exerc. 1995;27:1602–6.

119. Inuzuka R, Kuwata S, Kurishima C, et al. Influence of cardiac function and loading conditions on the myocardial performance index – theoretical analysis based on a mathematical model. Circulation. 2016;80:148–56.
120. Carrick-Ranson G, Doughty RN, Whalley GA, et al. The larger exercise stroke volume in endurance-trained men does not result from increased left ventricular early or late inflow or tissue velocities. Acta Physiol (Oxf). 2012;205:520–31.
121. Uemura K, Kawada T, Zheng C, et al. Myocardial performance index is sensitive to changes in cardiac contractility, but is also affected by vascular load condition. Conf Proc IEEE Eng Med Biol Soc. 2013:695–8. https://doi.org/10.1109/EMBC.2013.6609595.
122. Seydi V, Pancholy S, Johnson J, et al. Comparison of left ventricular performance in healthy young women and men during exercise. J Nucl Cardiol. 1996;3:415–21.
123. Cacciapuoti F, Marfella R, Paolisso G, et al. Is the aging heart similar to the diabetic heart? Evaluation of LV function of the aging heart with Tissue Doppler Imaging. J Aging Clin Exp Res. 2009;21:22–6.
124. Fujimoto N, Prasad A, Hastings JL, et al. Cardiovascular effects of 1 year of progressive and vigorous exercise training in previously sedentary individuals older than 65 years of age. Circulation. 2010;122:1797–805.
125. Carrick-Ranson G, Hastings JL, Bhella PS, et al. The effect of lifelong exercise dose on cardiovascular function during exercise. J Appl Physiol (1985). 2014;116:736–45.
126. Sagiv M. Exercise cardiopulmonary function in cardiac patients. London: Springer; 2012. p. 171–94.
127. Hambrecht R, Gielen S, Linke A, et al. Effects of exercise training on left ventricular function and peripheral resistance in patients with chronic heart failure: a randomized trial. Hambrecht R J Am Med Assoc. 2000;283:3095–101.
128. Saghiv M, Ben-Sira D, Sagiv M. Physical condition for the coronary artery disease patients. J Cardiol Cardiovasc Therapy. 4(2):555633. https://doi.org/10.19080/JOCCT.2017.04.555633.
129. Saghiv M, Sagiv M. Response of left ventricular volumes and ejection fraction during different modes of exercise in health and CAD patients. Int J Clin Cardiol Res. 2017;1:051–6.
130. Sagiv M, Hanson P, Goldhammer E, et al. Left ventricular and hemodynamic responses during upright isometric exercise in normal young and elderly men. Gerontology. 1988;34:165–70.
131. Sagiv M, Metrany R, Fisher N, et al. Comparison of hemodynamic and left ventricular responses to increased after-load in healthy males and females. Int J Sports Med. 1991;12:41–5.
132. Haykowsky MJ, Dressendorfer R, Taylor D, et al. Resistance training and cardiac hypertrophy: unravelling the training effect. J Sports Med. 2002;32:837–49.
133. Sagiv M, Ben-Sira D, Goldhammer E, et al. Left ventricular function during combined isometric and dynamic exercise-induced ST depression in coronary artery disease. Am J Cardiol. 2000;85:1359–61.
134. Kalam K, Otahal P, Marwick TH. Prognostic implications of global LV dysfunction: a systematic review and meta-analysis of global longitudinal strain and ejection fraction. Heart. 2014;100:1673–80.
135. Sagiv M, Amir O, Ben-Sira D, et al. Oxygen delivery of men and women at peak Wingate Test. Braz J Biomotricity. 2007;3:65–75.
136. Sagiv M, Ben-Sira D, Goldhammer E, et al. Left ventricular contractility and function at peak aerobic and anaerobic exercises. J Med Sci Sports Exerc. 2000;32:1197–201.
137. Sagiv M, Ben-Sira D, Sagiv M, et al. Left ventricular function at peak all-out anaerobic exercise in older men. Gerontology. 2005;51:122–5.
138. Subramanian R, Nayar S, Meyyappan C, et al. Effect of enhanced external counter pulsation treatment on aortic blood pressure, arterial stiffness and ejection fraction in patients with coronary artery disease. J Clin Diagn Res. 2016;10:OC30–4.
139. Saghiv M, Goldhammer E, Ben-Sira D, et al. Effects of all out-cycling bouts on left ventricular function in master cyclists. JJ Gerontol. 2017;3(1):025.

140. Ben-Sira D, Saghiv M. The effect of gender on left ventricular systolic function at Peak Wingate Anaerobic Test. Eur J Appl Physiol. 1977;75:549–53.
141. Scott JM, Esch BT, Haykowsky MJ, et al. Sex differences in left ventricular function and beta-receptor responsiveness following prolonged strenuous exercise. J Appl Physiol (1985). 2007;102:681–7.
142. Thijssen DH, Steendijk S, Hopman MT. Blood redistribution during exercise in subjects with spinal cord injury and controls. Med Sci Sports Exerc. 2009;41:1249–54.
143. Rizzo L, Thompson MW. Cardiovascular adjustments to heat stress during prolonged exercise. J Sports Med Phys Fitness. 2018;58:727–43.
144. Frisbee JC, Lewis MT, Wiseman RW. Skeletal muscle performance in metabolic disease: microvascular or mitochondrial limitation or both? Microcirculation. 2019;26(5):e12517. https://doi.org/10.1111/micc.12517.. Epub 2018 Dec 23
145. Malte H, Lykkeboe G. The Bohr/Haldane effect: a model-based uncovering of the full extent of its impact on O2 delivery to and CO2 removal from tissues. J Appl Physiol. 2018;125:916–22.
146. Boron WF, Boulpaep EL. Medical physiology: a cellular and molecular approach. Philadelphia, PA: Saunders/Elsevier; 2009.
147. Motohiro M, Yuasa F, Hattori T, et al. Cardiovascular adaptations to exercise training after uncomplicated acute myocardial infarction. Am J Phys Med Rehabil. 2005;84:684–91.
148. Goodman JM, Liu PP, Green HJ. Left ventricular adaptations following short-term endurance training. J Appl Physiol. 2005;98:454–60.
149. Capelli C, Antonutto G, Kenfack MA, et al. Factors determining the time course of VO 2(max) decay during bedrest: implications for VO 2(max) limitation. Eur J Appl Physiol. 2006;98:152–60.
150. Ekblom B, Astrand PO, Saltin B, et al. Effect of training on circulatory response to exercise. J Appl Physiol. 1968;24:518–22.
151. Gati S, Sharma S, Pennell D. The role of cardiovascular magnetic resonance imaging in the assessment of highly trained athletes. JACC Cardiovasc Imaging. 2018;11:247–59.
152. Paterick TE, Gordon T, Spiegel D. Echocardiography: profiling of the athlete's heart. J Am Soc Echocardiogr. 2014;27:940–8.
153. Vinereanu D, Florescu N, Sculthorpe N, et al. Left ventricular long-axis diastolic function is augmented in the hearts of endurance-trained compared with strength-trained athletes. Clin Sci. (Lond). 2002;103:249–57.
154. Sagiv M, Ben-Sira D, Sagiv A, et al. Left ventricular responses during prolonged treadmill walking with heavy load carriage. Med Sci Sports Exerc. 1994;26:285–8.
155. McArdle WD, Katch FI, Katch VL. Essentials of exercise physiology. 2nd ed. Philadelphia, PA: Lippincott Williams & Wilkins; 2000.
156. Sagiv M, Sagiv M, Ben-Sira D. Weight lifting training and left ventricular function in adolescent subjects. J Sports Med Phys Fitness. 2007;47:329–34.
157. Rowland T, Heffernan K, Jae SY, et al. Cardiovascular responses to static exercise in boys: insights from tissue Doppler imaging. Eur J Appl Physiol. 2006;97:637–42.
158. Sagiv M, Hanson P, Besozzi M, et al. Left ventricular responses to upright isometric handgrip and dead-lift in men with coronary artery disease. Am J Cardiol. 1985;55:1298–302.
159. Friedmann B, Frese F, Menold E, et al. Effects of acute moderate hypoxia on anaerobic capacity in endurance-trained runners. Eur J Appl Physiol. 2007;101:67–73.
160. James DV, Sandals LE, Draper SB, et al. Relationship between maximal oxygen uptake and oxygen uptake attained during treadmill middle-distance running. J Sports Sci. 2007;25:851–8.
161. Sagiv M, Sagiv M, Meckel Y, et al. Effects of different sprint cycling bouts on left ventricular function in top cyclists. J Sports Med Phys Fitness. 2008;48:360–5.
162. Sagiv M, Ben-Gal S, Ben-Sira D. Effects of gradient and load carried on human haemodynamic responses during treadmill walking. Eur J Appl Physiol. 2000;83:47–50.
163. Castello-Simoes V, Minatel V, Karsten M, et al. Circulatory and ventilatory power: characterization in patients with coronary artery disease. Arq Bras Cardiol. 2015;104:476–85.

164. Takehana K, Sugiura T, Nagahama Y, et al. Cardiovascular response to combined static-dynamic exercise of patients with myocardial infarction. Coron Artery Dis. 2000;11:35–40.
165. Sagiv M, Amir O, Goldhammer E, et al. Left ventricular contractility in response to upright isometric exercise in heart transplant recipients and healthy men. J Cardiopulm Rehabil Prev. 2008;28:17–23.
166. Sagiv M, Hanson P, Besozzi M. Left ventricular responses to upright isometric handgrip and deadlift in men with coronary artery disease. Am J Cardiol. 1985;55:1298–302.
167. do Prado DML, Rocco EA. The benefits of exercise training on aerobic capacity in patients with heart failure and preserved ejection fraction. Adv Exp Med Biol. 2017;1000:51–64.
168. Norrving B. Global atlas on cardiovascular disease prevention and control (PDF). World Health Organization in collaboration with the World Heart Federation and the World Stroke Organization. 2011:3–18.
169. Paffenbarger RS Jr, Hale WE, Brand RJ, et al. Work-energy level, personal characteristics, and fatal heart attack: a birth-cohort effect. Am J Epidemiol. 1977;105:200–13.
170. Aubourg P, Wanders R. Peroxisomal disorders. Handb Clin Neurol. 2013;113:1593–609.
171. Ward JPT. From physiological redox signaling to oxidant stress. Adv Exp Med Biol. 2017;967:335–42.
172. Imanaka T. Biogenesis and function of peroxisomes in human disease with a focus on the ABC transporter. Biol Pharm Bull. 2019;42:649–65.
173. Duncker DJ, Merkus D. Regulation of coronary blood flow. Effect of coronary artery stenosis. Arch Mal Coeur Vaiss. 2004;97:1244–50.
174. Shier D. Hole's Human anatomy and physiology. McGraw- Hill; 2002
175. Ramanathan T, Skinner H. Coronary blood flow. Contin Educ Anaesth Crit Care Pain. 2005;5:61–4.
176. Opie LH, Lopaschuk GD. Fuels: aerobic and anaerobic metabolism. In: Weinberg RW, Bersin J, Aversa F, editors. Heart physiology, from cell to circulation. Philadelphia, PA: Lippincott Williams & Wilkins; 2004. p. 306–54.
177. Tune JD, Gorman MW, Feigl EO. Matching coronary blood flow to myocardial oxygen consumption. J Appl Physiol. 2004;97:404–15.
178. Gute D, Fraga C, Laughlin MH, et al. Regional changes in capillary supply in skeletal muscle of high-intensity endurance-trained rats. J Appl Physiol. 1996;81:619–26.
179. Machado MV, Vieira AB, et al. Exercise training dose differentially alters muscle and heart capillary density and metabolic functions in an obese rat with metabolic syndrome. Exp Physiol. 2017;102:1716–28.
180. Laughlin MH, Korthuis RJ, Duncker DJ, et al. Regulation of blood flow to cardiac and skeletal muscle during exercise. In: Handbook of physiology. Exercise: regulation and integration of multiple systems. Bethesda, MD: Am Physiol Soc; 1996. p. 705–769.
181. Richards SH, Anderson L, Jenkinson CE, et al. Psychological interventions or coronary heart disease. Cochrane Database Syst Rev. 2017;4:CD002902. https://doi.org/10.1002/14651858.CD002902.pub4.
182. Bruning RS, Sturek M. Benefits of exercise training on coronary blood flow in coronary artery disease patients. Prog Cardiovasc Dis. 2015;57:443–53.
183. McPherson R, Tybjaerg-Hansen A. Genetics of coronary artery disease. Circ Res. 2016;118:564–78.
184. Khera AV, Kathiresan S. Genetics of coronary artery disease: discovery, biology and clinical translation. Nat Revs Genet. 2017;18:331–44.
185. Betts JG, Desaix P, Johnson E, et al. Anatomy and physiology. 1st ed., Kindle Edition; 2014.
186. Wong ND. Epidemiological studies of CHD and the evolution of preventive cardiology. Nat Rev Cardiol. 2014;11:276–89.
187. Ardehali A, Ports TA. Myocardial oxygen supply and demand. Chest. 1990;98:699–705.
188. Forjaz CLM, Matsudaira Y, Rodrigues FB, et al. Post-exercise changes in blood pressure, heart rate and rate pressure product at different exercise intensities in normotensive humans. Braz J Med Biol Res. 1998;31:1247–55.

189. Sagiv M, Goldhammer E, Ben-Sira D, et al. What maintains the metabolic cost at maximal exercise in heart transplant recipients and coronary artery disease patients? J Heart Lung Transplant. 2007;26:617–21.
190. Kronenberg MW, Konstam MA, Edens TR, et al. Factors influencing exercise performance in patients with left ventricular dysfunction. SOLVD Investigators. Studies of Left Ventricular Dysfunction. J Card Fail. 1998;4:159–67.
191. Colao A, Cuocolo A, Marzullo P, et al. Impact of patient's age and disease duration on cardiac performance in acromegaly: a radionuclide angiography study. J Clin Endocrinol Metab. 1999;84:1518–23.
192. Turakhia M, Tseng ZH. Sudden cardiac death: epidemiology, mechanisms, and therapy. Curr Probl Cardiol. 2007;32:501–46.
193. Irfan M, Asif IM, Harmon KG. Incidence and etiology of sudden cardiac death: new updates for athletic departments. Sports Health. 2017;9:268–79.
194. Varro A, Baczko I. Possible mechanisms of sudden cardiac death in top athletes: a basic cardiac electrophysiological point of view. Pflugers Arch. 2010;460:31–40.
195. Maron BJ. Hypertrophic cardiomyopathy: a systematic review. JAMA. 2002;287:1308–20.
196. Wisten A, Börjesson M, Krantz P, et al. Exercise related sudden cardiac death (SCD) in the young – pre-mortal characterization of a Swedish nationwide cohort, showing a decline in SCD among athletes. Resuscitation. 2019;144:99–105.
197. Schunkert H, Jackson B, Tang SS, et al. Distribution and functional significance of cardiac angiotensin converting enzyme in hypertrophied hearts. Circulation. 1993;87:1328–39.
198. Iwai N, Ohmichi N, Nakamura Y, et al. DD genotype of the angiotensin-converting enzyme gene is a risk factor for left. Circulation. 1994;90:2622–8.
199. Frey N, Luedde M, Katus HA. Mechanisms of disease: hypertrophic cardiomyopathy. Nat Rev Cardiol. 2011;25(9):91–10.
200. Wigle ED, Rakowski H, Kimball BP, et al. Hypertrophic cardiomyopathy. Clinical spectrum and treatment. Circulation. 1995;92:1680–92.
201. Maron BJ, Shirani J, Poliac LC, et al. Sudden death in young competitive athletes. Clinical, demographic, and pathological profiles. JAMA. 1996;276:199–204.
202. Sasson Z, Rakowski H, Wigle ED. Hypertrophic cardiomyopathy. Cardiol Clin. 1988;6:233–88.
203. Goldhammer E, Sagiv M, Abinader G. Asymptomatic hypertrophic apical caediomyopathy in a world-class athlete. J Cardiopulm Rehabil. 1987;7:303–6.
204. van der Werf C, van Langen IM, Wilde AA. Sudden death in the young: what do we know about it and how to prevent? Circ Arrhythmia Electrophysiol. 2010;3:96–104.
205. Hastings JL, Levine BD. Syncope in the athletic patient. Prog Cardiovasc Dis. 2012;54:438–44.
206. Edwards CP, Yavari A, Sheppard MN, et al. Anomalous coronary origin: the challenge in preventing exercise-related sudden cardiac death. Br J Sports Med. 2010;44:895–7.
207. Basso C, Maron BJ, Corrado D, et al. Clinical profile of congenital coronary artery anomalies with origin from the wrong aortic sinus leading to sudden death in young competitive athletes. J Am Coll Cardiol. 2000;35:1493–501.
208. Eckardt KU. Anemia in end-stage renal disease: pathophysiological considerations. Nephrol Dial Transpl. 2001;16(Suppl 7):2–8.
209. Koskolou MD, Roach RC, Calbet JA, et al. Cardiovascular responses to dynamic exercise with acute anemia in humans. Am J Physiol. 1997;273:H1787–93.
210. Mancini DM, Kunavarapu C. Effect of erythropoietin on exercise capacity in anemic patients with advanced heart failure. Kidney Int. 2003;64:S48–52.
211. Anderson L, Thompson DR, Oldridge N, et al. Exercise-based cardiac rehabilitation for coronary heart disease. Cochrane Database Syst Rev. 2016;2016(1):CD001800. https://doi.org/10.1002/14651858.CD001800.pub3.
212. Mampuya WM. Cardiac rehabilitation past, present and future: an overview. Cardiovasc Diagn Ther. 2012;2:38–49.

213. Saeidi A, Mostafavi S, Heidari H, et al. Effects of a comprehensive cardiac rehabilitation program on quality of life in patients with coronary artery disease. ARYA Atheroscler. 2013;9:179–85.
214. Williams MA, Haskell WL, Ades PA, et al. Resistance exercise in individuals with and without cardiovascular disease: 2007 update. A scientific statement from the American Heart Association Council on Clinical Cardiology and Council on Nutrition, Physical Activity, and Metabolism. Circulation. 2007;116:572–84.
215. Pollock ML, Franklin BA, Balady GJ, et al. Resistance exercise in individuals with and without cardiovascular disease. Benefits, rationale, safety, and prescription. An advisory from the Committee on Exercise, Rehabilitation, and Prevention, Council on Clinical Cardiology, American Heart Association. Circulation. 2000;101:828–33.
216. Noel M, Jobin J, Marcoux A, et al. Comparison of myocardial ischemia on the ergocycle versus the treadmill in patients with coronary heart disease. Am J Cardiol. 2010;105:633–9.
217. Hung MJ, Kao YC. Follow-up exercise electrocardiography can confirm the appropriateness of treatment for exercise-induced myocardial ischemia and life-threatening cardiac arrhythmias due to coronary vasospasm. Chin Med J (Engl). 2017;130:2003–4.
218. Sarma S, Levine BD. Beyond the Bruce protocol. Cardiol Clin. 2016;34:603–8.
219. Colquhuon D, Freedman B, Cross D, et al. Clinical exercise stress testing in adults. Heart Lung Circ. 2015;24:831–7.
220. Keles ME, Samadi A, Isikhan SY, et al. The effect of effort test on the levels of ischemia modified albumin, 7-ketocholesterol and cholestan-3β, 5α, 6β -triol and their role in the diagnosis of coronary artery disease. J Med Biochem. 2019;38:249–55.
221. Beltz NM, Gibson AL, Janot JM, et al. Graded exercise testing protocols for the determination of VO2max: historical perspectives, progress, and future considerations. J Sports Med. 2016:3968393, 12 pages. https://doi.org/10.1155/2016/3968393.
222. Lukaski H, Bolonchuk WW, Klevay LM. Comparison of metabolic responses and oxygen cost during maximal exercise using three treadmill protocols. J Sports Med Phys Fitness. 1989;9:223–9.
223. Borg GA. Psychophysical bases of perceived exertion. Med Sci Sports Exerc. 1982;14:377–81.

Chapter 7
Aging

Aging causes structural and functional decrements in skeletal muscles (sarcopenia), bone mineral density, cardiovascular, and VO_2. Overall cardiovascular function at rest in most healthy elderly individuals is adequate to meet the body's need for pressure and flow. The resting HR is unchanged. Heart size is essentially not different in younger vs. older adults, but heart wall thickness increases modestly, due largely to an increase in myocyte size [1]. Atherosclerosis and arteriosclerosis progressively decrease the tissue oxygen supply, cell constituents to change with aging, as a consequence, most biological functions show a progressive, age-related deterioration [2]. Progress of many cardiovascular diseases as atherosclerosis, endothelial dysfunction, and hypertension with aging suggests that aging should be seen as a risk factor. This is causing an imbalance between vasodilator–vasoconstriction substances ratio formed by the endothelium resulting in a significant reduction in nitric oxide production. Nitric oxide decrement causes some anomalies in blood artery function along with an increase in oxidative stress molecules' production, which, in turn, increases the production of ROS and nitrogen species. Secreted klotho regulates nitric oxide production which, in turn, benefits endothelial function: membrane klotho acts through (FGF)-23, while secreted klotho regulates nitric oxide produced in the endothelium. Exercise inactivity accelerates aging and its consequences; it is suggested as a major reason for increased disease and mortality. In recent years, however, there are enough studies about the effect of aerobic exercise on blood circulating s-Klotho. Recently, it has been demonstrated that the response of circulating s-Klotho depends on aerobic fitness level: values of s-Klotho were significantly higher in trained individuals compared to untrained once, suggesting that aerobic exercise training is a suitable model for mechanistically probing the role of physical activity on s-Klotho expression. In conclusion, the present chapter suggests that aerobic exercise delays the aging process by increasing Klotho gene expression, which, in turn, reduces ROS damages to the cell. The relationships between the circulating s-Klotho levels, a potential anti-aging factor, and aerobic exercise are beneficial to physically active aged individuals. Endothelial

M. S. Saghiv, M. S. Sagiv, *Basic Exercise Physiology*,
https://doi.org/10.1007/978-3-030-48806-2_7

dysfunction improvement depends on increased circulating s-Klotho levels through aerobic exercise. The purpose of this review is to discuss factors such as oxidative stress and pro-inflammatory cytokines involved in the progress of aging.

Aging Considerations

Physiological changes occur with aging in all organ systems. In response to exercise stress, the age-related reduction in physiological reserves causes a loss of homeostatic balance, the primary aging process, which has first a genetic component, occurs in the absence of disease and independent of lifestyle [3]. The Q decreases, BP increases, and arteriosclerosis develops. The lungs show impaired gas exchange, a decrease in VC and slower expiratory flow rates. Progressive elevation of blood glucose occurs with age on a multifactorial basis and osteoporosis is frequently seen due to a linear decline in bone mass after the fourth decade. Lean body mass declines with age and this is primarily due to loss of muscle mass (sarcopenia), which significantly reduces elderly movement.

The aging progression happens at dissimilar rates between different body's organs. As of today, the knowledge of how the aging process is controlled at the molecular and cellular levels is not fully understood. Tissues' deterioration that occurs during the aging process, which is not reversible except for the long-lasting aerobic exercise, intervention that improves elderly's functional capacity, including increased physiological beneficial effects at the mitochondrial level.

Aging is a usual life course that includes all physiological systems, seen as an advanced biological degeneration and decrease in physical capacities that raises the possibility of disease. Aging is a gradual decline of various organ functions of the body, deteriorating phenomenon defined by relations between the genetic expressions and external factors [4]. It is a complex process involving many variables: genetics, lifestyle factors, and chronic diseases that interact with one another, greatly influencing the manner in which the individual ages. Ecological epigenetics defines how natural factors affect cellular epigenetics and, thus, human performance [5]. Aging is a multifaceted process that not only involves the natural biological processes of aging, but also the increased risk for different diseases such as coronary heart disease, diabetes, and cancer [6]. Overall cardiovascular function at rest in most healthy elderly individuals is adequate to meet the body's need for pressure and flow. The resting HR is unchanged. Heart size is essentially not different in younger vs. older adults, but heart wall thickness increases modestly, due largely to an increase in myocyte size [1]. Atherosclerosis and arteriosclerosis progressively decrease the tissue oxygen supply, cell constituents to change with aging, as a consequence, most biological functions show a progressive, age-related deterioration [2]. Aging is a complex process involving many variables: genetics, lifestyle factors, and chronic diseases that interact with one another, greatly influencing the manner in which the individual ages. Participation in regular physical activity both aerobic and resistive exercises elicits a number of favorable responses that contribute to healthy aging. The primary aging course, itself genetically related, occurs both independently of lifestyle and in the absence of disease [1]. Aging causes structural changes and functional decrements in skeletal muscles (sarcopenia), bone mineral density, cardiovascular, and the ability for aged skeletal muscle to respond to physiological loads such as muscle loading or acute injury. Indeed, older adults often exhibit an age-related reduction in the number and size of muscle fibers known as sarcopenia [7]. Much has been learned recently regarding the adaptability of various biological systems, as well as the ways that regular exercise can influence them. The functional capacity of the cardiopulmonary system, as assessed by VO_{2max}, declines with advancing age in trained and untrained individuals. Accordingly, at maximal effort, decreases in maximal Q, $(a\text{-}v)O_2$, and VO_{2max} are noted [8]. These age modifications are the balance between cellular impairment due to metabolic measures happening inside the cell and offsetting molecular responses that can restore the damage. It is widely accepted that aging is an important risk factor for the development of hypertension and atherosclerosis with its vascular complications. The underlying mechanism whereby aging elevates the probability of disease onset remains obscure; however, it is likely closely related to underlying mechanisms of atherogenesis, the most accepted being the oxidation hypothesis [9]. In brief, an online vital phase in the progress of atherosclerosis is oxidative alteration of low-density lipoprotein. The oxidation of low-density lipoprotein is a free radical-driven lipid peroxidation process and the aldehyde products of lipid hydroperoxide breakdown are responsible for the modification of the low-density lipoprotein apoprotein [10]. Successful aging is a function of both genetic and environmental factors [11]. The primary aging process, which is genetically associated, occurs both independently of lifestyle and in the absence of disease [8]. Aging-related changes occur mainly in the cardiopulmonary, arteries, and skeletal muscles, bringing about a reduction in physical performance [12]. Two main factors influence the physical performance with aging: (a) the decrease in cardiopulmonary function and (b) a progressive decline in muscle mass and strength that occurs with aging [13]. Consequently, the VO_{2max} and power ability decrease, irrespective of lifestyle because of genetic factors, thereby severely limiting body function, quality of life, and longevity [14]. However, structured endurance training after 40 years of age is powerful enough to induce beneficial cardiovascular adaptations in later life [15].

Many cross-sectional studies have demonstrated a significant relationship between age and aortic stiffness, although the age-related changes observed in peripheral arteries appear to be less marked [16]. Age is the main clinical determinant of large artery stiffness, seen in elderly persons with isolated systolic hypertension characterized by fissuring and fracturing of the elastin protein, collagen proliferation, and calcium deposition, frequently associated with a widened and torturous aorta [17].

As large arteries dilate, wall tension and pulsatile stresses increase and exacerbate artery wall degeneration, thus initiating feedback whereby increased hypertension leads to further degeneration. There are several possible explanations for the influence of aging on the loss of elasticity of central arteries. The most likely explanation appears to be age-associated structural changes in the arterial wall. Aging is associated with a decrease in elastin and an increase in collagen and connective tissues in the arterial wall [18].

The aorta stiffens during senescence is indicated by an increase in arterial pulse wave velocity [19]. Pulse pressure rises markedly after the fifth decade due to arterial stiffening with age, which results in a progressive rise in SBP and fall in DBP as the elastic capacity of the aorta diminishes.

Maximal work capacity is decreased regardless of lifestyle because of genetic factors. Such consequences contribute to the geriatric syndrome of frailty, thereby severely limiting the function, quality of life, and longevity [20]. Muscle mass decreased significantly in aged individuals, after the sixth or seventh decades of life [21]. This reduction in muscle mass is a major factor in the decline of VO_{2max}, indicating the cardiopulmonary fitness of the individual [22], and therefore, bringing about a reduction in physical work capacity [12]. Such myocardial and peripheral functional changes include a decline in the maximal HR, SV, and left ventricular contractility, and an increase in TPR. Consequently, oxygen delivery to the working muscle decreases. The human population aged from 0 to 91 years screened previously by ELISA revealed that the level of s-Klotho was related to human physiological declines with aging.

Aging of the Arteries

Changes in the blood vessels normally occur with aging, yet several other changes if not treated may lead to heart complications. With age, baroreceptors' sensitive level decreases, which may partially explain orthostatic hypotension in the elderly, occurring when an individual goes from lying or sitting to standing. This may reduce brain blood flow, consequently, causing dizziness. In addition, capillary walls thickness increases slightly, lessening oxygen diffusion capacity. The aorta turns out to be thicker and stiffer, due to changes in the aorta connective tissue, which, in turn, increases BP that may result in myocardial hypertrophy, and this occurs in the other arteries as well.

Blood volume decreases slightly with age due to the reduction in body water volume. In addition, red blood cells' production rate is reduced, that sometimes ends up in anemia. Infection resistance decreases due to reduction in neutrophils (immunity), in spite that a great portion of the white blood cells remains unchanged with age.

Important changes occur in the cardiovascular system with advancing age, even in apparently healthy individuals. Thickening and stiffening of the large arteries develop due to collagen and calcium deposition and loss of elastic fibers in the medial layer. These arterial changes cause systolic blood pressure to rise with age, while diastolic blood pressure generally declines after the sixth decade [23]. However, as large arteries dilate, wall tension and pulsatile stresses increase and exacerbate artery wall degeneration, thus initiating feedback whereby increased hypertension leads to further degeneration [24]. Many cross-sectional studies have demonstrated a significant relationship between age and aortic stiffness, although the age-related changes observed in peripheral arteries appear to be less marked [16]. Age is the main clinical determinant of large artery stiffness, seen in elderly persons with isolated systolic hypertension characterized by fissuring and fracturing of the elastin protein, collagen proliferation, and calcium deposition, frequently associated with a widened and torturous aorta [24].

In response to a stress, the age-related reduction in physiological reserves causes a loss of regulatory or homeostatic balance. This, combined with another consequence of age-related changes, namely an increased perception of effort associated with sub-maximal work, sets, especially to a patient unbeknown to the above, a vicious cycle leading to decreased exercise capacity, resulting in an elevated perception of effort, subsequently causing avoidance of activity, and finally feeding back to an exacerbation of the age-related declines secondary to disuse.

Aging-related, environmental, and genetic factors are responsible for structural and functional changes of the arterial wall, leading to decreased elasticity and increased stiffness [25]. Aging is also associated with structural and functional changes of the vessel wall, which result in decreased vascular distensibility and elevated arterial stiffness [26]. There are several possible explanations for the influence of aging on the loss of elasticity of central arteries. The most likely explanation appears to be age-associated structural changes in the arterial wall. Aging is associated with a decrease in elastin and an increase in collagen and connective tissues in the arterial wall [27].

In patients with aortic stenosis, senescence profoundly influences the pattern of hypertrophic growth and diastolic function. Patients with pressure overload are characterized by more severe hypertrophy and interstitial fibrosis, as well as more severe impairment of relaxation, myocardial stiffness, and filling indices [28]. The aorta stiffens during senescence, as indicated by an increase in arterial pulse wave velocity [29]. Pulse pressure, measured as the difference between SBP and DBP, rises markedly after the fifth decade due to arterial stiffening with age, which results in a progressive rise in SBP and fall in DBP as the elastic capacity of the aorta diminishes [30]. Pulse pressure may be increased because of a larger forward pressure wave or an earlier or larger wave reflection. [31].

The Baltimore Longitudinal Study of Aging [29] found that aortic arterial-pulse wave velocity increased progressively with age in 50 healthy females (26–96 years) in whom only modest age-related increases in blood pressure were observed. This is noteworthy since elastic properties of arteries are not necessarily uniform [32], and aging has been reported to have different effects on the stiffness of peripheral arteries such as brachial and radial arteries and central arteries in men [33]. This may increase cardiovascular morbidity and mortality because of an elevation of systolic blood pressure, which raises left ventricular afterload, and because of a decrease in DBP, which alters coronary perfusion [34].

The loss of elasticity of the great arteries results in an increase of impedance at ejection. Thus, aging is associated with MABP. Aging-related, environmental, and genetic factors are responsible for structural and functional changes of the arterial wall, leading to decreased elasticity and increased stiffness [35]. There are several possible explanations for the influence of aging on the loss of elasticity of central arteries. The most likely explanation appears to be age-associated structural changes in the arterial wall.

Many cross-sectional studies have demonstrated a significant relationship between age and aortic stiffness, although the age-related changes observed in peripheral arteries appear to be less marked [16]. Aging is also associated with structural and functional changes of the vessel wall, which result in decreased vascular distensibility and elevated arterial stiffness [26]. Age is the main clinical determinant of large artery changes, seen in elderly persons with isolated systolic hypertension characterized by fissuring and fracturing of the elastin protein, collagen proliferation, and calcium deposition, frequently associated with a widened and torturous aorta [17]. As large arteries dilate, wall tension and pulsatile stresses increase and exacerbate artery wall degeneration, thus initiating feedback whereby increased hypertension leads to further degeneration [24].

Pulse pressure rises markedly after the fifth decade of life due to arterial stiffening with age, which results in a progressive rise in SBP and fall in DBP as the elastic capacity of the aorta diminishes [30]. Pulse pressure may be increased because of a larger forward pressure wave or an earlier or larger wave reflection [31]. In addition, arterial-pulse wave velocity is increasing progressively with age in healthy females aged 26–96 years [29].

At systole, time pressure rises in the whole cardiovascular system, so that the left ventricle is subjected to an increase of parietal tension to which it adapts itself by hypertrophy which normalizes this tension. The aorta stiffens during senescence, as indicated by an increase in arterial pulse wave velocity [29]. Arterial compliance is an important therapeutic target in older individuals in whom stiffening of the proximal aortic circulation is thought to underlie systolic hypertension and increased cardiac work. This indicates a positive association between systemic arterial compliance and fitness level in healthy older people and an inverse association between systemic arterial compliance and systolic blood pressure. These findings are consistent with either acquisition of a more compliant circulation and lower blood pressure due to enhanced physical activity or that a more compliant arterial circulation and lower blood pressure permit greater athletic performance [36]. For healthy subjects between the ages of 25 and 65 years, there is an interactive effect between age and gender and an independent effect of physical training on peripheral vascular function [37]. It was suggested that interventions to improve aerobic capacity alleviates the stiffening of the arterial tree that accompanies normative aging [29].

The Renin-Angiotensin System

Clinical and experimental studies have demonstrated a major role of the rennin-angiotensin system (RAS) in the functional and structural changes of the large arteries in hypertension. The RAS is important for regulating blood pressure and extracellular fluid. The concept of the RAS has recently evolved from a classical systemic endocrine system to an appreciation of local RASs functioning in a procaine manner, including in the vascular wall [38]. Angiotensin-converting enzyme (ACE) is a key component in the RAS system, generating the vasoconstrictor angiotensin II, and degrading vasodilator Kinins [39]. ACE is widely expressed in human tissues, including skeletal muscle, and may play a metabolic role during exercise. Angiotensin II, the predominant biological product of RAS, has been known for its effects on metabolism and is a recognized growth factor necessary for the hypertrophy of skeletal muscle in response to mechanical load [40]. Most of its known physiological and pathophysiological activities are mediated through the angiotensin II type 1 receptor (AT1R), which is the dominant receptor in the cardiovascular system [41]. Higher levels of ACE have been observed in human subjects with increased thickness of the carotid wall. Clinical and experimental pharmacological studies have shown that ACE inhibitors can prevent and/or beneficially affect hypertension-induced structural and functional alterations of the arterial wall independent of BP changes [42].

Growing evidence supports a role for the involvement of genetic factors in the development of hypertension and aortic stiffness. Recent genetic studies focused on the RAS and on the identification of RAS candidate genes. These studies may help us to understand the mechanisms underlying the involvement of RAS in arterial regulation.

A functional polymorphism of the human ACE gene has been identified in which the absence (deletion—D allele), rather than the presence (insertion—I allele), of a 287-bp Alu repeat element in intron 16 is associated with higher enzyme activity in both serum and tissue [43], resulting in greater production of angiotensin II and aldosterone and a decreased half-life of bradykinin [44]. The most extensively studied polymorphism in the AT1R gene is the A1166C variant. This polymorphism has been linked to enhanced physiological responses of angiotensin II resulting in increased vasoconstrictor activity [45].

Because RAS activity plays a major role in the regulation of vascular tone, the ACE ID genotype associated with ACE activity could be a candidate gene for large-artery stiffness. [46]. It was suggested that in hypertensive but not normotensive subjects, the AT1R and ACE genotypes are involved in the regulation of aortic rigidity. ACE genotype has been shown to affect exercise and glucose load responses. AT1R genotype appears to predispose to favorable anthropometric and metabolic traits, relative to cardiovascular risk and has previously been associated with the development of hypertension and coronary disease.

Aging Work Capacity

Aging reduces work capacity and performance during exercise, due to changes in the myocardium, lung, and skeletal muscles. The aging influence on work capacity and physical performance during exercise has been widely investigated. It has been found that there are aging changes in the myocardium, lung, and skeletal muscles which reduce exercise performance with advancing age [47]. Myocardial functional changes with aging in a sedentary adult include a decline in maximal HR, SV, and left ventricular contractility, and an increase in TPR. The changes in the structure and function of the cardiovascular system result in VO_{2max} decline, which is the best single indicator of physical working capacity. Whether maximal Q actually decreases with primary aging depends on many interactive factors. First, the primary aging process, which has a genetic component, occurs in the absence of disease and independent of lifestyle [48]. If oxygen transport is indeed limited by maximal Q, this may be calculated as the product of maximal HR, $(a\text{-}v)O_2$, and maximal SV. The second possible cause that may reduce VO_{2max} with aging is the reduced $(a\text{-}v)O_2$ at maximal effort [49]. The elderly are not generally anemic and the red blood cell content is usually well maintained. Some studies have shown a correlation between VO_{2max} and muscle mass [50]. Mitochondrial density was found to be lower in skeletal muscle of older individuals, which could further diminish the capacity for endurance work. Thus, it may be that the aging-related atrophy of skeletal muscle plays some role in the aging-related decline in work endurance [51]. However, others [51] do not support a causal role for muscular atrophy in the decline in VO_{2max} during aging. It seems that the related changes in VO_{2max} are dependent on a number of factors including the onset of disease and level of physical activity. The lowest rates of decline in VO_{2max} with age are found in those who remain disease-free and continue to maintain high levels of physical activity. No significant decline in the VO_{2max} over a 10-year period was found in a group of competitive master's athletes who maintained their training intensity and continued to compete [50]. The annual rate of decay in VO_{2max} for the general population is estimated to be 0.4–0.5 $mLO_2 \cdot kg^{-1} \cdot min^{-1} \cdot year^{-1}$. The highest rates of decline in VO_{2max} are in those individuals that have reduced their levels of exercise training as they age.

Cardiopulmonary Function in the Elderly

The functional capacity of the cardiopulmonary system, as assessed by VO_{2max}, declines with advancing age in trained and untrained individuals (Fig. 7.1). This reduction results in a decrease in physiological functional capacity that would contribute to a loss of independence, increased incidence of disability, and reduced quality of life with age. The primary aging process, itself genetically associated, occurs both independently of lifestyle and in the absence of disease [8].

Accordingly, one may expect maximal Q to decrease with aging irrespective of lifestyle because of genetic factors.

Resting HR, EF, and peak ejection rate remain unchanged with aging [52]. The loss of elasticity of the great arteries, and in particular the aorta which becomes tortuous and wider, results in an increase of impedance at ejection. At systole, time pressure rises in the whole cardiovascular system, so that the left ventricle is subjected to an increase of parietal tension to which it adapts itself by hypertrophy which normalizes this tension. EF and ESV are thus preserved, and the systolic function at rest globally remains unmodified by age [53].

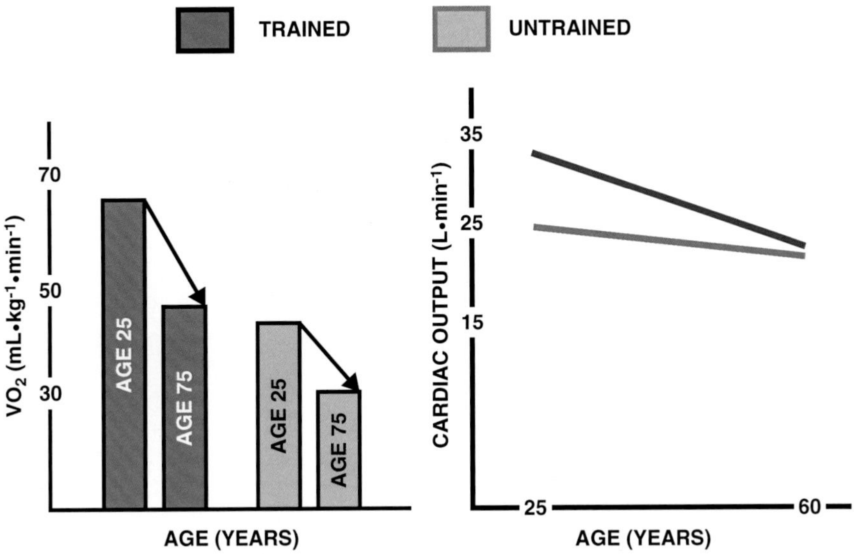

Fig. 7.1 In the elderly, the decrease in maximal oxygen uptake with age is due to the reduced cardiac output. The decline in trained individuals is significantly less compared to that in untrained individuals

Since metabolic demands during incremental exercise are increased over time, elderly may have different rate of reaction for energy supply between oxygen delivery and extraction [54]. The changes in the structure and function of the cardiovascular system result in VO_{2max} decline. Whether maximal Q actually decreases with primary aging depends on many interactive factors.

During short bursts of graded maximal aerobic exercise, reductions in peak work rate, HR, and left ventricular emptying occur, but higher BP and TPR occur in healthy older versus younger humans [55].

Reduced $(a\text{-}v)O_2$ at maximal effort is the second factor associated with aging which decreases VO_{2max} [56].

The evidence that VO_{2max} is limited by Q, the oxygen delivery capacity, and in some cases the pulmonary system, is nowadays undeniable. This statement pertains to healthy subjects performing whole-body, dynamic exercise [22].

Muscle mass influences myocardium function since both organs and lungs are connected by a closed-circuit artery. Therefore, any reduction in muscle mass (sarcopenia) that reduces its function, such as a decline in $(a\text{-}v)O_2$ ability during dynamic endurance exercise due to sarcopenia, will, in turn, increase HR in order to have the Q needed to deliver more oxygen to the working muscles. The increase in HR is reflected by the Fick equation. Thus, physical work capacity declines with advancing age, but the extent to which this can be attributed to a decrement in cardiac reserve is not certain.

Cardiopulmonary changes in the structure of the cardiopulmonary system appear to be influenced by age, associated with rising risk of cardiovascular disease and decreased functional capacity. Some changes appear to occur independently of gender, disease, and level of physical activity. Other structural changes such as an increase in left ventricular mass appear to be dependent on both gender and the level of physical activity. In the cardiovascular system, morphological and physiological changes have been identified in the elderly. The morphological changes typically do not produce clinical signs of cardiac dysfunction during life. In the elderly, there is a modest increase in left ventricular thickness, probably resulting from the observed increase in BP. Except for increased tortuosity, cross-sectional area, and degree of atherosclerosis, there are no clear age-associated changes in coronary arteries. However, studies in this area are limited. Aging is associated with a shift in the mechanism by which Q is maintained during submaximal exercise. In spite of the elderly having lower early diastolic filling rate during submaximal exercise, EDV is maintained and even increased as a function of age. As a result, older subjects appear to have a greater reliance on the Frank-Starling mechanism for the maintenance of adequate Q, during submaximal exercise. The decrease of maximal HR associated with advanced age is similar for the sedentary person, the average active person, and the top athlete. In humans, maximal HR peaks at around 10 years of age, and decreases by approximately 1 beat·min^{-1}·year^{-1}. This suggests that maximal HR does not adapt to chronic exercise [57]. HR dynamics, which depend on sympathetic nervous activity, are more sensitive to exercise stress, than to age, nonetheless, VO_{2peak} as estimated by the age-associated decline in maximum HR is unrelated to physical training status [58]. It is estimated that mean fall in Q is 0.23–0.35 L·min^{-1} per decade. Normalization of the Q for differences in body surface area does not eliminate the effects of aging. However, several of the changes noted with aging are related to disuse and normalized with increased activity [52]. Peak endurance performance is maintained until approximately 35 years of age, followed by modest decreases until 50–60 years of age, with progressively steeper declines thereafter.

Maximal HR decreases mainly because of a decreased responsiveness to circulating catecholamines. The classical equation peak HR = (220 − age in years) suggests a maximum of about 155 beats·min^{-1} at age 65 years. In a well-trained 65-year-old, maximal HR can be attained at 170 beats·min^{-1} and sometimes even or more [59]. The mechanism(s) underlying the decreased maximal HR is unknown. The changes may be attributed to the heart itself rather than to neural input. An example for the input ability of the nervous system in the elderly is the manifestation of ventricular arrhythmias. Ventricular tachycardia at a rate of 200 bpm was observed in elderly individuals who could not attain this HR by physiological means [60]. Moreover, variations in the number of pacemaker cells and volume of the sinoatrial-node result from aging [61]. Other investigations suggested that there is an age-related decline in maximal HR which could be related with decreased sympathetic drive [62], due to diminished response to adrenergic regulation of the heart and circulation [63]. In addition, the aging process reduces numerous hormonal control mechanisms which may result in early diabetes, may be resulting from obesity and physical inactivity.

Although older subjects rely on changes in SV, by dilatation of the left ventricle during intermediate and maximal exercise, to partially maintain Q, and to compensate for the reduction in HR [64], maximal SV is decreased due to reduced preload (circulating blood volume associated with physical inactivity) and increase in afterload. Thus, reducing the perfusion of the working muscles via many regulatory mechanisms. Age appears to be associated with an increase in the TPR independent of change in maximal limb blood flow. However, studies suggested that age is associated with a decrease in the maximal conductance of the vascular bed of the calf. Endurance training results in an increase in the maximal conductance independent of age. The mechanism for the change in maximal limb conduction with age is not well understood. Therefore, at peak exercise, older individuals tend to increase EDV and ESV compared with younger subjects [65]. The changes in left ventricular volumes during exercise may be due to reduced contractility, diastolic and systolic left ventricular functions, and loading conditions. The failure to augment contractility in the older subjects may be the cause of reduced diastolic and systolic functions. Contractility appears to be diminished with age. Compared to the young, older subjects have a greater ESV and a lower EF during submaximal exercise. The mechanisms for the decrease in contractility with age are unclear; however, increased afterload and/or a decreased responsiveness to adrenergic stimulation has been suggested. Maximal isometric contractile properties of left ventricular muscle are altered in old humans [60]. The contraction duration is prolonged and may result from a decrease in the rate of calcium transport by cardiac sarcoplasmic reticulum [66]. Similarly, the isotonic contractile properties are altered with aging [61], leading to change in two variables of isotonic contraction: velocity of shortening and relaxation time. In rats, these changes in isometrically and isotopically of the myocardial muscle contractions reduce elder myocardial peak filling. Sagiv et al. reported that diastolic left ventricular filling has been shown to be affected not only by ventricular property but also by the aging process [67]. Therefore, aging is associated with a shift of left ventricular filling from early phase to late phase, and thus, decreased passive/active ratio during diastole. Consequently, the heart relies more on atrial filling and in an increase of isovolumic relaxation time indicating diastolic dysfunction [68]. Left ventricular systolic performance is well-maintained with aging under resting conditions. However, during exercise, systolic function is reduced. This response may be mediated, at least in part by age-associated reductions in beta-adrenergic receptor sensitivity or density, reducing the inotropic response of the myocytes to a standard dose of catecholamine. Other manifestation of the decreased responsiveness of cardiovascular tissues to beta-adrenergic stimulation is the decreased ability of the atrium to dilate. The peripheral vasculature, however, does not show decreased vasodilatation by the vasodilatation stimuli of nitrates, indicating that the vascular changes are receptor-mediated rather than intrinsic. Thus, increasing BP and MABP progress with age, both at rest and during exercise. This rise in TPR may affect left ventricular loading conditions which, in turn, may mask or mimic changes in left ventricular volumes, diastolic, and systolic functions [66]. Numerous structural and functional changes in the lung, chest wall respiratory muscles, and vasculature occur with age [69]. However, in diverse populations and elderly, incremental exercise exposes the individuals to a relatively high degree of load, which may alter the left ventricular contractility and function [54].

This has the effect of placing a large load on the left ventricular, which might have significant effects on oxygen delivery to the working muscles.

Aging Effect on Pulmonary Function

Healthy aging causes reductions in lung elastic recoil, vital capacity, diffusion surface area, and chest wall compliance. Accordingly, in highly fit elderly individuals, significant expiratory flow limitation with an accompanying increase ventilatory work begins during sub-maximal exercise at VE values of the 70–80 $L \cdot min^{-1}$ range [70]. At rest and during exercise, coronary artery disease patients do not exhibit difficulties regarding the pulmonary capacity [71]. Aging is associated with a progressive deterioration in the structure and function of the pulmonary circulation. The older adult demonstrates several alterations in the ventilatory response to exercise relative to the young adult. These differences are primarily caused by the aging-induced loss of elastic recoil of the lung, which reduces expiratory airflow rates along the effort-independent portion of the maximal volitional flow-volume envelope. Following a 6-year period of aerobic training in active and highly trained older adults (mean age 67–73 years) suggests that habitual physical activity and high aerobic capacity modify neither the normal deterioration in resting lung function nor the increased levels of ventilatory work during exercise that occur with healthy aging over the sixth and seventh decades of life [72]. Despite these distinctions in the response to exercise of the aged adult, alveolar ventilation is adequate for CO_2 elimination even during maximal exercise, and arterial oxygen homeostasis is generally maintained [73]. Thus, elderly appear to be able to maintain alveolar ventilation at a level that allows maintenance of arterial blood gases within normal limits, even during heavy exercise. This ability to maintain adequate gas exchange likely occurs because age-related reductions in the maximal metabolic demand of exercise occur at a rate equal to or greater than the rate of deterioration in ventilatory reserve [74].

Mitochondrial Function with Aging

Mitochondrial density was found to be lower in skeletal muscle of older individuals, which could further diminish the capacity for endurance work [75]. Thus, it may be that the aging-related atrophy of skeletal muscle (Fig. 7.2) plays some role in the aging-related decline in aerobic capacity. The alterations in skeletal muscle structure and function after prolonged periods of physical inactivity are initiated by the chronic lack of mechanical stimulus of sufficient intensity, which is the result of a series of biochemical and metabolic interactions spanning from cellular to tissue/organ level [76]. Reduced activation of skeletal muscle alters the gene expression of myosin heavy chain isoforms to meet the functional demands of

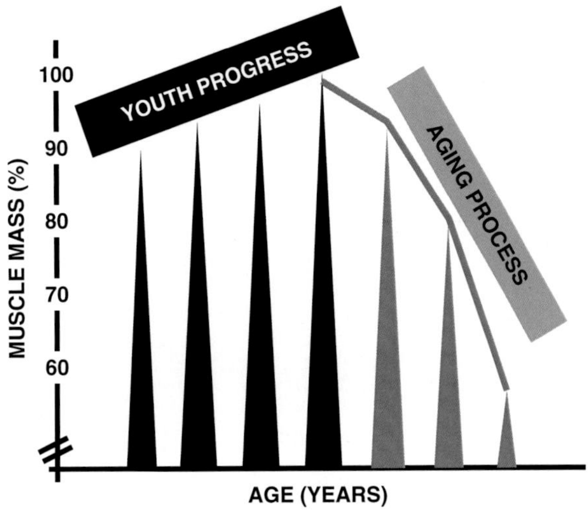

Fig. 7.2 For endurance performance, the key problem associated with aging is the gradual decline in muscle mass. The missing muscle makes it more difficult to perform a powerful physical activity. Muscles reach their maximum size during their 25th year of life, grow smaller by about 10% between the ages of 25 and 50, and then shrink by 45% over the next three decades

reduced mechanical load, which results in muscle atrophy and reduced capacity to process fatty acids.

In contrast, chronic loading results in the opposite pattern of adaptations [77].

A quarter of body muscle mass declines from the age of 50 to the age of 75 years; this is due to the impact of aging on fast and slow fiber atrophy [76], which translates into atrophy, mainly type II fibers and somehow a tendency to conserve type I fibers [78]. Consequently, muscle atrophy decreases muscle capacity to produce force by recruiting more motor units. This, in turn, increases the load on the cardiopulmonary system to provide more oxygen to the working muscles.

Age-Related Changes in ATP-Production

Aging goes with variability of alterations in muscle mass and strength decrements [79]. Aging may slowdown ATP synthesis rates from the metabolic pathways during maximal exercise, due to aging changes causing mitochondria function decline [80]. Thus, exercise work capacity in addition to bioenergetic alterations is reduced with age. The mechanisms by which mitochondrial function ability changes with age remains uncertain. Slight to moderate decreased oxidative capacity on the one hand, and increased oxidative damage to biomolecules due to the excessive production of ROS [81], which describes the best mitochondrial dysfunction. During the aging process, mitochondria are characterized by changes in oxidative stress, deterioration in mitochondrial DNA, a reduction in some enzyme activities, and alterations in mitochondrial respiration chain. ROS are considered to be unwanted and toxic by-products of the mitochondrial electron transport system. Although oxidative capacity is almost unchanged with age, peak glycolytic flux and overall ATP production from anaerobic glycolysis are lower in older men during a high-intensity contraction [82]. Mitochondrial DNA (mtDNA) has about 15-fold mutation rate, compared to nuclear DNA and declined repair machinery ability related to nuclear DNA [80]. Underlying causes of mitochondrial dysfunction and disruption in mitochondrial energetics that contribute to age-related phenotypes in aging are the result of both the buildup of mitochondrial DNA (mtDNA) alterations and the reduced mtDNA duplicate amount [83]. Therefore, in older individuals, the synthesis rate of muscle fibers and other mitochondrial proteins in skeletal muscle decays with advancing age, and thus, may modify muscle metabolic capacity [84].

Age Effect on Oxygen Uptake

VO_2 declines with aging and it is estimated that the rate of this decline is on average 1% per year or approximately 10% per decade after the age of 25 in sedentary subjects [85]. Total daily energy demand is defined mainly by resting metabolism that decreases with aging, around 10% compared to young adult, to the age of 65 years. One large cross-sectional study found the average decrease was 0.46 $mLO_2 \cdot kg^{-1} \cdot min^{-1} \cdot year^{-1}$ in men (1.2%) and 0.54 $mLO_2 \cdot kg^{-1} \cdot min^{-1}$ in women (1.7%) [86]. However, this deterioration might not be necessarily due to the aging process, since the trend to increase body weight constitutes by itself an explanation to the decrease in the relative VO_{2max} despite the maintenance of its absolute values. A reduction in the anaerobic threshold contributes to the decrease in endurance performance with aging although this may be secondary to decreases in VO_{2peak}.

Usually, the decline in age-related VO_2 can be accounted for by a reduction in maximal HR, maximal SV, i.e., Q and (a-v)O_2 [87]. Part of the age-related decline in VO_{2max} appears to be due to peripheral (a-v)O_2 ability rather than central circulatory factors, e.g., to a decrease in muscle mass during exercise, the ability to direct blood flow to muscles, and the ability of the muscle to utilize oxygen [88]. Previously it has been shown that a relationship exists between VO_2 and muscle mass [77].

Regarding the age influence on former elite athletes, it has been shown that VO_{2max} declines by almost 30% from ages 23 to 50 (from 70 to 50 $mLO_2 \cdot kg^{-1} \cdot min^{-1}$) when they stop training after their careers are over [89]. In some cases, the relative decline is greater than for the average population—as much as 15% per decade [90]. However, in comparison, master athletes who continue to keep training show a decrease of 5–6% per decade [90]. When they maintain the same relative intensity of training, a decrease of only 3.6% over 25 years has been reported and most of that was attributable to a small increase in body weight. It seems that training per se can slow down the age-induced decline rate of VO_2, becoming however less effective after the age of about 50 [91].

High level of VO_{2max} characterizes master athletes during their life span [92]. Long-lasting regular aerobic exercise training reduces slightly master athletes' VO_{2max}. However, a decline in muscle cell size and contractile performance is apparent; this may cause an aerobic performance decline [93].

In addition, due of being active for long time keeps body composition balance as at the youth period [94]. Greater fat mass is associated with a greater decline in lean body mass [95] With increasing age, body fat increases in untrained [95] and slightly in master athletes [96]. High-fat tissue does not allow the individual to achieve high levels of aerobic fitness [97]. Mechanisms underlying age-associated change in body composition such as an increase in body fat and a decrease in bone mass are not fully understood. A decrease in GH secretion and serum IGF-1 levels with aging may have some impact on these processes. However, due to their regular training, they show only slight changes in aerobic capacity and body composition across years [98]. In master athletes, regular aerobic exercise training preserves the age-related decrease in both VO_{2max} [94] and skeletal muscle mass [99]. At the age of 40 years, aerobic and resistance master athletes have declines in their performance levels, by almost 50% at the age of 80. Elderly who participated chronically in the weightlifting program increase more arm region bone mineral density [100]. Changes in body composition appeared to be less than those shown for the healthy sedentary population and were related to changes in training habits. The physically, success of the master athletes offers a strong sign that age-related atrophy can be significantly more gradually, by training. These physiological capacities of older athletes are reduced despite continued vigorous endurance exercise.

Aging Effect on the Balance Between Oxygen Delivery and Extraction

Untrained elderly compared to the trained elderly have a lesser response of the central factors, i.e., cardiopulmonary without a significant reduction in peripheral ability to extract oxygen at the muscle level. Differences arise between the trained and untrained in absolute VO_2 of the working muscles and peak power output at peak exercise test due to physical inactivity or physical activity. However, VO_2 in the untrained is compensated for the lower Q by a significant increase in $(a\text{-}v)O_2$ from rest to peak exercise as in the trained elderly.

This lack of differences between the two groups with regard to oxygen extraction is in agreement with that reported previously. It was found that in untrained elderly subjects, skeletal muscle mitochondrial capacity, tissue blood flow capacity, and oxygen exchange capacity appear to be well matched. It seems that intrinsic mitochondrial function and regulation are not altered significantly in the untrained elderly. The higher aerobic capacity in the trained elderly is related to increases in the abilities of cardiovascular factors and to the lesser extent to increases in muscle mitochondria concentration and capillarity. In addition, the different skeletal muscle fiber types and muscle remodeled, due to exercise training, may also explain in part the better aerobic capacities (VO_2) exhibited by the trained subjects.

Values for Q at peak aerobic exercise are low in the untrained compared to the trained at maximal aerobic exercise. The lower Q in the untrained elderly at peak exercise is related to the limited HR reserve, SV, left ventricular contractility, and to the inappropriate adjustment of the circulation [54]. The augmentation in maximal VO_2 in the trained elderly in response to exercise training is mainly due to the adaptive increases in Q and to the lesser extent to the $(a\text{-}v)O_2$ at maximal exercise [101]. The increase in maximal Q in the trained elderly is the result of a large increase in EDV and SV while ESV decreased [102]. These significant changes in left ventricular volumes at maximal exercise that are seen in the trained subjects are a result of an integrated mechanism that involves changes in contractility and afterload.

Exercise in Arterial Stiffness-Hypertension

The relationship between arterial stiffness and hypertension is more complex. Elastic artery stiffening, an age-related process, can be accelerated in the presence of hypertension. Hypertension may produce arterial stiffening by both functional and structural mechanisms. Therefore, the distending or MABP is an important confounder of measurements of arterial stiffness [16].

Physical activity has been recommended in the prevention and treatment of hypertension [103]. A large body of data demonstrates that changes toward a more physically active lifestyle positively affect blood pressure response in normotensive and hypertensive individuals. Meta-analysis of longitudinal aerobic training studies [104] in mild essential hypertensive subjects demonstrates an average reduction in resting SBP and DBP of 10.8 and 8.2 mmHg, respectively [105]. Therefore, participation in an exercise-training program may be viewed as a non-pharmacological approach for preventing and treating mild hypertension.

Previous studies have shown that a supervised exercise program leads to a decrease in BP in humans with essential arterial hypertension. Men in the high fitness group with resting systolic blood pressure above or equal to 140 mmHg had a lower death rate than their low-fitness, below 140 mmHg resting systolic-pressure counterparts. Nevertheless, even though the reduction in BP due to exercise training was detected in normotensive and hypertensive patients, the magnitude of reduction in SBP and DBP at rest and during submaximal exercise is greater in hypertensive subjects [106].

How does physical activity produce a decrease in BP in elderly and in borderline hypertensive patients? The mechanism(s) responsible for the reduction in arterial pressure has not been investigated, though a reduction in sympathetic nerve activity resultant from exercise training was suggested to lower arterial pressure [107]. However, the lack of change in muscle sympathetic nerve activity after training indicates that the lower arterial pressure is unrelated to a reduction in central sympathetic outflow. The failure of muscle sympathetic nerve activity to change at rest with isometric training is also typically observed with either forearm or leg dynamic exercise training [108]. Therefore, reductions in sympathetic outflow to skeletal muscle do not appear to be a prerequisite to lower arterial pressure in humans. However, it cannot be excluded that sympathetic outflow to other vascular beds (e.g., visceral regions) may be reduced and may contribute to the reduction in arterial pressure at rest [107].

Another possible mechanism for the reduction in arterial pressure is peripheral vascular adaptations. Because aerobic and isometric exercises elicit marked increases in muscle sympathetic nerve activity and norepinephrine release. Thus, vascular sensitivity to norepinephrine may be decreased with aerobic and isometric training.

Exercise Epigenetic in Aging

Epigenetic factors include DNA methylations, histone modifications, non-coding RNAs like lncRNA or microRNAs and can help to explain how cells with identical DNA can differentiate into different cell types with different phenotypes. Some interface of aging and energetic was identified: these included genes that accelerate aging or in contrast promote longevity in model organisms, DNA damage responses and telomeres, molecular mechanisms of life span extension by calorie restriction, and pharmacological interventions into aging [109]. Epigenetic mechanisms affected by physical exercise have also been seen to be involved in age-related processes. A major component of aging is a significant loss of DNA methylation over time [110]. Methyl deoxycytidine, which is a methylated cytosine on the 5′ carbon of cytosine, is involved in the process of cell differentiation and maintenance. Cell differentiation involves the methylation of different areas within the DNA of a cell, which can alter the transcription of genes. During cell differentiation, DNA methylation is important for establishing the identity and function of a cell because of its role in controlling gene expression. As one ages, the amount of DNA methylation slowly begins to decrease. A recent study [111] on genome epigenetic of children and older individuals suggested that the older individuals had significantly decreased overall DNA methylation and histone. These mechanisms include permanent structural changes to the organ caused by suboptimal levels of an important factor during a critical developmental period, changes in gene expression caused by epigenetic modifications (including DNA methylation, histone modification, non-coding RNAs like lncRNA or microRNAs) and permanent changes in cellular ageing. Studies have also looked at methyl deoxycytidine residues from tissues collected from rodents at various ages. These studies found that DNA methylation loss increased significantly as the rodent aged [110]. A major component of aging is a significant loss of DNA methylation over time [112], thus, aging rate is related to a reduction in DNA methylation activity [113]. This loss of DNA methylation appears to be slowed by physical exercise. Further studies have looked at the effects of aerobic exercise on DNA methylation and aging in humans, revealing that genome with DNA methylation in adult individuals who obtained 30 or more minutes of exercise a day had significantly more DNA methylation as compared to sedentary individuals [113, 114]. Such exercise has been shown to induce positive changes in DNA methylation within adipose tissue and regulate metabolism in both healthy and diseased individuals [115]. Increased DNA methylation of genes Hdac4 and Ncor2 has also shown to increase lipogenesis following exercise [114]. Consequently, aerobic exercise affects aging rate through slowing the pace of the DNA methylation loss over time. All these epigenetic alterations may have clinical relevance, thus, playing an important role in the prevention and confrontation of neurophysiological disorders, metabolic syndrome, cardiovascular diseases, and cancer. Acute exercise is associated with DNA hypomethylation of the entire genome in skeletal muscle cells of sedentary individuals and high-intensity exercise tends to cause reduction in promoter methylation of certain genes. In human skeletal muscle, the expression of many genes tends to increase or decrease between the third and seventh decades. The changes are modest when averaged over all of the cells in the tissue [116]. Exercise differentially influences sarcoplasmic reticulum calcium-ATPase isoform expression in type I fiber and type II fiber. Additionally, adenosine monophosphate-activated protein kinase-α2 influences the regulation of sarcoplasmic reticulum calcium-ATPase2a mRNA in type I skeletal muscle fibers following exercise training [117]. It is estimated that an average of 10 transcription units, the vast majority of which make long noncoding RNAs (lncRNAs), may overlap each traditional coding gene [117]. Noncoding RNAs (ncRNAs), including microRNAs (miRNAs) and long noncoding RNAs (lncRNAs), have emerged as a class of previously unrecognized transcripts. They are emerging as integral components of the gene regulatory networks in a broad range of biological processes, and dysregulation of their expression has been implicated in many human diseases [118]. Following aerobic exercise, skeletal muscle's myosin heavy chain gene expression (contractile ability) was altered. Muscle fiber type I and IIa myosin heavy chain mRNA was increased by 11%, while type IIb was decreased by 38% [119]. In addition, insights to myosin heavy chain regulatory mechanisms previously suggested that rats' lncRNA are conserved in humans. It has been discovered that MicroRNAs (miRNAs) play important roles in immune function and often act to attenuate or silence gene translation. miRNAs interfere with mRNA that is present and render it unusable and therefore decrease the product of that mRNA. miRNAs regulate many physiological processes, such as inflammation, angiogenesis, as well as ischemia. miRNA expression profiling of various human myopathies identified a common set of miRNA that were upregulated in each of the different muscular disorders [120]. Drummond and colleagues reported that let-7 miRNA expression was elevated in skeletal muscle of older humans [121]. The significance of this finding is that let7 expression may be a contributing factor to the alteration in satellite cell activity observed with age. Aerobic exercise training has been established as an important phenotype capable of altering the human skeletal muscle. In healthy subjects, aerobic exercise training can thus locally modulate some signaling ways in muscle regeneration. It reduces the overall number of various miRNAs within the skeletal muscle that produce negative effects and increase myomiRs expression with its positive effect on muscle reconstruction. Stimuli that cause the body to enter an anabolic, or constructive phase, such as resistance training as well as the correct diet, have also shown a reduction of miRNAs. This reduction may actually play a role in the growth of the muscle cell [122]. Exercise is also known to positively influence the expression patterns of miRNAs in leukocyte cells. It is associated with rapid changes in the profile of gene expression in circulating neutrophils [123].

Sarcopenia and Exercise

Compared to young individuals, aged subjects have sarcopenia, a degenerative loss of skeletal muscle mass, quality, and strength. Around the age of 30 years, adults lose 3% of their muscle strength every year, on average [124]. The reduction in lean body mass is lesser in the upper limbs and larger in the lower limbs, in both women and men [125]. This difference may be due to the greater loss of motor units in the lower limbs. Sarcopenia is a natural consequence, associated with aging universal, obvious, and a common condition that affects 10% of adults who are over 50 years old, and thus, increase the risk of disability, fragility, quality of life, and reduce longevity muscle synthesis signaling pathways. Neuropathic is the significant sarcopenia cause, since it is associated with α-motoneuron degeneration and muscle fiber denervation, consequently, it causes motor units loss [126]. Sarcopenia is different from weakness and wasting of the body due to severe chronic illness (cachexia), in which muscle is degraded through cytokine-mediated degradation, although both conditions may coexist. Sarcopenia is considered a component of the frailty syndrome [127], reducing muscle regeneration capacity. Consequently, it promotes further muscle wasting and raises the risk of metabolic syndrome progression. The problem defines the amount of lean body mass weight can be lost before death interrupts. It is hard to define sarcopenia, regarding lean body mass, since there is no critical muscle mass weight to define if sarcopenia is present. The rate of muscle loss depends on physical inactivity, inflammation, unbalanced diet, and sedentary lifestyle [128]. Muscle force and power decline more than muscle dimensions, it reduces functional capacity, it depends on progressive resistance training, which causes large increases in strength with relatively less increase in muscle mass. This demonstrates that muscle mass gain lags behind the increased strength following resistance training. Still it has been suggested that loss of more than about 40% of baseline lean mass is fatal, without differences between man and woman.

Sarcopenia reflects a progressive imbalance between decreased anabolism and increased catabolism, along with a reduced muscle renewal capability. Suggested mechanisms of muscle changes of sarcopenia include an imbalance between signals for muscle cell growth and signals for teardown, changes in satellite cell recruitment, changes in anabolic signaling, protein oxidation, inflammation, and developmental factors. The pathologic changes of sarcopenia include muscle atrophy along with a reduction in muscle tissue quality as reflected in the replacement of muscle fibers with fat, an increase in fibrosis, changes in muscle metabolism, oxidative stress, and degeneration of the neuromuscular junction [129]. Although basal protein synthesis and breakdown show little changes in old age, the proportion of muscles' fast to slow fibers, changes in sarcopenic muscle resulting in fast fiber atrophy, along with a slight decrease in slow fiber can be seen. Loss of motor units and an increase in hybrid fibers are typical findings of aging; however, a small amount of fast fibers that lost their motor nerve connection (innervated) are sometimes reinnervated by an extension from a slow type I fiber motor nerves [130], consequently, fast fibers are converted to type I fibers [131]. However, it appears that in the elderly, reinnervation quantity is less than that of denervation, bringing about a loss of motor units [132].

The failure to activate satellite cells upon injury or exercise is also thought to contribute to the pathophysiology of sarcopenia [129]. Additionally, oxidized proteins can lead to a buildup of lipofuscin and cross-linked proteins causing an accumulation of non-contractile material in the skeletal muscle and contribute to sarcopenic muscle [133].

Sarcopenia is preventable, a healthy diet and regular exercise can reverse it, increasing lifespan and quality of life. Exercise remains the primary intervention for sarcopenia treatment, but translation of research findings into clinical practice is challenging. The type, duration, and intensity of exercise are variable between studies, preventing a standardized exercise prescription for sarcopenia [134]. Lack of exercise is a significant risk factor for sarcopenia and exercise can dramatically slow the rate of muscle loss [135]. Exercise can be an effective intervention because aging skeletal muscle retains the ability to synthesize proteins in response to short-term resistance exercise [136]. Progressive resistance training in older adults can improve physical performance (gait speed) and muscular strength [137].

Exercise During Aging

During exercise that produces stress on the aged individual, the decreased muscle mass results in a loss of controlling or homeostatic balance. At rest, usually most physiological systems function normally, unless a pathological condition exists. Consequently, muscle strength declines after the age of 50 years at a rate of about 1.2–1.4% per year [138]. This loss of strength with age is due to many factors but is primarily attributed to a loss of muscle mass bringing to sarcopenia [139]. Because sarcopenia is related to a loss of functional abilities [140], dependency disability, physical inactivity, and muscle strength in older individuals, all increase risk of falls and fractures [141], and decreased bone mineral density. Thus, sarcopenia has negative consequences for the health status and functional abilities of older adults. Elevated perception of effort during submaximal effort is also an additional effect of sarcopenia that may cause evasion, physical activity, and disuse.

Level of endurance capacity [142] means exercising at intensities above the anaerobic threshold. There is growing evidence from longitudinal and cross-sectional studies that chronic endurance training attenuates the decline in VO_{2max}, associated with age [8]. In addition, the position of aerobic exercise as a tool to reduce the risk of numerous long-lasting deteriorating diseases has been convincingly recognized earlier [143] Also, improved physical work capacity following training may be considered as an accurate indicator of regular physical activity.

Improved work capacity includes different physiological system mechanisms in the cardiopulmonary capacity, body composition, muscular strength, and endurance capacity. Untrained and trained elderly can increase the response of the cardiopulmonary without a significant reduction in peripheral ability to extract oxygen at the muscle level. It was found that in elderly subjects, skeletal muscle mitochondrial capacity, tissue blood flow capacity, and oxygen exchange capacity appear to be well matched. It seems that intrinsic mitochondrial function and regulation are not altered significantly. Therefore, the higher aerobic capacity in the trained elderly is related to increases in the abilities of cardiovascular factors following training and to the lesser extent to increases in muscle mitochondria concentration and capillarity [144]. Both anaerobic exercise and aerobic exercise generate ROS differently. Increased aerobic exercise intensity along with oxygen uptake results in ROS rise affecting, muscle contractile, and decreases oxidative damage, as long as exercise intensity is up to 75% [145]. One has to imagine the potential of 2% of an individual to produce peroxisomal following a fitness session lasting 60 min run with an oxygen uptake of 180 LO_2. During aerobic exercise, ATP production is in the Krebs cycle with oxygen presence, where free fatty acids are the main source of energy production pathway producing ATP and lactate. Aerobic exercise compared to anaerobic exercise produces a larger ROS amount; however, anaerobic exercise induces continually ROS production during recovery. High blood lactate levels possibly have an effect on the increase in ROS production by equine polymorphonuclear leukocytes involved in the pathogenesis, which relates to alterations in essential immune functions within a short period after all-out-intensity exercise [146]. Although more oxygen is consumed during aerobic exercise, the generated ROS does not induce significant oxidative damage. Oxygen uptake as such may not be the major cause of exercise-induced oxidative damage. Metabolite peroxisome produced by the mitochondria during the energy production process is modified by aerobic exercise in the redox setting. The peroxisome proliferator-activated receptor (PPAR) γ coactivator 1-α (PGC-1α) is a lipid catabolism and mitochondrial function regulator [147]. PGC-1α activation could result in a decreased oxidative challenge, either by upregulation of antioxidant enzymes or by an increased number of mitochondria that allows lower levels of respiratory activity for the same degree of ATP generation. Usually, Oleic acid, a monounsaturated fatty acid, initiates oxygen-production through NADPH oxidase. However, oleic acid also can alter cell's reactive oxygen species (ROS) production. Oleic acid slows down electron's sequence of reactions between the cytochromes within the Krebs respiratory chain and thus, due to their protonophoric act on the inner mitochondrial membrane by translocating protons across lipid bilayers so it significantly reduces ROS production in the converse mode of electron transport. These effects of oleic acid can modulate signaling functions of ROS, and thus decrease the apoptotic process of cells [148]. The oxidative activity of ROS is a normal function of cellular respiration but can impair the normal functioning of other molecules and systems if present in high amounts [149]. The body has natural antioxidant systems that help to keep ROS levels normal; however, these antioxidant systems can be overwhelmed, causing an acute accumulation of ROS known as oxidative stress [150]. Study by Saghiv et al. [151]

tested the hypothesis that long-lasting aerobic exercise training could prevent the age-associated reduction in s-Klotho serum levels, in 30 healthy sportsmen: 15 young aerobically well-trained elite athletes and 15 aerobically well-trained master athletes. This study demonstrated that circulating s-Klotho levels are similar for young healthy well-trained elite runners and elite master athletes. This suggests that the response of s-Klotho depends on the aerobic fitness level [7]. In addition, levels of s-Klotho were significantly higher in both exercising groups when compared with age-matched untrained subjects reported earlier by [2], suggesting that long-lasting aerobic training may be an appropriate model for mechanistically probing the role of physical activity and s-Klotho expression.

Anaerobic Exercise in the Elderly

Aerobic pathway capacity is reduced in older subjects compared to young ones, thus maximal energy provided at peak endurance exercise is limited [152]. Anaerobic metabolic pathways have received less attention in old subjects: anaerobic energetic release during exercise has not been evaluated previously. Therefore, data in the elderly are limited on anaerobic energetic yield during all-out strenuous exercise. More importantly, the interplay of both aerobic and anaerobic pathways may enable to calculate the aerobic/anaerobic relative contribution to total energy release at peak all-out vigorous exercise [153]. It seems that in the well-trained older men, the lower aerobic ability compared to young individuals, at peak all-out vigorous exercise, is due at least partially to the aged decreased capacity to produce ATP by oxidative phosphorylation owing it to dysfunctional mitochondria. The impairment of mitochondrial function is owed to decreased rates of electron transfer by the selectively diminished activities of the four-electron transfer or respiratory complexes (complexes I–IV) [154]. In the aged individuals, compared to young subjects, mitochondria do not only produce less ATP but also increase the production of ROS as by-products of aerobic metabolism in the aging tissues. Moreover, the activities of free radical-scavenging enzymes are altered in the aging process. The concurrent age-related changes of these two systems result in the elevation of oxidative stress in aged tissues [155].

Safety of Resistance Training in the Aged

Disuse of skeletal muscle is often cited as a potentially reversible cause of ill-health in elderly people. It can lead to a serious decline in functional health and increases the risk of outpatient care, hospitalization, nursing home admission, and death. The loss of muscle mass accounts for most of the observed loss of strength. The loss of muscle tissue is due to a decrease in the number of muscle fibers and atrophy of the type II muscle fibers. The subsequent decline in strength reduces the capacity to carry out basic activities of daily life and puts people at risk for falls, impaired mobility, and dependence [156]. Although sedentary lifestyle, and aging itself, may contribute to muscle weakness and loss of skeletal muscle mass, that conditions are potentially preventable or reversible with targeted interventions [157]. It is suggested that high-intensity resistance exercise training is a feasible and effective means of counteracting muscle weakness and physical frailty in elderly people [158].

The principal goal of resistive training in the elderly is to restore physical, psychological, and vocational function. However, resistance exercise has previously been regarded as thermodynamically hazardous, due to exaggerated BP responses causing higher double pressure products leading to more ischemic events. Traditionally, training programs for older subjects have emphasized lower extremity aerobic exercise.

Various vocational, recreational, and other activities of daily living actually resemble heavy resistance exercise more than aerobic exercise. Many elderly people lack the ability to perform such common activities as carrying groceries or luggage or doing yard work. Therefore, it is important to recognize that the elderly require a minimum level of muscle strength for activities of daily living. Evidence has been accumulated which suggests that resistance exercise training may be less hazardous than is presumed. The benefits of resistance training for the elderly are improved muscle strength and endurance, enhanced bone mineral density, weight reduction, maintaining an improved self-image, and gaining confidence.

Resistance training can produce a slight increase in aerobic capacity, which is associated with an increase in muscle mass. In the elderly, exercise capacity can be limited by leg fatigue resulting in termination of exercise without coexisting evidence of cardiopulmonary limitation. Studies conducted on the elderly have demonstrated increased strength with enhanced treadmill performance in circuit weight training trainees. Therefore, those with weak leg muscles will gain additional benefits if conventional endurance training is combined with resistance training.

Electrocardiographic and direct arterial BP responses during single-arm and single- and double-leg lifting at 20, 40, 60, and 80% of one maximal repetition do not cause clinically significant ST-segment depression, angina, or ventricular arrhythmias, except when subjects work against high resistance: 80% of 1-RM, double pressure product values exceeded those attained during cycle ergometer testing at 85% of maximal aerobic capacity. Thus, light-to-moderate resistive training exercise can be considered safe for low-risk cardiac patients. Even the risk of developing compromised left ventricular function is less during a workout against weight resistance compared to that in conventional clinical aerobic exercise test [159].

Circuit weight training is an approach for strength development especially designed for the elderly. It incorporates a series of selected resistive training exercises that are performed in a sequence. With circuit weight training, one performs approximately 10–12 repetitions using about 50–55% of 1-RM, on specialized weight machines. Individuals move from one weight machine to another with a short rest (30–60 s) between stations. The fact that circuit weight training can improve strength, body composition, and, in a modest way, cardiorespiratory endurance makes this form of exercise appealing for training programs for elderly [160].

A large amount of evidence on the safety and efficacy of this kind of resistive training in the elderly has been gathered in the past two decades [161]. BP measurements in elderly subjects collected before, during, and after circuit weight training demonstrated a slight increase or no change compared to resting values. In comparison to BP values measured during dynamic exercise at 85% of VO_{2peak}, circuit weight training BP values were lower or slightly higher than those achieved at 85% of VO_{2peak} [162].

Weight training programs result in an increase in muscle strength, which may reduce the relative stress imposed by activities of daily living. Thus, resistance exercise can contribute to better health by preventing musculoskeletal disorders, helping to maintain desirable body composition, and improving self-image and self-efficacy. Also, engaging in resistance exercise may reduce some risk factors for CAD. Improvements are seen in physical fitness and quality of life, as well as in the modification of classical risk factors for coronary artery disease. In addition, the aging musculoskeletal system retains its responsiveness to progressive resistance training, and, most importantly, the correction of disuse is accompanied by a significant improvement in the level of functional mobility and the overall ability to conduct activities safety [163].

S-Klotho Effects on Aging

Klotho is a transmembrane protein that also effects and controls over the connection of humans to insulin and seems to be associated to an anti-aging function [164]. The α-Klotho gene is highly preserved in humans and circulates in blood as s-Klotho which acts by way of a humoral factor [165]. In humans, serum levels of s-Klotho decrease with aging progress [166] that may be observed in patients with several aging-related diseases such as coronary artery disease, cancer, hypertension, and kidney disease [167]. The klotho gene encodes a single-pass transmembrane protein that binds to multiple fibroblast growth factor receptors and functions as a co-receptor for FGF23, a bone-derived hormone that suppresses phosphate reabsorption and vitamin D biosynthesis in the kidney [168]. In addition, the extracellular domain of s-Klotho protein is shed and secreted [169]. It has previously been demonstrated that Klotho protein increases the ability of cells to remove damaging ROS by increasing the cell functions that detoxify harmful ROS [170]; however, the precise mechanism of klotho's action is not fully understood, but it changes cellular calcium homeostasis, by increasing the expression and activity of transient receptor potential cation channels, vanilloid subfamily, member 5 (TRPV5) which plays a key role in active Ca^{++} reabsorption in the kidney and decreasing that of transient receptor potential canonical 6 (TRPC6), a subtype of calcium-permeable channel [171]. Additionally, Klotho increases membrane expression of the inward rectifier ROMK1 channels, i.e., renal outer medullary K^+ [171, 172]. Klotho-deficient mice show increased production of vitamin D and altered mineral-ion homeostasis is suggested to be a cause of premature aging like phenotypes because the lowering of vitamin D activity by dietary restriction reverses the premature aging like phenotypes and prolongs survival in these mutants. These results suggest that aging like phenotypes were due to Klotho-associated vitamin D metabolic abnormalities [168]. The Klotho protein deficiency is known to participate in premature aging. Klotho protein contributes in numerous pathways that control aging, for example, phosphate homeostasis, insulin signaling, and Wnt signaling regulation. As an aging suppressor, Klotho is an important molecule in aging processes and its overexpression results in longevity [173]. The Klotho gene encodes a transmembrane protein that after cleavage is also found as a secreted protein. However, previously it has been shown that the circulating s-Klotho levels are significantly higher than the membrane-bound form in humans and mice and the latter is significantly higher than the former [173]. Importantly, its overexpression suppresses insulin/IGF-I signaling and thus extends the lifespan. In addition, Klotho participates in the regulation of several other intracellular signaling pathways, including regulation of FGF23 signaling, cAMP, PKC, transforming growth factor-β (TGF-β), p53/p21, and Wnt signaling [174]. A population-based longitudinal study revealed that low plasma levels of Klotho are associated with decreased activities of daily living in older individuals. In older community-dwelling adults, plasma Klotho is an independent predictor of all-cause mortality [174]. Low plasma Klotho levels have also been associated with poor muscle strength in older community-dwelling adults [175].

Safety Exercising in Aging

Disuse of skeletal muscle is often cited as a potentially reversible cause of ill health in elderly people. It can lead to a serious decline in functional health and increases the risk of outpatient care, hospitalization, nursing home admission, and death. The loss of muscle mass accounts for most of the observed loss of strength. The loss of muscle tissue is due to a decrease in the number of muscle fibers and atrophy of the type II muscle fibers. The subsequent decline in strength reduces the capacity to carry out basic activities of daily life and puts people at risk for falls, impaired mobility, and dependence [176]. Although sedentary lifestyle and aging itself may contribute to muscle weakness and loss of skeletal-muscle mass, that conditions are potentially preventable or reversible with targeted interventions [157]. It has been suggested that high-intensity resistance exercise training is a feasible and effective means of counteracting muscle weakness and physical frailty in elderly people [177]. The principal goal of resistance training in the elderly is to restore physical, psychological, and vocational function. However, resistance exercise has previously been regarded as thermodynamically hazardous due to exaggerated BP responses causing higher double pressure products leading to more ischemic events. Traditionally, training programs for older subjects have emphasized lower extremity aerobic exercise (i.e., walking, stationary cycle ergometer, and stair climbing). Various vocational, recreational, and other activities of daily living actually resemble heavy resistance exercise more than aerobic exercise. Many elderly people lack the ability to perform such common activities as carrying groceries or luggage or doing yard work. Therefore, it is important to recognize that the elderly require a minimum level of muscle strength for activities of daily living. Evidence has been accumulated, which suggests that resistive exercise training may be less hazardous than was once presumed. The benefits of resistance training for the elderly are improved muscle strength and endurance, enhanced bone mineral density, weight reduction, maintaining an improved self-image, and gaining confidence. Resistance training can produce a slight increase in aerobic capacity, which is associated with an increase in muscle mass. In the elderly, exercise capacity can be limited by leg fatigue resulting in termination of exercise without coexisting evidence of cardiorespiratory limitation. Studies conducted on the elderly have demonstrated increased strength with enhanced treadmill performance in circuit weight training trainees. Therefore, those with weak leg muscles will gain additional benefits if conventional endurance training is combined with resistance training. Electrocardiographic and direct arterial blood pressure responses during single-arm and single- and double-leg lifting at 20%, 40%, 60%, and 80% of one-repetition maximum (1-RM) did not reveal any clinically significant ST-segment depression, angina, or ventricular arrhythmias, except when subjects were working against high resistance: 80% of 1-RM, double pressure product values exceeded those attained during cycle ergometer testing at 85% of maximal aerobic capacity. Thus, light-to-moderate resistance training exercise can be considered safe for low-risk cardiac patients. Even the risk of developing compromised left ventricular function is less during a workout against weight resistance compared to that in conventional clinical aerobic exercise test [159]. Circuit weight training is an approach for strength development specially designed for the elderly. It incorporates a series of selected resistive training exercises that are performed in a sequence. With circuit weight training, one performs approximately 10–12 repetitions using about 50–55% of 1-RM on specialized weight machines. The individual moves from one weight machine to another with a short rest (30–60 s) between stations. The fact that circuit weight training can improve strength, body composition, and, in a modest way, cardiorespiratory endurance makes this form of exercise appealing for training programs for the elderly [160]. A large amount of evidence on the safety and efficacy of this kind of resistive training in the elderly has been gathered in the past two decades [161]. BP measurements in elderly subjects collected before, during, and after circuit weight training demonstrated a slight increase or no change compared to resting values. In comparison to BP values measured during dynamic exercise at 85% of peak oxygen uptake, circuit weight training blood pressure values were lower or slightly higher than those achieved at 85% of peak oxygen uptake [178]. Weight training programs result in an increase in muscle strength. A stronger musculature may reduce the relative stress imposed by activities of daily living. Thus, resistance exercise can contribute to better health by preventing musculoskeletal disorders, helping to maintain desirable body composition and improving self-image and self-efficacy. Also, engaging in resistance exercise may reduce some risk factors for CAD. Improvements are

Table 7.1 Endurance training-induced changes in cardiovascular and hemodynamic variables in elderly subjects

Variable	Rest	Submaximal exercise	Maximal exercise
Oxygen uptake	Unchanged	Unchanged	Increases
Systolic blood pressure	Decreases	Decreases	Decreases
Diastolic blood pressure	Decreases	Decreases	Decreases
Stroke volume	Decreases	Increases	Increases
Heart rate	Decreases	Decreases	Unchanged
Cardiac output	Unchanged	Unchanged	Increases
Contractility	Unchanged	Unchanged	Unchanged/increases
Total peripheral resistance	Unchanged	Unchanged	Decreases
Arteriovenous 02 difference	Unchanged	Unchanged	Increases

Table 7.2 Endurance training-induced changes in pulmonary function in elderly subjects

Variable	Rest	Submaximal exercise	Maximal exercise
Maximal voluntary ventilation	Unchanged	–	–
Maximal oxygen uptake	Unchanged	Unchanged	Increases
Minute ventilation	Unchanged	Decreases	Decreases
Ventilation/oxygen uptake	Unchanged	Unchanged	Decreases
Ventilation/volume CO_2	Unchanged	Unchanged	Decreases
Lactic acid	Unchanged	Decreases	Decreases

seen in physical fitness and quality of life, as well as in the modification of classical risk factors for CAD. In addition, the aging musculoskeletal system retains its responsiveness to progressive resistance training, and most importantly, the correction of disuse is accompanied by a significant improvement in the level of functional mobility and the overall ability to conduct activities safely.

Table 7.1 summarizes some of the changes associated with aging and training on cardiovascular and hemodynamic variables, while Table 7.2 reveals that efficiency of breathing is improved in the elderly following an endurance training program. These changes include a reduction in blood lactate levels and increased maximal ventilation.

Laboratory on Resistance

Equipment:

- Hand dynamometer
- Deadlift dynamometer
- Isokinetic device
- Borg table

Testing stages:

Experimental Procedure:

In this laboratory, each four individuals are included in a group, out of them two will be tested. The other two will serve as testers.

Pre-test instructions for subjects:

(a) Subjects must refrain from exercise 2 h prior to test.
(b) Subjects should minimize activity on the day of the test, if possible.

Test procedures:

- The tester will measure three times with hand dynamometer the maximal voluntary contraction (MVC).
- The highest value will be the subject's MVC.
- The same procedure will be followed by the deadlift dynamometer.
- The same procedure will be followed by the chest-press maneuver.
- The subject will perform with the isokinetic device knee flexion and knee extension.

Assignments:

- All maneuver's results will be reported as follows:
- Compare males with females.
- Do the same for lean body mass.
- Determined lean body mass by caliper as suggested earlier.
- Explain the results with scientific background.

Laboratory report

Name ____________ Weight __________

Isometric handgrip test

Trial 1________kg Trial 1_______kg Trial 1_______kg

maximal strength________________

Isometric deadlift

Trial 1________kg Trial 1_______kg Trial 1________kg

Chest press

Trial 1________kg Trial 1_______kg Trial 1________kg

Isokinetic test

Knee flexion maximal momentum value _____ N•m-1

Optimal degree____ °

Knee extension maximal momentum value _____ N•m-1

Optimal degree____ °

References

1. Dasari V, Srivastava S, Khan S, et al. Epigenetic factors Polycomb (Pc) and Suppressor of zeste (Su(z)2) negatively regulate longevity in Drosophila melanogaster. Biogerontology. 2018;19:33–45.
2. Shephard RJ. Physical activity and aging. 2nd ed. London: Croom Helm; 1987.
3. Reddy S, Comai L. Lamin A, farnesylation and aging. Exp Cell Res. 2011. [Epub ahead of print].
4. Santos Lozano A, Santamarina A, Pareja Galeano H, et al. The genetics of exceptional longevity: insights from centenarians. Maturitas. 2016;90:49–57.
5. Tiffon C. The impact of nutrition and environmental epigenetics on human health and disease. Int J Mol Sci. 2018;19(11).
6. Gault ML, Willems ME. Aging, functional capacity and eccentric exercise training. Aging Dis. 2013;4:351–63.
7. Avin KG, Coen PM, Huang W, et al. Skeletal muscle as a regulator of the longevity protein. Klotho Front Physiol. 2014;5:189.
8. Hawkins S, Wiswell R. Rate and mechanism of maximal oxygen consumption decline with aging: implications for exercise training. Sports Med. 2003;33:877–88.
9. Witztum JL. The oxidation hypothesis of atherosclerosis. Lancet. 1994;344:793–5.
10. Burkitt MJ. A critical overview of the chemistry of copper dependent low-density lipoprotein oxidation: roles of lipid hydroperoxides, alpha-tocopherol, thiols, and ceruloplasmin. Arch Biochem Biophys. 2001;394:117–35.
11. Sebastiani P, Solovieff N, Puca A, et al. Genetic signatures of exceptional longevity in humans. Science. 2010;333(6041):404.
12. Thomas L, Levett K, Boyd A, et al. Changes in regional left atrial function with aging: evaluation by Doppler tissue imaging. Eur J Echocardiogr. 2003;4:92–100.
13. Gregorich ZR, Peng Y, Cai W, et al. Top-down targeted proteomics reveals decrease in myosin regulatory light-chain phosphorylation that contributes to sarcopenic muscle dysfunction. J Proteome Res. 2016;15:2706–16.
14. Baumann CW, Kwak D, Liu HM, et al. Age-induced oxidative stress: how does it influence skeletal muscle quantity and quality? J Appl Physiol (1985). 2016;121:1047–52.
15. Matelot D, Schnell F, Kervio G, et al. Cardiovascular benefits of endurance training in seniors: 40 is not too late to start. Int J Sports Med. 2016;37:625–32.
16. McEniery CM, Wilkinson IB, Avolio AP. Age, hypertension and arterial function. Clin Exp Pharmacol Physiol. 2007;34:665–71.
17. Seals DR, Desouza CA, Donato AJ, et al. Habitual exercise and arterial aging. Appl Phys. 2008;105:1323–32.
18. Pang MYC, Yang FZH, Jones AYM. Vascular elasticity and grip strength Are associated with bone health of the hemiparetic radius in people with chronic stroke: implications for rehabilitation. Physic Ther. 2013;93:774–85.
19. Sethi S, Rivera O, Oliveros R, et al. Aortic stiffness: pathophysiology, clinical implications, and approach to treatment. Integr Blood Press Control. 2014;23(7):29–34.
20. Afilalo J, Karunananthan S, Eisenberg MJ, et al. Role of frailty in patients with cardiovascular disease. Am J Cardiol. 2009;103:1616–21.
21. Chien MY, Kuo HK, Wu YT. Sarcopenia, cardiopulmonary fitness, and physical disability in community-dwelling elderly people. Phys Ther. 2010;90:1277–87.
22. Bassett DR Jr, Howley ET. Limiting factors for maximum oxygen uptake and determinants of endurance performance. Med Sci Sports Exerc. 2000;32:70–84.
23. Fleg JL, Strait J. Age-associated changes in cardiovascular structure and function: a fertile milieu for future disease. Heart Fail Rev. 2011. [Epub ahead of print].
24. Nichols WW, O'Rourke MF. MacDonald's blood flow in arteries. 4th ed. London: Arnold, Hodder Headline Group; 1998.

25. Laurent S, Katsahian S, Fassot C, et al. Aortic stiffness is an independent predictor of fatal stroke in essential hypertension. Stroke. 2003;34:1203–6.
26. Avolio AP, Chen SG, Wang RP, et al. Effects of aging on changing arterial compliance and left ventricular load in a northern Chinese urban community. Circulation. 1983;68:50–8.
27. Tanaka H, DeSouza CA, Seals DR. Absence of age-related increase in central arterial stiffness in physically active women. Arterioscler Thromb Vasc Biol. 1998;18:127–32.
28. Lorell BH, Carabello BA. Left ventricular hypertrophy pathogenesis, detection, and prognosis. Circulation. 2000;102:470–9.
29. Vaitkevicius PV, Fleg JL, Engel JH, et al. Effects of age and aerobic capacity on arterial stiffness in healthy adults. Circulation. 1993;88:1456–62.
30. Franklin SS, Gustin W IV, Wong ND, et al. Hemodynamic patterns of age-related changes in blood pressure. Circulation. 1997;96:308–15.
31. Mitchell GF, Lacourciere Y, Ouellet JP, et al. Determinants of elevated pulse pressure in middle-aged and older subjects with uncomplicated systolic hypertension: the role of proximal aortic diameter and the aortic pressure-flow relationship. Circulation. 2003;108:1592–8.
32. Kelly R, Hayward C, Avolio A, et al. Noninvasive determination of age-related changes in the human arterial pulse. Circulation. 1989;80:1652–9.
33. O'Rourke M. Arterial haemodynamics and ventricular-vascular interaction in hypertension. Blood Press. 1994;3:33–7.
34. Mattace-Raso FU, van der Cammen TJ, Hofman A, et al. Arterial stiffness and risk of coronary heart disease and stroke: the Rotterdam Study. Circulation. 2006;113:657–63.
35. Laurent S, Kingwell B, Bank A, et al. Clinical applications of arterial stiffness: therapeutics and pharmacology. Am J Hypertens. 2002;5:453–8.
36. Cameron JD, Rajkumar C, Kingwell BA, et al. Higher systemic arterial compliance is associated with greater exercise time and lower blood pressure in a young older population. J Am Geriatr Soc. 1999;47:653–6.
37. Martin WH III, Ogawa T, Kohrt WM, et al. Effects of aging, gender, and physical training on peripheral vascular function. Circulation. 1991;84:654–64.
38. Rush JW, Aultman CD. Vascular biology of angiotensin and the impact of physical activity. Appl Physiol Nutr Metab. 2008;33:162–72.
39. Coates D. The angiotensin converting enzyme. Int J Biochem Cell Biol. 2003;35:769–73.
40. Gordon NF, Scott CB, Wilkinson WJ, et al. Exercise and mild essential hypertension. Recomm Adults Sports Med. 2001;10:390–404.
41. Guyton AC, Hall JE, editors. Textbook of medical physiology. 9th ed. Philadelphia, PA: WB Saunders; 2000. p. 227–30.
42. Benetos A, Gautier S, Ricard S, et al. Influence of angiotensin-converting enzyme and angiotensin II type 1 receptor gene polymorphisms on aortic stiffness in normotensive and hypertensive patients. Circulation. 1996;94:698–703.
43. Avila-Vanzzini N, Espínola-Zavaleta N, Masso-Rojas F, et al. ACE gene polymorphism correlation (I/D) with the ventricular function in patients with ischemic and idiopathic dilated cardiomyopathy. (Article in Spanish). Rev Investig Clin. 2006;58:39–46.
44. Puthucheary Z, Skipworth JR, Rawal J, et al. The ACE gene and human performance: 12 years on. Sports Med. 2011;41:433–48.
45. Van Geel PP, Pinto YM, Voors AA, et al. Angiotensin II type 1 receptor A1166C gene polymorphism is associated with an increased response to angiotensin II in human arteries. Hypertension. 2000;35:717–21.
46. Tanriverdi H, Evrengul H, Kaftan A, et al. Effects of angiotensin-converting enzyme polymorphism on aortic elastic parameters in athletes. Cardiology. 2005;104:113–9.
47. Sagiv M, Amir O, Goldhammer E, et al. Left ventricular contractility in response to upright isometric exercise in heart transplant recipients and healthy men. J Cardiopulm Rehabil Prev. 2008;28:17–23.
48. Camell CD, Sander J, Spadaro O, et al. Inflammasome-driven catecholamine catabolism in macrophages blunts lipolysis during ageing. Nature. 2017;550(7674):119–23. https://doi.org/10.1038/nature24022.

49. Strait JB, Lakatta EG. Aging-associated cardiovascular changes and their relationship to heart failure. Heart Fail Clin. 2012;8:143–64.
50. Saghiv M, Ben-Sira D, Goldhammer E, et al. What maintains the metabolic cost at peak anaerobic test in elite young and master male cyclists? Exerc Health Dis. 2018;2:1–4.
51. Reaburn P, Dascombe B. Endurance performance in masters athletes. Eur Rev Aging Phys Act. 2008;5:31–42.
52. Bortz WM. Disuse and aging. JAMA. 1982;248:1203–8.
53. Swine C. Aging of heart function in man. (French). Presse Med. 1992;21:1216–21.
54. Sagiv M, Ben-Sira D, Goldhammer E, et al. Left ventricular contractility and function at peak aerobic and anaerobic exercises. Med Sci Sports Exerc. 2000;32:1197–201.
55. Correia LC, Lakatta EG, O'Connor FC, et al. Attenuated cardiovascular reserve during prolonged submaximal cycle exercise in healthy older subjects. J Am Coll Cardiol. 2002;40:1290–7.
56. Tanaka H, Seals DR. Endurance exercise performance in masters athletes: age-associated changes and underlying physiological mechanisms. J Physiol. 2008;586:55–63.
57. Iemitsu M, Miyauchi T, Maeda S, et al. Exercise training improves cardiac function-related gene levels through thyroid hormone receptor signaling in aged rats. Am J Physiol Heart Circ Physiol. 2004;286:H1696–705.
58. Fukuoka Y, Nakagawa Y, Ogoh K, et al. Dynamics of the heart rate response to sinusoidal work in humans: influence of physical activity and age. Clin Sci (Lond). 2002;102:31–8.
59. Shephard RJ. Health and aerobic fitness. Champaign, IL: Human Kinetics Publishers; 1993.
60. Capasso JM, Puntillo E, Olivetti G, et al. Difference in load dependence of relaxation between the left and right ventricular myocardium as a function of age in rats. Circ Res. 1989;65:1499–507.
61. Izzo C, Carrizzo A, Alfano A, et al. The impact of aging on cardio and cerebrovascular diseases. Int J Mol Sci. 2018;19(2). pii:E481. https://doi.org/10.3390/ijms19020481.
62. Cruickshank JM. The role of beta-blockers in the treatment of hypertension. Adv Exp Med Biol. 2017;956:149–66.
63. Sagiv N, Sagiv M, Ben-Sira D. Weight lifting training and left ventricular function in adolescent subjects. J Sports Med Phys Fitness. 2007;47:329–34.
64. de Oliveira SN, Pereira Moro AR, et al. Effects of concurrent training with elastic tubes in hypertensive patients: a blind controlled randomized clinical trial. Exp Aging Res. 2019:1–15. https://doi.org/10.1080/0361073X.2019.1693030.
65. Peterson LR, Morton RR, Schechtman KB, et al. Peak exercise stroke volume: associations with cardiac structure and diastolic function. J Appl Physiol. 2003;94:1108–14.
66. Biering-Sorensen T. Cardiac time intervals by tissue Doppler imaging M-mode echocardiography: reproducibility, reference values, association with clinical characteristics and prognostic implications. Dan Med J. 2016;63(8). pii: B5279.
67. Sagiv M, Goldhammer E, Ben-Sira D. Effect of increased after-load on left ventricular filling properties in healthy elderly and young subjects. Med Exerc Nutr Health. 1992;1:48–53.
68. Dugan SA, Gabriel KP, Lange-Maia BS, et al. Physical activity and physical function: moving and aging. Obstet Gynecol Clin N Am. 2018;45:723–36.
69. Ketata W, Rekik WK, Ayadi H, et al. Aging of the respiratory system: anatomical changes and physiological consequences. [Article in French]. Rev Pneumol Clin. 2012;68(5):282–9. https://doi.org/10.1016/j.pneumo.2012.06.003.
70. Guenette JA, Witt JD, McKenzie DC, et al. Respiratory mechanics during exercise in endurance-trained men and women. J Physiol. 2007;581:1309–22.
71. Dufour Doiron M, Prud'homme D, et al. Time-of-day variation in cardiovascular response to maximal exercise testing in coronary heart disease patients taking a beta-blocker. Appl Physiol Nutr Metab. 2007;32:664–9.
72. McClaran SR, Babcock MA, Pegelow DF, et al. Longitudinal effects of aging on lung function at rest and exercise in healthy active fit elderly adults. J Appl Physiol. 1995;78:1957–68.

73. Johnson BD, Badr MS, Dempsey JA. Impact of the aging pulmonary system on the response to exercise. Clin Chest Med. 1994;15:229–46.
74. Taylor BJ, Johnson BD. The pulmonary circulation and exercise responses in the elderly. Semin Respir Crit Care Med. 2010;31:528–38.
75. Eynon N, Yamin C, Ben-Sira D, et al. Optimal health and function among the elderly: lessening severity of ADL disability. Eur Rev Aging Phys Act. 2009;6:5–6.
76. Porter C, Hurren NM, Cotter MV, et al. Mitochondrial respiratory capacity and coupling control decline with age in human skeletal muscle. Am J Physiol Endocirnol Metab. 2015;309:E224–32.
77. Dash RK, Dibella JA II, Cabrera ME. A computational model of skeletal muscle metabolism linking cellular adaptations induced by altered loading states to metabolic responses during exercise. Biomed Eng Online. 2007;20:14–42.
78. Kimyagarov S, Levenkron S, Shabi A, et al. Changes of skeletal muscle mass among disabled elderly. (Article in Hebrew). Harefuah. 2010;149:67–70.
79. Papa EV, Dong X, Hassan M. Resistance training for activity limitations in older adults with skeletal muscle function deficits: a systematic review. Clin Interv Aging. 2017;12:955–61.
80. Chistiakov DA, Sobenin IA, Revin VV, et al. Mitochondrial aging and age-related dysfunction of mitochondria. BioMed Res Int. 2014; https://doi.org/10.1155/2014/238463.
81. Reiter RJ, Tan DX, Rosales-Corral S, et al. Mitochondria: central organelles for melatonin's antioxidant and anti-aging actions. Molecules. 2018;23(2). pii: E509. https://doi.org/10.3390/molecules23020509.
82. Braganza A, Corey CG, Santanasto AJ, et al. Platelet bioenergetics correlate with muscle energetics and are altered in older adults. JCI Insight. 2019;5. pii: 128248. https://doi.org/10.1172/jci.insight.128248.
83. Zhang R, Wang Y, Ye K, et al. Independent impacts of aging on mitochondrial DNA quantity and quality in humans. BMC Genom. 2017;18(1):890. https://doi.org/10.1186/s12864-017-4287-0.
84. Marini M, Sarchielli E, Brogi L, et al. Role of adapted physical activity to prevent the adverse effects of the sarcopenia. A pilot study. Ital J Anat Embryol. 2008;113:217–25.
85. Fleg JL, Morrell CH, Bos AG, et al. Accelerated longitudinal decline of aerobic capacity in healthy older adults. Circulation. 2005;112:674–82.
86. Ishida K, Katayama K, Akima H, et al. Effects of deconditioning on the initial ventilatory and circulatory responses at the onset of exercise in man. Adv Exp Med Biol. 2010;669:319–22.
87. Rossiter HB, Kowalchuk JM, Whipp BJ. A test to establish maximum O2 uptake despite no plateau in the O2 uptake response to ramp incremental exercise. J Appl Physiol. 2006;100:764–70.
88. Lakatta EG. Changes in cardiovascular function with aging. Eur Heart J. 1990;Suppl C:22–9.
89. Huang YC, O'Brien SR, MacIntyre NR. Intrabreath diffusing capacity of the lung in healthy individuals at rest and during exercise. Chest. 2002;122:177–85.
90. Tolle J, Waxman A, Systrom D. Impaired systemic oxygen extraction at aximum exercise in pulmonary hypertension. Med Sci Sports Exec. 2008;40:3–8.
91. Wilmore JH, Costill DL. Physiology of sport and exercise: rate of reaction. 3rd ed. Champaign, IL: Human Kinetics; 2005.
92. Faulkner JA, Larkin LM, Claflin DR, et al. Age-related changes in the structure and function of skeletal muscles. Clin Exp Pharmacol Physiol. 2007;34:1091–6.
93. Trappe S. Marathon runners: how do they age? Sports Med. 2007;37:302–5.
94. Pollock ML, Mengelkoch LJ, Graves JE, et al. Twenty-year follow-up of aerobic power and body composition of older track athletes. J Appl Physiol. 1997;82:1508–16.
95. Koster A, Ding J, Stenholm S, et al. Health ABC study. Does the amount of fat mass predict age-related loss of lean mass, muscle strength, and muscle quality in older adults? J Gerontol A Biol Sci Med Sci. 2011;66:888–95.
96. Wroblewski AP, Amati F, Smiley MA, et al. Chronic exercise preserves lean muscle mass in masters athletes. Phys Sports Med. 2011;39:172–8.

97. Knechtle B, Knechtle P, Rosemann T. Upper body skinfold thickness is related to race performance in male Ironman triathletes. Int J Sports Med. 2011;32:20–7.
98. Knechtle B, Rust CA, Knechtle P, et al. Does muscle mass affect running times in male long-distance master runners? Asian J Sports Med. 2012;3:247–56.
99. Lepers R, Cattagni T. Do older athletes reach limits in their performance during marathon running? Age (Dordr). 2012;34:773–81.
100. American College of Sports Medicine Position Stand. Quantity and quality of exercise for developing and maintaining cardiorespiratory, musculoskeletal, and neuromotor fitness in apparently healthy adults: guidance for prescribing exercise. Med Sci Sports Exerc. 2011;43:1334–59.
101. Farinatti P, Monteiro W, Oliveira R, et al. Cardiorespiratory responses and myocardial function within incremental exercise in healthy unmedicated older vs. young men and women. Aging Clin Exp Res. 2018;30:341–9.
102. Montero D, Díaz-Canestro C. Endurance training and maximal oxygen consumption with ageing: role of maximal cardiac output and oxygen extraction. Eur J Prev Cardiol. 2016;23:733–43.
103. Fletcher GF, Balady G, Blair SN, et al. Statement on exercise: benefits and recommendations for physical activity programs for all Americans. A statement for health professionals by the Committee on Exercise and Cardiac Rehabilitation of the Council on Clinical Cardiology, American Heart Association. Circulation. 1996;94:857–62.
104. Hackam DG, Khan NA, Hemmelgarn BR, et al. The 2010 Canadian Hypertension Education Program recommendations for the management of hypertension: part 2 – therapy. Can J Cardiol. 2010;26:249–58.
105. Wiles JD, Taylor K, Coleman D, et al. The safety of isometric exercise: rethinking the exercise prescription paradigm for those with stage 1 hypertension. Medicine (Baltimore). 2018;97(10):e0105. https://doi.org/10.1097/MD.0000000000010105.
106. Oliveira A, Monteiro A, Jacome C, et al. Effects of group sports on health-related physical fitness of overweight youth: a systematic review and meta-analysis. Scand J Med Sci Sports. 2017;27(6):604–11. https://doi.org/10.1111/sms.12784.
107. Ray CA, Carrasco DI. Isometric handgrip training reduces arterial pressure at rest without changes in sympathetic nerve activity. Am J Physiol Heart Circ Physiol. 2000;279:H245–9.
108. Katayama K, Saito M. Muscle sympathetic nerve activity during exercise. J Physiol Sci. 2019;69:589–98.
109. Blagosklonny MV, Campisi J, Sinclair DA, et al. Impact papers on aging in 2009. Aging (Albany, NY). 2010;2:111–21.
110. Wilson VL, Smith RA, Ma S, et al. Genomic 5-methyldeoxycytidine decreases with age. J Biol Chem. 1987;262:9948–51.
111. Vickers MH. Early life nutrition, epigenetics and programming of later life disease. Nutrients. 2014;6:21652178.
112. Heyn H, Li N, Ferreira HJ, et al. Distinct DNA methylomes of newborns and centenarians. Proc Natl Acad Sci. 2012;109:10522–7.
113. Zhang FF, Cardarelli R, Carroll J, et al. Physical activity and global genomic DNA methylation in a cancer-free population. Epigenetics. 2011;6:293–9.
114. Ronn T, Volkov P, Davegardh C, et al. A six months exercise intervention influences the genome-wide DNA methylation pattern in human adipose tissue. PLoS Genet. 2013;9(6):e1003572.
115. O'Donovan A, Lin J, Dhabhar FS, et al. Pessimism correlates with leukocyte telomere shortness and elevated interleukin-6 in post-menopausal women. Brain Behav Immun. 2009;23:446–9.
116. Welle S, Brooks AI, Delehanty JM, et al. Gene expression profile of aging in human muscle. Physiol Genomics. 2003;14:149–59.

117. Morissette MP, Susser SE, Stammers AN, et al. Differential regulation of the fiber type-specific gene expression of the sarcoplasmic reticulum calcium ATPase isoforms induced by exercise training. J Appl Physiol (1985). 2014;117:544–55.
118. Lee JT. Epigenetic regulation by long noncoding RNAs. Science. 2012;338:1435–9.
119. Rinn JL, Chang HY. Genome regulation by long noncoding RNAs. Annu Rev Biochem. 2012;81:145–66.
120. Pandorf CE, Haddad F, Owerkowicz T, et al. The presence and regulation of antisense long non-coding RNA with altered myosin expression in exercising human muscle. (Abstract). FASEB J. 2012;26:1086.
121. Drummond MJ, McCarthy JJ, Sinha M, et al. Aging and microRNA expression in human skeletal muscle: a microarray and bioinformatics analysis. Physiol Genomics. 2010;43:595–603.
122. Ntanasis-Stathopoulos J, Tzanninis JG, et al. Epigenetic regulation on gene expression induced by physical exercise. J Musculoskelet Neuronal Interact. 2013;13:133–46.
123. Radom-Aizik S, Zaldivar F Jr, Oliver S, et al. Evidence for microRNA involvement in exercise-associated neutrophil gene expression changes. J Appl Physiol. 2010;109:252–61.
124. Wilson D, Jackson T, Sapey E, et al. Frailty and sarcopenia: the potential role of an aged immune system. Ageing Res Rev. 2017;36:1–10.
125. Janssen I, Heymsfield SB, Wang ZM, et al. Skeletal muscle mass and distribution in 468 men and women aged 18–88 yr. J Appl Physiol. 2000;89:81–8.
126. Wee AS. Carpal tunnel syndrome: comparison of the compound muscle action potentials recorded at the thenar region from ulnar and median nerve stimulation. Electromyogr Clin Neurophysiol. 2006;46:123–6.
127. Peterson SJ, Mozer M. Differentiating sarcopenia and cachexia among patients with cancer. Nutr Clin Pract. 2017;32:30–9.
128. Oliveira LF, Verneque D, Menegaldo LL. The influence of aging on the isometric torque sharing patterns among the plantar flexor muscles. Acta Bioeng Biomech. 2017;19:41–5.
129. Ryall JG, Schertzer JD, Lynch GS. Cellular and molecular mechanisms underlying age-related skeletal muscle wasting and weakness. Biogerontology. 2008;9:213–28.
130. Shibuya K. Cortical motor nuron hyperexcitability and motor neuron death in ALS: dying forward hypothesis. (Article in Japanese). Brain Nerve. 2017;69:565–9.
131. Doherty TJ. Invited review: Aging and sarcopenia. J Appl Physiol. 2003;95:1717–27.
132. Narici MV, Maffulli N. Sarcopenia: characteristics, mechanisms and functional significance. Br Med Bull. 2010;95:139–59.
133. Marcell TJ. Sarcopenia: causes, consequences, and preventions. J Gerontol Series A, Biol Sci Med Sci. 2003;58:M911–6.
134. Sayer AA. Sarcopenia the new geriatric giant: time to translate research findings into clinical practice. Age Ageing. 2014;43:736–7.
135. Abate M, Di Iorio A, Di Renzo D, et al. Frailty in the elderly: the physical dimension. Eur Medicophys. 2007;43:407–15.
136. Yarasheski KE. Exercise, aging, and muscle protein metabolism. J Gerontol Series A, Biol Sci Med Sci. 2003;58:M918–22.
137. Liu CJ, Latham NK. Progressive resistance strength training for improving physical function in older adults. Cochrane Datab Sys Rev. CD002759. https://doi.org/10.1002/14651858.cd002759.
138. Kim KM, Lim S, Oh TJ, et al. Longitudinal changes in muscle mass and strength, and bone mass in older adults: gender-specific associations between muscle loss. Gerontol A Biol Sci Med Sci. 2018;73:1062–9.
139. Chen L, Nelson DR, Zhao Y, et al. Relationship between muscle mass and muscle strength, and the impact of comorbidities: a population-based, cross-sectional study of older adults in the United States. BMC Geriatr. 2013;13:74. https://doi.org/10.1186/1471-2318-13-74.
140. Dey DK, Bosaeus I, Lissner L, et al. Changes in body composition and its relation to muscle strength in 75-year-old men and women: a 5-year prospective follow-up study of the NORA cohort in Göteborg, Sweden. Nutrition. 2009;25:613–9.

141. Sherrington C, Fairhall NJ, Wallbank GK, et al. Exercise for preventing falls in older people living in the community. Cochrane Database Syst Rev. 2019;1:CD012424. https://doi.org/10.1002/14651858.CD012424.pub2.
142. Baldwin KM, Haddad F. Research in the exercise sciences: where we are and where do we go from here–Part II. Exerc Sport Sci Rev. 2010;38:42–50.
143. Naichun J, Luan J, Hu F, Rubinek T, Modan Moses D. Klotho and the growth hormone/insulin-like growth factor 1 axis: novel insights into complex interactions. Vitam Horm. 2016;101:85–118.
144. Matsakas A, Narkar VA. Endurance exercise mimetics in skeletal muscle. Curr Sports Med Rep. 2010;9:227–32.
145. Matos N, Winsley RJ. Trainability of young athletes and overtraining. J Sports Sci Med. 2007;6:353–67.
146. Echigoya Y, Morita S, Itou T, et al. Effects of extracellular lactate on production of reactive oxygen species by equine polymorphonuclear leukocytes in vitro. Am J Vet Res. 2012;73:1290–8.
147. Suwa M, Nakano H, Kumagai S. Effects of chronic AICAR treatment on fiber composition, enzyme activity, UCP3, and PGC-1 in rat muscles. J Appl Physiol. 2003;95:960–8.
148. Mishra S, Chattopadhyay A, Naaz S, et al. Oleic acid ameliorates adrenaline induced dysfunction of rat heart mitochondria by binding with adrenaline: an isothermal titration calorimetry study. Life Sci. 2018;218:96–111.
149. McBride JM, Kraemer WJ. Free radicals, exercise, and antioxidants. J Strength Cond Res. 1999;13:175–83.
150. Park SY, Kwak YS. Impact of aerobic and anaerobic exercise training on oxidative stress and antioxidant defense in athletes. J Exerc Rehabil. 2016;12:113–7.
151. Saghiv M, Goldhammer E, Sagiv M, Ben Sira D. Klotho gene expression responses to long lasting aerobic training and aging. J Gerontol. 2015;1(2):009.
152. Betik AC, Hepple RT. Determinants of VO2 max decline with aging: an integrated perspective. Appl Physiol Nutr Metab. 2008;33:130–40.
153. Li Y, Niessen M, Chen X, et al. Overestimate of relative aerobic contribution with maximal accumulated oxygen deficit: a review. J Sports Med Phys Fitness. 2015;55:377–82.
154. Navarro A, Boveris A. The mitochondrial energy transduction system and the aging process. Am J Phys Cell Phys. 2007;292:C670–86.
155. Wei YH, Wu SB, Ma YS, et al. Respiratory function decline and DNA mutation in mitochondria, oxidative stress and altered gene expression during aging. Chang Gung Med J. 2009;32:113–32.
156. Ticinesi A, Meschi T, Narici MV, et al. Muscle ultrasound and sarcopenia in older individuals: a clinical perspective. J Am Med Dir Assoc. 2017;18:290–300.
157. Lovering RM, Brooks SV. Eccentric exercise in aging and diseased skeletal muscle: good or bad? J Appl Physiol (1985). 2014;116:1439–45.
158. Hernandez Morante JJ, Gomez Martínez C, Morillas-Ruiz JM. Dietary factors associated with frailty in old adults: a review of nutritional interventions to prevent frailty development. Nutrients. 2019;11(1). pii: E102. https://doi.org/10.3390/nu11010102.
159. Sagiv M, Sagiv A, Ben-Sira D, et al. Effects of chronic overload training and aging on left ventricular systolic function. Gerontology. 1997;43:307–15.
160. Monteiro AG, Alveno DA, Prado M, et al. Acute physiological responses to different circuit training protocols. J Sports Med Phys Fitness. 2008;48:438–42.
161. Karlsdottir AE, Foster C, Porcari JP, et al. Hemodynamic responses during aerobic and resistance exercise. J Cardpulm Rehabil. 2002;22:170–7.
162. Lichtenberg T, von Stengel S, Sieber C, et al. The favorable effects of a high-intensity resistance training on sarcopenia in older community-dwelling men with osteosarcopenia: the randomized controlled FrOST study. Clin Interv Aging. 2019;14:2173–86. https://doi.org/10.2147/CIA.S225618.
163. Sagiv M. Safety of resistance training in the elderly. Eur Rev Aging Phys Act. 2009;6:1–2.

164. Naichun J, Luan J, Hu F, et al. Aerobic exercise-stimulated Klotho upregulation extends life span by attenuating the excess production of reactive oxygen species in the brain and kidney. Exp Ther Med. 2018;16:3511–7.
165. Rubinek T, Modan Moses D. Klotho and the growth hormone/insulin-like growth factor 1 axis: novel insights into complex interactions. Vitam Horm. 2016;101:85–118.
166. Zuo Z, Lei H, Wang X, et al. Aging-related kidney damage is associated with a decrease in klotho expression and an increase in superoxide production. Age (Dordr). 2011;33:261–74.
167. Wang Y, Sun Z. Antiaging gene Klotho regulates endothelin-1 levels and endothelin receptor subtype B expression in kidneys of spontaneously hypertensive rats. J Hypertens. 2014;32:16291636.
168. Kuro OM. Klotho as a regulator of oxidative stress and senescence. Biol Chem. 2008;389:233–41.
169. Xiao NM, Zhang YM, Zheng Q, et al. Klotho is a serum factor related to human aging. Chin Med J (Engl). 2004;117:742–7.
170. Guo Y, Zhuang X, Huang Z, et al. Klotho protects the heart from hyperglycemia-induced injury by inactivating ROS and NFκB-mediated inflammation both in vitro and in vivo. Biochim Biophys Acta Mol basis Dis. 1864;2018:238–51.
171. Huang CL. Regulation of ion channels by secreted Klotho: mechanisms and implications. Kidney Int. 2010;77:855–60.
172. Liu BC, Yang LL, Lu XY, et al. Lovastatin-induced phosphatidylinositol-4-phosphate 5-kinase diffusion from microvilli stimulates ROMK channels. J Am Soc Nephrol. 2015;26:1576–87.
173. Mencke R, Hillebrands JL. NIGRAM consortium. The role of the anti-ageing protein Klotho in vascular physiology and pathophysiology. Ageing Res Rev. 2017;35:124–46.
174. Sopjani M, Rinnerthaler M, Kruja J, et al. Intracellular signaling of the aging suppressor protein Klotho. Curr Mol Med. 2015;15:27–37.
175. Semba RD, Cappola AR, Sun K, et al. Plasma Klotho and mortality risk in older community-dwelling adults. J Gerontol A Biol Sci Med Sci. 2011;66:794–800.
176. Tornbom K, Sunnerhagen KS, Danielsson A. Perceptions of physical activity and walking in an early stage after stroke or acquired brain injury. PLoS One. 2017;12(3):e0173463. https://doi.org/10.1371/journal.pone.0173463.
177. Park Y, Choi JE, Hwang HS. Protein supplementation improves muscle mass and physical performance in undernourished prefrail and frail elderly subjects: a randomized, double-blind, placebo-controlled trial. Am J Clin Nutr. 2018;108:1026–33.
178. Tavoian D, Russ DW, Law TD, et al. A randomized clinical trial comparing three different exercise strategies for optimizing aerobic capacity and skeletal muscle performance in older adults: protocol for the DART study. Front Med (Lausanne). 2019;6:236. https://doi.org/10.3389/fmed.2019.00236.

Chapter 8
Skeletal Muscles

Skeletal Muscle Anatomy

Skeletal muscle is one of the three major muscle types, the others being cardiac muscle and smooth muscle. It is a form of striated muscle tissue, which is under the voluntary control of the somatic nervous system [1]. Most skeletal muscles are attached to bones by bundles of collagen fibers known as tendons.

A skeletal muscle refers to multiple bundles (fascicles) of cells joined together called muscle fibers. The fibers and muscles are surrounded by connective tissue layers called fasciae. Muscle fibers, or muscle cells, are formed from the fusion of developmental myoblasts in a process known as myogenesis. Muscle fibers are cylindrical and have more than one nucleus. They also have multiple mitochondria to meet energy needs.

Muscle fibers are in turn composed of myofibrils. The myofibrils are composed of actin and myosin filaments, repeated in units called sarcomeres, which are the basic functional units of the muscle fiber. The sarcomere is responsible for the striated appearance of skeletal muscle and forms the basic machinery necessary for muscle contraction.

Although a skeletal muscle appears to function as a single unit, this is far more multifaceted than that. Cutting through the first muscle stage is the outer connective tissue (epimysium) covering the entire muscle and holding it together. Following the epimysium, there are small packs of fibers enveloped by a connective tissue covering (perimysium).

Beyond the perimysium, there are the muscle fibers, where each fiber is the muscle cells, covered by endomysium. Each fiber is composed of many sarcomeres, containing about 500,000 sarcomeres (Fig. 8.1). In fact, sarcomeres are the elementary functional units of the myofibril. Muscle fiber number varies between different muscles ranging from several hundred up to more than a million. Muscle fibers vary in diameter from 10 to 120 μm.

M. S. Saghiv, M. S. Sagiv, *Basic Exercise Physiology*,
https://doi.org/10.1007/978-3-030-48806-2_8

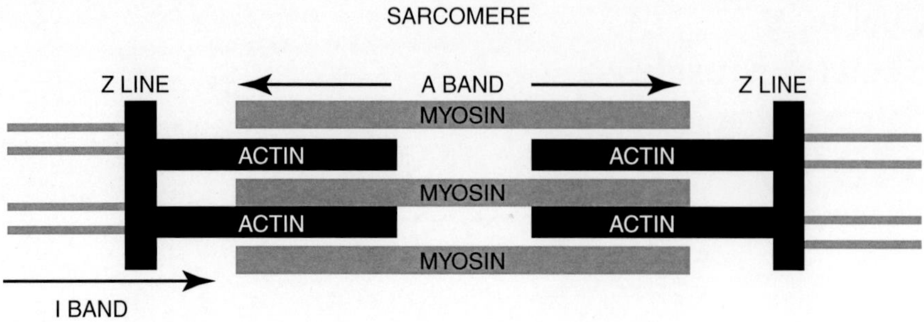

Fig. 8.1 Scheme of sarcomere

An individual muscle fiber is surrounded by a plasma membrane (plasmalemma) that is a part of a larger unit (sarcolemma). The plasmalemma is another name for the plasma membrane of a cell. Usually, each cell has one of these membranes, which functions as a containing wall to keep the contents separate from the surrounding environment. The plasmalemma is constructed from lipids, more commonly known as fats and proteins. Some of the proteins are fixed and some can move about within the cell membrane. A number of proteins are involved in transporting molecules into and out of the cell, while others help maintain the cell structure, act as receptors, or take part in chemical reactions as enzymes.

The sarcolemma contains the plasmalemma and the basement membrane. Each muscle fiber at its end, its plasmalemma rages with the tendon, made of fibrous cords of connective tissue that connect the fiber to the bone. Force produced by muscle fiber contraction is conveyed by the tendons generating motion.

The plasmalemma has a series of thin folds lengthwise the external of the fiber and when the fiber stretches, the folds disappear.

In addition, the plasmalemma preserves acid-base balance and conveyances metabolites from the capillary blood into the muscle fiber.

The growing and progress of skeletal muscle are completed by satellite cells situated among the plasmalemma and the basement membrane. Satellite cells are also responding to muscle's variation to injury and training. Satellite glial cells are glial cells that cover the surface of nerve cell bodies in sensory, sympathetic, and parasympathetic ganglia [2].

Both satellite glial cells and Schwann cells are derived from the neural crest of the embryo during development [3]. Satellite glial cells have been found to play a variety of roles, including control over the microenvironment of sympathetic ganglia [2]. They are thought to have a similar role to astrocytes in the central nervous system (CNS) [2].

Inside the plasmalemma, a muscle fiber contains successively smaller subunits. The largest of these are myofibrils being rod-like structures running the length of the muscle fibers. A gelatin-like substance fills the spaces within and between the myofibrils. This is the sarcoplasm. It is the fluid part of the muscle fiber—its cytoplasm. The sarcoplasm contains mainly dissolved proteins, minerals, glycogen, fats, and the necessary organelles. It differs from the cytoplasm of most cells because it contains a large quantity of stored glycogen as well as the oxygen-binding compound myoglobin, which is quite similar to hemoglobin.

The sarcoplasm also houses an extensive network of transverse tubules (T-tubules) which are extensions of the plasmalemma that pass laterally through the muscle fiber. With membranes that contain large concentrations of ion channels, transporters, and pumps, T-tubules permit rapid transmission of the action potential into the cell and also play an important role in regulating cellular calcium concentration. Through these mechanisms, T-tubules allow heart muscle cells to contract more forcefully by synchronizing calcium release throughout the cell. These tubules are interconnected as they pass among the myofibrils, allowing nerve impulses received by the plasmalemma to be transmitted rapidly to individual myofibrils. The tubules also provide pathways from outside the fiber to its interior, enabling substances to enter the cell and waste products to leave the fibers.

A longitudinal network of tubules, known as the sarcoplasmic reticulum, is also found within the muscle fiber. The sarcoplasmic reticulum is a membrane-bound structure found within muscle cells that is similar to the endoplasmic reticulum in other cells. The main function of the sarcoplasmic reticulum is to store calcium ions. Calcium ion levels are kept relatively constant, with the concentration of calcium ions within a cell being 100,000 times smaller than the concentration of calcium ions outside the cell [4]. These membranous channels parallel the myofibrils and loop around them. The sarcoplasmic reticulum serves as a storage site for calcium, which is essential for muscle contraction. The figure shown depicts the T-tubules and the sarcoplasmic reticulum.

Neural Control of Muscles

Neural regulation is the procedure utilized by the nervous system to regulate the whole thing from motion to physiological developments. The nervous system includes two main parts: central and peripheral (Fig. 8.2). Each system splits into two subdivisions, and altogether, the neural system includes four separate but highly team-up sub-systems, each of which gives an exclusive influence to motor control.

The nervous system is stimulated by sensory nerves and it is built in such a way that it passes on the electrical impulse to specific receptors. In the nervous system, there are neurons that are the basic structural units of the nervous system. Each neuron cell body has three major regions: A cranial nerve nucleus is a collection of neurons (gray matter) in the brain stem that is associated with one or more cranial nerves (Fig. 8.3).

Axons carrying information to and from the cranial nerves form a synapse first at these nuclei, body (soma), dendrites, and axon. Neurons are nerve cells that connect between sensors and the muscle; some are able to move fast the electrical impulse and others are slower. The neuron's exclusive reaction is the cell's ability to be

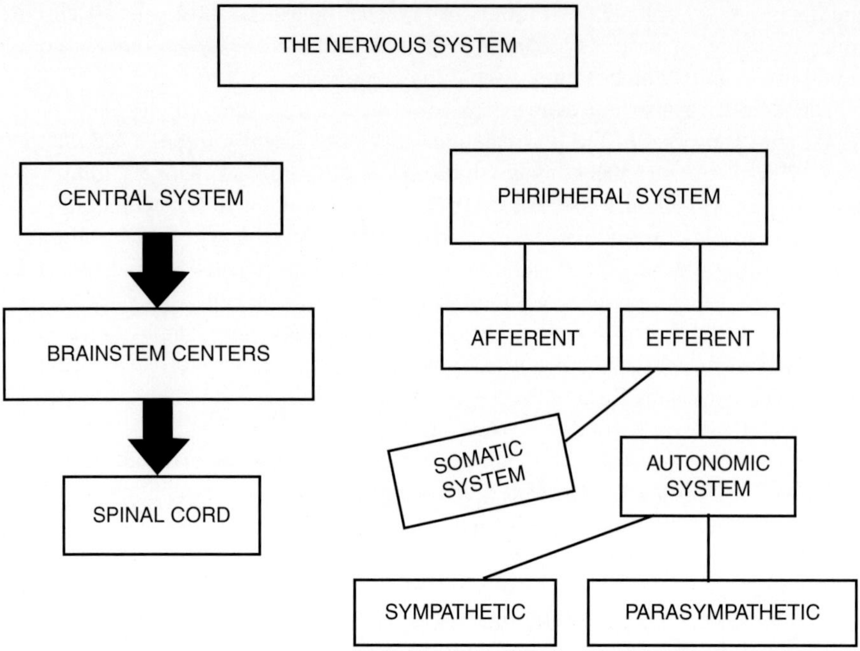

Fig. 8.2 The nervous system scheme: The neural courses consist of two types of nerves in the peripheral nervous system: sensory (afferent) and motor (efferent) nerves and are accountable for the regulation of movement. The motor nerves split into two categories: autonomic and somatic systems. The autonomic system also has two subdivisions: sympathetic and parasympathetic

Fig. 8.3 Each nerve cell includes the central part (nucleus or stoma), dendrites that are involved in stimulation, and an axon that is important for its involvement in the progression of the impulse from the point generated to other cells

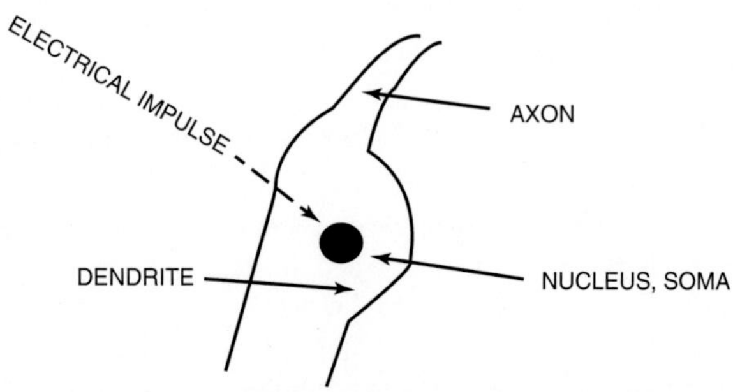

stimulated and to pass on the information from one part of the body to another part. Neurons are organized in a form of bundles called ganglia.

The applicable cells comprise inferior motor neurons and local circuit neurons that are the key base of synaptic contribution to the lower motor neurons. Instructions for motion are eventually carried to the muscles by the action of the lower motor neurons. The local circuit neurons attain sensory inputs as well as descending projections from more developed centers. Thus, the circuits coordinate between different muscles, which are significant for the beginning of any movement.

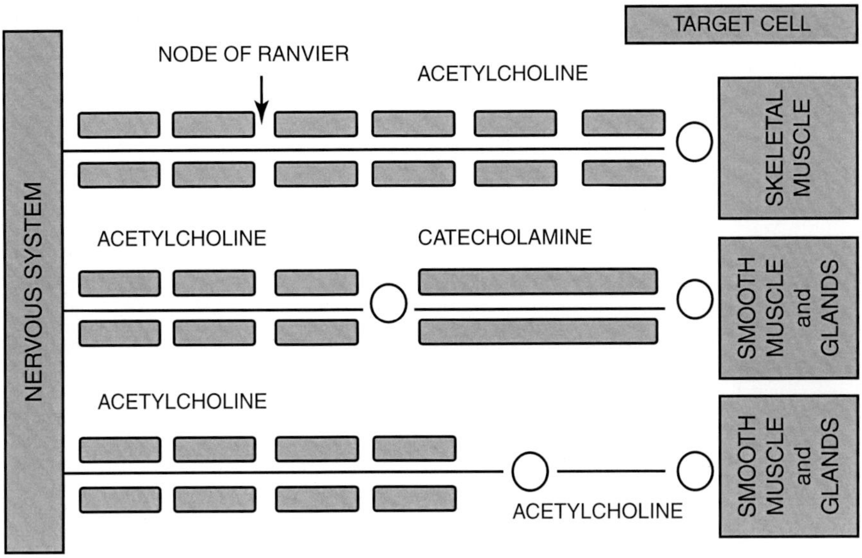

Fig. 8.4 Impulse generation by different chemicals in different nervous systems

The nervous system uses neurotransmitters frequently mentioned as the body's chemical messengers. They are the molecules utilized by the nervous system to convey communications between neurons or from neurons to muscles. The message between two neurons occurs in the small gap between the synapses of neurons (synaptic cleft). At this point, electrical signals (action potential) that have moved along the axon are briefly changed into chemical ones by the discharge of neurotransmitters, triggering a precise response in the receiving neuron. Different nervous systems use different neurotransmitters, namely catecholamine or acetylcholine (Fig. 8.4).

Muscle Function

Skeletal muscle shows notable adaptive abilities in response to a number of physiological and pathophysiological conditions, specifically, alterations occurring in the mitochondria in response to exercise or chronic contractile activity. Muscle tissues are derived from the mesodermal layer of embryonic germ cells in a process known as myogenesis. Skeletal muscles are covered by a hard layer of connective tissue (epimysium), which anchors muscle tissue to tendons at each end, where the epimysium becomes thicker and collagenous. Muscles are predominantly powered by the oxidation of fats and carbohydrates, but anaerobic chemical reactions are also used, particularly by fast twitch fibers. These chemical reactions produce ATP molecules that are used to power the movement of the myosin heads. Skeletal muscles adjust to chronic long-lasting exercise by significant functional improvements and structural changes in the mitochondria, capillary supply, and in myosin heavy chain (MHC) isoforms [5, 6].

Aerobic exercise training furthers the progress of an increase in mitochondrial volume, density, and mitochondrial proteins. These changes are seen in all muscle fiber types. The specific effects of increased contractile activity are caused by several factors, including nerve-dependent parameters, acute and sustained changes in intracellular free calcium concentration, metabolic factors such as hypoxia, and or weight loads [5]. A low mitochondrial functional capacity, for oxidative metabolism is related to insulin resistance and type 2 diabetes mellitus [7]. Exercise enhances mitochondrial enzymatic activity in skeletal muscle, which is likely related to the increases in mitochondrial biogenesis [8].

Muscular metabolic efficiency, defined as the ratio of work accomplished to energy expended or, as the ratio of mechanical work output to the total energy expenditure, calculated from VO_2, revealed 20–30% efficiency. When comparing individuals with low metabolic efficiency with those with high metabolic efficiency, one has to keep in mind mainly the training status [9] without significant gender differences. This low efficiency is the result of energy conversion to a different kind of energy or heat (second law of thermodynamics). When forming ATP from food and then by converting chemical energy into mechanical energy, a large portion of the energy turns into heat. During exercise, the amount of energy lost depends on the exercise mode and the type of muscle fibers being used (fast-twitch or slow-twitch). However, mitochondrial efficiency was not different in endurance trained versus untrained individuals. This suggests that the large differences among individuals during exercise seem to be due to the efficiency of transferring the chemical energy from ATP hydrolysis into physical work.

Skeletal muscle contractions are under the voluntary nervous system control, and they need synaptic contribution coming from motor neurons to initiate muscle contractions. One motor neuron is able to innervate numerous muscle fibers, causing the muscle fibers to contract at the same time. When the impulse passes the synapse, the actin and myosin filaments within each sarcomere of the skeletal muscle the thin filaments slide over the thick filaments. Actin and myosin filaments slide past each other to initiate a contraction. The rate of action potentials initiating contraction may be described as a twitch, summation, or tetanus. Contractions of smooth and cardiac muscles differ significantly from that of the skeletal muscle. Cardiac and smooth muscle cell contractions start by themselves (myogenic) and not as skeletal muscles that are stimulated by an external nerve. However, it is important to note that cardiac and smooth muscles can be controlled by the autonomic nervous system. Once the smooth and cardiac muscles contract, the mechanisms of contraction do not differ from that of the skeletal muscle.

The factor that starts and accelerates the movement is calcium (Ca^{++}), released from the sarcoplasmic reticulum. The electrical impulse passes the synapse to initiate the action potential in the muscle, which releases Ca^{++} from the sarcoplasmic reticulum. Figure 8.5 describes the five stages of impulse progression from the motoric fiber nerve to the motor unit:

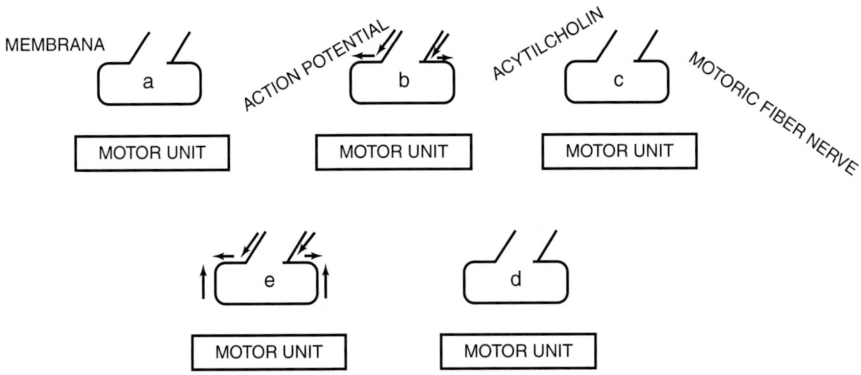

Fig. 8.5 The electrical impulse progression to contraction

(a) No contact between the motoric fiber nerve (causing muscle contraction) to the motor unit.
(b) Action potential progresses through the motoric fiber nerve.
(c) Acetylcholine, a neurotransmitter, is released when the impulse arrives at the motoric fiber nerve end.
(d) Acetylcholine connects to the receptors on the muscle, the membrane's motor unit.
(e) The last stage describes the action potential along the muscle fiber as a response to the Acetylcholine connection to the membrane's motor unit.

The end of muscle contraction is followed by muscle relaxation, which is a coming back of the muscle fibers to their small tension-generating state.

Sliding Filament Theory

Muscle contraction has a number of theories. The acceptable approach is the filament sliding (Fig. 8.6).

According to the sliding filament theory, the actin (thin) muscles' fiber filament slides over the myosin (thick) filament during muscle contraction, while the two groups of filaments remain at a comparatively constant length. The active contraction occurs when myosin reaches forward, binds to actin bridge, then myosin pulls the actin filament inward toward the center of the A band, and thus, the sarcomere (Z-to-Z line), the functional unit of striated muscle, is shortened, and muscle contracts. Relaxation takes place following contraction, and myosin releases actin and then grasps onward again to bind actin in a new cycle. This process is known as the cross-bridge cycle.

Figure 8.7, reveals that the connection between actin and myosin during muscle contraction and relaxation has eight stages [10]:

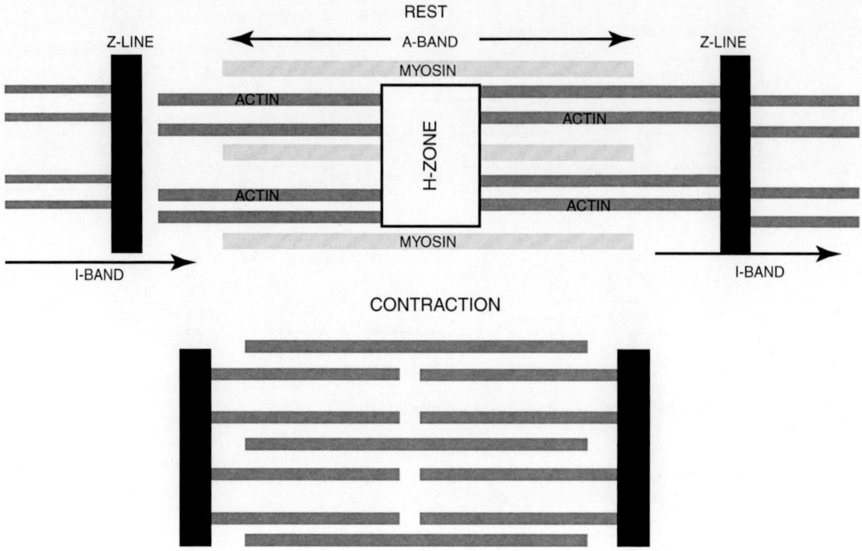

Fig. 8.6 Sliding filament theory: sarcomere Z-line to Z-line in relaxation and contraction stages

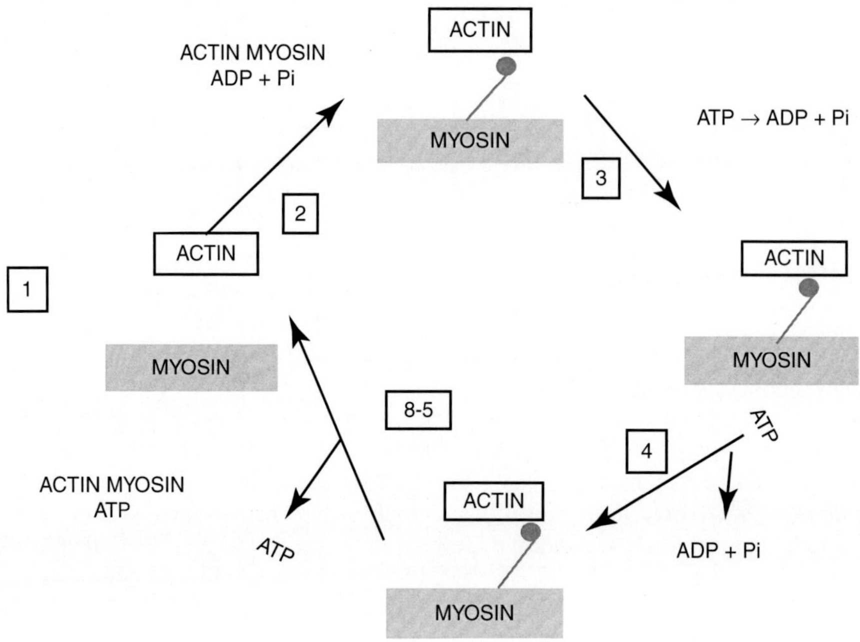

Fig. 8.7 Eight stages of cross-bridge attachment and detachment

In the first stage, there is no connection between myosin and actin.

1. Ca^{++} exits the sarcoplasmic reticulum to connect to troponin and tropomyosin proteins that cover the connecting site on actin to myosin.
2. The myosin cross-bridge is connected to actin facilitated by magnesium (Mg) and ATPase enzyme or water to split ATP to ADP and Pi along with energy release.
3. Actin slides toward myosin by the bridge rotation while creating momentum. The actin filament is connected at its middle part to a myosin cross-bridge, by the myosin's bridgehead that rotates clockwise, causing the actin filament to move to the right.
4. Caseation of the electrical impulse effect.
5. Detachment between Ca^{++} and troponin and tropomyosin proteins.
6. Ca^{++} returns to the sarcoplasmic reticulum.
7. Myosin bridge detaches from actin with an ATP cost.

Total energy cost of the cross-bridge attachment and the detachment cycle is 3–4 ATP.

Kinds of Muscle Contraction

Muscle contraction is the activation of tension-generating sites within muscle fibers [11]. Muscle contraction produces tension without changes in muscle length such as during isometric contraction [12]. The termination of muscle contraction is followed by muscle relaxation, which is a return of the muscle fibers to their low tension-generating state. As a result of the nerve impulse for contraction, muscles produce tension that depends on the impulse derivation. Therefore, muscle contraction does not essentially mean muscle shortening. Muscle contraction is the initiation of tension-generating spots within muscle fibers and may be defined based on two variables: tension and length. There are four different ways that muscle can contract:

(a) muscle tension rises but muscle length remains unchanged, and it is designated as isometric contraction.
(b) During isotonic contraction, muscle tension remains unchanged while muscle fibers are shortening.
(c) When muscle length shortens, it is a concentric contraction.
(d) when muscle length extends, it is eccentric contraction.

Force-Velocity Relationship

There are three main features affecting skeletal muscle force potential: velocity, length, and time; when the velocity of shortening increases the force ability decreases. When achieving maximal strength, the force-velocity relationship becomes the greatest significant mechanical characteristic because of the association between most athletic events and power. Muscle force is proportional to the physiological cross-sectional area, and muscle velocity is proportional to muscle fiber length. When muscle velocity shortening increases, force ability decreases. Therefore, rapid movements are linked to low force output and vice versa. The mechanical force-velocity relationship is the greatest imperative for strength athletes due to its variability during eccentric, isometric, and concentric contractions. It relies totally on movement speed, and therefore, the relative load is lifted. The force–velocity relationship has a physiological basis; when muscles contract, actin and myosin filaments create cross-bridges in the myofibrils. These cross-bridges take a limited amount of time to occur. As the velocity of shortening increases the filaments slide past one another faster and faster causing fewer cross bridges to be connected [13]. During muscle contraction, skeletal muscles increase in thickness or size. These form changes take part in moderating the skeletal muscle performance by increasing its range of motion velocities. Thus far, the underlying mechanisms related to muscle form change are still unknown [13].

Figure 8.8, explains the relationship between the opposing force that the muscle has to contract (load) against and power that the muscle produces. Force-velocity and muscle–force relationships are inversely related. It means that when an individual performs weightlifting, high levels of force are essential. However, muscle contraction velocity is low. On the contrary, once muscle velocity declines, time for the cross-bridges is enough to attach and to create force, and thus, force increases. Muscles increase their strength based on the force located crossways the muscle. Following resistance, training muscles develop forces that increase muscle strength. The force-velocity relationship is the observation that muscle force and contraction

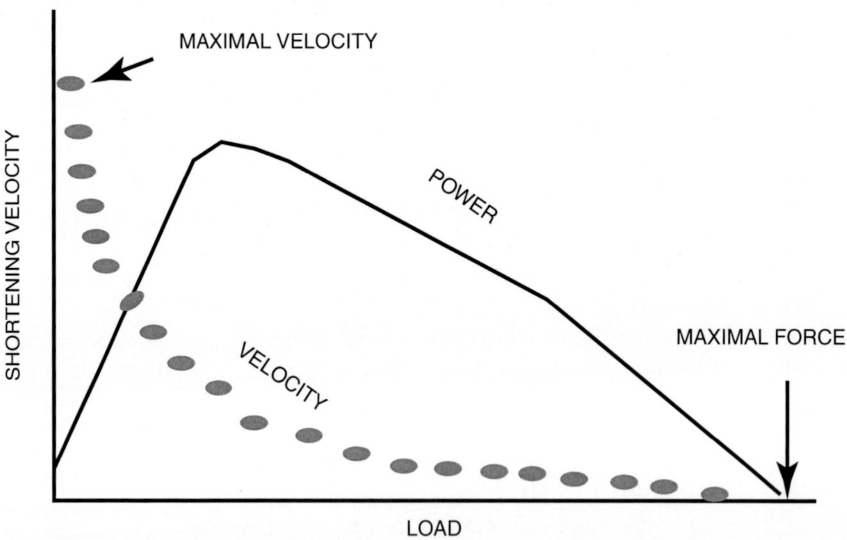

Fig. 8.8 Force-velocity relationship

Fig. 8.9 Sarcomere-length and tension relationship

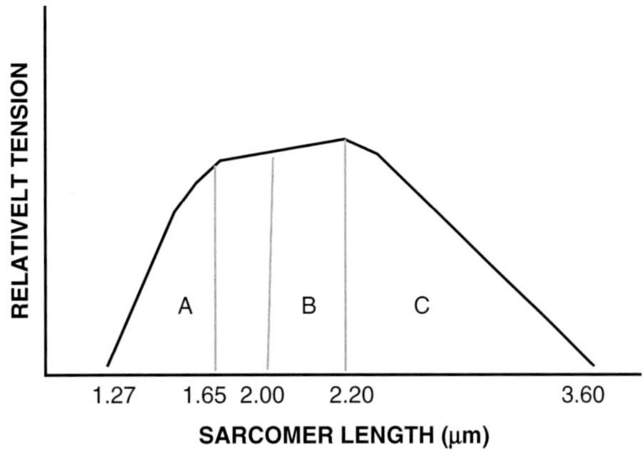

velocity are contrarily associated. In shortening concentric contraction, where contraction velocity is high, muscle force is low.

Figure 8.9, reveals that when the sarcomere length is shorter than 1.25–1.65 μm (A-zone), the distance between the actin and myosin is in such a distance that the sliding ability of the filaments is small and therefore contraction force is limited.

Where maximal contraction force is developed when the distance of the filaments is 2–2.2 μm (B-zone), the sliding ability is larger and therefore contraction force is maximal. One has to remember that contraction force depends on the amount of Ca^{++} released from the sarcoplasmic reticulum: a large amount of Ca^{++} was released, freeing more contacting sites between actin and myosin and thus stronger contraction. Situation C describes a sarcomere that the distance between the filaments is better than 2–2.2 μm, where the cross-bridge number connected to actin is low and therefore the sliding distance is limited, and consequently, contraction force is limited. As an example, a situation described in the C-zone is when a cardiac patient develops eccentric cardiomyopathy. Therefore, myocardial contraction is weak.

Muscle Fibers

Three kinds of muscles exist: skeletal muscle which is voluntary and involuntary such as the heart and smooth muscle. Skeletal muscle is a form of striated muscle tissue, which is under the voluntary control of the somatic nervous system [14]. Connective tissue is present in all muscles as fascia. Enclosing each muscle is a layer of connective tissue known as the epimysium; enclosing each fascicle is a layer called the perimysium, and enclosing each muscle fiber is a layer of connective tissue called the endomysium. Most skeletal muscles are attached to bones by bundles of collagen fibers known as tendons.

Many muscles are named by the action the muscle performs. These include: A flexor muscle decreases the anterior angle at a joint, an extensor increases the anterior angle at a joint, an abductor moves a bone away from the midline, an adductor moves a bone closer to the midline, an elevator raises a structure, a depressor moves a structure down, a supinator turns the palm of the hand up, a pronator turns the palm down, and a sphincter decreases the size of an opening; a tensor tenses a body part; and a rotator turns a bone around its axis [15].

The majority of muscles are built from different muscle fibers; muscle cells do not have a homogenous tissue. Muscle fibers are composed of myofibrils. The myofibrils are composed of actin and myosin filaments, repeated in units called sarcomeres, which are the basic functional units of the muscle fiber. The sarcomere is responsible for the striated appearance of the skeletal muscle and forms the basic machinery necessary for muscle contraction. Muscle fibers are the individual contractile units within a muscle. A single muscle contains many muscle fibers. Different clusters of cells are the myosatellite cells located between the basement membrane and the sarcolemma of muscle fibers [16]. These cells are usually inactive and turn to be active during exercise or pathological stage, to deliver nucleus to the muscle cell, for muscle evolution or restoration.

In muscles, there are two kinds of muscle fibers: slow twitch fibers that are mainly power-driven by the oxidation of fats and carbohydrates; nevertheless, anaerobic chemical reactions are also utilized, mainly by fast twitch fibers. Slow twitch fiber or red fiber or fiber I does not have subdivision, while the fast twitch or fiber II or white fiber has three subdivisions, fiber IIa, fiber IIb, and fiber IIc. The red fiber has high aerobic enzyme concentration for oxygenation and low anaerobic enzymatic ability. The white fiber has high anaerobic enzymes and very low aerobic enzymatic capacity. The subdivision of white fiber IIb is the fastest and classical in the body, with low aerobic enzyme concentration; however, it has the highest anaerobic enzymes of all fibers. Fiber IIa has low-moderate levels of aerobic capacity with high anaerobic ability. Fiber IIa is faster than the red fiber, still, slower than fiber IIb and fiber IIc. Fiber IIc is faster than fiber IIa and slower than fiber IIb. Aerobic exercise training influences fast twitch fibers by shifting them toward fiber IIa, while anaerobic exercise training changes fast twitch fibers to be more IIb. Muscle fiber composition is defined genetically, and under normal conditions, there is no way to switch between them. The red fiber differs from the white fiber in its mechanical and chemical responses. The red fiber differs from the white fiber in isometric ability and kinetic ability. The best evidence for that claim is the differences in the amount of ATPase, an enzyme that acts to split ATP on the cross-bridges. High ATPase levels as in fast twitch fibers increase ATP splitting, and thus, fiber activity is increased. In addition, skeletal muscle fibers also include two other vital governing proteins, troponin, and tropomyosin, needed for muscle contraction to occur. These proteins are preventing actin's contact with myosin. Skeletal muscle cells are excitable and are subject to depolarization by the neurotransmitter acetylcholine, released at the neuromuscular junction by motor neurons [17].

Muscular tension is influenced by the rate and characteristic of the muscle fiber recruitment rate of the electrical impulse movement along the nerve to the muscle, integration of the motor unit's activation, and fiber type: type I or type II with a higher ability of fiber II to generate greater muscular tension. Type I muscle fibers are mainly utilized during endurance exercise and compose the myocardium. Type I fibers yield slight degrees of force for lasting exercises such as aerobic activity (marathon runners). Type II fibers are utilized during anaerobic exercises, producing a high degree of force lasting short bursts of time such as weightlifting.

Most females and males do not differ in their mixture of both fiber I and fiber II. However, gender does affect the amount but not the superiority of the muscle, although the muscle fibers of females and males are typically similar (Fig. 8.10), males usually have extra muscle mass than females due to testosterone activity, and therefore, males are stronger than females. In males as in females, a significant difference exists between individuals in fiber compositions; some individuals are different genetically and may have a high amount of fibers I and a low percentage of fibers II; others have the opposite fiber composition. Comparing males and females reveals that males are stronger in any muscle group.

These differences are more pronounced in the upper limbs where female's strength is 50% less than males, while the differences in the lower limbs are about 30%. This might be the result of differences in lean body mass scattering and body size between females and males.

After the age of 30, the aging process muscle mass and bone mass are decreased and thus also strength and physical capacity. Muscle strength drops in individuals aged 40 years and above compared to subjects younger than 40 years up to 40%. With increasing age, body composition is changing with a loss of muscle mass and bone mass and a reduction of physical capacity over the years. Maximal work capacity is between the ages of 20 and 30 years. Marked changes with aging take place after the age of 50 years. As an example, at the age of 50 years, the leg lean body mass reduction rate is 2% per year and a strength loss of 1.5–5% per year. Although aging is a factor that humans do not have control on, muscle mass and strength can be increased following resistance training, with different strength rate gains, which are less with aging. However, resistance training still, delays the aging-related process of sarcopenia following resistance training.

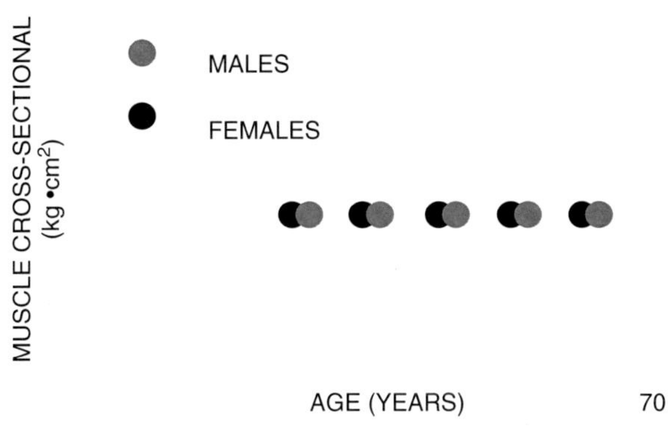

Fig. 8.10 Force generation for one unit of the biceps cross-section in males and females

Muscle Strength and Force

Muscle strength exertion depends on three factors: (a) physiological strength, resulting from muscle size, cross-sectional area, available cross-bridge availability, and responses to exercise training, (b) neurological strength depends on, the nerve signal strength and (c) the mechanical strength related to, muscle's leverage angel, arm length momentum and joint capabilities. Strength is the ability to generate force, and this is expressed during each body movement of the whole body or part of it. Force is a daily action, and most of it is during sports events. During several different power sports that need explosive force. Muscle strength is defined as the maximal force that a single muscle or group of muscles generate. The expression of muscular strength is specific to the test employed, the device utilized for peak force or peak torque. Consequently, the test applied should include the movements that are similar to the performance of interest. For instance, some tests of muscular strength can be performed using either machine weights, in which the movement is inhibited to follow an immovable trail, or free weights, in which the movement is quite unimpeded. Yet, a test done with machine weights not essentially yield the same result as the same test done with free weights. Additional factors that influence muscular strength are:

(a) the best day time to test an individual for muscular strength is in the early evening, and the mechanism for that is unclear yet, and,
(b) warm-up, prior to the test, optimizes performance, modifies the expression of muscular strength, and decreases the danger of injury.

Muscle force is influenced by motion speed. When the speed is 0, it is an isometric contraction and not dynamic contraction. The definition of isometric contraction is an increase in muscle tension without a change in its length. When the movement is done at different velocities, a dynamic force is measured. Directions of the dynamic contractions can be eccentric, which increase muscle's length or concentric, which shortens the muscle length. Lifting capacity is frequently measured by maximum static or dynamic physical efforts. Regularly, in sporting and ordinary activities, both velocity and acceleration characteristics alter through the movement. Velocity is defined as the amount at which the location of a body changes per unit of time, while acceleration is the rate at which the velocity changes per unit of time.

There are three main types of strength: The first kind is Static strength required during gymnastics maneuvers performance when the force execrated is steady for a period of time. The second type of strength is the Explosive strength that is expressed when performing a collision with the ball game when athletes meet the ball hard and fast, utilizing maximum energy. The third form is the Dynamic strength such during sprint and cycling, for example, dynamic strength is employed to work continuously with major leg muscles. However, a low to moderate relations between isometric strength measurements and lifting capacity suggest that estimates of the functional lifting capacity should not be based on static measurements alone. Dynamic lift tests are often a better simulation of the task being assessed and may be more appropriate for a back-injured population.

Resistance training is a mode of exercise that forces the muscles to contract against an exterior opposing force with the anticipation to increase strength, tone, muscle mass, and muscle endurance. Resistance exercise against external force relates to an extensive variety of activities leading to muscle contractions [18]. The exterior resistance can be anything that forces the muscles to contract. A major public health problem is physical inactivity that results in muscle mass loss and deficiency of muscle functions. This may reduce in the elderly especially glucose removal. Muscle mass loss and muscle dysfunction may reduce ATP resynthesize consequently, a reduction in chemical energy availability for muscle utilization. Although resistance exercise does not have an impact on aerobic and anaerobic capacities, yet, it has been widely recommended as a significant tool to improve muscle function. As consequences of training overload muscle, cross-sectional area, muscular endurance, and skeletal muscle size increase as well as strength and thus, may reduce the occurrence of sports injuries. These adaptations to resistance training alter body composition, cardiovascular function, and neuroendocrine responses. Training workloads suggest that resistance bouts to bring about physiological alterations and depend on sets number and repetitions. Thus, it is important to pay care when advising resistive-training workloads to elderly, sedentary individuals, and athletes. The reasons driving skeletal muscle atrophy and dysfunction are physical inactivity and aging. These were previously investigated intensively, suggesting that osteoporosis can result due to chronic inactivity that declines in bone density. In addition to physical inactivity, epigenetic changes with aging, mitochondrial dysfunction, and oxidative stress are factors affecting muscle atrophy. Resistance exercise training is a very nonpharmacological powerful anabolic stimulus for skeletal muscle regeneration and function in healthy individuals across the life span. However, the gains in lean body mass are modest in the elderly compared to young healthy individuals. The progression of muscle hypertrophy involves the increased synthesis of actin and myosin and the number of myofibrils.

Compared to aerobic exercise training, following resistance training, the number of capillaries is unchanged, the mitochondrial mass relative to the increase in fiber size is reduced, while quantities of fast heavy-chain myosin increase. Bones likewise muscles respond when load is applied creating strain on the tendons and muscles, which increase bone density. Consequently, the principle of progressive overload training places higher than normal demands on the working musculature, applies not only on muscles but bones as well. In addition, resistance training increases lean body mass along with a decrease in total body fat and percentage.

Resistance training induces tiny injury or tears (catabolism) to the muscle cells; however, the muscles renew (anabolism) and repair the cells, which in turn increases strength. Many biological processes of expansion in the body need some catabolism before anabolism prevails. Catabolism and healing of the muscle occur only under resting conditions. This is a good reason not to train every day without recovery time between workouts. The beneficial effects of resistance training are important since aging reduces muscle fiber number by 5 pounds of muscle every decade after the age of 30 years. Therefore, resistance training can reduce the pace of losing and sometimes may attenuate the aging process accordingly and increase skeletal muscle mass and strength. Resistance exercise training increases bone density in women and elderly Individuals suffering from bone loss (osteoporosis). In addition, resistance training may raise the metabolic rate, consequently maintaining the right lean body weight.

Cardiovascular Adaptations

During resistance training exercises, the immediate response of the HR increases up to120–150 beats·min^{-1} following a workout affected by the length of contraction, intensity (1 Rm%), and the muscle mass involved (handgrip vs. deadlift maneuvers). An abrupt and sharp increase in SBP, DBP, and MABP also is noted. This pressor response may be dangerous for sedentary individuals; however, in well-trained subjects, this hazard is reduced. Effects of resistance training on the cardiovascular system include left ventricular hypertrophy, almost without a change in left ventricular internal dimensions, resting HR, resting BP, and left ventricular systolic and diastolic functions.

Interestingly, in terms of chronic adaptations, there seems to be a drop or no changes in HR following resistance training. These results may be explained by the differences in intensity, repetition, muscle mass utilized, and duration that the training sessions last. Dynamic approaches to resistance training involve modest loads and high repetitions (such as circuit training) and are related to reductions in BP. The effects of other resistance modes of exercise training on BP revealed large variations in BP adaptations. Different investigations show a reduction in DBP values, unchanged SBP and DBP, and declines in SBP. This is not the case with competitive strength-trained athletes; they reduce their left ventricular cavity and thus volume due to an increase in wall thickness, septal thickness, and wall mass. This is different compared to aerobically trained athletes with increases in the left ventricular cavity. These changes reduce SV and Q, and therefore, as a compensation, HR is increased at rest and during exercise.

Muscle Elasticity

Muscle elasticity is a natural capability of the muscle, to return to its original muscle length when a tensile force is removed. The collagen present in the elastic element of the muscle has two types of performance when subjected to a tensile force: elasticity and plasticity. The collagen present in the elastic element of the muscle has two types of performance when exposed to an elastic force: elasticity and plasticity. Elastic performance in the muscle fiber returns the tissue to its original length when a stretchy force is removed [19].

Another type of performance is plasticity, which allows the muscle to adapt to the kind of overload that is imposed on it beyond its elastic range, so structural changes occur in the tissue, such as the tissue does not fit and returns to the original length when force is removed. In the case that muscle shortening takes place and comes before a concentric contraction, then the concentric contraction is more forceful resulting from muscle lengthening. Muscle changes its features by raising its tension through the elastic energy stored in the sequential elastic element of the muscle. This form of eccentric-concentric contractions is known as the muscle shortening-elongation cycle [20].

One of the important factors in the stretch-shortening cycle is the types of fibers used in it. Slow and fast-twitch fibers give a response to pre-stretching differently, and the stretching activity depends on them.

Muscles that possess fast fibers have more elastic energy as compared to slow fibers [21]. In a short distance, they can pre-stretch at an extremely high speed. As in the fast-twitch fibers cross-bridging of myosin takes place quickly, they can control the rapid elongation taking place, whereas in slow-twitch fibers, cross-bridging occurs slowly and energy doesn't get stored. Nonetheless, a benefit from the slow-twitch fiber is that of a vast range of motions from slow pre-stretching.

Following exercise, skeletal muscle has marked metabolic and morphological adaptations in response to the contractions. Skeletal muscles are highly capable to adapt in response to numerous biological, physical, and pathophysiological conditions [22]. Exercise alters the expression of a variety of nuclear and mitochondrial DNA gene products, resulting in phenotypic adaptations. This leads to an increase in muscle mitochondrial volume and changes in organelle composition, referred to as mitochondrial biogenesis [23]. The impact importance of biogenesis is enhanced resistance to fatigue. Signals initiated by the exercise bout involve changes in intracellular Ca^{++} as well as alterations in the ATP/ADP ratio and thus activation of kinases, such as AMP kinase and Ca^{2+}-calmodulin-activated kinases. These kinases activate transcription factors that bind DNA to affect the transcription of genes, the most evident manifestation of which occurs during the post-exercise recovery period when energy metabolism is directed toward anabolism, rather than contractile activity.

Skeletal muscle mass loss due to aging seems to be an important cause for the decreased mitochondrial quality and functional capacity. However, during aging, systematic exercise training can result in valuable physiological changes that improve cellular ATP production and metabolic process in skeletal muscle, including alterations to mitochondrial content, protein, and biogenesis [24].

Clinical Significance in Muscles

Skeletal muscle diseases (myopathies) and nerve diseases (neuropathies) affect muscle functions such as muscle hypertrophy, muscle atrophy, delayed onset muscle soreness, and mitochondrial myopathy.

Muscle Hypertrophy and Atrophy

Hypertrophy can be positive or negative. The positive hypertrophy is physiological hypertrophy, while the negative hypertrophy is pathological hypertrophy. Skeletal muscle mass and muscle fiber size differ according to physiological and pathological conditions. Hypertrophy occurs when the overall rates of protein turnover exceed the rates of protein degradation leading to muscle hypertrophy. Muscular hypertrophy is an increase in muscle mass and cross-sectional area [25], in response to contractile activity, such as strength exercise, aging, and specific hormones, such as androgens such as testosterone and β-adrenergic agonists contractile activity [26], and are biological factors affecting hypertrophy, until the late teens. Supplementation of additional testosterone or other anabolic steroids increases muscular hypertrophy.

Muscle can produce hypertrophy, independent of strength and muscular contraction, influenced by hormonal signaling, strength training, and disease. Mechanical overload applies stress on the skeletal muscle, forcing it, to adapt by increasing the size and amount of contractile proteins, which comprise the myofibrils within each muscle fiber, leading to an increase in the size of the individual muscle fibers and their consequent force production [27]. Enlargement of individual muscle fibers is a result of an increase in fiber size without a change in length, achieved without increasing muscle fiber number following exercise; rather, muscles grow larger through a combination of muscle cell growth as new protein filaments are added along with additional mass provided by undifferentiated satellite cells alongside the existing muscle cells [28].

Biological factors such as age and hormone levels can affect muscle hypertrophy. Two major signaling pathways regulate skeletal muscle growth: the IGF1–Akt–mTOR pathway, acting as a positive regulator, and the myostatin–Smad2/3 pathway, acting as a negative regulator [29], and additional pathways have recently been identified [27]. During puberty in males, hypertrophy occurs at an accelerated rate as the levels of growth-stimulating hormones produced by the body increase. Natural hypertrophy normally stops at full growth in the late teens. As testosterone is one of the body's major growth hormones, on average, men find hypertrophy much easier to achieve than women. Taking additional testosterone or other anabolic steroids will increase muscular hypertrophy.

Proliferation and fusion of satellite cells, leading to an increase in the number of myonuclei, may also contribute to muscle growth during early but not late stages of postnatal development and in some forms of muscle hypertrophy in the adult. Muscle atrophy occurs when protein degradation rates exceed protein synthesis and may be induced in adult skeletal muscle in a variety of conditions, including starvation, denervation, cancer cachexia, heart failure, and aging. Two major protein degradation pathways, the proteasomal and the autophagic–lysosomal pathways, are activated during muscle atrophy and variably contribute to the loss of muscle mass. These pathways involve a variety of atrophy-related genes or aerogenes, which are controlled by specific transcription factors, such as $FoxO_3$, which is negatively regulated by Akt and NF-κB, which is activated by inflammatory cytokines.

IGF-1 is a primary mediator of the effects of the growth hormone. The growth hormone is made in the anterior pituitary gland, is released into the bloodstream and then stimulates the liver to produce IGF-1. IGF-1 then stimulates systemic body growth and has growth-promoting effects on almost every cell in the body, especially skeletal muscle, cartilage, bone, liver, kidneys, nerve, skin, hematopoietic, and lung cells. In addition to the insulin-like effects, IGF-1 can also regulate cellular DNA synthesis [30]. For example, muscle-specific inactivation of the IGF1 receptor impairs muscle growth due to reduced muscle fiber number and size [31]. Conversely, muscle-specific overexpression of IGF1 causes muscle hypertrophy [29]. IGF1 is known to activate both the mitogen-activated protein kinase and extracellular signal-regulated kinase (MAPK/ERK), and the PI3K-Akt pathways result in hypertrophy of transfected muscle fibers [29]. FoxO factors have been shown to inhibit mTORC1 and activate Akt by inducing the expression of sestrin3 and Rictor in cultured mammalian cells [32]; nonetheless, this was not been seen in vivo.

Muscle atrophy includes the reduction of myofibers as a result of protein degradation rates exceeding protein synthesis, organelles, and cytoplasm induced, by immobility, malnutrition, aging, physical inactivity, or a wide range of injuries or diseases that impact the musculoskeletal or nervous system [29]. Myostatin causes upregulation of the E3 ubiquitin ligases MuRF1 and MAFbx, mediating skeletal muscle atrophy, myostatin decreases expression of these atrophy markers in differentiated myotubes, as well as other genes normally upregulated during differentiation [33]. Muscle atrophy leads to muscle weakness and causes disability. Daily activities break down 1–2% of muscle; however, the body is able to rebuild and recover these losses. The normal decline rate in muscle mass throughout the lifespan is 0.37% per year in women and 0.47% per year in men [34].

Inactivity and starvation in mammals lead to atrophy of skeletal muscle, a decrease in muscle mass that may be accompanied by a smaller number and size of the muscle cells as well as lower protein content.

Left Ventricular Physiological Hypertrophy

Cardiac growth or hypertrophy observed in trained athletes is referred to as "physiological," and it is characterized by normal or enhanced contractile function and normal architecture and organization of cardiac structure [35]. In recent years, several investigations have demonstrated an association between left ventricular remodeling and the angiotensin-converting enzyme (ACE I/D) and/or angiotensinogen polymorphisms, in athletes with the DD and/or TT genotypes, respectively. However, the impact of the known genetic determinants on left ventricular remodeling is at present incomplete, with genetic determinants explaining up to one-quarter of the overall variability of left ventricular dimensions [36]. The basic mechanism by which exercise activates genes involves a stimulus signal to the DNA, then transcription via messenger RNA, and finally translation into protein [37]. However, if the genetic predisposition is set for a lower level range, changes due to physical activity can be trained up to a certain extent.

Adaptation of the myocardium to strength training with a slight increase in left ventricular internal diameter and a large increase in left ventricular wall thickness can be explained on the basis of blood pressor response and low Q during weight lifting [38].

Hypertrophy is a physiological adaptation by which the heart compensates for an increased workload. At some point, this process can be defined as an athletic heart in healthy subjects or as pathological in cardiac patients. An important problem is determining when left ventricular hypertrophy ceases to be physiological and begins to be pathological. Physiological hypertrophy is defined as hypertrophy accompanied by a normal or augmented contractile rate in which the maximum rate at which myosin hydrolyzes ATP and the maximal velocity of muscle shortening is either normal or elevated [39]. Pathological hypertrophy, on the other hand, is associated with depressed contractility not necessarily accompanied by concomitant heart failure, in which case the rate of myosin ATPase activity and the velocity of muscle shortening are decreased [40]. To differentiate between physiological and pathological myocardial changes, the relationship between heart size and ergometric performance, as well as the echocardiographic measured ratio between left ventricular myocardial thickness and volume, is useful.

Pathological Ventricle Hypertrophy

Pathological hypertrophy is the response to stress or diseases such as hypertension, heart muscle injury (myocardial infarction), or neurohormones. Valvular heart disease is another cause of pathological hypertrophy. Pathological hypertrophy is induced by factors such as prolonged and abnormal hemodynamic stress, high BP, and myocardial infarction. Pathological hypertrophy differs from physiological hypertrophy in their underlying molecular mechanisms; cardiac phenotype [40]; prognosis; and signaling pathways that cause hypertrophy progression, interact cardiomyocytes with non-cardiomyocytes that induce cardiomyocyte hypertrophy, and progress myocyte death and fibrotic remodeling [41]. With birth, cardiomyocytes depart the myocardial cell cycle and differentiates immediately. Consequently, in the mature myocardium, cardiomyocytes grow in size as a substitute to a rise in cardiomyocyte number, and thus, the myocardium progresses hypertrophy to counter ventricular wall stress and increased workload. Physiological hypertrophy is associated with normal cardiac function, while pathological one relates to cardiac dysfunction. Left ventricular hypertrophy resulting from cardiovascular disease and/or hypertension is accompanied by significant morbidity and mortality and in advanced stages may be irreversible. Pathological left ventricular hypertrophy also leads to an increase in myocardium mass, but the myocardium does not increase its pumping ability and instead accumulates myocardial scarring (collagen). In pathological left ventricular hypertrophy, the heart can increase its mass by up to

150%. It is related to fibrosis, reduced capillary density, increased cytokines levels, and waning of signaling, autophagy destruction, and abnormal cardiomyocyte/non-cardiomyocyte relations. In addition, epigenetic changes, resulting in an inappropriate cardiac remodeling adjustment that generally progresses to heart failure.

Cardiac and/or hypertensive patients have a markedly lower average rapid left ventricular filling rate. Thus, in those patients with left ventricular hypertrophy, abnormal left ventricular diastolic filling rate is more closely related to impaired myocardial contractility than to left ventricular chamber ejection fraction [42]. Although abnormal left ventricular diastolic filling is more closely related to impaired myocardial contractility than to left ventricular chamber ejection fraction, coronary artery disease patients and/or hypertensive patients with reduced ejection fraction and left ventricular filling pattern have higher left ventricular mass, eccentric left ventricular geometry (with thin left ventricular wall thickness), or concentric with wall thicknesses of 12–18 mm, prolonged left ventricular isovolumic relaxation time, atrial filling fraction, and heart rate compared to those patients with normal ejection fraction [43].

Athlete's Heart

Physiological hypertrophy of the heart occurs in response to the normal growth of children or during pregnancy, as well as in athletes [41]. Cardiovascular adaptations to exercise have been systematically defined and differ according to the type of conditioning: aerobic training such as long-distance running and swimming; and strength training (also referred to as static, isometric, power, or anaerobic) such as wrestling, weightlifting, or throwing heavy objects. Long-lasting cardiovascular adaptation to dynamic training produces an increase in Q and $(a\text{-}v)O_2$. Physical exercise is associated with hemodynamic changes that alter the loading conditions of the myocardium. Thus, the predominant effect of endurance exercise is to produce volume overload on the left ventricle, while strength exercise causes a pressure overload for the most part. Left ventricular eccentric hypertrophy forms, where ventricular wall thickness and cavity size are increased, as a consequence of the volume overload produced during exercise [44], while left ventricular concentric hypertrophy forms, where ventricular wall thickness increases and cavity size is unchanged or decreased, as a consequence of the pressure overload produced during resistance exercise [45]. The increase in wall thickness induced by pressure overload results from an increase in the myocyte cross-sectional area [46].

Athlete's heart is a collection of morphological cardiac changes, including increased left ventricular volume, increased left ventricular mass, increased left atrial volume, and right ventricular structural changes as physiological responses to exercise training. These structural changes fall within the normal reference ranges of appropriately matched control subjects for most trained individuals; however, there are significant numbers of athletes who have "abnormal" measurements. The ability to distinguish between physiological changes associated with athlete's heart and structural abnormalities that may represent underlying cardiac disease is of paramount importance [47].

In addition, athletes demonstrate a spectrum of alterations in the 12-lead electrocardiogram (ECG) pattern, including a marked increase in pre-cordial R-wave or S-wave voltages, ST segment or T-wave changes, and deep Q waves, suggestive of left ventricular hypertrophy, which may raise the possibility of a pathologic heart condition but have also been viewed as a consequence of morphological cardiac remodeling induced by athletic conditioning.

Even though the morphology of athlete's heart and the impact of different sports on cardiac structure have been investigated previously by several authors [48], they have not been able to satisfactorily resolve the question of whether two types of athlete's heart exist.

The Renin-Angiotensin System

Studies of left ventricular hypertrophy causes center on the renin-angiotensin system and identification of candidate genes. Data show that different molecular mechanisms are responsible for the different types of left ventricular mass increases seen in athletes and cardiac patients.

Puthucheary et al. [49] have shown the association between renin-angiotensin system genes (angiotensin-converting enzyme gene ACE) with changes in left ventricular mass in response to army physical training, which is a mix of static and endurance exercise. Training-induced left ventricular mass changes are significantly associated with the presence of angiotensin-converting enzyme D allele.

Other studies [50], suggested that exercise-induced left ventricular growth in young males is strongly associated with the angiotensin-converting enzyme I/D polymorphism. It was found that the absence of the D allele from the 287-base pair marker in the angiotensin-converting enzyme gene resulted in a higher than normal angiotensin-converting enzyme level.

However, the role of angiotensin-converting enzyme gene in power athletes who developed concentric cardiac hypertrophy is unclear. It was hypothesized that ACE genotypic modulation characteristics may affect left ventricular mass in power athletes [51].

Duchenne's Muscle Dystrophy

Duchenne's muscle dystrophy is a genetic disorder that gradually weakens the body's muscles; thus, immobility ultimately occurs. It is caused by unseemly or absence of genetic information that stops the body from producing the proteins desired to construct and preserve vigorous muscles. It is a genetic condition that is inborn through abnormal genes passed from one parent. Duchenne's dystrophy only occurs in males, but it may be passed on by a mother who is a carrier of the abnormal gene.

Even though some benefits of muscle exercise were consistently reported across studies, the benefits might be due to the small size of studies and other biases [52]. Therefore, whether muscle exercise is beneficial or damaging for patients with muscle dystrophy has been a dispute for a long time and is divisive [53]. Strength training or aerobic exercise programs might optimize muscle and cardiorespiratory functions and prevent additional disuse atrophy and deconditioning in people with muscle dystrophy. Muscle weakness is the key problem and exercise could be of help if it counterbalances the loss of muscle tissue and strength [54]. Although resistance and aerobic exercises at moderate intensity in muscle dystrophy seem to be unharmed, there is an inadequate indication to accomplish that those exercises are helping. On the other hand, in mitochondrial myopathy, aerobic exercise combined with strength training appears to be safe and may be effective in increasing submaximal endurance capacity [54]. Muscle dystrophy depends on the type of muscular dystrophy, and in general, it can appear around the age of 40. The deterioration period is slow, although sometimes it is very severe that may shorten life expectancy. Unfortunately, no treatment or cure for muscular dystrophy exists.

Delayed Onset Muscle Soreness

Delayed onset muscle soreness is one symptom of exercise-induced muscle damage. It is prevalent subsequently to strenuous exercise, which typically means that muscles are being stronger. Symptoms can vary from muscle sensitivity to Spartan incapacitating pain [55]. Delayed onset muscle soreness is the pain and stiffness arising from muscle weakness, numerous hours to days following unaccustomed exercise or all-out exercise. Soreness is theoretical as gloomy and painful in the injured muscle, often combined with tenderness and stiffness. Usually, the pain is felt only when the muscle contracts or stretched; however, at rest, there is no pain. This soreness, a characteristic symptom of delayed onset muscle soreness (hyperalgesia), is referred to as muscular mechanical soreness [56]. Sometimes, delayed onset muscle soreness is an acute muscle soreness, which is felt during exercise and nearly post-exercise. Delayed onset muscle soreness is most common following a period of decreased activity (rest) and at the beginning of the competitive season [55]. Even though there is an alteration between modes of exercises and individuals, it is utmost strongly felt within 24–72 h peaking up to 2 days later, following the exercise bout [57], then lessens and fades away within 7 days following exercise. It seems that delayed onset muscle soreness is caused by lengthening the muscle-tendon complex during eccentric exercise, which causes microtrauma (micro-injury) to the muscle fibers, lesions at the Z-line of the muscle sarcomere. Exercise-induced muscle damage, reduce muscle strength, and power has a substantial effect on endurance running performance, mediated by modifications in the sense of effort [58], and this reduction in performance is impermanent. It is an exercise with an intensity that is above and beyond what the individual is used to perform. Exertional rhabdomyolysis seems to be the risky form of delayed onset muscle soreness.

Regardless of an augmented number of research studies concentrating on the eccentric contraction, a substantial difference yet remains in the knowledge of the cellular and molecular mechanisms, causing the early micro-injury response and successive adaptations to eccentric exercise [59]. Following such an exercise, the muscle treatment is rest, since the muscle adjusts fast to avoid muscle damage and thus soreness [57].

In order to avoid these contrary effects, individuals should, in addition to rest, repeat sessions at submaximal eccentric exercises, with an increasing intensity along the training [59].

Measuring Isometric and Dynamic Strength

The most important tool utilized to measure isometric strength is the cable tensiometer (Fig. 8.11). This device enables us to measure a variety of muscle groups and a range of motions in different angles. Another device to measure isometric strength is the dynamometer: handgrip (Fig. 8.12) computing hands' strength and deadlift (Fig. 8.13) measuring legs' strength. The maximal strength that a single or a group of muscle contractions can generate at a constant speed in a given direction is

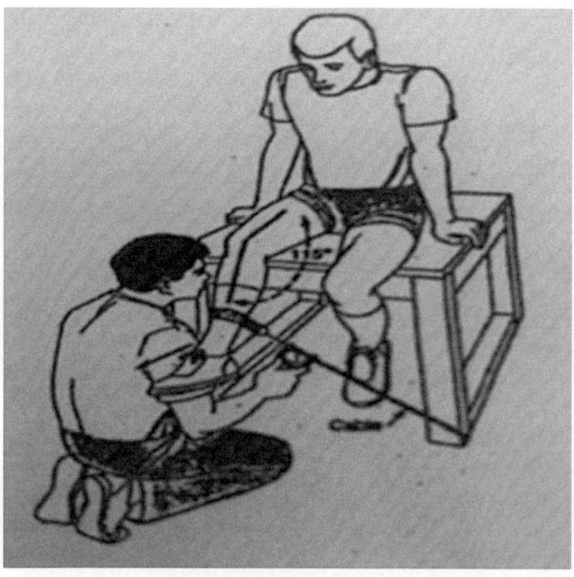

Fig. 8.11 Tensiometer device

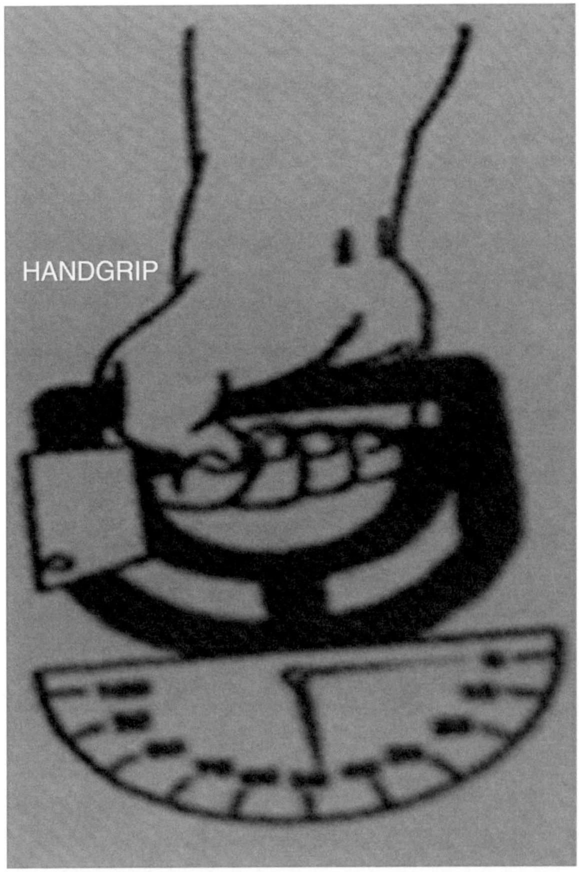

Fig. 8.12 Handgrip dynamometer

Maximal dynamic strength. It can be measured by 1-Rm, which means the strength generated in one maximal repetition.

Isokinetic contraction is measured in constant speed along the full range of motion. Isokinetic force is measured with computerized dynamometer electromechanical strength (Fig. 8.14).

Laboratory on Resistance

Equipment:

- Hand dynamometer
- Deadlift dynamometer
- Isokinetic device
- Borg table

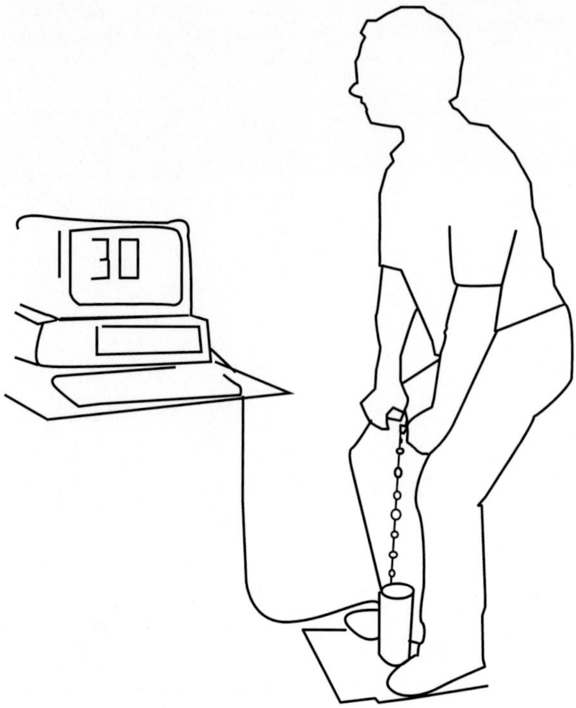

Fig. 8.13 Deadlift dynamometer

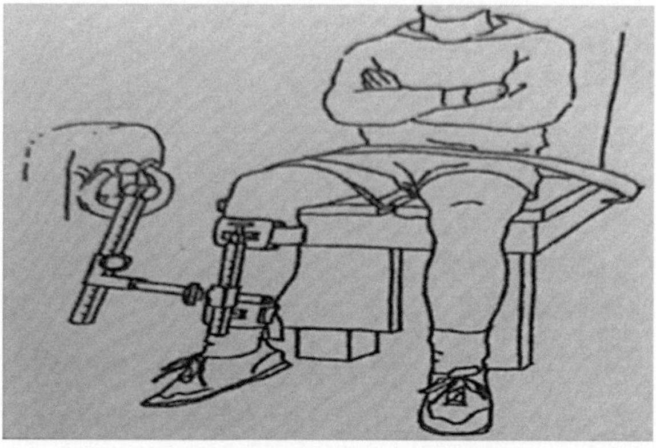

Fig. 8.14 Isokinetic device measuring maximal strength with constant speed along the full range of motion

Testing stages:

Experimental Procedure:

In this laboratory, four individuals, out of them, two will be tested. The other two will serve as testers.

Pre-test instructions for subjects:

(a) Subjects must refrain from exercise 2 h prior to the test.
(b) Subjects should minimize activity on the day of the test, if possible.

Test procedures:

- The tester will measure three times with hand-dynamometer maximal voluntary contraction (MVC).
- The highest value will be the subject's MVC.
- The same procedure will be followed by the deadlift dynamometer.
- The same procedure will be followed by the chest-press maneuver.
- The subject will perform with the isokinetic device knee flexion and knee extension.

Assignments:

- All maneuver's results will be reported as follows:
- Compare males with females.
- Do the same for lean body mass.
- Determine the lean body mass by a Caliper as suggested earlier.
- Explain the results with a scientific background.

Laboratory report

Name ____________ Weight __________

Isometric handgrip test

Trial 1__________kg Trial 1__________kg Trial 1_________kg

maximal strength________________

Isometric deadlift

Trial 1__________kg Trial 1__________kg Trial 1__________kg

Chest press

Trial 1__________k g Trial 1__________kg Trial 1__________kg

Isokinetic test

Knee flexion maximal momentum value _____ N•m^{-1}
Optimal degree____°

Knee extension maximal momentum value _____ N•m^{-1}
Optimal degree____°

References

1. Birbrair A, Zhang T, Wang Z-M, et al. Role of pericytes in skeletal muscle regeneration and fat accumulation. Stem Cells Dev. 2013;22:2298–314.
2. Hanani M. Satellite glial cells in sympathetic and parasympathetic ganglia: in search of function. Brain Res Rev. 2010;64:304–27.
3. Gotz M, Hartfuss E, Malatesta P. Radial glial cells as neuronal precursors: a new perspective on the correlation of morphology and lineage restriction in the developing cerebral cortex of mice. Brain Res Bull. 2002;57:777–88.

4. Bronner F. Extracellular and intracellular regulation of calcium homeostasis. Sci World J. 2003;1:919–25.
5. Fluck M, Hoppeler H. Molecular basis of skeletal muscle plasticity–from gene to form and function. Rev Physiol Biochem Pharmacol. 2003;146:159–216.
6. Hood DA, Irrcher I, Liubicic V, et al. Coordination of metabolic plasticity in skeletal muscle. J Exp Biol. 2006;209:2265–75.
7. Petersen KF, Morino K, Alves TC, et al. Effect of aging on muscle mitochondrial substrate utilization in humans. Proc Natl Acad Sci U S A. 2015;112:11330–4.
8. Simoneau JA, Colberg SR, Thaete FL, et al. Skeletal muscle glycolytic and oxidative enzyme capacities are determinants of insulin sensitivity and muscle composition in obese women. FASEB J. 1995;9:273–8.
9. Hopker JG, Coleman DC, Passfield L. Changes in cycling efficiency during a competitive season. Med Sci Sports Exerc. 2009;41:912–9.
10. Fitts RH. The cross-bridge cycle and skeletal muscle fatigue. J Appl Physiol. 2007;104:551–8.
11. Silverthorn DU. Muscles. Human physiology: an integrated approach. 7th ed. San Francisco, CA: Pearson; 2016. p. 377–416.
12. Widmaier EP, Hersel R, Strang KT. Muscle. Vander's human physiology: the mechanisms of body function. 12th ed. New York, NY: McGraw-Hill; 2010. p. 250–91.
13. Dick TJM, Wakeling JM. Shifting gears: dynamic muscle shape changes and force-velocity behavior in the medial gastrocnemius. J Appl Physiol (1985). 2017;123:1433–42.
14. Birbrair A, Zhang T, Zhong-Min W, et al. Role of pericytes in skeletal muscle regeneration and fat accumulation. Stem Cells Dev. 2013;22:2298–314.
15. Tortora G, Anagnostakos N. Principles of anatomy and physiology. 5th Harper international ed: Harper & Row; 1987. p. 219.
16. Zammit PS, Partridge TA, Yablonka-Reuveni Z. The skeletal muscle satellite cell: the stem cell that came in from the cold. J Histochem Cytochem. 2006;54:1177–91.
17. Costanzo LS. Physiology. 2nd ed. Philadelphia: Saunders; 2002. p. 23.
18. American College of Sports medicine. American College of Sports medicine position stand. Progression models in resistance training for healthy adults. Med Sci Sports Exerc. 2009;41:687–708.
19. Lieber RL, Roberts TJ, Blemker SS, et al. Skeletal muscle mechanics, energetics and plasticity. J Neuroeng Rehabil. 2017;14(1):108. https://doi.org/10.1186/s12984-017-0318-y.
20. Rassier DE. Sarcomere mechanics in striated muscles: from molecules to sarcomeres to cells. Am J Phys Cell Physiol. 2017;313:C134–45.
21. Fitts RH, Widrick JJ. Muscle mechanics: adaptations with exercise-training. Exerc Sport Sci Rev. 1996;24:427–73.
22. Huertas JR, Casuso RA, Agustín PH, et al. Stay fit, stay young: mitochondria in movement: the role of exercise in the new mitochondrial paradigm. Oxidative Med Cell Longev. 2019;2019:7058350. https://doi.org/10.1155/2019/7058350.
23. Popov DV. Adaptation of skeletal muscles to contractile activity of varying duration and intensity: The role of PGC-1α. Biochemistry (Mosc). 2018;83:613–28.
24. Seo DY, Lee SR, Kim N, et al. Age-related changes in skeletal muscle mitochondria: the role of exercise. Integr Med Res. 2016;5:182–6.
25. Iorga B, Neacsu CD, Neiss WF, et al. Micromechanical function of myofibrils isolated from skeletal and cardiac muscles of the zebrafish. J Gen Physiol. 2011;137:255–70.
26. Schiaffino S, Dyar KA, Ciciliot T. Mechanisms regulating skeletal muscle growth and atrophy. FEBS J. 2013;280:4294–314.
27. Scheid LM, Weber C, Bopp N, et al. Extraction protocols for individual zebrafish's ventricle myosin and skeletal muscle actin for *in vitro* motility assays. Front Physiol. 2017;8:367. https://doi.org/10.3389/fphys.2017.00367.
28. Poole RM, editor. The incredible machine. Washington, DC: Nat Geograp Soc; 1986. p. 307–11.

29. Schiaffino S, Mammucari C. Regulation of skeletal muscle growth by the IGF1-Akt/PKB pathway: insights from genetic models. Skelet Muscle. 2011;1(1):4. https://doi.org/10.1186/2044-5040-1-4.
30. Yakar S, Rosen CJ, Beamer WG, et al. Circulating levels of IGF-1 directly regulate bone growth and density. J Clin Invest. 2002;110:771–81.
31. Mavalli MD, DiGirolamo DJ, Fan Y, et al. Distinct growth hormone receptor signaling modes regulate skeletal muscle development and insulin sensitivity in mice. J Clin Invest. 2010;120:4007–20.
32. Chen CC, Jeon SM, Bhaskar PT, et al. FoxOs inhibit mTORC1 and activate Akt by inducing the expression of Sestrin3 and Rictor. Dev Cell. 2010;18:592–604.
33. Glass DJ. PI3 kinase regulation of skeletal muscle hypertrophy and atrophy. Curr Top Microbiol Immunol. 2010;346:267–78.
34. Mitchell WK, Williams J, Atherton P, et al. Sarcopenia, dynapenia, and the impact of advancing age on human skeletal muscle size and strength; a quantitative review. Front Physiol. 2012;3:260.
35. Shiojima I, Walsh K. Regulation of cardiac growth and coronary angiogenesis by the Akt/PKB signaling pathway. Genes Dev. 2006;20:3347–65.
36. Di Mauro M, Izzicupo P, Santarelli F, et al. ACE and AGTR1 polymorphisms and left ventricular hypertrophy in endurance athletes. Med Sci Sports Exerc. 2010;42:915–21.
37. Morucci G, Punzi T, Innocenti G, et al. New frontiers in sport training: genetics and artistic gymnastics. J Strength Cond Res. 2014;28:459–66.
38. Sagiv M, Amir O, Goldhammer E, et al. Left ventricular contractility in response to upright isometric exercise in heart transplant recipients and healthy men. J Cardiopulm Rehabil Prev. 2008;28:17–2.
39. Olah A, Nemeth BT, Matyas C, et al. Physiological and pathological left ventricular hypertrophy of comparable degree is associated with characteristic differences of in vivo hemodynamics. Am J Physiol Heart Circ Physiol. 2016;310:H587–97.
40. Nakamura M, Sadoshima J. Mechanisms of physiological and pathological cardiac hypertrophy. Nat Rev Cardiol. 2018;15:387–407.
41. Shimizu I. Physiological and pathological cardiac hypertrophy. J Mol Cell Cardiol. 2016;97:245–62.
42. Palmieri V, Bella JN, DeQuattro V, et al. Relations of diastolic left ventricular filling to systolic chamber and myocardial contractility in hypertensive patients with left ventricular hypertrophy (The PRESERVE Study). Am J Cardiol. 1999;84:558–62.
43. Wachtell K, Papademetriou V, Smith G, et al. Relation of impaired left ventricular filling to systolic midwall mechanics in hypertensive patients with normal left ventricular systolic chamber function: the Losartan Intervention for Endpoint Reduction in Hypertension (LIFE) study. Am Heart J. 2004;148:538–44.
44. Turner MJ, Spina RJ, Kohrt WM, et al. Effect of endurance exercise training on left ventricular size and remodeling in older adults with hypertension. J Gerontol Ser A. 2000;55:M245–M25.
45. Barauna VG, Rosa KT, Irigoyen MC, et al. Effects of resistance training on ventricular function and hypertrophy in a rat model. Clin Med Res. 2007;5:114–20.
46. Gaasch WH, Zile MR. Left ventricular structural remodeling in health and disease: with special emphasis on volume, mass, and geometry. J Am Coll Cardiol. 2011;58:1733–40.
47. Atchley AE Jr, Douglas PS. Left ventricular hypertrophy in athletes: morphologic features and clinical correlates. Cardiol Clin. 2007;25:371–82.
48. Utomi V, Oxborough D, Whyte GP, et al. Systematic review and meta-analysis of training mode, imaging modality and body size influences on the morphology and function of the male athlete's heart. Heart. 2013;99:1727–33.
49. Puthucheary Z, Skipworth JR, Rawal J, et al. The ACE gene and human performance: 12 years on. Sports Med. 2011;41:433–48.

50. Gharavi AG, Lipkowitz MS, Diamond JA, et al. Deletion polymorphism of the angiotensin-converting enzyme gene is independently associated with left ventricular mass and geometric remodeling in systemic hypertension. Am J Cardiol. 1996;77:1315–9.
51. Kasikcioglu E, Kayserilioglu A, Ciloglu F, et al. Angiotensin-converting enzyme gene polymorphism, left ventricular remodeling, and exercise capacity in strength-trained athletes. Heart Vessel. 2004;19:287–93.
52. Gianola S, Pecoraro V, Lambiase S, et al. Efficacy of muscle exercise in patients with muscular dystrophy: a systematic review showing a missed opportunity to improve outcomes. PLoS One. 2013;8(6):e65414. https://doi.org/10.1371/journal.pone.0065414.
53. Ansved T. Muscle training in muscular dystrophies. Acta Physiol Scand. 2001;171:359–66.
54. Voet NBM, van der Kooi EL, Riphagen II, et al. Strength training and aerobic exercise training for muscle disease. Cochrane Data Sys Rev. 2010;CD003907. https://doi.org/10.1002/14651858.CD003907.pub3.
55. Cheung K, Hume P, Maxwell L. Delayed onset muscle soreness: treatment strategies and performance factors. Sports Med. 2003;33:145–64.
56. Taguchi T, Matsuda T, Tamura R, Sato J, Mizumura K. Muscular mechanical hyperalgesia revealed by behavioural pain test and c-Fos expression in the spinal dorsal horn after eccentric contraction in rats. J Physiol. 2005;564:259–68.
57. Nosaka K. Muscle soreness and damage and the repeated-bout effect. In: Tiidus PM, editor. Skeletal muscle damage and repair. Human kinetics. 2008. p. 59–76.
58. Marcora SM, Bosio A. Effect of exercise-induced muscle damage on endurance running performance in humans. Scand J Med Sci Sports. 2007;17:662–71.
59. Hody S, Croisier JL, Bury T, et al. Eccentric muscle contractions: risks and benefits. Front Physiol. 2019;10:536. https://doi.org/10.3389/fphys.2019.00536.

Chapter 9
Thermoregulation

Body Water

Water is the main chemical component assembling the human body. In a young male, water amount is 50–57% of body weight. Water keeps equal heat between intracellular and extracellular fluids throughout the entire cell. In case that water does not keep equal heat between intracellular and extracellular fluids, cell temperature increases sharply due to metabolic heat production [1]. Thus, that heat increase may break down the cells. Sweat is all salt water with some other constituents. When an individual sweats, the sweat glands filter out water from the blood and it appears as sweat all over the skin [2], consequently increasing sweat rate and skin blood flow.

Water level is critically influential on electrolyte concentration: Na, K, and Ca; therefore, core temperature regulation is important. Changes in those electrolytes' concentration have a significant impact on the electrical impulse progression.

The physiological reason and advantages that the body uses water for temperature regulation are as follows:

(a) Adding solvent to the water changes frizzing and boiling temperatures.
(b) Water is able to keep ion's solvent which is very important for electrical impulse progression.
(c) For the Van der Waals forces, it takes a long time to increase water temperature. In order to increase water temperature, it needs a large amount of energy to change water's physical state. This energy absorption by water can be seen as heat storage without increasing the core temperature.

Note, cell's water is a medium for chemical reactions. This medium facilitates cell's material exchange and right blood volume.

The primary sweating source for water is the intracellular fluid followed by the within tissue extracellular fluid. In extreme conditions the cardiovascular system can be the water source for sweating. Fluid movements between water

M. S. Saghiv, M. S. Sagiv, *Basic Exercise Physiology*,
https://doi.org/10.1007/978-3-030-48806-2_9

compartments are done by osmosis. Under harsh conditions, when water loss is significant in addition to hot surroundings, a significant secretion of aldosterone and antidiuretics (ADH) takes place to increase kidney water absorption [3]. Following acclimatization and training, water compartments are stabilized especially blood plasma and improve water exchange between compartments. In addition, sweat mechanism is improved along with an increase in plasma volume.

During high-intensity exercise, an individual may lose water up to 2 $L{\cdot}h^{-1}$ [4]; under this condition, signs of dehydration appear, including dark-colored urine, muscle cramps, decreased sweat rate, and increased fatigue. However, usually, water loss is around to 1 $L{\cdot}h^{-1}$, which can be replaced almost immediately. Environmental temperature, exercise intensity, humidity, VO2max level, and environment acclimation are factors affecting sweat rate. Replacing fluids during and after exercise is very important for staying hydrated and preventing dehydration.

Thermoregulation

Exercise heat acclimation induces physiological adaptations that improve thermoregulation, attenuate physiological strain, reduce the risk of serious heat illness, and improve aerobic performance in warm-hot environments and potentially in moderate environments [5]. The adaptations include improved sweating, improved skin blood flow, lowered core temperatures, reduced cardiovascular strain, improved fluid balance, altered metabolism, and enhanced cellular protection. The magnitudes of adaptations are determined by the intensity, duration, frequency, and number of heat exposures, as well as the environmental conditions (i.e., dry or humid heat). Evidence is emerging that controlled hyperthermia regimens where a target core temperature is maintained enable more rapid and complete adaptations relative to the traditional constant work rate exercise heat acclimation regimens. Furthermore, inducing heat acclimation outdoors in a natural field setting may provide more specific adaptations based on direct exposure to the exact environmental and exercise conditions to be encountered during competition.

Ambient temperature has an important role in the ability of humans during their workout, because core temperature regulation is vital for physiological function [6]. In addition to heat generated by the contracting muscles, ambient temperature and humidity impose excessive load on the thermoregulatory system. Under these circumstances, heat interchange from the body to the atmosphere is reduced considerably. Humans are able to keep core temperature around 37 °C (98.6 °F), while normal ranges are 36.1 °C (97.0 °F) to 37.8 °C (100.0 °F); even when the temperature is close in range, the thermoregulation system acts. This thermoregulation capacity is a state of dynamic stability or temperature homeostasis. Increased core temperature above 45 °C or decrement below 18 °C changes body's metabolic enzymatic activity during chemical reaction that might result in death. Humans are tropical animals that can stand cold weather than hot environments. Thermoregulation is the preservation of a comparatively continuous core temperature. The range in which life can exist is 26–42 °C. That means that humans can increase their core temperature by only 5 °C while reduce it by 11 °C (Fig. 9.1).

Muscular contraction produced heat at rest and during exercise due to the breakdown of muscle fat, glucose, and glycogen to produce ATP and mechanical energy [7]. Metabolic processes are inefficienct since, during muscle contraction, more than 75% of the oxidized molecule's energy potential is converted into heat, which must be dissipated. Rate heat production is linear with muscle metabolic level. This heat production varies during exercise and is linear to exercise intensity. At rest and during exercise, many biological mechanisms are stimulated to balance heat gain–loss (Fig. 9.2) to control body core temperature within the normal range.

In spite of the existence of cooling factors, there are situations due to high heat loads when the body does not function normally: hyperthermia, reduction in substrate delivery to produce ATP in the cells [8], and the loss of the ability to regulate BP due to a large vasodilatation.

Core temperature is regulated and steadied mainly by the hypothalamus, linking the endocrine system to the nervous system. When the core temperature differs from the normal temperature range, endocrine response recruits different mechanisms to elevate or reduce energy production or dissipation as needed to return to the normal temperature range toward the predetermined set point. When exercise commences, heat gradient between the working muscles and core temperature is created, which is transferred to the body core through conductive and convective.

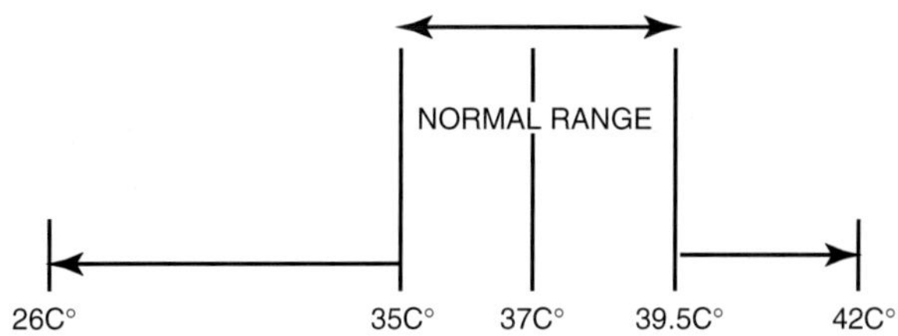

Fig. 9.1 Core temperature boundaries within the normal range

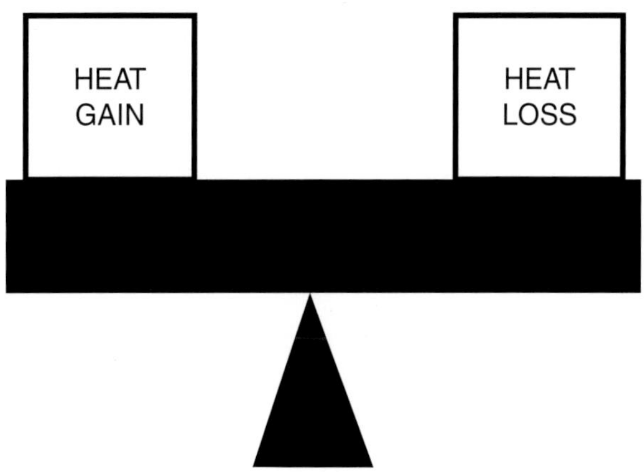

Fig. 9.2 Heat-gain balance

Factors Affecting Heat Gain–Loss

Factors that elevate core temperature at rest and during exercise are external heat, exercise intensity, muscle heat production, and shivering (Fig. 9.3). Factors that elevate ambient temperature include radiation directly from the sun and radiation from the surrounding (external) heat; thus, even in the shade the body is exposed to heat and ground radiation as an example is the radiation from snow while skiing.

The amount of heat gain or loss through radiation depends on the surroundings and clothes worn. On a rainy day, with bright clothing radiation to the body is low while with dark clothing and clear sky radiation is high and accordingly heat gain. In addition, clothes interfere withradiation, conduction, convection, and evaporation. Exercise intensity determines muscle metabolic rate, higher the intensity the more heat production. Working muscles by shivering are able to produce heat when the body's internal temperature declines; this is thermogenesis. Under cold conditions, muscles involuntarily start to contract and produce heat. Muscles can produce fast and bulky amounts of heat by thermogenesis.

At rest and during exercise, numerous physiological mechanisms are stimulated to avoid an extreme increase in body core temperature. Heat body loss (Fig. 9.4) is achieved by radiation of the body's heat to the surroundings and loss of heat through the respiratory system is best when the ambient temperatures are low. Therefore, heat loss by radiation and respiration stops to be a factor in temperatures above 30–31 °C. Radiation is the transfer of heat through electromagnetic waves through space, without the need for molecules or atoms, unlike convection or conduction.

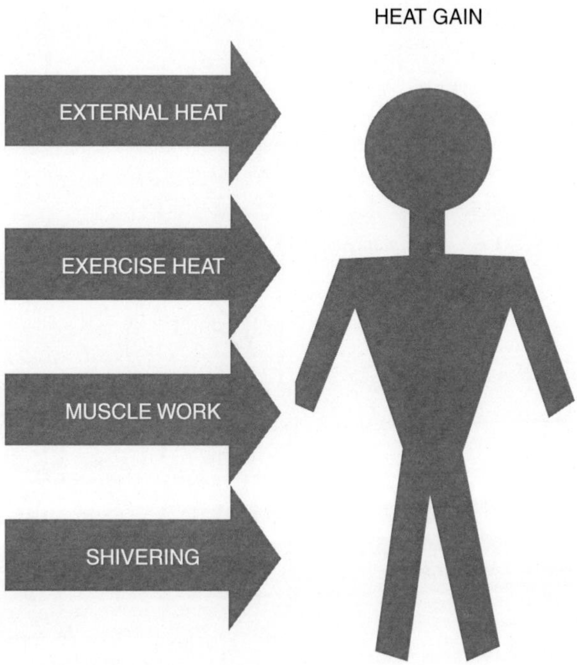

Fig. 9.3 Factors affecting heat gain

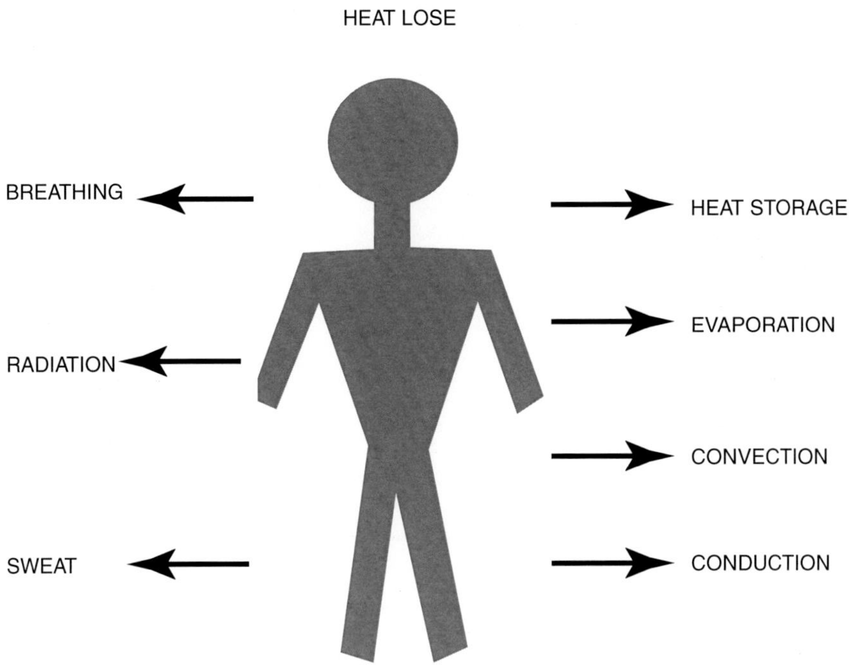

Fig. 9.4 Factors affecting heat loss

In cases of extreme ambient temperatures (above 40 °C) the respiratory system may add heat to the body. When the surrounding temperature reaches the value of 40 °C, the respiratory system as a cooling factor is no more effective. Conduction and convection are two factors with high importance in low surrounding temperatures. Heat is carried by the blood to the skin where it evaporates. Once the ambient temperature exceeds that of core temperature, sweat is the only physiological means for individuals to dissipate heat. The increase in core temperature, at some point, activates skin vasodilation and sweating. Eccrine sweat glands under the skin secrete sweat, on the skin exterior, causing heat loss by evaporation. The only situation that influences the rate and amount of sweat secretion is core temperature, which heats the blood. Arteriolar vasodilation allows increased blood flow through superficial capillaries in the skin increasing heat loss by convection and conduction. Conduction is a process by which heat is transferred from the hot area of a solid object to the cool area of a solid object by the collisions of particles. Energy is transferred to the nearest atom or molecule by physical contact or collision. Convection is a heat transfer by molecules such as air or liquid from a solid body to another by the movement of the air or liquid. The thin air layer next to the skin is heated by bloodstream coming from the body. Then the heat is carried away from the body in the ambient air flows. The amount of heat loss by convection depends on the skin surface area, ambient air flow velocity, and temperature difference between the surrounding's air and the skin surface.

Control of Skin Circulation

During aerobic exercise, a key role in heat dissipation processes is performed by skin temperature. Whenever aerobic exercise commences, at any given surrounding and skin temperatures, body core temperature increases along with skin blood flow [9]. Sympathetic neural control of skin blood flow includes the catecholamine vasoconstrictors, which increase their activity during cold exposure, to decrease skin blood flow, and a sympathetic active vasodilator occurs via cholinergic nerve transmission and has been shown to include potential roles for nitric oxide, that is, responsible for up to 90% of the cutaneous vasodilation that occurs during body heat stress [10]. Yet, there is no agreement on which is the single mediator leading to cutaneous vasodilation [10]. During environmental heat exposure and during aerobic exercise, there is a massive increase in skin blood flow due, to cutaneous vasodilatation, thus, body heat lose is increased, while during cold exposure, cutaneous vasoconstriction, avoids unnecessary heat dissipation [11]. At very high environmental temperatures, skin temperature rise, and cutaneous blood flow increases meaningfully. Due to the rise in skin temperature. The increased cutaneous blood flow transfers metabolic heat production from the body core to the skin. In the skin, heat is lost to the surroundings by convective, radiation, and evaporative mechanisms. High levels of skin blood flow increase fluid losses by filtration leading to reduction in central venous pressure [12]. Consequently, venous return declines, which lowers SV; consequently, HR increases to keep an adequate Q. Following long-lasting aerobic exercise, heat acclimation increases sweat rate during thermal stress, resulting in a significant reduction in skin and core temperatures during exercise, which in turn lessen skin blood flow.

Sweat and Evaporation

Sweating is the production of fluids secreted by the sweat glands in the skin. The sweat glands are under the influence of the sympathetic nervous system connected to the hypothalamus by an electrical impulse generated following the increase in blood heat. Generally, the sympathetic nervous system acts by means of catecholamines while on the sweat glands the neurotransmitter is acetylcholine. Body sweating areas differ between individuals; usually, in untrained subjects the active sweat glands are in the back. In trained individuals the active sweat glands are in the front meeting air while in motion. The forehead is the place where most people lose significant heat.

Sweating is the most powerful autonomic thermo-effector. The evaporation of sweat provides by far the greatest potential for heat loss, and it represents the only means of heat loss when air temperature exceeds skin temperature. Sweat production results from the integration of afferent neural information from peripheral and central thermoreceptors which leads to an increase in skin sympathetic nerve activity. At the neuroglandular junction, acetylcholine is released and binds to muscarinic receptors which stimulate the secretion of a primary fluid by the secretory coil of eccrine glands. The primary fluid subsequently travels through a duct where ions are reabsorbed. The end result is the expulsion of hypotonic sweat on to the skin surface. Sweating increases in proportion with the intensity of the thermal challenge in an attempt by the body to attain heat balance and maintain a stable internal body temperature. The control of sweating can be modified by biophysical factors, heat acclimation, dehydration, and nonthermal factors. The purpose of this chapter is to review the role of sweating as a heat loss thermo-effector in humans.

The warm blood reaches the hypothalamus which in turn increases the rate and amount of sweat. Maximal amount of sweat during exercise may be 4 $L{\cdot}h^{-1}$. During strenuous aerobic exercise the rate can increase to 1 Lometer 2 of body surface area. Exercising in hot environment costs an additional of 12% in sweating amount. In hot ambient conditions when workload is equal to 3 $LO^2{\cdot}min^{-1}$, loss of body water can be 1.5–1.75 $L{\cdot}min^{-1}{\cdot}h^{-1}$. Children, adolescents, and older (above 50 years) individuals are restricted in their capacity to cool down by perspiration due to the limited ability to increase sweating rate. During exercise in high temperatures, sweat secretion exceeds water drinking, and consequently, hypohydration stage may be reached. Factors influencing the rate of evaporation include the following: (a) whenever body core temperature rises, (b) with the increase in skin surface area, (c) evaporation declines with an increase in humidity, and (d) increases when wind velocity rises (Fig. 9.5).

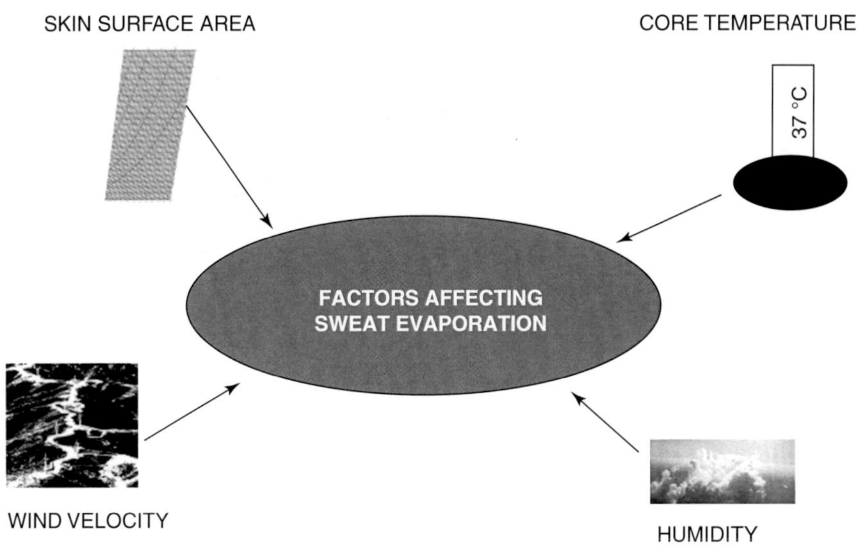

Fig. 9.5 Factors affecting evaporation

Table 9.1 Skin moister level in different air humidity percentages in ambient temperature of 31 °C

Air humidity (%)	Skin water (%)
20	20
40	60
55	80
75	100

Evaporating rate depends on the difference between air humidity pressure and skin water pressure as expressed in Table 9.1 and the following equation:

$$(\text{Ps} - \text{Pa}) \cdot \text{wet skin area}$$

where, Pa = air water pressure; Ps = skin water pressure.

Blood Flow During Exercise

The human ability to perform prolonged exercise is decreased in warm surroundings. Hypohydration or hypovolemia causes larger heat storage and reduces an individual's ability to bear heat stress. In addition, hypovolemia and the shift of blood from the working muscles to the skin make it tough to preserve central venous pressure and thus an acceptable Q in addition to backing muscular metabolism and thermoregulation during exercise-heat stress.

During aerobic exercise, there is a heat storage, namely that when heat production reaches a steady-state level, body core temperature remains elevated above resting values during the exercise bout. In a well-trained individual, heat storage increases his core temperature up to 38 °C while in untrained it reaches 38.5 °C. This temperature differences affect the post-exercise recovery time for the untrained subject. Heat storage increases enzymatic activity which increases in turn the efficiency of the metabolic process and gives time for the thermoregulation system to become active.

In dry and hot ambient heat dissipation during exercise is done mainly by sweat evaporation, because of large differences between skin-air-water pressures; thus, in hot and saturated surroundings loss of heat by sweating is very limited. Under these conditions, it is preferable to be in ambient temperature of 40 °C with humidity less than 25% than ambient temperature of 32 °C with 85% humidity. When ambient temperature is 34 °C with 100% humidity (sauna), there is no heat loss. The result is metabolic heat accumulation and rapid core temperature increase, which can be fatal. Increase in air humidity decreases the rate of heat dissipation. On the other hand, when desert ambient temperature with very low humidity sweat evaporation is very high and unnoticeable, this may cause an imbalance in body liquids. Wind is an additional factor that facilitates heat dissipation. More wind molecules contact the skin and by that heat dissipation rate is increased. Wind increases body heat loss to the surroundings by convective heat transfer even when the ambient temperature is higher than skin temperature. In high surrounding temperatures, convective heat transfer is increased by 50%; however, sweat rate is reduced only by 15% since convection has a minor role in the cooling process. In 10 °C ambient temperature and wind of 15 km h^{-1} heat loss from the body is done by convection to such a level that there is no need for sweat. In addition, this also decreases skin temperature, and thus, increases gradient between core temperature and skin temperature.

Personal factors such as metabolic rate that increase linearly body heat production rate have an important role during exercise. Acclimatization following aerobic training in warm environments increases plasma volume by 10–12%, sweat starts during exercise earlier, and its rate increases threefolds. In addition, heat storage is lower, electrolyte loss is less, and blood flow to the skin is reduced. All these factors together decrease core temperature during exercise compared to untrained individuals. Acclimatization process following training and exposure to warm ambient is short and is completed within 7–14 days. However, without exercising acclimatization process may be very long (months). Physical fitness level is an important factor that influences core temperature. A well-trained individual is able to perform physical activity in warm conditions better than untrained due to higher sweat rate, watery sweat, and a significant higher Q that enables him to divert more blood toward the skin for cooling and at the same time delivers enough oxygen to the working muscles.

Heat dissipation from the body depends on its transfer to the skin. At rest, this is achieved mainly by conduction and convection, due to the different temperature gradient between core temperature (37 °C) and 30–32 °C skin temperature (30–32 °C) in a wide range of ambient temperatures (Fig. 9.6).

In trained subjects during exercise, gradient between the temperatures is increased for the rise in core temperature to 38 °C. Heat dissipation from the skin to the surroundings depends (as mentioned earlier) on the differences in water pressure between skin to atmosphere. This is not the case for radiation, conduction, and

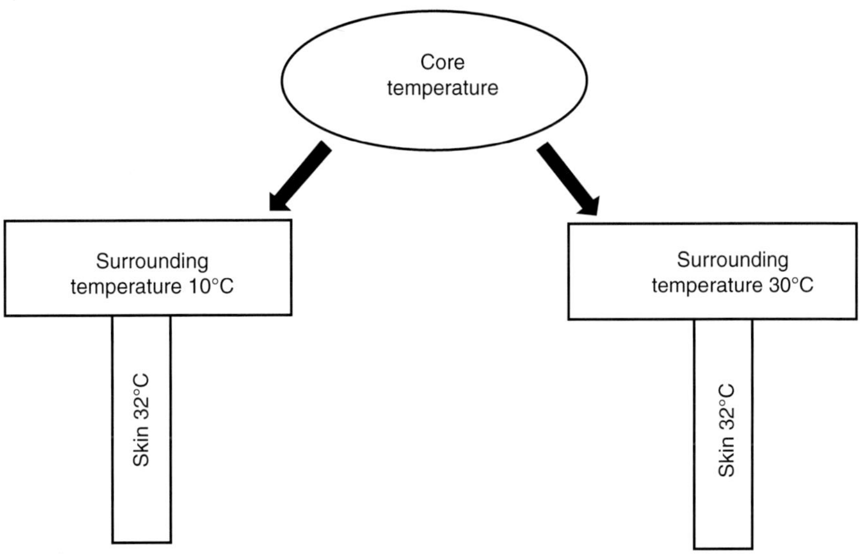

Fig. 9.6 Temperature gradients between core body temperature and skin and from the skin to the surroundings

convection in which heat transfer depends on temperature gradients. It is worth to note that muscle heat production does not depend on ambient temperature, while it depends only on the work intensity. As an example, a well-trained individual that runs once at a speed of 12 km·h^{-1} at ambient temperature of 10 °C and once at ambient temperature 30 °C at the same speed, his core temperature during both events is the same (38 °C). Figure 9.7 summarizes the different ways for heat dissipation. At ambient temperature of 30 °C, 95% of the cooling is done through sweating (Fig. 9.8) and 5% by respiration, conduction, convection, and radiation. However, at

Fig. 9.7 Different ways for heat dissipation at ambient temperatures of 10 °C and 30 °C

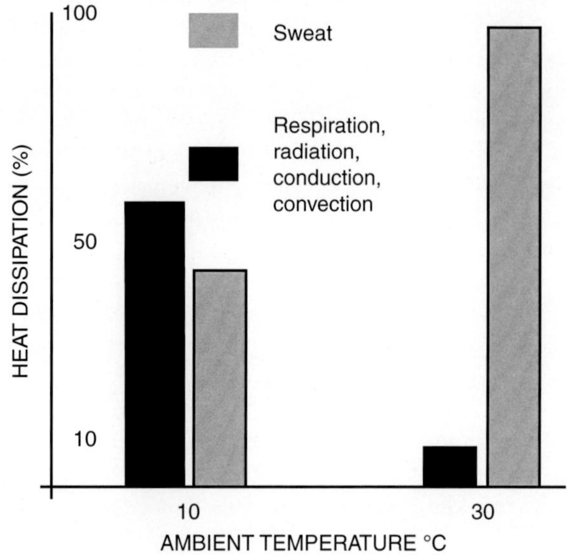

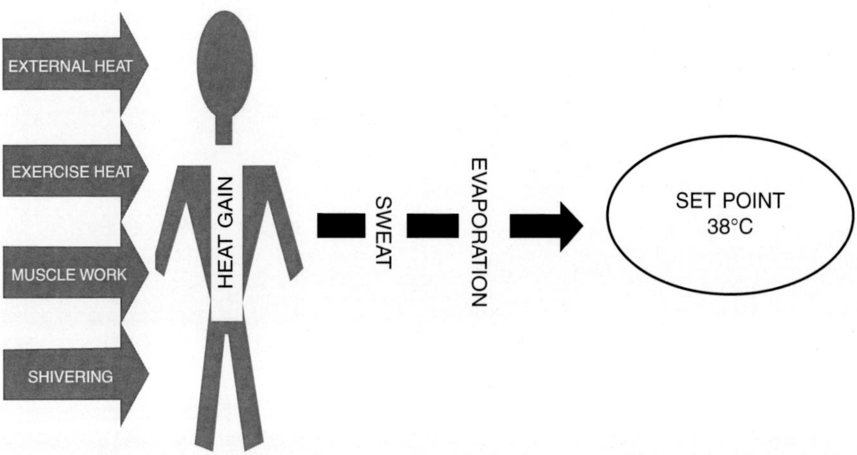

Fig. 9.8 At high surrounding temperature 95% of the heat removal is done by sweating to keep body core temperature at 37 °C

ambient temperature of 10 °C, 40% of the cooling is done through sweating and 60% by respiration, conduction, convection, and radiation. If the ambient temperatures will be above 30 °C, sweat share for cooling is increased, whereas, in ambient temperatures below 10 °C, importance of respiration, conduction, convection, and radiation increase for cooling.

Heat Regulation

The thermoregulation system is capable of keeping core temperature at rest and during aerobic exercise within normal range in a very wide variety of temperatures. Difficulty arises in heat regulation when an individual exercises intensively in very hot surroundings. Metabolic heat production increases during intense exercise, by 20–25 folds of that at rest. Under these conditions when the heat regulation system is not active, core temperature increases 0.1 °C each 5–7 min, until it is stabilized. In ambient temperature of 8–28 °C, when the thermoregulation system is active, it is possible to perform intensive bouts lasting for a long time. This is due to the fact that core temperature depends mainly on the metabolic rate and not on the surrounding temperatures. There is no general agreement between researches about the final means of thermoregulation at rest and during exercise. One assumption is that within the hypothalamus exists a sort of "thermostat" that activates thermoregulation when heat is out of the normal range.

The "thermostat" is rested during exercise to 38 °C in well-trained individuals, due to the heat storage. The new thermostat setup is regulated by the sympathetic nervous system (Fig. 9.9). Few nervous fibers in the sympathetic nervous system are sensitive to the changes in blood heat levels. These fibers are

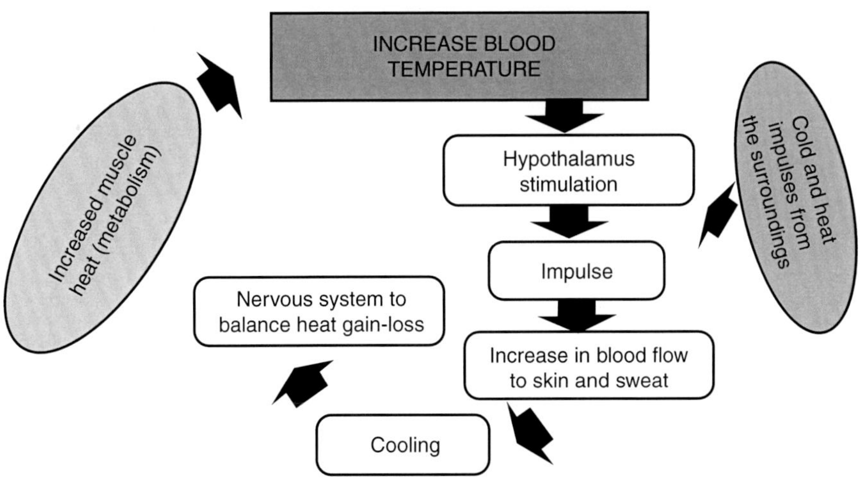

Fig. 9.9 Factors affecting hypothalamus for thermoregulation

connected to the hypothalamus that is the main center for thermoregulation. When core temperature increases, the hypothalamus activates physiological systems to reduce core temperature.

Cardiac Output During Exercise in Warm Environments

In the surroundings with temperatures of 20–22 °C, Q increases linearly with VO2 and with blood flow to the working muscles. With the increase in core temperature, blood flow to the skin is increased for cooling, and therefore, Q is higher compared to the same submaximal exercise intensity in cold climate (Fig. 9.10).

As an example, when running at a speed of 12 km·h^{-1}, in ambient temperature of 35 °C, HR can reach 170 beats·min^{-1}. At the same workload, in surrounding temperature of 10 °C, HR is about 130 beats·min^{-1}. The higher 40 beats·min^{-1} of HR in the hot environment are due to the increased blood flow to the skin for cooling. One has to keep in mind that the significant HR increase in the hot environment in spite of cardiac redistribution occurring during aerobic exercise.

Control mechanisms effect preload and afterload, and thus, SV is maintained or slightly decreased, influenced by variations in heat and cold temperatures [13]. Left-ventricular compliance does not change during acute thermal stress, measured at pulmonary capillary wedge pressure, to left-ventricular EDV [14]. During whole-body heating, indices of left-ventricular diastolic function are well-maintained in spite of preload reduction in preload, therefore indicating that indirect improvements in diastolic function may be occurring [15].

Q redistribution during exercise in hot ambient temperatures sometimes is not enough to deal with metabolic and thermoregulatory demands; it increases from 4 KJ·min^{-1} at rest up to 2000 KJ·min^{-1} at peak exercise, or from 80 W to several

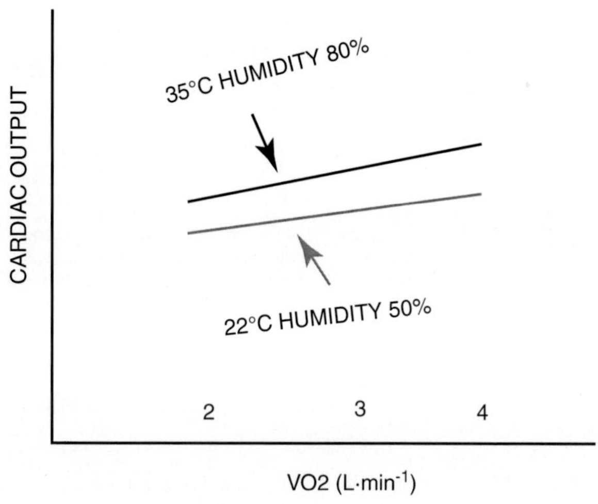

Fig. 9.10 Cardiac output responses during submaximal exercise in different surrounding temperatures

hundred watts. The heat produced set an extraordinary stress on the circulation. Consequently, Q cannot meet heat dissipation demands. Generally, during aerobic exercise, blood flow to working muscles is not reduced by heat stress; however, Q is insufficient to meet competing blood flow demands from skin and muscle [12]. During exercise, under these conditions, temperature controlling mechanisms act in order to avoid tissue damage. In hot compared to cool environments, exercise capacity is reduced. Because Q to the exercising muscles is reduced, contributing to the reduced performance in hot conditions. However, heat acclimation lowers the heart rate [16] and increases plasma and stroke volumes and thus maximal cardiac output in hot conditions [17]. The mechanism for the increased plasma volume may be an isosmotic volume expansion caused by an influx of protein to the vascular compartment and a sodium retention induced by a significant increase in aldosterone [18]. Environmental heat stress can challenge the limits of human cardiovascular and temperature regulation, body fluid balance, and thus aerobic performance. Hyperthermia reduces VO2max [19, 20], which leads to higher relative exercise intensity and an exponential decline in aerobic performance at any given exercise workload. That local hyperthermia also induces vasodilatation of the skeletal muscle microvasculature, thereby contributing to heat stress and exercise hyperemia. The increased muscle vasodilatation in the face of elevated muscle sympathetic vasoconstrictor activity is closely related to the rise in arterial plasma ATP and local tissue temperature [21].

Performance may be impaired by the development of hyperthermia and concomitant rise in cardiovascular strain increasing relative exercise intensity. A thermoregulatory-mediated rise in cardiovascular strain is associated with reductions in supportable power output, VO2 peak, and maximal power output during prolonged exercise in the heat [22]. Other possible cause for the reduced physical performance is a decrease in hemoglobin concentration from the vascular volume. This in turn will increase relatively exercise intensity which increases cardiovascular strain; as a consequence, an earlier fatigue ensues [23]. Severe heat stress has been shown to markedly suppress VO2max and work capacity without altering the initial rate of rise in whole-body VO2 [24]. These results can be reversed following heat acclimatization.

Heat stress increases muscles blood flow and Q, based on thermoregulatory demand of the skin circulation. In ambient temperatures of 20–28 °C, VO2 increases linearly with Q and can increase blood flow to the working muscles; however, in temperatures above 28 °C, Q does not increase linearly with VO2.

During aerobic exercise, the myocardium and contracting muscle are required to meet the increased metabolic energy demands, while blood flow to the skin is required to dissipate heat. This blood flow competition results in blood flow [25]. If Q is not large enough to supply blood to the working muscles and to the skin, anaerobic metabolism and lactate production follows, while the core temperature rises. In that case, one selection is to increase blood flow to the working muscles and less to the skin. Under this condition, lactate production levels are normal. However, core temperature will rise to dangerous levels. The other choice, which is better than the first one, is reduction in blood flow to the working muscles and increased blood flow to the skin. With this condition, lactate production will increase, and thus, an imbalance in acid–base is reached bringing to a halt in exercising, before the core temperature reaches high and dangerous levels [26].

Therefore, vasoconstriction decreases blood flow to the kidneys, liver, pancreas, spleen, and digestive track (cardiac redistribution). In thermally stressful environments, renal and splanchnic flows are further reduced, with this additional fraction of the available Q redirected to the skin to facilitate heat dissipation [27]. Aerobic exercise causes cutaneous vasoconstriction at exercise onset through increased vasoconstrictor activity. As exercise continues, internal temperature reaches a threshold for increased active vasodilator activity that is elevated by exercise; this vasodilator function compromises temperature regulation [12]. It is unclear at present how these two mechanisms leading to augmented skin blood flow, i.e., increased Q and visceral vasoconstriction, are affected by an individual's VO2max level. Additional blood flow to the skin is not a concern. Instead, circulatory control factors prevent muscle vascular conductance from reaching such high levels. Otherwise, BP could be severely threatening. Such a compromised BP regulation is in fact seen in conditions wherein sympathetic vasoconstrictor function is compromised [28].

The heat transfer to the skin by the circulation during exercise is possible following the opening of closed blood vessels in the skin. In warm environment compared to normal ambient temperatures, during submaximal dynamic exercise at the same absolute load, HR is higher while SV and Q are lower (Fig. 9.11). Core temperature has been shown to increase HR and reduce SV when skin blood flow, blood volume, and Q are constant [29].

It is worth to point out that the (a-v)O2 and Q are not influenced by ambient temperatures below 280 °C. In any case at any given ambient temperature, Q increases. In the higher ambient temperature, the additional increased 40 beats·min^{-1} is due to Q augmentation, in order to increase blood flow to the skin for heat dissipation.

Additional factor that broadens Q reduction during dynamic exercise in warm environment is dehydration. It reduces substantially plasma volume that progresses

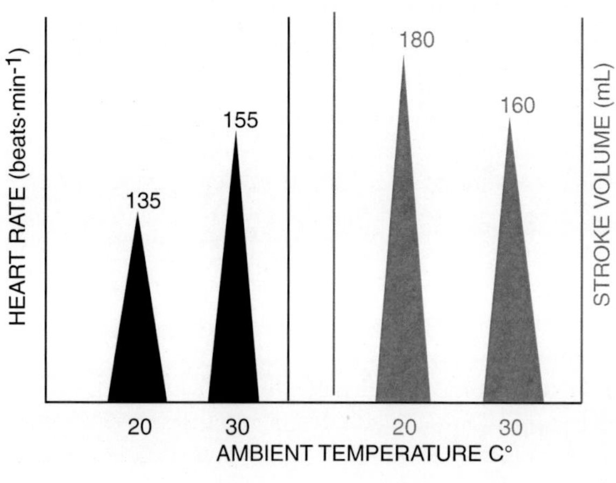

Fig. 9.11 At the same, absolute workload heart rate and stroke volume are significantly higher at ambient of 30 °C compared to environment temperature of 20 °C

during exercise and, thus, lessens venous return and SV. These effects are worsened when the exercise is performed in an extreme hot environment. Mild hypohydration impairs cardiovascular and thermoregulatory strain and tended to impair endurance performance; nonetheless, aerobic fitness attenuates the physiological effects. Yet, athletes who often dehydrate during exercise are hypervolemic and have less cardiovascular sensitivity to acute hypervolemia. This, however, may have negative consequences in CAD patients exercising in hot environment. Dehydration reduces aerobic endurance performance and results in increased body temperature, HR, perceived exertion, and possibly increased reliance on anaerobic metabolic pathway as an energy source [30]. In hot environments, dehydration coupled with the relatively reduced Q may result in a decrement of sweating rate and increase in body core temperature, hence discontinuing exercise.

During prolonged exercise, venous return decreases due to dehydration which further decreases SV and Q and increases HR (Fig. 9.12). Under normal conditions, in hot environment, the water movement between the body compartments assists in the maintainance of adequate level of sweating without an influence on Q. However, the loss of water through sweat interferes with the movement of water from the extracellular compartment to the cardiovascular compartment and, thus, reduces venous return.

Figure 9.12 reveals that in ambient temperatures of 22–28 °C (mild intensity) or in ambient temperatures of 32–40 °C (high intensity), any loss of water above 3–4% body weight will bring to a significant decrease in Q. At that stage, HR cannot be further increased, i.e., the reduction in SV is so significant that any increase in HR cannot compensate for the Q reduction.

Acclimatization to warm environment and the benefits of aerobic training improve physical performance in trained and untrained persons. In trained

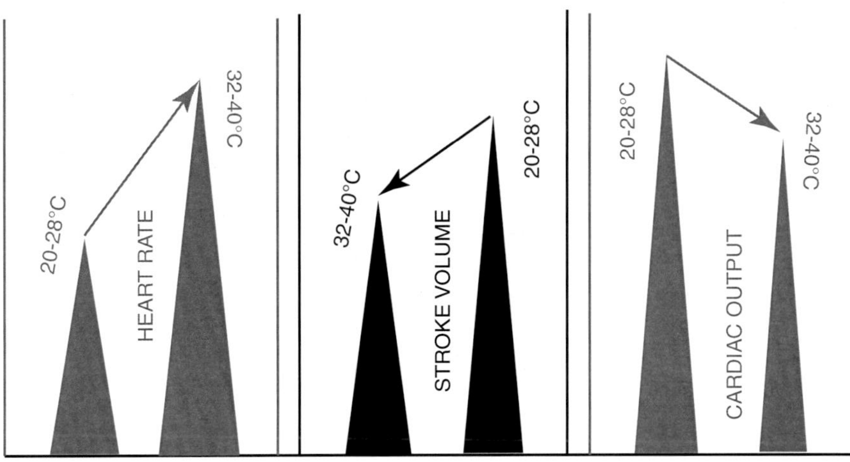

Fig. 9.12 With the loss of body's water (sweat) above 3–4% of the individual's weight in warm environment, cardiac output decreases significantly due to reduced stroke volume and as a consequence heart rate increases significantly

Table 9.2 Summary of hemodynamic responses in different environment temperatures during dynamic aerobic exercise

Variable	At 22 °C	At 40 °C
Heart rate	Increase	Increase
Stroke volume	Maximal	Reduce
Cardiac output	Increase	Increase
Skin flow	Increase	Increase
Plasma volume	Normal	Reduce
VO2	Increase	Increase
Minute ventilation	Normal	Reduce

individuals the improvement is greater than that of the untrained. This is due to the larger amount of body water in the trained individuals following training. This in turn facilitates temperature regulation by keeping core temperature low in the warm ambient. Cardiac output is reduced less under the same environment condition following acclimatization. The reduction in venous return is attenuated so stroke volume is reduced not as much as before acclimatization.

The influence of the acclimatization on cardiac output is still controversial. One approach holds that acclimatization period should be 12 weeks, while the second thought is that 6-week training is sufficient. Table 9.2 compares physiological responses to endurance exercise in different ambient temperatures.

Under high extremely ambient temperatures VO2max in untrained is lower than that achieved by them in normal temperatures (20–22 °C). However, in few well-trained athletes there is no significant difference in VO2max in both temperatures.

Cardiovascular Response in Cold Environment

Exercising in the cold is not a striking choice for many sedentary individuals and athletes, yet, the meaning of cold is hard to define and differs between different athletic events. As an example, exercising at moderate intensity, in the surrounding temperatures of 10–12 °C, is an advantage for the exercising individual since almost the entire Q goes to the contracting muscle, while heat dissipation is achieved, through conduction and convection [31]. When the ambient temperatures are between 2 °C and 22 °C, Q increases linearly with the increase in VO2 and blood flow to the muscles. In ambient temperatures below 2 °C, Q does not increase in a linear fashion with VO2 [32]. In cold temperatures, most of the Q goes to the working muscles and not to the skin. This increases moderately blood flow to the internal organs. The decrease in heat transfer to the skin, during endurance exercise, in the cold environment (water and air alike) is possible from the vasoconstriction of blood vessels [33]. Vasoconstriction is an effective approach for the prevention of heat loss from the body through radiation and convection, thus decreasing blood flow to the skin. This in turn increasees blood volume in the chest, and therefore, venous return is increased as well as SV. Consequently, during submaximal dynamic endurance exercise in cold surroundings, HR is lower and SV larger compared to the same workload at normal temperature (Fig. 9.13). During efforts between 40

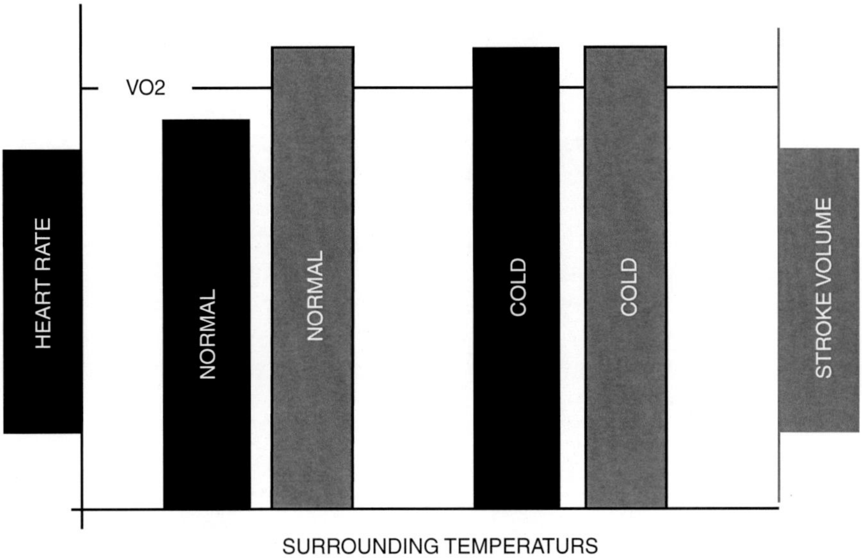

Fig. 9.13 Differences in heart rate and stroke volume responses for the same oxygen uptake, in normal and cold temperatures

and 85% VO2max, there is no linear relationship between HR and VO2 as in normal ambient temperatures. Therefore, the submaximal PWC-170 cycle test is not valid with these cold temperatures. The response to cold compared to heat is that in the cold behavior plays a central role in body core temperature preservation, such as wearing appropriate clothing. Additional difference is catecholamines and thyroid hormones secretion during cold stress.

In humans, this is accomplished by decreasing blood flow to the skin—that is, by skin vasoconstriction. Additional mechanism for the increase in core temperature is shivering. Shivering reflex is an involuntary somatic motor thermogenic response, involving involuntary contractions in response to cold [34]. Any reduction in body core temperature triggers the shivering reflex and results in the generation of heat as a by-product, in order to sustain homeostasis. Skeletal muscles begin to contract in low intensity, which do not produce any work but produce heat from the utilization of the ATP molecule [35]. When drops in body core temperature are significant, shivering becomes fierce in order to reach temperature homeostasis .

Renal Role During Exercise in the Heat

In the heat two target organs—the sweat gland and the kidney—are of interest. While sweat can be improved following heat acclimation, renal function is under hormonal restricted control [36]. Renal balances body fluid tonicity and effective circulating volume. Hypohydration resulting in hypovolemia, due to exercising in the heat, depends on renal perfusion capacity. Fluid supplement for maintaining hydration, during aerobic exercise in hot surroundings, is important due to the large amount of fluid lost over the period of exercise. Therefore, the effects of hydration status on the kidneys while exercising aerobically in the heat are also significant. Dehydration decreases blood volume, reducing renal blood flow, and subsequent reductions in GFR are increased when dehydration is associated with exercise in a hot environment [37]. Even though that total fluid loss is replaced during strenuous exercise, renal blood flow is reduced to 0.25 L out of 1 L resting values, as a result of renal blood vessel vasoconstriction during Q redistribution [38]. Hypohydration, reduction in body weight above 2%, resulted from extreme sweat loss starts to limit aerobic performance warm–hot environments [39]. Hypohydration may occur from voluntary fluid restriction, insufficient fluid availability, or thermoregulatory sweating [40]. The reduced amount of renal blood flow is further reduced, following dehydration, when exercising in the heat [41]. If water lost is not replaced, hypohydration reduces plasma volume and, thus, increases plasma osmotic pressure, which reduces the sensitivity of sweating and blood flow responses. Vigorous exercise results in body fluid losses from the lungs, when exhaling air as well as from sweating. A larger amount of water during aerobic exercise suggests greater dehydration. During a marathon race in hot environment, runners may lose 3–4 kg body water, while running a marathon in the cold they lose 2–2.5 kg [42]. This finding suggests that exercising in the heat can lead to a greater reduction of renal blood flow than exercising in the cold.

In response to exercise in the heat, there are numerous alterations in fluid and electrolyte homeostasis. The endocrine system plays an important role in the regulation of fluid and electrolyte homeostasis that must occur with exercise. Arginine vasopressin is the main regulator of water to maintain plasma osmolality [43]. Osmotic regulation of plasma tonicity generally occurs through activation of the vasopressin type 2 receptor in the kidney, which enables water reabsorption along osmotic gradients and associated antidiuresis when plasma tonicities are high [44].

Heat Acclimatization and Training

Acclimatization and training in hot environment improve physical performance of trained and untrained individuals. However, the improvement in the trained individuals is greater, due to the fact that they have a higher body water value compared to that of the untrained. Compared to Q responses prior to acclimatization and following training, Q increases because of the relatively smaller reduction in venous return and, consequently, a relatively small SV reduction, and thus, HR increases moderately. Acclimatization and training generate earlier sweating, due to an increase in reaction time of the autonomic nervous system. This reduces the amount of heat stored, and therefore, core temperature is kept within the normal range. It is possible that aerobic training increases sweat gland sensitivity in addition to the redistribution of body's sweating areas. As mentioned before, trained individuals have sweating glands in their front compared to untrained individuals in whom

Table 9.3 Physiological responses to acclimatization and training

Variables	Pre-acclimatization and training	Post-acclimatization and training
Core temperature	High	Very low
Temp. Core-skin gradient	Low	High
Skin blood flow	High	Lower
Sweating rate	High	Higher
Time to start sweating	Long	Short
Heart rate	High	Lower
Cardiac output	High	Higher
Stroke volume	Low	Higher
Na removal by kidney	Low	Lower
Na in sweat	Low	Very low

sweating area is in the back. The general agreement for the time length needed for acclimatization and training is 12 weeks.

Others deliberate that aerobic training shortens the period to 6 weeks. Table 9.3 reveals the physiological improvements following acclimatization and training.

Fluid Regulation During Exercise

The capacity to control plasma osmolality that is the body's electrolyte–water balance [45] is very limited and with close physiological range. This emphasizes the high standing of conserving body water and sodium balance at rest and during exercise. Cells are freely permeable to water; the osmolality of the extracellular fluid is about equivalent to that of the intracellular fluid. Consequently, plasma osmolality is a controller to intracellular osmolality. This is important, as it shows that variations in extracellular fluid osmolality have a great effect on intracellular fluid osmolality—changes that can result in problems with normal cell function and volume. If the extracellular fluid is too hypotonic, water would readily fill surrounding cells, increasing their volume and potentially dissoluting them (cytolysis). Many poisons, medications, and diseases affect the balance between the intracellular fluid and extracellular fluid, affecting individual cells and homeostasis as a whole [46].

During aerobic exercise, there are several adjustments in fluid and electrolyte homeostasis. These occur nearly when exercise commences and can continue for a very long time (up to days), following the exercise termination. The endocrine system plays an important role in the regulation of fluid and electrolyte homeostasis that must occur with exercise. Dysregulation of the endocrine system may limit exercise activity and, in some incidences, result in debilitating morbidities or death [47]. Dehydration above 2% of body weight alters at least one of the following systems: cardiovascular, thermoregulatory, central nervous system, and metabolic functions, which in turn will reduce physical performance during endurance exercise. This work capacity decrement is heightened by heat stress. To diminish the contrary significances of body water shortages on endurance exercise act, it is suggested that fluid consumption has to be adequate to lessen dehydration to a smaller amount than 2% of body weight loss [23]. Since inadequate fluid intake during exercise in the heat reduces tolerance to central hypovolemia, independent of increases in body temperature [48], non-osmotic arginine vasopressin secretion, combined with over fluid replacement, which exceeds fluid loss due to sweat, results in low blood sodium concentrations and ease fluid retention and sodium loss, thus probably assimilating both the water intoxication and sodium loss [49]. Therefore, it is recommended to consume fluid amount of $1L{\cdot}h^{-1}$ [50].

The posterior pituitary hormone, antidiuretic arginine vasopressin, is the main fluid homeostasis endocrine regulator [51]. Unsuitable arginine vasopressin secretion may result in pathophysiology stage, morbidity, and occasional mortality. Antidiuretic hormone arginine vasopressin secretion at rest is mainly defined by variations in osmotic regulation. During exercise, osmotic control of arginine vasopressin secretion can be improved by numerous potential non-osmotic factors simultaneously stimulated, particularly during strenuous or prolonged exercise.

Exercise-associated hyponatremia has appeared as an important complication of prolonged endurance physical activities [52]. In collapsed subjects with increased cardiac troponin T levels, at the 2006 Boston marathon, abnormal sodium value, hypernatremia, hyponatremia, and ionized calcium level was low as well as ionized magnesium level, while, blood lactate level was elevated in 95% of subjects. The frequency of elevated postrace cardiac troponin T levels in collapsed athletes after endurance exercise is similar to that in asymptomatic runners [53].

Sweat evaporation has a critical key role in thermoregulation, and this is most seen when the capacity to sweat is not at the best, during long-lasting strenuous exercise in hot surroundings [54], which may lead to heat exhaustion or heat stroke.

Temperature Regulation in the Cold

Physical activity, leisure activity, and work in cold ambient expose the body to different conditions than that of the hot surroundings. Physiological responses such as shivering mechanism and vessels vasoconstriction, clothing, and behavioral strategies help to conserve body heat and body temperature. Thermogenesis restores body heat losses, while peripheral vessels vasoconstriction expands thermal padding of the body and thus reduce the amount of heat loss. Body heat is kept by:

- Aerobic exercise increases metabolic heat production more than shivering. While exercise increases metabolic heat production, it also facilitates heat loss from the body by increasing blood flow to the skin and active muscles.
- Lying in the wrapped position reduces body surface area.
- Head to be covered since heat loss is the highest from forehead.
- Body fat is a good insulator; each centimeter of fat layer increases core temperature by half Celsius degree.

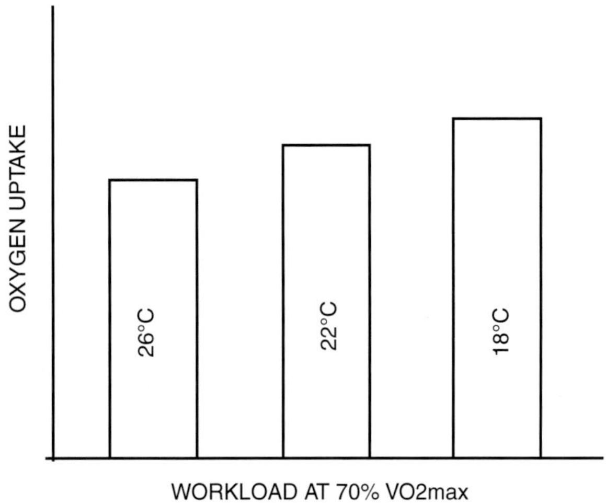

Fig. 9.14 As water temperature becomes colder, VO2 being higher during similar submaximal aerobic exercises

- Avoiding sweat since heat is lost through water 20-folds higher than through air.
- Physiological shivering mechanism and vessel vasoconstriction through the sympathetic nervous system may increase core temperature by half Celsius degree.

During similar submaximal aerobic exercise, different water temperatures affect VO2; in water temperature of 18 °C VO2 is higher than that of 22 °C which is higher than water temperature of 26 °C (Fig. 9.14).

The increased VO2 in the cold is a result of the higher blood catecholamine levels. In cold water muscles partially lose their contractility ability, thus, they recruit more muscle fibers to accomplish the physical task which in turn increase VO2. In the case where core temperature decreases, chemical reactions that oxygen is involved with become slower, and therefore, the muscle utilizes the glycolytic pathway for energy production to contract through the white muscle fibers. Under low ambient temperatures when core temperature is below 35 °C, VO2max is significantly lower than that achieved under normal temperatures. The VO2max decline in the cold is due to the reduction in muscle's contraction strength and movement limitations.

In cold conditions, BP increases mainly SBP due to vasoconstriction from the cold and to the high catecholamine levels. DBP does not increase since not all the blood vessels under the skin are constricted. Therefore, during prolong physical activity in cold water, catecholamine level increases and thus core temperature. When core temperature reaches 25 °C, there is a failure of motor-unit recruitment, to contract along with a significant increase in SBP. Individuals participating in long-lasting exercise challenging cold-surrounding activities are at risk to develop thermoregulatory fatigue, which reduces muscle working capacity, movement efficiency, and performance.

Normal body temperature averages 37 °C (98.6 °F), physiologically; hypothermia, a potentially dangerous drop in body temperature, is defined as body core temperature under 36.8 °C. However, clinically, it is defined as a core temperature below 35 °C (95 °F). Hypothermia symptoms for adults include shivering, slow breathing, confusion, sleepiness, exhaustion, unclear speech, loss of coordination, and a sluggish frail heartbeat.

Usually hypothermia is caused by prolonged exposure to cold temperatures mainly during the winter periods. It progresses when heat losses exceed metabolic heat production. The cardiovascular system is the main controller of body temperature, compared to young individuals; elderly are particularly susceptible to hypothermia; it is likely for elder individuals to lose heat due to cardiovascular changes with aging. In addition to vascular changes with aging, other features can contribute to hypothermia such as malnutrition and changes in the central nervous system or brain.

Hypothermia is categorized into three different levels: mild, moderate, and severe. In severe hypothermia, core body temperature may be under 28 °C (82 °F) or lower. At body core temperature below 18–12 °C, there is an electrical silence. To determine correctly hypothermia, temperature should be measured in the rectum or esophagus.

Normally, when a person chills, the response includes shivering, which is a kind of exercise of the muscles that promotes warming. Shivering is often absent in elderly patients with hypothermia. These patients feel cold not just in the extremities, which can be normal, but also in the trunk. They appear pale. As the cold progresses, it depresses bodily functions. Finally, coma may occur.

Clothing is the best way in preventing chillness and hypothermia injury and responding to weather changes.

Declines in body core temperature may ultimately affect exercise functional and work capacity. Each athlete, exercising in the cold, responds differently depending on body composition and underlying medical risk factors. An understanding of wind-chill temperatures, wetness, and the weather forecast allows athletes and event coordinators to properly respond to changing weather conditions. At the first sign of a freezing cold injury, ensure warm, dry clothes and move to a protected environment. Although cold weather activities are in general safe for athletes, cold injuries can be prevented with proper education, preparation, and response to changing weather conditions.

Heat Disorders When Exercising in Hot Environment

Hyperthermia can result in dry skin, dehydration, nausea, vomiting, headache, low BP, heart tachycardia, shortness of breath, confusion, fainting, dizziness, and in extreme situation, even death. Heat is produced from cell metabolism, due to the chemical reactions. With the help of radiation and sweating, the body accomplishes heat loss. Under normal conditions, sweat and evaporation keep skin temperature below that of the surroundings. Radiation is not an effective tool for heat dissipation, when the surrounding temperatures are higher than that of the body skin. In high surrounding temperatures, when humidity is also high, sweating efficacy is reduced due to the decline in evaporation rate. This is why the body has difficulties in dissipating heat in hot and humid surroundings. Imbalance between heat production and ineffective heat loss can lead to exertional heat-related illness heat. Exertional heat-related illness contains numerous phases that is a problem for exercising individuals, in high environmental heat stress [55]. Convinced forms of exertional heat-related illness may develop an intimidating and very dangerous condition if not taken care, such as heat stroke, which is a situation where body core temperature is above 40 °C and causes unstable mental standing changes. This condition is the most severe form of exertional heat-related illness. Exertional heat stroke needs an immediate treatment action that includes rapid body cooling to avoid complications and even life threatening. Many exertional heat-related illnesses events such as heat exhaustion, exercise-related failure, exercise-related muscle spasms and pain, exercise-associated hyponatremia, and exertional rhabdomyolysis are in general avoidable by avoiding strenuous exertion in very hot environment, heat acclimatization, and shortening the length and intensity of exercise sessions during high environmental heat stress. In addition, maintaining adequate hydration, avoiding exercising in the heat when ill, and seeking medical personal assistance.

Laboratory on Thermoregulation

Laboratory Purpose To investigate the physiological response to prolonged exercise in the heat

Equipment
1. Warm sports outfits
2. Computerized heat measurement device
3. Ambient temperature measuring device
4. Humidity measuring device
5. Bicycle ergometry
6. HR monitor
7. Metabolic cart
8. Weight scale
9. Watch stopper
10. Borg scale

Experiment Stages
Measure room's temperature

- Weigh the subject.
- For measuring skin temperature: (1) connect to the right and left between the hands and elbows, (2) connect to the right and left thighs, and (3) measure core temperature by putting the thermometer deep in the rectum (8–10 cm).
- Put the HR monitor on the subject.

- The subject will dress the warm sports outfits covering all body except the head and right forearm.
- Connect the subject to the metabolic cart.
- The subject will be sited on the bicycle in full relaxation for 5 min to collect resting values of: temperatures of 4 sites, HR, and VO2.
- During 5 min rest, the subject will be explained how to use the Borg scale.
- Subject will paddle in an intensity correspondent to 75% VO2 for 30–45 min: every 5 min 4 sites of temperature, HR, Borg rating, and VO2 will be measured. When the first sweat drops (sweat threshold) will appear, HR, rectal temperature, and skin temperature will be measured.
- With exercise secession, the subject will be weighted again for sweat loss.

Assignments

(a) Plot 4 graphs to describe the changes of the following variables with time.

- Rectal temperature, uncovered right forearm, and covered left forearm.
- Rectal temperature and Borg rating values.
- HR, VO2, and Borg Rating values.
- Rectal temperature and HR.

(b) Did you find any temperature differences between the uncovered right forearm and covered left forearm?

(c) Find if there are temperature differences between the thighs and left forearm. If there are differences answer why?

(d) How the above variables score was if the subject's workout was uncovered?

(e) What is the impact of cloth on the heat load?

(f) Why in general the body does not sweat immediately?

(g) What variables' score will be different when humidity is 100% once when the subject is covered and once uncovered?

Laboratory Report

Name_______ Room Temp._____ (°C) Humidity____ (%)

Pre-test Weight____ (kg) Post-test weight_____ (kg)

	Time minutes	HR	VO2	Rectal Temp.	Skin Temp.	Forearm Temp.	Thigh Temp.
Rest	5						
Load (w)	10						
	15						
	20						
	25						
	30						
	35						
	40						
	45						

References

1. Hamley IW. Introduction to soft matter. Hoboken, NJ: John Wiley; 2007.
2. Shibasaki M, Crandall CG. Mechanisms and controllers of eccrine sweating in humans. Front Biosci (Schol Ed). 2010;2:685–96.
3. Bankir L. Antidiuretic action of vasopressin: quantitative aspects and interaction between V1a and V2 receptor-mediated effects. Cardiovasc Res. 2001;51:372–90.
4. Sousa CV, da S Aguiar S, dos R Olher R, et al. Hydration status after an ironman triathlon: a meta-analysis. J Hum Kinet. 2019;70:93–102.
5. Hein A. M. Daanen, SR, Periard JD, et al. Heat acclimation decay and re-induction: A systematic review and meta-analysis. Sports Med. 2018;48:409–30.
6. Racinais S, Cocking S, Periard JD. Sports and environmental temperature: From warming-up to heating-up. Temperature (Austin). 2017;4:227–57.
7. Flouris AD, Webb P, Kenny GP. Noninvasive assessment of muscle temperature during rest, exercise, and postexercise recovery in different environments. J Appl Physiol (1985). 2015;118:1310–20.
8. Nybo L. Hyperthermia and fatigue. J Appl Physiol. 2008;104:871–8.
9. Galan-Carracedo J, Suarez-Segade A, Guerra-Balic M, et al. The dynamic and correlation of skin temperature and cardiorespiratory fitness in male endurance runners. Int J Environ Res Public Health. 2019;16(16):E2869. https://doi.org/10.3390/ijerph16162869.
10. Johnson JM, Minson CT, Kellogg DL Jr. Cutaneous vasodilator and vasoconstrictor mechanisms in temperature regulation. Compr Physiol. 2014;4(1):33–89. https://doi.org/10.1002/cphy.c130015.
11. Charkoudian N. Mechanisms and modifiers of reflex induced cutaneous vasodilation and vasoconstriction in humans. J Appl Physiol (1985). 2010;109:1221–8.
12. Johnson JM. Exercise in a hot environment: the skin circulation. Scand J Med Sci Sports. 2010;20:29–39.
13. Wilson TE, Crandall CG. Effect of thermal stress on cardiac function. Exerc Sport Sci Rev. 2011;39:12–7.
14. Wilson TE, Brothers RM, Tollund C, et al. Effect of thermal stress on Frank-Starling relations in humans. J Physiol. 2009;587:3383–92.
15. Stohr EJ, Gonzalez-Alonso J, Pearson J, et al. Effects of graded heat stress on global left ventricular function and twist mechanics at rest and during exercise in healthy humans. Exp Physiol. 2011;96:114–24.
16. Nielsen B, Strange S, Christensen NJ, et al. Acute and adaptive responses in humans to exercise in a warm, humid environment. Pflugers Arch. 1997;434:49–56.
17. Lorenzo S, Halliwill JR, Sawka MN, Minson CT. Heat acclimation improves exercise performance. J Appl Physiol. 2010;109:1140–7.
18. Nielsen B, Hales JR, Strange S, et al. Human circulatory and thermoregulatory adaptations with heat acclimation and exercise in a hot, dry environment. J Physiol. 1993;460:4674–85.
19. Gonzalez-Alonso J, Calbet JA. Reductions in systemic and skeletal muscle blood flow and oxygen delivery limit maximal aerobic capacity in humans. Circulation. 2003;107:824–30.
20. Gonzalez-Alonso J, Crandall CG, Johnson JM. The cardiovascular challenge of exercising in the heat. J Physiol. 2008;586:45–53.
21. Gonzalez-Alonso J. Hyperthermia impairs brain, heart and muscle function in exercising humans. Sports Med. 2007;37:371–3.
22. Periard JD, Cramer MN, Chapman PG, et al. Cardiovascular strain impairs prolonged self-paced exercise in the heat. Exp Physiol. 2011;96:134–44.
23. Cheuvront SN, Kenefick RW, Montain SJ, et al. Mechanisms of aerobic performance impairment with heat stress and dehydration. J Appl Physiol. 2010;109:1989–95.
24. Nybo L, Jensen T, Nielsen B, et al. Effects of marked hyperthermia without dehydration on Vo2 kinetics during intense exercise. J Appl Physiol. 2001;90:1057–64.

25. Lynn BM, Minson CT, Halliwill JR. Fluid replacement and heat stress during exercise alter post-exercise cardiac haemodynamics in endurance exercise-trained men. J Physiol. 2009;587:3605–17.
26. Drust B, Rasmussen P, Mohr M, et al. Elevations in core and muscle temperature impair repeated sprint performance. Acta Physiol Scand. 2005;183:181–90.
27. Pearson J, Low DA, Stohr E, et al. Hemodynamic responses to heat stress in the resting and exercising human leg: insight into the effect of temperature on skeletal muscle blood flow. Am J Physiol Regul Integr Comp Physiol. 2011;300:R663–73.
28. Puvi-Rajasingham S, Smith GD, Akinola A, et al. Abnormal regional blood flow responses during and after exercise in human sympathetic denervation. J Physiol. 1997;505:841–9.
29. Gonzalez-Alonso J, Rodriguez M, Coyle EF. Stroke volume during exercise: interaction of environment and hydration. Am J Physiol Heart Circ Physiol. 2000;278:H321–30.
30. Merry TL, Ainslie PN, Cotter JD. Effects of aerobic fitness on hypohydration-induced physiological strain and exercise impairment. Acta Physiol (Oxf). 2010;198:179–90.
31. Nimmo MA. Exercise in the cold. J Sports Sci. 2004;22:898–915.
32. Shiojiri T, Shibasaki M, Aoki K, et al. Effects of reduced muscle temperature on the oxygen uptake kinetics at the start of exercise. Acta Physiol Scan. 1997;159:327–33.
33. Rintamaki H. Human responses to cold. Alaska Med. 2007;49:29–31.
34. Jubrias SA, Vollestad NK, Gronka RK, et al. Contraction coupling efficiency of human first dorsal interosseous muscle. J Physiol. 2008;586:1993–2002.
35. Loiselle DS, Johnston CM, Han JC, et al. Muscle heat: a window into the thermodynamics of a molecular machine. Am J Physiol Heart Circ Physiol. 2016;310:H311–25.
36. Buono MJ, Kolding M, Leslie E, et al. Heat acclimation causes a linear decrease in sodium ion concentration. J Therm Biol. 2018;71:237–40.
37. Racinais S, Alonso JM, Coutts AJ, et al. Consensus recommendations on training and competing in the heat. Scand J Med Sci Sports. 2015;25:S6–19.
38. Oani H, Kaya M, Tsujita J. Effect of the volume of fluid ingested on urine concentrating ability during prolonged heavy exercise in a hot environment. J Sports Sci Med. 2013;12:197–204.
39. Sawka MN, Cheuvront SN, Kenefick RW. Hypohydration and human performance: impact of environment and physiological mechanisms. Sports Med. 2015;4:S51–60.
40. Kenefick RW, Cheuvront SN. Physiological adjustments to hypohydration: impact on thermoregulation. Auton Neurosci. 2016;196:47–51.
41. Shirreffs SM, Sawka MN, Stone M. Water and electrolyte needs for football training and match-play. J Sports Sci. 2006;24:699–707.
42. Neviackas JA, Bauer JH. Renal function abnormalities induced by marathon running. South Med J. 1981;74:1457–60.
43. Carroll HA, James LJ. Hydration, arginine vasopressin, and Glucoregulatory health in humans: a critical perspective. Nutrients. 2019;11(6):E1201. https://doi.org/10.3390/nu11061201.
44. Harrois A, Anstey JR. Diabetes insipidus and syndrome of inappropriate antidiuretic hormone in critically ill patients. Crit Care Clin. 2019;35:187–200.
45. Erstad BL. Osmolality and osmolarity: narrowing the terminology gap. Pharmacotherapy. 2003;23:1085–6.
46. Sureda-Vives M, Morell-Garcia D, Rubio-Alaejos A, et al. Stability of serum, plasma and urine osmolality in different storage conditions: relevance of temperature and centrifugation. Clin Biochem. 2017;50:772–6.
47. Wade CE. Hormonal regulation of fluid and electrolyte homeostasis during exercise. In: Endocrinology of physical activity and sport. Contemporary endocrinology. Cham: Springer Nature; 2013. p. 221–44.
48. Schlader ZJ, Gagnon D, Rivas E, et al. Fluid restriction during exercise in the heat reduces tolerance to progressive central hypovolaemia. Exp Physiol. 2015;100:926–34.
49. Hew-Butler T. Arginine vasopressin, fluid balance and exercise: is exercise-associated hyponatraemia a disorder of arginine vasopressin secretion? Sports Med. 2010;40:459–79.

50. Cheuvront SN, Carter R III, Sawka MN. Fluid balance and endurance exercise performance. Curr Sports Med Rep. 2003;2:202–8.
51. Hew-Butler T, Noakes TD, Soldin SJ, et al. Acute changes in arginine vasopressin, sweat, urine and serum sodium concentrations in exercising humans: does a coordinated homeostatic relationship exist? Br J Sports Med. 2010;44:710–5.
52. Verbalis JG. Renal function and vasopressin during marathon running. Sports Med. 2007;37:455–8.
53. Siegel AJ, Januzzi J, Sluss P, et al. Cardiac biomarkers, electrolytes, and other analytes in collapsed marathon runners: implications for the evaluation of runners following competition. Am J Clin Pathol. 2008;129:948–51.
54. Casa DJ, Cheuvront SN, Galloway SD, et al. Fluid needs for training, competition, and recovery in track-and-field athletes. Int J Sport Nutr Exerc Metabol. 2019;29:175–80.
55. Nichols AW. Heat-related illness in sports and exercise. Curr Rev Musculoskelet Med. 2014;7:355–65.

Chapter 10
The Immune System

Introduction

The environment contains a wide variety of potentially harmful pathogens, such as bacteria, viruses, fungi, protozoa, and multicellular parasites, which will cause disease if they enter the body and are allowed to multiply. Immunology is a scientific field of research dedicated to studying all aspects of the immune system. Although there have been many studies on physical activity and the immune system response, information collected is little about on its linkage to illness.

In rats, moderate exercise has been related with a one-third decrement in the occurrence of upper respiratory tract infections (URTI) [1]. It also reduces the risk of infection and increases other immunological variables, compared with a sedentary individual. In addition, epidemiological evidence suggests that moderate exercise has a beneficial effect on the human immune system, while extreme exercise impairs it, an effect which is modeled in a J curve. Nonetheless, elite athletes following very prolonged and heavy exertion bouts for long-lasting training times, or participation in acute or in very demanding sports events, influence the systemic immunity by increasing the risk of infection [2]. In marathon runners, it has been found that their long-lasting prolonged high-intensity exercise bout was related to an increased risk of an infection, although others did not find that result. Strenuous exercise during continuing infection and fever may be risky and should always be avoided [3]. Because myocarditis may or may not be accompanied by fever, athletes should be informed about the symptoms of this dangerous condition. In athletes, pulmonary dysfunctional capacity and illness are common during competitions events, which weakens significantly physical performance. Long-lasting vigorous exercise training affects the white blood cell [4], which results in a decreased host protection, during which viruses and bacteria can gain a grip, thus developing an infection. These immune changes happen in some sections of the immune system and body such as the skin, upper respiratory tract mucosal tissue, lung, blood, and muscle. Other factors such as psychological stress, lack of sleep, and malnutrition

M. S. Saghiv, M. S. Sagiv, *Basic Exercise Physiology*,
https://doi.org/10.1007/978-3-030-48806-2_10

can also depress immunity and lead to increased risk of infection. Inadequate recovery time between training sessions leads to an impermanent condition of immunological depression. This condition is reversible if the athlete takes a few days of rest.

Immune cell functions are impaired following acute sessions of prolonged high-intensity exercise, and some studies have found that athletes are at a higher risk for infections. The immune systems of athletes and nonathletes are generally similar. Compared to sedentary individuals, athletes may have slightly raised cytolytic action and natural killer cell count at rest; however, these are questionable to be clinically important. Compared to sedentary individuals, active individuals demonstrate reductions in C-reactive protein, which is related to enduring sicknesses and the helpful effects of exercise as an anti-inflammatory tool.

The immune system is a host defense system including many biological structures and processes within the human body, which protects against disease. The immune system protects through physical barriers organisms from pathogens such as bacteria and viruses from entering the organism. It utilizes multiple and diverse tools, cells, and tissues to decrease any possible harm to the body. In principle, the immune system will react to anything that is in the wrong place at the wrong time or has the potential to harm the body for any reason [5]. To function properly, an immune system must detect a wide variety of agents, known as pathogens, from viruses to parasitic worms, and distinguish them from the organism's own healthy tissue [6]. In many species, there are two major subsystems of the immune system: the innate immune system and the adaptive immune system. If a pathogen breaches these barriers, the innate immune system provides an immediate but non-specific response [7]. If pathogens successfully evade the innate response, vertebrates possess a second layer of protection, the adaptive immune system, which is activated by the innate response. Both subsystems use humoral immunity and cell-mediated immunity to perform their functions [6]. These subsystems of the immune system collaborate, yet have different triggers at times and different biological tools, to respond to their triggers. In humans, the blood–brain barrier, blood–cerebrospinal fluid barrier, and similar fluid–brain barriers separate the peripheral immune system from the neuroimmune system, which protects the brain.

Chronologically and on an evolutionary timeline, the innate immune system existed prior to the development of the adaptive immune system. Thus, the adaptive immune system is considered a more advanced system in comparison to the innate immune system. While the innate immune system is found in all plants, insects, fungi, and other multicellular organisms, the adaptive immune system is found in advanced organisms and creations only. The main goals of the innate immune system can be simplified as follows:

- Existing as a physical and chemical barrier against foreign agents.
- Activation of the adaptive immune system via the antigen presentation process.
- The release of chemical factors alerting nonspecific and specific immune cells to the specific locations, including immune mediators such as cytokines.
- Identification and removal of foreign threats present in cells, tissues, organs, blood, and lymph, by specialized white blood cells (WBC).

- Activating of the Complement Cascade resulting in bacteria identification, cell activation, and dead cells and/or antibody complexes clearance promotion.

Pathogens can rapidly evolve and adapt and thereby avoid detection and neutralization by the immune system; however, multiple defense mechanisms have also evolved to recognize and neutralize pathogens [8]. Humans have sophisticated defense mechanisms, including the ability to adapt over time to recognize specific pathogens more efficiently. Adaptive immunity creates immunological memory after an initial response to a specific pathogen, leading to an enhanced response to subsequent encounters with that same pathogen [5]. Other basic immune mechanisms include phagocytosis, antimicrobial peptides called defensins, and the complement system, including the ability to adapt over time to recognize specific pathogens more efficiently. This process of acquired immunity is the basis of vaccination.

Disorders of the immune system can result in autoimmune diseases, inflammatory diseases, and cancer [9]. Immunodeficiency occurs when the immune system is less active than normal. In humans, immunodeficiency can either be the result of a genetic disease such as severe combined immunodeficiency, acquired conditions such as HIV/AIDS, or the use of immunosuppressive medication. In contrast, autoimmunity results from a hyperactive immune system attacking normal tissues as if they were foreign organisms. Common autoimmune diseases include Hashimoto's thyroiditis, rheumatoid arthritis, diabetes mellitus type 1, and systemic lupus erythematosus. Immunology covers the study of all aspects of the immune system [10].

For the most part, the immune system deals with almost all the implications of being a creation that has multiple direct connections to the environment (i.e., the respiratory system, ears, eyes).

The innate immune system is mostly located in areas that allow the entrance to our body. In this sense, the innate immune system is the first immune line of defense [11], utilizing physical barriers, mechanical production of force, chemical changes, and non-selective cells as the main ways to protect the body. Physical barriers may include sticky and condensed solutions that physically either capture the unwanted factor or delay its entrance to the body as much as possible to improve the chance of a biological defense reaction (by the innate immune system, the adaptive immune system, or both).

While there are several different physical barriers utilized in nature, examples of physical barriers in humans mainly include the skin itself, skin oils, hair in multiple locations (head, skin, nose etc.), ear wax, tears, urine, mucus, stomach acids, vaginal secretions, semen, breast milk, and saliva in the mouth, which is rich in lysozyme—an enzyme that destroys bacteria by digesting their cell walls. Mechanical production of force processes includes sneezing, coughing, puking, and passing gas [12]. These processes incorporate the work of muscles triggered by the immune system to create force that physically pushes the trigger toward the outer parts of the body and hopefully completely outside of the body. Such processes will occur as either natural responses or forced responses. For example, sneezing is not forced, yet a person can be externally stimulated to vomit and can superficially cough. Areas of the body where the chemical makeup of the environment can be changed (i.e., pH, Iron concentration) assist in battling invaders by creating a hostile chemical environment. Core body temperature can be used and adjusted to assist battling threats as another measure of immunity. Since proteins change their 3D confirmation (i.e., denaturation) at certain temperatures, and thus lose their function, the increase in body core temperature may cause the biological threat to become dysfunctional. Additionally, heat may accelerate defensive chemical reactions in favor of our body. As a generalization, it is said that the innate immune system is unable to learn (adapt) and it produces no systematic memory of prior encounters.

The adaptive immune system is regarded as more sophisticated due to the fact that it is able to adapt to changing threats and create a systematic memory of prior encounters (immunological memory). The adaptive immune system is able to create a stronger immune response that is additionally more specific to every threat [13]. The adaptive immune system has a biological method of identifying unique biological features of threats and "tagging" them accordingly. These unique features are used to "remember" them for future encounters which in turn improve the quality and time of response. In other words, the adaptive system "learns" from every encounter with a threat.

Mechanically, processes that produce a fair amount of friction endanger adjacent cells since the vast majority of friction turns into heat which in turn may burn and kill cells. Responding to the heat created, fluids are used to absorb the heat and reduce the risk of killing more cells. Fluids that cannot be condensed often increase the mechanical pressure on adjacent nerves causing pain. The combination of dead cells and pain triggers the immune system to respond even more, resulting in either a controlled or uncontrolled inflammation. Certain types of immune cells are charged with the task of removing dead cells or part of them and eliminating the possibility that others may use the dead cells for their own purposes. Additionally, dead cells interfere with the normal processes in the area. In some cases, the immune system may attack its own cells if the immune response is out of control or it results in an overall better outcome for the body as a whole. For example, the immune system may attack its own cell or molecules if it is in the wrong place in the body. It may also kill its own cells to prevent from cancer cells to utilize its resources to spread. Attacks upon the body's own cells may also occur due to a mistake in identification and/or a genetic disease, disorder, or syndrome.

Another common generalization regarding the immune system is that it is one of the first systems in the body to suffer when overall energy is low. Thus, any process or state of existence that increases the energy cost of maintaining such a state will come at the expense of the immune system to an extensive scale. Such is the case during and after exercise. During exercise, most physiological processes address the metabolic needs of the skeletal muscles, leaving very little energy for the immune system. As a result, the immune system is operating while depleted of energy compared to the optimal possible.

Harmful agents that can lead to the occurrence of a disease are called pathogens. Pathogens can include viruses, parasite worms, bacteria, fungi, etc. The body utilizes both hormonal immunity and cell-mediated immunity to fight the pathogens [14]. In humans, the brain's immune system is separated by the blood–brain barrier, blood–cerebrospinal fluid barrier, and other similar fluid–brain barriers.

Once a pathogen has been successful in entering the body, the immune system utilizes diverse tools and processes to eliminate the threat and prevent its evolution or hiding within the body. Some of the defense mechanisms may include engulfing the pathogen, cutting it off of energy resources, digesting it (breaking it apart), disabling it, and disposing of it. However, some pathogens have changed tactics to limit immune discovery or to use the immune response to their gain [14].

Immune Deficiency and Autoimmunity

Immunodeficiency is a condition in which the immune system's capacity to function against infectious disease is limited or completely vague. An individual who has an immunodeficiency of any kind is said to be immunocompromised. Most situations of immunodeficiency are assimilated due to extrinsic factors such as nutrition that affects the immunity functional capacity [14]. Following organ transplant, some drugs may cause immunosuppression, while it can be either an antagonistic effect or being the right action. The number of immunodeficiency syndrome possibilities is large, which represent autoimmunity condition. The inability of the immune system to cure some patients from infections may be accountable for triggering autoimmunity through continuous immune system activation [15]. The cause of immunodeficiency varies depending on the nature of the disorder. The cause can be either genetic or acquired by malnutrition and poor sanitary conditions. Only for some genetic causes, the exact genes are known [13].

Autoimmunity results in a more than normal reactive immune system, attacking normal tissues and cells as if they are a foreign threat [16]. Common autoimmune diseases include Hashimoto's thyroiditis, rheumatoid arthritis, diabetes mellitus type 1, and systemic lupus erythematosus. Allo-immunity is regarded as part of autoimmunity where the body rejects any tissue that is contributed and/or transplanted into the host.

Antigens

Antigens (antibody generators) are structures specifically bound by antibodies or a cell surface version of antibodies: B-cell antigen receptor. Antigens are non-self (foreign) molecules that are able to bind with a specific immune receptor resulting in an immune response. Native antigen molecule after processing it can be recognized by T-cell receptor. Both T cells and B cells are cellular components of adaptive immunity [17]. T cells are a type of lymphocyte which matures in the thymus gland and plays a key role in the immune response. T cells can be distinguished from other lymphocytes by the presence of a T-cell receptor on the cell surface. While B cells (B lymphocytes) are a type of white blood cell of the small lymphocyte subtype [18].

Theoretically, newborns do not have any immune encounters with pathogens; thus, their adaptive immune system has no prior experience nor memory cells. One may say that newborns are at high risk for infections at their early stages of life on the one hand and that adaptive immunity is built and strengthened one encounter at a time. Being able to control the "new encounter" of the adaptive immune system is preferred as practiced by immunizations (weakened pathogens meant to create immune memory cells yet at a low risk of a disease). At first, passive immunity is provided by the mother as she passes on to the baby IgG antibodies. It represents approximately 75% of serum antibodies in humans; IgG is the most common type of antibody found in blood circulation. IgG molecules are created and released by plasma B cells [19]. They are transferred to the baby via the placenta giving the baby the same range of antibodies as the mother has. Antibodies are further passed on to the baby's gut through the mother's breast milk and colostrum until it can synthesize its own. Such passive immunity is short term (usually up to a few months only). Passive immunity can also be achieved by the infusion of antibody-rich serum (of external origins) and is used in modern medicine.

Lines of Defense

The body prevents infection via a number of non-specific and specific mechanisms working on their own or together. The body's first lines of defense are external barriers that prevent germs from entering. The innate immune at the beginning reacts through physical barriers that stop pathogens from entering the body, extinguish them following their entry, or move them away from the body, before they can damage body's soft tissues. Barrier defenses are the most basic defense mechanisms. Barrier defenses, protection, do not take care of the infections themselves, which the adaptive immune system takes care of them [20]. The largest physical barrier is the skin, where very few organisms are able to penetrate undamaged skin. Skin immunity is a property of the skin that allows it to resist infections from pathogens. Skin and mucus membrane are the examples of anatomical barriers that provides immunity. These barriers are commonly divided into physical barriers, chemical barriers, and biological barriers (Fig. 10.1). There are other physical barriers and a variety of chemical defenses.

Examples of these non-specific defenses are (a) mucus, a sticky trap secreted by all the surfaces inside the body that are directly linked to the outside, also contains antibodies and enzymes; (b) cilia—microscopic hairs in the airways that move to pass debris and mucus up away from the lungs; (c) lysozyme—a chemical (enzyme) present in tears and mucus that damages bacteria; (d) phagocytes—various cells that scavenge and engulf debris and invading organisms, which form part of the surveillance system to alert the immune system of attack; (e) commensal bacteria—bacteria on skin and gut that compete with potentially harmful bacteria for space and nutrients; (f) acid, in stomach and urine, makes it hard for any germs to survive; and (g) fever—elevated body temperature making conditions unfavorable for pathogens to survive.

Fig. 10.1 Anatomical barriers; physical, chemical, and biological

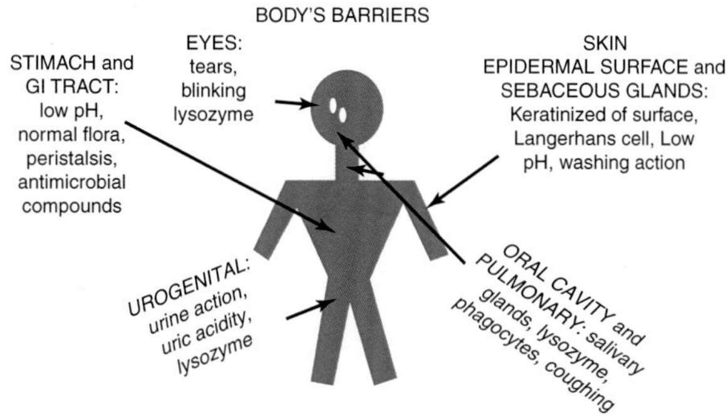

Physically, most infections are not capable of penetrating the epithelial surfaces created by the body [21]. These act as the very first physical defense of the body. Yet, the existence of such surfaces does not necessarily prevent the threat from existing on the surface itself. Thus, other defensive options are utilized in addition to the mere existence of the surfaces.

Shedding of one's skin (termed Desquamation) is one of the strategies to prevent bacteria and other forms of threats from existing on such surfaces for long. Flushing, sweat, organic acids, the lack of blood vessels, sebaceous glands, and the lack of moisture increase the hardship of surviving on an individual's skin [22].

Threats that still find their way into more internal areas such as the respiratory tract and gastrointestinal tract encounter additional barriers such as hair, gastric and bile acids, digestive enzymes, flushing, gut flora, defensins, thiocyanate, peristalsis, cilia, mucociliary escalator, and surfactants. In the nasopharynx they will encounter lysozymes, saliva and mucus, tears in the eyes, and endothelial cells as well as P-glycoprotein (mediated active transportation) in the blood–brain barrier.

Hair helps trap and delay pathogens from reaching the actual skin as well as exposes them to wind, water, and heat that might affect their function or kill them while stranded on the hair. Saliva and mucus physically trap the pathogen within their slimy solution and are either removed by the body or treated by the immune system. Acids oppose a chemical challenge by decreasing the pH of the environment, making it harder and perhaps impossible for the pathogen to survive [23]. Digestive enzymes engulf the pathogen, digest, and break it down to its basic components, which in some cases can actually be repurposed for the body's benefit. Flushing uses natural sources of fluids to wash away the pathogens such as tears and urine. The gut flora are regarded as "friendly bacteria" that reside in one's gut and prevent colonization of pathogens. By the secretion of toxins, competing over the same nutrients and occupation of surfaces, gut flora decreases the pathogens' chances of success, especially that of bacterial pathogens [24]. Muscles that are stimulated to contract create a wavelike effect that removes pathogens and/or makes it harder for them to stay put. This phenomenon is called peristalsis. In a similar way, cells that resemble grass or hair called cilia move in wavelike motion to trap pathogens or prevent them from staying put. Constantly on the move, peristalsis, cilia, and the mucociliary escalator process decrease the pathogens' chances of affecting the body on the one hand, while increasing the time for them to be detected and attacked by the immune system on the other hand. Defensins are small cysteine-rich cationic proteins that function as host defense peptides. They are active against bacteria, fungi, and many enveloped viruses and non-enveloped viruses. Cells of the immune system contain these peptides to assist in killing phagocytosed (engulfed) bacteria [25].

Most defensins bind to the pathogen's cell membrane forming a pore within the pathogen's outer membrane allowing ions and nutrients to escape or harmful molecules and cells to enter the threatening pathogen.

Defensins disrupt the structure of bacterial cell membranes and play a central role in the defense against pathogens. Recent research suggests the possibility of defensins being upregulated by biological activating factors as found (for example) in breast milk.

Thiocyanate is analogous to the cyanate ion, [OCN], whereas oxygen is replaced by sulfur and thiocyanate is a potent competitive inhibitor of the thyroid sodium-iodide symporter. Research shows that reduced concentrations of thiocyanate cause low host defense immunity.

Surface-active agents (surfactants) decrease the surface tension between different materials of different sorts (gases, fluids, and solids). These are organic amphiphilic compounds [26] that may act as wetting agents, emulsifiers, detergents, foaming agents, and dispersants. While they assist the lungs mechanically during gross respiration, their properties also increase the hardship of pathogens colonizing the lungs. Lysozymes are antimicrobial enzymes that function as part of the innate immune system. It specializes in the breaking down of certain part of the gram-positive bacterial cell's wall. With the cell wall broken, lysis of the bacteria may occur or be initiated. They are commonly found in saliva, tears, mucus, and breast milk.

P-glycoprotein is an important protein that disposes of molecules through the cell's membrane. For this reason, it creates an opening on the cytoplasmic side of the membrane that can be exploited from an immune standpoint to attack the pathogen cell.

Inflammation

Inflammation is associated with the innate immune system since it is considered a generic response of a non-specific nature. Inflammations are a normal standard immune reaction to cell injury meant to prevent the spread of an infection, prevent further damage and death of the body's cells and tissues, and assist in the healing process to follow [13]. While inflammation has a negative reputation, it is actually a protective process of positive nature. The source of this bad reputation is most likely the few inflammatory reactions that the body loses control over, thus becoming very lengthy and probably painful. While an infection of sort is the cause, the inflammation process is the response; thus, they do not represent the same idea. Inflammatory responses are associated with infections since they are part of a wide range of responses to an infection. It is important to remember that not every infection results in an inflammatory response (such would be the case with eosinophilia or parasitosis) and that an inflammation can be the chosen response, whereas the trigger is not an infection (such as in autoimmune diseases, trauma, atherosclerosis, and/or ischemia). The vast majority of inflammations are kept under control and do not induce the need for medical treatment. From an immune aspect, the earlier the onset of inflammation and the more accurate it is, the less the potential harm to the body [27].

Common causes of an inflammatory response may include:

- An infection by pathogens
- Stress
- Toxins (gas, fluid and solid)
- Chemical irritants
- Burns
- Frostbite
- Immune-hypersensitivity
- Ionizing radiation
- Alcohol consumption
- Psychological excitement
- Entry of a physical foreign body (splinters, nails, etc.)
- Trauma
- Actual cell/tissue injury

Inflammation commonly presents with five symptoms that include pain, fever, swelling, redness, and function loss (usually partial function loss). Pain can be the result of the release of chemical agents such as histamine and bradykinin [28] increasing the stimulation of the endings of nerves, or the increased mechanical pressure on the nerves themselves is commonly the result of the accumulation of fluids in the area of injury which in turn also increases the mechanical pressure in the area as fluids cannot be condensed. Since fluids cannot be condensed and in accordance with Newton's law, trying to forcefully condense fluids will result in the fluids applying an equal and opposite force on the environment that results in pressure on the nearby nerves. Redness is the result of increased blood flow to an area and local heat production. The importance of "getting right" the extent of the inflammation response lies in the fact that either too little or too much of an inflammatory response may result in harm being caused to the body's cells and tissues. Chronic (long-lasting) inflammation is associated with diseases such as atherosclerosis and osteoarthritis. [29]. The initial inflammatory process (acute inflammation) involves somewhat different cells and processes compared to those of the prolonged inflammatory response (chronic inflammation). Changes to plasma distribution and the activations and migration of leukocytes (mainly granulocytes) from the bloodstream to the injured tissues are characteristics of an acute inflammation. The chronic inflammatory response includes the utilization of mononuclear cells and activation of healing almost parallel to the attacking of the pathogens and injured cells/tissues. Inflammations are caused due to three common dynamics; the first dynamic is biological in nature such as the reaction to a pathogen; the second dynamic is chemical in nature such as the reaction to a toxin inhaled or secreted; and the third dynamic is physical or thermodynamical in nature such as the reaction to a physical injury. In all cases, the common denominator is cell injury being one of the outcomes that induces the immune system to respond. The biological injury typically includes changes to molecules important to cell function and exploitation of the cell's metabolic resources, whereas the intention of the invader is to use our body's resources in order to survive or flourish/multiply. Chemically based cell injury is the result of the effect of chemical irritants and toxins that either manipulate the function of the cell or disrupt it at the molecular level or electrical level. Chemically induced inflammation can also originate from day-to-day activities, whereas chemicals and molecules are the result of the body's metabolism, yet the high concentration presents a threat [30]. For example, research has shown that an immune response may be triggered by elevated levels of lactic acid due to exercise. While the creation of lactic acid is a normal metabolic process associated with exercise, high levels may be perceived by the immune system as harmful, eliciting an immune response [31]. Physical injuries demand that actual physical contact be applied to our body in order to induce the injury, thus leading to not only the physical disruption of the cell (physical injury) but also the creation of friction as part of the process. Friction is often transformed mostly into heat that in turn can cause internal or external burns which are harmful to living cells. For example, two tissues or organs of the body that are in motion and in contact with each other create friction that in turn will transform into heat causing neighboring cells to be injured or die, acting as the trigger for the immune system to respond. Such is the case with inflammatory reactions to exercising, especially with heavy mechanical loads (weights; resistance training).

Decreasing the time of reaction during the initiation of an acute inflammation is possible due to the fact that immune cells are already located in the affected cells (resident macrophage cells). The vast majority of these resident immune cells include macrophages, dendritic cells, histiocytes, Kupffer cells, and mast cells characterized by their ability to utilize pattern recognition via pattern recognition receptors (PRRs), which recognize pathogen-associated molecular patterns (PAMPs), and damage-associated molecular patterns (DAMPs) further explained below [32].

At the onset of cell injury, the resident immune cells are activated as a result of a PRR identifying a PAMP or DAMP releasing chemical mediators that cause the signs of an inflammation [33]. The acute increase in the blood flow both increases the availability of immune cells located within the bloodstream and the creation of edema due to the increased permeability of blood vessels on the one hand and the body's tissues on the other hand. Thus, plasma fluids and proteins enter the tissue.

Immediate alterations to the environment include increased permeability to white blood cells, increased sensitivity to pain, increased local temperature, increased local mechanical pressure, and decreased function on a neurological basis (a natural response of the nervous system to pain). Parallelly, other systems are called to action such as the complement system, the coagulation system, the fibrinolysis system, and the adaptive immune system [34]. The response of the innate immune system is a "first-line immune response" and is relatively short-lived due to the fact that its mediators are short-lived as well.

The vascular components of the acute inflammatory response include vasodilation (increasing the diameter of blood vessels) which immediately results in increased blood flow and increased permeability of cells and tissues to certain molecules and fluids. The vascular phase of the acute immune response also precedes the cellular phase based on specific immune cells (commonly myeloid granulocytes).

Vasodilation is initiated by mastocytes and macrophages that release vasoactive components such as serotonin, leukotriene B4, histamine, and prostaglandin E2 that create changes in nearby blood vessels. Blood plasma volume in the nearby tissues increases as nitric oxide (NO) is released resulting in the dilation of blood vessels and increased permeability. The fluids that enter the tissue contain antimicrobial factors that immediately engage the microbes and fight them. Such antimicrobial factors may include antibodies and lysozymes. When the injury induces bleeding, platelets, plasmin, kinins, and coagulants react to stop the bleeding. These clotting mediators create a fibrin lattice which aids phagocytic debridement and wound repair in later phases [35]. Parallel to the previously described processes, the lymphatic system is used to transport some of the exuded fluids to nearby lymph nodes to both remove bacteria from the original site and allow the adaptive immune system to investigate and initiate a response.

The progression of vasodilation and increase in the amount of blood is from arteriole to capillary levels resulting in stasis. Stasis is characterized by enlarged blood vessels full of cells. Stasis assists leukocytes to be recruited into tissue by allowing them to advance along the endothelium, which is prevented by normal blood flow that directs blood flow toward the middle of the blood vessel.

The complement system (the Complement Cascade) improves antibodies and phagocytic cells' ability to confront microbes and injured cells, promote inflammation, and ultimately attack the pathogen's cell membrane. While regarded as part of the innate immune system, the adaptive immune system has means of its own to activate the complement system if needed. The term cascade is used to describe a multilevel process or a multiphase process where a certain order of processes must occur one before the other. Cascades also result in certain responses dependent on the mediator or activator utilized prior. Thus, inactive components or responses can be activated at a certain phase of the whole cascade; they can be activated in sequence or in parallel. Often, immune cascades relay on a few specific factors to either deactivate the cascade or advance it. The complement system utilizes small proteins (serum proteins and cell membrane receptors) that are synthesized by the liver and released into the blood where they circulate in their inactive form [36]. When stimulated, proteases activate the proteins by cleaving their deactivating molecular component. The immediate result is the release of cytokines that in turn activate more components of the cascade needed downstream (later processes of the cascade). After the complement cascade has been fully utilized, phagocytes remove foreign cells, injured cells, and debris, initiate inflammation to attract more phagocytes, and activate the attack on pathogens' membrane in order to kill it. The complement immune cascade can be activated via different immune pathways.

The Classic Pathway is triggered by the activation of the C1 complex which occurs when subcomponents of the C1 complex bind with antigens [37]. The activation can be the result of one or more of the complex's subcomponents encountering an antigen or binding directly to the surface of the pathogen itself. The binding initiates 3D conformational changes in the C1q subcomponent, activating the C1r subcomponent, allowing both to cleave a third subcomponent called C1s. The cascade continues via the cleaving and activating of more molecules resulting in the creation and activation of C3 convertase. C3 convertase then cleaves C3 into C3a and C3b. While the anaphylatoxin C3a interacts with its C3a receptor (C3aR) to recruit leukocytes, C3b contributes to further downstream complement activation.

C3b then binds to the C3 convertase to form C5 convertase that then cleaves C5 into C5a and C5b. C5a is also an anaphylatoxin that interacts with its cognate C5a receptor to recruit and attract leukocytes. Additional interactions between C5b and other terminal components (C6–C9) form a complex that attacks pathogen membranes by forming pores in the target cell membranes.

The complement system is associated with immune inflammatory responses in adipose tissues implicated in the development of obesity [38]. High levels of the C1 component are one characteristic of obesity resulting in tissue inflammation as well as insulin resistance. The classic pathway has clinical significance as the activation of the classic complement pathway is used as a means to fight HIV-infected cells, methicillin-resistant Staphylococcus Aureus, cancer, and tumors. The activation of the classic pathway is achieved by the C1-inhibitor, whereas the inhibition of the C1 component prevents the occurrence of all other reactions downstream in the cascade. The lack of C1-inhibitor has clinical implications of itself; in the absence of enough C1-inhibitor (acquired or hereditary), the regulation of vascular permeability is lost leading to angioedema.

Deficiency in the C1q protein subcomponent may result in the development of systemic lupus erythematosus. The C1q protein subcomponent activates the binding between immune complexes and apoptotic cells directly onto phagocytes [13]. Thus, in the face of insufficient amount of the C1q protein subcomponent autoantibodies and apoptotic cells accumulate as part of systemic lupus erythematosus.

Since C3 is somewhat unstable in aqueous solutions (environments) on the one hand, and the alternative pathway reacts to C3 hydrolysis on the other hand, the alternative pathway is constantly active to a mild extent [39]. The alternative pathway is unique in comparison to the other pathways since it does not rely on pathogen-binding antibodies as part of its function. While C3b is inactivated by factors H and I on the one hand, when the internal thioester of C3 reacts with a pathogen's surface hydroxyl group or amino group of a molecule, the C3b is protected from the effects of factor H on the other hand. C3b binds to factor B to form the C3bB complex. The C3bB complex is cleaved by factor D into Ba and Bb. Bb remains connected to C3b forming C3bBb, which is the alternative pathway's form of C3 convertase.

Properdin (factor P) binds to the C3bBb complex allowing a greater extent of C3 cleavage that amplifies the complement activation several folds [39]. Endogenous complement regulatory proteins (CD35, CD46, CD55 and CD59) limit complement activation. The alternative complement pathway distinguishes self from non-self-based due to the fact that foreign cells, pathogens, and abnormal surfaces present great amounts of C3b and iC3b on their surfaces.

The lectin complement pathway functions in a similar way and structure to the classic complement pathway since after activation, C4 and C2 activate complement proteins needed further down the cascade (downstream). The lectin pathway's initial reactions involve mannose-binding lectin (MBL) and/or ficolin binding to certain carbohydrates/sugars. MBL is a protein that is part of the collecting family that is produced by the liver and can initiate the complement cascade by binding to pathogen surfaces. MBL binds to Salmonella, Listeria, Neisseria strains, Candida Albicans, Cryptococcus Neoformans, HIV-1, and respiratory syncytial virus (RSV).

MBL deficiency results mainly in recurrent upper respiratory tract infections and of other body systems such as pneumonia and meningitis. Infants and young children with MBL deficiency are more susceptible to infections as well as those with suppressed immunity such as people with cancer subjected to chemotherapy or low immunity. A key conclusion regarding the complement immune system as a whole is that while it is an important part of the immune response in general and the innate immune response in particular, the partial or complete lack of its regulation withholds extremely harmful implications often associated with or suspected as contributing to the cause or severity of diseases such as Barraquer–Simons syndrome, asthma, lupus erythematosus, glomerulonephritis, arthritis, autoimmune heart disease, multiple sclerosis, inflammatory bowel disease, paroxysmal nocturnal hemoglobinuria, atypical hemolytic uremic syndrome, ischemia-reperfusion injuries, rejection of transplanted organs, Alzheimer's disease, and spinal cord injuries.

Pattern Recognition

The initiation of the inflammation process (acute inflammation) is achieved by immune cells that are present in all cells of the body, macrophages, dendritic cells, histiocytes, Kupffer cells, and mast cells to name a few [40]. These cells contain pattern recognition receptors (PRRs), which recognize molecules that are common for pathogens but can be distinguished from host molecules known as pathogen-associated molecular patterns (PAMPs) [41]. When a PAMP is recognized by a PRR at the onset of cell injury, these cells are activated releasing inflammatory mediators.

Chemical signals, also known as chemical factors, such as bradykinin, serotonin, histamine, leukotrienes, and prostaglandins further elicit changes such as increased blood flow (the result of the dilation of blood vessels; vasodilation) and increased presence of neutrophils-phagocyte immune cells [7]. Neutrophils then release chemical factors acting as the trigger for additional parts of the immune system such as leukocytes and lymphocytes. Additionally, cytokines such as TNF, HMGB1, and IL-1 are produced by macrophages and other cells of the innate immune system in order to mediate the inflammatory response [42].

Inflammation is associated with common symptoms such as increased local blood circulation resulting in redness of the skin, increased core body temperature (a fever; which helps chemical reactions accelerate, weakens the pathogen, and indirectly causes the sick person to rest, increasing the energy available for the immune system), localized or whole body pain, increased production of mucus, productive cough and/or a runny nose, swelling, rheumatoid arthritis, and possible dysfunctional organs and tissues.

Toll-Like Receptors

Pathogen-associated molecular patterns (PAMPs) that are external to the cell are recognized via the mediation of transmembrane proteins known as Toll-like receptors (TLRs). Toll-like receptors all include a similar structural component known as the leucine-rich repeats (LRR) that allows them their specific functionality. Toll-like receptors trigger the synthesis and secretion of cytokines and activation of other host defense options necessary for both innate and/or adaptive immunity [43].

Cytosolic Receptors

Cytosolic receptors pertain to the pattern recognition receptors that detect infection or cell injury in the cytosol. Cytosolic receptors are mainly divided into three major classes: NOD–like receptors, RIG (retinoic acid-inducible gene)-like receptors (RLRs), and cytosolic DNA sensors [44].

Nucleotide-binding oligomerization domain-like receptors or NOD-like receptors (NLRs) are sensors within the cell (intracellular) of pathogen-associated molecular patterns (PAMPs) that enter the cell via phagocytosis or pores and damage-associated molecular patterns (DAMPs) that are associated with cell stress. They are pattern recognition receptors (PRRs) that regulate the innate immune response. The regulation of inflammatory and apoptotic responses can result from the collaboration of NLRs and TLRs. They are components of lymphocytes, macrophages, dendritic cells, and the epithelium. RIG-like receptors are virus recognizing intracellular pattern recognition receptors of the innate immune system. They specialize in cytoplasmic pattern recognition of intracellular viral RNA that is a product of a viral infection, making them the frontline defense against viral infections in most body tissues. The cGAS–STING pathway is a component of the innate immune system that detects cytosolic DNA and initiates inflammatory defense mechanisms. DNA localized in the cytosol of the cell instead of the cell nucleus is characteristic of tumorigenesis (tumor creating agents), viral infections, and intracellular bacteria [44].

Cellular Components

The cellular components of immunity rely on actual cells of the body with specificity of function. A wide range of such cells exist. While cellular components of immunity can be found in multiple tissues of the body, the bloodstream is the perfect tissue to have the majority of immune cellular components as it circulates throughout the body and is a prime avenue for threats to travel through the body.

Leukocytes (WBC; white blood cells) are independent single-celled organisms often perceived as the second branch of the innate immune system. The innate immune system's leukocytes include several cells such as natural killer cells (NKs), eosinophils, mast cells, innate lymphoid cells, basophils, and phagocytes [45]. Phagocytes can further be divided into macrophages, neutrophils, and dendritic cells. These cells are entrusted with identifying and eliminating larger pathogens via engulfment and contact leading to the killing of microorganisms.

Cellular innate immunity commonly utilizes phagocytosis (engulfment or eating) as an important component of its function [46]. Phagocytes that engulf whole pathogens or part of them are carried within the bloodstream in order to search for pathogens increasing their chance of being encountered. They may also be chemically summoned to specific locations by cytokines. The phagocyte engulfs the pathogen, trapping it in an intracellular vesicle called a phagosome, which then combines a lysosome to form a phagolysosome. Digestive enzymes within the phagolysosome kill the pathogen by means of respiratory burst (the release of free radicals into the phagolysosome). In some cases, phagocytosis can be utilized as a means of acquiring nutrients that can be used to the body's advantage. Often, it is regarded as one of the earliest forms of immuno-defense.

Phagocytes

Phagocytes include neutrophils, macrophages, and dendritic cells (DCs). Neutrophils represent up to 60% of all circulating leukocytes and are abundantly found, thus making them the most common sort of phagocyte cells, normally found in the bloodstream. Neutrophils can further be divided into neutrophil-killers and neutrophil-cager cells [47]. During the preliminary phase of inflammation (the acute phase) caused mainly by bacterial infection, neutrophils migrate toward the site of inflammation via chemotaxis as the first cells to respond. Macrophages are known as the "scavengers" of the body that get rid of dead cells, parts of cells (debris), and the antigen-presenting cells (that activate the adaptive immune system). Macrophages are located within tissues and produce a diversity of complement proteins, enzymes, and cytokines.

Dendritic cells (DCs) specialize in contact with the external environment [48]. Thus, they are located mainly in locations that have an immediate physical connection with the environment such as the skin, nose, lungs, stomach, and intestines. Contrary to their name, phagocyte dendritic cells are not connected to the nervous system. Phagocyte dendritic cells mediate the immune responses of both the innate and adaptive immune systems via mediation of T Cells that play a key role in the immune response [49].

Granulocytes

Neutrophils, mast cells, basophils, and eosinophils are all considered as granulocytes [50]. They are part of the leukocyte family characterized by granules in their cytoplasm. Mast cells (MCs) are located in connective tissue and mucous membranes in order to regulate the inflammatory response to allergies and anaphylaxis. Basophils and eosinophils secrete chemical mediators that are involved in the immune response to parasites and allergic reactions [51]. Asthma may act as one possible example.

Innate Lymphoid Cells

Innate lymphoid cells (ILCs) are innate immune cells that are derived from a common lymphoid lineage. These cells are often characterized by the lack of antigen-specific B- or T-cell receptor because of their lack of the recombination activating gene (RAG).

The lymphocyte family of cells includes natural killer (NK) cells that destroy compromised host cells [52]. NK cells identify such compromised cells via the "missing self" component, commonly including tumor cells and virus-infected cells. Cells characterized as "missing self" cells can be recognized via the low levels of a cell surface marker called major histocompatibility complex (MHC I) caused when a host cell is infected. NK cells are activated directly due to the "missing self" change to the surface of the infected cells, which acts as an advantage due to shortening of the time to their activation [13]. Cells that express intact self MHC antigens are not attacked by NK cells since MHC antigens are recognized by killer cell immunoglobulin receptors (KIR) deactivating the NK cells.

The Adaptive Immune System

Also known as the acquired immune system and/or the specific immune system, the adaptive immune system is structured to be accurate and specific to certain stimuli killing them and preventing their growth via colonization.

The adaptive immune system distinguishes itself from the innate immune system via two main characteristics: (1) the ability to learn, adapt, and create memory cells that act as a memory of past biological encounters and (2) it utilizes a highly specific extent of immune tools/cells.

Memory cells are the tool utilized to establish immunological memory which in turn improves and accelerates the immune response of the adaptive immune system the second time it encounters the same threat that has previously created a memory cell.

The operating principle of the adaptive immune system is the basis for the clinical implementation in medicine of vaccinations. Similar to the innate immune system, the adaptive immune system utilizes humoral immunity components and cell-mediated immunity components as well. Though it does not apply to all possible cases, immunological memory can provide our body with long-term immunity thereafter the initial encounter with a pathogen.

Distinguishing harmful from harmless foreign molecules is cardinal to the success of the adaptive immune system's function yet the system is sometimes unable to perfectly distinguish one from the other. White blood cells called lymphocytes carry out the acquired immune response; these responses are further divided into antibody responses and cell-mediated immune responses. Activated B cells secrete antibodies (Immunoglobulins). Antibodies bind to the foreign antigen resulting in its deactivation.

The adaptive immune system exploits two genetic processes [53] to its favor, allowing it to clone immune system cells relatively fast and in great numbers:

1. *Somatic hypermutation (SHM)*—involves a programmed process of mutation affecting the variable regions of immunoglobulin genes. Unlike germline mutation, SHM affects only an organism's individual immune cells, and the mutations are not transmitted to the organism's offspring.

   B cells proliferate/divide when they recognize an antigen [11], whereas the receptor locus undergoes repeated somatic mutations that is ≥105 times greater than the normal rate by means of genetic single-base substitutions, insertions, and deletions. Hypervariable regions of DNA known as "hotspots" and usually involved in antigen recognition on the immunoglobulin are commonly where such mutations occur. Such locations are also termed "complementarity-determining" DNA regions. This directed hypermutation allows for the selection of B cells that express immunoglobulin receptors possessing an enhanced ability to recognize and bind a specific foreign antigen. As the process is repeated thousands of times, including the mutations, an increase in specificity variation is achieved that results in multiple different variations of surface receptors on the B cells. The variation with the highest affinity to the antigen can then be selected for action. The B cells with the greatest affinity will then differentiate into plasma cells producing antibody and long-lived memory B cells [54]. Mechanisms/models explaining somatic hypermutation are also divided into two: (a) The DNA deamination model—it is focused completely on DNA substrates such as the Neuberger "DNA deamination model" based on activation-induced cytidine deaminase (AID) and short-patch error-prone DNA repair by DNA polymerase-eta operating around AID C-to-U lesions [55]. While this model explains somatic mutations at A:T and G:C base pairs, it does not explain how strand-biased mutations are generated. (b) The reverse transcriptase model—is an RNA/RT-based mechanism which explains in part the production of the full spectrum of strand-biased mutations at A:T and G:C base pairs. In such cases, mutations of A and G exceed mutations of T and C. An RNA-dependent DNA polymerase copying the base-modified Ig pre-mRNA template and integrating the now error-filled cDNA copy back into the normal chromosomal site creates error-prone cDNA synthesis. The errors in the Ig pre-mRNA are a combination of adenosine-to-inosine (A-to-I) RNA editing and RNA polymerase II transcription elongation complex copying uracil and AID-mediated lesions into the nascent pre-mRNA using the transcribed (TS) DNA as the copying template strand.

2. V(D)J Recombination—an irreversible genetic recombination of antigen receptor gene segments allows a small number of genes to generate a vast number of different antigen receptors uniquely expressed on each individual lymphocyte cell. Since the genetic change to DNA is irreversible, all progeny cells inherit the same receptor specificity, including the memory cells (B and/or T) [56]. V(D)J recombination occurs in bone marrow for B cells and the thymus for T cells. Almost completely randomly, it rearranges variable (V), diversity (D), and joining (J) gene segments, giving the process its name of V(D)J recombination. The process produces new amino acid sequences in the antigen-binding regions of immunoglobulins and TCRs that recognize all antigens of bacteria, parasites, pollen and other allergens, worms, viruses, and "altered self-cells" as such is the case with cancer. Such recognition can also lead to autoimmunity problems if it matches host tissues.

Immunoglobulins

The production of antibodies is the main function of the humoral immune system. An immunoglobulin (Ig) also known as an antibody (Ab) is a large, Y-shaped protein produced mainly by plasma cells that is used to neutralize pathogens such as bacteria and viruses [57]. The antibody recognizes a unique molecule of the pathogen, called an antigen, via the fragment antigen-binding (Fab) variable region. Each tip of an antibody contains a paratope that is specific for one epitope on an antigen, allowing these two structures to bind together with precision. Using this binding mechanism, an antibody can "tag" a microbe or an infected cell for future attack by immune cells. In some cases, the binding is enough to neutralize its target directly if the binding region is essential to the invaders' ability to function.

The ability of an antibody to communicate with the other components of the immune system is mediated by a region located at the base of the immunoglobin's "Y" shape known as the Fc region, which contains a conserved glycosylation site involved in such interactions. Antibodies are secreted by B cells of the adaptive immune system, mostly plasma cells [58]. Antibodies exist in both a soluble form and a membrane-bound form. The soluble form is secreted into the blood plasma and then attaches to the surface of a B-cell to create the B-cell receptor (BCR). BCRs are unique to the surface of B cells and are part of the process of the activation of these cells themselves and thereafter their differentiation into either plasma cells (which produce antibodies) or memory B cells. T helper cells are needed in most cases in order to activate the B cell, antibody generation, and antigen binding [11]. Soluble antibodies are released into the blood and tissue fluids as well as many secretions to continue to survey of invading microorganisms.

Antibodies are glycoproteins belonging to the immunoglobulin superfamily. They constitute most of the gamma globulin fraction of the blood proteins. They are typically made of basic structural units—each with two large heavy chains and two small light chains. There are several different types of antibody heavy chains that define the five different types of crystallizable fragments (Fc) that may be attached to the antigen-binding fragments. The five different types of Fc regions allow antibodies to be grouped into five isotypes. Each Fc region of a particular antibody isotype is able to bind to its specific Fc receptor (except for IgD, which is essentially the BCR), thus allowing the antigen–antibody complex to mediate different roles depending on which FcR it binds. The ability of an antibody to bind to its corresponding FcR is further modulated by the structure of the glycans present at conserved sites within its Fc region. The ability of antibodies to bind to FcRs helps to direct the appropriate immune response for each different type of foreign object they encounter. For example, IgE is responsible for an allergic response consisting of mast cell degranulation and histamine release [59]. IgE's Fab paratope binds to allergic antigen, for example house dust mite particles, while its Fc region binds to Fc receptor ε. The allergen–IgE–FcRε interaction mediates allergic signal transduction to induce conditions such as asthma.

Though the general structure of all antibodies is very similar, a small region at the tip of the protein is extremely variable, allowing millions of antibodies with slightly different tip structures, or antigen-binding sites, to exist. This region is known as the hypervariable region. Each of these variants can bind to a different antigen. This enormous diversity of antibody paratopes on the antigen-binding fragments allows the immune system to recognize an equally wide variety of antigens. The large and diverse population of antibody paratope is generated by random recombination events of a set of gene segments that encode different antigen-binding sites (or paratopes), followed by random mutations in this area of the antibody gene, which create further diversity. This recombinational process that produces clonal antibody paratope diversity is called V(D)J or VJ recombination. Basically, the antibody paratope is polygenic, made up of three genes, V, D, and J. Each paratope locus is also polymorphic, such that during antibody production, one allele of V, one of D, and one of J are chosen [59]. These gene segments are then joined together using random genetic recombination to produce the paratope. The regions where the genes are randomly recombined together is the hypervariable region used to recognize different antigens on a clonal basis.

Antibody genes also reorganize in a process called class switching that changes the one type of heavy chain Fc fragment to another, creating a different isotype of the antibody that retains the antigen-specific variable region. This allows a single antibody to be used by different types of Fc receptors, expressed on different parts of the immune system [11].

The Immunoglobulin Heavy Chain

In B cells, the first recombination event occurs between one D and one J gene segment of the heavy chain locus. As a result, DNA between these two gene segments is deleted. One V gene segment joins the D-J recombination forming a rearranged VDJ gene segment. Similar to before, all other gene segments between the V and D segments are deleted [13]. Primary and non-spliced RNA transcript is created containing the VDJ region of the heavy chain and the mu and delta chains (V-D-J-Cμ-Cδ). A polyadenylated (poly-A) tail is added after the Cμ chain and translation of this mRNA leads to the production of the Ig M heavy chain protein.

The Immunoglobulin Light Chain

In the rearrangement of the immunoglobins' light chain, the kappa (κ) and lambda (λ) chains rearrange in a similar way, excluding the involvement of a D segment. Thus, recombination involves joining the V and J chains, creating the VJ complex prior to adding the constant chain gene during primary transcription. Translation of the spliced mRNA for either the kappa or lambda chains results in the formation of the Ig κ or Ig λ light chain protein [60]. The combination of the Ig μ heavy chain and one of the light chains results in the formation of the membrane-bound form of the immunoglobulin IgM that is expressed on the surface of the immature B cell.

Effects of Exercise on the Immune System

Exercise results in a vigorous inflammatory response mostly categorized by the mobilization of leukocytes and a rise in circulating inflammatory mediators shaped by immune cells and right from the contracting skeletal muscles. Constructive and undesirable possessions on immune function and defenselessness to negligible illness have been seen after different training modes. While training in moderate exercise intensity may improve immune functional abilities in sedentary individuals, extreme volumes of long, high-intensity exercise may reduce immune practical capacity [4]. Thirty minutes following high-intensity exercise, white blood cells increased significantly above resting values [61] and were elevated for almost 3 h into the recovery period. The elevated white blood cells are owed to neutrophil and macrophage activity. Total lymphocyte count increased 30 min from the exercise cessation [62].

White blood cells increased following moderate resistance exercise training, while after intense resistance exercise training did not result in white blood cell changes after intense exercise [63]. Natural killer cell numbers and natural killer cell cytolytic activity were increased following exercise, with significantly higher increase following intense exercise compared to moderate exercise [4].

Exercise training may affect the immune response by increasing levels of pro-inflammatory cytokines (e.g., TNF-α, IL-6, IL-1) and stress proteins (heat shock protein [64]) and by altering circulating populations of peripheral blood mononuclear cells [1]. The effects of exercise on cytokine concentration in blood (IL-6, IL-8, IL-1β, IL-10, and TNF-α) are the stimulation of both pro- and anti-inflammatory responses. However, these responses are temporary, with levels returning to baseline, which takes 5–24 h following the exercise session [65].

Exercise has an adjustable effect on the immune system due to factors such as training intensity, type of sport, neuroendocrine, and infectors [66]. The outcomes of different investigations suggested that the intensity and bulk of aerobic exercise training impact the responses of the immune system [67]. Thus, aerobic exercise modulates the essential and detailed arms of the immune system, by VO2max percentage-dependent response and ventilatory threshold, determined during a treadmill stress test [68]. However, few ways are known to enhance the immune system and sickness by exercise [5]. Direct result of aerobic exercise on the immune system following intensities of 60–75% of VO2max appears to apply a defensive outcome, while frequent sessions of prolonged strenuous exercise are associated with an increased risk for immune dysfunction [69]. In addition, environmental and the type of sport involved contribute to infectious risk such as upper pulmonary tract infection [70]. The association between exercise and immunity has significant importance for public health and for medical body. Indications suggest that moderate exercise has positive effects on the human immune system function [71], which is expressed in a J-curve hypothesis (Fig. 10.2), describing the association between exercise intensity and the risk of getting infections [64]. Different studies did not confirm the intensity effect [72].

The immune response to exercise has focused on aerobic exercise, while the effect of an all-out exercise bout such as anaerobic exercise has received relatively minute

Fig. 10.2 Association between exercise intensity and the effect on the immunity system

attention. An acute bout of aerobic exercise induces mobilization of all lymphocyte subsets to the blood [73]; the neutrophil and lymphocyte concentration increases and falls below values found before exercise after intense long aerobic exercise, but is not suppressed after moderate aerobic exercise [74]. The effector functions of the immune system are in need of the suitable circulation of immune cells in the body organs. An acute all-out exercise bout redistributes immune cells between different body organs, which may improve the immune system efficiency [75].

Immune cell functions are impaired following heavy prolonged exercise sessions that result in noticeable immunodepression [76], such as leukocytosis described in athletes performing all-out strenuous exercise [77]. In addition, studies have suggested that athletes may develop risk for infections [78]. Furthermore, strenuous exercise that lasts long durations may decrease the lymphocytes levels [79]. Relative to sedentary individuals, the positive effects of exercise may be due to its anti-inflammatory effects. In individuals with heart disease, exercise interventions lower blood levels of fibrinogen and C-reactive protein, an important cardiovascular risk marker and an increased risk of diabetes [80]. The depression in the immune system following acute bouts of exercise may be one of the mechanisms for this anti-inflammatory effect.

Immune Response to All-Out Anaerobic Exercise

The immune response to exercise has received increased attention in the last decade. Most of this attention has focused on aerobic exercise, whereas the effect of anaerobic exercise has received comparatively little notice. An acute bout of aerobic exercise induces mobilization of all lymphocyte subsets to the blood [81]. Following intense long aerobic exercise, the neutrophil and lymphocyte concentration increases and falls below values of pre-exercise, while they are not suppressed after moderate aerobic exercise [82]. However, leukocytosis has been described in athletes submitted to acute strenuous exercise.

Two different bouts of exercise, aerobic and anaerobic, are normal training procedures for elite athletes. Such training with different energy sources has been used for many years to improve athlete's performance in order to compete at an international level.

Thus, changes in lymphocyte levels are workload-dependent, with different stress-induced immune system functions between maximal aerobic and all-out anaerobic efforts.

Differences in the lymphocyte's subsets response between maximal aerobic and all-out anaerobic efforts are explained by the differential influence brought by the exercise mode and workloads [83]. It has been suggested that the physiological events associated with the onset of exercise result in different increases in capillary hydrostatic pressure and capillary filtration pressure. The capillary hydrostatic pressure and capillary filtration pressure differ with the mode of exercise, and thus, the stage is set for different amount of water movement out of the vascular system resulting in a different acute decrease in plasma volume [84]. Since during plasma shifts immune cells do not leave the vascular system, the increase in lymphocyte subset levels during exercise in the circulating blood may reflect the level of the decrease in plasma volume [84]. To elicit this notion regarding the plasma volume as a cause for the differences found between the two modes of exercise suits the reestablishment of plasma volume and immune measured variables to pre-exercise values following one hour into recovery [85].

Strenuous Exercise Effects on Immunity in Elderly

The anaerobic system responds to high-intensity training with biochemical, neural, and anatomic adaptations. Unlike the aerobic system, this response tends to be primarily a local phenomenon with little systemic adaptation [86]. This kind of effort brings about oxygen debt, which increases blood lactate concentration, in direct association with increase in maximum performance.

Longitudinal data indicate that after more than two decades from not competing anymore, the physiological capacities of aged athletes are compromised, regardless of training [87]. However, the elderly respond to anaerobic training, and as a result, their independence and quality of life can be improved. Thus, appropriate application of anaerobic system assessments and training principles is an important aspect of sports medicine practice.

The aging process is accompanied by numerous limitations on the magnitude and quality of motor activity, imposed in part by functional changes in the neuromuscular apparatus.

Resting immune function, increased risk of infection, tumor development, and autoimmune diseases. Since cyclists' training regimes include speed endurance performance to volitional exhaustion [88], they may alter immune system function. Catecholamine appears to play an important role in the activation of the changes in lymphocyte levels which are dependent on β-adrenergic receptor expression on the respective cell types and also on cyclic adenosine monophosphate (cAMP) which activates protein kinase A [89].

In response to acute exercise, there is a rapid interchange of immune cells between peripheral lymphoid tissues and the circulation system. The response depends on many factors, including exercise intensity, duration, and mode. However, extremely little is known about the interaction of exercise, aging, and the immune system.

Master athletes have a preserved ability to recruit T lymphocytes and natural killer cells in response to an acute bout of exercise [90].

Acute strenuous exercise evokes immediate responses in the immune system, in particular the number and activity of natural killer lymphocytes. Of all immune cells, neutrophil natural killer cells, lymphocyte subsets, and leukocytes appear to be most responsive to the effects of strenuous exercise, both in terms of numbers and function.

The CD4 and CD8 cells represent the lymphocyte subsets and play a key role in inducing an adequate immune response to an abnormal stimulation of antigen. They increase the binding of antibodies to antigen and hence improve the body's ability to combat foreign agents carrying the antigen [91]. In epidemiological studies, CD4/CD8 ratio predicts morbidity and mortality in the elderly [92]. These changes in CD4/CD8 ratio may suggest that CD8 cells are preferentially recruited into circulation during exercise. Exercise-induced mobilization of these cells seems to depend on β-adrenergic receptor expression on the respective cell types and also on cAMP second messenger system, which works less well in B cells [93].

It has been suggested that stress-induced changes in the immune system are controlled by beta-adrenergic mechanisms. Exercise-induced catecholamine modulates the expression of adhesion molecules on immune cells, resulting in increased neutrophil count concomitant with the recruitment of natural killer cells and T cells into the circulation as reflected by the elevation of the total lymphocyte count [94].

The lymphocyte subsets increase could be explained by the liberation of cells from the endothelial vessel wall after catecholamine interaction with the beta-adrenoceptors and by mobilization from lymph nodes and spleen after beta-adrenergic stimulation [95].

Resting natural killer cell number and percentage of lymphocytes appear to be higher in master athletes compared with untrained elderly [96]. Changes in natural killer cells are also in relation to the schedule sequence. In addition, aged individuals have similar natural killer cell response as in younger individuals, perhaps because of a low initial proliferative capacity. Previously, aged individuals demonstrated less stimulation of lymphocyte proliferation by moderate physical activity and less suppression with exhausting exercise, possibly because resting immune function is lower than in the young. Moreover, resting natural killer cell activity of sedentary elderly individuals seems to be increased by training [97]. Vigorous exercise bout improves the cytotoxicity of natural killer cells to NPC and at the same time promotes the platelet-impeded apoptosis of NPC induced by natural killer cells. However, warm-up exercise reduces the resistance of platelet to natural killer cells—nasopharyngeal carcinoma cells interaction, increasing the efficiency of anti-nasopharyngeal carcinoma cell cytotoxicity by natural killer cell following vigorous exercise activity. Thus, natural killer cells have the magnitude of changes in number, which depend on exercise intensity [98].

During exercise, changes in plasma catecholamine concentration, namely epinephrine, correlate with changes in natural killer cell activity and number in peripheral blood mononuclear cells and with lactic acid levels in the blood [99].

Several studies noting large increases in natural killer cell number immediately following exercise have also reported a significant rise in catecholamine concentration. Moreover, exercise and epinephrine infusion induce similar changes in CD16 cell number [74].

Oxidative Burst

Oxidative burst or respiratory burst is the quick release of reactive oxygen species (ROS), namely superoxide anion (O2.−) and hydrogen peroxide (H_2O_2), from different types of cells. ROS are produced in aerobic cells either as by-products during mitochondrial electron transport or by several oxidoreductases and metal-catalyzed oxidation of metabolites [100]; it is produced regardless of the sort of pathogen. Usually it means the release of these chemicals from immune cells, e.g., neutrophils and monocyte macrophages. When neutrophils and monocyte macrophages are challenged by different bacteria or fungi, ROS kill the pathogens by engulfing them by phagocytosis. Oxygen burst plays an important role in the immune system, as a vital response occurring in phagocytes to destroy affected particles and bacteria. To combat infection, immune cells use nicotinamide-adenine dinucleotide phosphate-reduced (NADPH) oxidase, to reduce O_2– to an and then hydrogen peroxide (H_2O_2) [100]. NADPH oxidase, an enzyme family in the vasculature, produces superoxide, which impulsively recombines with other molecules to produce reactive free radicals. The superoxide reacts with nitric oxide, leading to the construction of peroxynitrite, reducing the bioactive nitric oxide needed to dilate terminal arterioles and feed arteries and resistance arteries. Superoxide anion, peroxynitrite, and other ROS result in pathology through peroxidation of proteins and lipids and by activation of redox-sensitive signaling cascades and protein nitrosylation. NADPH oxidase activation has been suggested to depend on prior protein kinase C activation [101]. Myeloperoxidase enzyme is part of neutrophils; it uses the ROS hydrogen peroxide to produce hypochlorous acid (HOCl), in the presence of chloride ions. To combat infections, immune cells use NADPH oxidase to reduce oxygen to a superoxide free radical (O^{-}_2) and then H_2O_2. Absence of NADPH oxidase will prevent the formation of ROS and will result in chronic granulomatous disease.

The increase in superoxide and H_2O_2 that is caused by the stimulus of the NADPH oxidase is temporary, in part due to the presence of the antioxidant enzymes, which return their concentrations to the required steady-state level. Thus, the antioxidant enzymes may function in the "turn-off" phase of signal transduction by ROS. During its transient elevation, H_2O_2 may act as a modifier of key signaling enzymes through reversible oxidation of critical thiols. The rapid reaction of thiols with H_2O_2 when in their unprotonated state would provide a potential mechanism for the specificity that is necessary for physiological cell signaling.

References

1. Gholamnezhad Z, Boskabady MH, Hosseini M, et al. Evaluation of immune response after moderate and overtraining exercise in wistar rat. Iran J Basic Med Sci. 2014;17:1–8.
2. Wentz LM, Ward MD, Potter C, et al. Increased risk of upper respiratory infection in military recruits who report sleeping less than 6 h per night. Mil Med. 2018;183:e699–704.

3. Scharhag J, Meyer T. Return to play after acute infectious disease in football players. J Sports Sci. 2014;32:1–6.
4. Cerqueira E, Marinho DA, Neiva HP, et al. Inflammatory effects of high and moderate intensity exercise: a systematic review. Front Physiol. 2020;10:1550. https://doi.org/10.3389/fphys.2019.01550.
5. Beck G, Habitat GS. Immunity and the invertebrates. Sci Am. 1996;275:60–6.
6. Litman GW, Cannon JP, Dishaw LJ. Reconstructing immune phylogeny: new perspectives. Nature Rev Immunol. 2005;5:866–79.
7. Kumar H, Kawai T, Akira S. Pathogen recognition by the innate immune system. Intl Rev Immunol. 2011;30:16–34.
8. Restifo NP, Gattinoni L. Lineage relationship of effector and memory T cells. Curr Opin Immunol. 2013;25:556–63.
9. O'Byrne KJ, Dalgleish AG. Chronic immune activation and inflammation as the cause of malignancy. Br J Cancer. 2001;85:473–83.
10. He W, Wang B, Li Q, et al. Aberrant expressions of co-stimulatory and co-inhibitory molecules in autoimmune diseases. Front Immunol. 2019;10:261. https://doi.org/10.3389/fimmu.2019.00261.
11. Alberts B, Johnson A, Lewis J, Raff M, Roberts K, Walters P. Molecular biology of the cell. Fourth ed. New York and London: Garland Science; 2002.
12. Boyton RJ, Openshaw PJ. Pulmonary defences to acute respiratory infection. Br Med Bull. 2002;61:1–12.
13. Janeway CA, Travers P, Walport M, et al. Immunobiology. 5th ed. New York and London: Garland Science; 2001.
14. Cobey S. Pathogen evolution and the immunological niche. Ann N Y Acad Sci. 2014;1320:1–15.
15. Grammatikos AP, Tsokos GC. Immunodeficiency and autoimmunity: lessons from systemic lupus erythematosus. Tren Mol Med. 2012;18:101–8.
16. Patt H, Bandgar T, Lila A, et al. Management issues with exogenous steroid therapy. Indian J Endocrinol Metabol. 2013;17:S612–7.
17. Abul KA, Lichtman A, Shiv P. Cellular and molecular immunology. 9th ed. Philadelphia: Elsevier; 2018. p. 97.
18. Kenneth M. Janeway's immunobiology. 8th ed. New York: Garland Science; 2012.
19. Gestur V, Gillian D, Theo R. IgG subclasses and allotypes: From structure to effector functions. Front Immunol. 2014;5:520.
20. Sperandio B, Fischer N, Sansonetti PJ. Mucosal physical and chemical innate barriers: lessons from microbial evasion strategies. Semin Immunol. 2015;27:111–8.
21. Evans DJ, Fleiszig SMJ. Why does the healthy cornea resist Pseudomonas aeruginosa infection? Am J Ophthalmol. 2013;155:961–70.
22. Evans DJ, Fleisziga SMJ. Why does the healthy cornea resist Pseudomonas aeruginosa infection? Am J Ophthalmol. 2013;155:961–70.
23. Kupper TS, Fuhlbrigge RC. Immune surveillance in the skin: mechanisms and clinical consequences. Nature Rev Immunol. 2004;4:211–22.
24. Chai LY, Netea MG, Vonk AG, et al. Fungal strategies for overcoming host innate immune response. Med Mycol. 2009;47:227–36.
25. Becker KL, Ifrim DC, Quintin J, et al. Antifungal innate immunity: recognition and inflammatory networks. Semin Immunopathol. 2015;37:107–16.
26. Marcos CM, de Oliveira HC, de Cassia VM, et al. Anti-immune strategies of pathogenic fungi. Front Cell Infect Microbiol. 2016;6:142. https://doi.org/10.3389/fcimb.2016.00142.
27. Schubert C, Behl R, Lesley A, et al. Amyloid peptides are toxic via a common oxidative mechanism. Proc Natl Acad Sci. 1990;92:1989–93.
28. Janeway CA, Travers P, Walport M, et al. Immunobiology. 6th ed. New York: Garland Science; 2005.

29. Mutschler E, Schäfer-Korting M. Arzneimittelwirkungen (in German). 7th ed. Stuttgart: Wissenschaftliche Verlagsgesellschaft; 1997.
30. Anderson HC, Mulhall D, Garimella R. Role of extracellular membrane vesicles in the pathogenesis of various diseases, including cancer, renal diseases, atherosclerosis, and arthritis. Lab Investig. 2010;90:1549–57.
31. Wirtz S, Neufert C, Weigmann B, Neurath MF. Chemically induced mouse models of intestinal inflammation. Nat Protoc. 2007;2:541–6.
32. Munford RS, Suffredini AF. Mandell, Douglas, and Bennett's principles and practice of infectious diseases (Eighth Edition). ScienceDirect. 2015;1:914–34.
33. Filiano AJ, Gadani SP, Kipnis J. Interactions of innate and adaptive immunity in brain development and function. Brain Res. 2015;1617:18–27.
34. Muralidharan S, Mandrekar P. Cellular stress response and innate immune signaling: integrating pathways in host defense and inflammation. J Leukoc Biol. 2013;94:1167–84.
35. Oikonomopoulou K, Ricklin D, Ward PA. Interactions between coagulation and complement-their role in inflammation. Semin Immunopathol. 2012;34:151–65.
36. Collins CK. Robbins pathologic basis of disease. Philadelphia: W.B Saunders Company; 1998.
37. Carpanini SM, Torvelland M, Morgan BP. Therapeutic inhibition of the complement system in diseases of the central nervous system. Front Immunol. 2019;10:362. https://doi.org/10.3389/fimmu.2019.00362.
38. Mortensen SA, Sander B, Jensen RK, et al. Structure and activation of C1, the complex initiating the classical pathway of the complement cascade. PNAS. 2017;114:986–99.
39. Stolarczyk E. Adipose tissue inflammation in obesity: a metabolic or immune response? Curr Opin Pharmacol. 2017;37:35–40.
40. Harboe M, Mollnes TE. The alternative complement pathway revisited. J Cell Mol Med. 2008;12:1074–84.
41. Galli SJ, Borregaard N, Wynn TA. Phenotypic and functional plasticity of cells of innate immunity: macrophages, mast cells and neutrophils. Nature Immunol. 2011;12:1035–44.
42. Thompson MR, Kaminski JJ, Kurt-Jones EA, et al. Pattern recognition receptors and the innate immune response to viral infection. Viruses. 2011;3:920–40.
43. Lotze MT, Tracey KJ. High-mobility group box 1 protein (HMGB1): nuclear weapon in the immune arsenal. Nat Rev Immunol. 2005;5:331–42.
44. Delneste Y, Beauvillain C, Jeannin P. Innate immunity: structure and function of TLRs. (article in French). Med Sci (Paris). 2007;23:67–73.
45. Tracey KJ. Physiology and immunology of the cholinergic antiinflammatory pathway. J Clin Invest. 2007;117:289–96.
46. Arai S, Meagher R, Swearingen M, et al. Infusion of the allogeneic cell line NK-92 in patients with advanced renal cell cancer or melanoma: a phase I trial. Cytotherapy. 2008;10:625–32.
47. Mittal R, Kodiyan J, Gerring R, et al. Role of innate immunity in the pathogenesis of otitis media. Int J Infect Dis. 2014;29:259–67. https://doi.org/10.1016/j.ijid.2014.10.015.
48. Guilliams M, Ginhoux F, Jakubzick C, et al. Dendritic cells, monocytes and macrophages: a unified nomenclature based on ontogeny. Nat Rev Immunol. 2014;14:571–8.
49. Boukhaled GM, Corrado M, Guak H, et al. Chromatin architecture as an essential determinant of dendritic cell function. Front Immunol. 2019;10:1119. https://doi.org/10.3389/fimmu.2019.01119.
50. Chang Y-J, Kim HY, Albacker LA, et al. Innate lymphoid cells mediate influenza-induced airway hyper-reactivity independently of adaptive immunity. Nature Immunol. 2011;12:631–8.
51. Web MD. Leukocyte, Polymorphonuclear. In: Webster's New World medical dictionary. 3rd ed. Boston, MA: Houghton Mifflin Harcourt; 2009. p. 244.
52. Stone KD, Prussin C, Metcalfe DD. IgE, mast cells, basophils, and eosinophils. J Allergy Clin Immunol. 2010;125:S73–80.
53. Artis D, Spits H. The biology of innate lymphoid cells. Nature. 2015;517:293–301.
54. Nicholson LB. The immune system. Essays Biochem. 2016;60:275–301.

55. Hoffman W, Lakkis FG, Chalasani G. B cells, antibodies, and more. Clin J Am Soc Nephrol. 2016;11:137–54.
56. Neuberger MS, Harris RS, Di NoiaSvend J, et al. Immunity through DNA deamination. Trends Biochem Sci. 2003;28:305–12.
57. Gellert M. V(D)J recombination: RAG proteins, repair factors, and regulation. Annu Rev Biochem. 2002;71:101–32.
58. Fagarasan S, Honjo T. Intestinal IgA synthesis: regulation of front-line body defences. Nature Rev Immunol. 2003;3:63–72.
59. Dorner T, Radbruch A. Antibodies and B cell memory in viral immunity. Immunity. 2007;27:384–92.
60. Duarte J, Deshpande P, Guiyedi V, et al. Total and functional parasite specific IgE responses in Plasmodium falciparum-infected patients exhibiting different clinical status. Malar J. 2007;6:1. https://doi.org/10.1186/1475-2875-6-1.
61. Brebner JA, Stockley RA. Polyclonal free light chains: a biomarker of inflammatory disease or treatment target? F1000 Med Rep. 2013;5:4. https://doi.org/10.3410/M5-4.
62. Stelzer I, Kropfl JM, Fuchs R, et al. Ultra-endurance exercise induces stress and inflammation and affects circulating hematopoietic progenitor cell function. Scand J Med Sci Sport. 2015;25:e442–50.
63. Abbasi A, Hauth M, Walter M, et al. Exhaustive exercise modifies different gene expression profiles and pathways in LPS-stimulated and un-stimulated whole blood cultures. Brain Behav Immun. 2013;39:130–41.
64. Sarin HV, Gudelj I, Honkanen J, et al. Molecular pathways mediating immunosuppression in response to prolonged intensive physical training, low-energy availability, and intensive weight loss. Front Immunol. 2019;10:907. https://doi.org/10.3389/fimmu.2019.00907.
65. Draganidis D, Chatzinikolaou A, Jamurtas AZ, et al. The time-frame of acute resistance exercise effects on football skill performance: the impact of exercise intensity. J Sports Sci. 2013;31:714–22.
66. Gonzalo-Calvo D, de Davalos A, Montero A, et al. Circulating inflammatory miRNA signature in response to different doses of aerobic exercise. J Appl Physiol. 2015;119:124–34.
67. Khakroo Abkenar I, Rahmani-Nia F, et al. The effects of acute and chronic aerobic activity on the signaling pathway of the inflammasome NLRP3 complex in young men. Medicina (Kaunas). 2019;55(4):E105. https://doi.org/10.3390/medicina55040105.
68. San Juan AF, Chamorro-Vina C, Mate-Munoz JL, et al. Functional capacity of children with leukemia. Int J Sports Med. 2008;29:163–7.
69. Gleeson M. Immune function in sport and exercise. J Appl Physiol. 2007;103:693–9.
70. WHo report. https://www.who.int/mediacentre/news/releases/2006/pr32/en/.
71. Martin SA, Pence MB, Woods JA. Exercise and respiratory tract viral infections. Exerc Sport Sci Rev. 2009;37:157–64.
72. MatthTopel ML, Sandesara PB, Stahl EP, et al. Mechanisms underlying the J-curve for diastolic blood pressure: subclinical myocardial injury and immune activation. Int J Cardiol. 2019;276:255–60.
73. Kostka T, Berthouze SE, Lacour J, et al. The symptomatology of upper respiratory tract infections and exercise in elderly people. Med Sci Sports Exerc. 2000;32:46–51.
74. Brown FF, Campbell JP, Wadley AJ, et al. Acute aerobic exercise induces a preferential mobilisation of plasmacytoid dendritic cells into the peripheral blood in man. Physiol Behav. 2018;194:191–8.
75. Dhabhar FS, Malarkey WB, Neri E, et al. Stress-induced redistribution of immune cells--from barracks to boulevards to battlefields: a tale of three hormones--Curt Richter award winner. Psychoneuroendocrinology. 2012;37:1345–68.
76. Brunelli DT, Rodrigues A, Lopes WA, et al. Monitoring of immunological parameters in adolescent basketball athletes during and after a sports season. J Sports Sci. 2014;32:1050–9.

77. Moreira A, Delgado L, Moreira P, et al. Does exercise increase the risk of upper respiratory tract infections? Br Med Bull. 2009;90:111–31.
78. Saghiv MS, Sherve C, Sagiv M, et al. Effect of different exercise lactic acid levels on the immune system response. J Aller Immunol. 2017;4(1):021.
79. Wentz LM, Ward MD, Potter C, et al. Increased risk of upper respiratory infection in military recruits who report sleeping less than 6 h per night. Mil Med. 2018;183:e699–704.
80. Niemela M, Kangastupa P, Niemela O, et al. Acute changes in inflammatory biomarker levels in recreational runners participating in a marathon or half-marathon. Sports Med open. 2016;2(1):21. Epub 2016 Mar 2
81. Ursini F, Abenavoli L. The emerging role of complement C3 as a biomarker of insulin resistance and cardiometabolic diseases: preclinical and clinical evidence. Rev Recent Clin Trials. 2018;13:61–8.
82. Pedersen BK, Toft AD. Effects of exercise on lymphocytes and cytokines. Br J Sports Med. 2000;34:246–51.
83. McCarthy D, Dale MM. The leukocytosis of exercise: a review and model. Sports Med. 1998;6:333–63.
84. Coker RH, Krishna MG, Lacy DB, et al. Role of hepatic alpha- and beta-adrenergic receptor stimulation on hepatic glucose production during heavy exercise. Am J Phys. 1997;273:E831–8.
85. Neumayr G, Pfister R, Mitterbauer G, et al. Short-term effects of prolonged strenuous endurance exercise on the level of haematocrit in amateur cyclists. Int J Sports Med. 2002;23:158–61.
86. Lobigs LM, Peeling P, Dawson B, et al. The need for an alternative method to determine intravascular volumes. Eur J Sport Sci. 2018;18:1255–63.
87. Laursen PB. Training for intense exercise performance: high-intensity or high-volume training? Scand J Med Sci Sports. 2010;20(Suppl 2):1–10. https://doi.org/10.1111/j.1600-0838.2010.01184.x.
88. Schneider J. Age dependency of oxygen uptake and related parameters in exercise testing: an expert opinion on reference values suitable for adults. Lung. 2013;191:449–58.
89. Campbell JP, Turner JE. Debunking the myth of exercise-induced immune suppression: redefining the impact of exercise on immunological health across the lifespan. Front Immunol. 2018;9:648. https://doi.org/10.3389/fimmu.2018.00648.
90. Sagiv M, Ruddoy J, Sagiv A, et al. Effect of changes in load carriage while walking on the left ventricular function in highly trained elderly subjects. Gerontology. 2002;48:289–92.
91. Cao Dinh H, Beyer I, Mets T, et al. Effects of physical exercise on markers of cellular immunosenescence: a systematic review. Calcif Tissue Int. 2017;100:193–215.
92. Malm C. Exercise immunology: a skeletal muscle perspective. Exerc Immunol Rev. 2002;8:116–67.
93. Sato H, Adachi E, Lim LA, et al. CD4/CD8 ratio predicts the cellular immune response to acute hepatitis C in HIV-coinfected adults. J Infect Chemother. 2019;25:646–8.
94. Kakanis MW, Peake J, Brenu EW, et al. The open window of susceptibility to infection after acute exercise in healthy young male elite athletes. Exerc Immunol Rev. 2010;16:119–37.
95. Pedersen BK, Bruunsgaard H, Kloker M, et al. Exercise-induced immunomodulation-possible roles of neuroendocrine and metabolic factors. Int J Sports Med. 1997;18:S2–7.
96. Weicker H, Werle E. Interaction between hormones and the immune system. Int J Sports Med. 1991;12:S30–7.
97. Pedersen BK. Exercise-induced myokines and their role in chronic diseases. Brain Behav Immun. 2011;25(5):811–6. https://doi.org/10.1016/j.bbi.2011.02.010.
98. Buyukyazi G, Kutukculer N, Kutlu N, et al. Differences in the cellular and humoral immune system between middle-aged men with different intensity and duration of physically training. J Sports Med Phys Fitness. 2004;44:207–14.
99. Wang JS, Chung Y, Chow SE. Exercise affects platelet-impeded antitumor cytotoxicity of natural killer cell. Med Sci Sports Exerc. 2009;41:115–22.

100. Nagoa F, Suzui M, Takeda K, et al. Mobilization of NK cells by exercise: downmodulation of adhesion molecules on NK cells by catecholamines. Am J Physiol Regul Integr Comp Physiol. 2000;279:R1251–6.
101. Kitagawa M, Sugiyama H, Morinaga H, et al. A decreased level of serum soluble Klotho is an independent biomarker associated with arterial stiffness in patients with chronic kidney disease. PLoS One. 2013;8(2):e56695.

Chapter 11
Exercise in Hostile Environment

Introduction

High altitude is considered to be between 1500 and 3500 m above sea level. Very high altitude is at heights between 3599 and 5500, while extreme altitude is above 5550 m (Fig. 11.1). Oxygen percentage at sea level and in high altitude stays the same at 20.9%.

However, what changes is the barometric pressure (a measurement of air's force against a surface) decline at high altitude. Accordingly, as barometric pressure decreases, the volume needed to contain one million air molecules increases. Although the volume of air is affected by temperature and pressure, the concentration of CO_2 is not affected. If a climber started with 1000 part(s) per million of CO_2, then he finishes with 1000 parts per million of CO_2 despite the changes in the air volume. Oxygen partial pressure (PO_2) is reduced; as an example, at a height of 4000 m barometric pressure is 480 mm Hg and PO_2 is 100 mm Hg compared to sea level where barometric pressure is 760 mm Hg and PO_2 is 159 mm Hg (Table 11.1).

Therefore, the molecule number of oxygen and other gases to a given volume is the highest at sea level. Limited compensation at altitude for the reduced oxygen molecule number is:

(a) Significant increase in VE
(b) Increase in hemoglobin production

Decreased barometric pressure at altitude decreases the ability to move oxygen from the atmosphere to the lung and therefore VE increases significantly. VE is increased due to increases in VT and f.

The body's reaction to long-lasting acclimatization to altitude is an increase in red blood cell production. Red blood cells improve oxygen carriage by the blood. In addition, the characteristic of hemoglobin molecule changes physiologically and chemically, thus facilitating oxygen attachment and detachment compared to sea level.

M. S. Saghiv, M. S. Sagiv, *Basic Exercise Physiology*,
https://doi.org/10.1007/978-3-030-48806-2_11

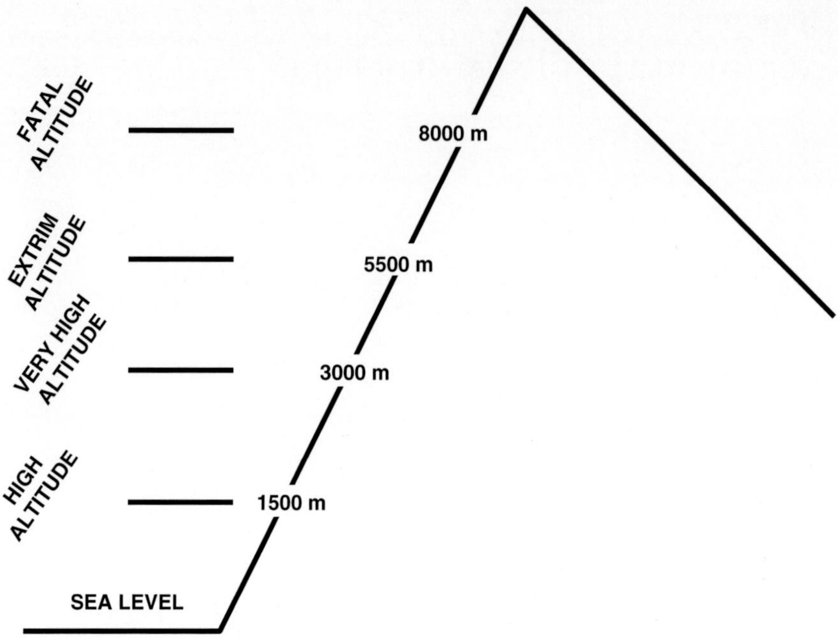

Fig. 11.1 Different altitude definitions based on height

Table 11.1 Effect of barometric pressure on PO_2 at different heights

Height (m)	Barometric pressure (mm Hg)	Oxygen partial pressure (mm Hg)
0	760	159
1500	620	130.2
3000	523	109.8
6100	346	72.7
9150	224	47

At 4000 m, hemoglobin level continues to increase for another 3 months; above 6000 m hemoglobin increment can last 38 weeks. However, an increase in blood hematocrit and blood density imposes load on the left ventricle ejection. In addition, to these physiological variables, Q increases also mainly by HR. Following 8–9 weeks at high altitude, VE decreases however, still significantly above sea level, while Q decreases to sea level and hemoglobin levels stay the same.

Altitude/Hypoxic Training Effects

Meta-analysis study, of randomized controlled trials, where low oxygen partial pressure applies a disadvantageous effect on performance, compared the usefulness of altitude/hypoxic training with sea-level training on the cardiopulmonary system function for oxygen delivery during aerobic exercise capacity of elite athletes. It has been suggested that altitude/ hypoxic training seems a more efficient method for enhancement of oxygen delivery capacity and thus increases aerobic exercise capacity [1]. It becomes clear that altitude/hypoxic training plays an important role in preparing elite and well-trained individuals. Hypoxic training techniques are usually used by endurance athletes to improve exercise performance at altitude and at sea level in normoxic [2]. Adaptation to environmental hypoxia produces a sequence of metabolic, cardiopulmonary, and muscular adaptations that effect oxygen delivery and extraction [3]. In addition, training in hypoxia significantly increases VO2max and nitric oxide facilitated by a significant increase in the serum concentrations of proangiogenic factors and serum nitric oxide levels [4].

When the training volume at altitude is comparable to the quantity at sea level, altitude/hypoxic training induces enhancements in the muscle oxidative enzymes, myoglobin, and molecular adaptations and improves muscle capillarization [5, 6].

Mitochondrial oxidative phosphorylation produces mainly cellular ATP from substrates such as glucose and fatty acids. In order to preserve ATP homeostasis, the mitochondria need a continuous supply of fuels and oxygen. In healthy individuals at altitude, tissue oxygen levels decrease and a loss of skeletal muscle mitochondrial density occurs [7], even in the muscles of Himalayan Sherpas than those of unacclimated lowlanders [8]. Under these conditions, hypoxic challenge forces the cell to somehow keep energetics and limit oxidative stress. Studying on cultured cells and genetic mouse models [9] revealed some adaptations that allow cells with a diminished mitochondrial population to function effectively in hypoxia. These include the glucose transporter, GLUT1, which mediates non-insulin-stimulated glucose uptake by heart and skeletal muscle [10], and protein levels of both GLUT1 and the insulin-stimulated glucose transporter, GLUT4, are increased in hypoxic rat skeletal muscle.

Cardiovascular improvements following exposure to hypoxia and hypoxic training, for 3 weeks, lasting 2–3 h each session, at heights of 2700–4200 m, enhance blood flow through the myocardial muscle and exercise work capacity, while 3 weeks of intermittent hypoxic exposure treatment may improve baroreflex sensitivity and HR variability [11]. In untrained individuals, submaximal exercise ability is improved significantly, time to exhaustion is increased, and HR variability is enhanced. Hematological adaptations include increases in hemoglobin mass; even sportspersons with high initial hemoglobin mass may expect hemoglobin mass gain, following altitude/hypoxic training [12]. Changes in hemoglobin levels in the blood during exercise are increased, due to the fluid movement from the cardiovascular system to the extracellular compartment, following the increased BP. Consequently, oxygen delivery capacity is increased.

Exercise at Altitude

Exercising at high elevation is a strain that triggers the sympathetic system and the adrenal system, which affect body's responses to critical altitude introduction and advance adaptations during chronic altitude exposure. High altitude is a hazardous environment in which it provokes a response that affects exercise ability and physical performance. Exercising at high altitudes decreases oxygen availability; therefore, oxygen demand-supply reaches imbalance (hypoxia); consequently, physical performance is reduced especially maximal exercise. Compared to sea level, oxygen obtainability during aerobic exercise at 4000 m is 66% and at 6000 m 50%; therefore, at 6000 m VE is doubled and tripled at 8000 m. Hyperventilation is attained largely by expressive increase in breathing rate (Fig. 11.2).

Hypoxia affects, in different degrees, all body organs, systems, and functions. An attempt by untrained individual to exercise at altitude 6000 m reveals that its work capacity to perform submaximal load is only 50% of his ability at sea level (Fig. 11.3). In addition, at altitude hypoxia reduces the ability to oxygenate the

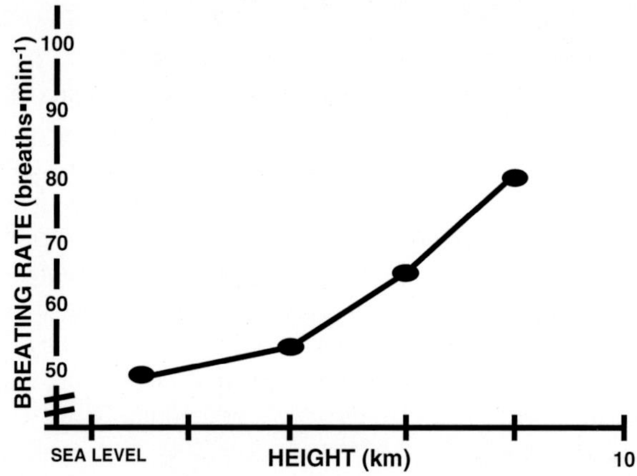

Fig. 11.2 Breathing rate during exercise at different heights

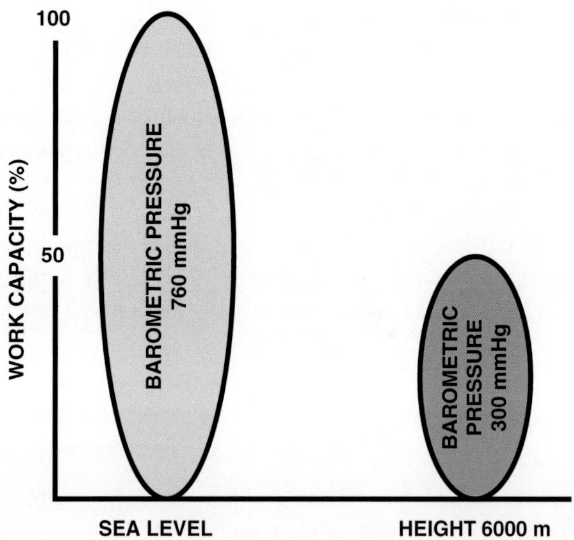

Fig. 11.3 Work capacity at 6000 m

tissues owing to the decreased diffusion capacity between the oxygenated arterial blood and the tissue. Consequently, the lungs hyperventilate; however, unfortunately, hyperventilation eliminates CO_2 in large quantities.

In addition to the inability to produce oxygen deficit, skeletal muscle's lactate tolerance is also reduced. This is why mountain climbers need to stop every few steps for long recovery. Limited VE is seen when at 5200 m hyperventilation is 120 L·min^{-1} which yields VO2 value of only 2 LO2·min^{-1}. Unfortunately, hyperventilation has a negative impact on CO_2 by eliminating large CO_2 quantities bringing about an acid/base imbalance and may cause altitude sickness. In addition, air at high altitude is dry; therefore, when the lungs hyperventilate it causes a significant water loss, resulting in dehydration. Previously it has been suggested that different body organs react differently at high altitude. Research at the Everest height revealed that the brain is the most vulnerable organ, and therefore, it seems that pulmonary blood vessel regulation is damaged. On the contrary, the skeletal muscles, respiratory muscles, myocardium, and nervous system function normal at maximal exercise. In addition, following acclimatization the main energy for metabolism is fat due to the dramatic decrement in carbohydrates metabolism due to:

(a) Changes in the autonomic nervous system that increase BMR, causing transfer from glycolytic metabolism to fat metabolism.
(b) Reduced oxygen supply initiates muscle anaerobic metabolism that produces high lactate levels. At a certain point, where hypoxia is reached the glycolitic pathway is interrupted and therefore, fat becomes the only energy source.

The Cardiovascular System

Climbing to 4000 m the exposure to low barometric pressure forces the heart to compensate for the created hypoxic situation. As compensation, Q increases by 20% at rest and during submaximal exercise, while maximal Q cannot increase above the values at sea level during all-out aerobic exercise. Under hypoxic state at 4000 m height, during aerobic exercise, oxygen arterial blood content decreases by 25%; thus, muscle (a-v)O_2 is reduced thus, VO2max is reduced by 25%. During submaximal aerobic exercise, muscles increase (a-v)O_2 up to a certain limited level. The shift from increased Q to increase (a-v)O_2 is achieved after 1 week of height exposure. Thereafter, Q decreases to a lower value than its value at sea level. This reduction in Q points out that (a-v)O_2, capillary density, and tissue myoglobin concentration increased significantly. as compensation for a higher value than at sea level. Changes in Q are the result of the interplay between HR and SV. Maximal HR up to elevations of 4000 m does not differ from its values at sea level; however, at heights above 6000 m HR is suppressed and may reach maximal HR of 145 beats·min^{-1} (Fig. 11.4). This has a negative effect on muscle strength and power generation due to a decrease in muscle mass owing to the reduction in training intensity. Compared to sea level, at a given submaximal aerobic exercise at altitude,

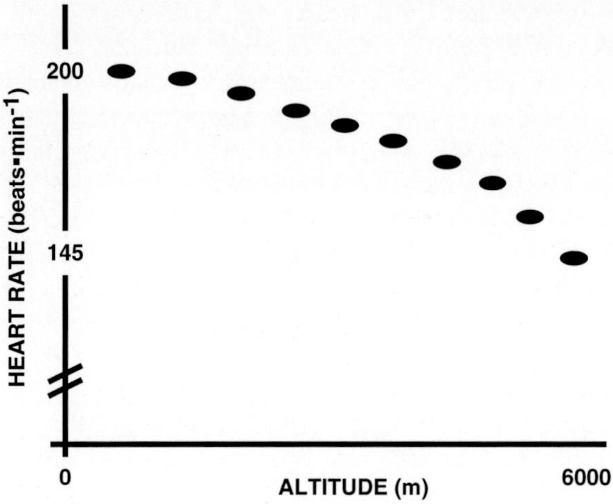

Fig. 11.4 Heart rate decreases with increased altitude

a sharp increase in HR is noted, as a reaction to hypoxia and due to the activation of the chemical sensors. This increased HR may stay at its level for 1 month more, while others reported that HR decreases after 1 week to sea-level values.

The conflicting data reported may the result of different sample size, protocols used, and individuals' fitness degree. The second Q component SV decreases within one hour from arrival and stables after 1 week at the high altitude. The decline in SV is due to the reduction in plasma volume due to water loss in altitude; therefore, HR increases to keep appropriate Q. BP remains unchanged pointing to the reduction in TPR due to the autoregulation mechanism. Following 1 week, BP increases to its previous sea-level values. It seems that these BP variations are connected to the Q variations.

Preferably acclimatization should be progressive. Above 3000 m individuals should climb no more than 300–400 m each day. An individual that lives in high altitude needs to be acclimatized and adapt to physiological changes in order to overcome the hypoxic effects on physical tasks. Following 3 days of climbing, a 24–48-h rest is a must. When symptoms of acute mountain sickness appear climbing should be ceased and the climber should turn downhill. The most important feature of acclimatization to altitudes up to about 5000 m is the increase in VE due to hypoxia. The steady decrease in oxygen partial pressure causes individual's acclimatization to physiological strain to depend on the extent of hypoxia associated with altitude, time of acclimatization, exercise intensity, and individual basal fitness. With time, the lungs overcome the low oxygen partial pressure and the lower oxygen saturation level of hemoglobin in the arterial blood; thus, oxygen tissue levels are also lower. Partial pressure is vital in defining the way and how much gas can enter the respiratory membrane and cross the pulmonary capillaries blood. A gap in oxygen partial pressures between the alveoli and the blood causes a decrease in oxygen amount to pass the respiratory membrane. Thus, less oxygen molecules are tied by hemoglobin; however, the muscles receive an adequate quantity of oxygen during rest at high altitudes. Owing it to two main mechanisms:

(a) At altitude, the number of oxygen molecules entering the muscles is almost identical as at sea level. However, at sea level, hemoglobin saturation is higher than that of at altitude; therefore, less oxygen molecules enters the muscles, while at high altitudes
(b) Production of the hormone erythropoietin is increased following the kidney stimulation which in turn increases the production of erythrocytes, which increase 2,3-bisphosphoglyceric acid (2,3-BPG) production and thus improve oxygen release from hemoglobin.

Low quantities of oxygen in the arterial blood, during strenuous exercise, may cause altitude sickness. Yet, although hemoglobin molecule is poorly saturated due to the low oxygen partial pressure, the significant increase of red blood cells and thus hemoglobin facilitates the oxygen delivery. With time, this permits the individual to participate in physical effort at efforts between 50 and 60%.

Renal Regulation of Fluid at Altitude

Variations in body fluid, electrolyte, and acid–base homeostasis, in high-altitude climbers, are controlled mainly, by the kidneys' function during the acclimatization processes and in mountain sickness syndromes [13]. Altitiude-related loss of fluid (Hypovolemia) echoses excessive fluid losses due to humidity and tachypnea, reduced fluid intake due to decreased thirst [13].

Additional physiological response to altitude is diuresis, which happens during the early phase of adaptation to altitude which is obligatory. This initial phase is followed by a norepinephrine, renin, and ADH [14], triggered by hypoxia-induced chemoreceptor stimulation. Under these conditions, sympathetic nervous is activated by hypoxemia; consequently, the cardiovascular system increases HR and Q. Usually, most individuals respond by hyperventilation resulting in hypocapnia and respiratory alkalosis. Under these conditions of hypoxia and alkalosis, water diuresis and fluid shift away from intravascular space. Within a few hours of exposure to hypoxia, the hypoxic diuretic response occurs [15], but prolonged hypoxia leads to restoration of renal sodium and fluid excretion rates at or below the sea-level baseline [16]. Acute hypoxemia depresses aldosterone secretion secondary to a direct effect on adrenal cells [14]. Also, plasma renin is decreased in resting hypoxemic conditions, but the mechanism remains unknown.

Renal variations include renal plasma flow and glomerular enlargement, and conserved glomerular filtration rate, due to augmented filtration fraction, is found in high-altitude populations [17, 18]. However, the effect of altitude on renal plasma flow and glomerular filtration rate in mountain climbers, exposed for a short time, is unclear. In some reports, high altitude did not change effective renal plasma flow and glomerular filtration rate [19], whereas other studies demonstrate a significant change. It has been assumed that hypocapnia and hypoxia may elevate renal blood flow significantly; however, glomerular filtration rate is decreased [20]. Actual renal plasma flow decreases one-third at a height of 6000 m, while blood viscosity is increased also significantly [21]. The phenomenon of hypoxia-induced diuresis is still in question, since diuresis is attributed completely to respiratory alkalosis, which in turn improves renal sodium excretion and extravascular fluid shift. The mechanisms for the improvements are independent of hypoxia.

Health Risks of Acute Exposure to Altitude

Altitude illness affects mountain climbers at high altitudes; the rate of sickness depends on the rate of climbing up and individual's vulnerability and other existing risk factors. Acute altitude mountain sickness is the main sickness due to ascending to heights, which characteristically results in headache and malaise within few hours of reaching altitude. Acute altitude mountain sickness may develop to high-altitude cerebral edema in some individuals. The onset of cerebral edema is indicated by declining symptoms of acute mountain sickness; however, it develops to ataxia, and if not taken care, sooner or later it ends in loss of consciousness and death. High-altitude pulmonary edema is rare, but nonetheless is the foremost reason of altitude illness-associated death. Slow ascent is the most important measure to prevent the onset of altitude illness. If this is not possible, or if symptoms occur despite slow ascent, medications may be used. Descent is required for all individuals with high-altitude cerebral edema or pulmonary edema.

Altitude Mountain Sickness

The number of travelers visiting high-altitude regions at altitudes over 3000 m above sea level is increasing. In addition, during athletic events at high altitudes, such as mountain races, athletes may practice very rapid ascent and thus may develop altitude illness. Separately from harsh surrounding conditions at high altitude, solar radiation, extreme cold, and strong winds, mountain climbers may have sympthoms of altitude mountain sikness, the most common syndrome [22]. In numerous conditions the elderly ascending to heights may have underlined chronic diseases, because of which altitude and hypoxic condition might be a risk factor, due to a short time for adaptation and acclimatization [23]. Although not life threatening, acute altitude mountain sickness shows within a few hours of ascent and, usually, consists of pain, loss of appetite, headache (the most common symptom), nausea, vomiting, difficult breathing, sleeplessness, fatigue, and dizziness [24], causing incapacitating for several days or longer. More progressive forms of altitude mountain sickness may go together with peripheral edema, periorbital edema, a change in mental status, ataxia, or rales [24]. These may start between 6 and 48 h from arrival at high altitude and can be more severe on the second and third days. In some cases, the condition can worsen to pulmonary edema or high-altitude cerebral edema.

Some early climbers stated of facing symptoms that nowadays are described as altitude mountain sickness. However, the exact mechanism of altitude illness is not fully understood. Those individuals who experience the greatest distress at altitude suffer from a low VE response to hypoxia. Consequently, this insufficient VE that reduces PO_2 further causes CO_2 accumulation in the tissues, which in turn induce most of the symptoms related with altitude sickness.

The incidence of acute altitude sickness varies with the altitude, the rate of ascent, and the individual's susceptibility. Several studies have been conducted to determine the incidence of acute altitude sickness in groups of hikers and climbers [25]. Results vary widely; at elevations of 2500–3500 m heights, about 7% men and 22% for women of mountain climbers usually experienced acute altitude sickness; however, the reason for this gender difference is undefined. At heights of 3000–5500 m, acute altitude sickness occurrence rate varies between 1 and 53%.

Training regimes, at high altitude, are part of preparation for high-level athletes competing in long-lasting events. However, the question is: if athlete's high fitness level can avoid symptoms of acute altitude sickness? Unfortunately, no clear answer exists for this question; highly aerobically trained individuals appear to have little defense against the effects of hypoxia.

High-Altitude Cerebral Edema

Although the pathophysiology of acute mountain sickness and high-altitude cerebral edema is not completely understood, current evidence suggests that hypobaric hypoxemia leads to relative fluid retention, in contrast to the usual response of marked diuresis in no affected climbers. This may lead to mild cerebral edema that occurs in individuals with moderate to severe acute mountain sickness and high-altitude cerebral edema. Suggested mediators of this response include vascular endothelial growth factor, nitric oxide synthase, and bradykinin [26]. This neurological high-altitude cerebral edema causes mental and/or ataxia (difficulty walking) alteration in an individual with acute mountain sickness, or without acute mountain sickness, this sickness may progress to unconsciousness and death. Rare cases of high-altitude cerebral edema, which is fluid accumulation in the cranial cavity, have been reported. Symptoms of acute mountain sickness and early high-altitude cerebral edema include headache and at least one of the following: anorexia, nausea or vomiting, dizziness or lightheadedness, difficulty sleeping, and fatigue or weakness. These symptoms can easily be misinterpreted as a viral illness [27]. The key criteria, however, are a recent gain in altitude and (although not specifically stated) the absence of other causes of the symptoms. Symptoms of acute mountain sickness typically occur within six to 12 h of gaining altitude and range from mild with spontaneous resolution (especially at altitudes less than 3500 m to severe with progression to high-altitude cerebral edema [28]. High-altitude cerebral edema is considered end-stage acute mountain sickness and is defined by ataxia or altered mental status, usually in an individual with acute mountain sickness [29]. High-altitude cerebral edema can progress in a matter of hours from mild ataxia to coma and death [30].

High-Altitude Pulmonary Edema

In unacclimated individuals who rapidly ascend to altitudes above 2500 m, it is life threating. High-altitude pulmonary edema differs from acute mountain sickness: high-altitude pulmonary edema is the buildup of fluids in the lungs; however, the cause of this condition is unknown. It may be due to the hypoxic effect, which causes pulmonary vasoconstriction, resulting in the formation of blood clots in the pulmonary. The fluid buildup restricts minute ventilation, resulting in reduced blood oxygenation, causing cyanosis. High-altitude pulmonary edema is treated via administration of supplemental oxygen and movement of the individual to a lower altitude [31].

Scuba Diving

Descending to deep water increases barometric pressure (hyperbaric). Diving is an issue that attracted physiologist's attention: with snorkel or SCUBA (self-contained underwater breathing apparatus) diving. Human experience in an aquatic environment is more limited compared to the knowledge about altitude. Humans are very limited compared to marine animals, in survival at depth deeper than 25–30 m. Humans cannot be acclimatized to the depth. At rest, VO2 is about 250 $mLO_2 \cdot min^{-1}$ and the body's ability to accumulate oxygen is 2.2 L; this enables theoretically to remain in the deep water around 9 min. However, the body is not able to utilize the full amount of 2.2 L; consequently, the time decreases to 3–6 min. At the same time, the increase in CO_2 forces the diver to renew respiration before he utilizes the full accumulated oxygen; this is more pronounced during exercise.

At sea level barometric pressure is 760 mm Hg that equals to one atmosphere. With every 10 m of descending in seawater, barometric pressure increases by additional one atmosphere, meaning that diving at a depth of 10 the pressure will be 2 atmospheres. Table 11.2 emphasizes the relationship between the descending in

Table 11.2 The relations between depth barometric pressure and lungs volume

Depth (meters)	Atmospheric (pressure)	Lungs volume (mL)
Sea level	1	5000
10	2	2500
20	3	1666
30	4	1250
40	5	1000
90	10	500

seawater and the changes in the atmospheric pressure and the effect of this relation on pulmonary volumes.

This sharp increase with the descent to the deep water generates physiological and medical problems. This is very noticeable when diving; it includes the following:

(a) Exposure to the surrounding water pressure around the body generates local pressure differences on different parts of the body. Examples are the local pressure differences between the eardrum, sinuses in the nose, and lungs. When the surrounding pressure is two atmospheres, lungs volume equals half of that at sea level. This increase by one atmosphere (reduced lungs volume) represents different kind of underwater strain. Additional view of hyperbaric is the gases exiting the lungs into the blood where they dissolve with the body liquids, especially nitrogen. Pressure gradient of oxygen and nitrogen between the lungs and blood changes linearly with the atmospheric pressure variations. However, CO_2 pressure gradient between blood and lungs is affected by CO_2 metabolism production, while depth does not affect the CO_2 amount released. In one atmosphere, one nitrogen liter is dissolved and two liters at two atmospheres and so on. Whenever diving deeper, more nitrogen dissolves in the body liquids; this still is not a problem, but, when ascending from the deep fast, it causes joint bends. This is caused by the reduction in ambient pressure (decompression) that the diver experiences during ascent from depth. In addition, it is the removal of dissolved inactive gases from the diver's body, occurring during ascent. Therefore, while ascending, the diver must stop though ascending known as decompression stops. A diver who only breathes gas at atmospheric pressure when or snorkeling does not regularly need to decompress.

(b) According to Boyel's law, the pressure and volume of a gas have an inverse relationship: when temperature is held constant, if one increases, the other decreases. Therefore, the product of pressure (P) and volume (V) is constant (C), as follows:

$$P \cdot V = C.$$

As an example, one liter at sea level turns half a liter in 10 m depth. Lungs in underwater, for example, shrink as its volume is compressed due to descending; equally, when the individual ascends from the underwater depth, the lung expands and may be erupted as it rises and its volume expands. This principle is the foundation of SCUBA safety training to prevent barotrauma, or injury to the body from changing pressure. Based on Boyel's law as a result of gas compression, the number of molecules increases in a given volume (corrected for STPD). Consequently, when an individual breathes at sea level, VT is 500 mL containing (6.022 o 1023)/2 (Avogadro's number) molecules; in a depth of 10 m, VT is still 500 mL; however, molecule number is doubled and so on.

Partial pressure of gas is linearly related to its molecule number in a given gas volume. The sum of the partial pressures of all gases in the mixture gives the total pressure. Gases dissolve, diffuse, and react according to their partial pressures, and not according to their concentrations in gas mixtures or liquids. That means that every time that the diver inhales at 10 m depth, oxygen molecule number will be twice of that at sea level and so on. Increase in oxygen partial pressure in the depth can cause oxygen toxicity. Cells normally deactivate free radicals as soon as they are formed; however, once an individual inhales high oxygen partial pressure, free radicals' amount in the cells increases significantly due to slower elimination ability; consequently, oxygen turns out to be toxic. Oxygen toxicity is a medical disorder following an exposure to oxygen at high pressure. Oxygen toxicity is a concern for SCUBA divers who are exposed to high oxygen concentrations diving deep or by using mixed gases. Leisure divers who dive on air do not have to face oxygen toxicity if they follow the guidelines of safe diving. Pulmonary oxygen toxicity occurs, when divers inhale a partial pressure of oxygen at atmospheric pressure above two atmospheres and beyond 14 h. Pulmonary oxygen toxicity damages diver's lungs or cells dies.

Central nervous system oxygen toxicity is caused by exposure to a high oxygen partial pressure greater than two atmospheres. CNS oxygen toxicity typically is manifested as uncontainable spasms and unconsciousness, which happens within a moment; this may result in drowning.

(c) In addition, it can result in spasms, pulmonary barotrauma, and death. On the same bases as oxygen, nitrogen gas with an increase in its partial pressure causes nitrogen narcosis, caused by the anesthetic effect of certain gases at high pressure. Nitrogen narcosis usually becomes noticeable at depths of 30 m and reaches its peak at 100 m. Narcosis produces a state similar to drunkenness. Rarely, it can occur also during shallow dives, but does not usually become noticeable at depths less than 30 m. Narcosis may be fully inverted by ascending to a lower depth, without long-term effects.

(d) Pressure on the central nervous system happens when an individual descends to 200 m and more. Abnormal nerve-muscle conduction due to the high pressure on the nerves has nothing to do with shrinking and spreading gas and it is related to hydrostatic pressure.

Underwater Physical Activity

Under water physical activity at a depth of 15 m, SV is decreased due to the water pressure on the chest cavity outside and inside. The compensation to keep desired Q is a significant increase in HR. In the deep water, an increase in core temperature facilitates HR increment. Significant deep immersion where surrounding temperature is low brings about a reduction in the heart's reserve capacity, namely that SV decreases while HR compensation is very limited. Compared to sea level, paddling on a bicycle vigorously at a depth of 30 m decrease significantly, VE, lung diffussion capacity and VT. Additional factors affecting cardiopulmonary function are external surrounding factors such as cold, hypothermia, and murky water and internal factors such as gender, age, and fitness level.

Introduction to Space Physiology

Gravity, also called gravitation, is a force that exists among all material objects in the universe. The force of attraction that exists between any two masses, any two bodies, and any two particles defines gravity. It is not just the attraction between objects and the Earth. The force of gravity keeps all of the planets in orbit around the sun. Gravity is the force by which a planet or other body draws objects toward its center. The force of gravity keeps all of the planets in orbit around the sun. Albert Einstein defined gravity as a curve in space that wraps around an object, such as a star or a planet. If another object is nearby, it is pulled into the curve. Gravity is an invisible force that pulls objects toward each other. Anything that has mass also has gravity. Objects with more mass have more gravity. Earth's gravity is what keeps us on the ground and what makes things fall.

It has been demonstrated mathematically that gravitational fields cause alteration in space and time. When gravity is not felt by humans, the sense of weightlessness or zero gravity exists. Force that attracts two bodies to each other in the universe means gravity. Nonetheless in a gravity-free environment, such as in space with weightlessness, astronauts regularly do not feel its effects. When space station moves in space its movement forward almost equals the speed of its direction fall to the universe. Thus, space stations are held in orbit by the gravity of the planet which they are orbiting. Therefore, astronauts traveling inside space station do not have any pulling toward a specific direction; therefore, they float. However, this weightlessness in the long run affects numerous health problems related to it.

At normal gravity the body systems are forced to work against it. However, weightlessness causes these systems to relax, since it is no longer working against the gravity pulling. In addition, sometimes astronauts are disoriented to realize the process by which the body can vary muscle contraction in immediate response to incoming information regarding external forces, by utilizing stretch receptors in the muscles, to keep track of the joint position in the body. This may cause astronauts to become unsettled and nervous. Therefore, space station has extra handholds and footholds on the exterior of their spacecraft so that they can anchor. Long-lasting periods in space impose medical problems such as loss of bone calcium through urine; thus, astronauts are more vulnerable to bone fractures, in addition to muscle atrophy and reduction in VO2max. But then, it has been found that the combination of resistance exercise and aerobic exercise is linked to better outcomes in bone density muscle strength and cardiopulmonary function. Nevertheless, despite exercising during a typical 6-month space mission, yet, astronauts need additional few months of physiological rehabilitation to adjust to Earth conditions.

Earth's gravity is what keeps the body on the ground and what makes things fall. Consequently, the closer objects are to each other, the stronger their gravitational pull is. Earth's gravity comes from all its mass. All its mass makes a combined gravitational pull on all the mass of the body.

Humans practice their own body weight as a result of the supporting force of gravity, which results in a normal force applied to an individual by the surface of a supporting object, on which the person is standing or sitting. In the absence of this force, a person would be in free fall and would experience weightlessness. It is the transmission of this reaction force through the human body, and the resultant compression and tension of the body's tissues, that results in the sensation of weight.

Because of the distribution of mass throughout an individual's body, the extent of the force response differs between an individual's feet and head. At any horizontal cross section of an individual's body, the extent of the compressive force being resisted by the tissues below the cross-section is equal to the weight of the portion of the body above the cross section. As an example, the shoulders carry the weight of the outstretched arms and are subject to a considerable torque. The term microgravity, also, μg is an expression for weightlessness and a very small zero-G is a very weak gravity, as in an orbiting spacecraft.

A real micro-g environment relatively to Earth or sun requires traveling into deep space to reduce the effect of gravity by attenuation to almost zero (Fig. 11.5). This is the simplest in conception but requires traveling an enormous distance, rendering it highly impractical. For example, to reduce the gravity of the Earth by a factor of one million, one needs to be at a distance of six million kilometers from the Earth, but to reduce the gravity of the sun to this amount, one has to be at a distance of 3.7 billion kilometers. Thus, it is not impossible, but it has only been achieved so far by four interstellar probes (Voyager 1 and 2 of the Voyager program, and Pioneer 10 and 11 of the Pioneer program) and they did not return to Earth. If travel at the light were possible for a spacecraft, it would take roughly three and a half hours to reach this microgravity environment (a region of space where the acceleration due to gravity is one-millionth of that experienced on the

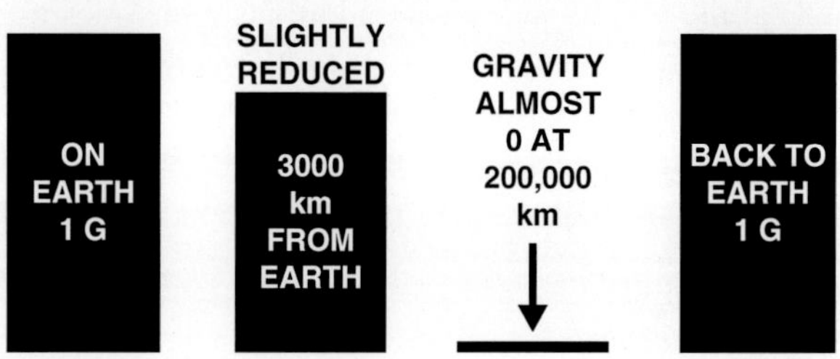

Fig. 11.5 describes the effect of the distance travelled from earth on gravity

Earth's surface). To reduce the gravity to one-thousandth of that on Earth's surface, one needs to be at a distance of 200,000 km.

Arterial Gradient

During spaceflight, arterial pressure gradients do not exist compared to Earth where arterial pressure gradients within the body are significant during upright posture (Fig. 11.6). The fluid pressure factors in the Starling–Landis equilibrium are affected by the lack of gravity, and only colloid osmotic pressure of the blood remains to oppose fluid flow into upper body tissue [32]. During spaceflight, microgravity alters the Starling–Landis Equilibrium to larger transcapillary filtration into tissues, resulting in facial edema and elevated intracranial pressure. The Starling–Landis equilibrium is a hypothesis in physiology, stating that the flow of fluids across capillary walls depends on the equilibrium between the force of BP on the walls, which tends to force fluids out, and the osmotic pressure across the walls, which tends to force them in due to the larger concentration of dissolved substances in the blood. In microgravity, where gravitational acceleration drops to zero; the weight of the fluid and body tissues is essentially zero, so that interstitial fluid pressure also falls, while intracranial pressure is elevated [33].

In addition, microgravity also brings about orthostatic intolerance. The body responds by increasing HR and BP to increase venous blood returning to the myocardium. This decline in venous return may have serious physiological significances to left-ventricular function and contractility, and therefore, during spaceflight, with less blood volume, astronauts are more susceptible to black out.

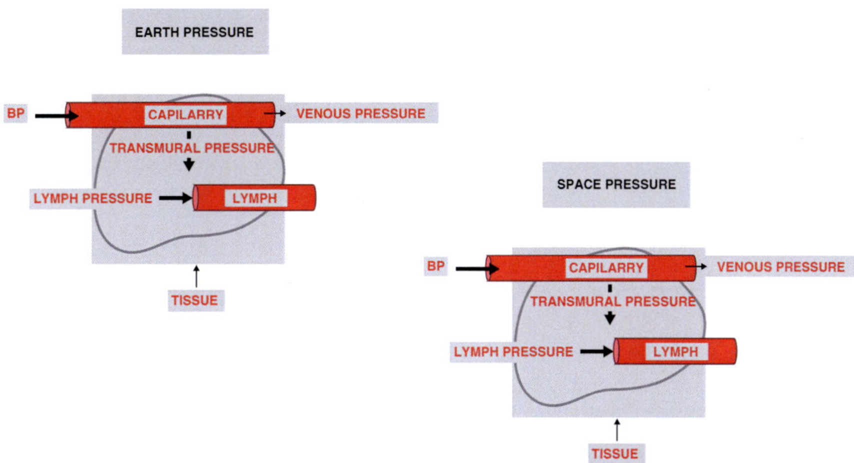

Fig. 11.6 Increased pressure from the capillary to the tissue (transmural pressure) during spaceflight. Smaller arrows indicate lower pressure gradients

Blood Depletion in Spaceflight

The total quantity of fluid circulating within the arteries, capillaries, veins, venules, and chambers of the heart at any time composes the blood volume. For astronauts' blood volume, regulation during spaceflight, is an important issue, since it may be life threatening. Two main components compose the blood volume: plasma which equals almost 60% of total blood volume, while red blood cells (erythrocytes), white blood cells (leukocytes), and platelet make up the rest 40% [33]. Diuresis is often a side effect of heart failure or kidney failure. It can develop as a response to high BP, extreme consumption of liquids, a kidney obstruction, and diabetes. During spaceflight, diuresis is increased, within 24–48 h of exposure, to the level of hypovolemia, due to an increase in central blood volume. Due to excessive elimination of fluid by the kidneys, there is also a decrease in red blood cells formation. Sometimes anemia is seen within 4 days of spaceflight. This central blood volume increment is a result of microgravity condition, in which blood from the lower limbs returns at a higher rate to the upper body part [34]. Under these conditions, the kidneys eliminate the extra chest blood volume. Hypovolemia in space is the result of a negative balance of decreased fluid intake and smaller reduction of urine output; fast fluid shifts between the body compartments, such as fluid movement from the intravascular to interstitial, due to the lower transmural pressure particularly of the chest and reduced muscle tone allow the fluid movement from the intravascular. Additionally, loss of erythrocytes reduces blood volume. When blood is lost, the body quickly pulls water from tissues outside the bloodstream in an attempt to keep the blood vessels filled. As a result, the blood is diluted, and the hematocrit is reduced [35]. The attenuated diuresis during space flight can be explained by increased retention after stress-mediated sympathetic activation during initial phase of space flight, stimulation caused by reduced red cell mass, and activation after fast blood volume contraction. Additionally, the relation between plasma osmolarity and vasopressin release might be disturbed in microgravity.

Endurance exercise increases blood volume over time, even following a single bout of exercise, and intensifies with weeks of training. Improved blood volume increases exercise performance, work capacity, and cardiovascular function. Factors affecting blood volume increment after exercise training, cardiopulmonary exercise mode, and release of antidiuretic hormone and aldosterone hormone, affecting the kidneys to retain or reabsorb water, thus, increase blood plasma levels. Since plasma is a major component of blood volume, any increase in plasma will increase blood volume amount. In addition, aerobic exercise raises proteins mainly albumin amount within the blood plasma. The increase in plasma protein content results in water retention within blood plasma, which in turn increases. In addition to the improvement in oxygen transportation to the contracting muscles during exercise, individuals who exercise regularly increase their total number of red blood cells. A well-trained individual has greater red blood cells values compared to a sedentary individual. Spaceflight has an effect on the cardiopulmonary system, like detraining, in which the individual's fitness level decreases; thus, blood volume returns to a pre-training level [36]. However, it can be regained by exercising during the spaceflight and upon landing. However, it can be regained again by exercising during the spaceflight and in Earth.

The Impact of μg on the Skeletal

Rats that were flown to space exhibited muscle atrophy, protein content loss, and impaired muscle function within 1 week [37], losing one-third of their muscle mass and peak force. Atrophy is significantly found more in slow muscle myofibers containing the slow myosin heavy chain protein and less in fiber II muscle [38]. It seems that fiber II types muscle are not as sensitive to the unloading provocation as the fiber I types of muscle are. Myosin heavy chain with protein actin derives the force, work, and power to accomplish movement and stabilizing body posture. Nonetheless, when spaceflight duration is above 6 months fiber I and fiber II muscles exhibit similar reductions of around one-third in isokinetic strength [39]. In some experiments, it has been reported that the myofibrillar fraction can be degraded to a greater extent than other muscle fractions [40]. These atrophies may result in incapacitated astronaut members, when returning to Earth due to the increased gravity surroundings [41]. Skeletal muscles, mainly extensor more than flexors muscles, are utilized broadly for postural support and motion. Extensor muscles of the lower limb suffer from a rapid loss of cell mass (atrophy), structural changes, and metabolic alterations during spaceflight [42]. This muscle atrophy is also associated with decline in peak force and power [43]. In contrast to the contractile apparatus, studies on skeletal muscle metabolic enzymes in rodent demonstrated a diversity response without a definite adaptive variation in oxidative enzyme expression [42].

Low-orbit spaceflight induces bone breakability at weight-bearing skeletal sites and increases bone suction. Mechanical load on the skeletal structure, by muscular contraction, has a key role in the maintenance, preservation, function, and growth of bone tissue [44].

Bone loss affected by μg conditions is a considerable obstacle to astronauts in long-lasting spaceflight, specifically bone metabolism imbalance [45]. In addition, total body calcium, released from the skeleton of astronauts, increases the danger of renal stone progress [46].

Additional factor affecting bone loss in space is Wnt-β-catenin signaling pathway. It controls the quantity of the transcriptional co-activator β-catenin which regulates main evolving gene expression [47]. Wnt-β-catenin signaling pathway is involved in the adult's tissue homeostasis [46]. Alterations in the Wnt pathway are frequently related to cancer and other diseases [48]. In studies on animals, the image of bone to spaceflight was found to be related with decrement in both cortical and trabecular bone formation, modification in mineralization shapes, inefficiency of collagen, and non-collagenous protein metabolism [44]. Yet, the machinery affecting bone mass and calcium depletion during spaceflight is still not fully understood. Bone that is lost during long spaceflight mission is not completely recovered, and bone density can last to worsen after landing, maybe due to osteocyte death. In spite of exercising in space, astronauts still return to Earth with substantial musculoskeletal deficits [49]. Exercise and other interventions designed to reduce the loss of bone in space are not completely effective. The explanation for the limited intervention efficiency in space is the loading forces applied on the astronaut in space, which are less and not comparable to those in Earth [50]. Bone mineral density is decreased by about 1%·month^{-1} in the proximal femur; this lost amount is equal to that lost by postmenopausal women in 1 year [51]. During long-lasting spaceflight above a month, there is a significant reduction in bone mass and density, specifically in the spinal and lower limbs. This decrement is due to the lack of loads on those body parts in μg conditions [52]. Effective countermeasure regime is needed to adapt again, to gravity in Earth upon landing, to recover from cartilaginous tissues, such as intervertebral disks, and loss of structure and function in space, as much as possible. Subsequent to a time period of physical inactivity in microgravity during spaceflight, once astronaut lands on Earth, muscles atrophy and bones deteriorate, turning any exercise or tasks of daily living nearly unbearable. Although muscle atrophy and bone deterioration have medical consequences, muscles can be constructed back up with the right exercise training regime, moderately in short time; however, regaining lost bone takes more time and sometimes it does not recover fully.

The most used parameter to describe mass loss and bone strength is the measurement of bone mineral density. This measurement does not reflect density which is an engineering definition that describes general mass divided by the objective volume (gm·cm^3); rather, it stands for mineral density (gm·cm^2). To avoid mineral loss in space, astronaut utilizes ultimately the DXA technique. The bone mineral density is measured by dual-energy X-ray absorptiometry (DXA). It is bidimensional, divided by numerous areas where in a specific area the bone mineral content is calculated by dividing the amount observed by the X-ray by the size of the specific area.

The Impact of μg on the Cardiovascular System

Long-lasting spaceflight effects on the cardiovascular system are well documented [53]. The environment of space and long-term weightlessness have negative effects on the cardiovascular system functions and lessen creation of red blood cells. Astronauts returning to Earth usually demonstrate reduced orthostatic tolerance when assessed on a tilt table or quiet standing. Under normal Earth conditions, when an individual lies down, stands up quickly, and feels light-headed, that is orthostatic intolerance. Nevertheless, there was no space postflight assessment of astronauts on orthostatic tolerance during activities of daily living. Before taking part in 6 months of space mission and with landing back to Earth, BP measurements of astronauts did not show orthostatic intolerance or hypotension during activities of daily living [54]. Similarly, during spaceflight, SBP decreased in space by 12–14 mmHg, while DBP remains unchanged. Thus, long-lasting experience to spaceflight has in general no effect on SBP and DBP variability and its distribution and remained similar before, during, and after spaceflight [55]. Definitely, there are changes in cardiac structure during spaceflight, plasma volume is decreased, left-ventricular mass decreases, and the autonomic nervous system adjustments that may alter myocardial electrical conduction with minor changes in QT interval were detected in a small number of astronauts after long duration in spaceflight. Cardiac arrhythmias disorders were seen in some astronauts, during long duration of spaceflight. These were associated with cardiovascular disease; nonetheless, no one knows if the arrhythmias existed before boarding the spaceflight, while atrial fibrillation can progress with time. Although this has not been detected, worry exists that numerous astronauts may develop cardiac rhythm disturbances due to long-lasting experience to microgravity.

The Immune System in Space

The immune system is a structure composed of biological assemblies and developments. Any change such as decreased activity in one component leads to a disease risk within the human body. During flights to space, there has been a focus on the health effects of spaceflight-related environmental stressors and circadian rhythm [56]. Immune system dysfunction during spaceflight signifies a considerable risk for clinical, cellular, and genetic basis of immune system. Studies have shown in microgravity that there are immune system modifications. In spite of the immune system dysregulation, an adaptation reaction of the immune system to microgravity environment starts after 2 weeks and continuing until the mission is over for the T-cell and macrophages.

This may create an environment where, in some astronauts, rashes, unusual allergies, and latent virus reactivation may present themselves. Evidence of virus DNA has been found in astronauts' blood, urine, and saliva samples. This can occur when microgravity weakens the immune system or under traumatic circumstances on Earth. NASA is trying to define the reduction of the severity of immune matters prior to boarding for long-lasting space missions, where microgravity conditions for longer periods of time will affect the immune system.

Some elementary immune preventative counteractions, which must be developed in the future, may include vaccines, balanced nutrition, and exercise, and in addition, steps such as isolation of astronauts before taking off and filtered air in order to avoid diseases, bacteria, and viruses that may cause medical problems for the astronaut.

Pulmonary Function in Space

The pulmonary does not have structural variations in microgravity, and so there is no apparent degradation in lung function upon return to earth, even after 6 months in space. Gravity has an effect that matches VE to Q, by the gradients in VE, Q, and gas exchange [57]. This matching is available, due to the pulmonary construction, with its network of alveoli and the capillary connection. Gravity on Earth offers appreciated inevitable effects that affect pulmonary function. Due to gravity, VE goes mainly to the lungs bottom, where it matches lung blood perfusion for VE/Q that equals one at rest. This VE/Q ratio keeps gas exchange effectiveness. Under weightlessness conditions, VE and Q are more uniform; consequently, edema is not present in the lungs and gas exchange is well maintained [58]. Lung diffusing capacity increased significantly (by 28%), due to an increase in the diffusing capacity of the alveolar membrane not seen with gravity conditions. These changes in the diffusion capacity may be the result of the effects of uniform filling of the pulmonary capillary bed [59]. Q increases also significantly (by 18%) during spaceflight; this is in the face of hypovolemia and reduced central venous pressure [59]. Consequently, the VE/Q ratio is maintained well. In addition, in space, obstructive sleep apnea is not seen, since the main reason for it is the gravity effect on the upper airway tracks.

Exercising in Space

Space is a very hostile dangerous environment for humans; during long-lasting missions, weightlessness and high radioactivity characterize main intimidations to astronaut's well-being [60]. Interestingly, spaceflight enforces on the body of healthy fit astronauts, sort of a faster aging course or disease [61]. Subsequent to few weeks of spaceflight, the cardiopulmonary system function is significantly reduced and myocardial deterioration and pulmonary dysfunction may limit myocardial function. While following months of weightlessness, loss of bone density and muscle atrophy are noticed. However, exercise lessens the burden during spaceflight and it can be alterable subsequently to landing [60]. These findings may offer an understanding tool about the pathophysiological developments, understanding mechanisms of physiological deconditioning, and developing countermeasures to alleviate this development, which is an interesting issue for medical personal.

Head-down tilt bed rest has the same adaptations and responses as related to space microgravity. Response such as a fast reduction in VO2max resulting from the decrease in body circulating blood volume occurs in response to the head-ward body fluid shift [62]. Healthy individuals in microgravity of space have negative adaptive physiological alterations similar to accelerated aging. The atrophy of the cardiovascular system in space and continuous bed rest is tenfolds sooner than in the natural aging progression [63]. Long-duration head-down tilt bed rest is analogous to microgravity and affects organs such as bone damage, muscle atrophy myocardial dysfunction, nervous system function and coordination deterioration, orthostatic intolerance, and low physical work and motion capacity [64]. In addition to effects of 6 months spaceflight on the international space station, some astronauts developed neuro-ophthalmic and neuro-cerebral abnormalities [65]. Yet, those established abnormalities are revocable, though the inquiry is what will happen following longer exposures. The dislocation of humans from Earth to interplanetary results in deconditioning due to transitional gravitational forces [66]. This space deconditioning due to microgravity has applications as a medical risk when returning to Earth as in the inactive elderly [63].

Being in environment of space and in microgravity (μG) carries out a specific physiologic variation on healthy astronauts. When returning to Earth from space, these variations seem as physical incapability which requires a period of restoration [67]. The main problems arising from being in spaceflght and in μG environments are muscular atrophy, contractile strength, functional capabilities, bone demineralization, neuro-vestibular symptoms, and orthostatic intolerance [68]. Exercise counteraction is designed to lessen the harmful properties of μG on skeletal muscle structure, function, and contractile strength and cardiopulmonary function [69]. Exercise is the main counteraction utilized during spaceflight exposures as an effective tool to preserve or reestablish cardiopulmonary, musculoskeletal structure, and orthostatic function [69]. During the 96 days of Soyuz 26-Salyut-6 mission, exercise training was irregular for 25 days, thus decreasing work capacity. Renewal of regular exercise training, for the rest 70 days of the spaceflight, enhanced physiological work capacity variables; this, clearly, confirmed the significance of lasting regular combined exercise during spaceflight [70]. In addition, introduction to μG environments affects strength and VO2max. Long-lasting period of physical inactivity and conditions in space affect the astronaut, when returning to Earth; muscle atrophy and loss in bone mass result in significantly reduced daily living functional capacity. Since, in physical fitness, there is no one single exercise mode that restores all of the abovementioned physiological problems; therefore, for those experiencing μG, it is suggested to combine aerobic exercise and resistance exercise training, with strength coming first and then aerobic stimulus [71]. Duration, frequency, intensity, and mode of exercise can affect the general course of adaptation to space microgravity. Efficient exercise program for the astronauts on board the space station include exercising for 75–120 min·day^{-1} using three different modes of exercise: cycle ergometer, treadmill (both aerobic), and eccentric resistance exercise device [72]. Weightlifting device is more powerful than that used on Earth, since any mass in space would weigh significantly less for microgravity. This, occupies significant part of the astronaut's daily routine. The combined aerobic and resistance training has proven to be effective in reestablishing cardiopulmonary, bone, and musculoskeletal functions in spaceflights with long duration [73]. The exercise apparatus is attached on an elevated platform for the lack of gravity which has the tendency to bring astronauts to float; binds are attached to the astronauts to hold them to the bicycle and treadmill to avoid the effect of microgravity and to diminish the noise the equipment does (Figs. 11.7 and 11.8). Cycle ergometer is utilized to gain and measure work capacity and HR response and to determine labor done. Treadmill: Walking or jogging on the treadmill is like walking under normal conditions in Earth. This mode of exercise is the most important means to keep bones density and muscles mass.

Fig. 11.7 Harnesses are attached to the astronauts to hold them to the bicycle to avoid the effect of microgravity

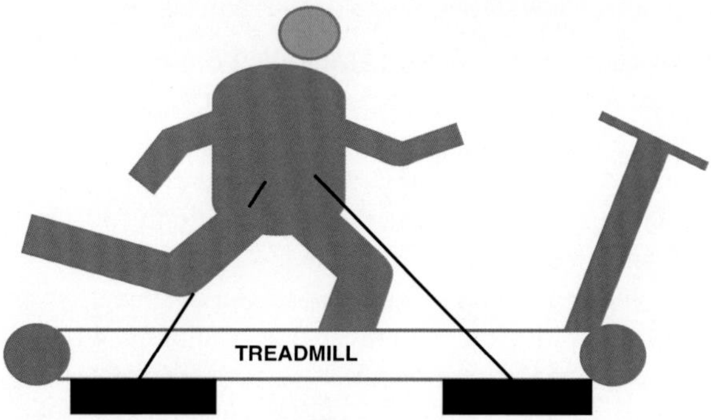

Fig. 11.8 Harnesses are attached to the astronauts to hold them to the treadmill to avoid the effect of microgravity

Fig. 11.9 Harnesses are attached to the astronauts to hold them to the resistance device to avoid the effect of space

The resistance exercise device (Fig. 11.9) utilized in space is similar to multipurpose weightlifting equipment on Earth which enables the astronauts to be active against load in different positions for different muscle groups. A heavy object on Earth loses its weight in space; therefore, resistance exercise equipment requires to be modified for utilization in space.

References

1. Park H-Y, Hwang H, Park J-H, et al. The effects of altitude/hypoxic training on oxygen delivery capacity of the blood and aerobic exercise capacity in elite athletes – a meta-analysis. J Exerc Nutr Biochem. 2016;20:15–22.
2. McLean BD, Gore CJ, Kemp J. Application of 'live low-train high' for enhancing normoxic exercise performance in team sport athletes. Sports Med. 2014;44:1275–87.
3. Bailey DM, Davies B. Physiological implications of altitude training for endurance performance at sea level: a review. Br J Sports Med. 1997;31:183–90.
4. Zebrowska A, Jastrzębski D, Sadowska-Krępa E, et al. Comparison of the effectiveness of high-intensity interval training in hypoxia and normoxia in healthy male volunteers: a pilot study. Biomed Res Int. 2019;2019:7315714. https://doi.org/10.1155/2019/7315714.
5. Brocherie F, Millet GP, D'Hulst G, et al. Repeated maximal-intensity hypoxic exercise superimposed to hypoxic residence boosts skeletal muscle transcriptional responses in elite team-sport athletes. Acta Physiol (Oxf). 2018;222(1):e12851. https://doi.org/10.1111/apha.12851.
6. Rizo-Roca D, Bonet JB, Inal B, et al. Contractile activity is necessary to trigger intermittent hypobaric hypoxia-induced fiber size and vascular adaptations in skeletal muscle. Front Physiol. 2018;9:481. https://doi.org/10.3389/fphys.2018.00481.
7. Ferretti G. Limiting factors to oxygen transport on Mount Everest 30 years after: a critique of Paolo Cerretelli's contribution to the study of altitude physiology. Eur J Appl Physiol. 2003;90:344–50.
8. Howald H, Hoppeler H. Performing at extreme altitude: muscle cellular and subcellular adaptations. Eur J Appl Physiol. 2003;90:360–4.
9. Murray AJ. Metabolic adaptation of skeletal muscle to high altitude hypoxia: how new technologies could resolve the controversies. Genome Med. 2009;1(12):117. https://doi.org/10.1186/gm117.
10. Semenza GL. Expression of hypoxia-inducible factor 1: mechanisms and consequences. Biochem Pharmacol. 2000;59:47–53.
11. Lizamore CA, Hamlin MJ. The use of simulated altitude techniques for beneficial cardiovascular health outcomes in nonathletic, sedentary, and clinical populations: a literature review. High Alt Med Biol. 2017;18:305–21.
12. Hauser A, Troesch S, Steiner T, et al. Do male athletes with already high initial haemoglobin mass benefit from 'live high-train low' altitude training? Exp Physiol. 2018;103:68–76.
13. Goldfarb-Rumyantzev AS, Alper SL. Short-term responses of the kidney to high altitude in mountain climbers. Nephrol Dial Transplant. 2014;29:497–506.
14. Ramirez G, Hammond M, Agosti SJ, et al. Effects of hypoxemia at sea level and high altitude on sodium excretion and hormonal levels. Aviat Space Environ Med. 1992;63:891–8.
15. Hildebrandt W, Ottenbacher A, Schuster M, et al. Diuretic effect of hypoxia, hypocapnia, and hyperpnea in humans: relation to hormones and O2 chemosensitivity. J Appl Physiol. 2000;88:599–610.
16. Steinback CD, Poulin MJ. Cardiovascular and cerebrovascular responses to acute isocapnic and poikilocapnic hypoxia in humans. J Appl Physiol (1985). 2008;104:482–9.
17. Naeye RL. Children at high altitude: pulmonary and renal abnormalities. Circ Res. 1965;16:33–8.
18. Bestle MH, Olsen NV, Poulsen TD, et al. Prolonged hypobaric hypoxemia attenuates vasopressin secretion and renal response to osmo-stimulation in men. J Appl Physiol. 2002;92:1911–22.
19. Olsen NV, Kanstrup IL, Richalet JP, et al. Effects of acute hypoxia on renal and endocrine function at rest and during graded exercise in hydrated subjects. J Appl Physiol. 1992;73:2036–43.
20. Pichler J, Risch L, Hefti U, et al. Glomerular filtration rate estimates decrease during high altitude expedition but increase with Lake Louise acute mountain sickness scores. Acta Physiol (Oxf). 2008;192:443–50.
21. Singh MV, Salhan AK, Rawal SB, et al. Blood gases, hematology, and renal blood flow during prolonged mountain sojourns at 3500 and 5800 m. Aviat Space Environ Med. 2003;74:533–6.

22. Schneider M, Bernasch D, Weymann J, et al. Acute mountain sickness: influence of susceptibility, preexposure and ascent rate. Med Sci Sports Exerc. 2002;34:1886–91.
23. Navot Mintzer D, Leshem E, Chazan B, et al. High altitude exposure in travelers with preexisting medical condition. [article in Hebrew]. Harefuah. 2015;154:725–9.
24. Taylor AT. High-altitude illnesses: physiology, risk factors, prevention, and treatment. Rambam Maimonides Med J. 2011;2(1):e0022. https://doi.org/10.5041/RMMJ.10022.
25. West JB, Schoene RB, Milledge JS. High altitude medicine and physiology. 4th ed. Hodder Arnold: London, Great Britain; 2007.
26. Khodaee M, Grothe HL, Seyfert JH, et al. Athletes at high altitude. Sports Health. 2016;8:126–32.
27. Sutton JR, Coates G, Houston CS. Hypoxia and mountain medicine. Vt. Queen City Printers: Burlington; 1992.
28. Basnyat B, Murdoch DR. High-altitude illness. Lancet. 2003;361:1967–74.
29. Wu T, Ding S, Liu J, et al. Ataxia: an early indicator in high altitude cerebral edema. High Alt Med Biol. 2006;7:275–80.
30. Hackett PH, Roach RC. High-altitude illness. N Engl J Med. 2001;345:107–14.
31. Luks AM. Do we have a "best practice" for treating high altitude pulmonary edema? High Alt Med Biol. 2008;9:111–4.
32. Kashirina DN, Percy AJ, Pastushkova LK, et al. The molecular mechanisms driving physiological changes after long duration space flights revealed by quantitative analysis of human blood proteins. BMC Med Genet. 2019;12(Suppl 2):45. https://doi.org/10.1186/s12920-019-0490-y.
33. Leach CS, Alfrey CP, Suki WN, et al. Regulation of body fluid compartments during short-term spaceflight. J Appl Physiol (1985). 1996;81:105–16.
34. Goswami N, Blaber AP, Hinghofer-Szalkay H, et al. Lower body negative pressure: physiological effects, applications, and implementation. Physiol Rev. 2019;99:807–51.
35. Diedrich A, Paranjape SY, Robertson D. Plasma and blood volume in space. Am J e Med Sci. 2007;334:80–6.
36. Tomilovskaya E, Shigueva T, Sayenko D, et al. Dry immersion as a ground-based model of microgravity physiological effects. Front Physiol. 2019;10:284. https://doi.org/10.3389/fphys.2019.00284.
37. Tyganov SA, Mochalova EP, Belova SP, et al. Effects of plantar mechanical stimulation on anabolic and catabolic signaling in rat postural muscle under short-term simulated gravitational unloading. Front Physiol. 2019;10:1252. https://doi.org/10.3389/fphys.2019.01252.
38. Tischler ME, Henriksen EJ, Munoz KA, et al. Spaceflight on STS-48 and earth-based unweighting produce similar effects on skeletal muscle of young rats. J Appl Physiol (1985). 1993;74:2161–5.
39. Hackney KJ, Scott JM, Hanson AM, et al. The astronaut-athlete: optimizing human performance in space. J Strength Cond Res. 2015;29:3531–45.
40. Riley DA, Bain JL, Thompson JL, et al. Thin filament diversity and physiological properties of fast and slow fiber types in astronaut leg muscles. J Appl Physiol (1985). 2002;92:817–25.
41. Adams GR, Caiozzo VJ, Baldwin KM. Skeletal muscle unweighting: spaceflight and ground-based models. J Appl Physiol (1985). 2003;95:2185–201.
42. Fitts RH, Riley DR, Widrick JJ. Functional and structural adaptations of skeletal muscle to microgravity. J Exp Biol. 2001;204:3201–8.
43. Riley DA, Bain JL, Romatowski JG, et al. Skeletal muscle fiber atrophy: altered thin filament density changes slow fiber force and shortening velocity. Am J Physiol Cell Physiol. 2005;288(2):C360–5.
44. Zerath E. Effects of microgravity on bone and calcium homeostasis. Adv Space Res. 1998;21:1049–58.
45. Thirsk R, Kuipers A, Mukai C, et al. The space-flight environment: the international Space Station and beyond. CMAJ. 2009;180:1216–20.

46. Grimm D, Grosse J, Wehland M, et al. The impact of microgravity on bone in humans. Bone. 2016;87:44–56.
47. MacDonald BT, Tamai K, He X. Wnt/β-catenin signaling: components, mechanisms, and diseases. Dev Cell. 2009;17:9–26.
48. Clevers H. Wnt/beta-catenin signaling in development and disease. Cell. 2006;127:469–80.
49. Rittweger J, Albracht K, Fluck M, et al. Sarcolab pilot study into skeletal muscle's adaptation to long-term spaceflight. NPJ Microgravity. 2018;4:18. https://doi.org/10.1038/s41526-018-0052-1.
50. Genc KO, Gopalakrishnan R, Kuklis MM, et al. Foot forces during exercise on the international Space Station. J Biomech. 2010;43:3020–7.
51. Lang T, LeBlanc A, Evans H, et al. Cortical and trabecular bone mineral loss from the spine and hip in long-duration spaceflight. J Bone Miner Res. 2004;19:1006–12.
52. He JP, Feng X, Wang JF, et al. Icariin prevents bone loss by inhibiting bone resorption and stabilizing bone biological apatite in a hindlimb suspension rodent model. Acta Pharmacol Sin. 2018;39:1760–7.
53. Convertino VA. Mechanisms of microgravity-induced orthostatic intolerance and implications of effective countermeasures: overview and future directions. J Gravit Physiol. 2002;9:1–12.
54. Lee SMC, Feiveson AH, Stein S, et al. Orthostatic intolerance after ISS and space shuttle missions. Aerosp Med Hum Perform. 2015;86(suppl):A54–67.
55. Fu Q, Shibata S, Hastings JL, et al. Impact of prolonged spaceflight on orthostatic tolerance during ambulation and blood pressure profiles in astronauts. Circulation. 2019;140:729–38.
56. Sonnenfeld G. The immune system in space, including earth-based benefits of space-based research. Curr Pharm Biotechnol. 2005;6:343–9.
57. Prisk GK. Microgravity and the respiratory system. Eur Respir J. 2014;43:1459–71.
58. Verbanck S, Larsson H, Linnarsson D, et al. Pulmonary tissue volume, cardiac output, and diffusing capacity in sustained microgravity. J Appl Physiol (1985). 1997;83:810–6.
59. West JB, Elliott AR, Guy HJ, et al. Pulmonary function in space. JAMA. 1997;25(277):1957–61.
60. Demontis GC, Germani MM, Caiani EG, et al. Human pathophysiological adaptations to the space environment. Front Physiol. 2017;8:547. https://doi.org/10.3389/fphys.2017.00547.
61. Hargens AR, Bhattacharya R, Schneider SM. Space physiology VI: exercise, artificial gravity, and countermeasure development for prolonged space flight. Eur J Appl Physiol. 2013;113:2183–92. https://doi.org/10.1007/s00421-012-2523-2525.
62. Convertino VA. Exercise and adaptation to microgravity environments. In: Handbook of physiology; 1996. p. 815–43.
63. Vernikos J, Schneider VS. Space, gravity and the physiology of aging: parallel or convergent disciplines? A mini-review. Gerontology. 2010;56:157–66.
64. Hargens AR, Vico L. Long-duration bed rest as an analog to microgravity. J Appl Physiol (1985). 2016;120:891–903.
65. Marshall-Goebel K, Mulder E, Donoviel D, et al. An international collaboration studying the physiological and anatomical cerebral effects of carbon dioxide during head-down tilt bed rest: the SPACECOT study. J Appl Physiol (1985). 2017;122:1398–405.
66. Cardus D, McTaggart WG. Artificial gravity as a countermeasure of physiological deconditioning in space. Adv Space Res. 1994;14:409–14.
67. Walsh NP, Whitham N. Exercising in environmental extremes. Sports Med. 2006;36:941–76.
68. Elmann-Larsen B, Schmitt D. Staying in bed to benefit ESA's astronauts and Europe's citizens. ESA Bull. 2003;113:34–9.
69. Jones TW, Petersen N, Howatson G. Optimization of exercise countermeasures for human space flight: operational considerations for concurrent strength and aerobic training. Front Physiol. 2019;10:584. https://doi.org/10.3389/fphys.2019.00584.
70. Convertino VA, Sandler H. Exercise countermeasures for spaceflight. Acta Astronaut. 1995;35:253–70. https://doi.org/10.1016/0094-5765(95)98731-N.

71. Steele J, Androulakis-Korakakis P, Perrin C, et al. Comparisons of resistance training and "cardio" exercise modalities as countermeasures to microgravity-induced physical deconditioning: new perspectives and lessons learned from terrestrial studies. Front Physiol. 2019;10:1150. https://doi.org/10.3389/fphys.2019.01150.
72. Convertino VA. Physiological adaptations to weightlessness: effects on exercise and work performance. Exerc Sports Sci Rev. 1990;18:119–65.
73. Rittweger J, Beller G, Armbrecht G, et al. Prevention of bone loss during 56 days of strict bed rest by side-alternating resistive vibration exercise. Bone. 2010;46:137–47.

Chapter 12
Epigenetics in Exercise

Genes

Genetics is a branch of biology that is related to the study of genes, genetic variation, and heredity in organisms. Organisms receive characteristics by way of distinct [1].

The molecular basis for genes is deoxyribonucleic acid (DNA), which normally exists as a double-stranded molecule, coiled into the shape of a double helix [2]. Both strands of double-stranded DNA store the same biological information. A large part of DNA (more than 98% for humans) is noncoding, meaning that these sections do not serve as patterns for protein sequences. The two strands of DNA run in opposite directions to each other and are thus antiparallel. Attached to each sugar is one of the four types of nucleobases (informally, bases). It is the sequence of these four nucleobases along the backbone that encodes genetic information. RNA strands are created using DNA strands as a template in a process called transcription, where DNA bases are exchanged for their corresponding bases except in the case of thymine (T), which RNA substitutes for uracil (U) [3]. Under the genetic code, these RNA strands specify the sequence of amino acids within proteins in a process called translation.

The two DNA strands are also known as polynucleotides as they are composed of simpler monomeric units called nucleotides. Each nucleotide is composed of one of the four nitrogen-containing nucleobases cytosine (C), guanine (G), adenine (A), and thymine (T); a sugar called deoxyribose; and a phosphate group (Fig. 12.1).

The nucleotides are joined to one another in a chain by covalent bonds between the sugar of one nucleotide and the phosphate of the next, resulting in an alternating sugar-phosphate backbone. Genetic information exists in the sequence of these nucleotides, and genes exist as stretches of sequence along the DNA chain [4]. Viruses are the only exception to this rule—sometimes viruses use a very similar

M. S. Saghiv, M. S. Sagiv, *Basic Exercise Physiology*,
https://doi.org/10.1007/978-3-030-48806-2_12

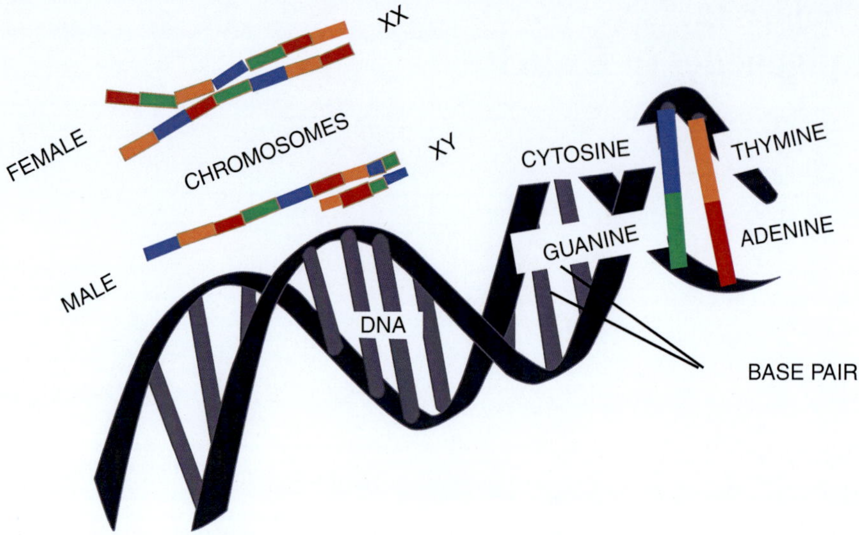

Fig. 12.1 Chromosome and DNA structures. The complementary nitrogenous bases are divided into two groups, pyrimidines and purines. In DNA, the pyrimidines are thymine and cytosine; the purines are adenine and guanine

RNA molecule instead of DNA as their genetic material [5]. Viruses cannot reproduce without a host and are unaffected by many genetic processes, so they cannot be considered as living organisms.

The complementary nitrogenous bases are divided into two groups, pyrimidines and purines. In DNA, the pyrimidines are thymine and cytosine and the purines are adenine and guanine. Thus, in its two-stranded form, each strand effectively contains all necessary information. This DNA structure is the physical basis for inheritance: DNA replication duplicates the genetic information when the two strands separate, by splitting the strands and using each strand as a template for the synthesis of a new partner strand [1]. This information is replicated as and when the two strands separate. Genes are arranged linearly along the long chains of DNA base-pair sequences. In bacteria, each cell usually contains a single circular gonophore, while humans have their DNA arranged in multiple linear chromosomes. DNA is organized into long structures called chromosomes. These DNA strands are often extremely long and genes vary in size; the largest human chromosome, for example, is about 247 million base pairs in length [6]. The DNA of a chromosome is related to structural proteins that organize, compact, and control access to the DNA, forming a material called chromatin; in humans, chromatin is usually composed of nucleosomes, segments of DNA wound around cores of histone proteins. Before typical cell division, these chromosomes are duplicated in the process of DNA replication, providing a complete set of chromosomes for each daughter cell. The full set of hereditary material in an organism (usually the combined DNA sequences of all chromosomes) is called the genome.

Humans have two of each chromosome and thus two copies of every gene [1]. A nucleus contains 46 molecules called chromosomes—23 are donated by the female and 23 by the male. The two alleles for a gene are located on identical loci of the two homologous chromosomes, each allele inherited from a different parent.

Many species have so-called sex chromosomes that determine the gender of each organism [1]. In humans and many other animals, the Y chromosome contains the gene that triggers the development of specifically male characteristics. In evolution, this chromosome has lost most of its content and also most of its genes, while the X chromosome is similar to the other chromosomes and contains many genes. The X and Y chromosomes form a strongly heterogeneous pair.

Regulation of Gene Expression

The genome of a given organism contains thousands of genes, but not all these genes need to be active at any given moment. A gene is expressed when it is being transcribed into mRNA, and there exist many cellular methods of controlling the expression of genes such that proteins are produced only when needed by the cell. Transcription factors are regulatory proteins that bind to DNA, either promoting or inhibiting the transcription of a gene [7]. Within a genome, there exists a series of genes essential for the production of the amino acid tryptophan. However, when the amino acid tryptophan is now obtainable to the cell, these genes for amino acid synthesis are no longer needed. The presence of amino acid directly affects the activity of the genes—tryptophan molecules bind to the tryptophan repressor, which is a transcription factor, changing the repressor's structure such that the repressor binds to the genes. The tryptophan repressor blocks the transcription and expression of the genes, thereby creating negative feedback regulation of the tryptophan synthesis process [8]. Transcription factors bind to DNA, influencing the transcription of associated genes.

Differences in gene expression are especially clear within multicellular organisms, where cells all contain the same genome but have very different structures and behaviors due to the expression of different sets of genes. All the cells in a multicellular organism derive from a single cell, differentiating into variant cell types in response to external and intercellular signals and gradually establishing different patterns of gene expression to create different behaviors. As no single gene is responsible for the development of structures within multicellular organisms, these patterns arise from the complex interactions between many cells.

Gene is a sequence of DNA that encodes a directly functional RNA that is an elementary physical and practical unit of inheritance, acting as a template for the production of proteins. In humans, genes differ in magnitude from a few hundred DNA bases to more than two million bases, and these genes make up different DNA sequences called genotypes. The Human Genome Project estimated that humans have between 20,000 to 25,000 genes. The transmission of genes to human's offspring is the basis of the inheritance of phenotypic traits [4]. Most human characteristics are under the effect of various genes as well as the environment. Some genetic traits are promptly noticeable, such as hair color while others are not, such as muscle fiber composition. Genes may obtain alterations in their arrangement, leading to different variations (alleles) in the population. These alleles encode slightly different versions of a protein, which cause different traits. Genes develop due to the discrepancy existence of alleles. One percent of genes differ between individuals, and alleles contain the same gene with minor differences in their order of DNA bases. These contribute to each individual's exceptional physical abilities. Each individual has two replicas of each gene, one came from each parent [9]. The entire accompaniment of genes in an animal or cell is identified as its genome, which is stored on one or more chromosomes. Each gene occupies a specific position on a chromosome. Because genes provide instructions for making proteins and proteins determine the structure and function of each cell in the body, it follows that genes are responsible for all the characteristics an individual inherits [10].

Epigenetics

Epigenetic mechanisms control developments beyond DNA-stored information. The definition of epigenetics and its specific mechanisms including DNA methylation and gene imprinting, modifications of nucleosome histones associated with silencing or activation of gene transcription, RNA interference, chromosomal silencing, and the role of mobile elements are discussed [11]. There are several pathways, including histone tail modifications, histone variant incorporation, nucleosome remodeling, DNA methylation and noncoding RNAs that altogether contribute to the dynamic ensemble of chromatin under distinct developmental options. Recent clues suggest a potential link between caffeoyl phenylethanoid glycoside (CpG) methylation and other histone modifications [12]. Complex interactions have been discovered among various components of DNA methylation and histone methylation pathways, proteins involved in the formation of higher-order chromatin structure, chromatin remodeling activities, and RNA interference [13]. The histone tail modifications are most variable and over 50 marks have by now been mapped. While the majority of these modifications are transient, histone lysine methylation and, in particular, a histone lysine tri-methyl state has been regarded as a more robust signal, consistent with proposed roles to impart long-term epigenetic memory [12]. Histone lysine methylation plays a fundamental role in chromatin organization and function. This epigenetic mark is involved in many biological processes such as heterochromatin formation, chromosome X inactivation, genomic imprinting, and transcriptional regulation [14]. A complex picture is emerging in which DNA methylation and histone modifications are parts of an epigenetic program that integrates gene-silencing networks within the cell [15]. Histone deacetylation and methylation at lysine 9 of H3 might also contribute to the establishment of DNA methylation patterns, a long-standing mystery in epigenetics. Heritable patterns of gene expression and gene silencing are determined by chromatin states that either permit or restrict transcription. Restrictive heterochromatin in most eukaryotes is characterized by high levels of DNA methylation and histone H3 methylation at lysine 9 [13].

Many of these changes occur through epigenetic alterations to DNA, such as histone modifications, DNA methylations, expression of microRNAs, and changes of the chromatin structure. All these epigenetic alterations may have clinical relevance, thus playing an important role in the prevention and confrontation of metabolic syndrome, cardiovascular diseases, and cancer [16].

DNA methylation, histone deacetylation, and methylation of histone H3 at lysine 9 are the three best-characterized covalent modifications associated with a repressed chromatin state [15]. Complex interactions have been discovered among various components of DNA methylation and histone methylation pathways, proteins involved in the formation of higher-order chromatin structure, chromatin remodeling activities, and RNA interference [13]. Heterochromatin mediates various nuclear processes including centromere function, gene silencing, and nuclear organization. Recent reports have demonstrated that distinct and novel chromatin-associated factors, including DNA, RNA, and histone modifications, are involved in each of these events. These new findings define a novel conserved mechanism of heterochromatin formation that is likely to have an impact on all eukaryotic silencing pathways [17].

The role of epigenetic processes in the control of gene expression has been known for a number of years. Epigenetics is a process by which the operation of genes is changed but not the DNA itself. It occurs on the outside of the gene, mainly through a process called methylation. DNA methylation is a major form of epigenetic modification and plays essential roles in physiology and disease processes. In the human genome, about 80% of cytosines in the 56 million CpG sites are methylated to 5-methylcytosines. The methylation pattern of DNA is highly variable among cell types and developmental stages and influenced by disease processes and genetic factors [18]. DNA methylation at cytosine residues is of particular interest for epigenetic studies as it has been demonstrated to be both a long-lasting and a dynamic regulator of gene expression [18]. Sequencing-based DNA methylation analysis applies the coverage density and flexibility enabled by next-generation sequencing for whole genome analysis of DNA methylation patterns [19]. Epigenetic modifications are heritable alterations that are not due to changes in the sequence of nucleotides, which are organic molecules that serve as monomers, or subunits, of nucleic acids like DNA and RNA [20]. Epigenetic modifications, such as histones that are highly alkaline proteins found in eukaryotic cell nuclei that package and order the DNA into structural units called nucleosomes modifications [21] and DNA methylation, alter the accessibility to DNA and change the chromatin structure a complex of macromolecules found in cells, consisting of DNA, protein, and RNA, thereby regulating patterns of gene expression: the process by which information from a gene is used in the synthesis of a functional gene product [20]. Methylated histones can act as binding sites for certain transcription factors due to their bromodomains and chromodomains. Methylated histones can also prevent the binding of transcription factors by hiding the transcription factor's recognition site, which is usually found on the major groove of DNA. The methyl groups bound to the cytosine residues lie in the major groove of DNA, the same region most transcription factors use to read a DNA sequence. A common epigenetic tag found in DNA is the covalent attachment of a methyl group to the C5 position of the cytosine found in CpG dinucleotide sequences [20]. CpG methylation is an important mechanism of transcriptional silencing (a term used to describe the epigenetic regulation of gene expression). Methylation of CpG islands is shown to reduce gene expression by the formation of tightly condensed heterochromatin that is transcriptionally inactive. CpG sites in a gene are most commonly found in the promoter regions of a gene while also being present in nonpromoter regions. The CpG sites in nonpromoter regions tend to be constitutively methylated, causing transcription machinery to ignore them as possible promoters. The CpG sites near promoter regions are mostly left unmethylated until a cell decides to methylate them and repress transcription. Methylation of CpGs in promoter regions results in the transcriptional silencing of a gene. Environmental factors including physical exercise have been shown to have a beneficial influence on epigenetic modifications.

Chronic aerobic exercise training decreases numerous main cardiovascular risk factors [22]. Following long-lasting aerobic training active athletes increase their life expectancy by an additional 5–6 years compared to healthy inactive young adults [23]. Recently, it has been suggested that following long-lasting aerobic training, gene expression is modified (epigenetic), which may be accounted for the reduced mortality [24]. Thus, being engaged in long-lasting athletic activities is the most promising way to increase longevity [25, 26]. In addition, epigenetics following chronic aerobic exercise training include various genetic factors related to the progressions at the DNA level, for a variety of physiological responses beginning with inflammation up to tissue repair [27].

Epigenetics in Metabolism

VO_2max is one of the main variables in the field of exercise physiology and is frequently used to indicate the fitness level of an individual [28]. At maximal exertion, $(a–v)O_2$ can reach values of 140–180 mLO_2 L^{-1} [29]. Age causes structural and functional changes in cardiovascular $(a–v)O_2$ ability and skeletal muscles in humans, bringing about a reduction in aerobic power and physical performance [30]. First, the primary aging process, which has a genetic component, occurs in the absence of disease and independent of lifestyle [31].

Aerobic exercise induces several metabolic adaptations to meet increased energy requirements. Promoter DNA methylation, histone post-translational modifications, noncoding RNAs like lncRNA or microRNA expression are involved in the gene expression changes implicated in metabolic adaptation after exercise. Although there is no uniform definition for epigenetics, it has the potential to explain various biological phenomena. It is a series of chemical tags that modify DNA and its associated structures constitute the epigenome and include any genetic expression modifier independent of the DNA sequence of a gene. However, it does not involve changes in the nucleotide sequences [32]. Epigenetic modifications and many epigenetic enzymes are potentially dependent on changes in the levels of metabolites, such as oxygen, tricarboxylic acid cycle intermediates, 2-oxoglutarate, 2-hydroxyglutarate, and β-hydroxybutyrate, and are therefore susceptible to the changes induced by exercise in a tissue-dependent manner. Most of these changes are regulated by important epigenetic modifiers that control DNA methylation (DNA methyl transferases and ten-eleven-translocation proteins) and post-translational modifications in histone tails controlled by histone acetyltransferases, histone deacetylases, and histone demethylases [33]. In addition to restructuring the muscular and skeletal system to better handle mechanical stress during effort, aerobic exercise also affects gene expression with respect to metabolism. The effects are widespread and can affect anything from muscle growth to aerobic stamina to diabetes and other metabolic disorders [16]. In general, even a small amount of exercise can induce hypomethylation of the whole genome within muscle cells. This means that many regulatory genes can be turned on for pathways like muscle repair and growth. The intensity of the exercise directly correlates with the amount of promoter demethylation, so more strenuous exercise activates more genes [16].

Class IIa Histone deacetyltransferases (HDACs) are highly expressed within human skeletal muscles. HDACs and acetyltransferases control the epigenetic regulation of gene expression through modification of histone marks. Chromatin remodeling by the post-translational modification of histones is thought to be a central mechanism for the epigenetic regulation of gene expression [34]. Exercise helps to reduce HDACs' activity, especially at promoters, which affects gene expression. In mice, this regulation of HDACs has been shown to increase the amount of type I fibers in muscle. The amount of type I fibers is positively correlated with the maximal aerobic capacity and fat metabolism. Type I fibers are slow twitch, endurance fibers that aging (57–76 years of age range) do not impair the muscle angiogenic response to exercise training [35].

Angiotensin-Converting Enzyme

Measuring genetic markers centered on the renin-angiotensin system (ACE) and identification of candidate genes revealed that different molecular mechanisms are responsible for the different types of left ventricular mass increases seen in athletes and in cardiac patients [36]. Kasikcioglu and colleagues [37] suggested that exercise-induced left ventricular growth in young males is strongly associated with ACE I/D polymorphism. In addition, the absence of the D allele from the 287-base pair marker in the angiotensin-converting enzyme gene resulted in a higher than normal ACE level. Following a 10-week physical training period in male military recruits, echocardiographic measurements of left ventricular mass revealed that persons with D allele in the 287-base pair marker in the ACE had a mixed cardiac response to military basic training, with both eccentric and concentric left ventricular hypertrophy, while individuals with ACE of 11 genotype did not have a cardiac morphology response to training. These two last studies [36] may have implications for which the exercise mode will best serve athletes or cardiac patients in terms of modulating the cardiovascular function during exercise training.

Despite exercise's general benefit, individual responses to exercise vary [38]. The basis for this is unclear, but there appears to be a strong genetic component [39]. An insertion (I)/deletion (D) polymorphism in intron 16 of the ACE gene has been identified as a potential marker for the differential response to exercise. In the field of hypertension, the D allele is associated with significantly higher ACE. In response to exercise, the D allele has been associated with increased muscle strength and power while I allele has been associated with better muscular endurance although data are not entirely consistent. Since both muscle strength and endurance are determinants of physical function in older adults, maintenance of physical function could be related to the ACE I/D genotype. However, studies involving younger individuals suggest that a genotype effect would be seen primarily in response to high physical activity levels [40].

Epigenetics in Aging

Aging is a complex multifactorial process that not only involves the natural processes of aging but also the increased risk for different diseases—coronary heart disease, diabetes, and cancer [41]. Studies have demonstrated that genetics can play a major role in aging. When researchers adjust the genes in certain mice, yeast cells, and other organisms, they can almost double the lifespan of these creatures. Successful aging is a function of both genetic and environmental factors [42]. The primary aging process that is genetically associated, occurs both, independent of lifestyle and in the absence of disease [43]. Aging-related changes occur mainly in the cardiopulmonary and skeletal muscles, bringing about a reduction in physical performance [44]. Maximal work capacity is decreased with aging regardless of lifestyle because of genetic factors. Consequently, the VO_2max decreases. Such consequences contribute to the geriatric syndrome of frailty, thereby severely limiting the function, quality of life, and longevity [45]. Apart from genetic endowment, an individual must also interact with environmental factors associated with longevity. One of these factors includes maintaining a high level of physical activity [46]. Chronic endurance training will attenuate the decline in VO_2max associated with age [47].

Data further suggest that the epigenetic pattern may change during aging, affecting key genes [48] and targeting age-related DNA methylation. While DNA methylation is increased in the skeletal muscle of elderly compared with young, the opposite pattern is found for gene expression. Additionally, the transcript level of DNA methylation in the skeletal muscle is associated with increased VO_2max in vivo [49]. These data demonstrate how age can influence DNA methylation, gene expression, and subsequently in vivo metabolism. However, chronic aerobic exercise training significantly impacts positively DNA methylation and left ventricular growth in young males associated with ACE I/D polymorphism. The role of genetics in determining exercise performance is best viewed at different levels of influence. The very nature of genetics is based on the expression of genetic information (i.e. from genes) that directs the cellular development of an organism. In humans, it is possible to illustrate cellular genetic regulation in the type and concentration of certain enzymes in skeletal muscle, which in turn influences the metabolic capacity of the muscle, influences adaptability to training, and determines the potential to excel during exercise. However, this review will focus on the influence of exercise on genetics and thus aging.

Epigenetics and Longevity

The aging process results in a decline of numerous systematic functions including tissue regeneration ability and compensatory capacity. Consequently, a homeostasis imbalance is reached, without the capability to balance it over [50]. Therefore, the body organs turn out to be more vulnerable to stress, diseases, and injuries, leading to age-related pathologies, which at times result in death (Fig. 12.2).

The following insulin/IGF-I, PI3K, TOR, MAPK, AMPK, PKC, NF-κB, TGF-β, Notch, and WNT are signaling pathways that facilitate regulation of the aging process. These signaling waterfalls regulate body homeostasis, growth, and replica [51]. Ecological epigenetics defines how natural factors affect cellular epigenetics and, thus, human performance. The alterable abilities of epigenetic mechanisms offer a hopeful path for healing and delay of age-associated reduction and sickness [52]. The genetic variations and somatic mutations cause longevity; nonetheless, other factors besides genetics play too a key role [53].

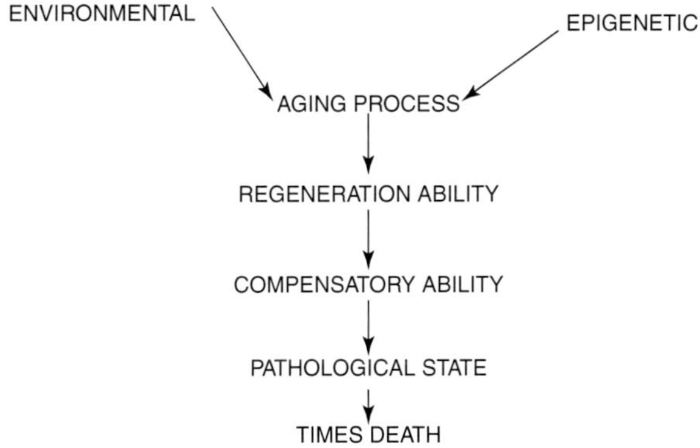

Fig. 12.2 Environmental and epigenetic effects on the aging process and their consequences

The Klotho gene affects longevity, while disorders in s-Klotho values in the bloodstream are related to numerous aging phenotypes [54]. Klotho is a transmembrane protein that also affects and controls over the connection of humans to insulin and seems to be associated with an anti-aging function [55]. The α-Klotho gene is highly preserved in humans and circulates in blood as s-Klotho, which acts by way of a humoral factor [56]. In humans, serum levels of s-Klotho decrease with the aging process [57] that may be observed in patients with several aging-related diseases such as coronary artery disease, cancer, hypertension, and kidney disease [58]. The klotho gene encodes a single-pass transmembrane protein that binds to multiple fibroblast growth factor receptors and functions as a co-receptor for FGF23, a bone-derived hormone that suppresses phosphate reabsorption and vitamin D biosynthesis in the kidney [59]. In addition, the extracellular domain of s-Klotho protein is shed and secreted [60]. It has previously been demonstrated that Klotho protein increases the ability of cells to remove damaging ROS by increasing the cell functions that detoxify harmful ROS [61]; however, the precise mechanism of klotho's action is not fully understood, but it changes cellular calcium homeostasis, by both increasing the expression and activity of transient receptor potential cation channels, vanilloid subfamily, member 5 (TRPV5), which plays a key role in active Ca++ reabsorption in the kidneys and decreasing that of transient receptor potential canonical 6 (TRPC6), a subtype of the calcium-permeable channel [62]. Additionally, Klotho increases membrane expression of the inward rectifier ROMK1 channels i.e. renal outer medullary K+ [62]. Klotho-deficient mice showed increased production of vitamin D, and altered mineral-ion homeostasis is suggested to be a cause of premature aging-like phenotypes because the lowering of vitamin D activity by dietary restriction reverses the premature aging-like phenotypes and prolongs survival in these mutants. These results suggest that aging-like phenotypes were due to Klotho-associated vitamin D metabolic abnormalities [63]. The Klotho protein deficiency is known to participate in premature aging. Klotho protein contributes in numerous pathways that control aging, for example, phosphate homeostasis, insulin signaling, and Wnt signaling regulation [64]. As an aging suppressor, Klotho is an important molecule in aging processes and its overexpression results in longevity. The Klotho gene encodes a transmembrane protein that after cleavage is also found as a secreted protein. However, previously it has been shown that the circulating s-Klotho levels are significantly higher than the membrane-bound form in humans and mice and the latter is significantly higher than the former. Importantly, its overexpression suppresses insulin/IGF-I signaling and thus extends the lifespan. In addition, Klotho participates in the regulation of several other intracellular signaling pathways, including regulation of FGF23 signaling, cAMP, PKC, transforming growth factor-β (TGF-β), p53/p21, and Wnt signaling [65].

The study by Crasto et al. [66], a population-based longitudinal study, revealed that low plasma levels of Klotho are associated with decreased activities of daily living in older individuals. In older community-dwelling adults, plasma klotho is an independent predictor of all-cause mortality [67]. Low plasma Klotho levels have also been associated with poor muscle strength in older, community-dwelling adults [68].

Telomere Shortening

Another component of aging is the gradual shortening of telomeres located at the end of chromosomes. Telomere length regulates gene expression long before telomeres become short enough to produce a DNA damage response, and thus, the main role of telomeres is a tumor suppressor mechanism [69]. Rapid telomere shortening may indicate a cellular hyperactivation, hyperproliferation, and/or hypersecretary phenotype often associated with cellular senescence, stem cell exhaustion, and diseases of aging [69]. Endogenous human genes can be regulated by the length of telomeres prior to the onset of DNA damage signals and suggest the possibility that cell turnover/telomere shortening may provide a mechanism for adjusting cellular physiology. The upregulation of ISG15 with telomere shortening may contribute to chronic inflammatory states associated with human aging [70]. Mammalian telomeres are formed by tandem repeats of the TTAGGG sequence, which are progressively lost with each round of cell division. Aging and age-related diseases are associated with the significant shortening of these sequences. The shrinking of telomeres occurs in somatic cells where telomerase, the enzyme in control of telomere lengthening, is not expressed [71]. However, it has been seen that telomeres can transcribe noncoding RNA or functional RNAs that do not get translated into protein. These noncoding RNAs can be positively impacted by physical exercise. A study found that mice exposed to short-term running phases had increased noncoding RNA transcription at telomeres as compared to sedentary controls [72]. This increase in noncoding RNA transcription aided telomere stability, making the exercise group's telomeres less likely to be as affected by aging over time.

Exercise Epigenetic

Epigenetic modifications are heritable alterations that are not due to changes in the sequence of nucleotides [20]. Nucleotides are molecules consisting of a nucleoside and a phosphate group. They are the basic building blocks of DNA and RNA and are the study of epigenetic modifications resulting from exercise training to the genome of cells. It is well established that engaging in regular exercise prevents many chronic diseases such as the reduction of the risk for type 2 diabetes and cardiovascular disease, modulates the function of many physiological systems such as the musculoskeletal and nervous systems, decreases the mortality risk, and increases the longevity [73]. It is well established that mechanical load and contracting metabolism effects modulate the function of the nervous system, skeletal muscles, and the cardiopulmonary system. Epigenetics influences numerous changes in the above-mentioned systems, by altering DNA. Alterations include histone modifications, DNA methylations, expression of microRNAs, and changes of the chromatin structure. These changes may play a key role in preventing neurological disorders, metabolic syndrome, and cardiovascular diseases [74]. However, researchers have searched for molecular mechanisms explaining the health benefits of regular exercise for decades, and it is well established that exercise alters the gene expression pattern in multiple tissues [75]. Recent advances highlight an essential, intricate web of interactions among these mechanisms, generating a self-reinforcing, self-perpetuating cycle of epigenetic events that lead to long-term transcriptional repression.

Exercise positively influences epigenetic mechanisms and improves health. However, several issues remain unclear concerning the links between exercise and epigenetics [76].

Exercise can give some estimation of the different body system range and abilities. There is growing concern about the negative influence of excessive and persistent exercise on health. How an individual physically adapts to the prevailing environmental conditions might influence epigenetic mechanisms and modulate gene expression. The modulating effect of aerobic exercise and resistance exercise on gene expression has been known for some time now and has provided us with an understanding of the biological responses to exercise. Emerging research data suggest that epigenetic modifications are extremely important for both development and disease in humans. In data suggested, epigenetic modifications: DNA methylation and histone acetylation) and microRNAs (miRNAs) are responsive to acute aerobic exercise and resistance exercise in brain, blood, skeletal muscle and cardiac muscle, adipose tissue and even buccal cells [73]. Aerobic exercise, especially long-term repetitive above 60% of VO_2max, positively affects health and reduces the aging process, through induced stress and epigenetic mechanisms, in turn modulating the link between the environment, human lifestyle factors, and genes. Six months of aerobic exercise alters whole-genome DNA methylation in skeletal muscle and adipose tissue and directly influences lipogenesis. Some miRNAs are related to VO_2max and trainability that are differentially expressed amongst individuals with high and low VO_2max. Remarkably, miRNA expression profiles discriminate between low and high responders to resistance exercise (miR-378, 26a, 29a and 451) and correlate with gains in lean body mass [73].

In general, alterations in the skeletal muscle phenotype following progressive resistance exercise training include gains in strength, increases in lean body mass, and muscle fiber hypertrophy, thus causing gains in skeletal muscle mass and enhanced insulin sensitivity. MicroRNAs (miRNA), small noncoding RNA molecules, may regulate protein synthesis, while resistance exercise training is an efficient strategy for stimulating muscle protein synthesis in vivo. In addition, resistance exercise training increases muscle mass, with a very wide range of effectiveness in humans. It has been suggested that resistance exercise training induces hypertrophy in human skeletal muscles associated with selected changes in miRNA abundance. Also, miRNAs may play a role in the phenotypic change and pronounced intergroup variation in the resistance exercise training response [77].

In methylation, groups of atoms, called methyl groups, attach to the outside of a gene and make the gene more or less able to receive and respond to biochemical signals. Far less has been known about the effect of exercise on methylation. A few studies have found that a single bout of exercise leads to immediate changes in the methylation patterns of certain genes in muscle cells. Nonetheless, whether longer-term regular exercise training affects methylation or how it does has been unclear. The basic mechanism by which exercise activates genes (epigenetic) involves a stimulus signal to the DNA, then transcription via messenger RNA, and finally translation into proteins [78]. Additionally, the transcript level of DNA methylation in skeletal muscle is associated with increased VO_2max in vivo [49].

Interestingly, following exercise, many of the methylation changes are on portions of the genome known as enhancers that can amplify the expression of proteins by genes. Thus, gene expression is noticeably increased or changed in thousands of muscle cells. These data demonstrate how age can influence DNA methylation, gene expression, and subsequently in vivo metabolism. Chronic aerobic exercise training significantly impacts positively DNA methylation and left ventricular growth in young males associated with ACE I/D polymorphism. The role of genetics in determining exercise performance is best viewed at different levels of influence. The very nature of genetics is based on the expression of genetic information (i.e. from genes) that directs the cellular development of an organism. In humans, it is possible to illustrate cellular genetic regulation in the mode of exercise and concentration of certain enzymes in skeletal muscle, which in turn influences the metabolic capacity of the muscle, influences adaptability to training, and determines the potential to excel during exercise [74].

Different types of exercise and physical activities induce mechanical and hormonal activities and improve metabolic characteristics, leading to adaptions of almost every tissue and cell type. Both acute and chronic exercises are epigenetic modifiers, changing the functional genome in gametes. Exercise that induced epigenetic modifications may counter disease changes of the epigenome, for example, in metabolic and cardiovascular diseases. It is hypothesized that positive preventive and rehabilitative effects of exercise are affected by epigenetic modifications, which reduce systemic inflammation. Also, response to resistance training shows different epigenetic effects compared to aerobic exercise. Additionally, different types and intensities of exercise exhibit specific epigenetic responses, measured in muscle tissue and circulation system functions. Consequently, exercise indicators may be reflected from the epigenetic changes following exercise [79].

Aging and Exercise Epigenetics

Epigenetic factors including DNA methylations, histone modifications, and non-coding RNAs like lncRNA or microRNAs can help to explain how cells with identical DNA can differentiate into different cell types with different phenotypes. Some interfaces of aging and energetics were identified; these included genes that accelerate aging or in contrast promote longevity in model organisms, DNA damage responses and telomeres, molecular mechanisms of life span extension by calorie restriction and pharmacological interventions into aging [80].

Epigenetic mechanisms affected by physical exercise have also been seen to be involved in age-related processes. A major component of aging is a significant loss of DNA methylation over time [81]. Methyl deoxycytidine, which is a methylated cytosine on the 5′ carbon of cytosine, is involved in the process of cell differentiation and maintenance. Cell differentiation involves methylation of different areas within the DNA of a cell, which can alter the transcription of genes. During cell differentiation, DNA methylation is important for establishing the identity and function of a cell because of its role in controlling gene expression. As one ages, the amount of DNA methylation slowly begins to decrease. A recent study [82] looking at genome epigenetic of children and older individuals suggests that the older individuals had significantly decreased overall DNA methylation and histone. These mechanisms include permanent structural changes to the organ caused by suboptimal levels of an important factor during a critical developmental period, changes in gene expression caused by epigenetic modifications (including DNA methylation, histone modification, noncoding RNAs like lncRNA or microRNAs) and permanent changes in cellular ageing. Studies have also looked at methyl deoxycytidine residues from tissues collected from rodents at various ages. These studies found that DNA methylation loss increased significantly as the rodent aged [81]. A major component of aging is a significant loss of DNA methylation over time [83], and thus, the aging rate is related to a reduction in DNA methylation activity [84].

This loss of DNA methylation appears to be slowed by physical exercise. Further studies have looked at the effects of aerobic exercise on DNA methylation and aging in humans, revealing that genome with DNA methylation in adult individuals who obtained thirty or more minutes of exercise a day had significantly more DNA methylation as compared to sedentary individuals [85]. Such exercise has been shown to induce positive changes in DNA methylation within adipose tissue and regulate metabolism in both healthy and diseased individuals [70]. Increased DNA methylation of genes Hdac4 and Ncor2 has also shown to increase lipogenesis following exercise [70]. Consequently, aerobic exercise affects the aging rate through slowing the pace of the DNA methylation loss over time.

All these epigenetic alterations may have clinical relevance, thus playing an important role in the prevention and confrontation of neurophysiological disorders, metabolic syndrome, cardiovascular diseases, and cancer. Acute exercise is associated with DNA hypomethylation of the entire genome in skeletal muscle cells of sedentary individuals and high-intensity exercise tends to cause a reduction in promoter methylation of certain genes [86].

In human skeletal muscle, the expression of many genes tends to increase or decrease between the third and seventh decades. The changes are modest when averaged over all of the cells in the tissue [87]. Exercise differentially influences sarcoplasmic reticulum calcium-ATPase isoform expression in type I and type II fibers. Additionally, adenosine monophosphate-activated protein kinase $\alpha 2$ influences the regulation of sarcoplasmic reticulum calcium-ATPase2a mRNA in type I skeletal muscle fibers following exercise training [88]. It is estimated that an average of 10 transcription units, the vast majority of which make long noncoding RNAs (lncRNAs), may overlap each traditional coding gene [89]. Noncoding RNAs (ncRNAs), including microRNAs (miRNAs) and long noncoding RNAs (lncRNAs), have emerged as a class of previously unrecognized transcripts. They are emerging as integral components of the gene regulatory networks in a broad range of biological processes, and dysregulation of their expression has been implicated in many human diseases [90]. Following aerobic exercise, skeletal muscle's myosin heavy chain gene expression (contractile ability) was altered. Muscle fiber type I and IIa myosin heavy chain mRNA was increased by 11%, while type IIb was decreased by 38%. In addition, insights into myosin heavy chain regulatory mechanisms previously suggested that in rats, lncRNA are more conserved than in humans [91]. It has been discovered that MicroRNAs (miRNAs) play important roles in immune function and often act to attenuate or silence gene translation. miRNAs interfere with mRNA that is present and render it unusable and therefore decrease the product of that mRNA. miRNAs regulate many physiological processes, such as inflammation, angiogenesis, and ischemia. miRNA expression profiling of various human myopathies identified a common set of miRNAs that were upregulated in each of the different muscular disorders [92]. Drummond and colleagues reported that let-7 miRNA expression was elevated in the skeletal muscle of older humans [93]. The significance of this finding is that let7 expression may be a contributing factor to the alteration in satellite cell activity observed with age.

Aerobic exercise training has been established as an important phenotype capable of altering the human skeletal muscle. In healthy subjects, aerobic exercise training can thus locally modulate some signaling ways in muscle regeneration [94]. It reduces the overall number of various miRNAs within the skeletal muscle that produces negative effects and increases myomiR expression with a positive effect on muscle reconstruction [92]. Stimuli that cause the body to enter an anabolic, or constructive, phase, such as resistance training as well as the correct diet, have also shown a reduction of miRNAs. This reduction may actually play a role in the growth of muscle cells [16]. Exercise is also known to positively influence the expression patterns of miRNAs in leukocyte cells. It is associated with rapid changes in the profile of gene expression in circulating neutrophils [95].

References

1. Griffiths AGF, Miller JH, Suzuki DT, et al., editors. Genetics and the organism: introduction. An introduction to genetic analysis. 7th ed. New York: W.H. Freeman; 2000.
2. Alberts B, Johnson A, Lewis J, et al. Chapter 4: DNA, chromosomes and genomes. In: Garland P, editor. Molecular biology of the cell. 6th ed. New York: Garland Science; 2014.
3. Somerville RL. Thymine. Encyclopedia of genetics. ScienceDirect. 2001, pp 1965–1966. https://doi.org/10.1006/rwgn.2001.1292
4. Pearson H. Genetics: what is a gene? Nature. 2006;441:398–401.
5. Prescott L. Microbiology. Dubuque: Wm. C. Brown Publishers; 1993.
6. Gregory SG, Barlow KF, McLay KE, et al. The DNA sequence and biological annotation of human chromosome 1. Nature. 2006;441:315–21.
7. Brivanlou AH, Darnell JE Jr. Signal transduction and the control of gene expression. Science. 2002;295:813–8.

8. Holmqvist E, Unoson C, Reimegard J, et al. A mixed double negative feedback loop between the sRNA MicF and the global regulator Lrp. Mol Microbiol. 84:414–27. https://doi.org/10.1111/j.1365-2958.2012.07994.x.
9. Maher B. ENCODE: the human encyclopedia. Nature. 2012;489:46–8.
10. Pertea M. The human transcriptome: an unfinished story. Genes (Basel). 2012;3:344–60.
11. Tchurikov NA. Molecular mechanisms of epigenetics. Biochemistry (Mosc). 2005;70:406–23.
12. Jenuwein T. The epigenetic magic of histone lysine methylation. FEBS J. 2006;273:3121–35.
13. Stancheva I. Caught in conspiracy: cooperation between DNA methylation and histone H3K9 methylation in the establishment and maintenance of heterochromatin. Biochem Cell Biol. 2005;83:385–95.
14. Voolkel P, Angrand PO. The control of histone lysine methylation in epigenetic regulation. Biochimie. 2007;89:1–20.
15. Fuks F. DNA methylation and histone modifications: teaming up to silence genes. Curr Opin Genet Dev. 2005;15:490–5.
16. Ntanasis-Stathopoulos J, Tzanninis JG, Philippou A, et al. Epigenetic regulation on gene expression induced by physical exercise. J Musculoskelet Neuronal Interact. 2013;13:133–46.
17. Grewal SI, Rice JC. Regulation of heterochromatin by histone methylation and small RNAs. Curr Opin Cell Biol. 2004;16:230–8.
18. Zhang Y, Jeltsch A. The application of next generation sequencing in DNA methylation analysis. Genes. 2010;1:85–101.
19. Masser DR, Stanford DR, Freeman WM. Targeted DNA methylation analysis by next-generation sequencing. J Vis Exp. 2015;24:96. https://doi.org/10.3791/52488.
20. Handy DE, Castro R, Loscalzo J. Genetics primer for the general cardiologist. Circulation. 2011;123:2145–56.
21. Cox M, Nelson DR, Lehninger AL. Lehninger principles of biochemistry. San Francisco: W.H. Freeman; 2005.
22. Reimers CD, Knapp G, Reimers AK. Does physical activity increase life expectancy? A review of the literature. J Aging Res. 2012;2012:243958. https://doi.org/10.1155/2012/243958.
23. Kettunen JA, Kujala UM, Kaprio J, et al. All-cause and disease-specific mortality among male, former elite athletes: an average 50-year follow-up. Br J Sports Med. 2015;49:893–7.
24. Kontro TK, Sarna S, Kaprio J, et al. Mortality and health-related habits in 900 Finnish former elite athletes and their brothers. Br J Sports Med. 2018;52:89–95.
25. Lemez S, Baker J. Do elite athletes live longer? A systematic review of mortality and longevity in elite athletes. Sports Med Open. 2015;1(1):16.
26. Lemez S, Wattie N, Baker J. The end game: Mortality outcomes in North American professional athletes. Scand J Med Sci Sports. 2018;28:1722–30.
27. Radom-Aizik S, Zaldivar F Jr, Leu SY, et al. A brief bout of exercise alters gene expression and distinct gene pathways in peripheral blood mononuclear cells of early- and late-pubertal females. J Appl Physiol (1985). 2009;107:168–75.
28. Basset DR Jr, Howley ET. Limiting factors for maximum oxygen uptake and determinants of endurance performance. Med Sci Sports Exerc. 2000;32:70–84.
29. Beck KC, Randolph LN, Bailey KR, et al. Relationship between cardiac output and oxygen consumption during upright cycle exercise in healthy humans. J Appl Physiol. 2006;101:1474–180.
30. Hayes LD, Grace FM, Sculthorpe N, et al. Does chronic exercise attenuate age-related physiological decline in males? Res Sports Med. 2013;21:343–54.
31. Oeltmann T, Carson R, Shannon JR, et al. Assessment of O-methylated catecholamine levels in plasma and urine for diagnosis of autonomic disorders. Auton Neurosci. 2004;116:1–10.
32. Kanherkar RR, Bhatia-Dey N, Csoka AB. Epigenetics across the human lifespan. Front Cell Dev Biol. 2014;2:49. https://doi.org/10.3389/fcell.2014.00049.
33. Pareja-Galeano H, Sanchis-Gomar F, García-Gimenez JL. Physical exercise and epigenetic modulation: elucidating intricate mechanisms. Sports Med. 2014;44:429–36.
34. Shahbazian MD, Grunstein M. Functions of site-specific histone acetylation and deacetylation. Annu Rev Biochem. 2007;76:75–100.

35. Gavin TP, Kraus RM, Carrithers JA, et al. Aging and the skeletal muscle angiogenic response to exercise in women. J Gerontol A Biol Sci Med Sci. 2015;70(10):1189–97.
36. D'Andrea A, Caso P, Scarafile R, et al. Biventricular myocardial adaptation to different training protocols in competitive master athletes. Intern J Cardiol. 2007;115:342–9.
37. Kasikcioglu E, Kayserilioglu A, Ciloglu F, et al. Angiotensin-converting enzyme gene polymorphism, left ventricular remodeling, and exercise capacity in strength-trained athletes. Heart Vessel. 2004;19:287–93.
38. Perusse L, Rankinen T, Rauramaa R, et al. The human gene map for performance and health-related fitness phenotypes: the 2002 update. Med Sci Sports Exerc. 2003;35:1248–64.
39. An P, Borecki IB, Rankinen T, et al. Evidence of major genes for exercise heart rate and blood pressure at baseline and in response to 20 weeks of endurance training: the HERITAGE Family Study. Int J Sports Med. 2003;24:492–8.
40. Ettinger WH Jr, Burns R, Messier SP, et al. The Fitness Arthritis and Seniors Trial (FAST): a randomized trial comparing aerobic exercise and resistance exercise to a health education program on physical disability in older people with knee osteoarthritis. JAMA. 1997;277:25–31.
41. Gault ML, Willems MET. Aging, functional capacity and eccentric exercise training. Aging Dis. 2013;4:351–63.
42. Sebastiani P, Solovieff N, Puca A, et al. Genetic signatures of exceptional longevity in humans. Science. 2010; https://doi.org/10.1126/science.1190532.
43. Hawkins S, Wiswell R. Rate and mechanism of maximal oxygen consumption decline with aging: implications for exercise training. Sports Med. 2003;33:877–88.
44. Thomas L, Levett K, Boyd A, et al. Changes in regional left atrial function with aging: evaluation by Doppler tissue imaging. Eur J Echocardiogr. 2003;4:92–100.
45. Afilalo J, Karunananthan S, Eisenberg MJ, et al. Role of frailty in patients with cardiovascular disease. Am J Cardiol. 2009;103:1616–21.
46. Baldwin KM, Haddad F. Research in the exercise sciences: where we are and where do we go from here--Part II. Exerc Sport Sci Rev. 2010;38:42–50.
47. Buford TW, MacNeil RG, Clough LG, et al. Active muscle regeneration following eccentric contraction-induced injury is similar between healthy young and older adults. J Appl Physiol (1985). 2014;116:1481–90.
48. Ling C, Groop L. Epigenetics: a molecular link between environmental factors and type 2 diabetes. Diabetes. 2009;58:2718–25.
49. Ronn T, Poulsen P, Hansson O, et al. Age influences DNA methylation and gene expression of COX7A1 in human skeletal muscle. Diabetologia. 2008;51:1159–68.
50. Moskalev AA, Aliper AM, Smit-McBride Z, et al. Genetics and epigenetics of aging and longevity. Cell Cycle. 2014;13:1063–77.
51. Barzilai N, Huffman DM, Muzumdar RH, et al. The critical role of metabolic pathways in aging. Diabetes. 2012;61:1315–22.
52. Sen P, Shah PP, Nativio R, et al. Epigenetic mechanisms regulating longevity and aging. Cell. 2016;166:822–39.
53. Deepashree S, Shivanandappa T, Ramesh SR. Is longevity a heritable trait? evidence for non-genomic influence from an extended longevity phenotype of drosophila melanogaster. Curr Aging Sci. 2018;11:24–32.
54. Kuro-o M. Klotho as a regulator of oxidative stress and senescence. Biol Chem. 2008;389:233–41.
55. Rubinek T, Modan-Moses D. Klotho and the growth hormone/insulin-like growth factor 1 axis: novel insights into complex interactions. Vitam Horm. 2016;101:85–118.
56. Zuo Z, Lei H, Wang X, et al. Aging-related kidney damage is associated with a decrease in klotho expression and an increase in superoxide production. Age (Dordr). 2011;33:261–74.
57. Wang Y, Sun Z. Current understanding of klotho. Ageing Res Rev. 2009;8:43–51.
58. Wang Y, Sun Z. Antiaging gene Klotho regulates endothelin-1 levels and endothelin receptor subtype B expression in kidneys of spontaneously hypertensive rats. J Hypertens. 2014;32:1629–36.

59. Xiao NM, Zhang YM, Zheng Q, et al. Klotho is a serum factor related to human aging. Chin Med J (Engl). 2004;117:742–7.
60. Baluchnejadmojarad T, Eftekhari SM, Jamali-Raeufy N, et al. The anti-aging protein klotho alleviates injury of nigrostriatal dopaminergic pathway in 6-hydroxydopamine rat model of Parkinson's disease: Involvement of PKA/CaMKII/CREB signaling. Exp Gerontol. 2017;10:70–6.
61. Guo Y, Zhuang X, Huang Z, et al. Klotho protects the heart from hyperglycemia-induced injury by inactivating ROS and NF-κB-mediated inflammation both in vitro and in vivo. Biochim Biophys Acta Mol Basis Dis. 1864;2018:238–51.
62. Liu BC, Yang LL, Lu XY, et al. Lovastatin-induced phosphatidylinositol-4-phosphate 5-kinase diffusion from microvilli stimulates ROMK channels. J Am Soc Nephrol. 2015;26:1576–87.
63. Mencke R, Hillebrands JL. NIGRAM consortium. The role of the anti-ageing protein Klotho in vascular physiology and pathophysiology. Ageing Res Rev. 2017;35:124–46.
64. Sopjani M, Rinnerthaler M, Kruja J, Dermaku-Sopjani M. Intracellular signaling of the aging suppressor protein Klotho. Curr Mol Med. 2015;15:27–37.
65. Semba RD, Cappola AR, Sun K, et al. Plasma klotho and mortality risk in older community-dwelling adults. J Gerontol A Biol Sci Med Sci. 2011;66:794–800.
66. Crasto CL, Semba RD, Sun K, et al. Relationship of low-circulating "anti-aging" klotho hormone with disability in activities of daily living among older community-dwelling adults. Rejuvenation Res. 2012;15:295–301.
67. Semba RD, Cappola AR, Sun K, et al. Relationship of low plasma klotho with poor grip strength in older community-dwelling adults: the InCHIANTI study. Eur J Appl Physiol. 2012;112:1215–20.
68. Skulachev MV, Skulachev VP. Programmed aging of mammals: proof of concept and prospects of biochemical approaches for anti-aging therapy. Biochemistry (Mosc). 2017;82:1403–22.
69. Robin JD, Ludlow AT, Batten K, et al. Telomere position effect: regulation of gene expression with progressive telomere shortening over long distances. Genes Dev. 2014;28:2464–76.
70. O'Donovan A, Lin J, Dhabhar FS, et al. Pessimism correlates with leukocyte telomere shortness and elevated interleukin-6 in post-menopausal women. Brain Behav Immun. 2009;23:446–9.
71. Henriques CM, Ferreira MG. Consequences of telomere shortening during lifespan. Curr Opin Cell Biol. 2012;24:804–8.
72. Werner C, Hanhoun M, Widmann T, et al. Effects of physical exercise on myocardial telomere-regulating proteins, survival pathways, and apoptosis. J Am Coll Cardiol. 2008;52:470–82.
73. Denham J, Marques FZ, O'Brien BJ, et al. Exercise: putting action into our epigenome. Sports Med. 2014;44:189–209.
74. McGee SL, Hargreaves M. Epigenetics and exercise. Trends Endocrinol Metab. 2019;30:636–45.
75. Ling C, Ronn T. Epigenetic adaptation to regular exercise in humans. Drug Discov Today. 2014;19:1015–8.
76. Sanchis-Gomar F, Garcia-Gimenez JL, Perez-Quilis C, et al. Physical exercise as an epigenetic modulator: eustress, the "positive stress" as an effector of gene expression. J Strength Cond Res. 2012;26:3469–72.
77. Davidsen PK, Gallagher IJ, Hartman JW, et al. High responders to resistance exercise training demonstrate differential regulation of skeletal muscle microRNA expression. J Appl Physiol (1985). 2011;110:309–17.
78. Macarthur DG, North KN. Genes and human elite athletic performance. Hum Genet. 2005;116:331–9.
79. Zimmer P, Schenk A, Bloch W. Medical epigenetics. Epigenetics in exercise science and sports medicine. ScienceDirect. 2016:515–530. https://doi.org/10.1016/B978-0-12-803239-8.00029-6. Get rights and content
80. Blagosklonny MV, Campisi J, Sinclair DA, et al. Impact papers on aging in 2009. Aging [Albany NY]. 2010;2:111–21.
81. Wilson VL, Smith RA, Ma S, et al. Genomic 5-methyldeoxycytidine decreases with age. J Biol Chem. 1987;262:9948–51.

82. Vickers MH. Early life nutrition, epigenetics and programming of later life disease. Nutrients. 2014;6:2165–78.
83. Heyn H, Li N, Ferreira HJ, et al. Distinct DNA methylomes of newborns and centenarians. Proc Natl Acad Sci U S A. 2012;109:10522–7.
84. Zhang FF, Cardarelli R, Carroll J, et al. Physical activity and global genomic DNA methylation in a cancer-free population. Epigenetics. 2011;6:293–9.
85. Ronn T, Volkov P, Davegardh C, et al. Six months exercise intervention influences the genome-wide DNA methylation pattern in human adipose tissue. PLoS Genet. 2013;9:e1003572. https://doi.org/10.1371/journal.pgen.1003572.
86. Walter CA, Zhou ZQ, Manguino D, et al. Health span and life span in transgenic mice with modulated DNA repair. Ann N Y Acad Sci. 2001;928:132–40.
87. Welle S, Brooks AI, Delehanty JM, et al. Gene expression profile of aging in human muscle. Physiol Genomics. 2003;14:149–59.
88. Morissette MP, Susser SE, Stammers AN, et al. Differential regulation of the fiber type-specific gene expression of the sarcoplasmic reticulum calcium-ATPase isoforms induced by exercise training. J Appl Physiol (1985). 2014;117:544–155.
89. Lee JT. Epigenetic regulation by long noncoding RNAs. Science. 2012;338:1435–9.
90. Rinn JL, Chang HY. Genome regulation by long noncoding RNAs. Annu Rev Biochem. 2012;81:145–66.
91. Weiss K, Antoniou A, Schratt G. Non-coding mechanisms of local mRNA translation in neuronal dendrites. Eur J Cell Biol. 2015. pii: S0171-9335(15)00057-6. https://doi.org/10.1016/j.ejcb.2015.05.011
92. McCarthy JJ. microRNA and skeletal muscle function: novel potential roles in exercise, diseases, and aging. Front Physiol. 2014; https://doi.org/10.3389/fphys.2014.00290.
93. Drummond MJ, McCarthy JJ, Sinha M, et al. Aging and microRNA expression in human skeletal muscle: a microarray and bioinformatics analysis. Physiol Genomics. 2010;43:595–603.
94. Beiter T, Hoene M, Prenzler F, et al. Exercise, skeletal muscle and inflammation: ARE-binding proteins as key regulators in inflammatory and adaptive networks. Exerc Immunol Rev. 2015;21:42–57.
95. Radom-Aizik S, Zaldivar F Jr, Oliver S, et al. Evidence for microRNA involvement in exercise-associated neutrophil gene expression changes. J Appl Physiol. 2010;109:252–161.

Chapter 13
Exercise Equipment

Introduction

An important index of exercise tolerance is the amount of oxygen used up by the working muscles at peak exercise [1]. When it is not possible to measure VO_2, it can be measured by means of an ergometer; in Greek ergon = work and metry = measure [2]. Equations related to the mechanical work rate are helpful in estimating the metabolic equivalent of the physical effort done [3].

These metabolic measurements are appropriate to be used for clinical and general population [4]. The idea behind the utilization of commercial fitness equipment is that it increases fitness, improves cardiovascular health, lowers blood pressure, regulates body weight, and improves the immune system efficiency. The cardiopulmonary exercise test examines the physiological responses to different work intensities. The data from graded exercise testing can be practically used to improve athletic performance and medical screening. The cardiopulmonary exercise test is utilized to measure true VO_{2max} or VO_{2peak} [5].

The ergometer gives good control to the estimated increase in metabolic demand during different exercise modes and intensities [6]. Each metabolic equation includes the resting metabolic cost, the horizontal component, and, if exists, the vertical one. The evaluation of exercise tolerance is important in assessing cardiopulmonary health status or in identifying the potential for endurance performance [7].

The ability for reasonable measurement of VO_2 (aerobic), intensity, and power (anaerobic) exercises requires stationary equipment at the laboratory. Maximal aerobic power and related variables during submaximal work can be determined on a bicycle, an arm cranking ergometer, and a treadmill [6, 8]. These are necessary items to increase the metabolic demand and to generate physiological changes due to exercise. Work and power are measured during exercise tests, the physical performance test, training, and physical rehabilitation [9]. Calibrating the ergometers is essential for having the precise power output performed when demonstrating and

M. S. Saghiv, M. S. Sagiv, *Basic Exercise Physiology*,
https://doi.org/10.1007/978-3-030-48806-2_13

for investigation purposes. Work upon an object is the result of a force acting on a body that causes movement of the body. When a force acts upon a body to cause motion, it is said that work was done upon the body. Nevertheless, both force and movement are vector quantities, work has no direction due to the nature of a dot product in vector mathematics. This description is steady with the correct meaning since a constant force participates purely in the product of force and distance.

The elementary calculation of work is as follows:

$$W = F \cdot d$$

where W = work, F = force, and d = distance the body moved.

Mathematically, work can be expressed by the following equation

$$W = F \cdot d \cdot \cos \text{theta}$$

where F is the force, d is the displacement, and the angle (theta) is defined as the angle between the force and the movement vector. When defining the position in the exertion equation, it is significant to understand that the angle is the position between the force and the movement route. The equation for work lists three variables—each variable is associated with one of the three keywords mentioned in the definition of work (force, displacement, and cause). The angle theta in the equation is associated with the amount of force that causes a displacement. A force is a push or pull upon a body resulting from the body's interaction with another body. Whenever there is an interaction between two bodies, there is a force upon each of the bodies. Forces only exist as a result of an interaction.

Every time a new sum is presented in physics, the metric units are linked with that amount. In work or energy, the metric unit is the Joule (abbreviated J). One Joule is equivalent to one Newton of force causing an object to move 1 m.

$$1\,\text{J} = 1\,\text{N} - \text{m}$$

or

$$1\,\text{J} = 1\,\text{N} - \text{m}$$

Measuring functional capacity, energy expenditure, and power output has also applications in different areas of exercise [10]. Typically, a graded exercise test performed on a bicycle or a treadmill is used to determine VO_2 which is an important parameter of aerobic function [5] and energy expenditure. VO_{2max} is an outstanding predictor of chronic disease morbidity and mortality risk [11]. Advanced exercise physiology utilizes regularly these ergometers to establish individuals' fitness and health status. As an example, for coaches, the knowledge of the energy cost needed for any physical activity such as running, walking, biking, etc. will help them to plan training sessions for physicians to complete clinical diagnosis and/or rehabilitation. Thus, understanding processes with energy cost and the ability to measure such a cost is vital for training programs in health and disease [12].

Muscle Mass Influence on VO_2

At a given percent of maximal workload, active muscle mass determines VO_2; at the same relative workload, the larger the active muscle mass involved, the higher will be VO_2 [13]. It will be lower on an arm cranking ergometer compared to a bicycle ergometer and values for the bicycle ergometer will be lower than that attained on the treadmill for both the trained and untrained individuals. When comparing trained individuals with untrained individuals during submaximal effort at a given percentage of their VO_{2max}, HR and pulmonary VE will be similar on arm cranking, bicycle, and treadmill. However, in trained subjects, HR will be lower and pulmonary VE will be more efficient compared to sedentary subjects [14]. During maximal exercise, sedentary and trained subjects reached significantly lower values on the arm-cranking ergometer than on the bicycle ergometer for VO_{2max}, while values for the bicycle ergometer will be lower than that attained on the treadmill for both the trained and untrained subjects. However, values of VO_{2max} will be significantly higher in the trained subjects. The lower VO_{2max} values and the arm-cranking ergometer, cycling ergometer, and treadmill in both groups are due to the amount of muscle mass involved during exercise. Larger muscle mass means higher VO_{2max}. Also, the differences between the results obtained in untrained subjects and in trained ones support the concept of specific training. In general, arm-cranking exercise elicits a VO_{2max} corresponding to approximately 70% of the value reached during leg exercise [15, 16]. The scale of this alteration seems to be reduced in elderly and sedentary subjects [17]. However, in arm-trained subjects such as rowers, cross-country skiers, and swimmers, the arms' VO_{2max} approaches or surpasses the leg value. During exercise with large muscle mass, VO_{2max} is mainly limited by oxygen delivery, which in turn depends on blood flow and the oxygen content of the arterial blood [18]. However, during exercise with small muscle mass, VO_{2max} is not limited by Q since when reaching fatigue stage, Q has not reached maximal values [19].

Stair climbing can be a vigorous activity and can lead to cardiopulmonary physiological benefits [20]. One approach that has been used to promote routine physical activity is encouraging stair use in multilevel environments. Stair promotion has been attempted in various public settings, including subway stations and shopping malls.

Stair Climbing

Stair climbing is one mode of daily living activity that can improve work capacity. Stair-Climb Test is submaximal in its intensity. Stair-climbing test is considered as a tool that can be used for cardiopulmonary exercise testing, without leading the individual to maximal exertion [21]. Symptom-limited stair climbing test offers a simple and inexpensive means to predict VO_{2max} and can be applied for all ages. To date, the stair-climbing test has not been properly standardized [22]. The stair-climbing test physiological parameters, correlated with VO_{2max}, demonstrate that the results were comparable in predicting postoperative complications although there was still some difficulty in determining which stair-climbing test [23].

When it is not possible to measure VO_2 during exercise, it can be calculated by means of an ergometer. Equations related to the mechanical work rate are helpful for estimating the metabolic equivalent of the physical effort done [24]. These metabolic measurements are appropriate for clinical and general population use. The ergometer gives good control to the estimated increase in metabolic demand during different exercise modes and intensities. Laboratory ergometry equipment such as stairs, arm cranking, bicycle, and treadmill are essential items to increase metabolic demand and to measure changes of work done and power output, during exercise stress tests, physical performance test, training, and physical rehabilitation. Maximal aerobic power and related variables during submaximal work can be determined on the bicycle, arm cranking ergometer, and treadmill [25, 26].

Stepwise Test

Aerobic fitness can be evaluated using a special test that measures the VO_2 of a person running on a treadmill and bicycle for a prescribed distance. However, it would be more economical to find a formula that uses simpler measurements that evaluate fitness and predict VO_2. Utilization of stair climbing is inexpensive, obtainable, and easy to operate.

The stepwise test allows predicting work, power, and energy expenditure. To achieve cardiopulmonary benefits, stair climbing can be used, although stair climbing provides a minimal intensity for cardiovascular workout. In addition, the estimation that ascending of 15 cm during stair-climbing costs 11% of kcal, while descending costs almost half of it (5–6 kcal).

Comparisons between stair test, bicycle, and treadmill for power outputs can be calculated from the measured distance, time, and the individual's weight. Therefore, knowing the stairs height, number of repetitions of up and down, and time of test lasting enables the calculation of power. Staircase correlates reasonably with VO_{2max} that is the gold standard. During the test, the individual performs a series of steps or flights of steps for passing from one level to another (Fig. 13.1).

The advantages of the stepwise test are its affordability and accessibility to all kind of populations, as well as the fact that it can be carried out in any location that has a staircase of the minimum height, low cost, and the ability to test a large group of subjects at the same time. However, according to the literature, in most stair-climbing test studies, the height of the staircase is not suitably determined,

Fig. 13.1 Flights of steps for passing from one level to another

the stair-climbing time is not standardized, and the variable power is not properly considered. Consequently, there is no clearly defined stair-climbing test or can be drawn.

Stair-Treadmill Ergometer

Although a stair-treadmill ergometer sounds like a high-tech piece of exercise equipment, it is actually another name for a stair climber or step-mill. A stair-treadmill is the combination of a standard stepper and a treadmill. A regular stepper consists of two hydraulic pedals that push up and down as if climbing stairs. A stair-treadmill combines the stepping action of a stepper and the tread belt of a treadmill. The tread belt on a step-mill covers a set of stairs that rotate through the machine. The individual walks up the stairs as they travel downward. It is similar to walking up an escalator that is traveling down.

Stair-treadmill ergometers provide an intense cardiovascular workout. The individual can adjust the speed at which the steps rotate to increase or decrease the intensity of the individual's workout. If the individual usually uses a treadmill, elliptical or stationary bike for his cardiovascular exercise, a stair-treadmill can offer him a more challenging cardiovascular session and an intense lower-body workout.

A high-tech ergometer might measure the individual's BP and VO_2, but a basic ergometer is a piece of fitness equipment that provides basic workout data, such as the individual's exercise time, distance traveled, calories expenditure, and speed. A stair-treadmill ergometer is with a console that displays information about the individual's workout. Some machines also include an HR monitor and a fitness assessment program.

Calculating Stepwise Oxygen Uptake

Work done (or VO_2) on a stepwise (Fig. 13.2) can be calculated by multiplying the individual's weight by the distance achieved during the test. As an example, such calculations for a man weighing 70 kg, working on a stepwise with 0.5 m height, steps up and down 30 times per minute and lasting work for 10 min.

Distance completed from stairs up and down is calculated as follows:

$$D = 30t \cdot 0.5\,\mathrm{m\,h} \cdot 10\,\mathrm{min} = 150\,\mathrm{m}.$$

where D = distance; t = up and down repetitions; h = height; min = working time in minutes; and m = meter.

The calculated work done on stairs is as follows:

$$\mathrm{Work} = 70\,\mathrm{kg} \cdot 150\,\mathrm{m} = 10{,}500\,\mathrm{kgm}.$$

where Person's weight kg = kilogram; m = distance; and kgm = kilogram-meter.

Calculating power on stair:

$$\mathrm{Power} = 10{,}500\,\mathrm{kgm} / 10\,\mathrm{min} = 1050\,\mathrm{kgm\,min}^{-1}$$

where kgm = kilogram-meter, min = minute, and kgm min^{-1} = kilogram-meter per minute;

$$\mathrm{SI\,units} = 1050\,\mathrm{kgm\,min}^{-1} / 6.12\,\mathrm{kgm} = 171.6\,\mathrm{W}$$

Fig. 13.2 Stepwise test; up and down

where kgm min^{-1} = kilogram-meter per minute, kgm = kilogram-meter, 6.12 kgm = 1 W, and W = watts;
or

$$\text{SI units} = 1050\,\text{kgm}\,\text{min}^{-1} / 0.101\,\text{kgm} = 10{,}396.04\,\text{J}$$

where kgm min^{-1} = kilogram-meter per minute, kgm = kilogram-meter, and J = Joules.

Calculating metabolic cost of stepping up and down

Calculation of oxygen uptake is from four components: horizontal, slopped, climbing up, and climbing down:

$$VO_2 = \left[\left(\#\,\text{of steps}\,\text{min} \cdot 1.8\,\text{height}\,\text{m} \cdot 1.33\right) + \left(\#\,\text{of steps}\,\text{min}/10\right)\right] + 3.5$$

where 1.8 = oxygen cost (mLO_2 kg^{-1} min^{-1}) of moving vertically up during stepping, 1.33 = oxygen cost (mLO_2 kg^{-1} min^{-1}) of the vertically down component, steps #/10 = is a correction factor for positive (up) component of stepping (10 is a constant #), and 3.5 = VO_2 at rest (mLO_2 kg^{-1} min^{-1}).

In essence, VO_2 is the best indicator of exercise intensity because it is tied closely to energy expenditure.

Example

What will be the VO_2 cost of a person weighing 70 kg that goes up and down 40 times per minute at a stair with a height of 0.5 m for 10 min?

Answer: VO_2 = [(40 · 1.8 · 0.5 · 1.33) + (40/10)] + 3.5 = 55.38 mLO_2 kg^{-1} min^{-1}

Question: How many METs are 55.38 mLO_2 kg^{-1} min^{-1}?

Answer: 55.38/3.5 = 15.8 METs

Question: What is the total VO_2 (whole body) for the 10 min?

Answer: 55.38 mLO_2 kg^{-1} min^{-1} · 70 kg · 10 min = 38,766 mLO_2

Question: How many liters of oxygen equal 38,766 mLO_2?

Answer: 38,766/1000 = 38.766 LO_2

Question: How many kcal equal 38.766 LO_2?

Answer: Each LO_2 equals 4.83 kcal, and therefore, 38.766 LO_2 = 187.24 kcal

Stationary Bicycle

A fixed cycling machine (Fig. 13.3) is used in fitness testing to, estimate the exercise intensity (power output) from, the revolution per minute (rpm) and the resistance to pedaling, which can be adjusted to vary the intensity.

Most exercise bikes provide a mechanism for applying resistance to the pedals that increases the intensity of the exercise. Resistance mechanisms include magnets, fans, and friction.

A stationary bicycle is a device with saddle, pedals, and some form of handlebars on the bicycle but used as exercise equipment rather than transportation. An exercise bicycle is usually a special-purpose exercise machine resembling a bicycle

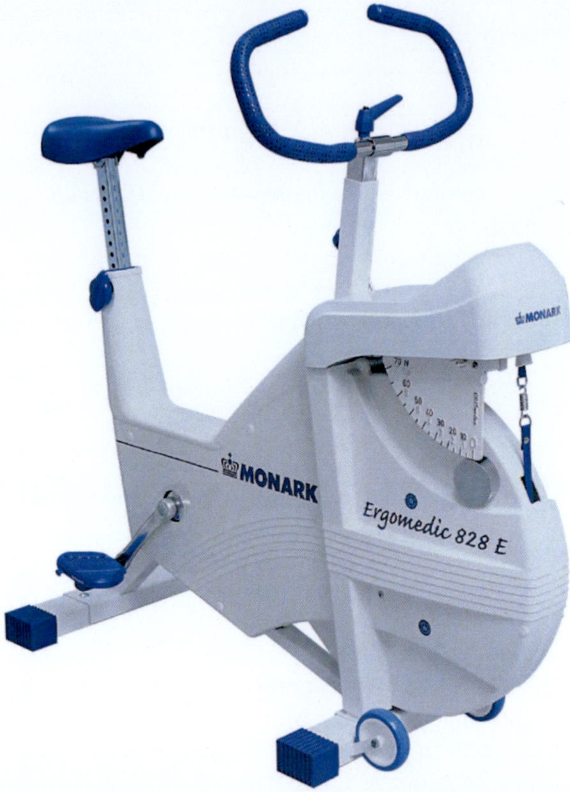

Fig. 13.3 Mechanical bicycle ergometry

Fig. 13.4 A bicycle mounted on a 'Fluid 2' trainer

without true wheels. A stationary bicycle is often used by racing cyclists to warm up before racing or to train on their own machines indoors (Fig. 13.4). A bicycle ergometer is a stationary bicycle with an ergometer to measure the work done by the exerciser. Some models feature handlebars connected to the pedals so that the upper body exercise along the lower body.

Exercise bikes are used, to increase general fitness, and for training on the road to cycle events and weight loss. Biking has long been used for physical therapy because of its low impact, safety, and effective properties on the cardiopulmonary system.

The low-impact movement involved in operating an exercise bike does not put much stress on joints and does not involve sporadic motions.

Calculating Bicycle's Work and Power

During the bicycle test, the individual tested will paddle against resistance. Calculating work done on a bicycle is by multiplying paddling revolution per minute (rpm) with the resistance in kg.

Example
Calculations for a man weighing 70 kg, paddling on a bicycle at a rate of 60 rpm and a resistance of 1.5 kg for 10 min.

Therefore,

$$\text{Distance} = 60\,\text{rpm} \cdot 6\,\text{m pre} - \text{revolution} \cdot 10\,\text{min} = 3600\,\text{m min}^{-1},$$

$$\text{Work} = 1.5\,\text{kg resistance} \cdot 3600\,\text{m min}^{-1}\,\text{distance} = 5400\,\text{kgm},$$

$$\text{Power} = 5400\,\text{kgm work} / 10\,\text{min} = 540\,\text{kgm min}^{-1},$$

$$\text{SI units} = 540\,\text{kgm min}^{-1} / 6.12\,\text{kgm} = 88.24\,\text{W},$$

or

$$\text{SI units} = 540\,\text{kgm min}^{-1} / 0.101\,\text{kgm} = 5346.53\,\text{J}.$$

Calculating the metabolic cost of bicycle ergometry

Calculating oxygen uptake for whole body's energy cost:

$$\text{VO}_2 = \left(\text{rpm} \cdot 6\,\text{m} \cdot 2\,\text{mLO}_2 \cdot \text{resistance}\right) + \left(\text{body weight} \cdot 3.5\,\text{mLO}_2\,\text{kg}^{-1}\,\text{min}^{-1}\right)$$

where rpm = revolutions per minute, 6 m = distance covered for each revolution of the wheel, 2 mLO_2 = oxygen cost for each revolution, resistance = the friction on the wheel (kg), whole body resting VO_2 = body weight (kg) · 3.5 $mLO_2\ kg^{-1}\ min^{-1}$

Note: calculations on the bicycle ergometry are for whole body $mLO_2\ min^{-1}$ and not $mLO_2\ kg^{-1}\ min^{-1}$

Example
What will be the oxygen uptake cost of a person weighing 70 kg that paddles 60 rpm with a resistance of 5 kg and exercise for 10 min?

Note: for a person weighing 70 kg, his whole-body oxygen uptake at rest will be equal to 245 $mLO_2\ min^{-1}$.

Answer: $VO_2 = (60 \cdot 6 \cdot 2 \cdot 5) + (70 \cdot 3.5) = 3845\ mLO_2\ min^{-1}$
Question: How many METs equal 3845 $mLO_2\ min^{-1}$?
Answer: 3845 $mLO_2\ min^{-1}$/245 $mLO_2\ min^{-1}$ = 11 METs
$mLO_2\ min^{-1}$
Question: What was the total VO_2 (whole body) for the 10 min?
Answer: 3845 $mLO_2\ min^{-1} \cdot$ 10 min = 38,450 $mLO_2\ min^{-1}$
Question: How many liters of oxygen equal 38,450 $mLO_2\ min^{-1}$?
Answer: 38450/1000 = 38.45 LO_2
Question: How many kcal equal 38.450 LO_2?
Answer: Each LO_2 equals 4.83 kcal, and therefore, 38.450 LO_2 = 185.71 kcal

Treadmill Ergometer

The best fitness equipment that gives maximal results is the treadmill. It is designed to improve fitness by means of running and walking. Treadmill is built on the principle that by exercising at higher workload intensity, the gain is higher. Working on a treadmill is comfort without stress on the defined target HR. Treadmill is also ideal for increasing cardiopulmonary function by walking or running, particularly when the surrounding ambient temperatures are unbearable. Workloads on a treadmill can be adjusted by training HR by adjusting speed and grade.

Compared to bicycle effort, body balance and weight-bearing factors are present while running on the treadmill, which increases muscle mass involved in the effort and thus could influence VO_2, force–velocity relationship, and power. In addition, the sizes of muscle mass and joints engaged in the movement are more important than body balance and weight-bearing. Cardiopulmonary variables at standardized percentages of VO_{2max} (between 50 and 100%), tested on a treadmill or a cycle ergometer revealed that VO_{2max} and HR were significantly higher on the treadmill than that of the bicycle due to larger muscle mass involved. These results demonstrated that athletes could use the relationship between HR and percentage VO_{2max} measured during a treadmill or a cycle ergometer test to obtain interchangeable reference HRs for planning their exercise intensity and running and cycling training sessions.

Calculating treadmill's work and power: During a treadmill test (Fig. 13.5), the tested individual is asked to walk/run on a motor drive belt. Work done on a treadmill is calculated by multiplying working time with the horizontal and vertical components.

Example
Man weighing 70 kg, running on a treadmill at a speed of 12 km h^{-1} for 10 min.

$$\text{Distance} = (12{,}000 / 60\,\text{min}) \cdot 10\,\text{min} = 200\,\text{m}\,\text{min}^{-1}.$$

$$\text{Work} = 70\,\text{kg weight} \cdot 200\,\text{m}\,\text{min}^{-1}\,\text{distance} = 14{,}000\,\text{kgm}.$$

Fig. 13.5 Motor drive treadmill ergometer with handrails

$$\text{Power} = 140{,}000\,\text{kgm work} / 10\,\text{min} = 1400\,\text{kgm min}^{-1}.$$

$$\text{SI units} = 14{,}000\,\text{kgm min}^{-1} / 6.12\,\text{kgm} = 228.76\,\text{W}.$$

or

$$\text{SI units} = 1400\,\text{kgm min}^{-1} / 0.101\,\text{kgm} = 1386.14\,\text{J}.$$

Treadmill while walking: VO_2 ($mLO_2\ kg^{-1}\ min^{-1}$) of the horizontal component (without elevation) during walking on a treadmill can be calculated as follows:

$$VO_2\ \text{horizontal} = \left[\left(\text{speed in m} \cdot 0.1O_2\right) / \left(60\,\text{min}\right)\right] + 3.5\text{mLO}_2\ \text{kg}^{-1}\ \text{min}^{-1}$$

where 0.1 = oxygen cost per meter of walking, dividing by 60 min is to have the walking speed in meters per minute and the additional 3.5 $mLO_2\ kg^{-1}\ min^{-1}$ is the resting VO_2 required at rest per kg body per minute.

Calculating the metabolic cost of the horizontal component on a treadmill

Example

Question: what will be the energy cost of a man weighing 70 kg that walks on a treadmill without elevation at a speed of 6 km h^{-1} (6000 m h^{-1}) for 35 min?

Answer: VO_2 horizontal = (6000/60 · 0.1) + 3.5 = 13.5 $mLO_2\ kg^{-1}\ min^{-1}$

Question: How many METs equal 13.5 $mLO_2\ kg^{-1}\ min^{-1}$?

Answer: METs = 13.5 $mLO_2\ kg^{-1}\ min^{-1}$/3.5 $mLO_2\ kg^{-1}\ min^{-1}$ = 3.86

Meaning that the subject's energy cost during exercise is 3.86 higher than that at rest.

Question: What was the total energy cost for the 35 min of exercise?
Answer: $VO_2 = 13.5\ mLO_2\ kg^{-1}\ min^{-1} \cdot 70\ kg \cdot 35\ min = 33{,}075\ mLO_2$
Question: How many liters of oxygen equal 33,075 mLO_2?
Answer: 33075/1000 = 33.075 LO_2
Question: How many Kcal are in 33.075 LO_2?
Answer: If 1 L of oxygen equals 5 Kcal, then 33.075 LO_2 equals 165,375 Kcal:

Calculating the energy cost of walking on a sloped treadmill: During walking on a sloped treadmill, VO_2 will be higher compared to that without elevation. VO_2 will be the sum of the vertical and horizontal components. The horizontal component is calculated as follows:

$$VO_2\ \text{horizontal} = \left[\left(\text{speed in m} \cdot 0.1 O_2\right)/60\,\text{min}\right)] + 3.5\,mLO_2\ kg^{-1}\ min^{-1}$$

The vertical component is calculated as follows:

$$VO_2\ \text{vertical}\, mLO_2\ kg^{-1}\ min^{-1} = \left(\text{speed in m} \cdot 1.8\, O_2 \cdot \text{grade\%}\right) / \left(60\,\text{min} \cdot 100\%\right)$$

where speed – m h^{-1}; 1.8 = oxygen cost per meter of climbing uphill; grade = % of inclination; dividing by 60 min is to have the walking speed in meters per minute, and 100%.

Example

Question: what will be the energy cost of a man weighing 70 kg that walks on a treadmill with 12% elevation at a speed of 6 km h^{-1} (6000 m h^{-1}) for 35 min?

Answer: calculated as before VO_2 horizontal = (6000 · 0.1 O_2)/60 + 3.5 = 13.5 m $LO_2\ kg^{-1}\ min^{-1}$

$$VO_2\ \text{vertical} = (6000 \cdot 1.8 \cdot 12)/(60 \cdot 100) = 21.6\,mLO_2\ kg^{-1}\ min^{-1}$$

Question: What was the overall metabolic cost?
Answer: VO_2 horizontal + VO_2 vertical = 13.5 + 21.6 = 35.1 $mLO_2\ kg^{-1}\ min^{-1}$
Question: How many METs equal 35.1 $mLO_2\ kg^{-1}\ min^{-1}$?
Answer: METs = 35.1 $mLO_2\ kg^{-1}\ min^{-1}$/3.5 $mLO_2\ kg^{-1}\ min^{-1}$ = 10 METs
Question: What was the total energy cost for the 35 min of exercise?
Answer: $VO_2 = 35.1\ mLO_2\ kg^{-1}\ min^{-1} \cdot 70\ kg \cdot 35\ min = 85{,}995\ mLO_2$
Question: How many liters of oxygen equal 85,995 $mLO_2\ min^{-1}$?
Answer: 85995/1000 = 85.995 LO_2
Question: How many Kcal are in 85.995 LO_2?
Answer: If 1 L of oxygen equals 5 Kcal, then 85.995 LO_2 equals 430 Kcal

Calculating the total metabolic cost of the horizontal component on a treadmill during running: During running on a sloped treadmill, VO_2 will be higher compared to that without elevation. VO_2 will be the sum of the vertical and horizontal components. The VO_2 ($mLO_2\ kg^{-1}\ min^{-1}$) of the horizontal component (without elevation) during running on a treadmill can be calculated as follows:

$$VO_2\ \text{horizontal} = \left[\left(\text{speed in m} \cdot 0.2\, O_2\right)/\left(60\,\text{min}\right)\right] + 3.5\,mLO_2\ kg^{-1}\ min^{-1}$$

where 0.2 = oxygen cost per meter of running since at a certain point both legs are in the air; dividing by 60 min is to have the running speed in meters per minute; and the additional 3.5 $mLO_2\ kg^{-1}\ min^{-1}$ is the resting VO_2 required at rest per kg per minute.

The vertical component is calculated as follows:

$$VO_2 \text{ vertical } mLO_2\ kg^{-1}\ min^{-1} = (\text{speed in m} \cdot 1.8\, O_2 \cdot \text{grade\%}) / (60\,\text{min} \cdot 100\%)$$

where speed – m h^{-1}; 1.8 = oxygen cost per meter of climbing uphill; grade = % of inclination; dividing by 60 min is to have the walking speed in meters.

Example

Question: what will be the energy cost of a man weighing 70 kg that runs on a treadmill with 3% elevation at a speed of 12 km h^{-1} (12,000 m h^{-1}) for 50 min?

Answer: VO_2 horizontal = (12,000 · 0.2) + 3.5 = 43.5 $mLO_2\ kg^{-1}\ min^{-1}$

$$VO_2 \text{ vertical} = (12{,}000\,\text{m}\,h^{-1} \cdot 1.8\, O_2 \cdot 3\%) / (60\,\text{min} \cdot 100\%) = 10.8\, mLO_2\ kg^{-1}\ min^{-1}$$

Question: What was the total VO_2 for 1 min?

Answer: 43.5 + 10.8 = 54.3 $mLO_2\ kg^{-1}\ min^{-1}$

Question: How many METs it equals?

Answer: METs = 54.3 $mLO_2\ kg^{-1}\ min^{-1}$/3.5 $mLO_2\ kg^{-1}\ min^{-1}$ = 15.5 METs

Question: What was the total energy cost for 50 min of exercise?

Answer: VO_2 = 54.3 $mLO_2\ kg^{-1}\ min^{-1}$ · 70 kg · 50 min = 190,050 mLO_2

Question: How many liters of oxygen equal 190,050 $mLO_2\ min^{-1}$?

Answer: 190,050/1000 = 190.050 LO_2

Question: How many Kcal are in 190.050 LO_2?

Answer: If 1 L of oxygen equals 5 Kcal, then 190.050 LO_2 equals 950.25 Kcal:

Laboratory Assignments

Stepwise: Two subjects, one female (60 kg) will work on a stepwise with 0.45 m height, steps up and down 15 min^{-1}, and one male (75 kg) will work on a stepwise with 0.6 m height, steps up and down 18 min^{-1}. Calculate the following for each subject: (a) Distance (m), (b) Work done (kgm), (c) Power (kgm min^{-1}), (d) SI units = (W). (e) Metabolic cost ($mLO_2\ kg^{-1}\ min^{-1}$), (f) How many METs was the effort intensity? and (g) How many kcal the effort costs?

Stationary bicycle: Two subjects (females or males) will exercise on a stationary bicycle, one weighing 60 kg for 15 min with 3 kg resistance and 60 rpm the other one weighing 70 kg for 17 min with 3.5 kg resistance and 60 rpm. Calculate the following for each subject: (a) Distance (m), (b) Work (kgm), (c) Power = (kgm min^{-1}), (d) SI units (watts) or (e) SI units (J), (f) Metabolic cost ($mLO_2\ kg^{-1}\ min^{-1}$), (g) How many METs was the effort intensity? and (h) How many kcal the effort cost?

Treadmill: Two subjects (males or females) will run on a treadmill for 15 min one 10 km min^{-1} and 6% grade weighing 70 kg and the other at 12 km min^{-1} with 2% weighing 80 kg. Calculate the following for each subject:

(a) Horizontal energy cost VO_2 (mLO_2 kg^{-1} min^{-1}), (b) vertical energy cost VO_2 (mLO_2 kg^{-1} min^{-1}), (c) Total energy cost VO_2 (mLO_2 kg^{-1} min^{-1}), (d) How many METs it equals? And (e) How many kcal was expended for the total bout?

Data of the stepwise test

Name ______________ Height ______________

Weight ______________ Date______________

Resting HR (beats min^{-1})

Peak HR (beats min^{-1})

Resting BP (mmHg)

Peak BP (mmHg)

Repetition # ups and downs

Stair's height

Bout's time

Calculate: (a) Distance (m) ____________, (b) Work done (kgm) ____________, (c) Power (kgm min^{-1}) ____________, (d) SI units = (W) ____________, (e) Metabolic cost (mLO_2 kg^{-1} min^{-1}) ____________, (f) How many METs was the effort intensity? ____________ and (g) How many kcal the effort cost? ____________

Data of a bicycle ergometer

Name ______________ Height ______________ Weight ______________ Date ______________

Resting HR (beats min^{-1})

Peak HR (beats min^{-1})

Resting BP (mmHg)

Peak BP (mmHg)

Revolution per minute

Resistance in kg

Bout's time

Calculate: (a) Distance (m), (b) Work (kgm), (c) Power = (kgm min^{-1}), (d) SI units (watts) or (e) SI units (J), (f) Metabolic cost (mLO_2 kg^{-1} min^{-1}), (g) How many METs was the effort intensity? and (h) How many kcal the effort cost?

Data of a treadmill ergometer

Name ______________ Height______________

Weight______________ Date______________

Resting HR (beats min^{-1})

Peak HR (beats min^{-1})

Resting BP (mmHg)

Peak BP (mmHg)
Speed and % elevation
Bout's time

Calculate: (a) Horizontal energy cost VO_2 ($mLO_2\ kg^{-1}\ min^{-1}$), (b) vertical energy cost VO_2 ($mLO_2\ kg^{-1}\ min^{-1}$), (c) Total energy cost VO_2 ($mLO_2\ kg^{-1}\ min^{-1}$), (d) How many METs it equals? and (e) How many Kcal were expended for the total bout?

References

1. Nelson N, Asplund CA. Exercise testing: who, when, and why? PM&R. 2016;8:S16–23.
2. Smirmaul BCP, Bertucci DR, Teixeira IP. Is the VO2max that we measure really maximal? Front Physiol. 2013;4:203. https://doi.org/10.3389/fphys.2013.00203.
3. Kramer PA, Sarton-Miller I. The energetics of human walking: is Froude number (Fr) useful for metabolic comparisons? Gait Posture. 2008;27:209–15.
4. Psota T, Chen KY. Measuring energy expenditure in clinical populations: rewards and challenges. Eur J Clin Nutr. 2013;67:436–42.
5. Beltz NM, Gibson AL, Janot JM, et al. Graded exercise testing protocols for the determination of VO2max: historical perspectives, progress, and future considerations. J Sports Med (Hindawi Publ Corp). 2016;2016:3968393.
6. Driss T, Vandewalle H. The measurement of maximal (anaerobic) power output on a cycle ergometer: a critical review. Biomed Res Int. 2013;2013:589361. https://doi.org/10.1155/2013/589361.
7. Duruturk N, Tonga E, Karatas M, et al. Activity performance problems of patients with cardiac diseases and their impact on quality of life. J Phys Ther Sci. 2015;27:2023–8.
8. Orr JL, Williamson P, Anderson W, Ross R, McCafferty S, Fettes P. Cardiopulmonary exercise testing: arm crank vs cycle ergometry. Anaesthesia. 2013;68:497–501.
9. Fletcher GF, Balady GJ, Amsterdam EA, et al. Exercise standards for testing and training. Circulation. 2001;104:1694–740.
10. Hills AP, Mokhtar N, Byrne NM. Assessment of physical activity and energy expenditure: an overview of objective measures. Front Nutr. 2014;1:5. https://doi.org/10.3389/fnut.2014.00005.
11. Williams CJ, Williams MG, Eynon N, et al. Genes to predict VO2max trainability: a systematic review. BMC Genomics. 2017;18(Suppl 8):831. https://doi.org/10.1186/s12864-017-4192-6.
12. Hochsmann C, SteWalz SP, Schafer J, Holopainen J, Hanssen H, Schmidt-Trucksass A. Mobile exergaming for health-effects of a serious game application for smartphones on physical activity and exercise adherence in type 2 diabetes mellitus—study protocol for a randomized controlled trial. Trials. 2017;18:103. https://doi.org/10.1186/s13063-017-1853-3.
13. Maciejczyk M, Więcek M, Szymura J, et al. The influence of increased body fat or lean body mass on aerobic performance. PLoS One. 2014;9(4):e95797.
14. Ramos-Jimenez A, Hernandez-Torres RP, et al. The respiratory exchange ratio is associated with fitness indicators both in trained and untrained men: a possible application for people with reduced exercise tolerance. Clin Med Circ Respirat Pulm Med. 2008;2:1–9.
15. Loughney L, West M, Pintus S, et al. Comparison of oxygen uptake during arm or leg cardiopulmonary exercise testing in vascular surgery patients and control subjects. Br J Anaesth. 2014;112:57–65.
16. Larsen RT, Christensen J, Tang LH, et al. A systematic review and meta-analysis comparing cardiopulmonary exercise test values obtained from the arm cycle and the leg cycle respectively in healthy adults. Int J Sports Phys Ther. 2016;11:1006–39.

17. Kim C-H, Wheatley CM, Behnia M, et al. The effect of aging on relationships between lean body mass and VO2max in rowers. PLoS One. 2016;11(8):e0160275. https://doi.org/10.1371/journal.pone.0160275.
18. Bassett DR Jr, Howley ET. Limiting factors for maximum oxygen uptake and determinants of endurance performance. Med Sci Sports Exerc. 2000;32:70–84.
19. Cardinale DA, Larsen FJ, Jensen-Urstad M, et al. Muscle mass and inspired oxygen influence oxygen extraction at maximal exercise: role of mitochondrial oxygen affinity. Acta Physiol (Oxf). 2019;225(1):e13110. https://doi.org/10.1111/apha.13110.
20. Avitsland A, Solbraa AK, Riiser A. Promoting workplace stair climbing: sometimes, not interfering is the best. Arch Public Health. 2017;75:2.
21. Brunelli A, Refai M, Xiume F, et al. Performance at symptom-limited stair-climbing test is associated with increased cardiopulmonary complications, mortality, and costs after major lung resection. Ann Thorac Surg. 2008;86:240–7.
22. Higueras-Fresnillo S, Esteban-Cornejo I, Gasque P, et al. Criterion-related validity of self-reported stair climbing in older adults. Aging Clin Exp Res. 2018;30:199–203.
23. Dong J, Mao Y, Li J, et al. Stair-climbing test predicts postoperative cardiopulmonary complications and hospital stay in patients with non-small cell lung cancer. Med Sci Monit. 2017;23:1436–41.
24. Zelik KE, Kuo AD. Mechanical work as an indirect measure of subjective costs influencing human movement. PLoS one. 2012;7(2):e31143. https://doi.org/10.1371/journal.pone.0031143.
25. Vrijens J, Van Cauter C. Cardiorespiratory response to treadmill and bicycle exercise in runners. Eur J Appl Physiol Occup Physiol. 1980;43:243–51.
26. Smith PM, Doherty M, Price MJ. The effect of crank rate strategy on peak aerobic power and peak physiological responses during arm crank ergometry. J Sports Sci. 2007;25:711–8. https://doi.org/10.1080/02640410600831955.

Appendix A: Measurements in Exercise Physiology

The purpose of this Laboratory is to explain and demonstrate standard metric units and Laboratory equipment measuring physiological variables. The metric system is now often used as a synonym for the International System of Units (SI)-the official system of measurement in almost every country in the world (Table A.1). The metric system is officially not used in the United States, and it remains one of the few industrialized countries that have not adopted the metric system as its official system of measurement. Although the United Kingdom uses the metric system for most official purposes, the use of the imperial system of measurement, particularly for use at home, is widespread and is permitted by the law.

From its beginning, the main features of the metric system were the standard set of inter-related base units and a standard set of prefixes in powers of ten. These base units are used to derive larger and smaller units that could replace a huge number of other units of measure in existence. Although the system was first developed for commercial use, the development of coherent units of measure made it particularly suitable for science. A common set of decimal-based prefixes that have the effect of multiplication or division by an integer power of ten can be applied to units that are too large or too small for practical use. The prefix kilo, for example, is used to

Table A.1 SI units

Variable	SI units
Mass	Kilogram (kg), gram (g)
Time	Second (s)
Distance	Meter (m)
Force	Newton (N)
Energy	Joules (J)
Work	Kilogram-meter (kgm)
Power	Watts (W)
Velocity	Meter-second (m s^{-1}), Newton-meter (N m^{-1}), J
Torque	(N m^{-1})
Volume	Liter (L), milliliter (mL)

M. S. Saghiv, M. S. Sagiv, *Basic Exercise Physiology*,
https://doi.org/10.1007/978-3-030-48806-2

multiply the unit by 1000, and the prefix milli is to indicate a one-thousandth part of the unit. Thus, a kilogram and kilometer are a thousand grams and meters, respectively, and a milligram and millimeter are one thousandth of a gram and meter, respectively. In the original version of the metric system, the base units could be derived from a specified length (meter) and the weight (mass) of a specified volume (1/1000 of a cubic meter) of pure water (Fig. A.1). Tables A.2, A.3, A.4, A.5, and A.6 present different values for units in the metric system.

During its evolution, the metric system has adopted many units of measure. The introduction of SI rationalized both the way in which units of measure were defined and also the list of units in use. These are now catalogued in the official SI Brochure. The table below lists the units of measure in this catalogue and shows the conversion factors connecting them with the equivalent units that were in use on the eve of the adoption of SI. The SI brochure also catalogues certain non-SI units that are

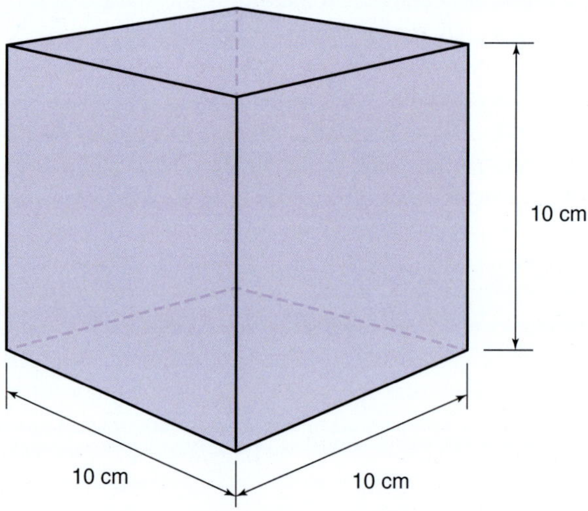

Fig. A.1 One liter is equivalent to the volume of a cube with edges of 10 cm, and the kilogram was originally designed to be one liter of water at the melting point of ice

Table A.2 Length units

Metric units	Equivalent in metric units
1 meter (m)	100 cm or 1000 mm
1 centimeter (cm)	0.01 m or 10 mm
1 millimeter (mm)	0.001 m or 0.1 cm

Table A.3 Mass units (weight)

Metric units	Equivalent in metric units
1 ton (kg)	1000 kg
1 kilogram (kg)	1000 g or 1,000,000 mg
1 gram (g)	0.001 kg or 1000 mg
1 milligram (mm)	0.000001 kg or 0.001 g

Table A.4 Volume units (weight)

Metric units	Equivalent in metric units
1 liter (L)	1000 mL
1 milliliter (mL)	0.001 L

Table A.5 Time and pressure units

Variable	Equivalent units
Time (t)	Minute (min), 60 seconds (s) hour (h), 60 min day (d) 24 h
Pressure (P) Atmosphere (atm)	1 atm (760 torr) 1 torr (1 mmHg) 1 bar (0.98692 atmosphere)

widely used with the SI in matters of everyday life or units that are exactly defined values in terms of SI units and are used in particular circumstances to satisfy the needs of commercial, legal, or specialized scientific interests (Table A.5). These units include:

Definitions of Relevant Terms

- Mass: a body of coherent matter, usually of indefinite shape and often of considerable size: a mass of muscles or a collection of incoherent particles, parts, or objects regarded as forming one body: weight of the whole body.
- Force: forces resulting from action-at-a-distance are those types of forces that physical activity uses by being able to exert a push or pull. Force units are Newton (N) and it is defined as the force needed to accelerate 1 kg of mass to a distance of 1 m in a second.

Newton's Laws of Motion

Newton's laws of motion are three physical laws that together laid the foundation for classical mechanics. They describe the relationship between a body and the forces acting upon it and its motion in response to the said forces. They have been expressed in several different ways over nearly three centuries and can be summarized as follows:

- First law: When viewed in an inertial reference frame, an object either remains at rest or continues to move at a constant velocity, unless acted upon by an external force.
- Second law: The vector sum of the forces F on an object is equal to the mass m of that object multiplied by the acceleration vector a of the object: $F = ma$.
- Third law: When one body exerts a force on a second body, the second body simultaneously exerts a force equal in magnitude and opposite in direction on the first body.

- Torque: moment or moment of force is the tendency of a force to rotate an object about an axis, fulcrum, or pivot. Just as a force is a push or a pull, a torque can be thought of as a twist to an object. A force applied at a right angle to a lever multiplied by its distance from the lever's fulcrum (the length of the lever arm). Mathematically, torque is defined as the cross product of the lever-arm distance vector and the force vector, which tends to produce rotation.
- Energy: it is the capacity of the body to perform work. Energy exists in several forms such as heat, nuclear, mechanical energy, light, chemical energy, electrical, or other forms. In exercise physiology, we are mainly concerned with chemical and mechanical energies. According to the law of conservation of energy, the total energy of a system remains constant, though energy may transform into another form. In biological systems, chemical energy is transformed into mechanical energy which turn is changed to heat production by the muscle (Table A.6).
- Work: in physics, a force is said to do work if, when acting on a body, there is a displacement of the point of application in the direction of the force. For example, when a ball is held above the ground and then dropped, the work done on the ball as it falls is equal to the weight of the ball (a force) multiplied by the distance to the ground (a displacement). The SI unit of work is the Newton-meter or joule (J), and therefore:

$$W(\text{work}) = F(\text{force}) \cdot D(\text{distance})$$

- Power or work rate describes exercise intensity. It is defined as the amount of work performed per unit of time. It is a term that describes the rate in which an action or movement, especially one requiring skill is completed, and therefore:

$$\text{Power}(P) = \text{work}(W) / \text{time}(t)$$

or

$$P = W / t$$

Power is an element of skill-related fitness that needs to excel in physical performance. Increased strength does not always translate into increased power. For example, a strong upper body lifts a high amount of weight. However, a strong upper body does not always have the ability to throw a shot put very far if enough workload cannot be generated.

Table A.6 Units expressing work, energy, and force

Term	Short term	Value
kilogram-meter	kgm	1 kgm = 9.81 J
kilopond-meter	kpm	1 kgm = 1 kpm
kilocalories	Kcal	1 kcal = 4.186 J
Joule	J	1 J = 0.101 kgm
Watt	W	1 W = 6.12 kgm min^{-1}

- Velocity: it is the time needed to move the body from one place to another, such as rapidity of motion or operation; swiftness; and workload: a high body motion velocity. Time-motion analysis has been used extensively to investigate the movement patterns and physiological demands of high-intensity intermittent field sports.
- MET: it is defined as the amount of oxygen consumed while sitting at rest and is equal to 3.5 mLO_2 kg^{-1} min^{-1}. Using MET can quantify the energy cost of activities. METS are also routinely utilized to describe the functional capacity or aerobic power of an individual and to provide list of activities in which the individual can safely participate. Transformation of VO_2 measurements may be obtained by dividing the value of the VO_2 in mLO_2 kg^{-1} min^{-1}; measurement of 42 mLO_2 kg^{-1} min^{-1} is equal to 12 METs.

In spite of its limitations, the MET concept provides a convenient method to describe the functional capacity or exercise tolerance of an individual as determined from progressive exercise testing and to define a range of physical activities in which a person may participate safely, without exceeding a prescribed intensity level. One metabolic unit (MET) equals the VO_2 at rest.

Other conversion units are:

1 kg = 2.2 pounds (lbs).
1 Mile min^{-1} = 26.8 m min^{-1}
1 Mile h^{-1} = 1.609 km h^{-1}
1 lb of fat = 3500 kcal
1 MET = 3.5 mLokg^{-1} min^{-1}
1 LO_2 = 4.83 kcal
1 inch = 2.54 cm

Appendix B: Report Formats

General Guidelines for All Laboratories

Laboratory Report

Data of the stepwise test
Name ______________ Height____ (cm) Weight___ (kg)
Date______________

Resting heart rate (beats min^{-1})	
Peak heart rate (beats min^{-1})	
Resting blood pressure (mmHg)	
Peak blood pressure (mmHg)	
Repetition (# of ups and downs)	
Stair's height (cm)	
Bout's time (min)	

Calculate:

Distance: _______ (m)
Work: ________ (VO_2)
Power:
SI units ______(watts)
SI units ______(joules)

M. S. Saghiv, M. S. Sagiv, *Basic Exercise Physiology*,
https://doi.org/10.1007/978-3-030-48806-2

Data of the bicycle Ergometer
Name ______________ Height____(cm) Weight___(kg)
Date________

Resting heart rate (beats min^{-1})	
Peak heart rate (beats min^{-1})	
Resting blood pressure (mmHg)	
Peak blood pressure (mmHg)	
Revolution (per minute)	
Resistance (kg)	
Bout's time (min)	

Calculate:

Distance: ______(km or miles)
Work: _________(VO_2)
Power: ______________
SI units ____(watts)
SI units ____(joules)

Data of the treadmill Ergometer
Name ______________ Height____(cm) Weight____(kg)
Date___________

Resting HR (beats min^{-1})	
Peak HR (beats min^{-1})	
Resting BP (mmHg)	
Peak BP (mmHg)	
Elevation (%)	
Bout's time (min)	
Bout's time (min)	
Workload (VO_2)	
Elevation (%)	
Bout's time (min)	

Calculate:

Distance: ______ (km or miles)
Work: _______(VO_2)
Power:
SI units ____(watts)
SI units ____(joules)

General Notes

- Write in the past tense, since your report follows the laboratory experiance.
- Avoid the use of "I, me, or my". Refer to the subject as "the subject" or "the participant".
- Use "he, him, his", or "she, her, her" (as appropriate) instead of "they", "them", "their", which are all plural.
- Laboratory report consists of Introduction, Methods, Results, and Discussion.
- Reports should be typed (double spaced, with 1 inch margins) and follow the format.

Introduction:

- Provide an introduction to the concept being studied.
- Provide background information of the test.
- Justify the importance of this laboratory to learn this concept.
- State the purpose of the laboratory presented in laboratory session.

Method:

- Provide step by step detailing of the procedures for data collection and analysis.

Results:

- Give subject's information (height, weight, age, sex, etc.).
- Provide results (may refer to appendix containing chart)

Discussion:
In this section, you have to convince that you understand the topic studied.

- Discuss the meaning of the results.
- Compare the results to norms.
- Discuss possible sources of error.

Conclusion:

- This section consists of a bottom-line statement describing the value of the laboratory.

Listed organizations can be useful in choosing a profession both before and after graduation!

- American College of Sports Medicine http://www.acsm.org/
- American Council on Exercise http://www.acefitness.org/
- American Association of Cardiovascular and Pulmonary Rehabilitation http://www.aacvpr.org/
- IDEA Health and Fitness Association http://www.ideafit.com/
- National Strength and Conditioning Association http://www.nsca-lift.org/

Index

M. S. Saghiv, M. S. Sagiv, *Basic Exercise Physiology*,
https://doi.org/10.1007/978-3-030-48806-2

P

R

The manufacturer's authorised representative in the EU is Springer
Nature Customer Service Centre GmbH, Europaplatz 3, 69115 Heidelberg,
Germany. If you have any concerns regarding our products, please
contact ProductSafety@springernature.com

Printed and bound by CPI Group (UK) Ltd, Croydon, CR0 4YY
30/09/2026
02224059-0011